Principles of Biotechnology and Genetic Engineering

PRINCIPLES OF BIOTECHNOLOGY AND GENETIC ENGINEERING

By

Dr. A. JAYAKUMARAN NAIR
M.Sc., Ph.D.
Sr. Lecturer, Department of Biotechnology
University of Kerala,
Thiruvananthapuram, Kerala

UNIVERSITY SCIENCE PRESS

(An Imprint of Laxmi Publications Pvt. Ltd.)

BANGALORE • CHENNAI • COCHIN • GUWAHATI • HYDERABAD
JALANDHAR • KOLKATA • LUCKNOW • MUMBAI • RANCHI
NEW DELHI

Published by :

UNIVERSITY SCIENCE PRESS

(An Imprint of Laxmi Publications Pvt. Ltd.)

113, Golden House, Daryaganj,
New Delhi-110002

Phone : 011-43 53 25 00
Fax : 011-43 53 25 28

www.laxmipublications.com
info@laxmipublications.com

First Edition : 2007
Second Edition : 2010

OFFICES

✆ **Bangalore**	080-26 61 15 61	✆ **Jalandhar**	0181-222 12 72
✆ **Chennai**	044-24 34 47 26	✆ **Kolkata**	033-22 27 43 84
✆ **Cochin**	0484-237 70 04, 405 13 03	✆ **Lucknow**	0522-220 95 78
✆ **Guwahati**	0361-254 36 69, 251 38 81	✆ **Mumbai**	022-24 91 54 15, 24 92 78 69
✆ **Hyderabad**	040-24 65 23 33	✆ **Ranchi**	0651-221 47 64

UPB-9462-395-PRIN BIOTEC & GENE ENGG-NAI **C—593/09/09**

Typeset at : Goswami Associates, Delhi *Printed at :* Repro Knowledgecast Limited, Thane

CONTENTS

PREFACE

Biotechnology as a fast developing technology as well as a science has already shown its impact on different aspects of day-to-day human life such as public health, pharmaceuticals, food and agriculture, industry, bioenergetics and information technology. Now it is very clear that biotechnology will be a key technology for the 21st century and the science of the future. It has the potential to ensure food security, dramatically reduce hunger and malnutrition, and reduce rural poverty, particularly in developing countries. Considering its commercial potential and its possible impact on the economy, the government has taken a number of measures to build up trained human resources in biotechnology and promote research and development and its commercial aspects. The introduction of biotechnology as a subject discipline by various universities is such an initiation.

This book covers all the fundamental aspects of biotechnology. It has been written in a very simple manner and explains the fundamental concepts and techniques in detail so that they are very easily understood, even by those without even a basic understanding of biology. Reading this textbook will give readers an idea of the relationship between biotechnology and health, nutrition, agriculture, environment, industry, etc., and will explain different applications of biotechnology in everyday life.

—Author

Part 1 INTRODUCTION TO BIOTECHNOLOGY

Chapter 1

OVERVIEW

In This Chapter

1.1 INTRODUCTION AND DEFINITION

The term 'biotechnology' was used before the twentieth century for traditional activities such as making dairy products such as cheese and curd, as well as bread, wine, beer, etc. But none of these could be considered biotechnology in the modern sense. Genetic alteration of organisms through selective breeding, plant cloning by grafting, etc. do not fall under biotechnology. The process of fermentation for the preparation and manufacturing of products such as alcohol, beer, wine, dairy products, various types of organic acids such as vinegar, citric acid, amino acids, and vitamins can be called classical biotechnology or traditional biotechnology. Fermentation is the process by which living organisms such as yeast or bacteria are employed to produce useful compounds or products.

Modern biotechnology is similar to classical biotechnology in utilizing living organisms. So what makes modern biotechnology modern? It is not modern in the sense of using various living organisms, but in the techniques for doing so. The introduction of a large number of new techniques has changed the face of classical biotechnology forever. These modern techniques, applied mainly to cells and molecules, make it possible to take advantage of the biological process in a very precise way. For example, genetic engineering has allowed us to transfer the property of a single gene from one organism to another. But before going into the details of biotechnology and the techniques that make it possible, let us first define biotechnology.

Definition of Biotechnology

There are several definitions for biotechnology. One simple definition is that it is the commercialization of cell and molecular biology. According to United States National Science Academy, biotechnology is the "controlled use of biological agents like cells or cellular components for beneficial use". It covers both classical as well as modern biotechnology. More generally, biotechnology can be defined as "the use of living organisms, cells or cellular components for the production of compounds or precise genetic improvement of living things for the benefit of man".

Even though biotechnology has been in practice for thousands of years, the technological explosion of the twentieth century, in the various branches of sciences—physics, chemistry, engineering, computer application, and information technology—revolutionized the development of life sciences, which ultimately resulted in the evolution of modern biotechnology.

Supported by an array of biochemical, biophysical, and molecular techniques besides engineering and information technology, life scientists were able to develop new drugs, diagnostics, vaccines, food products, cosmetics, and industrially useful chemicals. Genetically-altered crop plants, which can resist the stress of pests, diseases, and environmental extremes were developed. New tools and techniques to extend the studies on genomics and proteomics, not only of man but other organisms were also developed. The involvement of information technology and internet in biotechnology, particularly genomics and proteomics, has given birth to a new branch in biotechnology—the science of bioinformatics and computational biology. The skills of biotechnology, like any other modern science, are founded on the previous knowledge acquired through the ages. If one wants to understand biotechnology thoroughly, one should also know the history of its development.

1.2 HISTORICAL PERSPECTIVES

Biotechnology as a science is very new (about 200 years old) but as a technology it is very old. The word biotechnology, first used in 1917, refers to a large-scale fermentation process for the production of various types of industrial chemicals. But the roots of biotechnology can be traced back to pre-historical civilizations, such as Egyptian and Indus valley civilizations, when man learned to practice agriculture and animal domestication. Even before knowing about the existence of microorganisms, they had learned to practice biotechnology.

Biotechnology in Prehistoric Times

Primitive man became domesticated enough to breed plants and animals; gather and process herbs for medicine; make bread, wine and beer and create many fermented food products including yogurt, cheese, and various soy products; create septic systems to deal with digestive and excretory waste products; and to create vaccines to immunize themselves against diseases. Archeologists keep discovering earlier examples of the uses of microorganisms by man. Examples of most of these processes date back to 5000 BC. Ancient Indus people, for example, prepared and used various types of fermented foods, beverages, and medicines. The ancient Egyptians and Sumerians used yeast to brew wine and to bake bread as early as 4000 BC. People in Mesopotamia used bacteria to convert wine into vinegar. Many ancient civilizations exploited tiny organisms that live in the earth by rotating crops in the field to increase crop yields. The Greeks used crop rotation to maximize crop yield and also practiced various methods of food preservation such as drying, smoking, curing, salting, etc. All these techniques and processes were practiced in the Middle East and South East Asia including ancient India. The Egyptian art of mummification used the technique of dehydration using a mixture of salts.

Use of Genetic Resources

The ancient people were also aware of the role of natural genetic resources such as plants in the economic growth of a land. The rulers at those time used to send plant-collectors to gather prized exotic species of plants that produced valuable spices and medicines. Likewise, in modern times, colonial powers mounted huge plant-collecting expeditions across Latin America, Asia, and Africa, installing their findings in botanic gardens. These early 'gene banks' helped the colonial powers to establish agricultural monocultures around the globe.

Microorganisms and Fermentation

Although baking bread, brewing beer, and making cheese has been going on for centuries, the scientific study of these biochemical processes is less than 200 years old. Clues to understanding fermentation emerged in the seventeenth century when Dutch experimentalist Anton Van Leeuwenhoek discovered microorganisms using his microscope. He unraveled the chemical basis of the process of fermentation using analytical techniques for the estimation of carbon dioxide. Two centuries later, in 1857, a French scientist Louis Pasteur published his first report on lactic acid formation from sugar by fermentation. He published a detailed report on alcohol fermentation later in 1860. In this report, he revealed some of the complex physiological processes that happen during fermentation. He proved that fermentation is the consequence of anaerobic life and identified three types of fermentation:

- Fermentation, which generates gas;
- Fermentation that results in alcohol; and
- Fermentation, which results in acids.

At the end of the nineteenth century, Eduard Buchner observed the formation of ethanol and carbon dioxide when cell-free extract of yeast was added to an aqueous solution of sugars. Thus, he proved that cells are not essential for the fermentation process and the components responsible for the process are dissolved in the extract. He named that substance 'Zymase'. The fermentation process was modified in Germany during World War I to produce glycerine for making the explosive nitroglycerine. Similarly, military armament programs discovered new technologies in food and chemical industries, which helped them win battles in the First World War. For example, they used the bacteria that converts corn or molasses into acetone for making the explosive cordite. While biotechnology helped kill soldiers, it also cured them. Sir Alexander Fleming's discovery of penicillin, the first antibiotic, proved highly successful in treating wounded soldiers.

The Genesis of Genetics

In 1906, biotechnology took a leap forward when Gregor John Mendel announced the findings of his experiments as the 'laws of genetics'. He predicted the presence of 'units of heredity'—later called genes—which did not change their identity from generation to generation but only recombined. The science of genetics derived from the term 'genesis', which relates to the origin of something, tried to explain how organisms both resemble their parents and differ from them. It was believed that every gene directly corresponds to a specific trait. By the 1920s, genetics was helping plant breeders improve their crops. By the 1940s, genetics had transformed the agriculture sector, which led to the Green Revolution in the 1960s.

DNA and Genetic Engineering—The Beginning of Modern Biotechnology

The science of genetics was transformed by the discovery of DNA (deoxyribonucleic acid), which carries the hereditary information in the cells. The chemical DNA had already been discovered in 1869 by Friederich Miescher but was not taken seriously as the chemical basis of genes until the early 1950s. Two scientists, Francis Crick and James Watson along with Rosalind Franklin, in 1953, discovered that the DNA structure was a double helix: two strands twisted around each other like a spiral staircase with bars across like rings. The structure, function, and composition of DNA are virtually identical in all living organisms—from a blade of grass to an elephant. What differs—and makes each creature unique—is the precise ordering of the chemical base in the DNA molecule. This gave scientists the idea that they might change this ordering and so modify lifeforms. Marshall Nirenberg and H. Gobind Khorana carried out the deciphering of the genetic code in 1961.

Soon scientists and industrialists were seeking to alter the genetic make-up of living things by transferring specific genes from one organism to another. They could now modify lifeforms by altering the hereditary material at the molecular level. Walter Gilbert carried out the first recombinant DNA experiments in 1973, and the first hybridomas created in 1975. The production of monoclonal antibodies for diagnostics was carried out in 1982, and the first recombinant human therapeutic protein, insulin (humulin), was produced in 1982. In 1976, the U.S. company Genentech became the first biotech company to develop technologies to rearrange DNA. Commercial uses of recombinant-DNA-assisted biotechnology include the development of interferon, insulin, and a number of genetically-modified crop plants such as the high-solids-processing tomato that has 20% less water. Transgenic animals have been created such as the unfortunate onco-mouse designed to develop cancer ten months after birth to study cancer.

Companies have been assisted and encouraged in their research by the 1980 ruling of the U.S. Supreme Court allowing genetically-engineered microorganisms to be patented. This means that virtually any lifeform on this planet can theoretically become the private property of the company or person who 'creates' it. One of the greatest threats of the new biosciences is that life will become the monopoly of a few giant companies.

An estimated 600 pharmaceutical companies worldwide are conducting research and development into genetically-engineered products. Mistakes are bound to happen. And with something so powerful as genetic engineering, one mistake could have profound and wide-ranging effects. The whole gene revolution is on the verge of becoming the private property of a few multinationals. We must impose tough controls on the genetics supply industry and work to make sure that the new techniques are in the service of the global community.

Milestones in the History of Biotechnology

5000 BC	Indus and Indo-Aryan civilizations practiced biotechnology to produce fermented foods and medicines and to keep the environment clean.
4000 BC	Egyptians used yeasts to make wine and bread.
1750 BC	The Sumerians brewed beer.
250 BC	The Greeks used crop rotation to maximize crop fertility.
1500 AD	The Aztecs made cake from spirulina.
1663	Robert Hook first described cells.
1675	Microbes were first described by Anton Van Leeuwenhock.
1859	Darwin published his theory of evolution in *'The Origin of Species.'*
1866	Gregor John Mendel published the basic laws of genetics.
1869	DNA was isolated by Friederich Miescher.
1910	Genes were discovered to be present in chromosomes.
1917	The term 'biotechnology' was used to describe fermentation technology.
1928	The first antibiotic, penicillin, was discovered by Alexander Flemming.
1941	The term 'genetic engineering' was first used.
1944	Hereditary material was identified as DNA.
1953	Watson and Crick proposed the double helix structure of DNA.
1961	Deciphering of genetic code by M. Nirenberg and H.G. Khorana.
1969	The first gene was isolated.
1973	The first genetic engineering experiment was carried out by Walter Gilbert.
1975	Creation of the first hybridomas.
1976	The first biotech company.
1978	World's first 'test-tube baby,' Louise Brown, was born through *in vitro* fertilization.
1981	The first gene was synthesized. The first DNA synthesizer was developed.
1982	The first genetically engineered drug, human insulin, produced by bacteria, was manufactured and marketed by a U.S. company. Production of the first monoclonal antibodies for diagnostics.
1983	The first transgenic plant was created—a petunia plant was genetically engineered to be resistant to kanamycin, an antibiotic.
1983	The chromosomal location of the gene responsible for the genetic disorder, Huntington's disease, was discovered leading to the development of genetic screening test.
1985	DNA fingerprinting was first used in a criminal investigation.
1986	The first field tests of genetically-engineered plants (tobacco) were conducted.
1990	Chymosin, an enzyme used in cheese making, became the first product of genetic engineering to be introduced into the food supply.

1990	Human genome project was launched.
1990	The first human gene therapy trial was performed on a four-year-old girl with an immune disorder.
1991	The gene implicated in the inherited form of breast cancer was discovered.
1992	Techniques for testing embryos for inherited diseases were developed.
1994	First commercial approval for transgenic plant by the U.S. government.
1995	First successful xenotransplantation trial was conducted, transplanting a heart from a genetically-engineered pig into a baboon.
1996	First commercial introduction of a 'gene chip' designed to rapidly detect variances in the HIV virus and select the best drug treatment for patients.
1996	Dolly, the sheep was cloned from a cell of an adult sheep.
1998	Embryonic stem cells were grown successfully, opening new doors to cell- or tissue-based therapies.
1999	A U.S. company announced the successful cloning of human embryonic cells from an adult skin cell.
1999	Chinese scientists cloned a giant panda embryo.
1999	Indian scientists and companies started producing recombinant vaccines, hormones, and other drugs.
2002	The draft of human genome sequence was published.

1.3 SCOPE AND IMPORTANCE OF BIOTECHNOLOGY

In the past, biotechnology concentrated on the production of food and medicine. It also tried to solve environmental problems. In the nineteenth century, industries linked to the fermentation technology had grown tremendously because of the high demand for various chemicals such as ethanol, butanol, glycerine, acetone, etc. The advancement in fermentation process by its interaction with chemical engineering has given rise to a new area—the bioprocess technology. Large-scale production of proteins and enzymes can be carried out by applying bioprocess technology in fermentation. Applying the principles of biology, chemistry, and engineering sciences, processes are developed to create large quantities of chemicals, antibiotics, proteins, and enzymes in an economical manner. Bioprocess technology includes media and buffer preparation, upstream processing and downstream processing. Upstream processing provides the microorganism the media, substrate, and the correct chemical environment to carry out the required biochemical reactions to produce the product. Downstream processing is the separation method to harvest the pure product from the fermentation medium. Thus, fermentation technology changed into biotechnology, now known as classical biotechnology. Now if we look at biotechnology, we find its application in various fields such as food, agriculture, medicine, and in solving environmental problems. This has led to the division of biotechnology into different areas such as agricultural biotechnology, medical or pharmaceutical biotechnology, industrial biotechnology, and environmental biotechnology.

Modern biotechnology is mainly based on recombinant DNA (rDNA) and hybridoma technology in addition to bioprocess technology. rDNA technology is the main tool used to not only produce genetically-modified organisms, including plants, animals, and microbes, but also to address the fundamental questions in life sciences. In fact, modern biotechnology began when recombinant human insulin was produced and marketed in the United States in 1982. The effort leading up to this landmark event began in the early 1970s when research scientists developed protocols to construct vectors by cutting out and pasting pieces of DNA together to create a new piece of DNA (recombinant DNA) that could be inserted into the bacterium, *e. coli* (transformation). If one of the pieces of the new DNA includes a gene for insulin or any other therapeutic protein or enzyme, the bacterium would be able to produce that protein or enzyme in large quantities by applying bioprocess technology.

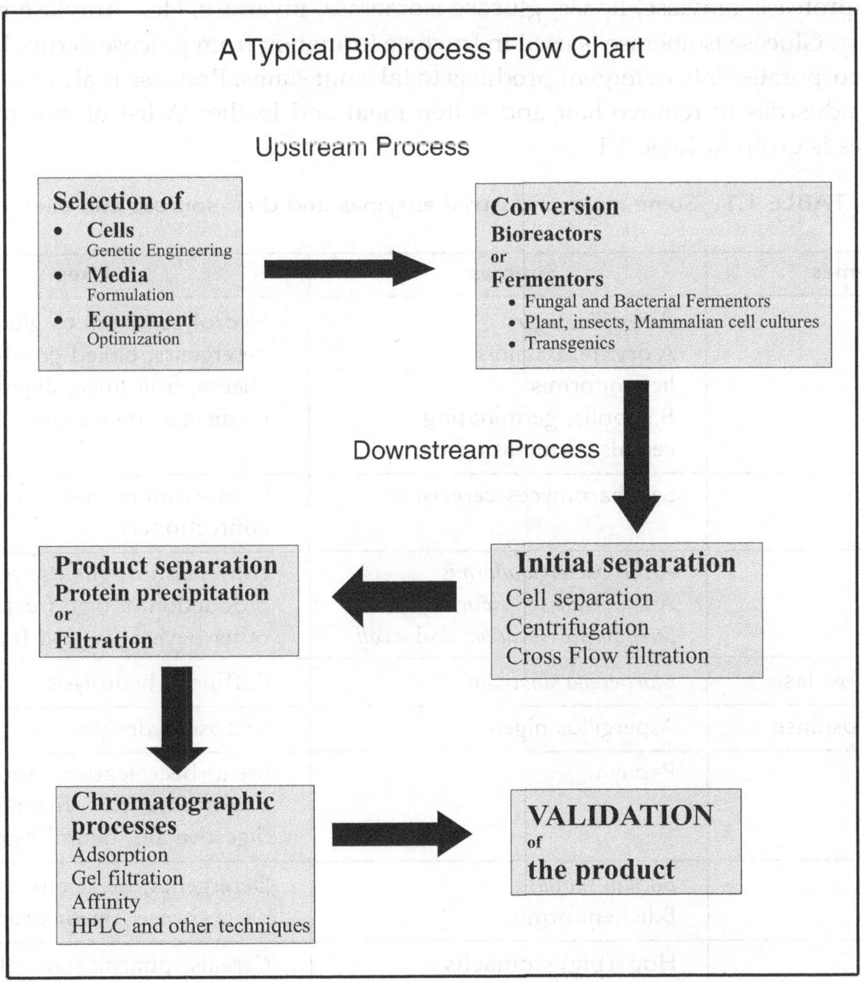

FIGURE 1.1 **An overview of modern bioprocessing.**

Another way of preparing human therapeutic proteins, vaccines, and diagnostic proteins is by hybridoma technology. The first hybridoma experiments were carried out in 1975. In hybridoma technology, a B-lymphocyte secreting antibody against a specific antigen is fused with a myeloma cell. The resulting (a cancerous B-lymphocyte) cell, if injected into a mouse's abdomen or if cultured in a bioreactor by applying bioprocess technology, will grow and divide indefinitely, producing large quantities of the antibody, which can then be harvested. The resulting proteins are called monoclonal antibodies (MAb) and are most often used in diagnostic kits. The most famous MAb-containing diagnostic kit is the pregnancy test.

In agriculture, rDNA technology can be used to produce new varieties of crop plants with improved agricultural and nutritive qualities. Transgenic plants, which are resistant to biotic and abiotic stresses such as salinity, drought, and disease, have been produced.

Recombinant microorganisms, plant cells, and animal cells can be cultivated and used for the large-scale production of industrially-important enzymes and chemicals. Examples of such enzymes are protease, amylase, lipase, glucose isomerase, invertase, etc. Amylase is used in the starch industry. Glucose isomerase is used in fructose formation from glucose syrup. Proteases and lipases are incorporated into detergent products to take out stains. Protease is also used in the meat and leather industries to remove hair and soften meat and leather. A list of enzymes and their industrial uses is given in Table 1.1.

TABLE 1.1 Some major industrial enzymes and their sources and uses.

Enzymes	Sources	Uses
Amylases	*Aspergillus niger* A.oryzae, Bacillus licheniformis B.subtilis, germinating cereals germinating barley	Hydrolyze starch to glucose, detergents, baked goods, milk, cheese, fruit juice, digestive medicines, dental care
Invertases	Saccharomyces cerevisiae	Production of invert sugar, confectionery
Glucose isomerase	*Arthrobacterglobiformis* *Actomoplanes missouriensis* *Streptomycesolivaceus* and *e.coli*	conversion of glucose to fructose production of high fructose syrup, other beverages, and food
α D-Galactosidase	*Mortierella vinacease*	Raffinose hydrolysis
β D-Galactosidase	Aspergillus niger	Lactose hydrolysis
Papain	Papaya	Meat, beer, leather, textiles, pharmaceuticals, meat industry, digestive aid, dental hygiene, etc.
Proteases	*Bacillus subtilis* B.licheniformis	Detergents, meat tenderizers, beer, cheese, flavor production
Pepsin	Hog (pig) stomachs	Cereals, pharmaceuticals
Trypsin	Hog and calf pancreases	Meat, pharmaceuticals
11-β-Hydroxylase	*Curvularia lunata*	Steroid conversion, bioconversion of organic chemicals
Ficin	Figs	Leather, meat, pharmaceuticals
Bromelain	Pineapple	Meat, beer, pharmaceuticals

Since the manufacturing of human insulin using recombinant *e. coli* began in 1982, many other proteins (for human and veterinary therapeutics, vaccines, and diagnostics) have been manufactured. Today, there are a large number of human therapeutic proteins or vaccines made by modern biotechnology methods, approved by the government and marketed in the country. Besides more than 200 other human therapeutic and vaccine proteins are at clinical trial stage. Products are being tested to target diseases such as cancer, AIDS, heart disease, multiple sclerosis, Lyme disease, herpes, rheumatoid arthritis, and viral diseases. Products are also being developed to reduce bleeding from surgical procedures, aid in wound healing, and prevent organ-transplant rejection.

It is difficult to predict the future of this exciting new field of modern biotechnology. There is no doubt about its ability to improve the human life and the economy of the world. But along with the advancement in the research and development of life sciences and biotechnology, arise several social, environmental, and ethical problems. Several organizations are looking into various issues and addressing the general concerns.

1.4 COMMERCIAL POTENTIAL

Biotechnology is considered the commercialization of life sciences. It has a significant impact on various applied sciences, manufacturing processes, on medicine and health, and agriculture and environmental sciences. With monitoring and diagnostic systems it made giant strides in the field of health and medicine. Biotechnology plays an important role in monitoring the use of both traditional and non-conventional energy resources.

Commercial biotechnology products are already available and include, for instance, new diagnostics, recombinant vaccines and therapeutic proteins, and biochips or DNA chips. The biochips or the DNA microarrays currently being produced are revolutionizing the design and output of gene analysis in the field of molecular medicine. Bioremediation technologies for the elimination of toxic factory effluents with the help of genetically-altered microorganisms, purification of rivers, fresh water ecosystems, and drinking water are now carried out commercially.

In its economic potential, biotechnology runs parallel with the computer industry. The biotech industry is waiting to explode in the consumer market. Consumers are going to see scores of new biotech products, such as foods that contain vaccines or super-nutritious foods that will change the way people view agriculture.

In addition to the similarities in their economical potential, there is also a resemblance in the technical side also. There's a parallel between genetic code and computer code. Computers and living organisms both organize their essential information in a similar fashion. Computers are directed by a series of ones and zeros, known as the binary code. All living organisms use a code made up of four parts, a quaternary code. Instead of ones and zeros, the information is conveyed by a series of four chemicals—adenine, thymine, guanine, and cytosine—which geneticists simply call A, T, G, and C. Like computer code, the arrangement of these four chemicals strung together form genes, which contain the information that tells the cells whether you are to be a linebacker-sized human or a lemming!

Scientists first learned that they could manipulate these four chemicals to form new genes in the mid-1970s. The recombinant DNA technique was first developed in 1974, and today even high school kids can cut and stitch genes together. The development of this science has been mind-boggling and so has been the rise of biotech industries all over the world.

1.5 AN INTERDISCIPLINARY CHALLENGE

Even though the basic sciences—physics, chemistry and biology—seem to be independent of each other, they are really not. The research and development in a particular discipline is not at all possible without the involvement of other scientific disciplines. By the middle of the twentieth century there was tremendous growth in every scientific discipline because of the very close interplay of physical, chemical, and biological sciences. The close interaction of these sciences has created a large number of hybrid disciplines. This has proved that at the higher levels of study, science is interdisciplinary. Some of the new disciplines are listed below.

- Anatomy
- Biochemistry
- Bioorganic chemistry
- Biophysics
- Bio statistics
- Bioprocess technology
- Cell biology
- Chemical evolution
- Computational biology
- Developmental biology
- Ecology
- Eco physiology
- Embryology
- Ethno biology
- Ethno botany
- Ethno pharmacology
- Evolution
- Genetics
- Genomics
- Human genetics
- Immunology
- Inorganic chemistry
- Medicinal chemistry
- Microbiology
- Molecular biology
- Molecular evolution
- Molecular taxonomy
- Organic chemistry
- Pathology
- Pharmaco genomics
- Pharmacology
- Photobiology
- Photochemistry
- Physical chemistry
- Physiology
- Polymer chemistry
- Population genetics
- Proteomics
- Taxonomy
- Thermo chemistry
- Toxicology
- Virology

Modern biotechnology is really an interdisciplinary science, which takes the fundamental principles of biological sciences and integrates it with all the other sciences including mathematics, statistics, and engineering. Recently, out of its interaction with Information Technology, a new branch has emerged—Bioinformatics.

1.6 A QUANTITATIVE APPROACH

All lifeforms, from a virus to human, are very complex in their organization and workings, even though there is a gradient in complexity from one lifeform to another. Despite the structural complexity and the thousands of biochemical reactions involved, all these lifeforms obey the fundamental laws of physics and chemistry in their growth and development. They obey the laws of thermodynamics, the law of conservation of matter, the law of mass action, etc. During the growth of an organism it consumes substrate or food materials as a source of energy and matter. It will be metabolized in the body and will be incorporated into the cells and tissues or will be secreted as products. The energy of the substrate or food material will be used for the building up process of the body or for the production of the product, which is a byproduct of its metabolic activities, to maintain its existence.

If we focus mainly on the consumption of certain compounds and the products that are produced by organisms, we can see that the product formation is directly proportional to the substrate consumption at a particular set of physical, chemical and biological conditions. The following is an equation that represents the aerobic cell growth, in which the final extra cellular products formed are only CO_2 and H_2O.

$$\underset{\text{(Substrate)}}{C_wH_xO_yN_z} + aCO_2 + \underset{\text{(Nitrogen source)}}{bH_gO_hN_i} \longrightarrow \underset{\text{(Dry biomass)}}{cCH_\alpha O_\beta N_\delta} + dCO_2 + eH_2O$$

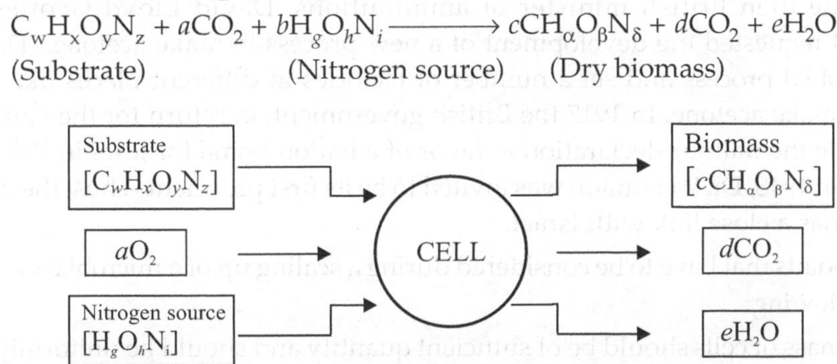

In the above equation, the substrate is $C_wH_xO_yN_z$, $bH_gO_hN_i$ represents the chemical formula of a nitrogen source and $cCH_\alpha O_\beta N_\delta$ is that of the dry biomass. a, b, c, d, and e are the coefficients representing the number of moles in each case. If the substrate is glucose, $w = 6$, $x = 12$, $y = 6$, and $z = 0$. In this equation it is shown that a mole of oxygen was consumed and a mole of carbon dioxide was released per substrate.

Now take the case of glucose utilization in the alcohol fermentation by yeast cells. The equation goes like this.

$$C_6H_{12}O_6 \xrightarrow[\text{(Yeast)}]{} 2CH_3\,CH_2\,OH + 2CO_2$$

This equation represents the anaerobic mode of life. During the process of alcohol fermentation by yeast in the absence of oxygen, one mole glucose is converted into two moles of ethyl alcohol and two moles of carbon dioxide. First, these experiments are conducted in laboratories in small volumes. A laboratory-scale process can be standardized with respect to the molar concentrations of the substrate utilized and the products formed. Other culture parameters such as media composition, pH, and temperature also have to be optimized. Once the experiment conditions are optimized for a small-scale volume like 500 ml or 1 liter cultures, it has to be scaled up to the industrial level. Conversion of a laboratory-scale process into a large-scale process suitable for an industry is not a simple thing. A large number of unforeseen problems can arise during this scaling up of process and these have to be addressed satisfactorily. The development of an industrial process from the laboratory-scale experiment by applying the principles of biochemical engineering is commonly known as **scaling up** or process development.

Process Development or Scale Up

The scaling up of a laboratory-scale experiment or process into an industrial production process mainly depends on the amount of the product that the industry wants to produce. Depending on the nature of the microbial or biochemical process a defined set of procedures has to be applied to convert a small-scale experiment to the level of an industry. The first large-scale fermentation process was developed during the time of First World War to meet the high demand of alcohol, acetone, glycerine, and butanol based chemicals. Acetone was required for the manufacturing of explosives such as cordite. The traditional method of acetone production by distilling wood could not meet the requirement. During that time (1912), Dr. Chaim Weizman discovered the bacterium *Clostrudium acetobutylicum*. The then British minister of ammunitions, David Lloyd George, contacted Dr. Weizmann and requested the development of a new process to make acetone. Dr. Weizmann developed a microbial process and set a number of factories in different places like Canada, the U.S., and India to make acetone. In 1917 the British government, in return for the valuable service of Weizmann, made the Balfour declaration in favor of a nation home for Jews in Palestine. When Israel was formed in 1948, Dr. Weizmann was invited to be its first president. Thus, the microbiology and fermentation has a close link with Israel.

The major points that have to be considered during a scaling up of a microbial or fermentation process are the following:

- The biomass of cells should be of sufficient quantity and should be uniformly distributed in the culture medium so that the cells will be able to multiply and grow freely.
- The substrate molecules present in the culture medium have to be in contact with the cell surface.
- The substrate has to be transported to the site of action within the cell.
- The concentration of the substrate molecules within the cell should be sufficient enough to get the maximum efficiency of the process. The transport of substrate and product molecules across the cell membrane is known as mass transfer.
- The byproducts formed during the process of reactions have to be removed from the site of action and should be transported out of cells.
- There should not be any localized holding up of products or localized accumulation of products within the reacting system. This can lead to the inhibition of cell growth and product formation.
- If aeration is a factor in the growth or product formation, there should be sufficient availability of oxygen in the culture medium. In many microbial systems, mass transfer and diffusion of oxygen to the microorganisms are important rate-limiting factors.

- The quality and efficiency of the process depends on the sterility maintained in the reacting system. Contamination by unwanted microorganisms can reduce the quality of the product as well as the efficiency of the process.
- The chemical and physical environment of the culture with respect to the pH and temperature has to be optimal.

All these parameters can be maintained in specially designed culture vessels called *fermentors or bioreactors,* which are used extensively in microbial process studies.

The Fermentor

Fermentor, in general terms, is something that, as its name suggests, ferments. The process of fermentation has been known for thousands of years, but has been mainly used to convert the glucose found in various fruits, seeds, and tubers into alcohol, later used for human consumption. In recent times, however, with increased knowledge of bacteria and fungi, fermentors have been put to a more productive use.

More scientific and recent uses of fermentors in biotechnology include growing large quantities of genetically-engineered organisms such as bacteria and yeast, and plant and animal cells. These bacteria, having had the genes that code for various proteins (human insulin, for example) spliced into them, will grow and reproduce and will express the inserted gene. This will result in the desired protein being released into the growth medium, where it can be harvested, purified, and then sold and used.

The general idea behind the fermentor is to provide a stable and optimal environment for microorganisms in which they can reproduce and do whatever they want. It is a specialized container where all the culture conditions that are optimized for the growth and product synthesis can be maintained continuously throughout the process of fermentation under sterile conditions. Fermentors are specially designed to suit industrial-scale fermentation for the production of antibiotics, hormones, vaccines, enzymes, and specialty chemicals. But now fermentation experiments can be conducted in laboratories using laboratory fermentors, which are also available in a range of volumes. Figure 1.2 shows an ordinary laboratory fermentor with a microprocess controller. The main components and features of an ordinary laboratory fermentor are explained below:

1. **Type of vessel:** A closed vessel made up of glass or steel, attached with an air inlet that has a filter to maintain sterility inside.
2. **Agitator:** An agitator or metallic blade to stir the medium for properly mixing the medium and cells to improve the material and oxygen transfer between the cells and the medium.
3. **Baffles:** This is a rectangular strip of metal attached to the vessel wall. It can improve the oxygen transfer by increasing the turbulence of the culture medium.
4. **Sensors:** The main sensors in the fermentor are the pH sensor, temperature sensor, and antifoam sensor.
   (*a*) pH regulator—When the cells grow and multiply by metabolizing the media constituents, the pH will change continuously. It can change the physiological state of the cells and thereby the efficiency of the product formation. The sensor constantly

monitors the pH change of the media; its specific pH will be maintained constant by adding acid or alkali.

(*b*) Temperature regulator—Microbial culture under active metabolic conditions will produce heat and that will increase the temperature of the medium beyond its optimum temperature. To counter this there is a heat exchange coil inside the vessel or a jacket around the vessel through which hot water or cold water circulates to maintain the temperature.

(*c*) Antifoam monitor—Agitation and aeration of the culture medium can result in excessive foaming. Foaming will increase at high cell density. There will be a sensor placed just above the culture medium, which can detect the formation of foam when it touches the sensor. When the foam touches the sensor, the electrical signals activate the pump attached to supply the antifoam agent into the medium.

5. **Addition Ports:** This is the inlet for adding culture medium and inoculum (microbial culture) required for the fermentation process.

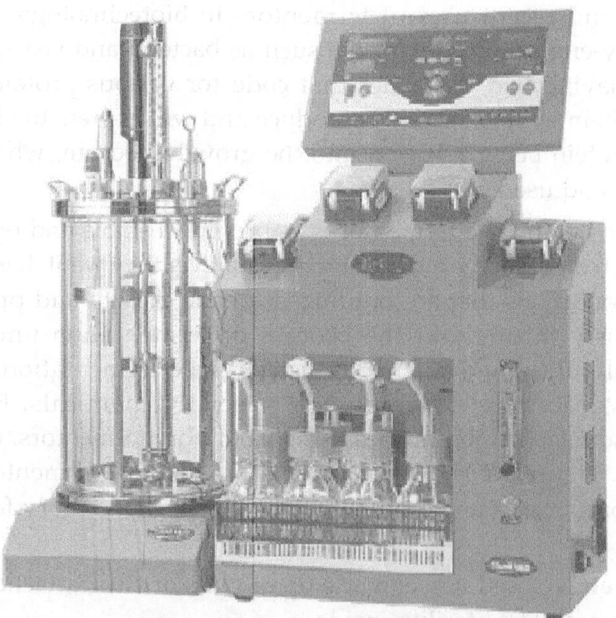

FIGURE 1.2 **A laboratory fermentor (BioG-Micom) is a precise and unique fermentation system with a microprocessor controller and display system. This system is designed for various types of fermentation processes.**

6. **Outlet port:** There may be a separate outlet for removing samples intermittently during the process and also at the end after the completion of the fermentation.
7. **Sterilization:** The entire process of fermentation depends on the sterility of the culture medium and other equipments involved in the process. Therefore the fermentor vessel and its accessories that come into contact with the culture media should be sterilized. Usually the fermentor vessel is sterilized by circulating boiled water followed by filling

it with steam for a specific period of time. For carrying out this process there is a steam inlet and outlet. Fermentor vessels of small volumes can also be sterilized by autoclaving.

Aseptic Operation of a Fermentor

A fermentor is a small bioreactor specially designed to cultivate microorganisms in aseptic conditions. There should not be any contamination by other microorganisms. This is very essential for the successful completion of fermentation. The entire fermentor and all other accessories of the fermentor and solutions such as growth medium used in the experiment are to be sterile. The air used in the fermentation process should be free of microbes. In short, the entire fermentation process has to be conducted under aseptic conditions.

The fermentor vessel in which the cells are growing has to be washed with hot water and then sterilized by conducting steam through it. Care should be taken to see that steam is reaching all parts of the fermentor assembly. If it is a small volume fermentor, it can be separated from the assembly and can be autoclaved separately. The media is normally steam sterilized separately in autoclave, but also can be sterilized in the fermentor vessel itself by passing steam through the jacket or cooling coil around the vessel. In addition to the media components, the other additives such as antifoam agents should also be sterilized. But there is no need to sterilize the acids or alkalis used to maintain the pH, if it is strong. The air used in the fermentor should be filtered using a bacterial filter or a better filter. By using ordinary bacterial filters it is possible to make the air free of bacteria and fungal spores. But it is not possible to avoid the presence of bacteriophages in the incoming air and this can cause serious damage to the culture. By choosing the right type of air filters and compressors and by regulating the speed of airflow, the contamination of cultures in the fermentor can be minimized. For example, some types of compressors can generate sufficient heat to kill the bacteriophages present in the air. In certain special cases, if cultivating a pathogenic microorganism or a genetically-engineered organism, the air leaving the fermentor also needs to be sterilized as a safety measure. This will also help in protecting a new strain of microorganism from becoming freely available to others.

1.7 CLASSICAL vs MODERN CONCEPTS

Though there are no differences in the principles, the technological advancement of utilizing living cells for the benefit of man differentiates between classical and modern biotechnology. The classical biotechnology that emerged during the early twentieth century was basically a microbial-based fermentation process in which the principles of biochemical engineering have been applied to change it into an industrial process. In short, it is a hybrid of fermentation and biochemical engineering. Modern biotechnology is based on the ability of recombinant DNA and Hybridoma technology to genetically alter the cells and organisms—microbes, plants, and animals—and to use them for different purposes.

Another important aspect of modern biotechnology is the ownership of the bioprocess technology, transgenic organisms, and the socio-political, ethical and economical consequences that accompany these experiments.

In plant biotechnology, the transgenic crop plants are used in agriculture to enhance productivity. With classical plant breeding techniques it is possible to transfer a trait from one plant to another plant. But there are limitations. It is highly non-specific in the sense that the hybrid plant produced may not have the desired trait. Gene transfer through hybridization is only possible between related plants. But genetic engineering made it possible to transfer a specific gene to a plant from any organism. A major use of transgenic plants is their application in non-agricultural sectors. For example, bioremediation of toxic wastes having heavy metals such as arsenic, mercury, etc. The genetically-altered organisms, both plants and animals, may also be used as bioreactors for the production of vaccines, hormones, and other therapeutic proteins. Recombinant DNA technology or genetic engineering, in addition to these, is a powerful tool in molecular biology. Along with genetic engineering, hybridoma technology has also made the modern biotechnology powerful in medical and pharmaceutical industries. Isolation and characterization of new genes, proteins, and genome sequencing has led to the formation of a huge volume of computerized data and a lot of associated problems, giving birth to the new science of bioinformatics and computational biology.

The completion of a "working draft" of the human genome—an important landmark in the history of biotechnology—was announced in June 2000 at a press conference at the White House and was published in the February 15, 2001 issue of the journal *Nature*. According to the article, there are only 32,000 genes, which is just about 2 to 3 % of the total genome sequence. The remaining 97% of the sequence does not contain code for any gene. It is not even known whether these sequences have any function. But with these 32,000 genes several things can be done by geneticists, molecular biologists, and pharmacologists. For example, with all these 32,000 genes on a microchip, it is possible to do a large throughput screening of new drugs for a specific group of the population and develop a population-specific or a tribe-specific designer drug. A newly developed drug to treat breast cancer is good for the treatment of those patients in whom one specific gene, Her-2 Neu, is over-expressed. The development of gene chips, (*i.e.*, complete genome or specific DNA sequences immobilized on a microscopic silica, glass, or nylon chip) facilitates rapid screening of genomes and proteomes for various purposes such as drug design and development and toxicological and pharmacological trials of drugs. Similarly, there are protein chips where protein molecules are immobilized on microscopic chips. The emergence of these protein and DNA chips has changed the proteomics and genomic researches and new fields such as pharmaco genomics and toxico genomics have emerged.

The excitement and optimism about biotechnology has encouraged both public and private sectors to make huge investments in research and technology developments. Biotech industries, started as university supported private enterprises, have undergone many changes. Many small industries have closed or merged with bigger pharmaceutical or chemical companies. But the market response to biotechnology products was not encouraging in the beginning. The main reason was the regulatory policies of the concerned governments and the negative approach of the public toward genetically-engineered products.

Several other factors also hindered the growth of biotechnology industries. In the agricultural sector, biotech products, particularly the edible materials, have to compete with the products from the conventional sources that people prefer. But in the case of life-saving medicines—vaccines, alcohol, pesticides, weedcides etc.—the products of conventional methods have been taken over by the biotech products.

The biotech industries also have the potential to replace several polluting industries, using environment-friendly manufacturing processes. The vaccines produced by genetic engineering are supposed to be safer than the vaccines made from conventional methods. The vaccine produced conventionally carries the inactive virus, which may become active at any time. However, the optimism about biotechnology products has to be supported by hard scientific evidences. Only hard scientific proofs can boost the morale of the public and industry. For example, there is no evidence yet to show that the *r*DNA product (a protein) has the correct folding of polypeptide chain to form the three-dimensional structure by the post-translational modifications as it happens in the actual system. Today, most of the drugs that are coming to the market or those in the process of development are from modern biotechnology. This is because the pharma industries are mainly focused on discovering new biotechnology-based drugs.

1.8 QUALITY CONTROL IN MANUFACTURING

The ultimate aim of any good manufacturing process is to bring a product of superior quality to the market. Therefore, quality control is of great importance in any manufacturing process.

A product has to undergo stringent quality control tests and procedures. **Quality control (QC)** ensures that a product is not released for use until its quality has been judged satisfactorily. Quality control is concerned with sampling, specification, and testing as well as organization, documentation, and release procedures. Each manufacturing unit has a separate quality control unit, which operates independently to ensure this. Quality control has to take care of all parameters, which directly or indirectly affect the quality of the product.

The parameters of quality control depend on the type and the final use of the product. For example, if the product is a therapeutic protein like a vaccine or hormone, it should be biologically active, pure, and chemically stable. The product has to be examined for its:

- Biological activity,
- Purity, chemical, and physical properties, and finally its
- Shelf life and stability.

The following are some of the major points to be taken into consideration to successfully incorporate quality control in a manufacturing unit.

- The procedure of the manufacturing process should be standardized and should be approved by a research laboratory.
- Enough trained manpower to carry out manufacturing, sampling, inspection, and testing of the materials at the different stages of the process—starting materials, intermediates, and final products.
- Proper packing with correct labeling, which includes some information about property and handling procedures.
- Samples should be taken in an approved manner to test its qualities.
- Records of sampling, inspection, and test procedures carried out should be maintained properly and should be made available for inspection.
- The final product should contain the active ingredients that comply with the qualitative and quantitative composition of the stated formulation.
- The products should be certified by a qualified person or by an approved committee before released for sale.

- Sufficient quantity of the starting materials, intermediates, and final products should be retained for future examinations if required.
- The packaging materials should also be of the right type, which would guard the product from any degradation or change. The impact of physical parameters on the property of the product should be recorded on the label.

1.9 PRODUCT SAFETY

Based on international guidelines, governments in each country frame rules to ensure the safety not only of the product but also of those who conduct the genetic engineering experiments. When the first rDNA experiments were carried out, scientists themselves imposed a moratorium on further genetic experiments until proper guidelines were structured. In 1975, the Asilomar Conference on Recombinant DNA formulated some guidelines and the major points are the following:

- Recombinant DNA experiments should be carried out in a laminar flow-chamber kept in a clean room so that the recombinant organisms do not escape.
- The host organism used to carry out the gene cloning or the cloning of rDNA should be specially made for such experiments. Even if the organisms escape the laboratory they will not survive and the recombinant DNA molecule will be lost. This led to the development of special safe vectors that can be used for rDNA experiment without fear. There are also non-pathogenic *e.coli* developed for use in the rDNA experiments as host cells.

In modern biotechnology, the product in most cases will be therapeutic proteins like vaccines, hormones, and enzymes. The proteins that are to be used as a drug should meet all the safety requirements and regulations imposed by the regulatory authorities. They are very strict about the data regarding the experiments on animal models, terminal patients, non-terminal patients, and independent clinical trials by separate agencies. If a transgenic organism is released into the environment, in addition to the ordinary safety norms, its environmental impact should also be monitored. The clinical trials of drugs produced by rDNA method have to be carried out in three phases at least.

Phase I: In the first phase of study, the side effects and dose of tolerance by the patients are systematically carried out on selected patients.

Phase II: In the second phase, collecting data from pharmacological, pharmacokinetic, metabolic, and toxicological studies on a selected number of patients optimizes the use of the drug.

Phase III: In the last phase, the studies focus on the safety aspects of the drugs. During these studies the harmful effects of the drug, if any, are monitored in addition to its effectiveness as a drug. In dosage range, the interactions with other drugs are also investigated.

A new biotechnology product, whether it is a drug or a food material, has to undergo trial studies, and only after the regulatory agencies are fully satisfied by the data, they give permission or license for its mass production and marketing.

1.10 GOOD MANUFACTURING PRACTICES (GMP)

All industries must have **good manufacturing practices (GMP).** Quality control actually forms a part of that. Since most of the products of biotechnology are directly used for human consumption, it is very important to realize these two aspects of manufacturing very stringently. A good manufacturing practice follows prescribed guidelines including the procedure for the manufacturing of a particular product intended for a specific use. It should follow a well-defined manufacturing process and should be provided with all the necessary facilities. This includes the space, suitable building and premises, equipments and instruments, in addition to the facilities for routine supply of the raw materials, storage, and transport of the products. Trained manpower is an essential requirement of GMP to carry out all the processes very promptly and accurately and to keep the records of the manufacturing processes. Finally, GMP will reflect in the quality of the product. The use of the correct type of raw materials, its quality and quantity, etc., determine the quality of the product. Therefore, it is very important to monitor the manufacturing process in all its different steps to ensure the superior quality of the product.

The following are some important points to be observed to achieve good manufacturing practices:

- The manufacturing unit should be housed in a compound that is well protected. It should be protected from insects, animals, and migration of extraneous materials from outside the building or from one region of the building to another.
- The building should be very strong to accommodate the machinery required for the manufacturing. It should meet all the required criteria with enough space, ventilation, and provisions to meet any urgent situation such as fire.
- The building should be constructed in such a way that it guards the machinery, raw materials, and products from contamination.
- Temperature and humidity should be controlled to protect both raw materials and products.
- There should be good sanitary practices. The toilet, rest room, dress-changing room, etc., should be separated from the manufacturing areas.
- The production area should be well maintained in terms of sterility and hygienity. The flow of personnel in the production area should be controlled to minimize contamination.
- The sophisticated instruments and analytical equipment should be well maintained and protected from vibrations, electrical impacts, moisture, heat, etc. If possible, separate rooms should be provided in such situations.
- Utilities and support systems like sterile water, compressed gas, nitrogen and liquid nitrogen, etc., should be of standard quality and should be supplied promptly.
- Storage and distribution of the product should be well organized. A well-maintained storage and packing facility is essential to prevent contamination and mixing of products.
- To carry out all these, the factory and its premises should be well maintained in a very good working state.

1.11 GOOD LABORATORY PRACTICES (GLP)

To carry out research projects and experiments including trial experiments, some basic guidelines and criteria are to be followed by laboratories. This is also applicable to manufacturing establishments. These guidelines are necessary to ensure the safety of the environment, the public, and the researcher. Moreover it is a must for the success of the experiments.

There are four components to **Good Laboratory Practices (GLP):**

1. **Management:** The administrative authority that coordinates the implementing and monitoring activity of different committees and organizations to monitor safety and ethical issues.
2. **Quality Assurance:** This is an internal body or committee that ensures the practice of GLP within the laboratory. Animal ethics committee, Bio ethics committee, and Bio safety committee are some examples.
3. **Study Director:** The person who is responsible for the overall conduct of the safety and study committees.
4. **National Compliance Monitoring Authority:** The approval of safety study related to health and environment conducted on a new biotechnology product by this authority is necessary for international recognition and mutual acceptance. A guideline or framework is provided by the Organization of Economic Cooperation and Development (OECD) to the National Compliance Monitoring Institution regarding GLP for product development and marketing.

The GLP related to the product development and its health and environment safety are compulsory for testing any product before it goes into the market for human consumption. The study and its data will be assessed by a national regulatory authority for the purpose of giving license or registration. These regulations in the trial study can prevent scientific misconduct and fraud, which can affect the safety of man and environment.

Good Laboratory Practice for Students

A certain code of conduct is to be followed in a laboratory to ensure maximum safety. Every person has a responsibility toward the health and safety of himself/herself and of all other persons who may be affected. As all laboratories are potentially dangerous, it is essential that students follow the safety instructions strictly. Eating and drinking in laboratory areas should be prohibited at all times. Students are advised to wear laboratory coats whenever they are in the laboratory.

Use fume cupboards and personal protective equipments (gloves, goggles, etc.) when advised to do so by a teacher or supervisor, if the work involves the use of hazardous materials like radioactive chemicals, pathogenic organisms, or UV radiations.

If using unfamiliar equipment, try to understand how it works with the help of the instruction manual or seek the help of a supervisor or a technician. Look out for instructions or warnings labeled on the equipment. Students should plan their experiments in such a way that it can be carried out during normal working hours. Avoid working alone.

Keep work within the bench area allotted. Do not leave equipment such as pH meters, laminar flow cabinets, fume cupboards, etc., in an untidy state. Clean all equipment, including balances, after use.

Always avoid mouth pipetting; use filling devices (rubber bulbs, Pi-pumps) or transfer pipettes. Rinse empty glassware as soon as possible after use and move to the designated location for washing. Pay particular attention to the removal of corrosive or toxic chemicals, plant material, microbial cultures, soil, agar, etc., and their proper disposal. Make sure all chemicals and media are clearly and unambiguously labeled.

Clean up spillages immediately using the appropriate procedure. For example, if it is a liquid, mop with cloth or paper towel. Do not try to suck them up using pipettes, etc. In the case of strong acids or alkalis, flush with water, neutralize, and then mop it up. In the case of solid chemicals, wipe with damp cloth or paper towel. Never blow or brush the dusty spills or other solid chemicals. This can only increase the risk from hazardous materials.

Do not put paper towels used for cleaning up chemical spillages into waste paper bins where cleaning staff may come into contact with the harmful substances. Use black waste sacks for disposing such waste material.

Students will need to keep a lab notebook in order to write down procedures, data, results, observations, etc. Keep notes organized because you will want to use them later for reference.

1.12 MARKETING

Marketing high-technology products and innovations like that of biotechnology is not the same as marketing more traditional products and services. The marketing of a familiar product is very different from marketing products with which customers are unfamiliar. The marketing of biotechnology products is also dependent on the normal criteria for marketing. Introducing a new product, particularly an unconventional product like that of modern biotechnology, is very difficult. Marketing is the process of planning and executing the conception, pricing, promotion, and distribution of ideas, goods, and services to create exchanges that satisfy individual and organizational objectives.

In 1995, sales of products derived from recombinant DNA-based technologies approached $8 billion, less than 4% of worldwide pharmaceutical sales. Modern biotechnology products started appearing in the market only in the late 1980s. Today, many products of modern biotechnology are available in the national as well as international markets.

The industry has its own strategies for getting into and competing in the market. New technologies and their demonstration among the consumers is vital to the future marketing success of protein-based drugs and other biotechnology products. The reasons include the following:

- Proteins are erratically absorbed or are inactivated when swallowed, thus they tend to be injected.
- Proteins are generally unstable and reactive.
- Most proteins have a short life in plasma after injection.

Some of the technologies that could appear in future markets include insulin that could be delivered by nasal sprays, or orally, self-regulated implants, and artificial pancreases with monoclonal insulin-producing cells. Protein drug doses could be administered in very small, controlled delivery systems over days, months, or even years on a single re-filling. In the case of drug targeting, the drug goes only to infected cells and not the systemic bloodstream. The use of liposomes as drug containers or vehicles to carry toxic solutions, labeling liposomes with antibodies or surface reactive chemicals will lead to better targeting and tissue-selective delivery systems. New technologies in the production of drugs by biotechnology can reduce the cost of producing drugs and other products, and then it will become more consumer friendly. Table 1.2 gives some of the therapeutic proteins and vaccines already on the market.

TABLE 1.2 New biotechnology products added to the global market or under development.

Serial Number	Name of the drug	On market	Under trial /development
1.	Vaccines	8	68
2.	Monoclonal antibodies	16	70
3.	Gene therapy	0	44
4.	Growth factor	4	26
5.	Interferons	7	21
6.	Interleukins	1	18
7.	Recombinant human proteins	12	8
8.	Human growth hormones	16	14
9.	Clotting factors	5	6
10.	Colony-stimulating factors	2	4
11.	Erythro Proteins	2	4
12.	Others	20	48

(Source: A combined survey conducted by Biotechnology Industry Organization and Pharmaceutical and Manufacturers of America).

Global Scenario

The United States is the forerunner in biotechnology research and development. Ever since the start of biotechnology research, the U.S. has promoted biotechnology enterprises by shaping the policies that promote biotechnology innovations as far as possible. The federal government has supported basic research in National Institute of Health, the prime institute of biotechnology research in the U.S., and other scientific agencies, by funding, accelerating the administrative processes for approving new medicines, encouraging private-sector research investment and small business development through tax incentives and promoting intellectual property protection and open international markets for biotechnology inventions and products. The U.S. has developed public databases and internet facilities that allow the scientists to coordinate their efforts and research findings for easy and systematic studies. In addition, the government also improves science education and promotes the freedom of scientific enquiries. The administration has provided guidelines to protect patients from misuse or abuse of new drugs and sensitive medical information and has also provided the federal regulatory agencies with sufficient resources to maintain sound, science-based review and regulations of biotechnology products.

REVIEW QUESTIONS

1. Define biotechnologies.
2. Compare modern biotechnology with classical biotechnology.
3. What is your justification in considering biotechnology an old technology?
4. Describe the contribution of Louis Pasteur toward development of classical biotechnology.
5. What are the different types of fermentation?
6. What is lactic acid bacillus? Explain its role in the formation of curd.
7. Name some of the fermented food products of India.
8. What is the percentage human genome coding for gene?
9. Name four recombinant DNA products available on the market.
10. Describe Good Laboratory Practices for industry and research laboratories.

Chapter 2

FUNDAMENTALS OF BIOCHEMICAL ENGINEERING

In This Chapter

2.1 INTRODUCTION

We are all aware of opportunities created by advances in molecular biology. Living cells and their components can be used to produce a large number of useful compounds such as therapeutics and other products. But to obtain significant benefits as a commercial operation, molecular biology needs the support of biochemical engineering. The vital area of biotechnology that is concerned with practical application of biological agents (whole cell systems and biocatalysts) and the methodologies and processes associated with it on an industrial scale is biochemical engineering. Biochemical engineering is applicable in different areas of biotechnology such as biochemical reactions, enzyme technology, environmental biotechnology, microbial manipulations, bioseparation technology, plant and animal cell cultures, and food technology. It consists of the development of new process technology, designing bioreactors, developing efficient, and economically feasible extraction and purification procedures (downstream processing).

A detailed understanding of the behavior of biological agents—living cells or its components—and an expertise in the principles of biochemical engineering is a prerequisite for developing a bioprocess technology to beneficially exploit the potential of the biological system. Thus, biochemical engineering has an important role in transforming a biological laboratory experiment into an economically-feasible industrial process.

It is very difficult to handle living cells in bioprocess development because the cells are very fragile and require a very specific type of chemical and physical environment, including nutritional requirements, pH, and temperature. The product formed in the cells may not be secreted into the medium. In such cases, the cells have to be ruptured to extract the products. In most cases, the product may be combined with other similar cellular products from which the actual product has to be separated and purified economically.

The basic problems associated with the development of a cost-effective bioprocess technology include:

- The low concentration of substrate in the culture media.
- The reaction must be conducted in optimal temperature and pH that facilitates both culture growth and maximum product output.
- The pressure build up in the bioreactor due to the generation of gas by the growing cells.

When a laboratory process is converted into an industrial-scale process a large number of similar problems can arise. These problems have to be addressed properly for the success of the technology.

The main areas of studies in biochemical engineering include:

- Operating considerations for bioreactors for different types of cell cultures and suspension and immobilized cell cultures of plant, animal, microbes, and genetically-engineered organisms.
- Selection, scale up, operation, and control of bioreactors (fermentors).
- Methods for recovery and purification of products.
- Medical applications of bioprocess engineering.

- Mixed cultures.
- Stoichiometry of microbial growth and product formation.
- Metabolic pathways and pathway engineering.

Applications of biochemical engineering include novel technologies for cheaper and cleaner alternatives for energy and chemical feedstocks. New biochemical engineering techniques are also available for cost-effective pollution control.

2.2 CONCEPT OF pH

The biochemical reactions that are necessary for life are profoundly influenced by the pH of cellular solutions. This part of the chapter discusses the chemical nature of pH, its measurement and its biological importance. Biomolecules such as proteins, both their structure and function, are deeply influenced by the chemical environment, the acidic or basic nature of the medium. The various biochemical reactions in the living system are also controlled by the state of pH in the biological medium.

An understanding of acid-base chemistry and dissociation of water is essential to understanding pH and buffer.

General Characteristics of Acids and Bases

Acids are generally a class of substances that taste sour (but do not use this method to identify a compound), such as vinegar, which is a dilute solution of acetic acid. Bases, or alkaline substances, are characterized by their bitter taste and slippery feel. The first precise definition of an acid and base was given by Svante Arrhenius and is referred to as the Arrhenius Theory.

Arrhenius Theory

In the 1830s, it was known that all acids contain hydrogen, but not all hydrogen-containing compounds are acids. In 1889, Swedish chemist Svante Arrhenius, connected acidic properties with the presence of hydrogen (H^+) ions and basic properties with presence of hydroxyl (OH^-) ions. Arrhenius defined an acid as a substance that ionizes in water to give hydrogen ions, and a base as a substance that ionizes in water to give hydroxide ions.

If a solution contains more H^+ than OH^- ions, then it is acidic.

If a solution contains more OH^- than H^+ ions, then it is basic.

If H^+ and OH^- ions are equal, then it is a neutral solution.

Hydrochloric acid (HCl) is a strong acid and is very soluble in water. It dissociates into its component ions in the following manner:

$$HCl\ (g) \longrightarrow H^+\ (aq) + Cl^-\ (aq)$$

The hydrogen ion interacts strongly with a lone pair of electrons on the oxygen of a water molecule. The resulting ion, H_3O^+, is called the hydronium ion.

$$H^+ + \underset{\displaystyle H}{\overset{}{O}}\!-\!H \longrightarrow \left[H\!-\!\underset{\displaystyle H}{O}\!-\!H \right]^+$$

Acidic solutions are formed when an acid transfers a proton to water. The reaction of HCl with water can be written in either of the following ways:

$$HCl\ (aq) + H_2O\ (l) \rightleftharpoons H_3O^+\ (aq) + Cl^-\ (aq)$$

$$HCl\ (aq) \rightleftharpoons H^+\ (aq) + Cl^-\ (aq)$$

The Bronsted-Lowry Concept of Acids and Bases

According to Bronsted and Lowry, "acids are substances that are capable of donating a proton, and bases are substances capable of accepting a proton." So, in the example above, HCl acts as a **Bronsted acid** by donating a proton in water, and water in turn acts as a **Bronsted base** by accepting a proton from HCl. Thus, water can act as an acid or a base.

Here is another example:

$$NH_3\ (aq) + H_2O(l) \rightleftharpoons NH_4^+(aq) + OH^-(aq)$$

Here, H_2O acts as a Bronsted acid by donating a proton to NH_3, which acts as a Bronsted base. Using the Arrhenius definition, we say that the resulting solution is basic because it contains OH^- ions, thus, we say that the NH_3 molecule is basic (a proton acceptor).

> **All Arrhenius acids are also Bronsted acids.**
> **All Arrhenius bases are also Bronsted bases.**

Conjugate Acid-Base Pairs

Let's look at the reaction of NH_3 and H_2O again:

$$(1)\ NH_3 + H_2O \rightleftharpoons NH_4^+ + OH^-$$

The reverse of this reaction is:

$$(2)\ NH_4 + OH^- \rightleftharpoons NH_3 + H_2O$$

In this case, NH_4^+ acts as an acid, which donates a proton to OH^-. OH^- acts as a base. An acid and a base that are related by the gain and loss of a proton are called a **conjugate acid-base pair**. For example, NH_4^+ is the **conjugate acid** of NH_3, and NH_3 is the **conjugate base** of NH_4^+.

> **Every acid has a conjugate base associated with it.**
> **Likewise, every base has a conjugate acid associated with it.**

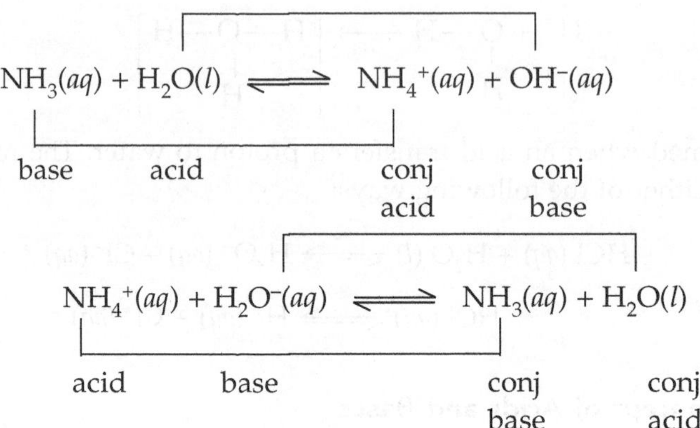

and

For any reaction:

$$HA + H_2O \rightleftharpoons H_3O^+ + A^-$$

If HA is a strong acid because it gives up its proton readily, then A^- is a weak base because it has little affinity for the proton.

If HA is a weak acid because it donates very few protons to the water, then A^- has a high affinity for a proton, and A^- is a stronger base than water.

The Dissociation of Water

Pure water is a poor conductor of electricity because it does not ionize to a great extent. So essentially it is a non-electrolyte. It consists almost entirely of H_2O molecule. A very small percentage of H_2O molecules undergo ionization and it can act as both a proton donor and acceptor for itself. A proton can be transferred from one water molecule to another, resulting in the formation of one hydroxide ion (OH^-) and one hydronium ion (H_3O^+).

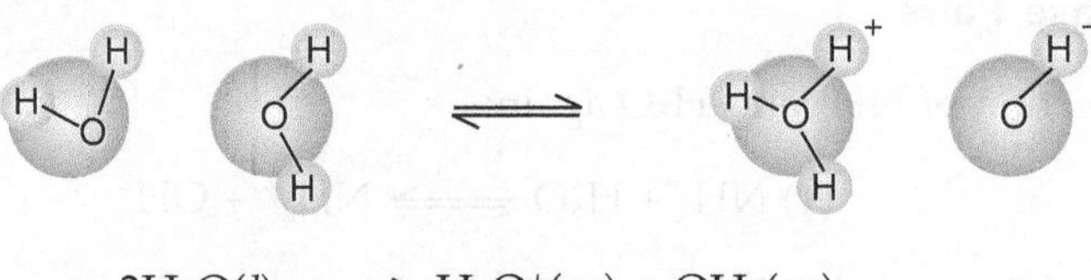

$$2H_2O(l) \rightleftharpoons H_3O^+(aq) + OH^-(aq)$$

This is called the **auto-ionization** or **dissociation** of water. This equilibrium can also be expressed as:

$$H_2O(l) \rightleftharpoons H^+(aq) + OH^-(aq)$$

In the above equilibrium, water acts as both an acid and a base. The ability of a species to act as either an acid or a base is known as **amphoterism.**

The concentrations of H_3O^+ and OH^- produced by the dissociation of water is equal. The corresponding equilibrium expression for the auto-ionization reaction is:

$$K = \frac{[H^+][OH^-]}{[H_2O]}$$

In pure water the concentration of H_2O at 25°C is 55. 6 M

$$[H_2O] = 55.6 \text{ M}$$

This value is relatively constant in relation to the very low concentration of H^+ and OH^-.

Molarity is one method of expressing concentration. It is the number of moles in one liter of the solution.

$$\textbf{Moles} = \frac{\textbf{Weight in grams}}{\textbf{Molecular weight}}$$

$$\textbf{Therefore, molarity of water} = \frac{1000}{18} = 55.6 \text{ M}$$

Therefore,

$$K = \frac{[H^+][OH^-]}{55.6 \text{ M}}$$

Rearranging gives:

$$K\ (55.6 \text{ M}) = [H^+][OH^-]$$

Now substitute a new constant K_W for K (55.6 M)

So now the equation becomes:

$$K_W = [H^+][OH^-]$$

where K_W designates the product **(55.6 M)K** and is called the **ion-product constant** for water. At 25°C, K_W is equal to 10^{-14}.

$$K_W = [H^+][OH^-] = 1 \times 10^{-14}$$

Therefore,

$$[H^+] = \frac{1 \times 10^{-14}}{[OH^-]}$$

$$[OH^-] = \frac{1 \times 10^{-14}}{[H^+]}$$

The ion-product constant *always* remains constant at equilibrium (as the name implies). Consequently, if the concentration of either H^+ or OH^- rises, then the other must fall to compensate. In acidic solutions, $[H^+] > [OH^-]$, and in basic solutions $[H^+] < [OH^-]$. A solution for which $[H^+] = [OH^-]$ is said to be neutral.

In pure water:

$$[H^+] = [OH^-] = 1 \times 10^{-7}$$

pH Scale

The **pH scale** is used as a measure of **acidity** or the concentration of H^+ ions in aqueous medium. The symbol pH is an abbreviation for *"pondus hydrogenil"*, of French (translated as potential hydrogen), meaning "hydrogen power." It is defined as the **negative logarithms** (10^{-n}) of the hydronium ions (H_3O^+) or hydrogen ions (H^+) concentration expressed in moles per liter of water. Dr. Sorensen has been credited as the founder of the modern pH concept.

$$pH = -\log([H_3O^+]) \quad \text{or} \quad ([H^+])$$

- **To obtain a mole of any substance, simply measure out its molecular weight in grams.**
- **One gram of hydrogen atoms is a mole.**

  44 grams of CO_2 is a mole (the atomic weight of C = 12, that of O = 16, and 12 + 2(16) = 44).
- **A mole of any substance contains the same number of particles, Avogadro's number, 6.02×10^{23}.**
- **The liter is a particular volume of water.**

  Thus, a pH of 0 means a hydrogen ion concentration of 10^0 molar or moles per liter, which is one mole.

TABLE 2.1 pH and its corresponding H^+ ion concentration.

pH	Concentration of H^+ ions in moles per liter			
0	10^0	=	1.0	molar
1	10^{-1}	=	0.1	molar
5	10^{-5}	=	0.00001	molar
7	10^{-7}	=	0.0000001	molar
10	10^{-10}	=	0.0000000001	molar

Since normal dissociation of water molecules into H^+ and OH^- yields 10^{-7} moles of H^+, pure water has a **pH of 7**, the **neutral** point in the scale. Since the pH scale is logarithmic, a difference in 1 unit is a 10-fold increase or decrease in concentration of hydrogen ions.

$$pH = -\log [H^+]$$

$$pH = -\log 10^{-7} = 7$$

From this it is obvious that at 25°C:

pH < 7-solution is acidic

pH = 7-solution is neutral

pH > 7-solution is basic

Notice that the pH of a solution measures the concentration of dissociated protons and not the total concentration of acid in a solution.

The negative log scale is useful for measuring other minute quantities, for example, to measure $[OH^-]$:

$$pOH = -\log [OH^-]$$

Knowing this, we obtain the following useful expression:

$$pH + pOH = -\log K_W = 14.00$$

TABLE 2.2 The relationship between pH and pOH.

pH	pOH	$[H^+]$ mol/L	$[OH^-]$ mol/L
0	14	1.0	10^{-14}
2	12	0.01	10^{-12}
4	10	0.0001	10^{-10}
6	8	10^{-6}	10^{-8}
8	6	10^{-8}	10^{-6}
10	4	10^{-10}	0.0001
12	2	10^{-12}	0.01
14	0	10^{-14}	1.0

Buffers

In any biological system, one of the most important parameters of an aqueous solution is the concentration of protons ($[H^+]$) or the pH. Although the $[H^+]$ is quite low, typically 10^{-6} to 10^{-8} M, it must be maintained within this range for life to exist. If an aqueous solution is able to maintain its pH near constant, it can be called a buffer. A buffer is a solution of a weak acid or a weak base and its corresponding salt that keeps pH constant under diverse conditions. Weak acids and bases are those that dissociate very little in solution (as opposed to strong acids and bases like HCl and NaOH which dissociate completely). The weak acid or base is in equilibrium with its dissociated salts. By the law of mass action, an increase in the concentration of either reactant or product forces the reaction in the opposite direction, thereby maintaining the original equilibrium. An acidic buffer (pH below 7) contains a weak acid and a salt of the acid (conjugate base). A basic buffer (pH above 7) contains a weak base and its conjugate acid (salt of the base).

Strong Acids and Bases

A strong acid undergoes complete dissociation into ions in aqueous solution. For example, when added to water HCl undergoes ionization as follows:

$$HCl + H_2O \longrightarrow H_3O^+ + Cl^-$$

NaOH is a strong base because it completely dissociates into Na^+ and OH^- ions in water. When there is complete ionization like this, that solution cannot act as a buffer.

Weak Acids and Bases

A buffer is a solution of a weak acid or a weak base and its corresponding salt that keeps the pH constant under diverse conditions. Weak acids and bases are those that dissociate very little in solution (as opposed to strong acids and bases like HCl and NaOH that dissociate completely). In water, a weak acid and base dissociate partially. The reaction can be represented as follows:

$$HA \rightleftharpoons A^- + H^+$$

The actual reaction in water is:

$$HA + H_2O \rightleftharpoons A^- + H_3O^+$$

Since the concentration of water is constant, it can be represented more precisely by the first equation.

The weak acid or base is in equilibrium with its dissociated salts. By the law of mass action, an increase in the concentration of either reactant or product forces the reaction in the opposite direction, thereby maintaining the original equilibrium.

Weak acids and bases can therefore serve as reservoirs that can react with any extra hydrogen or hydroxide ions that may enter the medium. For example, if excess acid is added to a buffer solution of a weak acid and its salt (left), the introduced hydrogen ions initially increase the pH (middle). But as the weak acid's anions (negatively charged ions) take up the extra H^+ the original pH becomes reinstated (right). If a base is added, the free protons combine to form additional salt. In both cases, the ratio of (buffer acid) to (buffer salt) remains constant and so does the pH.

The law of mass action can explain the above reversible reaction and the dissociation constant for this reaction is given by

$$K_a = \frac{[A^-][H^+]}{[HA].}$$

According to the **Henderson-Hasselbalch equation**

$$pH = pK_a + -\log \frac{[A^-]}{[HA]}$$

So if the pK_a of a weak acid or base is known, it is easy to make a buffer of the desired pH by taking appropriate proportion of acid and base. At equilibrium, when the concentration of the acid ([HA]) is equal to its conjugated base ($[A^-]$)

$$pH = pK_a$$

Example

Take the case of the weak acid. Acetic acid is a weak acid that dissociates incompletely into its ions as follows:

$$CH_3\,COOH + H_2O \rightleftharpoons CH_3\,COO^- + H_3O^+$$

According to the law of Mass action

$$\text{Equilibrium constant} \quad K = \frac{[CH_3\ COO^-][H_3O^+]}{[CH_3\ COOH][H_2O]}$$

In the case of a dilute solution, the concentration of H_2O is a constant and that of solute is always below 1M.

$$K.[H_2O] = \frac{[CH_3\ COO^-][H_3O^+]}{[CH_3\ COOH]}$$

$K.[H_2O] = K_a$, the dissociation constant.

$$K_a = \frac{[CH_3\ COO^-][H_3O^+]}{[CH_3\ COOH]}$$

Therefore, when applying the Henderson-Hasselbalch equation:

$$pH = pK_a + -\log\frac{[A]}{[HA]}$$

$$pH = pK_a + -\log\frac{[CH_3\ COO^-]}{[CH_3\ COOH]}$$

If we know the pK_a of a weak acid or base, it is very easy to make the buffer of any desired pH by weighing out appropriate concentrations of salt and acid. For making the acetate buffer we need sodium acetate and acetic acid in appropriate quantities.

pH of a Solution of Weak Acid or Base

In the case of a weak acid, a molecule dissociates and gives rise to both positive and negative ions equally.

$$HA \rightleftharpoons A^- + H^+$$

where $[A^-] = [H^+]$

Then the dissociation constant K_a can be expressed as follows:

$$K_a = \frac{[H^+]^2}{[HA]}$$

The hydrogen ion concentration of the solution is controlled by the dissociation of acid molecule and contribution by H_2O toward this is negligible.

Therefore,

$$[H^+]^2 = K_a.\ [HA]$$

$$[H^+] = \sqrt{K_a.}\ [HA] \qquad \text{since pH is negative logarithm of } [H^+]$$

$$-\log[H^+] = -\log(\sqrt{K_a.}\ [HA])$$

$$-\log [H^+] = -\frac{1}{2} (\log K_a + \log [HA])$$

$$pH = \frac{p K_a + p[HA]}{2}$$

A similar expression can also be derived for weak base also.

Buffering Capacity

For the dissociation reaction

$$[HA] \rightleftharpoons [A^-] + [H^+]$$

At equilibrium [HA] = [A^-]

According to the Henderson-Hasselbalch equation

$$pH = pK_a + -\log \frac{[A]}{[HA]}$$

$$pH = pK_a$$

Therefore, the pK_a is the pH at which the dissociation of the weak acid or weak base will be at an equilibrium point at which the ionized species and non-ionized species of molecules are equal. The maximum buffering capacity will be at pK_a or near the pK_a value of the weak acid or base. The dissociation constants of commonly used acids and bases are given in Table 2.3.

Henderson-Hasselbalch Equation

$$HA \rightleftharpoons A^- + H^+$$

$$K_a = \frac{[HA]}{[A^-][H^+]} \quad \text{i.e.,} \quad K_a [A^-] [H^+] = [KA]$$

On taking logarithms on both sides

$$(\log K_a) (\log [HA]) = \log [A^-] + \log [H^+]$$

$$\log K_a = \frac{\log [A^-] + \log [H^+]}{(\log [HA])}$$

On rearranging the above equation

$$-\log [H^+] = -\log K_a + \log \frac{[A^-]}{[HA]}$$

$$(-\log [H^+] = pH \text{ and } -\log K_a = pK_a)$$

Therefore, $$pH = pK_a + \log \frac{[A^-]}{[HA]}$$

TABLE 2.3 Dissociation constants for some weak acids. (K_a of "very large" indicates a strong acid)

Acid	Formula	Conjugate Base	K_a
Perchloric	$HClO_4$	ClO_4^-	Very large
Hydroiodic	HI	I^-	Very large
Hydrobromic	HBr	Br^-	Very large
Hydrochloric	HCl	Cl^-	Very large
Nitric	HNO_3	NO_3^-	Very large
Sulfuric	H_2SO_4	HSO_4^-	Very large
Hydronium ion	H_3O^+	H_2O	1.0
Iodic	HIO_3	IO_3^-	1.7×10^{-1}
Oxalic	$H_2C_2O_4$	$HC_2O_4^-$	5.9×10^{-2}
Sulfurous	H_2SO_3	HSO_3^-	1.5×10^{-2}
Hydrogen sulfate ion	HSO_4^-	SO_4^{2-}	1.2×10^{-2}
Phosphoric	H_3PO_4	$H_2PO_4^-$	7.5×10^{-3}
Citric	$H_3C_6H_5O_7$	$H_2C_6H_5O_7^-$	7.1×10^{-4}
Nitrous	HNO_2	NO_2^-	4.6×10^{-4}
Hydrofluoric	HF	F^-	3.5×10^{-4}
Formic	HCOOH	$HCOO^-$	1.8×10^{-4}
Benzoic	C_6H_5COOH	$C_6H_5COO^-$	6.5×10^{-5}
Acetic	CH_3COOH	CH_3COO^-	1.8×10^{-5}
Carbonic	H_2CO_3	HCO_3^-	4.3×10^{-7}
Hydrogen sulfite ion	HSO_3^-	SO_3^{2-}	1.0×10^{-7}
Hydrogen sulfide	H_2S	HS^-	9.1×10^{-8}
Hypochlorous	HClO	ClO^-	3.0×10^{-8}
Dihydrogen phosphate ion	$H_2PO_4^-$	HPO_4^{2-}	6.2×10^{-8}
Boric	H_3BO_3	$H_2BO_3^-$	7.3×10^{-10}
Ammonium ion	NH_4^+	NH_3	5.6×10^{-10}
Hydrocyanic	HCN	CN^-	4.9×10^{-10}
Phenol	C_6H_5OH	$C_6H_5O^-$	1.3×10^{-10}
Hydrogen carbonate ion	HCO_3^-	CO_3^{2-}	5.6×10^{-11}
Hydrogen peroxide	H_2O_2	HO_2^-	2.4×10^{-12}
Monohydrogen phosphate ion	HPO_4^{2-}	PO_4^{3-}	2.2×10^{-13}
Water	H_2O	OH^-	1.0×10^{-14}

2.3 PHYSICAL VARIABLES

Scientists measure all kinds of stuff in the laboratory. While many observations are qualitative—what color, what state, etc.—many observations are quantitative. Measuring the mass of a reactant, or the volume of a liquid, performing a titration, and other more sophisticated measurements require careful determination of value, which must be recorded along with the proper unit. Both the magnitude of the number and the unit are essential for communicating information to other chemists, wishing to repeat an experiment. The magnitudes of these measurable observations are called the **physical variables.**

Physical variables or the measurable observations can be divided into two groups.

- Substantial variables and
- Natural variables.

Substantial variables: These variables have a unit. They are measured against a precise physical standard. These standards are called units. Examples for substantial variables are: mass, length, volume, time, viscosity, heat, temperature, etc.

Natural variables: These are variables known as dimensionless numbers or groups. They do not require any units to express their measurement. Examples for unitless variables are refractive index, specific gravity, specific viscosity, etc.

There are some physical phenomena that do not have any units, such as the Reynolds number. **The Reynolds number (Re)** is a dimensionless measurement in fluid mechanics used to characterize the nature of flow of fluid through tunnels and pipes. Transition from laminar to turbulent flow can be expressed by the Reynolds number, which depends on the velocity, viscosity, and density of the fluid and also on the geometry of the pipe. The Reynolds number for full flow of a fluid through a tunnel or pipe with a circular cross section is given by

$$Re = \frac{Du\rho}{\mu}$$

where,

ρ = Fluid density
D = Diameter of the pipe or tunnel
u = Fluid velocity
μ = Fluid viscosity

2.4 DIMENSIONS AND UNITS

The physical variables that we use in physics and chemistry can be classified into two categories—**fundamental quantities** and **derived quantities**. There are some physical variables which form the basis of all measurements and quantities and are known as fundamental quantities or dimensions or base quantities. The units to express them are known as **base units**. Table 2.4 gives the seven basic quantities or dimensions and their units. All other quantities are derived from the fundamental quantities by their multiplication and/or division. The units to express them are also derived from base units. These units are called **derived units** and their dimensions are the combination of fundamental quantities.

For example, measurements such as area, volume, velocity, etc. are derived from base quantities or fundamental quantities.

Area = length × length
Volume = length × length × length
Velocity = distance/time

Units

Physical variables are measured against certain standards known as **units. Base units** are those used to express the dimensions or the fundamental quantities, and the derived units are those derived from the fundamental or base units. There are different systems of units such as MKS, CGS, SI, and FPS units. Units of one system can be converted into units of another system. SI units is the officially accepted system and is widely in use. There are two clusterings of **metric units** in science and engineering. One cluster, based on the centimeter, the gram, and the second, is called the **CGS system**. The other, based on the meter, kilogram, and second, is called the **MKS system**. Similarly, **FPS system** is the old British system that uses foot, pound, and second as the basic units.

SI Unit

All systems of weights and measurements, metric and non-metric, are linked through a network of international agreements supporting the **International System of Units**. The International System is called the **SI**, using the first two initials of its French name *Le Système International d'Unités*. It officially came into use in October 1960 and has been officially recognized and adopted by nearly all countries, though the amount of actual usage varies considerably. It is based upon seven principal units, one in each of seven different categories—length, mass, time, temperature, amount of substance, electric current and luminous intensity. The seven basic physical variables and their units as per the SI system are given in Table 2.4.

The SI is maintained by a small agency in Paris, the International Bureau of Weights and Measures (**BIPM**, for *Bureau International des Poids et Mesures*), and it is updated every few years by an international conference, the General Conference on Weights and Measures (**CGPM**, for *Conférence Générale des Poids et Mesures*), attended by representatives of all the industrial countries and international scientific and engineering organizations.

TABLE 2.4 Base Measurements and Base Units - SI (*Systeme International*)

Dimensions	Basic Units	
	Name	Symbol
Length	Meter	M
Mass	Kilogram	Kg
Time	Second	S
Electric current	Ampere	A
Temperature	Kelvin	K
Luminous intensity	Candela	cd
Amount of substance	Mole	Mol

The following are the official definitions of the seven base SI units, as given by the BIPM.

meter [m]

The meter is the basic unit of length. It is the distance light travels, in a vacuum, in 1/299792458th of a second.

kilogram [kg]

The kilogram is the basic unit of mass. It is the mass of an international prototype in the form of a platinum-iridium cylinder kept at Sevres in France. It is now the only basic unit still defined in terms of a material object, and also the only one with the prefix [kilo] already in place.

second [s]

The second is the basic unit of time. It is the length of time taken for 9192631770 periods of vibration of the caesium-133 atom to occur.

ampere [A]

The ampere is the basic unit of electrical current. It is the current that produces a specified force between two parallel wires, which are one meter apart in a vacuum. It is named after the French physicist Andre Ampere (1775-1836).

kelvin [K]

The kelvin is the basic unit of temperature. It is 1/273.16th of the thermodynamic temperature of the triple point of water. It is named after the Scottish mathematician and physicist William Thomson 1st Lord Kelvin (1824-1907).

mole [mol]

The mole is the basic unit of substance. It is the amount of substance that contains as many elementary units as there are atoms in 0.012 kg of carbon-12.

candela [cd]

The candela is the basic unit of luminous intensity. It is the intensity of a source of light of a specified frequency, which gives a specified amount of power in a given direction.

SI-derived Units

Other SI units, called **SI-derived units**, are defined algebraically in terms of these fundamental units. For example, the SI unit of force, the **newton**, is defined as the force that accelerates a mass of one kilogram at the rate of one meter per second. This means the newton is equal to one kilogram meter per second squared, so the algebraic relationship is $N = kg \cdot m \cdot s^{-2}$. Currently, there are 22 SI-derived units. They include:

- the **radian** and **steradian** for plane and solid angles, respectively;
- the **newton** for force and the **pascal** for pressure;

- the **joule** for energy and the **watt** for power;
- the **degree Celsius** for everyday measurement of temperature;
- units for measurement of electricity: the **coulomb** (charge), **volt** (potential), **farad** (capacitance), **ohm** (resistance), and **siemens** (conductance);
- units for measurement of magnetism: the **weber** (flux), **tesla** (flux density), and **henry** (inductance);
- the **lumen** for flux of light and the **lux** for illuminance;
- the **hertz** for frequency of regular events and the **becquerel** for rates of radioactivity and other random events;
- the **gray** and **sievert** for radiation dose; and
- the **katal**, a unit of catalytic activity used in biochemistry.

TABLE 2.5 Units derived from the basic units.

1. Length, Surface area, Volume

Size	SI-Unit		Further Units		Relationship
	Name	Symbol	Name	Symbol	
Length	Meter	m			
Surface (Area)	Square meter	m^2			
Volume	Cubic meter	m^3	Liter	l	$1\ l = 10^3\ m^3$

2. Mass

Size	SI-Unit		Further Units		Relationship
	Name	Symbol	Name	Symbol	
Mass	Kilogram	kg	Metric tonn	t	$1t = 10^3$ kg
			Atomic mass unit	u	1u = 1.66053. 10^{-27} kg
Density	Kilogram per cubic meter	kg . m^{-3}			
Specific Volume	Cubic meter per kilogram	$m^3.kg^{-1}$			

3. Amount of Substance

Size	SI-Unit		Further Units		Relationship
	Name	Symbol	Name	Symbol	
Amount of substance	Mole	mol			
Molar mass	Mass per amount of substance	Kg . mol^{-1}		g.mol^{-1}	
Concen-tration of a substance	Amount of substance in given volume of solvent	mol.n^{-3}		mol.l^{-1}	
Molality	Amount of substance per mass of solvent	mol.kg^{-1}		mol.g^{-1}	

4. Temperature

Size	SI-Unit		Further Units		Relationship
	Name	Symbol	Name	Symbol	
Temperature	Kelvin	K	Degree centigrade	°C	

5. Time

Size	SI-Unit		Further Units		Relationship
	Name	Symbol	Name	Symbol	
Time	Second	S			
Time interval			Minute Hour Day	min h d	1 min = 60 s 1 h = 60 min 1 d = 24 h
Frequency	Hertz	Hz			1 Hz = s^{-1}
Velocity	Meter per second	m. s^{-1}	Kilometer per second	Km. h^{-1}	1 km.h^{-1} = 1/3.6 m.s^{-1}

6. Force, Energy, Power

Size	SI-Unit		Further Units		Relationship
	Name	Symbol	Name	Symbol	
Force	Newton	N			1 N = 1 kg.m.s^{-2}
Pressure	Newton per square meter Pascal	N.m^{-2} Pa	Bar	Bar	1 Pa = 1N.m^{-2} 1 bar =10^5 Pa
Energy	Joule	J Kilowatt-hour	kW.h	1kW.h = 3.6 MJ	1 J = 1N.m = 1 W.s = 1 kg.m^2.s
Power	Watt	W			1W=1J.s^{-1} =1 Nm.s^{-1} = 1VA

Definition of the Important Derived Units of the SI System

farad [F]

The farad is the SI unit of the capacitance of an electrical system, that is, its capacity to store electricity. It is a rather large unit as defined and is more often used as a microfarad. It is named after the English chemist and physicist Michael Faraday (1791-1867).

hertz [Hz]

The hertz is the SI unit of the frequency of a periodic phenomenon. One hertz indicates that 1 cycle of the phenomenon occurs every second. For most work much higher frequencies are needed such as the kilohertz [kHz] and megahertz [MHz]. It is named after the German physicist Heinrich Rudolph Hertz (1857-94).

joule [J]

The joule is the SI unit of work or energy. One joule is the amount of work done when an applied force of 1 newton moves through a distance of 1 meter in the direction of the force. It is named after the English physicist James Prescott Joule (1818-89).

newton [N]

The newton is the SI unit of force. One newton is the force required to give a mass of 1 kilogram an acceleration of 1 meter per second per second. It is named after the English mathematician and physicist Sir Isaac Newton (1642-1727).

ohm [Ω]

The ohm is the SI unit of resistance of an electrical conductor. Its symbol is the capital Greek letter 'omega'. It is named after the German physicist Georg Simon Ohm (1789-1854).

pascal [Pa]

The pascal is the SI unit of pressure. One pascal is the pressure generated by a force of 1 newton acting on an area of 1 square meter. It is a rather small unit as defined and is more often used as a kilopascal [kPa]. It is named after the French mathematician, physicist, and philosopher Blaise Pascal (1623-62).

volt [V]

The volt is the SI unit of electric potential. One volt is the difference of potential between two points of an electrical conductor when a current of 1 ampere flowing between those points dissipates a power of 1 watt. It is named after the Italian physicist Count Alessandro Giuseppe Anastasio Volta (1745-1827).

watt [W]

The watt is used to measure power or the rate of doing work. One watt is a power of 1 joule per second. It is named after the Scottish engineer James Watt (1736-1819).

Dimensional Homogeneity in Equations

There are some formulas or equations derived to obtain the derived quantities from the base dimensions. Therefore, the units of the derived quantities are dependent on the units of the base quantities or dimensions. The equation represents the relationship between the physical variables or dimensions on either side, and they must be homogeneous dimensionally. That is, the units of the right side of the equation should be equal to the left side.

For example, the following is an equation for cell growth:

$$\mu t = \frac{[X]}{[X_0]}$$

where X = cell density at time t,
X_0 = initial cell density, and
μ = specific growth rate.

The same equation, when undergoing mathematical changes, should be on either side of the equation as shown below.

$$\log(\mu t) = \log\left(\frac{[X]}{[X_0]}\right)$$

2.5 MEASUREMENT CONVENTIONS

The development of a bioprocess in biotechnology requires the quantitative analysis of common physical variables having units or without units. The magnitude of these variables should be expressed with correct conventions for further analysis and understanding. It indicates the conditions of measurements and analysis and is essential mainly for comparing the magnitudes.

For example, the magnitudes of the substantial variables like density and pressure are always expressed with respect to temperature. Take the case of density and specific gravity. Density is defined as mass per volume at a fixed temperature, whereas the specific gravity is the ratio of density of a material with that of water. Therefore, it is a dimensionless variable.

Since the specific gravity is the relative density, it is always expressed along with the temperature of the substance and its reference material. The specific gravity of ethanol is represented as $0.789^{20°C}_{4°C}$. It means that the specific gravity of ethanol at 20°C is 0.789 referenced against that of water at 4°C. Density of water at 4°C is exactly 1.000 g. cm^{-3}, therefore we can say immediately that the density of ethanol is 0.789 $g.cm^{-3}$.

2.6 PHYSICAL AND CHEMICAL PROPERTY DATA

In all scientific studies, information about the properties of the material being studied is a prerequisite. Properties regarding the physical and chemical data are essential in the case of engineering studies for process development. The basic and essential physical and chemical data regarding elements and compounds are now available as handbooks. Therefore, the time-consuming measurements of such data can be avoided in all experiments. Some very important chemical, physical, and engineering handbooks include the following:

- Handbook of Chemistry
- Handbook of Chemistry and Physical Properties
- Biochemical Engineering and Biotechnology Handbook
- Chemical Engineering Handbook
- International Critical Table

2.7 STOICHIOMETRIC CALCULATIONS

Biotechnology basically depends on the life activities, or biochemical reactions, going on within the cells of living organisms. In chemical and biochemical reactions, atoms and molecules undergo rearrangements to form new groups and molecules. The groups of molecules or atoms that undergo rearrangements are called the **reactants** and the new types molecules and groups formed out of rearrangements are called the **products**. The determination of mass and molar relationships (the number of molecules consumed and the number of new molecules formed) in a chemical or biochemical reaction is known as **stoichiometric calculations**. The stoichiometric calculations can be carried out from molecular equations of the reactions and correct molecular and atomic weights of the reactants and products.

In the case of alcoholic fermentation the biochemical equation is as follows:

$$C_6H_{12}O_6 \rightarrow 2C_2H_5OH + 2CO_2$$

In this biochemical reaction the quantities on either side should be stable.

The total mass of the reactants = Total mass of products.

The total number of elements in the reactants = Total number of elements of the products.

The number of C, H, and O on either side of the equation is equal.

In biochemical engineering, the correct molar combination of the reactants is very important for the complete conversion of the reactants and thereby the maximum output of the products. Stoichiometric information regarding the biochemical reaction is a must for this operation. This will be of great help in controlling the speed of the process and preventing unnecessary wastage of reactants.

The following are some of the usual terms used in biochemical reactions:

- **Limiting factor:** The reactants or the components which are present in the smallest stoichiometric amount. It has a direct impact on the speed of the reaction.
- **Excess reactants:** The reactant or component that is present in large stoichiometric amounts, more than sufficient to combine with the limiting factors.
- **Conversion:** Indicates the percentage of reactants converted into the products.
- **Degree of completion:** The percentage of limiting reactant converted into products.
- **Selectivity:** The quantity of a particular product is expressed as the fraction of the amount that could have been produced if the entire amount of the reactant is converted to that particular product alone.
- **Yield:** The ratio of the mass or moles of the product formed to the mass or moles of reactant consumed.

2.8 ERRORS IN DATA AND CALCULATIONS

Measurements are prone to errors. Therefore, all techniques for data analysis must consider this error in measurements. Experimental errors while taking measurements are sometimes unavoidable and may depend on accuracy. For example, consider the measurement of length. Measure the length of a table as 5 meters. Here, we are actually comparing the length of the table with that of a standard that is 1 meter long. In this comparison, there is always some uncertainty regarding its accuracy. It depends on the accuracy of the scale that you have used for measuring the length. If the length measured is between 5 and 6 and the scale that was used did not have any subdivisions of meters marked on that, the measurement is not accurate. To get a more accurate measurement of the length, use a scale where the meter is subdivided into centimeters and the length of the table can be measured to the accuracy of centimeters, say 5 meters and 3 centimeters.

Experimentally-determined quantities always have errors to varying degrees. The reliability of the conclusions drawn from this data must take experimental errors into considerations for calculations. Minimization of errors by adopting accurate measurement scales, estimation of the errors and principles of error propagation in calculations are very important in all sciences to prevent deceptive and confusing interpretation of facts.

2.9 ABSOLUTE AND RELATIVE UNCERTAINTY

Experimental and measurement errors always create uncertainty in the final data. This problem can be solved by introducing the rules of significant figures. In this method, we specify the range of error by which each of the given values can be varied. Each of the readings will be uncertain within this range of error. This error value is known as **absolute error.** The same error can be represented in terms of percentage, and then it is called **relative error.**

For example, when representing the temperature of a solution it will be 37 ± 3°C. Here, ± 3°C represents the actual temperature range by which the reading is uncertain or can be varied and this is known as the absolute error. When the same error is represented as a percentage it is known as relative error.

37 ± 3°C can be represented as 37 ± 1.25%. Here, the error, 1.25 % is called relative error.

2.10 TYPES OF ERRORS

Experimental errors can be broadly classified into two categories:

- Systemic errors and
- Random errors

When an error affects all measurements in the same way it is called a **systemic error.** In most cases, the cause of this error is known and introducing a correction factor can minimize the error. For example, a watch showing an error of + five minutes (five minutes fast). In this case we can reduce five minutes from the time shown by the clock to get the correct time. A balance that shows an error of – 0.5 gm can be adjusted for that error effectively if the fact is known.

If an error occurs due to unknown reasons it is called a **random error** or an accidental error. This type of error can be detected by repeating the experiments under the same conditions. If different experimental values or results when repeating the experiments without changing the experimental conditions are found, then there are random errors. These errors can be quantified and minimized by applying methods of statistical analysis.

The results or data of an experiment should be reliable and reproducible. The term **precision** refers to the reliability and reproducibility of results. It also indicates the magnitude by which the data is free from random errors. We also use the term **accuracy** to refer to the quality of the data. When there is a minimum of both systemic and random errors or when it is almost zero and the results are reproducible, then we refer to the data as **accurate.**

2.11 STATISTICAL ANALYSIS

Data, Information, and Knowledge

Data is the set of results that is obtained from an experiment. Data makes a crude form of information. Information is the communication of knowledge. Knowledge is established or proved facts supported by evidence or data. But data is not knowledge. The data can be converted into knowledge systematically as per the sequence shown below.

DATA ⇒ INFORMATION ⇒ FACTS ⇒ KNOWLEDGE

Data becomes information when it becomes relevant to solve your specific problem. Information becomes facts when the data can support the information. Facts become knowledge when they are useful in the successful explanation of the problem, phenomenon, or process.

Statistics play an important role in the systematic conversion of data into knowledge. It is science that helps you in making decisions under uncertainties based on a numerical and measurable scale. This decision making should be based on the data, but not on personal views and belief. Statistical analysis of data involves the study of the laws of probability, collection, organization and presentation of data, data properties, relationships of data, etc.

Types of Data and Levels of Measurement

Data can be of two types—Qualitative data and quantitative data.

Data such as color, size, or any other attribute of a population is not computable by arithmetic relations and is considered **qualitative data.** They are the markers by which we can identify an individual, process, or to which group or class they belong. They are called **categorical variables.**

Quantitative data consists of measurements in the form of numerical values. The statistical analysis is applicable only in the case of this type of data. Quantitative data can be of discrete data or continuous data. Discrete data are countable data. For example, the number of unripe fruits present among the fruits of a basket or box. When the parameters are measurable and are expressed in a continuous scale, it is called continuous data. For example, the weight of tissues used in an experiment.

Statistical analysis of data includes a number of steps. The first thing in statistical analysis is to measure or to count. This measuring or counting is the connection between the reality and the data. A set of data is the representation of the reality in the form of a numerical or measurable scale. If the analyst is involved in collecting the data, it is called **primary type data** otherwise, it is called **secondary type data.**

Data, which is in discrete or continuous type, can be in any one of the following forms: **Nominal, Ordinal, Interval and Ratio–(NOIR)**

Under the conditions of uncertainty, decision making is largely dependent on the application of statistical analysis of data for probabilistic risk assessment of the decision. Figure 2.1 is the graphical representation constructing statistical models for decision making under the conditions of uncertainties.

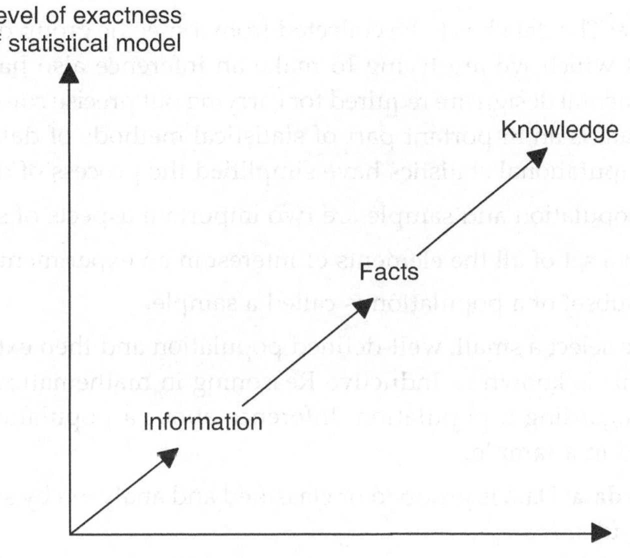

FIGURE 2.1 **The statistical thinking process in decision making under uncertainties.**

The Process of Statistical Analysis

Statistics are sets of mathematical methods used to collect, analyze, present, and interpret data to get to a conclusion about the problem. They are now used in a wide variety of professions to solve many complex experimental problems. The methods of statistical analysis are very helpful for decision makers, managers, and administrators of political, business, and economics to enable them to arrive at correct and better decisions about uncertain states of affairs.

The advancement in computer technology and software has greatly simplified statistical analysis, and a great number of statistical information is available in today's economic socio-political environments. New developments in software engineering have played an important role in statistical data analysis. There are very efficient software packages with extensive data-handling capabilities. They are ideal for handling various types of data from very small to very elaborate forms, which can be carried out routinely. Even though computers assist in the statistical analysis, the analysis mainly focuses on the outcome, in its ability to make correct predictions and decisions.

The statistical analysis of a data involves four basic steps:

- Definition of (understanding) the problem;
- Data collection or its compilation;
- Analyzing the data; and
- Final assessment and reporting of results.

Defining the problem: A clear vision of the problem is a prerequisite. The correct definition of the problem will help in collecting the exact type of data for analysis.

Collecting data: The data has to be collected from a specific group or population. Therefore, the population about which we are trying to make an inference also has to be clearly defined. Sampling and experimental design are required for carrying out precise collection of data. Designing the ways to collect data is an important part of statistical methods of data analysis, even though improvements in computational statistics have simplified the process of data collection.

Defining the population and sample are two important aspects of statistical analysis.

(*a*) Population: a set of all the elements of interest in an experiment or study.

(*b*) Sample: a subset of a population is called a sample.

In statistics, we select a small, well-defined population and then extend the inference to the whole population. This is known as Inductive Reasoning in mathematics. Its main purpose is to test the hypothesis regarding a population. Inference about a population is obtained from the information contained in a sample.

Analyzing the data: Data is grouped or classified and analyzed by suitable methods turning its conversion into results.

Reporting the results: Finally, the results are expressed in a suitable form such as tables, graphs, or a set of percentages. Since only a small collection or sample has been examined and not the entire population, the results should reflect the uncertainty condition through probability statements, intervals of values, and errors.

2.12 PRESENTATION OF EXPERIMENTAL DATA

Data has to be analyzed and converted into a result that tells the proper information or knowledge. The data that we obtain may be from small groups or samples, which represent the entire population. Samples are the only the realistic way to obtain data because of time and cost constraints. For the convenience of statistical analysis, data can be classified into two categories: cross-sectional and time series data:

Cross-sectional data	Data collected at the same time or approximately the same point of time.
Time series data	Data collected at different time intervals over a specific time period.

The data may be collected from existing sources or from a new observation of experiments designed to get new data. In experimental studies there will be a number of factors influencing the process. First, the variable of interest is identified and then the other variables or factors are controlled so that data can be collected on the influence of the variables. A survey is the most common type of observational study.

2.13 DATA ANALYSIS

In statistics, there are mainly two categories of data analysis—**exploratory methods and confirmatory methods**.

Simple arithmetic calculations are used to analyze data and easy-to-draw pictures are used to summarize the data in exploratory methods.

A **probability theory** is used in the confirmatory method of data analysis. Probability is important in decision making because it provides a means for measuring, expressing, and analyzing the uncertainties linked with future events.

Data Processing: Coding, Typing, and Editing

The data that is recorded on a data sheet will go through three stages:

- Coding: The data are transferred, if necessary, onto coded sheets.
- Typing: Data are typed and stored by at least two independent data- entry persons.
- Editing: The data is compared to the independently entered data to check for errors.

When the data is recorded or entered into the data sheet or computer, the following types of errors are possible :

- Recording errors
- Typing errors
- Transcription errors (incorrect copying)
- Inversion, (example- 123.45 is typed as 123.54) errors
- Repetition errors
- Deliberate errors.

2.14 TRENDS

Experimental data is displayed in a suitable graphical form to analyze the trends of variation among the variables. In certain cases it can be observed that the values are highly variable and fluctuate around a mean value. This type of phenomenon is called **scatter** and the distribution so obtained is called **Gaussian distribution**. For example, if we want to plot the variation of blood glucose levels as a function of time, we may get a scattered distribution. If we want to draw a line through all the values, it will result in a highly fluctuating line. In such cases we draw a line through the middle of the scattered values assuming that the system is smooth and continuous and the values of the data are without any experimental errors and would lie on the line, even though the scatter of points is considerable (Figure 2.2). In such cases, several questions will arise. To which points would the curve pass closer? Should all the data points be included or are some points clearly in the range of error? In statistics, there are special methods developed to deal with such situations.

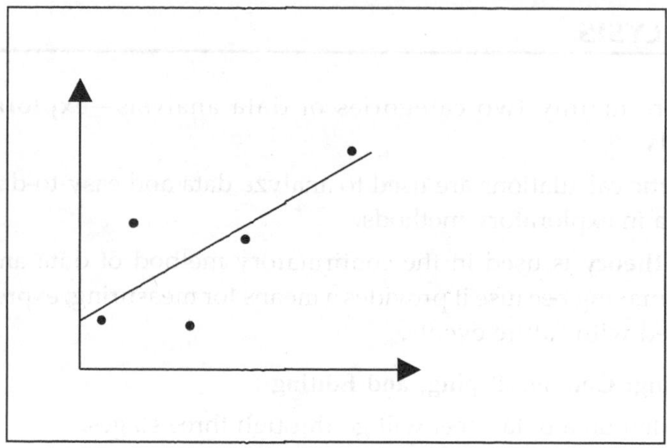

FIGURE 2.2 **Curve of best fit.**

2.15 TESTING MATHEMATICAL MODELS

The following are the main mathematical models used for testing the distribution of variables.

Normal

Application: It is a basic distribution of statistics and an appropriate model for many physical phenomena. Many applications arise from the central theorem—average of values of 'n' number of observations approach normal distribution, irrespective of form of original distribution under quite general conditions.

Example: Distribution of physical measurements, intelligence test scores, product dimensions, average temperatures, etc.

Many methods of statistical analysis presume to be normal distribution. The generalized Gaussian distribution has the following probability density function—(pdf).

$$A. \exp[-B|X|^n], \text{ where A, B, and } n \text{ are constants.}$$

If $n = 1$, it is Laplacian and if $n = 2$ it is Gaussian distribution. This distribution approximates reasonably good data in some image coding applications.

Slash distribution: The distribution of the ratio of a normal random variable to an independent uniform random variable.

Log-normal

Application: The representation of a random variable whose logarithm follows normal distribution. This is a model for processes arising from many small multiplicative errors and is appropriate when the value of an observed variable is a random proportion of the previously observed value.

Example: Distribution of various biological phenomena, distribution of sizes from breakage process, distribution of income size, life distribution of some transistor types, etc.

In cases where the data are log-normally distributed, the geometric mean acts as a better data descriptor than the mean. The more closely the data follows a log-normal distribution, the closer the geometric mean is to the median, and therefore log re-expression produces symmetrical distribution. The ratio of two log-normally distributed variables is known as log-normal.

Poisson

Application: It is usually used in quality control, reliability, queuing theory, etc. If the events take place independently at a constant rate, it gives a probability of exactly x independent occurrences during a given period of time. It may also represent the number of occurrences over constant areas or volumes. It is frequently used as approximation to binomial distribution.

Example: Used to represent distribution of a number of defects in a piece of material, customer arrivals, insurance claims, incoming telephone calls, radiation emitted, etc.

Geometric

Application: It gives probability of the number of binomial trials required before the first success is achieved.

Example: It can be used in quality control, reliability, and other industrial situations.

Binomial

Application: It gives probability of exact success in n number of independent trials, when probability of success p on single trial is a constant.

Used frequently in quality control, reliability, survey sampling, and other industrial problems.

A Short History of Statistics and Probability

The term statistics originated from the Italian word for 'state.' The original idea of statistics was derived for the collection of information about and for state. The origin of statistics happened in a church in Britain around the seventeenth century. A man named John Graunt reviewed a weekly church publication issued by the clerk of a local church community. It contained the lists of the number of births, deaths, and christenings in each community or church. The list of deaths carried the death reason and was named as the Bills of Mortality. He organized this data in a form now known as descriptive statistics. This was published as Natural and Political Observations made upon the Bills of Mortality. Thereafter he was elected as a member of the Royal Society. Thus, there are some concepts and terms from social and population sciences such as 'population'.

Probability is derived from the word 'to probe' (to find out). It was applied to those things not easily accessible or understandable. The term originated in the sixth century from the study of games of chance and gambling. It is a branch of mathematics studied by Blaise Pascal and Pierre de Ferment in the seventeenth century. Now the theories of probability and probabilistic modeling are used in almost all fields of industry, business, and science such as the study of flow of traffic through a highway, telephone exchanges, computer process, genetics, quality control insurance, investments, finance, software developments, etc. And the list still is increasing.

2.16 GOODNESS OF FIT (CHI-SQUARE DISTRIBUTION)

In **chi-square distribution**, the probability distribution curve stretches over the positive side of the line and has a long right tail. The form of the curve depends on the value of the degree of freedom. Chi-square distribution is mainly used in Chi-square tests for association. Chi-square tests are of statistical significance and widely used in bivariate tabular association analysis. The hypothesis is based on whether or not two different populations are different enough in some characteristic or aspect of their behavior based on two random samples. This procedure is also known as the **Pearson Chi-square test**. The Chi-square test is used to see if an observed distribution is in accordance to any particular distribution. This test is calculated by comparing the observed data with the expected data based on the particular distribution.

2.17 USE OF GRAPH PAPER WITH LOGARITHMIC COORDINATES

The use of graph paper in statistics and other experiments is essential for displaying and analyzing data. It is very easy to use ordinary graph paper for the display of results. But certain data becomes more informative when it is displayed on logarithmic graph paper. All logarithmic graph papers are designed in log in powers of ten.

How to Plot a Graph on Logarithmic Paper

Determine the range of your data. For example, consider data that ranges from 20 to 900 on the X-axis and from 0.01 to 8.0 on the Y-axis. The logarithmic paper is divided into decades on both axis in powers of ten. On the X-axis, the lowest power of ten of the data is 10 (10^1). So label the left-most end of X-axis 10. The next power of ten is 100 (10^2) and label the next decade 100. Thus, the next decade will be 1000 (10^3) followed by 10,000 (10^4) and so on until all values of X-axis are marked. The same process is followed for plotting on Y-axis. The smallest value 0.01 is represented as 10^{-2}, the next value is 0.1 (10^{-1}), followed by 1 (10^0), and then 10 (10^1). Thus, it will continue and cover the Y-axis.

If you want to plot the values 20 and 0.6 on the X-axis and Y-axis, respectively (2, 0.6), first find the value of 2 on the X-axis (between 10 and 100). This value represents 2×10 or 20. The value of Y is between 0.1 and 1 on the Y-axis. The value represents 6×0.1 or 0.6. All the values of the data can be plotted like this on logarithmic graph paper.

Since the logarithm of negative numbers and zero are not suitably defined, these numbers cannot be plotted on logarithmic graph paper. It is possible to construct logarithmic paper using standard linear graph sheets once you become familiar with logarithmic graph paper and plotting of values of data on it. When the range of data is small, for example, between 10,000 and 25,000 units, standard logarithmic paper is not available for plotting. In such situations you can make your own logarithmic graph paper.

How to Make Your Own Logarithmic Paper

If logarithmic graph paper is not available, you can make the graph on ordinary graph paper in a very simple way. First determine the range of the data to be graphed. Consider data having the range of 10 to 23, with the lowest value 10 and the highest 23. The log of 10 is 1, mark it on the left-most corner of the axis. The highest value is 23, and since it is not a round figure let us take 25 as the highest limit. The logarithm of 25 is 1.398 (1.40). Label the highest limit, 1.40, at the right-most corner of the axis. Now take the other values in your data, find out its logarithm, and plot it between 1 and 1.40. For example, the value 16 in your data has a logarithm of 1.20 and is therefore plotted at 1.20. Now finally, label the axis in the units of data so that the graph can be read easily. The left-most corner is 10 (log 1) and the right-most corner of the axis is 25 (log 1.40) and the distance between these two has to be divided and labeled. For example, take another two points, 15 and 20, and take the logarithm of 15 and 20 and plot this point on the axis, then label 15 and 20. In this way, you can plot a logarithmic graph on an ordinary graph paper.

2.18 PROCESS FLOW DIAGRAM

The bioprocess is the industrial application of biological pathways or reactions mediated by living whole cells of animals, plants, and microorganisms or enzymes under controlled conditions for the biotransformation of raw material into products. The product may be directly useful as food, medicine or industrial compounds, or the bioprocess may be without any direct product. In such cases the bioprocess may be used as a means to detoxify the industrial wastes or treatment of factory effluents with or without byproduct. The conversion of a laboratory scale biochemical reaction into an industrial bioprocess technology involves the application of biochemical engineering.

Biochemical engineering also includes the allied subjects such as Biochemistry, Molecular Biology, Microbiology, Genetics, Cell biology, Immunology, Biomedical, and Environmental sciences in addition to engineering principles. The need for biochemical engineering specifically developed because traditional chemical engineering deals primarily with petroleum and chemical industries.

The bioprocess technology involves mainly three parts:

- Upstream processes;
- Bioreaction and bioreactors; and
- Downstream processing.

Upstream processing: This part of biochemical engineering involves the development of processes for aseptic treatment of substrates or raw materials with the microorganism or the biocatalyst. It includes the media preparation, its sterilization, feeding into the bioreactor, regulation of temperature, pH and pressure, inoculation of the cell culture into the medium aseptically, etc.

Bioreactor: The designing of the appropriate type of bioreactor or fermentor is one of the important steps in the development of bioprocess technology. **Bioreactors** are vessels in which raw materials are biologically converted into specific products, using microorganisms, plants, animals, or human cells or individual enzymes. A bioreactor supports the natural process of cells by trying to maintain their environment to provide optimum growth conditions by providing appropriate temperature, pH, substrates, salts, vitamins, and oxygen. In most of the bioreaction processes the substrate of the biotransformation and the carbon source of the organisms will be the same. Table 2.6 gives some of the carbohydrates commonly used in the various fermentation processes as the carbon source and substrate for the reaction.

Bioreactors can be classified according to the type of biocatalysts and the type of bioreaction. The first classification is based on the type of biological agent used:

- microbial fermentors or
- enzyme (cell-free) reactors.

Further classification is possible based on biochemical reactions and process requirements.

Downstream processing: The recovery and purification of the required product from the growth medium through a set of separation and purification techniques is called downstream processing. Each stage in the overall separation procedure is strongly dependent on the history and quality of the biological production process. Maximization of production can lead to great difficulties in downstreaming and recently more attention is being paid to overall process optimization. It includes techniques such as filtration, centrifugation, sedimentation, various types of chromatographic techniques, electrophoresis, etc. Some fermentation parameter factors affecting the DSP are listed as follows.

- The properties of organisms or cells used in the fermentation process (safety, classification, morphology, thermal stability, tendency to flocculate, size, cell wall rigidity, etc.) influence the filtration, sedimentation and homogenization performance.
- The location of the product (intracellular, deposited in vacuoles or inclusion bodies, or excreted into the broth-or biomass itself) will define the initial separation steps and purification strategy.
- The stability of the product defines the need and kind of pre-treatment for inactivation or stabilization.
- The product, by-products, and impurities as well as any additions to the broth (antifoam) may form an interfacial layer in extraction steps, give peaks in chromatography, and block membranes in ultra-filtration and analytical equipment; also salts and trace elements often have to be removed prior to pharmaceutical use.
- Nutrient medium residues (pesticides, herbicides, etc.) present in the product.

TABLE 2.6 Substrates or carbon sources frequently used in the fermentation process.

Substrate/Carbon source	Type of Biocatalyst	Product
Sucrose Molasses Hydrolyzed starch	Saccharomyces cerviceae	Ethyl alcohol
Sucrose Molasses Hydrolyzed starch	Aspergillus	Citric acid and allied products
Sucrose Molasses Hydrolyzed starch	Pencillium sps	Penicillin
Sucrose/Glucose	Plant cells	Various secondary metabolites
Glucose	Escherichia coli	Various recombinant products
Glucose/whey	Lacobacillus	Lactic acid
Glucose	Hybridoma	Monoclonal antibodies
Glucose	Animal cells	Various products

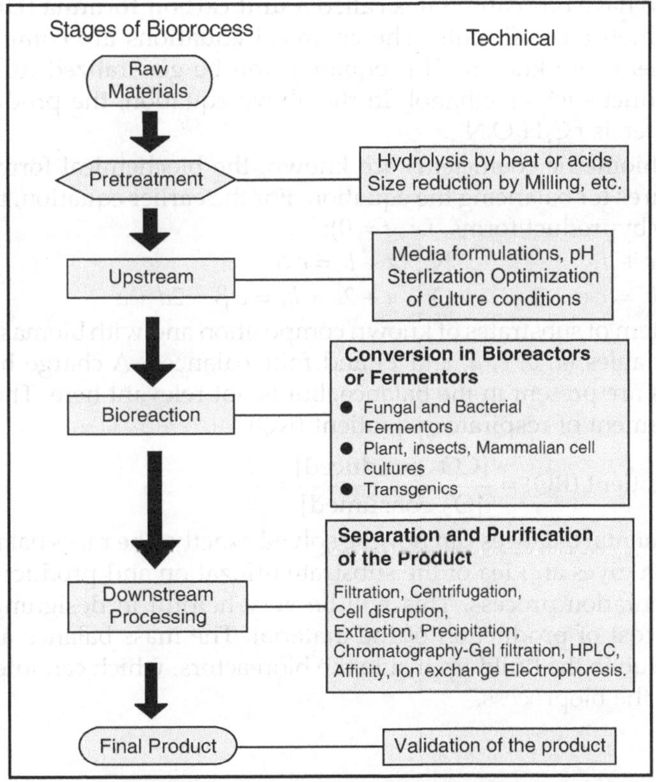

FIGURE 2.3 **Flow diagram of a bioprocess.**

2.19 MATERIAL AND ENERGY BALANCES

Material Balance

In bioprocess technology developments, stoichiometric equations indicating both material balance as well as energy balance are very important to determine the exact molar concentration of the substrates that facilitate the high rate of formation of the products and energy conservation in the biochemical process.

Consider the growth of microorganisms in a culture, which produces only carbon dioxide and water as the final products with increase in the biomass formation. This aerobic growth can be represented by the following equation.

$$aC_wH_xO_y + bO_2 + cN_iH_gO_h \longrightarrow C_\alpha H_\beta N_\gamma O_\delta + d\,H_2O + e\,CO_2$$

In the above equation $aC_wH_xO_y$ represents the chemical formula for the substrate or the carbon source such as glucose; $cN_iH_gO_h$ represents the formula of nitrogen source; and $C_\alpha H_\beta N_\gamma O_\delta$ indicates the dry biomass.

$$aC_wH_xO_yN_z + bO_2 + bH_gO_hN_i \rightarrow cCH_\alpha O_\beta N_\delta + dCO_2 + eH_2O + f\,C_jH_kO_lN_m$$

In the above biochemical equation, in addition to the biomass ($CH_\alpha O_\beta N_\delta$) formation, there is the product $f\,C_jH_kO_lN_m$. Note that compounds like ATP and NADH have been ignored. They may be part of metabolism, but are not exchanged with the environment. In addition, other compounds like vitamins, which are needed in very small quantities, have been neglected. When the formula for biomass is rewritten to have one carbon it is called a **unit carbon formula (UCF)**. The letters a, b, c, d, and e are stoichiometric coefficients. The chemical equations are complete only when these stoichiometric coefficients are known. This equation can be generalized further by including the formation of a byproduct such as ethanol. In the above equation, the product in addition to the biomass, CO_2 and water, is $f\,C_jH_kO_lN_m$.

Once the stoichiometric coefficients are known, the biochemical formula can be evaluated using normal procedures for balancing the equation. For the earlier equation, the elemental balances are (assuming that no byproduct forms, *i.e.*, $f = 0$):

C: $w = c + d$, N: $z + b_i = c\,\delta$

H: $x + b_g = c\,\alpha + 2e$, O: $y + 2a + b_h = c\,\beta + 2d + e$

For a given system of substrates of known composition and with biomass composition known, there will be five variables (*a, b, c, d,* and *e*) and four balances. A charge balance could provide additional data if ions are present in the balance, but is not relevant here. The solution is to use an experimental measurement of respiratory quotient (RQ).

$$\text{Respiratory quotient (RQ)} = \frac{[CO_2 \text{ produced}]}{[O_2 \text{ consumed}]} = \frac{d}{b}$$

Hence, the elemental balances can now be solved exactly. The mass balance of the cell growth and product formation gives an idea of the substrate utilization and product formation during the industrial scale fermentation process. This will be very helpful in designing the bioreactors and assessing the overall cost of production of the material. The mass balance analysis also gives the wastage of nutrients due to the build up inside the bioreactors, which can interfere with the overall reaction procedure of the bioprocess.

Energy Balances

Biological reactions, like any other phenomenon, always occur with a change in energy content and free energy, which can be explained by the principles of thermodynamics. The reaction can be either endergonic or exergonic. The product formed may be comparatively a high-energy state or low energy state, which is proportional to the degree of reduction or oxidation that the substrate molecule has undergone.

TABLE 2.7 Heat of combustion for various compounds at standard conditions (298°K and 1 atm) and pH 7.

Compound	Formula	γ	ΔH°_c (kJ/mole)	ΔH°_c (kJ/C-mole)	$\Delta H^\circ_c/\gamma$ (kJ/C-mole)
Formic acid	CH_2O_2	2	255	255	127.5
Acetic acid	$C_2H_4O_2$	4	875	437	109.5
Propionic acid	$C_3H_6O_2$	4.67	1527	509	109.0
Butyric acid	$C_4H_8O_2$	5.00	2184	546	109.2
Valeric acid	$C_5H_{10}O_2$	5.2	2841	568	109.2
Palmitic acid	$C_{16}H_{32}O_2$	5.75	9978	624	108.5
Lactic acid	$C_3H_6O_3$	4	1367	456	114.0
Gluconic acid	$C_6H_{12}O_7$	3.67			
Pyruvic acid	$C_3H_4O_3$	3.33			
Oxalic acid	$C_2H_2O_4$	1	246	123	123.0
Succinic acid	$C_4H_6O_4$	3.50	1491	373	106.6
Fumaric acid	$C_4H_4O_4$	3	1335	334	111.3
Malic acid	$C_4H_6O_5$	3	1328	332	110.7
Citric acid	$C_6H_8O_7$	3	1961	327	109.0
Glucose	$C_6H_{12}O_6$	4	2803	467	116.8
Fructose	$C_6H_{12}O_6$	4	2813	469	117.2
Galactose	$C_6H_{12}O_6$	4	2805	468	117.0
Sucrose	$C_{12}H_{22}O_{11}$	4	5644	470	117.5
Lactose	$C_{12}H_{22}O_{11}$	4	5651	471	117.8
Methane	CH_4	8	890	890	111.3
Ethane	C_2H_6	7	1560	780	111.4
Propane	C_3H_8	6.67	2220	740	110.9
Methanol	CH_4O	6	727	727	121.2
Ethanol	C_2H_6O	6	1367	683	113.8
iso-Propanol	C_3H_8O	6	2020	673	112.2
n-Butanol	$C_4H_{10}O$	6	2676	669	111.5
Ethylene glycol	$C_2H_6O_2$	5	1179	590	118.0
Glycerol	$C_3H_8O_3$	4.67	1661	554	118.6
Acetone	C_3H_6O	5.33	1790	597	112.0
Formaldehyde	CH_2O	4	571	571	142.8
Acetaldehyde	C_2H_4O	5	1166	583	116.6
Urea	CH_4ON_2	6	632	632	105.3
Ammonia	NH_3	3	383		127.7
Biomass	$CH_{1.8}O_{0.5}N_{0.2}$	4.8	560	560	116.7

The energy state of a molecule is given by the molar heat of combustion at standard conditions (Table 2.7) for that molecule. The heat of combustion of a molecule at standard conditions (normal temperature, pressure, and pH) is directly proportional to the degree of reduction. If the product formed is in a higher reduced state, the reaction has taken place with an input of energy in the form of ATP consumption, and if the product is in the oxidized state, there is the formation of heat or ATPs since it is an exergonic reaction.

ΔG, Gibb's free energy, and ΔH, the heat change, are the terms commonly used to show the energy balance of a biochemical reaction. Gibb's free energy is the usable form of energy in the biological system. Heat is not a usable form of energy as far as the biological system is concerned. Knowledge about the total energy change—energy consumption or heat formation in the biochemical reaction—is very important in designing the bioreactor for the bioprocess development.

2.20 FLUID FLOW AND MIXING

The proper mixing of the reactants is very important because it maintains the uniformity of the physical and chemical factors of the reactants and the reaction conditions in the bioreactor. It provides the correct uniform biological environment. The bioprocess has to be carried out at the precise range of physiological temperature and pH that facilitate the maximum reaction rate and maximum product output. A continuous supply of oxygen is also an important factor in the cell growth and biomass formation to increase the reaction rate. Therefore, fluid mixing is very important in maintaining homogeneous chemical and physical environments in the bioreactor.

The media that is used in the fermentation process may be Newtonian fluids such as bacterial broth media or non-Newtonian fluids such as semisolid media used for the production of certain enzymes and antibiotics such as penicillin.

Viscosity and fluidity of the culture media in large-scale fermentation processes is an important factor, which may interfere with the mass transfer in terms of nutrient mixing, oxygen supply, and marinating the proper temperature and pH.

There are different types of fluid-mixing devices in fermentors that allow the proper mixing of the media components depending on the viscosity of the fluid. There are different types of mixing arrangements in bioreactors. Depending on the type of mixing devices there are different types of fermentors such as stirred tank reactors, air-lift reactors, and bubble-column reactors for aerobic fermentations.

2.21 MASS TRANSFER

Mass transfer refers to the transfer of oxygen from bulk gas into the cells. The transfer of oxygen from the gas phase to the liquid phase through gas/liquid film is one of the rate-limiting steps in large-scale fermentation. The **oxygen transfer rate (OTR)** is dependent on a number of factors. There are some methods given below, by which the OTR of a reaction system can be increased and maintained for smooth operation and better product output.

- Increasing the pressure
- O_2 enrichment of the inlet air
- Increasing agitation
- Increasing airflow rate
- Reducing foaming and removing bubbles

2.22 HEAT TRANSFER

Heat transfer is another major factor that can affect the overall performance of a biochemical reaction in bioprocess technology. The reaction that is taking place in a fermentor can be exothermic or endothermic. Therefore, most of the fermentors need careful temperature control. The increase in temperature of the fermentor can be done by the exothermic reaction as well as by the agitation and the respiration of the growing organisms. Once the details about the heat transfer—the evolution of heat by fermentation and heat generated due to agitation and aeration—are known, the desired temperature can be calculated. Based on the overall heat production during the process of fermentation, heating and cooling devices can be designed to provide the optimal temperature needed for the smooth operation of the process. A well-designed jacket around the bioreactor and circulation of steam, hot water, or chilled water can very well guard the fluctuation of temperature of the bioreactor. This facility can also be used for the sterilization of the media and fermentor before starting the process and autoclaving and cleaning of the fermentor after the process.

2.23 BIOREACTOR DESIGNING

The designing of a bioreactor depends on a number of factors such as type of cells, details about the type of metabolic reaction, information about the mass transfer and heat transfer, fluid mixing, etc.

Types of Bioreactors

Bioreactors or fermentors can be classified in different ways. For example, classification is based on the type of biological agent, the type of metabolism or product formation, classification based on modes of operation, and another classification based on the mode of stirring or mixing. The first classification is based on the type of biological agent used. There are two types:

- Microbial fermentors, and
- Enzyme (cell-free) reactors.

Further classification is possible based on other process requirements.

For example, on the basis of mode of fermentation, bioreactors can be designed for:

- Aerobic and
- Anaerobic type or
- Submerged or surface type of fermentation, etc.

Based on the type of metabolism there are three types of bioreactors:

Type 1. Production of primary metabolites. Growth-associated production (e.g., ethanol production).

Type 2. Overproduction of primary metabolites. The product may be an intermediate (and not the end product of a pathway). The production phase can be distinguished from the growth phase (e.g., citric acid production, production of amino acids).

Type 3. Non-growth associated production. The product is not connecting to catabolism or energy metabolism (e.g., production of secondary metabolites such as antibiotics). There are distinct growth and production phases. Sometimes production only begins when the main carbon source is exhausted and a secondary carbon source is used.

Classification is also based on the mode of operation. There are two types of reactors:

1. **Batch mode:** no exchange of liquid medium. With addition of inoculum, culture grows uncontrolled until some nutrient is exhausted. This is easy to set up.
2. **Fed-batch:** additions of media, on demand. Some medium components may be continuously fed without taking out anything. Hence, reaction volume will increase. The substrate concentration must be kept low in the beginning to minimize byproduct formation. As biomass increases, the feed rate might have to be correspondingly increased. This requires that a feeding strategy be devised based on a predetermined rate or by a feedback control.

In another type of classification, the mechanism of mixing has been taken into consideration.

The type of mixing that is provided in the fermentor is an important factor that facilitates proper mass transfer and heat transfer. The mechanism of mixing should not have a negative impact on the growth and metabolism of the cells. For example, plant cells, animal cells, and hybridoma, which grow slowly and are fragile compared to bacterial cells need less oxygen for growth and metabolism. A mechanical stirring can cause damage to the cells. At the same time, mixing needs to maintain uniformity in the distribution of nutrients and other components in the medium. The following are some of the bioreactors that have different modes of agitation or mixing.

1. **Stirred tank reactors:** In this type impellers provide agitation.
2. **Bubble column reactors:** Mixing is provided by air flow through the medium at constant velocity.
3. **Airlift fermentors:** Mixing is provided by forced circulation of nutrients inside the reactor by air.

2.24 UNIT OPERATIONS

Large-scale fermentation or bioprocess operations involve mainly three steps: **upstream processing, bioreaction,** and finally, **downstream processing** and **product recovery.** Each of these three steps involves further individual steps and each individual step of operation is called **unit operation**.

Unit operations involved in upstream processing are:

- Milling
- Mixing
- Media preparation
- Sterilization
- Cooling
- Heating

Unit operations involved in bioreactions are:

- Mixing
- Handling of microorganism, plant, or animal cells
- Inoculation of the cells
- Heating and cooling

Unit operations involved in downstream processing and product recovery are:

- Filtration
- Centrifugation
- Dialysis
- Crystallization
- Membrane separation
- Chromatography
- Solvent extraction
- Evaporation
- Precipitation

2.25 HOMOGENEOUS REACTIONS

Microbial Growth Kinetics and Fermentation

In large-scale fermentation, information about the nature of growth and metabolism of the organism or cells is very important. In microorganisms usually the growth (increase in biomass) takes place by binary fission. Yeasts normally reproduce by budding. Fission, mycelial growth, and branching are also possible in the case of yeast.

After inoculation, the microbial culture starts growing. The growth of the microorganism follows a typical pattern of growth as shown in Figure 2.4. A microbial batch culture has four distinct phases of growth.

The growth cycle includes :

- lag phase
- exponential phase or log (logarithmic) phase
- stationary phase, and
- a death phase.

The lag phase: Inoculation of cells in the growth medium results in a period where there is no increase in cell number. This sets the stage for cell multiplication and growth. This period is known as the lag phase. The duration of the lag phase depends on the state of the inoculum and its growth history. The lag phase is not an inactive period, but it is a very active period preparing for a rapid growth phase.

The log phase: This is the logarithmic phase or the exponential growth phase, which follows the lag phase. During this phase the biomass or the number of cells increases exponentially with time. Cells undergo doubling several times and the specific growth rate of the culture remains constant. Assuming growth rate depends on the available biomass, and if μ = specific growth rate, the rate of increase in biomass (the increase in cell mass with time), dX/dt is the product of specific growth rate m and total biomass concentration X.

$$dX/dt = \mu X$$

Similarly, if N is the total number of cells, then the increase in cell number dN/dt is the product of specific growth rate and cell number N.

$$dN/dt = \mu N$$

The cell mass concentration X_0 at the beginning of the exponential phase of growth, the time required for doubling the biomass, could be calculated by the above formula

$$dX/dt = \mu t$$

$$\text{Ln } X/X_0 = \mu t$$

$$\text{Ln } 2 = \mu t, \text{ thus } t = \frac{\text{Ln}2}{\mu}$$

The generation time of a culture is the time taken by the cells in the culture to double the number of the biomass under defined conditions. It corresponds to the doubling time which ranges from 20 minutes to 8 hours. The doubling time and specific growth rate of the organisms are very important in the large-scale fermentation process because they determine the requirements of media and other growth-related factors, such as fermentation batch time for the production of the required biochemicals.

The specific growth rate of an organism is very characteristic of that organism and is dependent on the growth environment including ionic concentration, pH, temperature, media composition, and dissolved oxygen levels. The doubling time of some the microorganisms are given in the table below.

TABLE 2.8 Doubling time of some microorganisms under normal culture conditions.

Microorganisms	T_d (minutes)
E.coli	20
Bacteria	45
Yeasts	90
Moulds	160
Animal cells	630 -1260
Plant cells	3600 - 6600

The stationary phase: The logarithmic phase or the exponential phase of growth is followed by a stationary phase. During this phase, the growth rate and death rate of cells are almost equal. Depletion of a growth-limiting nutrient in the medium or accumulation of toxic byproduct or secondary metabolite may be the cause for the stationary phase.

The death phase: Following the stationary phase, finally the culture enters the death phase. Here, the cells lose their viability and the death rate overtakes the growth rate. The damage to the cells may be due to cell lysis or some other mechanisms.

Effect of Temperature on Growth

Growth of organisms is influenced by a number of factors and temperature is among them. All biochemical reactions of cells are controlled by the state of temperature and thus it influences the growth rate. Microorganisms are classified into three groups based on their optimum temperature for growth. They are:

Psychrophiles = grow around 15°C
Mesophiles = grow around 37°C
Thermophiles = grow around or above 55°C

Each organism has a specific range of optimum temperatures and exact value depends on the species or strains of the organisms and other growth conditions. The ability of an organism to utilize the carbon source for the production of biomass is temperature-dependent and it declines with increase in temperature in the case of normal organisms. Therefore, optimization of growth temperature for microbial cultures is essential to get maximum product output in bioprocess technology.

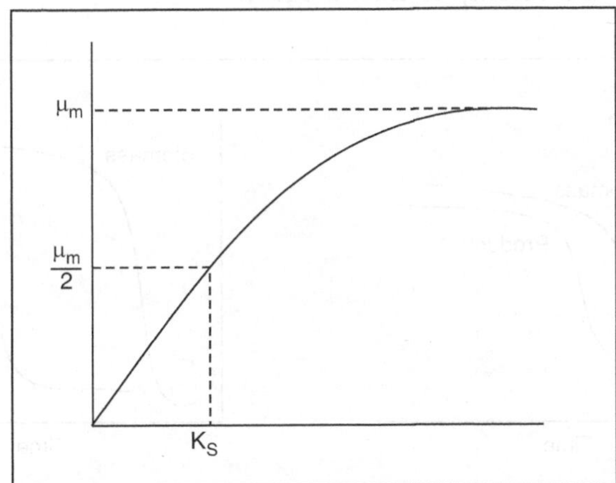

FIGURE 2.4 **Effect of substrate concentration on specific growth rate K_s is the substrate concentration at half μ_m.**

Effect of Substrate Concentration on Growth

Substrate concentration is another important factor that controls growth and rate of product formation in a microbial process. Usually, the carbon source is the rate-limiting component of the culture media and influence of such substrates can be described by the equation.

$$\mu = \mu_m S/(K_s + S),$$

where S is the concentration of the limiting nutrient, K_s is the substrate specific constant, and μ_m is the maximum specific growth rate. The value K_s indicates the affinity of the organism for the substrate or the nutrient. The effect of substrate concentration on growth of organism is represented in Figure 2.4. The specific growth rate, μ, increases with substrate concentrations until it becomes a rate-limiting factor. However, there can be inhibition to growth and product formation due to high substrate concentration in the culture media. K_s is the substrate concentration corresponding to the half of μ_m ($\mu_m/2$). Thus, the growth on a given substrate can be described by two constants—μ_m and K_s. Therefore, it is essential to know and evaluate the values of K_s, μ_m and μ of the organism with respect to the respective substrate to get an idea of the suitability of that material as substrate for growth and product formation.

The relation between growth and product formation is very important in the designing and development of a bioprocess technology. It is also very important in the designing of the bioreactor or fermentor. Based on this relationship the fermentation process can be classified into growth associated and non-growth associated fermentation.

In growth associated fermentation, product formation occurs along with the biomass increase. Both take place simultaneously. In such cases, the rate of product formation is directly proportional to rate of biomass formation (Figure 2.5). The product will be a primary metabolite and its production takes place primarily during the logarithmic phase of the growth curve. Most of the alcohol and acid fermentation belong to this class. In non-growth associated fermentation, the product formation is not directly related to the rate of growth of the cells. But it is proportional to the concentration of biomass. The product formation takes place at the stationary phase of growth. Production of secondary metabolites such as antibiotics belongs to this category.

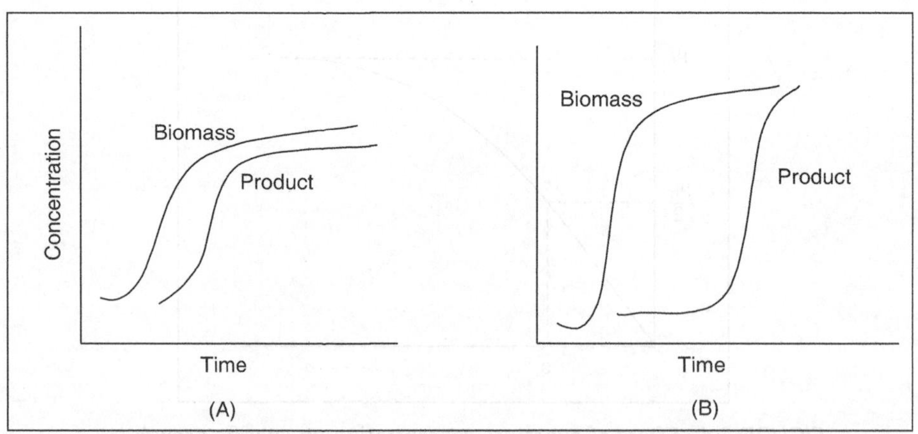

FIGURE 2.5 **Relationship between growth and product formation A. Growth Associated. B. Non-growth associated product formation.**

Therefore, information about the relationship between the growth kinetics and product formation kinetics is very crucial in the designing and development of an efficient bioprocess technology.

Selection of Growth Media

Selection of media and choice of media components is an important step in the process of fermentation. The growth media should be a balanced one supplemented with all required nutrients in sufficient quantities. The media should contain a suitable carbon source, nitrogen source, minerals, and other essential nutrients such as vitamins and hormones to produce a sufficient amount of biomass needed for the production of the product. Minimum amounts of nutrients are calculated from the stoichiometry of growth and product formation.

One should be very careful in the selection of media components. A number of factors have to be considered in this matter. The composition of the media is basically for maximizing the product yield. It should be economical in its cost. The selection of media also affects the different steps of the fermentation process in addition to the growth and metabolism of the organism or cells. It affects the upstream and downstream process of the fermentation. For example, the use of simple and pure carbon sources such as glucose can make the upstream and downstream process very simple. Whereas if you select complex mixtures such as molasses as the carbon source, special pre-treatment is needed for the growth of the microorganism in the medium, and the process of purification of the product will become difficult.

Therefore, selection and composition of the media should provide all the requirements for increasing the growth rate and product formation and also it should minimize or simplify the process operations. In certain cases of fermentation there may be two types of media, particularly in plant cell culturing.

Growth media: The media that facilitate growth of the cells.

Production media: The media meant for product synthesis.

The process is designed in such a way that when the cells grow to a sufficient amount of biomass in the growth media, they will be transferred to the production media where they start synthesizing the product.

2.26 REACTOR ENGINEERING

Reactor engineering includes all three stages of bioprocess engineering–pre-treatment or upstream process, bioprocess or the biochemical process, and the downstream process. While designing all these stages of the bioprocess, special care should be taken to bring down the cost of production.

Upstream processing equipment consists of a pre-treatment vessel. This vessel is specially designed to prepare the media of a specific volume by putting and mixing the correct components in correct quantity and volume, followed by sterilization.

The bioreactor or the fermentor is designed to carry out the specific biological reaction with the help of the biological agent—living cells or immobilized enzymes. The type of bioreactor depends on the type of cells, nature of growth, and the biochemical reactions. Huge bioreactors are needed for carrying out the bioreaction in controlled environments, with large volumes of the media and substrate.

Downstream processing includes the operations needed for the recovery of the product in a very simple and economic way. It includes a number of different filtration, separation, and purification methods. In general, large-scale fermentation and purification are carried out in a stepwise manner. A diagram representation of the process flow sheet of the fermentation process is given in Figure 2.6.

The entire operation starts with the media preparation and sterilization, and sterilization of the bioreactor or the fermentor with or without media. The microbial cultures are usually maintained in petri plates or in slants of culture tubes as stock cultures. These cells are in the dormant stage and need to be activated. First, the cells are inoculated in small culture volumes (5 to 10 ml) and grown overnight. Then they are inoculated into larger volumes (100 to 1000 ml) and then these are either transferred into the bioreactor directly or through a seed fermentor of 10 to 100 l (see Figure 2.6). This is very important to maintain the initial cell density as it can influence the rate of growth and product formation. The fermentor should be provided with the correct combination and volume of media with all its required components, sufficient aeration, control of pH, and antifoam and associated systems. The volume cell density and age of the inoculum are factors that control the exponential growth phase of the microbial growth.

Maintenance of sterile conditions during the process of fermentation, supply of sterile air for respiration, constant agitation and mixing to maintain the chemical uniformity within the fermentor are essential for the successful completion of the process.

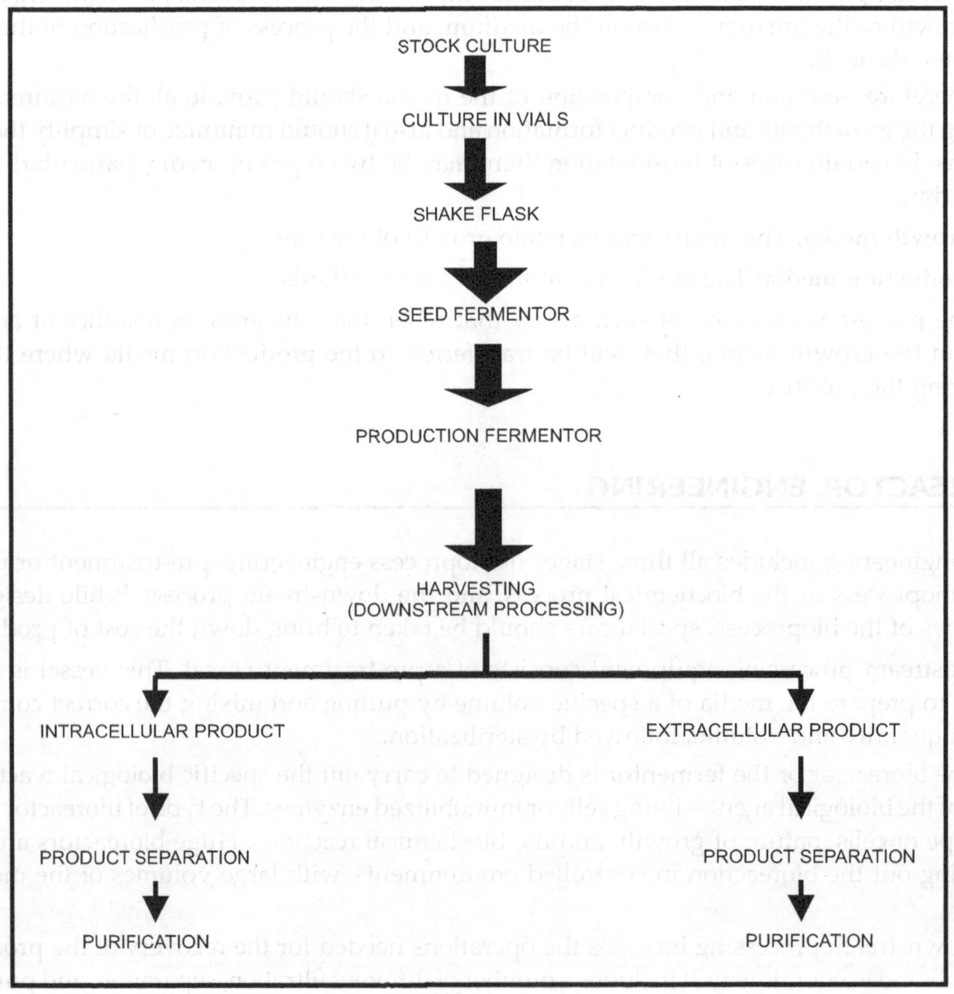

FIGURE 2.6 **Flow sheet of the fermentation process.**

Once the fermentation process is over, the cells are separated from the broth by filtration or centrifugation. If the product is extracellular like many acids, alcohols, antibiotics, and vitamins, the product can be purified from the cell, free media or supernatant. If the product is intracellular, the cells are separated from the culture and are disrupted by a suitable method. Then the product is purified from the mixture by following various separation methods that are part of downstream processing. Organization and different steps of the downstream processing are closely dependent on the type and nature of the product and the components of the media. Therefore, the success and efficiency of the bioprocess technology for the economical manufacturing of a biochemical is closely associated with the integrated designing and operation of all three important steps—upstream processing and bioreactor operations followed by downstream processing.

REVIEW QUESTIONS

1. Define pH. Derive the Henderson-Hasselbalch equation.
2. What is a buffer?
3. Explain the relationship between pK_a and pH.
4. What is a bioreactor?
5. What are the different steps of bioprocess technology development?
6. What are the factors that influence the designing of a fermentor?
7. Explain the following terms:
   (*a*) Reynolds number
   (*b*) Normal distribution
   (*c*) Poisson distribution
   (*d*) Mass transfer and material transfer

Chapter 3 BIOTECHNOLOGY AND SOCIETY

In This Chapter

3.1 PUBLIC PERCEPTION OF BIOTECHNOLOGY

Science and Society

Our perceptions or attitudes toward things are not always rational and are often culturally influenced. They are a combination of thoughts or the cognitive dimension, feelings, or the affective dimension, and the way we react—the behavioral dimension. The cognitive dimension consists of things we know, the affective dimension comprises of things we feel, and the behavioral dimension is how we will act on the attitudes we build. Attitudes help us to become socially acceptable; belonging to a group is very important, and it gives meaning to things we experience.

Advancements in science and technology have made our life very simple and fast. At the same time some of this advancement has caused great concern regarding the long-term impacts on environment and life. In 1985, the World Commission on Environment and Development (WCED), also known as Brundtland Commission appointed by United Nations (UN), recommended sustainable development preserving the environment without any degradation. The Commission defined sustainable development as 'the development that meets the needs of the present without compromising the ability of future generations to meet their own needs'. There are a lot of definitions and views regarding sustainable development for different nations. In all these views the common point on which all nations agree is that science and technology is portrayed as a double-edged sword.

There are two opposing ideologies regarding the use of science and technology:

- Holistic Ideology
- Reductionistic Ideology

Holistic ideology recommends the use of traditional methods on all fronts of life from agriculture to industry. This ideology argues that modern society gives more importance to formal knowledge and neglects informal and traditional types of knowledge. According to this ideology, there is no problem in the agriculture sector. Earth can produce sufficient food materials for the entire population, if we would return to consume some of the old or traditional grains, which can be easily grown. It also recommends the use of chemical manures, pesticides, herbicides, and the use of minimum tillage to conserve the land and to produce a crop. Whatever problem persists in the world is mainly due to the unequal distribution of food that is produced.

Reductionistic ideology argues for the use of new knowledge to improve the quality of agriculture and crop plants. It recommends continuous research and development studies to find out new solutions to problems.

Both these groups and their variant types have given different views regarding sustainable development with minimum impact on environment.

The Impact of Biotechnology on Society

At first glance, biotechnology may seem like simply a science and technology matter. Yet, biotechnology is valued and feared because of its potential impact on environment and our lives. Because of its power to affect social values, social relations, and environmental problems, biotechnology has become a lightening rod for the debate between competing socio-ethical perspectives.

There are significant differences in public perception of biotechnology according to country, sex, age, educational level, religious practices, and social groups. On the whole, attitudes are not nearly as positive as the industry would like them to be.

There are two things which dominate the public's perception of biotechnology: lack of trust and ignorance of science. Ignorance has both advantages and disadvantages. It is difficult to argue effectively with someone about a topic they do not know about in the first place. Ignorance means that many people will have anxieties about the technology. Unless people can understand the issues, their perceptions will easily be influenced by arguments that say that all biotechnology is morally dangerous, simply because of their in-built fear of the unknown.

The development of modern technologies is often the subject of heated and polarized debates. At first glance, biotechnology may seem like simply a science and technology matter, but it can change our social values and restructure social relations. Because of this there are several religious and social groups who view the developments in biotechnology in very different ways. For example, two important socially and ethically important issues are:

1. **Should man be allowed to play God?** Should man be allowed to genetically alter an organism by genetic engineering; i.e., clone animals and man?
2. **Who is the owner of the technology and creation**? In the case of genetic engineering, who is the owner of the genetically altered new organism? In the case of cloning, who are the parents of the cloned organism? Even in the case of test tube babies, who should be considered the real mother—the woman who is giving birth to the child or the contributor of the ovum.

In debates, the public, scientists, governments, and religious groups should discuss the social implications of biotechnology as a science, a technology, and a fast developing industry. If we want to address the obstacles to the successful development and commercialization of biotechnology products, we should develop a program that includes activities to harmonize and clarify biotechnology regulations, to understand consumer attitudes toward biotechnology, and to understand the social and ethical concerns raised by this technology. We should analyze and realize the competing social and ethical perspectives of biotechnology and discuss the competing pressures on governments to address socio-ethical and environmental issues.

Biotechnology is a method of production, the applications of which will impact virtually every aspect of human life. Both the method and the applications have socio-ethical implications. There is significant disagreement over whether biotechnology, and its many applications, will benefit or harm society. Biotechnology applications are so diverse that each instance must be judged individually.

In the debate, there are mainly three central levels of disagreement. First, there is disagreement over forecasts of the social consequences of this technology—what could happen. Second, there is disagreement over the ethics that should direct the consequences—what should happen. And finally, there is disagreement over the mechanism, which will be used to develop the ethics—who decides, what should and will happen? The debate over the development of biotechnology is, in fact, a debate over the way our social relations and social values will develop.

The application of biotechnology in society raises many social, ethical, and legal questions. To date, many countries have not developed any system or mechanism to manage the social and ethical issues raised by biotechnology. Instead, government allocation of public money to the research and development has determined the social consequences of biotechnology.

The commercialization process for biotechnology products has been uncertain and prolonged as products are forced to confront socio-ethical issues in hearings and parliamentary committees. Industrial development depends on the creation of a publicly derived socio-ethical framework to direct biotechnology's development.

If the ultimate goal is the successful commercialization of biotechnology, public support is crucial. Public support will, in turn, rest on the social applications of biotechnology. The ethics that will guide the social consequences of biotechnology must be representative of the public. To date, no organized investigation has been undertaken to explore the social and ethical guidelines with which the public would like to manage the development of biotechnology.

Public understanding of biotechnology as a science and technology is important because the products of biotechnology—and consequent benefits and risks—are going to affect everyone. Biotechnology holds great promise as a tool to preserve and enhance environmental quality, food problems, and human health. It can create environmentally safe methods and technologies to address these problems that affect humans and the environment. But without public understanding, acceptance, and support, the role that biotechnology could play in solving environmental and food production problems could be stymied.

3.2 PATENTING (INTELLECTUAL PROPERTY RIGHTS—IPR)

Intellectual property rights (IPR) is a collective term given to a number of different types of legal rights granted by each country. It can be considered as recognition for the contribution of the inventor, by the country, to the development of a new technology, process, or product. It is protected against the unauthorized exploitation for the industrial or commercial purpose.

Intellectual property rights is not a new concept in biotechnology or science. It is a product of the industrial world. It became famous among scientists and laymen equally due to the advancement in scientific researches and the marketing of those products. Some of the concerns about modern biotechnology have focused on the nature, impact, and legitimacy of IP rights as they are applied to gene technology, biodiversity, and to inventions that draw on genetic resources and associated traditional knowledge. Now most institutes and universities have their own experts or groups to deal with their own IP problems.

Types of Intellectual Property Rights

There are different types of intellectual property rights. They are mainly classified into five categories: Patents, trademarks, industrial designs, geographic indications of sources, plant species protection, and copyrights.

These rights are protected in each country by the laws enacted by the parliament or the governing body of the country.

1. **Patents:** This gives protection for inventions, manufacturing processes, or products. It gives its owner a monopoly right to the commercial utilization of the process or product for a limited period of time.
2. **Industrial design:** This relates to the shape and design of certain products such as ornaments and instruments.
3. **Trademark:** This relates to the distinctive words or symbols applied to products, services, or companies.
4. **Protection to plant varieties:** This is a type of IPR that falls under biotechnology. It is known as a plant variety right or plant breeder's right. It is given to new varieties of plants produced by hybridization procedures or by other genetic modifications.
5. **Copyright:** This IPR is for literary and artistic works and craftsmanship, engineering drawings, and software.

3.3 PATENTS

This is a system of laws giving protection to inventors and investors involved in an invention to protect the creation of the inventor for a period of 20 years against use for profit without their consent.

Patent Criteria

For an invention of a product or process to be patented, it should fulfill the following criteria:

- The invention or the information should have novelty. It should not already be in the possession of public.
- It should be inventive; it should not be a traditional technique or product. The technique or process should not be obvious to a person skilled in the particular art.
- It should be capable of industrial application.
- It should have process utility.

Economists have accepted the concept of patent as an encouragement for:

- Economic development through scientific discoveries.
- High investments in research and development.
- Investment in the production and marketing of new products and processes.
- Revelation of knowledge and technological information of inventions as against secrecy.

Discovery *vs* Invention

There is a sharp difference between discovery and invention. Discovery primarily relates to the acquiring of new knowledge by experiments, investigation, and thinking. It is intellectual in nature. The term itself indicates that it is the finding of a new principle, theory, or knowledge, which already exists or is concealed in nature. It adds to existing human knowledge. Example: Discovery of gravitational force by Sir Isaac Newton, elucidation of genetic code by Hargobind Khorana, etc.

Invention is actually the development of something new, which is not pre-existing. There is the application of pre-existing knowledge established by discovery. For example, electricity is a product of discovery, whereas an electric bulb is an invention. Similarly, the knowledge about the power of steam is a discovery and the steam engine is an invention. The techniques of gene cloning, PCR amplification, etc. are inventions. Invention also gives out new knowledge or information. A new procedure which resulted in an old product, a known procedure resulted in a new product, a new result or a new process or a new combination of material to produce a known product, etc., are some examples.

Generally speaking, an invention is patentable and a discovery is not patentable. Now there is confusion about materials that pre-existed and new ones. For example, a synthetic product is a new product, which did not exist previously. So it is definitely patentable. But when you consider a specific protein or gene or any other natural compound, which is isolated or purified, it is not a new compound. But it is patentable, Why? The reason is that these proteins or natural products never existed in previous times in isolated form or purified form. Similarly, a new strain of microorganisms isolated from nature will qualify for patenting if it exhibits a new property that is lacking in the previously known natural form of the bacteria. The patent offices of the U.S., Europe, and Japan have cleared these points through a joint statement of policy.

Product and Process Patents

A new product, which has been synthesized or isolated and purified from natural sources, is patentable according to the patent laws of many countries. Now the question is whether a process is patentable or not. A new method or procedure or process used for the synthesis, isolation, or purification of a known material—a compound or a microorganism—is patentable. The normal procedure is not patentable. If the product is a new one and the procedure used is a known procedure, then the procedure or the process is not patentable, only the product. If both process and product are new then both are patentable.

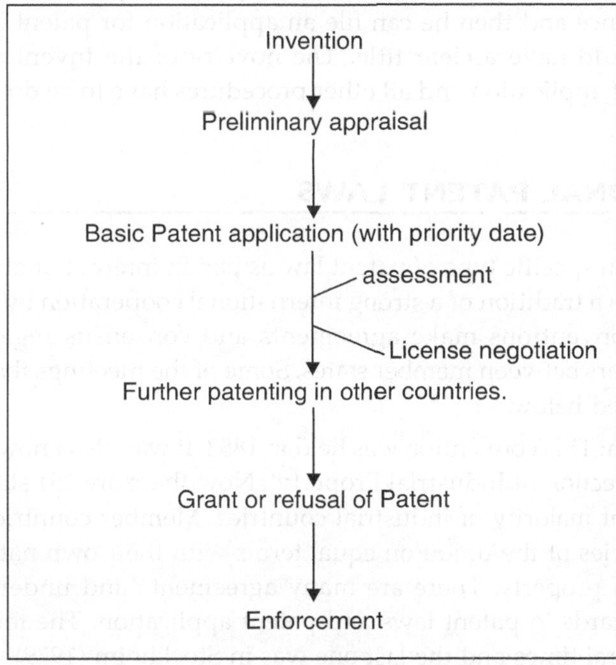

***FIGURE 3.1* The procedure of patenting.**

Reading a Patent

A patent document has a specific format as per international guidelines. The patent structure has two parts:

- Description and
- Claims.

Description: This technical part of the patent contains the following:

- Field of invention and its objective.
- Previous attempts to meet the objective, if any, with the results.
- The remaining part of the problem to be solved.
- The inventive solution and how to apply it.

The solution consists of a general discussion of the different parameters of the process and products with possible variations and substitutions. It will also have the description of some worked examples.

Claims: This pact is the legal component of the patent. This part defines the scope of legal protection required in terms of apparatus, process, products, use, and any other suitable part or category.

Patent Strategy

When someone invents something, the inventor first applies for a patent for that product or process in his country of residence and then he can file an application for patent in other countries. The patent application should have a clear title. The novelty of the invention should be checked thoroughly. The filing of application and all other procedures have to be done by a patent attorney.

3.4 INTERNATIONAL PATENT LAWS

Each country has its own specific type of patent law as per an international agreement or treaty or individually. But there is a tradition of a strong international cooperation by means of international conventions. These conventions make agreements and consensus regarding the formal and substantive patent matters between member states. Some of the meetings that shaped international patent laws are described below.

Paris Convention: This convention was held in 1883. It was also known as 'The International Convention for the Protection of Industrial Property'. Now there are 151 states in the Paris Union, which includes the great majority of industrial countries. Member countries must treat nationals of other member countries of the union on equal terms with their own nationals in regard to the protection of industrial property. There are many agreements and understanding between the member states with regards to patent laws and patent application. The text of Paris Convention has been modified several times and the last one was in Stockholm (1979).

Strasbourg Convention: This was held in 1963. This is the 'Convention on the Unification of Certain Points of Substantive Law on Patents for Invention'. This defines the common requirements for the patentability of an invention such as that it must be amenable to industrial application, must be novel, and must involve an inventive step, etc. Many of the features of this convention have been incorporated into the European Patent Convention. For example, the definition of 'the state of the art' against which the degree of novelty and inventiveness of the subject matter of a patent application must be assessed and judged. The exclusion of plant varieties and animal varieties from patent protection was also decided in this convention.

Patent Co-operation Treaty (PCT): This is considered the international patent body because it is administered by the world Intellectual Property Organization (WIPO) based in Geneva. This treaty was actually signed in 1970 and came into force in 1978 along with the European Patent Convention. It is the broadest international body and has 100 member states. The patent applications filed under PCT are considered 'international' because an international body (WIPO) initially processes them in an international phase before being formally introduced into a designated national system. The international phase is concerned with the formal preliminaries, a prior art search, and publication of the application.

European Patent Convention (EPC): This was held in 1973 and came into force in 1978. It established the European patent organization along with European Patent Office (EPO) and Administrative Council as its supportive organs. Each member country has its own regional patent office. A single patent application can be filed before EPO, which will be considered for its member states. The application may be filed in any of the regional offices or the national patent offices, but will be examined by the EPO in due course. When the patent is granted it will not be a single European Patent, but will be a bundle of national patents for each member state. For example, European Patent (UK), European Patent (Germany), European Patent (France), etc. Thus, a single patent application filed through the EPO will develop into a single unitary indivisible object of property covering the whole of the European economic community.

Budapest Treaty: This treaty known as 'The Budapest Treaty on the International Recognition of the Deposit of Microorganism for the Purposes of Patent Procedure' was signed in 1977 and came into force at the end of 1980. This treaty gives recognition for the microbial culture collections of the International Depository Authorities. A new strain of microorganisms can be deposited with any of the International Depositories for the purpose of a patent application in any member state.

The main functions of a microbial culture collection are to accept, maintain, and provide microbial cultures as a service to the scientific community. They accept new strains of microorganisms isolated by scientists, and its identity will be verified and properly classified. Then it is deposited into their culture collections and is maintained as a new strain. These depositories act as a source of microorganisms for scientific and industrial purposes. The new organism is allotted an accession number and date by which the strain is identified and referred in the future and in patent applications.

3.5 PATENTING IN BIOTECHNOLOGY

Biotechnology in one form or another has been part of human development since the dawn of agriculture. Human ingenuity has led to increased production and greater diversity and quality of livestock and varieties of crops. Today's food crops and domestic animals embody the benefits of many generations of selection and breeding.

Biotechnology continues to offer considerable potential for enhancing human health and well being. Modern biotechnology, including gene technology, is finding increasing application in health care and in a host of industrial and agricultural industries. Effectively applied, modern biotechnology may contribute to economic growth, technological development, and human welfare. Yet it has also raised concerns about ethical and moral issues, equitable sharing of the benefits of biotechnology, environmental impact, the accelerated pace of change, and regulatory challenges. Intellectual property (IP) rights are not new in the biotechnology domain, but some of the concerns about modern biotechnology have focused on the nature, impact, and legitimacy of IP rights as they are applied to gene technology and to inventions that draw on genetic resources and associated traditional knowledge.

Biotechnology patent law operates under the same general legal principles as other areas of patent law. However, the nature of biotechnology raises many unique patent law issues. Biotechnology law is becoming more complex in all industrialized nations of the world. Long-lasting, broad, international patent protection is vital to ensure a financial return after an invention travels the long pipeline from lab to marketplace. The complexity in biotechnology patenting involves patenting of living organisms and natural products, which are not patentable normally, in addition to the products and processes.

For example, a gene that causes a fatal neurological disease and another gene that dramatically improves crop yields. Even though these genes exist naturally and are not patentable, the gene is not occurring in isolated and pure form. In this sense genes and other naturally occurring materials are patentable and can be commercialized. The recent awareness of patents and their elaborated studies is an outcome of biotechnological inventions.

Examples of Inventions that May be Patented

Products

- Genes, including modified genes, expression vectors, and probes.
- Proteins, including modified proteins, monoclonal antibodies, receptors.
- Other chemical compounds such as small organic molecules.
- Cells and new strains or genetically modified organisms.

Methods/Processes

- A new and non-obvious method of using a known compound.
- Methods of medical treatment are not directly patentable in many countries, but they are patentable in the United States. However, methods of medical treatment can be protected indirectly.

Compositions

- Formulation of products for specific uses.

New Uses

- New uses for a previously known compound.
- Observing new abilities of known microorganisms for the synthesis of new compounds or known compounds like antibiotics.

New Methods of Treatment or Diagnosis

- New treatment or diagnosis procedures or methods for instruments, animals, plants, and humans.

Applying

The application process is the same as that of any other patent application. But biotechnology patenting includes a few unique twists. For example, applicants may have to make a cell or seed deposit with a culture collection before filing an application.

Patent Act

The *Patent Act* gives biotechnology patents a maximum term of 20 years from the filing date (patent applications filed before October 1, 1989 have a term of 17 years from the issue date).

3.6 VARIETAL PROTECTION

In conventional plant-breeding experiments for the production of desirable hybrids, there is the **Plant Breeder's Right** and **Farmer's Right**. The Plant Breeder's Right implies the protection of the new variety that has been created by the method of breeding procedures and selection. Farmer's Right means that farmers have the right to produce the seeds from a hybrid plant and raise the seedling for their agricultural purpose.

In the year 2001, two major steps were taken to protect the Farmer's Right in relation to the breeding of new varieties of crops. They are:

1. In the FAO's international treaty on plant genetic resources for food and agriculture there is a specific clause concerning the operation methods of the Farmer's Right.
2. The protection of plant varieties and the Farmer's Act 2001 gives concurrent attention to the rights of farmers, plant breeders, and researchers and also the protection of public interests. This include
   (*a*) Compulsory licensing of rights
   (*b*) Prevention of import of varieties incorporated with Genetic Use Restriction Technology (GURT) such as the seeds with the terminator gene (the terminator technology of Monsanto).
   In this technology, the plants are genetically engineered with a gene known as the terminator gene that makes the seeds non-viable. Therefore, at each time of cultivation the farmers have to purchase the seeds from the companies that produce the seedlings by tissue culture technology.

3.7 ETHICAL ISSUES IN BIOTECHNOLOGY—AGRICULTURE AND HEALTH CARE

The government's promotional activities surrounding biotechnology have inspired debate. The use of public funds to further biotechnology research supports the use of a method that has social implications and around which there exists significant social disagreement. The government's decision to allocate funds to certain niche areas determines the applications, which are most likely to be commercialized. This history has established a relationship between the government and the public. It has created trust and support among some, and distrust and opposition among others, because of the socio-ethical position implied in the policy.

Using public resources for biotechnology's development raises objections from stakeholders who do not believe this technology will benefit society, or who believe that its current course of development poses socio-ethical risks. The use of public resources for the development of biotechnology is supported by the biotech industry who believe that this technology offers significant health, environmental, and economic benefits to the public, and who believe that facilitating the development of this industry for those benefits is a government responsibility.

The key questions that arise when ethical issues are raised, are:

- Should government decide the ethics? and
- What should those ethics be?

The current debate on ethical issues related to biotechnology in agriculture and healthcare has raised certain debating points in the following fields:

- Release of genetically-modified plants and animals to the environment could cause disturbance in the existing ecological balance in an unpredictable manner.
- The release or use of genetically-engineered microorganisms for industrial purposes can lead to the generation of new infectious types of organisms.
- Introduction of artificial agents (such as genetically-engineered bacteria, fish, etc.) as biological control agents can cause serious imbalance in the environment and can lead to disastrous consequences.
- Genetic engineering of plants for herbicide resistance and enhanced photosynthesis could result in more tolerant weeds, as a result of cross-pollination with related plant species.
- Cloning of animals and humans caused much debate on ethical issues. There are a number of social and ethical issues that will be affected by biotechnology. For example, the social and ethical problems associated with new reproductive technologies, which include *in vitro* fertilization and human cloning. There are also technological risks associated with human and animal cloning. There is uncertainty regarding the current state of technology.
- The new reproductive technologies may be harmful for individual autonomy, equality, protection of the vulnerable, accountability, respect for human life and dignity, non-commercialization of reproduction, appropriate use of resources, and balancing individual and collective interests.

Proponents of biotechnology argue, correctly, that biotechnology is being held to standards that are not demanded of other advanced technologies such as computers or the information highway. There is considerable debate over whether biotechnology should be held to distinct standards because it is a power that raises special social questions. What is unquestionable is that biotechnology requires considerably more public trust than any other technology because of its potential power to transcend, quickly and intentionally, any God-given or natural limit to human activity.

REVIEW QUESTIONS

1. What is the public perception of biotechnology?
2. What is meant by sustainable development?
3. What is IP and IPR?
4. What is a patent? What are its advantages in scientific studies and industries?
5. What is the Farmer's Right and the Plant Breeder's Right?
6. Give the general structure of a patent. What are the criteria for a finding to be patented?
7. What are the special problems in the patent laws of biotechnology?
8. What is the justification for the patenting of natural materials like microbial strains, molecules like proteins, nucleic acids, and genes?
9. Expand the following:

   (*a*) PCT (*b*) GM Plants (*c*) GURT

   (*d*) WTO (*e*) EPC.

Part 2 BIOMOLECULES

Biomolecules are those compounds synthesized by living organisms. These groups of compounds have different sizes, shapes, chemical and physical properties, and biological functions. These biomolecules include different classes of compounds, which are broadly divided into two categories, depending on size and nature. Those molecules, which are polymers and bigger in size, are known as macromolecules and other molecules, which are simple and small in size, are biomolecules. There are four types of macromolecules in biological systems; namely, carbohydrates, proteins, lipids, and nucleic acids. Out of these four types three are polymers composed of monomers, or building blocks. Lipids are not polymers.

This part is divided into three chapters. In the first chapter we study the small molecules including the building blocks of macromolecules. This includes monosaccharides or sugars, amino acids, nucleotides, vitamins, coenzymes, and fatty acids. Some of these molecules form the building blocks of macromolecules. For example, amino acids are the building blocks of proteins. In biological systems, all these molecules, both macro and micro, are in a state of flux or in a dynamic state. That is, they are always subjected to chemical transformations in order to maintain the state of life. Actually, the sum total of all these biochemical changes forms the metabolism of an organism. These metabolic reactions are interconnected and arranged into specific metabolic pathways, and they may be linear or cyclic.

The second chapter mainly discusses macromolecules, such as protein, nucleic acids, carbohydrates, and lipids. In proteins, there is a special focus on enzymes, which mediate all cellular biochemical reactions. Nucleic acids are mainly genetic materials, carbohydrates form the primary energy sources, and lipids are the energy reservoirs, and they also form the barriers of cells and organelles. All these biomolecules organize in different manners and form supramolecular assemblies, and that gives rise to the organelles such as cell membranes, ribosomes, etc.

The third chapter deals with the various physico-chemical techniques for the isolation and purification of macromolecules such as protein, nucleic acid, lipids, etc. It includes techniques such as centrifugation and separation techniques such as electrophoresis and chromatography.

Chapter 4

BUILDING BLOCKS OF BIOMOLECULES—STRUCTURE AND DYNAMICS

In This Chapter

4.1 INTRODUCTION

The living cell of an organism is a chemical machine; its parts and its language are chemical. It is an assembly of various types of biomolecules having diverse physico-chemical and biological properties. All these small and large biomolecules take part in a variety of biochemical reactions and the phenomenon of life is an outcome of these activities. Biopolymers, or biological macromolecules, form the major class among biomolecules both structurally and functionally. All cellular structures such as cell walls, cell membranes, membranes of organelles, microtubules, flagella, and celia have various types of biomolecules, which are actively engaged in different metabolic reactions.

These macromolecules are also known as biopolymers. They can be classified into four major groups—carbohydrates, proteins, lipids, and nucleic acids. All these biopolymers are composed of small repeating units or the monomers of the same type or different types. These structural units of biopolymers are called building blocks. Even though lipids are not polymers they are macromolecules and put among the biopolymers. In this chapter we discuss the structure and dynamics of the building blocks of all major macromolecules of the biological system. The monomers of these biopolymers are characterized by the presence of certain specific functional groups. Therefore, before going into the details of these monomers, a basic understanding of the functional groups found in the biomolecules is required.

Monomers and Polymers

The basic principle of a cell is that it acquires small molecules (**monomers**), then assembles these into much larger molecules (**polymers**).

TABLE 4.1

Monomer	Polymer
amino acids	proteins
nucleotides	DNA, RNA
sugars	polysaccharides
fatty acids	complex lipids

Elements of Life

An element is a pure substance that is made up of only one kind of atom. There are 92 naturally occurring elements. All elements that are present on earth are not represented in the living system. Similarly, the abundance of elements on earth's crust is not mirrored in living tissue. Some of earth's major elements are not found in cells (e.g., aluminum) and some major elements in cells are not abundant on earth's crust (e.g., carbon). There are about 25 elements observed in the living cells.

The elements present in the cells can be identified as **macro** and **micro elements.**

Macro Elements—CHNOPS

These six elements are the predominant elements in all cells.

1. **C** = Carbon
2. **H** = Hydrogen
3. **N** = Nitrogen
4. **O** = Oxygen
5. **P** = Phosphorus
6. **S** = Sulfur

Some organisms have extra abundance of other elements (e.g., vertebrates have lots of calcium, mostly in bone).

Micro Elements and Trace Minerals

The remaining elements are required in comparatively fewer amounts, generally very small amounts relative to CHNOPS and are used in a variety of ways. Many of these minerals are critical for the function of certain enzymes (e.g., copper, molybdenum, zinc) and others are required for certain functions:

- Calcium for bone
- Silicon for glassy shell in certain protists
- Potassium and sodium for generation and transmission of nerve impulse in animals

Atoms and Isotopes

- An atom is the smallest unit of matter. It contains electrons (– charge), protons (+ charge), and neutrons (no charge), with electrons represented dynamically as "electron clouds."
- The number of protons and electrons in an atom is the same, unique to each **element.**

  Example: H = 1 proton; C = 6 protons; O = 8 protons. This is also called the **atomic number.**
- Electrons fit into specific geometrical patterns called **orbitals** (not described further here—learn about this in chemistry). Each orbital holds only two electrons.
- As orbitals closest to nucleus become filled, additional electrons must occupy orbitals further from the nucleus.
- Groups of orbitals can be approximated as **electron shells.**
- Some atoms lose or gain electrons and become charged **ions.**

  Examples: $Na \longrightarrow Na^+$; $Cl \longrightarrow Cl^-$
- Isotopes are the atoms having the same atomic number and different atomic weight. That is, they differ in neutron number. Example: isotopes of carbon.

  C^{12} = "normal" (99+ %)

  C^{13} = less common (<1%)

  C^{14} = "unstable" isotope, weakly radioactive; useful in research, in biology and medicine as tracer element.

Molecules

- Elements adjacent to noble gases tend to ionize and can form ionic bonds between them to form molecules.
- Elements further away from noble gases tend to form **covalent bonds** by electron sharing and become stably connected to **molecules.**
- Each atom in a molecule tries to complete an outer electron shell by sharing electrons.
- The number of electrons shared is called **valence.**

  Valences of C, O, N, H, S, P:

  H = 1, O = 2, N = 3, C = 4, P = 5, and S = 2

Some examples of molecules:

- 2H + O = H_2O **(water)**
- C + 4H = CH_4 **(methane)**
- 6C + 12H + 6O = $C_6H_{12}O_6$ **(glucose)**

The Nature of Chemical Bonds

Ionic Bonds

Ionic bonds are formed when there is a complete transfer of electrons from one atom to another, resulting in two ions, one positively charged and the other negatively charged. The electrostatic attractions between the positive and negative ions hold the compound together. For example, when a sodium atom (Na) donates one electron in its outer valence shell to a chlorine (Cl) atom, which needs one electron to fill its outer valence shell, NaCl (table salt) results. The symbol for sodium chloride is Na^+Cl^-. The energy value or strength of electronic bonds is often 4-7 kcal/mol.

Covalent Bonds

Covalent Bonds are the strongest chemical bonds and are formed by the sharing of a pair of electrons. These bonds are formed by elements such as C, H, N, O, P, S, which are usually not found immediately adjacent to noble gases in the periodic table. The energy of a typical single covalent bond is ~80 kilocalories per mole (kcal/mol). However, this bond energy can vary from ~50 kcal/mol to ~110 kcal/mol depending on the elements involved. Once formed, covalent bonds rarely break spontaneously. This is due to simple energetic considerations; the thermal energy of a molecule at room temperature (298 K) is only ~0.6 kcal/mol, which is much lower than the energy required to break a covalent bond.

There are single, double, and triple covalent bonds as shown in Table 4.2.

TABLE 4.2 Covalent bonds.

Bond Number	Example	Energy (kcal/mol)
Single	H │ H–C–H │ H	~80
Double	H H │ │ H–C = C–H │ │ H H	~150
Triple	H │ C = C │ H	~200

Note that carbon-carbon bonds are unusually strong and stable covalent bonds.

Usually covalent bonds are without any charges. But in certain cases they can also develop partial charges when the atoms involved have different electro negativities. Water is perhaps the most obvious example of a molecule with partial charges. The symbols delta+ and delta– are used to indicate partial charges.

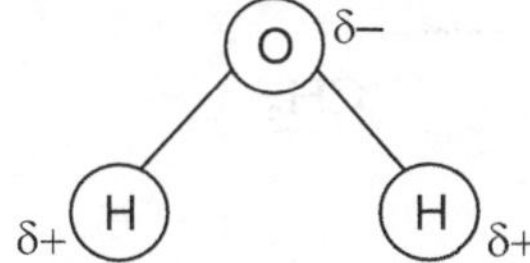

FIGURE 4.1 **Polarity of water molecule.**

Oxygen, because of its high electro-negativity, attracts the electrons away from the hydrogen atoms, resulting in a partial negative charge on the oxygen and a partial positive charge on each of the hydrogen atoms.

The possibility of hydrogen bonds (H-bonds) is a consequence of partial charges.

Hydrogen Bonds

A hydrogen bond is a weak interaction and is stable only for short periods of time. Its bond strength is ~ 3-5 kcal/mole.

In covalent linkages, if electrons are not evenly distributed, the result is **polarity,** which results in hydrogen bonding. A hydrogen atom covalently attached to a very electronegative atom (N, O, or P) shares its partial positive charge with a second electronegative atom (N, O, or P). In H_2O, H carries a fraction of positive charge and O fraction of negative charge. Therefore, water molecules can form hydrogen bonding with each other very easily. These bonds are frequently found in proteins and nucleic acids, and by reinforcing each other, serve to keep the protein (or nucleic acid) structure secure.

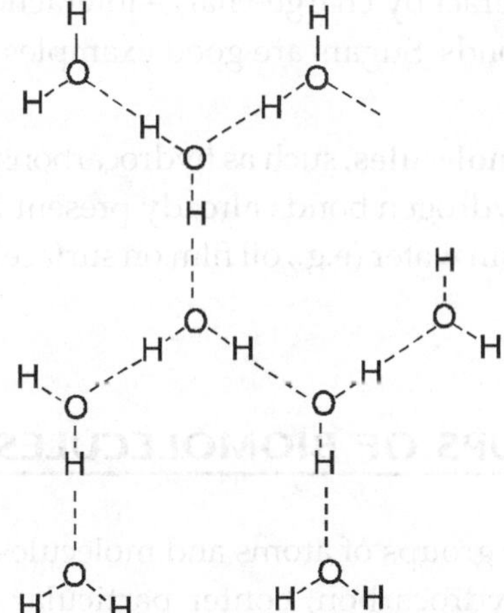

FIGURE 4.2 **Hydrogen bonding between water molecules.**

Van der Waal Bonds

Van der Waal interactions are very weak bonds (generally no greater than 1 kcal/mol) formed between non-polar molecules or non-polar parts of a molecule. The weak bond is created because a C-H bond can have a transient dipole and induce a transient dipole in another C-H bond.

```
 H            H
 |    ~~~~    |
CH3          CH3
```

Hydrophobic Interactions

Non-polar molecules cannot form H-bonds with H_2O, and are therefore insoluble in H_2O. These molecules are known as **hydrophobic** (water hating), as opposed to water loving hydrophilic molecules, which can form H-bonds with H_2O. Hydrophobic molecules tend to aggregate together in avoidance of H_2O molecules; hydrophobic interactions are clearly demonstrated when you put an oil drop on water. This attraction/repulsion is known as the hydrophobic (fear of water) force. To understand the energetics driving this interaction, visualize the H_2O molecules surrounding a "dissolved" molecule attempting to form the greatest number of hydrogen bonds with each other. The best energetic solution involves forcing all of the non-polar molecules together, thus reducing the total surface area that breaks up the H_2O H-bond matrix.

Polar and Non-polar Molecules

- Some molecules have electrons asymmetrically distributed that results in polar **molecules**. Because of polarity, molecules acquire a **dipole moment**, and will orient to one direction when placed in an electrical field.

  Example: **CH_3CH_2OH** (ethanol); the oxygen atom attracts more of the electrons than other atoms.
- All polar molecules interact by charge-charge interactions; interact well with water, and form good hydrogen bonds. Sugars are good examples of very soluble molecules, due to many **–OH** groups.
- By contrast, **non-polar molecules**, such as hydrocarbons, do not dissolve in water, because to do so would break hydrogen bonds already present in water. Instead, these molecules form a layer separate from water (e.g., oil film on surface) or form globules of fatty material inside water.

4.2 FUNCTIONAL GROUPS OF BIOMOLECULES

The functional groups are specific groups of atoms and molecules which, when substituted for one or more hydrogen atoms in a hydrocarbon, confer particular chemical properties to the new compound. The chemical property or the "behavior" of a molecule can be attributed to the functional group present in it. For example, all organic acids are characterized by the presence of –COOH group. The following are the important functional groups present in the biological molecules. Table 4.3 also gives a summary of the most common functional groups present in various biomolecules.

Hydrophobic Functional Groups

Methyl group Ethyl group Phenyl group

Polar Functional Groups

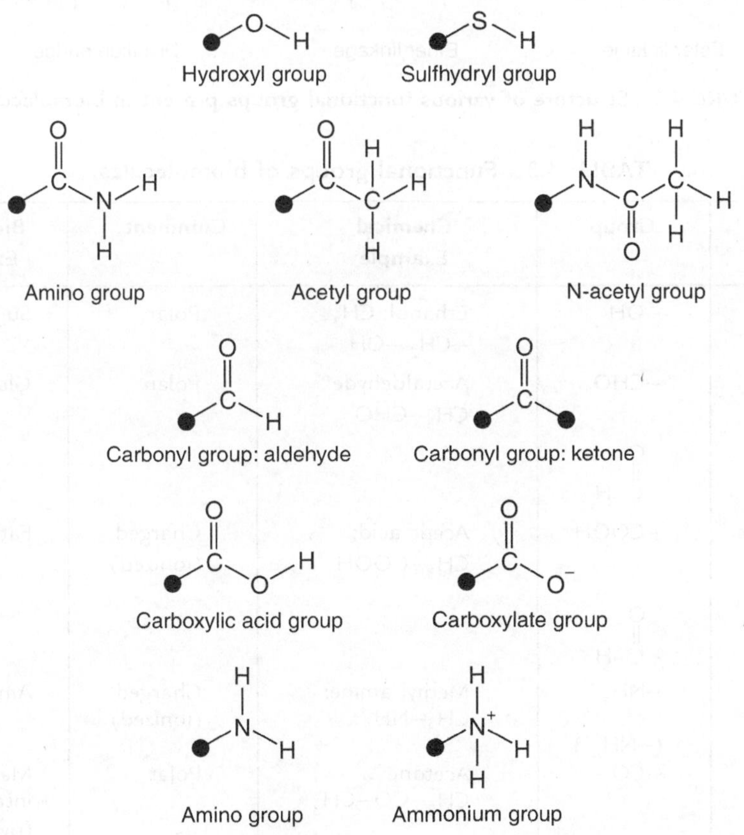

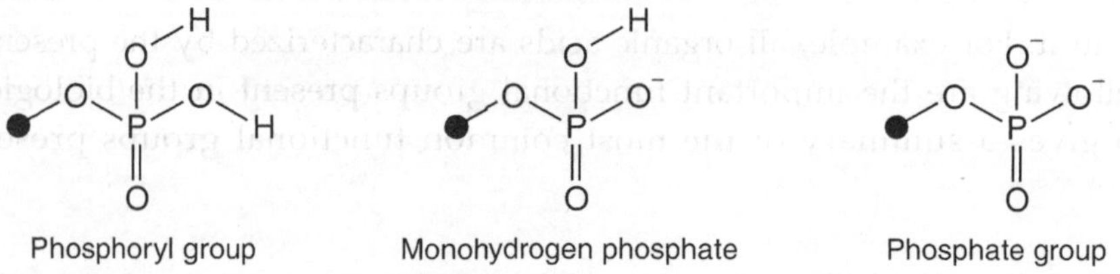

Linkages and Bridges

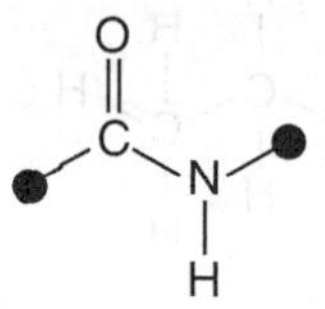

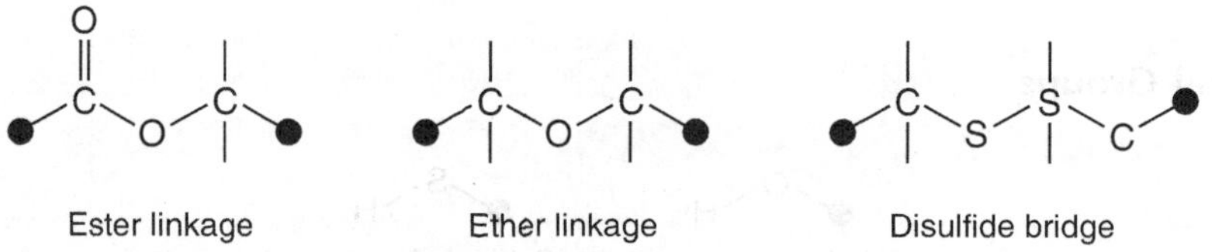

FIGURE 4.3 **Structure of various functional groups present in biomolecules.**

TABLE 4.3 Functional groups of biomolecules.

Name	Group	Chemical Example	Comment	Biological Example
Hydroxyl:	–OH	Ethanol: CH_3 $-CH_2-OH$	Polar	Sugars
Aldehyde:	–CHO	Acetaldehyde: CH_3-CHO	Polar	Glucose
Detail:	$-\overset{O}{\overset{\Vert}{C}}-H$			
Carboxylic acid:	–COOH	Acetic acid: CH_3-COOH	Charged (ionized)	Fatty acid
Detail:	$-\overset{O}{\overset{\Vert}{C}}-H$			
Amine:	$-NH_2$	Methyl amine: CH_3-NH_2	Charged (ionized)	Amino acid
Detail:	$(-NH_3^+)$			
ketone:	–CO–	Acetone: $CH_3-CO-CH_3$	Polar	Metabolic intermediate (pyruvic acid)

Detail:	O ‖ —C—			
Ether:	—O—	Ethyl ether: CH_3—CH_2—O—CH_2—CH_3	Not so polar (symmetric)	Some lipids
Ester:	—COOR	Methyl ester of acetic acid: CH_3—CO —OCH_3	Polar, not charged	Fats
Detail:	O ‖ —C—O—R			
Amide:	—$CONH_2$	acetamide: CH_3—$CONH_2$	Polar, not charged	Proteins (asparagine)
Detail:	O ‖ —C—NH_2			
Sulfhydryl:	—SH_2	2-mercaptoethanol: HO—CH_2 —CH_2—SH	Reducing agent	Protein (cysteine)

4.3 BUILDING BLOCKS OF CARBOHYDRATES

Carbohydrates are carbon compounds that contain large quantities of hydroxyl groups. The simplest carbohydrates, commonly known as sugars, also contain either an aldehyde group or a ketone group. The aldehyde-containing sugars are called **polyhydroxyaldehydes** and ketone-containing simple sugars are called **polyhydroxyketones.** All carbohydrates can be classified as **monosaccharides, oligosaccharides,** or **polysaccharides.** Monosaccharides are the monomers or the building blocks of carbohydrates. Anywhere from two to ten monosaccharide units, linked by glycosidic bonds, make up an oligosaccharide. Polysaccharides are much larger, containing hundreds of monosaccharide units. The presence of the hydroxyl groups allows carbohydrates to interact with the aqueous environment and to participate in hydrogen bonding, both within and between chains. Derivatives of the carbohydrates can contain nitrogens, phosphates, and sulfur compounds. Carbohydrates can also combine with lipids to form glycolipids or with protein to form glycoproteins.

Monosaccharides

The monosaccharides are commonly known as simple sugars such as glucose, fructose, galactose, etc. The various types of sugars present in nature are classified according to the number of carbons they contain in their backbone structures. Their basic molecular formula is $C_nH_2O_n$, and therefore are considered as the hydrates of carbon. The monosaccharides are classified into two categories based on the type of functional group they possess. Those sugars having aldehyde groups are called aldoses and those having keto groups are called ketoses. The major monosaccharides, both ketoses and aldoses, contain four to seven carbon atoms. Accordingly, monosaccharides are classified into seven groups depending on the number of carbon atoms. They are triose with three carbon atoms, tetrose with four carbon atoms, pentose with five carbon atoms, hexose with six carbon atoms, and heptulose with seven carbon atoms. Each of these groups includes both aldoses and ketoses.

The trioses-having aldehyde groups are called aldotriose and aldoketose. Similarly, tetrose may be aldotetrose or ketotetrose; pentose sugar includes aldopentose and ketopentose; and the hexose sugar includes aldohexose and ketohexose and so on. Heptulose or the seven-carbon sugar is not common and is not present in free form. Sedoheptulose, a keto form of heptose, is present in plants as an intermediate in the biosynthetic pathway of glucose in photosynthesis. Table 4.4 gives a summary of the classification of monosaccharides.

TABLE 4.4 Classifications of monosaccharides.

Number of carbons	Category name	Relevant examples
3	**Triose** Aldotriose Ketotriose	 Glyceraldehyde Dihydroxyacetone
4	**Tetrose** Aldotetrose Ketotetrose	 Erythrose Erythrulose
5	**Pentose** Aldopentose Aldoketose	 Ribose, Xylose Ribulose, Xylulose
6	**Hexose** Aldohexose Ketohexose	Glucose, Galactose, Mannose, Fructose
7	**Heptose** Aldoheptose Ketoheptose	Not naturally occurring aldoheptose, sedoheptulose

Nomenclature of Monosaccharides—D and L Forms

The predominant monosaccharides found in the body are structurally related to the aldotriose **glyceraldehyde** and to the ketotriose **dihydroxy-acetone**. All carbohydrates contain at least one asymmetrical (chiral) carbon and are, therefore, optically active. In addition, carbohydrates can exist in either of two conformations, as determined by the orientation of the hydroxyl group. With a few exceptions, those carbohydrates that are of physiological significance exist in the **D-conformation**. The mirror-image conformations, called **enantiomers**, are in the **L-conformation** (Figure 4.4).

CHO CHO

H — C — OH HO — C — H

CH_2OH CH_2OH

D–Glyceraldehyde L–Glyceraldehyde

FIGURE 4.4 **Structures of glyceraldehyde enantiomers.**

The simple sugars, or the monosaccharides, have two types of functional groups:

- —OH (Hydroxyl) and
- =O (Carbonyl) which includes both aldehyde and ketone.

The aldehyde and ketone moieties (carbonyl group) of the monosaccharides with five and six carbons will spontaneously react with alcohol groups present in neighboring carbons to produce intramolecular **hemiacetals** or **hemiketals**, respectively. This results in the formation of five- or six-membered rings. Because the five-membered ring structure resembles the organic molecule **furan**, derivatives with this structure are termed **furanoses.** Those with six-membered rings resemble the organic molecule **pyran** and are called **pyranoses**. Such structures can be depicted by either **Fischer** or **Haworth** style diagrams. The numbering of the carbons in monosaccharides proceeds from the carbonyl carbon, for aldoses, or the carbon nearest the carbonyl, for ketoses.

The rings can open and re-close, allowing rotation to occur about the carbon bearing the reactive carbonyl, yielding two distinct configurations (α and β) of the hemiacetals and hemiketals. The carbon about which this rotation occurs is the **anomeric carbon** (the carbon No. 1 in glucose and carbon No. 2 in fructose) and the two forms are called anomers. Carbohydrates can change spontaneously between α and β configurations—a process known as **mutarotation**. When drawn in the Fischer projection, the α configuration places the hydroxyl attached to the anomeric carbon to the right, toward the ring. When drawn in the Haworth projection, the α configuration places the hydroxyl downward as shown in Figure 4.5.

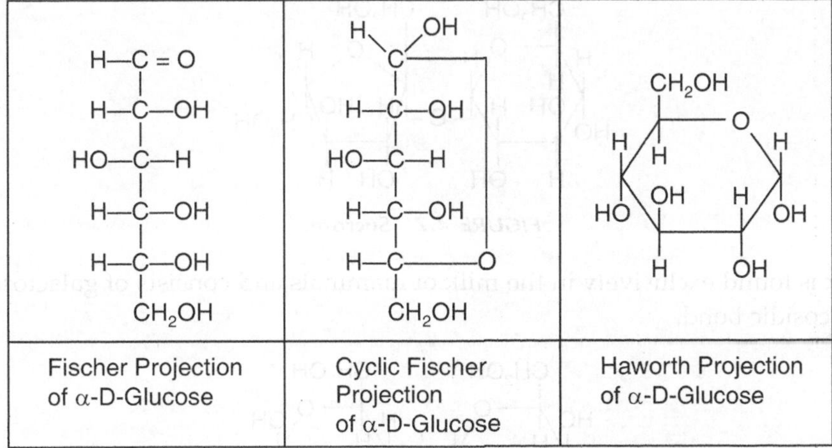

FIGURE 4.5 **Structure of glucose, three different structural representations.**

The spatial relationships of the atoms of the furanose and pyranose ring structures are more correctly described by the two conformations identified as the **chair form** (Figure 4.6) and the **boat form**. The chair form is the more stable of the two. Constituents of the ring that project above or below the plane of the ring are **axial** and those that project parallel to the plane are **equatorial**. In the chair conformation, the orientation of the hydroxyl group about the anomeric carbon of α-D-glucose is axial and equatorial in β-D-glucose.

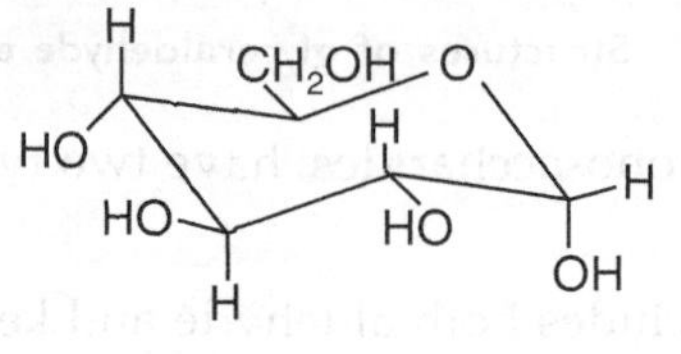

FIGURE 4.6 **Chair form of α-D-glucose.**

Disaccharides

The hydroxyl group of anomeric carbon of a cyclic monosaccharide can react with the hydroxyl group of another monosaccharide to form disaccharide. Covalent bonds between the anomeric hydroxyl of a cyclic sugar and the hydroxyl of a second sugar (or another alcohol-containing compound) are called **glycosidic bonds,** and the resultant molecules are **glycosides**. The linkage of two monosaccharides to form disaccharides involves a glycosidic bond. Various numbers of monosaccharides can be joined in this manner into a chain to form carbohydrate polymers (disaccharides, trisaccharides, oligosaccharides (a few) or polysaccharides (many)).

Several physiogically important disaccharides are sucrose, lactose, and maltose.

Sucrose: prevalent in sugar cane and sugar beets, is composed of glucose and fructose through an α-(1,2) β-glycosidic bond.

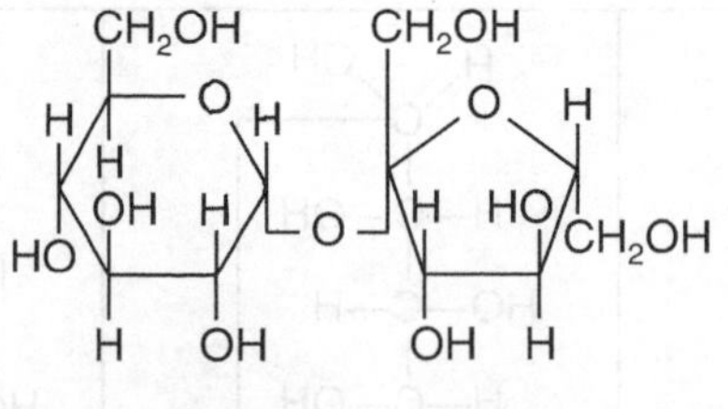

FIGURE 4.7 **Sucrose.**

Lactose: is found exclusively in the milk of mammals and consists of galactose and glucose in a β-(1,4) glycosidic bond.

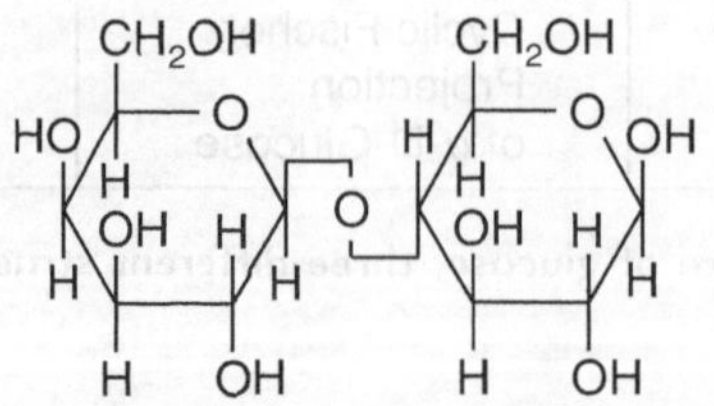

FIGURE 4.8 **Lactose.**

Maltose: the major degradation product of starch, is composed of two glucose monomers in an α-(1,4) glycosidic bond.

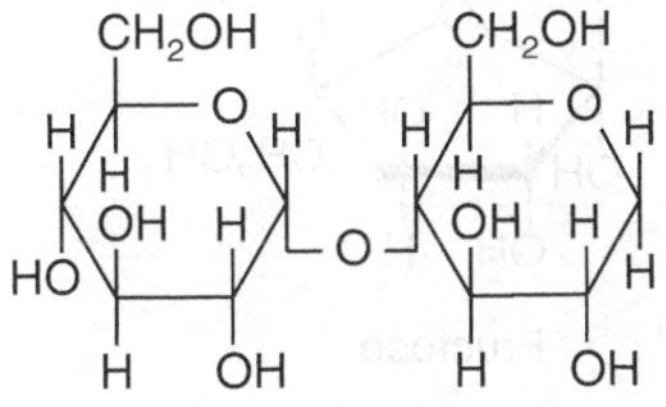

FIGURE 4.9 **Maltose.**

Polysaccharides

Most of the carbohydrates found in nature occur in the form of high molecular weight polymers called polysaccharides. The monomeric building blocks used to generate polysaccharides can be varied. In all cases, however, the predominant monosaccharide found in polysaccharides is D-glucose. When polysaccharides are composed of a single monosaccharide building block, they are called **homopolysaccharides.** Polysaccharides composed of more than one type of monosaccharide are called **heteropolysaccharides.**

Glycogen

Glycogen is the major form of stored carbohydrate in animals. This crucial molecule is a homopolymer of glucose in α-(1,4) linkage; it is also highly branched, with α-(1,6) branch linkages occurring every eight to ten residues. Glycogen is a very compact structure that results from the coiling of the polymer chains. This compactness allows large amounts of carbon energy to be stored in a small volume, with little effect on cellular osmolarity.

Starch

Starch is the major form of stored carbohydrate in plant cells. Its structure is identical to glycogen, except for a much lower degree of branching (about every 20-30 residues). Unbranched starch is called **amylose** (Figure 4.4), branched starch is called **amylopectin.**

Chemical Properties of Sugars

Monosaccharides are called reducing sugars. The presence of the hemiacetal group or the aldehyde group in monosaccharides makes them a reducing agent. It can reduce alkaline solutions of copper salts giving rise to red precipitate of cuprous oxide. This is the basic reaction of the qualitative tests such as Fehling's and Benedict's test used to identify the presence of glucose in various samples like urine, blood, etc. Both Fehling's and Benedict's reagents contain copper sulfate in an alkaline state. These tests are routinely conducted in pathological laboratories to find out the presence of sugar in clinical samples. All monosaccharides and disaccharides except sucrose are positive to these tests and are therefore reducing sugars. In sucrose, the glycosidic linkage is between the α-Glucose

Fructose

Sucrose
(glucose (α1 → 2) fructose)

FIGURE 4.10 **Glycosidic linkage in sucrose.**

anomeric carbon (C-1) of glucose and anomeric carbon (C-2) of fructose. Thus, both the anomeric carbons are involved in the glycosidic linkage and therefore there is no other active group to participate in the reaction (Figure 4.10). Hence, sucrose is a non-reducing sugar. Aldoses and ketoses are distinguished by Seliwanoff's test. Ketoses and aldoses undergo dehydration with concentrated sulfuric acid and form furfural derivatives. In the case of ketones, furfural derivatives condense with resorcinol and give a red complex.

Cellulose poly
(1, 4′-O-β-D-glucopyranoside)

FIGURE 4.11 **Starch and cellulose—the alpha and beta linkages.**

To distinguish between alpha and beta linkages, examine the position of the hydrogen on the first carbon molecule (Hemiacetal group). In an alpha linkage, the hydrogen is pointing up, and in a beta linkage it is pointing down.

4.4 BUILDING BLOCKS OF PROTEINS

Proteins are very large molecules composed of combinations of 20 different amino acids. The precise physical shape of a protein is very important for its function. There are many different proteins essential for the functioning of each cell in a living organism.

All proteins (peptides and polypeptides) are polymers of alpha amino acids. There are 20 α-amino acids that are relevant to the make-up of all proteins. Several other amino acids are found in the body free or in combined states (i.e., not associated with peptides or proteins). These non-protein associated amino acids perform specialized functions. Several of the amino acids found in proteins also perform functions distinct from the formation of peptides and proteins, e.g., tyrosine in the formation of thyroid hormones or glutamate acting as a neurotransmitter.

The α-amino acids in peptides and proteins consist of a carboxylic acid (**–COOH**) and an amino (**–NH_2**) functional group (except in proline, in which there is **–NH** group instead of **–NH_2** group) attached to the same tetrahedral carbon atom. This carbon is the α-carbon. Distinct R-groups, which distinguish one amino acid from another, are also attached to the alpha-carbon (except in the case of glycine where the R-group is hydrogen). The fourth substitution on the tetrahedral α-carbon of amino acids is hydrogen.

All 20 amino acids have the same general formula as given below.

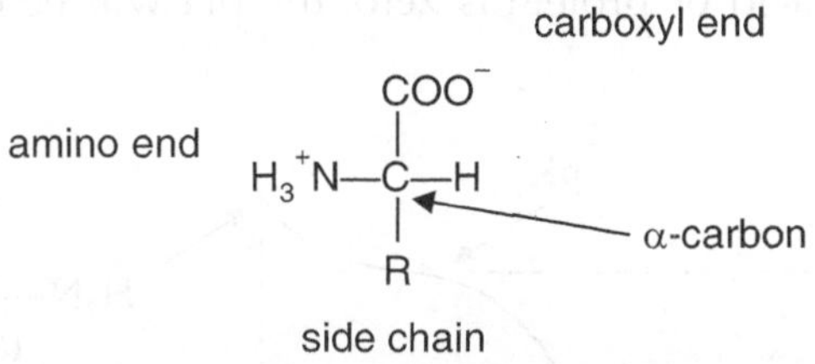

carboxyl end

$$\begin{array}{c} COO^- \\ | \\ H_3{}^+N\text{—}C\text{—}H \\ | \\ R \end{array}$$

amino end

α-carbon

side chain

FIGURE 4.12 **General structure of amino acid.**

The R-group in glycine is an H atom and in proline the α-carbon atom has an imino group (–NH) instead of an amino (–NH_2) group. All naturally occurring amino acids have L-configuration with respect to α-carbon position. There is no naturally occurring D-amino acid except in bacterial cell walls.

Properties of Amino Acids

Hydrophobicity and Hydrophilicity of Amino Acids

Each of the 20 α-amino acids found in proteins can be distinguished by the R-group substitution on the α-carbon atom. There are two broad classes of amino acids based upon whether the R-group is hydrophobic or hydrophilic. The hydrophobic amino acids tend to repel the aqueous environment and, therefore, reside predominantly in the interior of proteins. This class of amino acids does not ionize nor participates in the formation of H-bonds. The hydrophilic amino acids tend to interact with the aqueous environment, are often involved in the formation of H-bonds, and are predominantly found on the exterior surfaces of proteins or in the reactive centers of enzymes.

Acid-Base Properties of the Amino Acids

The α–COOH and α–NH_2 groups in amino acids are capable of ionizing (as are the acidic and basic R-groups of the amino acids). As a result of their ionizability, the following ionic equilibrium reactions may be written:

$$R\text{–}COOH \rightleftharpoons R\text{–}COO^- + H^+$$

$$R\text{–}NH_3^+ \rightleftharpoons R\text{–}NH_2 + H^+$$

The equilibrium reactions, as written, demonstrate that amino acids contain at least two weak acidic groups. However, the carboxyl group is a far stronger acid than the amino group. At physiological pH (around 7.4) the carboxyl group will be unprotonated and the amino group will be protonated. An amino acid with no ionizable R-group would be electrically neutral at this pH. This species is called a **zwitterion**.

Like typical organic acids, the acidic strength of the carboxyl, amino, and ionizable R-groups in amino acids can be defined by the dissociation constant, K_a or more commonly the negative logarithm of K_a, the pK_a. The **net charge** (the algebraic sum of all the charged groups present) of any amino acid, peptide, or protein, will depend upon the pH of the surrounding aqueous environment. As the pH of a solution of an amino acid or protein changes so too does the net charge. This phenomenon can be observed during the titration of any amino acid or protein. When the net charge of an amino acid or protein is zero, the pH will be equivalent to the **isoelectric point: pI**.

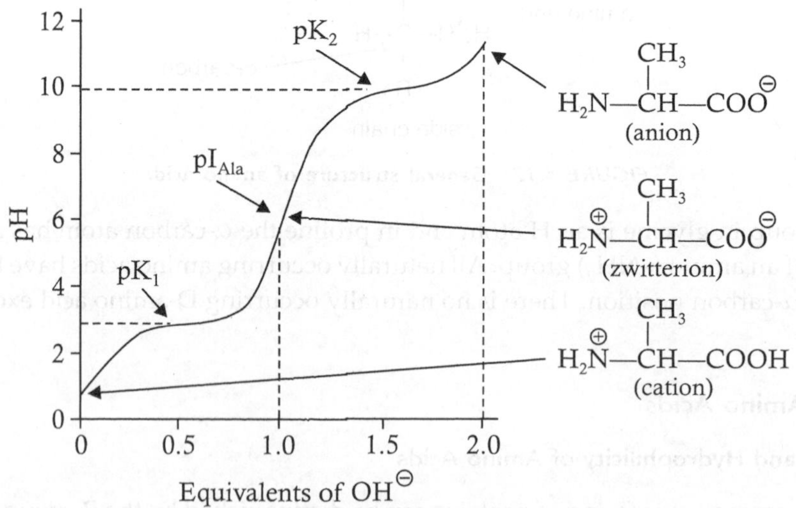

FIGURE 4.13 **Titration curve for alanine.**

Functional Significance of Amino Acid R-Groups

In solutions it is the nature of the amino acid R-groups that dictates structure-function relationships of peptides and proteins. The hydrophobic amino acids will generally be encountered in the interior of proteins shielded from direct contact with water. Conversely, the hydrophilic amino acids are generally found on the exterior of proteins as well as in the active centers of enzymatically active proteins. Indeed, it is the very nature of certain amino acid R-groups that allows enzyme reactions to occur.

The imidazole ring of histidine allows it to act as either a proton donor or acceptor at physiological pH. Hence, it is frequently found in the reactive center of enzymes. Equally important is the ability of histidines in hemoglobin to buffer the H^+ ions from carbonic acid ionization in red blood cells. It is this property of hemoglobin that allows it to exchange O_2 and CO_2 at the tissues or lungs, respectively.

The primary alcohols of serine and threonine as well as the thiol (–SH) of cysteine allow these amino acids to act as nucleophiles during enzymatic catalysis. Additionally, the thiol of cysteine is able to form a disulfide bond with other cysteines:

$$\textbf{Cysteine—SH +HS—Cysteine} \rightleftharpoons \textbf{Cysteine—S—S—Cysteine}$$

This simple disulfide is identified as cysteine. The formation of disulfide bonds between cysteines present within proteins is important to the formation of active structural domains in a large number of proteins. Disulfide bonding between cysteines in different polypeptide chains of oligomeric proteins plays a crucial role in ordering the structure of complex proteins, e.g., the insulin receptor.

Optical Properties of the Amino Acids

A tetrahedral carbon atom with four distinct constituents is said to be **chiral**. The one amino acid not exhibiting chirality is glycine since its "R-group" is a hydrogen atom. Chirality describes the handedness of a molecule that is observable by the ability of a molecule to rotate the plane of polarized light either to the right **(dextrorotatory)** or to the left **(levorotatory)**. All of the amino acids in proteins exhibit the same absolute steric configuration as **L-glyceraldehyde**. Therefore, they are all L-α-amino acids. D-amino acids are never found in proteins, although they exist in nature. D-amino acids are often found in polypeptide antibiotics.

The aromatic R-groups in amino acids absorb ultraviolet light with an absorbance maximum in the range of 280 nm. The ability of proteins to absorb ultraviolet light is predominantly due to the presence of the tryptophan, which strongly absorbs ultraviolet light.

Color Reaction of Amino Acids

One of the identification tests or the qualitative tests for amino acid is the 'Ninhydrin' test. The alpha amino group of amino acid reacts with the ninhydrin (triketohydrindene hydrate). Ninhydrin is a powerful oxidizing agent. In its presence, amino acids undergo oxidative deamination liberating ammonia, carbon dioxide, a corresponding aldehyde, and a reduced form of ninhydrin. The ammonia liberated from the alpha amino group of amino acid reacts with the ninhydrin and its reduced product to form a blue substance, diketohydrin or Ruhemann's purple. Proline is an exception. It produces a yellow color because of the presence of an α-imino group instead of α-amino group.

The Peptide Bond

Peptide-bond formation is a condensation reaction leading to the polymerization of amino acids into peptides and proteins. Peptides are small consisting of few amino acids. A number of hormones and neurotransmitters are peptides. Additionally, several antibiotics and antitumor agents are peptides. Proteins are polypeptides of greatly divergent length. The simplest peptide, a dipeptide, contains a single peptide bond formed by the condensation of the carboxyl group of one amino acid with the amino group of the second with the concomitant elimination of water. The presence of the carbonyl group in a peptide bond allows electron resonance stabilization to occur such that the peptide bond exhibits rigidity not unlike the typical —C = C— double bond. The peptide bond is, therefore, said to have partial double-bond character.

FIGURE 4.14 **Resonance stabilization forms of the peptide bond.**

TABLE 4.5 The 20 protein amino acids classified according to the nature of their R groups. The structures and names, both full names and three-letter abbreviations, and dissociation constants (pK_a Value) of each group (–COOH, –NH_2, and the R-group) in the amino acids are shown in separate columns.

Amino Acid	Symbol	Structure*	pK_1	pK_2	pK R Group
		Hydrophobic Amino Acids (with Aliphatic R-Groups)			
Glycine	Gly-G	$H-CH(NH_3^+)-COO^-$	2.4	9.8	
Alanine	Ala-A	$CH_3-CH(NH_3^+)-COO^-$	2.4	9.9	
Valine	Val-V	$(H_3C)_2CH-CH(NH_3^+)-COO^-$	2.2	9.7	
Leucine	Leu-L	$(H_3C)_2CH-CH_2-CH(NH_3^+)-COO^-$	2.3	9.7	
Isoleucine	Ile-I	$H_3C-CH_2-CH(H_3C)-CH(NH_3^+)-COO^-$	2.3	9.8	

Proline	Pro–P	Pyrrolidine ring with $\overset{+}{N}H_2$ and COO^-	2.0	10.6	
		Hydroxilated aliphatic Amino Acids (with Hydroxilated R-Groups)			
Serine	Ser-S	$HO{-}CH_2{-}\underset{NH_3^+}{CH}{-}COO^-$	2.2	9.2	~13
Threonine	Thr–T	$(H_3C)(HO)CH{-}\underset{NH_3^+}{CH}{-}COO^-$	2.1	9.1	~13
		Sulfur-Containing Amino Acids (R-Groups with Sulfur)			
Cysteine	Cys - C	$HS{-}CH_2{-}\underset{NH_3^+}{CH}{-}COO^-$	1.9	10.8	8.3
Methionine	Met–M	$\underset{\underset{CH_3}{S}}{CH_2}{-}CH_2{-}\underset{NH_3^+}{CH}{-}COO^-$	2.1	9.3	
		Acidic Amino Acids (Negatively charged R-groups) and their Amides			
Aspartic	Asp-D	$OOC{-}CH_2{-}\underset{NH_3^+}{CH}{-}CO$	2.0	9.9	3.9
Asparagine	Asn–N	$H_2N{-}\underset{O}{\overset{}{C}}{-}CH_2{-}\underset{NH_3^+}{CH}{-}\overset{O}{C}{-}O$	2.1	8.8	
Glutamic	Glu–E	$^-OOC{-}CH_2{-}CH_2{-}\underset{NH_3^+}{CH}{-}$	2.1	9.5	4.1
Glutamine	Gln-Q	$H_2N{-}\underset{O}{C}{-}CH_2{-}CH_2{-}\underset{NH}{CH}$	2.2	9.1	
		Basic Amino Acids (with positively charged R groups)			
Arginine	Arg-R	$HN{-}CH_2{-}CH_2{-}CH_2{-}\underset{N}{C}$; $\overset{+}{C}{=}NH_2$; H_2N	1.8	9.0	12.5

Lysine	Lys-K	$H_3\overset{+}{N}-CH_2-(CH_2)_3-CH(NH_3^+)-$	2.2	9.2	10.8
Histidine	His-H	imidazole $-CH_2-CH(NH_3^+)-COO^-$	1.8	9.2	6.0
		Aromatic Amino Acids (With aromatic R-groups)			
Phenylal anine	Phe-F	phenyl $-CH_2-CH(NH_3^+)-COO^-$	2.2	9.2	
Tyrosine	Tyr-Y	HO–phenyl $-CH_2-CH(NH_3^+)-COO^-$	2.2	9.1	10.1
Tryptophan	Trp-W	indole (N–H) $-CH_2-CH(NH_3^+)-COO^-$	2.4	9.4	

4.5 BUILDING BLOCKS OF NUCLEIC ACIDS: NUCLEOTIDES

Nucleic acids are the information-carrying molecules—our genetic material. As with many of our other compounds, the nucleic acids are composed of repeating units or monomers called **nucleotides**. Nucleotides are the building blocks of nucleic acids. As a class, the nucleotides may be considered one of the most important metabolites of cells. Nucleotides are found primarily as the monomeric units comprising the major nucleic acids of the cell, RNA, and DNA. However, they are also required for numerous other important and independent functions within the cell.

Functions of Nucleotides

- Serve as energy carrier molecules (ATP) and participate in energy transaction reactions. ATP or adenosine triphosphate predominantly carries out these reactions.
- Energy transport coenzymes, forming a portion of several important coenzymes such as NAD^+, $NADP^+$, FAD, and coenzyme A.
- Intra-cellular messengers, serving as mediators of numerous important cellular processes such as second messengers in signal-transduction events. The predominant second messenger is cyclic-AMP (cAMP), a cyclic derivative of AMP formed from ATP.
- Control numerous enzymatic reactions through allosteric effects on enzyme activity.
- Serve as activated intermediates in numerous biosynthetic reactions. These activated intermediates include S-adenosylmethionine (S-AdoMet) involved in methyl-transfer reactions as well as many sugar-coupled nucleotides involved in glycogen and glycoproteins synthesis.

Components of Nucleic Acids (which are long chains of nucleotides)

A nucleotide consists of three components:

- 5-carbon sugar or Pentose sugar
- Phosphate group
- Nitrogen base

The 5-carbon sugar in nucleic acid is ribose sugar in ribonucleic acids (RNA) and deoxyribose sugar in deoxyribose nucleic acids (DNA). The phosphoric acid in the form of phosphate residue is attached to the 5-carbon of the pentose sugar by an ester linkage. The nitrogen base is attached to the 1-carbon of the pentose sugar by N-glycosidic linkage. Both the reactions, the formation of ester link between pentose sugar and phosphoric acid and formation of N-glycosidic linkage between C-1 of the sugar and nitrogen bases are condensation reactions and result in the elimination of a molecule of water. The nucleotides found in cells are derivatives of the heterocyclic highly basic, compounds purine and pyrimidine.

Purine   Pyrimidine

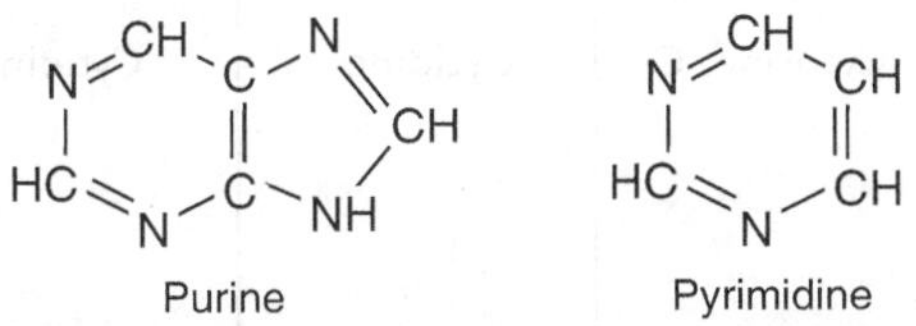

FIGURE 4.15 **Nitrogen bases—purine and pyrimidine.**

It is the chemical basicity of the nucleotides that has given them the common term "bases" as they are associated with nucleotides present in DNA and RNA. There are five major bases found in cells. The derivatives of purine are called adenine and guanine, and the derivatives of pyrimidine are called thymine, cytosine, and uracil. The common abbreviations used for these five bases are, **A, G, T, C,** and **U**. The purine and pyrimidine bases in cells are linked to D-ribose or 2'-deoxy-D-ribose through a β-N-glycosidic bond between the anomeric carbon (C-1) of the ribose and the N^9 of a purine or N^1 of a pyrimidine resulting in a nucleoside. The nucleoside condenses with phosphoric acid by forming an ester link with 5-terminal OH group resulting in a nucleotide. Since there are five types of nitrogen bases, there are five types of nucleoside and nucleotides as shown in Figure 4.16.

Nucleoside and Nucleotide Structure and Nomenclature

In nucleosides and nucleotides the base can exist in two distinct orientations about the N-glycosidic bond. These conformations are identified as, syn and anti. It is the anti-conformation that predominates in naturally occurring nucleotides.

Nucleosides are found in the cell primarily in their phosphorylated form. These are termed **nucleotides**. The most common site of phosphorylation of nucleotides found in cells is the hydroxyl group attached to the 5-carbon of the ribose. The carbon atoms of the ribose present in nucleotides are designated with a prime (′) mark to distinguish them from the backbone numbering in the bases. Nucleotides can exist in mono-, di-, or tri-phosphorylated forms.

Nucleotides are given distinct abbreviations to allow easy identification of their structure and state of phosphorylation. The monophosphorylated form of adenosine (adenosine-5-monophosphate) is written as, AMP. The di- and tri-phosphorylated forms are written as, ADP and ATP, respectively. The use of these abbreviations assumes that the nucleotide is in the 5-phosphorylated form. The di-and tri-phosphates of nucleotides are linked by acid anhydride bonds. Acid-anhydride bonds have a high free energy ($\Delta G^{0'}$) for hydrolysis imparting to them a high potential to transfer the phosphates to other molecules. It is this property of the nucleotides that results in their involvement in group-transfer reactions for energy transactions in the cell.

The nucleotides found in DNA are unique from those of RNA in that the ribose exists in the 2-deoxy form and the abbreviations of the nucleotides contain a *'d'* designation. The monophosphorylated form of adenosine found in DNA (deoxyadenosine-5-monophosphate) is written as dAMP.

Base Formula	Base	Nucleoside (with ribose or deoxyribose)	Deoxyribose or ribose Nucleotide phosphate
NH_2, N, O, N, X	Cytosine, C	Cytidine	Cytidine monophosphate CMP
O, H, N, O, N, X	Uracil, U	Uridine	Uridine monophosphate UMP
O, H, N, CH_3, O, N, dX	Thymine, T	Thymidine	Thymidine monophosphate TMP
NH_2, N, N, N, N, X	Adenine, A	Adenosine	Adenosine monophosphate AMP
O, H, N, N, H_2N, N, N, X	Guanine, G	Guanosine	Guanosine monophosphate GMP

FIGURE 4.16 Nitrogen bases and their corresponding nucleosides and nucleotides.

syn-Adenosine *anti*-Adenosine

FIGURE 4.17 ***syn*** **and** ***anti*** **conformations of nucleoside, using adenosine nucleoside as an example.** ***Anti*** **conformation predominates in nature.**

The Nucleotides of DNA

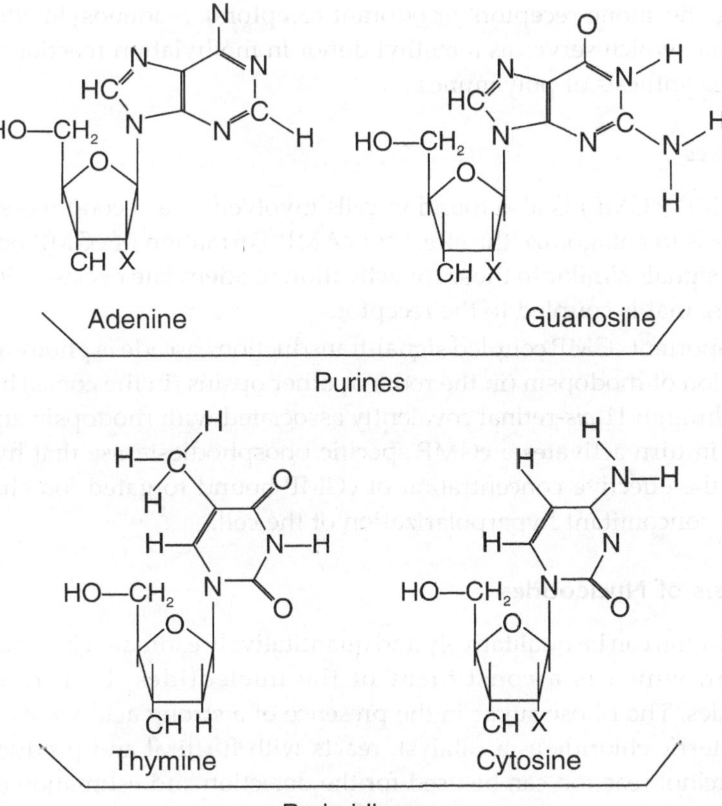

FIGURE 4.18 **Nucleotides of DNA.**

The nucleotide uridine is never found in DNA and thymine is almost exclusively found in DNA. Thymine is found in tRNAs but not rRNAs nor mRNAs. There are several less common bases found in DNA and RNA. The primary modified base in DNA is 5-methylcytosine. A variety of modified bases appear in the tRNAs. Many modified nucleotides are encountered outside the context of DNA and RNA that serve important biological functions.

Adenosine Derivatives

The most common adenosine derivative is the cyclic form, 3-5-cyclic adenosine monophosphate, cAMP. This compound is a very powerful second messenger involved in passing signal transduction events from the cell surface to internal proteins (e.g., cAMP–dependent protein kinase (PKA)). PKA phosphorylates a number of proteins, thereby, affecting their activity either positively or negatively.

Cyclic-AMP is also involved in the regulation of ion channels by direct interaction with the channel proteins (e.g., in the activation of odorant receptors by odorant molecules). Formation of cAMP occurs in response to activation of receptor-coupled adenylate cyclase. These receptors can be of any type (e.g., hormone receptors or odorant receptors). S-adenosylmethionine is a form of activated methionine, which serves as a methyl donor in methylation reactions and as a source of propylamine in the synthesis of polyamines.

Guanosine Derivatives

A cyclic form of GMP (cGMP) is also found in cells involved as a second messenger molecule. In many cases its' role is to antagonize the effects of cAMP. Formation of cGMP occurs in response to receptor-mediated signals similar to those for activation of adenylate cyclase. However, in this case it is *guanylate cyclase* that is coupled to the receptor.

The most important cGMP coupled signal-transduction cascade is photo-reception. However, in this case, activation of rhodopsin (in the rods) or other opsins (in the cones) by the absorption of a photon of light (through 11-*cis*-retinal covalently associated with rhodopsin and opsins) activates transducin, which in turn activates a cGMP specific phosphodiesterase that hydrolyzes cGMP to GMP. This lowers the effective concentration of cGMP bound to gated ion channels resulting in their closure and a concomitant hyperpolarization of the cell.

Qualitative Analysis of Nucleotides

Nucleotides in a solution can be qualitatively and quantitatively estimated by estimating or detecting the pentose sugar, which is a constituent of the nucleotides, both ribonucleotides and deoxyribonucleotides. The ribose sugar in the presence of a strong acid yields a furfural. Orcinol, in the presence of ferric chloride as a catalyst, reacts with furfural and produces a green-colored compound. This orcinol reaction can be used for the detection and estimation of ribonucleotides.

Diphenylamine can be used for the detection of deoxyribonucleotides. It reacts with diphenylamine under acidic conditions and gives a blue-colored complex. The reaction involves dehydration of deoxyribose of the nucleotide in the presence a strong acid to hydroxylevulinic aldehydes. These aldehydes react with diphenylamine and give the blue-colored complex.

4.6 BUILDING BLOCKS OF LIPIDS: FATTY ACIDS, GLYCEROL

Biological molecules that are insoluble in aqueous solutions and soluble in organic solvents are classified as lipids. The lipids of physiological importance for humans have four major functions:

1. They serve as structural components of biological membranes.
2. They provide energy reserves, predominantly in the form of triacylglycerols.
3. Both lipids and lipid derivatives serve as vitamins and hormones.
4. Lipophilic bile acids aid in lipid solubilization.

The building blocks or the chemical constituent of lipids are the long chain hydrocarbon acids known as fatty acids, the C_{18} amino alcohol or the sphingosine, glycerol, and cholesterol.

Major Types of Lipids

- Triglycerides commonly known as the fats and oils
- Waxes (very similar to triglycerides)
- Phospholipids—the main constituent of membranes
- Sterols (or steroids)
- Terpenes

Structure of Lipids

In chemical composition, lipids are similar to carbohydrates and have carbon, hydrogen, and oxygen as the main elemental components. However, the proportion of oxygen is low, so lipids are mostly hydrocarbons. The chemical structures of fats and oils, the most common lipids, are based on fatty acid building blocks and an alcohol, glycerol. The terms fats and oils are terms of convention. Fats are "hard" or solid at room temperature and oils are liquids at room temperature.

Fatty Acids

Fatty acids play two major roles in the body of plants and animals:

- As the components of more complex membrane lipids.
- As the major components of stored fat in the form of triacylglycerols.

Fatty acids are long-chain hydrocarbon molecules containing a carboxylic acid moiety at one end. The numbering of carbons in fatty acids begins with the carbon of the carboxylate group. At physiological pH, the carboxyl group is readily ionized, rendering a negative charge on to fatty acids.

Fatty acids are of two types: saturated fatty acids and unsaturated fatty acids.

Fatty acids that contain no carbon-carbon double bonds are called **saturated fatty acids**; those that contain double bonds are **unsaturated fatty acids**. The numeric designations used for fatty acids come from the number of carbon atoms, followed by the number of sites of unsaturation (e.g., palmitic acid is a 16-carbon fatty acid with no unsaturation and is designated by 16:0). The site of unsaturation in a fatty acid is indicated by the symbol Δ and the number of the first carbon of the double bond (e.g., palmitoleic acid is a 16-carbon fatty acid with one site of unsaturation between carbons 9 and 10, and is designated by $16{:}1^{\Delta 9}$).

Saturated fatty acids of less than eight carbon atoms are liquid at physiological temperature, whereas those containing more than ten are solid. The presence of double bonds in fatty acids significantly lowers the melting point relative to a saturated fatty acid.

The majority of body fatty acids are acquired through the diet. However, the lipid biosynthetic capacity of the body (fatty acid synthase and other fatty acid modifying enzymes) can supply the body with all the various fatty acid structures needed. Two key exceptions to this are the highly unsaturated fatty acids known as linoleic acid and linolenic acid, containing unsaturation sites beyond carbons 9 and 10. These two fatty acids cannot be synthesized from precursors in the body, and are thus considered the **essential fatty acids**; essential in the sense that they must be provided in the diet. Since plants are capable of synthesizing linoleic and linolenic acids humans can acquire these fats by consuming a variety of plants or else by eating the meat of animals that have consumed these plant fats.

TABLE 4.6 Physiologically relevant fatty acids.

Numerical Symbol	Common Name	Structure	Comments
14:0	Myristic acid	$CH_3(CH_2)_{12}COOH$	Often found attached to the N-term. of plasma membrane associated cytoplasmic proteins
16:0	Palmitic acid	$CH_3(CH_2)_{14}COOH$	End product of mammalian fatty acid synthesis
$16{:}1^{\Delta 9}$	Palmitoleic acid	$CH_3(CH_2)_5C{=}C(CH_2)_7$ COOH	
18:0	Stearic acid	$CH_3(CH_2)_{16}COOH$	
$18{:}1^{\Delta 9}$	Oleic acid	$CH_3(CH_2)_7C{=}C(CH_2)_7$ COOH	
$18{:}2^{\Delta 9,12}$	Linoleic acid	$CH_3(CH_2)_4C{=}CCH_2C$ $=C(CH_2)_7COOH$	Essential fatty acid
$18{:}3^{\Delta 9,12,15}$	Linolenic acid	$CH_3CH_2C{=}CCH_2C$ $=CCH_2C{=}C(CH_2)_7$ COOH	Essential fatty acid
$20{:}4^{\Delta 5,8,11,14}$	Arachidonic acid	$CH_3(CH_2)_3(CH_2C{=}C)_4$ $(CH_2)_3COOH$	Precursor for eicosanoid synthesis

Glycerol

Glycerol is a poly-hydroxy alcohol that forms a basic component of triglycerides, a major form of storage lipids. Triacylglycerides are composed of a glycerol backbone to which three fatty acids are esterified.

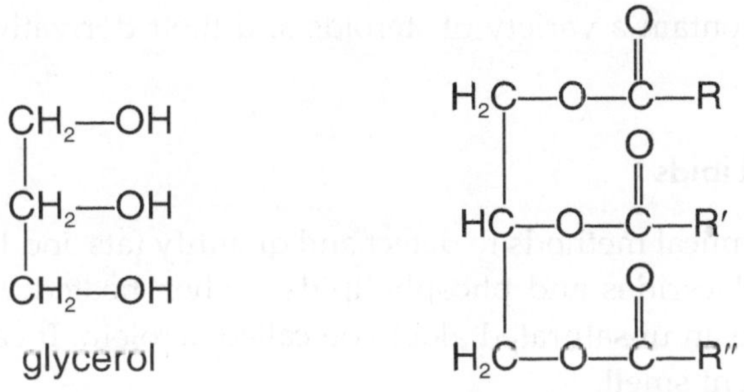

$$\begin{array}{l} CH_2\text{—}OH \\ | \\ CH_2\text{—}OH \\ | \\ CH_2\text{—}OH \end{array} \qquad \begin{array}{l} H_2C\text{—}O\text{—}\overset{O}{\overset{\|}{C}}\text{—}R \\ | \\ HC\text{—}O\text{—}\overset{O}{\overset{\|}{C}}\text{—}R' \\ | \\ H_2C\text{—}O\text{—}\overset{O}{\overset{\|}{C}}\text{—}R'' \end{array}$$

glycerol

FIGURE 4.19 **Basic composition of a triacylglycerides. R, R′, and R″ are the fatty acids.**

Basic Structure of Phospholipids

The basic structure of phospholipids is very similar to that of triacylglycerides except that C-3 of the glycerol backbone is esterified to phosphoric acid. The building block of the phospholipids is phosphatidic acid, which results when the X substitution in the basic structure (shown in Figure 4.20) is a hydrogen atom. Substitutions include ethanolamine (phosphatidylethanolamine), choline (phosphatidylcholine, also called lecithins), serine (phosphatidylserine), glycerol (phosphatidylglycerol), myo-inositol (phosphatidylinositol). These compounds can have a variety in the numbers of inositol alcohols that are phosphorylated generating polyphosphatidylinositols), and phosphatidylglycerol (diphosphatidyl-glycerol more commonly known as cardiolipins).

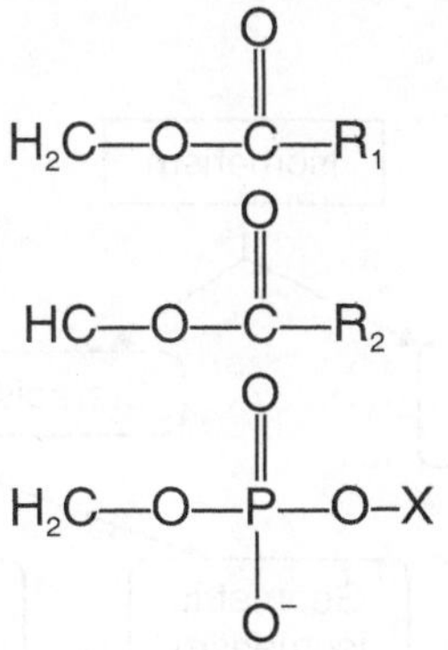

$$\begin{array}{l} H_2C\text{—}O\text{—}\overset{O}{\overset{\|}{C}}\text{—}R_1 \\ | \\ HC\text{—}O\text{—}\overset{O}{\overset{\|}{C}}\text{—}R_2 \\ | \\ H_2C\text{—}O\text{—}\underset{O^-}{\underset{|}{\overset{O}{\overset{\|}{P}}}}\text{—}O\text{—}X \end{array}$$

FIGURE 4.20 **Basic composition of a phospholipid; X can be a number of different substituents.**

Steroids

These are special type of lipids, derived from the building block cholesterol. Cholesterol is the most abundant steroid present in animals. A number of steroids particularly the steroid hormones like sex hormones—estrogens, testosterone, and other growth hormones and factors such as immuno-suppressive factors and vitamins such as Vitamin-D are derived from the basic molecule of cholesterols. Plants also contain a variety of steroids and their derivatives and carry out various biological functions.

Chemical Properties of Lipids

There are a number of chemical methods to detect and quantify fats and lipids. Glycerol, one of the main components of triglycerides and phospholipids when heated with potassium hydrogen sulphate, is dehydrated to an unsaturated aldehyde called acrolein. It can be very well identified by its characteristic pungent smell.

Unsaturated lipids can decolorize the colored solutions of halogens such as bromine, water. Unsaturated fatty acids add halogens (e.g., bromine) across their double bonds, and that leads to the discoloration of the solution.

4.7 OPTICAL ACTIVITY AND STEREOCHEMISTRY OF BIOMOLECULES

Isomerism

Isomerism occurs when two or more different compounds share the same molecular formula and different structural formula. When the atoms or groups are linked together in different ways the phenomenon is referred to as **structural isomerism.** When the atoms are linked together in the same way but the spatial arrangements are different, the type of isomerism is known as stereoisomerism. There are two distinct types of stereoisomerism; **geometric isomerism** and **optical isomerism.**

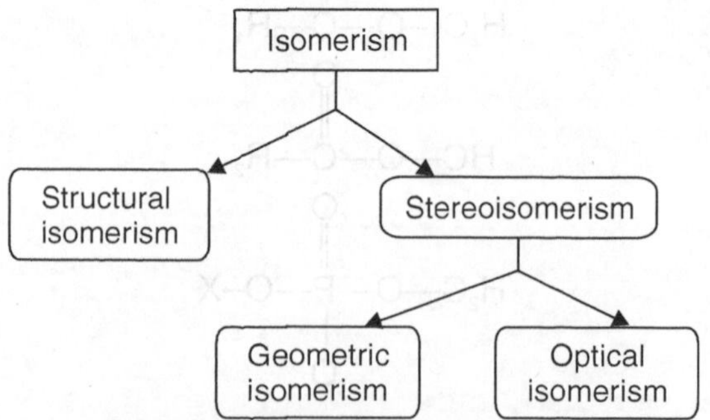

FIGURE 4.21 **Classification of Isomerism.**

Because so many types of isomers are possible, biological molecules can occur in an amazing variety of forms. A formula as simple as $C_6H_{12}O_6$ can refer to more than a dozen different molecules, each recognizably different from the others by cells and by chemists.

Structural Isomerism: These types of isomers differ in the arrangement or distribution of covalent bonds. The two substances A and B given in Figure 4.22 are isomers. They both have the same molecular formula, C_2H_6O. However, the atoms in each are connected in a different order. In structure A it is in the order C—O—C and in the structure B it is in the order C—C—O. Isomers like this are known as structural isomers. In the compounds C and D the arrangement and position of the functional group is different, and therefore C and D are structural isomers. Structural isomers have different physical and chemical properties and can even belong to different homologous series. But the compounds C and D, even though have different structural formulas, have identical functional groups. Hence, they may have similar chemical properties.

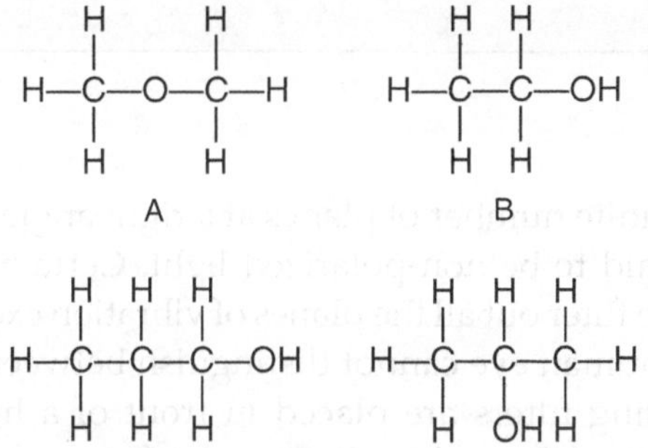

FIGURE 4.22 **Structural Isomerism.**

Stereoisomerism

Geometric Isomerism

Geometric isomerism always involves arrangement about a double bond. Double bonds are rigid and do not allow the joined atoms to rotate freely. Thus, the presence of a double bond results in some form of rigidity within the molecule. But this can also be caused by the presence of a ring. As a result, two isomers are always possible if different groups are attached around a double bond. For example, molecules (a) and (b) are not the same, even though they have the same functional groups attached to the same carbon atoms. Surprisingly, enzymes can easily recognize this difference and may bind tightly to one molecule and not at all to the other. Molecular shape is important in biology.

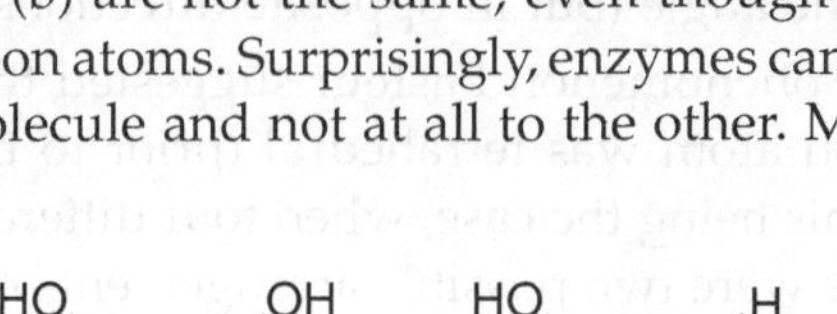

FIGURE 4.23 ***Cis*** **and** ***trans*** **isomerism.**

When both parts of the main chain of the compound are on the same side of the double bond the compound is called the **cis isomer** and when the parts of the main chain straddle the double bond the **trans isomer** results. The two geometric isomers are not superimposable.

The physical properties of geometric isomers are usually quite different. Although physically different, geometric isomers are normally chemically similar because they have the same functional groups in the same environment. In general, trans molecules can pack more closely than cis in the solid state and this results in the trans form having the highest melting point.

Geometric isomers caused by restricted rotation around a bond are distinguished using the terms : 'cis' and 'trans':

- **cis–both groups are on the same side of the double bond**
- **trans–the groups are on opposite sides of the double bond ('trans' means across)**

Optical Isomerism

Normal light vibrates in an infinite number of planes at a right angle to its direction of propagation. Under these conditions it is said to be non-polarized light. Certain materials, for example calcite and polaroid, have the ability to filter out all the planes of vibration except one. The result is polarized or plane-polarized light. The human eye cannot distinguish between polarized and non-polarized light; however, if two polarizing filters are placed in front of a light source and one is rotated relative to the other the intensity of the emitted light varies from maximum when the two filters are aligned to zero when they are out of phase.

It has been known for hundreds of years that solutions of many naturally occurring substances have the ability to rotate the plane of polarized light, which is known as **optical activity.** Louis Pasteur discovered that when 2,3-dihydroxy (tartaric) acid was recrystallized, crystals of two different forms were obtained. The two types of crystals were non-superimposable mirror images of each other.

When the two sets of crystals were dissolved in water it was found that one solution rotated the plane of polarized light to the right (clockwise) while the other rotated it to the left (anti-clockwise). When this phenomenon was investigated quantitatively it was found that under similar conditions solutions of the same concentration of the two optical isomers rotated the plane of polarized light through the same angle (but in opposite directions).

In order to explain this phenomenon Pasteur suggested that the arrangement of atoms or groups around a central carbon atom was tetrahedral (prior to this it was thought to be square planar, that is cross shaped). This being the case, when four different atoms or groups were bound to the central carbon atom there were two possible arrangements or configurations which give rise to the two optical isomers. Optical isomers or enantiomers consist of molecules, which are non-superimposable mirror images of each other.

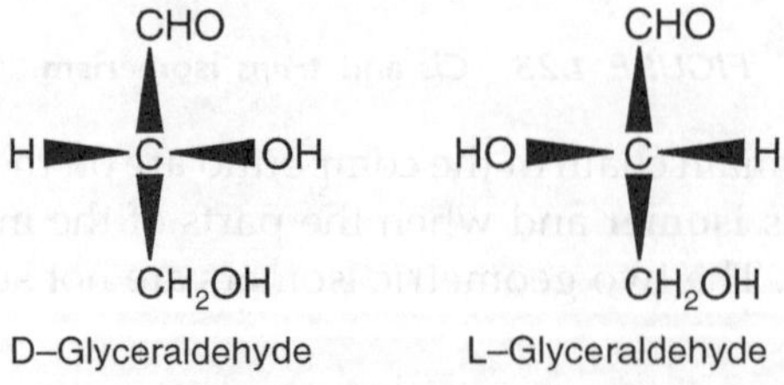

FIGURE 4.24 **Optical Isomers (enantiomers) of glyceraldehyde.**

Chirality

A molecule, which has four different groups or atoms joined to a central carbon atom, lacks symmetry; that is to say it is asymmetric. The central carbon atom is known as a chiral center. The word chiral comes from the Greek word for hand or handedness and this is quite appropriate as hands also lack symmetry and the right hand is a non-superimposable mirror image of the left hand.

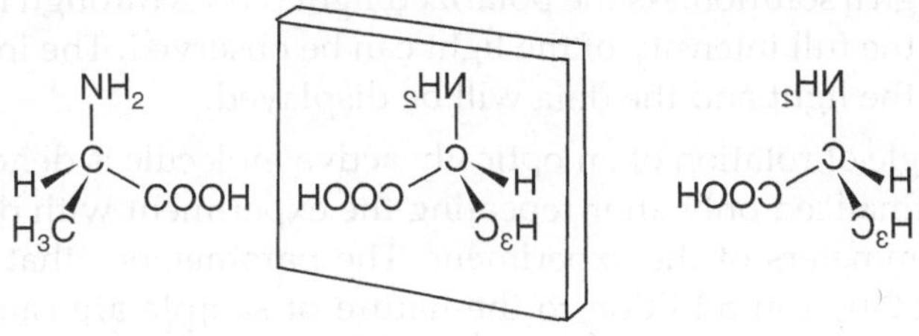

Structure 1 Imaginary mirror Structure 2

FIGURE 4.25 **Optical isomerism of alanine.**

Racemic Mixtures

Enantiomers are almost identical to each other. They contain the same functional groups in the same chemical environment. As a result, they are physically and chemically virtually identical to each other. The only differences between pairs of enantiomers are:

- Their effect on plane polarized light.
- Their crystals are non-superimposable mirror images of each other.
- Sometimes their effects on biological systems vary; for example, (–)–brucine is much more toxic than (+)–brucine.

This means that when a chiral compound is synthesized in the laboratory there is an equal chance of each enantiomeric form being obtained. As a result, the product is an equimolar mixture of enantiomers, which is optically inactive since the two forms cancel out each other's effect. This mixture is known as a **racemic mixture**, a **racemic modification** or a **racemate.**

Polarimetry

Polarimetry is the device used for studying the optical property of a molecule. Ordinary light is electromagnetic radiation of visible range of different wavelengths, vibrating in an infinite number of planes perpendicular to the direction of propagation. Monochromatic light, even though it has a specific wavelength, vibrates in many planes. Electromagnetic radiation has two vectors—a magnetic vector and an electric vector. Those radiations of the visible range, which can oscillate in a single plane, are called **plane-polarized light**. The French scientist Malus, in 1808, observed that light reflected from certain bodies at a particular angle had the special property of oscillation in plane. Such visible radiation, which can vibrate in a single plane, is plane-polarized light. Polarized light can be produced by passing ordinary light through a 'Nicol prism' (named after its discoverer, Nicol in 1828), which filters in only one of the polarized lights and the others are reflected out. The polarization of a monochromatic light transmitted through a Nicol prism can be easily detected by checking it through a second Nicol prism. When the second prism is placed parallel to the first one, which produced the plane-polarized light, and allowed the plane-polarized radiation to pass through, the second prism transmits the polarized light without any change in its intensity. But if the plane of polarization is perpendicular to the plane of prism, then it fails to pass through the second prism.

In a Polarimeter, a sample of solution is placed between two Nicol prisms. A polarized light is allowed to pass through a solution. As the polarized light passes through the solution, the second prism is rotated so that the full intensity of the light can be observed. The instrument will measure the angle of rotation of the light and the data will be displayed.

The observed angle of rotation of an optically active molecule is denoted by α. The result of optical rotation can be finalized only after repeating the experiment with different concentrations and changing other parameters of the experiment. The parameters that can directly affect the magnitude of optical rotation in addition to the nature of sample are sample thickness, sample concentration, solvent, temperature, and wavelength.

4.8 CONFORMATION AND CONFIGURATION

Configuration and conformation are two terms frequently used to represent the molecular state of a compound. Configuration commonly refers to the arrangement of groups or atoms about the carbon atoms or chain and that are absolute. That means the configuration of a molecule can be changed only by breaking the chemical link between atoms or groups. Conformation, on the other hand, is the arrangement of groups or atoms in the space of the whole molecule. Conformation of a molecule can change due to the rotation of C–C single bonds such as the chair and boat conformations of D-glycopyranose.

There are different methods to represent the configuration and conformation in organic molecules. For example, the projection formula was proposed by Emil Fischer in 1981 to represent the structure of organic molecules. In this representation, the asymmetric carbon is in the plane of projection, the groups at the top and bottom are inclined equally below the plane of projection and the groups on the left and right are inclined equally above the plane of projection. This is known as the **Fischer Projection Formula.** (Figure 4.26.)

In Fischer Projection Formula, the convention is that the most oxidized group is kept on top and the most reduced carbon is at the bottom. If hydrogen is to the left and hydroxide is to the right it is called **dextrorotatory** denoted by 'D' and if hydrogen is to the right and hydroxyl is to the left, the molecule is **levorotatory** denoted by 'L' as in the case of sugars. Emil Fischer established the complete stereochemistry of monosaccharides on the basis of the structure of the well-known compound phenylhydrazine. In the case of glucose there are four asymmetric carbon atoms. Accordingly, there will be a number of isomers. The general formula for the number of possible isomers is 2^n, where n is the number of asymmetric carbon atoms. Thus, the total number of possible isomers in glucose is $2^4 = 16$. In nature there is the preference for a particular isomer. In the case of carbohydrates, the D forms such as D-glucose, and D-ribose are more predominant, whereas the L forms are predominant natural amino acids. This is because the enzymes that metabolize these compounds have the active centers, which have the same configuration as that of the molecules.

In the case of cyclic form sugars—both pyranose and furanose structures—there is another method of representation known as the **Haworth Projection Formula**. In the ring structures, the C–C linkages are single bonds and therefore rotations about the bond are possible. But there can be steric hindrance because of bulky groups attached to the carbon. The steric hindrance can block the free rotation of bonds. Therefore, the ring structure can assume different conformations to avoid the steric hindrance. For example, the pyranose ring of hexose sugar is not planar in reality as the Haworth projection formula suggests. But it assumes 3-dimensional structure in space resembling a boat or a chair depending on the steric hindrance in the molecule (Figure 4.26). The various forms of 3-dimensional structures that a molecule can take depending on various surrounding molecular forces are called the **conformations of the molecule**. The individual form is called a **conformer**.

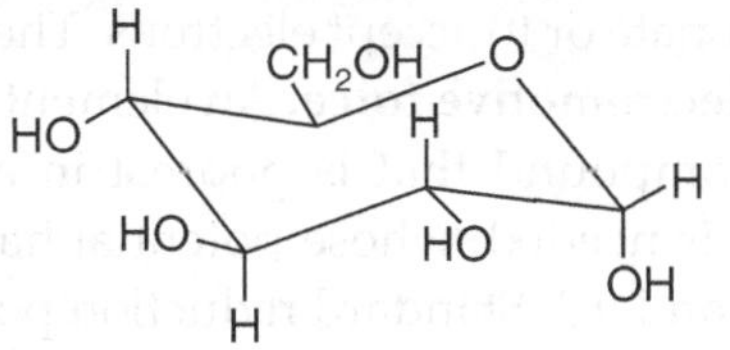

FIGURE 4.26 **The chair conformation of glucose.**

4.9 BIOCHEMICAL TRANSFORMATIONS

Every living cell continuously performs thousands of diverse chemical reactions. The nutrients taken into the cells are transformed into a multitude of new cell-specific biomolecules and components. In this way, sugars, amino acids and their precursors, organic acids, nucleotides, lipids, and other substances are synthesized. The totality of these reactions is summarized as the cell metabolism. In every biochemical reaction, a bond is either formed or broken. This process generally consumes or generates energy; consequently, the cell's metabolism and its energy balance have to be regarded as interdependent parameters. Every covalent bond of a molecule contains energy that is set free upon breaking down and that can subsequently be used for other purposes. It could, for example, be used for the formation of a new bond or it could be transformed into another form of energy such as movement, warmth, light, or electrical energy. Hence, activation energy is needed for the breakdown of a molecule. In chemical laboratories most of these reactions are possible, but under high temperatures and pressures with the help of certain inorganic or organic catalysts. The first two alternatives namely high temperature and pressure, are out of the question for a cell, which can only use catalysts. Biological catalysts are without exception enzymes (proteins). Their activity is often dependent on the presence of other molecules. Enzymes are highly specific and hardly any futile by-products are generated. The use of enzymes enables the cell to run thousands of thermodynamically possible and impossible reactions in parallel.

The biochemical reactions that take place in a living cell include oxidation-reduction reactions, group-transfer reactions, hydrolysis, breaking or formation of double bonds, transfer of functional groups within the molecule to form isomers, condensation reactions for the building of –C–C–, C–S–, –C–O–, and C–N–linkages by consuming energy from ATP hydrolysis.

Oxidations-Reductions-Redox Reactions

Redox reactions are among a cell's most important enzyme-catalyzed reactions. Oxidation and reduction refer to the transfer of one or more electrons from a donor to an acceptor, generally of another chemical species. The donor is oxidized, the acceptor reduced.

A substance that donates electrons is called a **reductant** or **reducing agent,** while the electron acceptor is called **oxidant** or **oxidizing agent**. Both, together, represent a redox couple:

$$\text{Electron donor} \rightarrow e^- + \text{electron acceptor}$$

Oxidation-reduction reactions are accompanied by a change in free energy. The free energy is a measure for the tendency to donate or to accept electrons. The flow of electrons can be measured and is called **redox potential** or **electromotive force.** An element is found to be in its highest degree of oxidation when it is in the compound that is poorest in energy. To depict redox reactions consistently, a common standard is needed, whose potential has arbitrarily been defined as zero. All other potentials refer to this standard. Standard reduction potentials are defined by convention with respect to the standard hydrogen half-reaction:

$$\tfrac{1}{2} H_2 <> H^+ + e^-$$

in which H^+ is in equilibrium with $H_2(g)$ at pH 0, 25°C and 1 atm. The redox potential of any redox pair can now be measured and related to the standard reduction potential. The dimension of the potential is V.

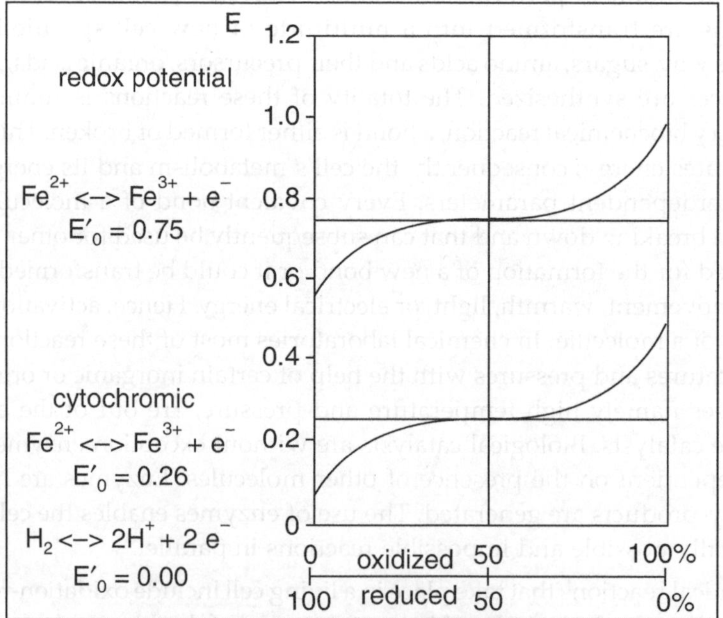

FIGURE 4.27 **The redox scale.**

The oxidation reduction reaction always involves the transfer of H^+ ions and electrons together or separately or in the form of hydride ion (H^-), followed by the conversion of C=O to C—OH, COOH to COH, etc., or vice versa. In biological systems, oxidation and reductions usually occur with the aid of a co-factor in the enzyme. The most important co-factors involved in the enzyme-mediated oxidation-reduction reactions are FAD/FADH (flavine adenine dinucleotide) and NAD/NADH (nicotinamide adenine nucleotide) that act as electron and proton acceptors and donors for these types of reactions. An example is the conversion of alcohol to acetic acid by alcohol dehydrogenase. The citric acid cycle, and the electron transport system of cellular respiration and the photo phosphorylation in the light reaction of photosynthesis, are rich in a series of oxidation-reduction reactions.

C-C Bond Formation or Cleavage

This one deals almost exclusively with C–C bonds. As we saw, carbons are not really active in reactions, except when they are activated. This usually means that we have heteroatoms or electron withdrawing groups (Z) directly attached to the carbons that make it (or the bond it is involved in) very reactive. Usually, these C–C bonds are weaker because the electron density of the C–C bond is distributed towards the heteroatoms bonded to the carbons. Examples are the formation of C–C bonds in the synthesis of fatty acids and sugars.

Condensation

Two groups come together, usually with the loss of a water molecule. Examples are the elongation of a protein by formation of a peptide bond, or an oligosaccharide by formation of a glycosidic bond. So, these reactions are the part of biosynthesis of macromolecules such as starch, glycogen, protein, and nucleic acids from their building blocks.

General formula: $A\text{—}OH + H\text{—}B \longrightarrow A\text{—}B + H_2O$

Hydrolysis

This reaction is just reverse of condensation. It is the substitution of a group or one of its components, by water. It involves adding water to split a covalent bond with the release of two smaller molecules. An example would be the hydrolysis of a peptide bond and release of amino acids or glycosidic bonds and release of monosaccharides or other subunits by the substitution of water molecules.

General formula: $A\text{—}B + H_2O \longrightarrow A\text{—}OH + H\text{—}B$

NOTE *A–B could be amino acids, sugars, etc.*

Rearrangement (Isomerization)

An isomerization reaction involves the transfer of groups or atoms from one location to another location within a molecule. The end result of these reactions is the relocation of a functional group or fragment in a molecule resulting in the formation of an isomer. Many times we can dissect them in other types of reactions occurring 'intramolecularly' (i.e., within the same molecule). An example is the conversion of glucose-6-phosphate into fructose-6-phosphate by an isomerase in the glycolysis part of glucose metabolism. As we will see, the names of most enzymes are associated with the reaction they catalyze.

Group-transfer Reactions

This is also involved in the biosynthesis of certain new molecules from the pre-existing molecule. The major types of group transfer reactions are transesterification, transamination, and transphosphorilation. A functional group or a part of a molecule attached to one molecule is transferred to another molecule. The best example is the transfer of a phosphate unit in adenosine triphosphate (ATP) to a sugar to form a sugar-phosphate and adenosine diphosphate (ADP), or synthesis of new amino acids by the transfer of an amine group from one amino acid to another carboxylic acid by transaminase.

ATP (Adenosine triphosphate)

ATP is a nucleotide that performs many essential roles in the cell. Some of the major roles of ATP in cellular metabolism are:

1. It is the major energy currency of the cell, providing the energy for most of the energy-consuming activities of the cell.
2. To keep on working, a cell must regenerate ATP. Cellular respiration provides the energy needed to drive the endergonic phosphorylation of ADP.
3. It is one of the monomers used in the synthesis of RNA and, after conversion to deoxyATP (dATP), DNA.
4. It regulates many biochemical pathways.

In mammals, it also functions outside cells.

- Its release from damaged cells can elicit pain, and
- Its release from the stretched wall of the urinary bladder signals when the bladder needs emptying!

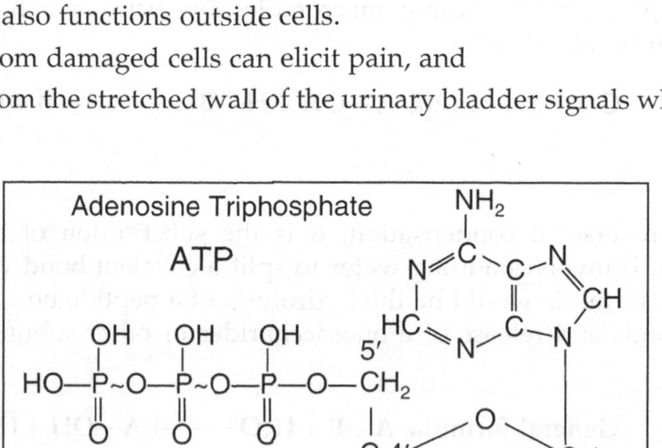

Adenosine Triphosphate

ATP

FIGURE 4.28 **Structure of adenosine triphosphate (ATP), the energy molecule of the cell.**

Energy of ATP

When the third phosphate group of ATP is removed by hydrolysis or by group transfer of inorganic phosphate (Pi), a substantial amount of free energy is released. The exact amount depends on the conditions, but we shall use a value of 7.3 kcal per mole under standard physiological conditions (at 25°C, 1 atmospheric pressure, and pH 7.0).

$$\textbf{ATP + H}_2\textbf{O} \longrightarrow \textbf{ADP + Pi + 7.3 kcal/mol}$$

ADP is adenosine diphosphate. P_i is inorganic phosphate.

For this reason, this bond is known as a "high-energy" bond and is depicted in the figure by a wavy red line. (The bond between the first and second phosphates is also "high-energy." But please note that the term is not being used in the same sense as the term "bond energy." In fact, these bonds are actually weak bonds with low bond energies.)

Nicotinamide Adenine Dinucleotide (NAD) and Nicotinamide Adenine Dinucleotide Phosphate (NADP)

Nicotinamide adenine dinucleotide (NAD) and its relative nicotinamide adenine dinucleotide phosphate (NADP) are two of the most important coenzymes in the cell. NADP is simply NAD with a third phosphate group attached as shown at the bottom of the figure.

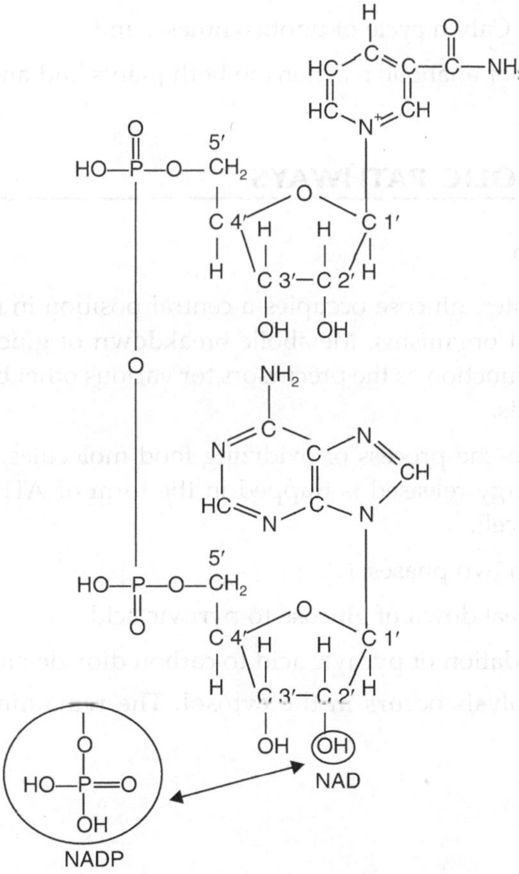

FIGURE 4.29 **Structure of NAD/NADP.**

Because of the positive charge on the nitrogen atom in the nicotinamide ring (upper right), the oxidized forms of these important redox reagents are often depicted as NAD^+ and $NADP^+$, respectively.

In cells, most oxidations are accomplished by the removal of hydrogen atoms. Both of these coenzymes play crucial roles in this. Each molecule of NAD^+ (or $NADP^+$) can acquire two electrons; that is, be reduced by two electrons. However, only one proton accompanies the reduction. The other proton, produced as two hydrogen atoms are removed from the molecule being oxidized, is liberated into the surrounding medium. For NAD, the reaction is thus:

$$\mathbf{NAD^+ + 2H \longrightarrow NADH + H^+}$$

NAD participates in many redox reactions in cells, including those:

- In glycolysis and most of those
- In the citric acid cycle of cellular respiration

NADP is the reducing agent

- Produced by the light reaction of photosynthesis
- Consumed in the Calvin cycle of photosynthesis and
- Used in many other anabolic reactions in both plants and animals

4.10 MAJOR METABOLIC PATHWAYS

Carbohydrate Metabolism

Among various carbohydrates, glucose occupies a central position in metabolism. Glucose is the major form of fuel for most organisms. Metabolic breakdown of glucose also generates various types of intermediates that function as the precursors for various other biomolecules such as amino acids, nucleotides, and lipids.

Cellular respiration is the process of oxidizing food molecules, such as glucose to carbon dioxide and water. The energy released is trapped in the form of ATP for use by all the energy-consuming activities of the cell.

The process occurs in two phases:

- Glycolysis, the breakdown of glucose to pyruvic acid
- The complete oxidation of pyruvic acid to carbon dioxide and water

In eukaryotes, glycolysis occurs in the cytosol. The remaining processes take place in mitochondria.

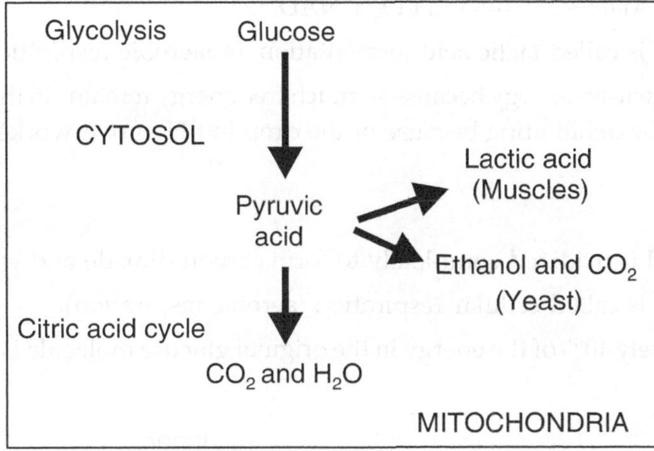

FIGURE 4.30 **Glycolysis and the fate of pyruvic acid.**

An almost universal pathway of glucose catabolism is known as glycolysis. It is defined as the anaerobic catabolism of glucose. It occurs in virtually all cells. In eukaryotes, it occurs in the cytosol. The overall reaction of glycolysis is

$$C_6H_{12}O_6 + 2NAD^+ \rightarrow 2C_3H_4O_3 + 2NADH + 2H^+$$

A glucose molecule is split into two molecules of pyruvic acid. The free energy stored in two molecules of pyruvic acid is somewhat less than that in the original glucose molecule. Some of this difference in energy is captured in two molecules of ATP.

The Fate of Pyruvic Acid

Pyruvic acid produced at the end of glycolysis can enter into any of the following metabolic pathways according to the type organism and availability of oxygen.

In Yeast

- Pyruvic acid is decarboxylated and reduced by NADH to form a molecule of carbon dioxide and one of ethanol.

$$C_3H_4O_3 + NADH + H^+ \Rightarrow CO_2 + C_2H_5OH + NAD^+$$

- This accounts for the bubbles and alcohol in, for example, beer and champagne.
- The process is called **alcoholic fermentation** (anaerobic respiration).
- The process wastes energy because so much of the free energy of glucose (some 95%) remains in the alcohol (a good fuel!).

In active Muscles

- Pyruvic acid is reduced by NADH forming a molecule of **lactic acid.**

- $C_3H_4O_3 + NADH + H^+ \longrightarrow C_3H_6O_3 + NAD^+$
- The process is called lactic acid fermentation. (anaerobic respiration)
- The process wastes energy because so much free energy remains in the lactic acid molecule. (It can also be debilitating because of the drop in pH of overworked muscles.)

In Mitochondria

- Pyruvic acid is oxidized completely to form carbon dioxide and water.
- The process is called cellular respiration (aerobic respiration).
- Approximately 40% of the energy in the original glucose molecule is trapped in molecules of ATP.

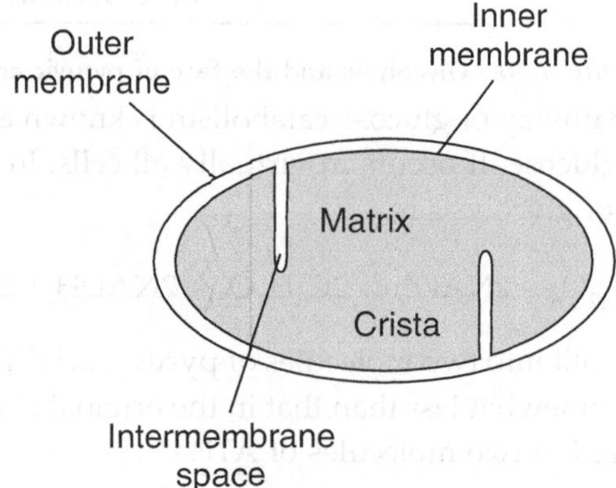

FIGURE 4.31 **Diagram sketch of mitochondria.**

The Citric-acid Cycle

The citric-acid cycle or the tricarboxylic-acid cycle (TCA cycle) is the central pathway of aerobic respiration. During this metabolic process the following reactions take place:

- The link between glycolysis and the krebs cycle is the conversion of pyruvic acid to acetyl CoA by a multienzyme complex in the matrix of the mitochondrion.
- The acetic acid of acetyl CoA joins a four-carbon molecule, oxaloacetic acid, to form the six-carbon citric acid molecule.
- The resulting molecule of citric acid (which gives its name to the process) undergoes the series of enzymatic steps shown in the diagram.
- Citric acid is subsequently degraded back to oxaloacetic acid in a series of steps constituting in one turn of the cycle.

During this process of citric acid cycle:

- Each of the three carbon atoms present in the pyruvic acid that entered the mitochondrion leaves as a molecule of carbon dioxide (CO_2).
- A pair of electrons ($2e^-$) is removed and transferred to NAD^+ reducing it to $NADH + H^+$.
- A pair of electrons is removed from succinic acid and reduces FAD to $FADH_2$.

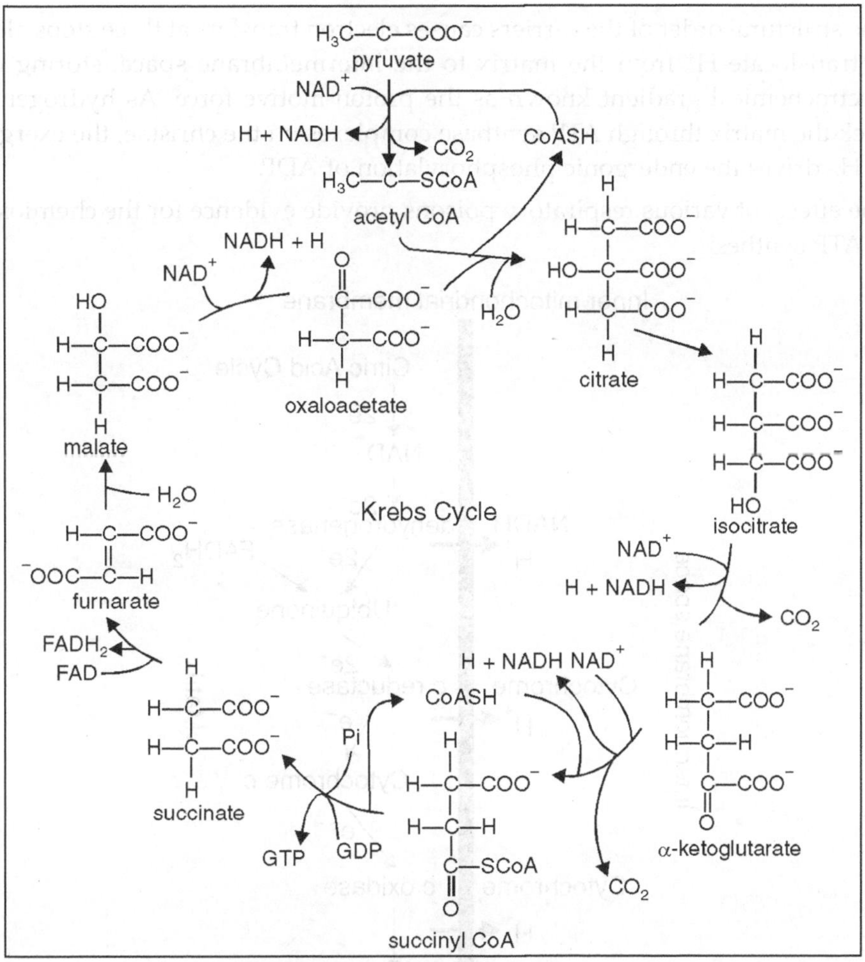

FIGURE 4.32 The TCA cycle.

Electron Transport Chain and Oxidative Phosphorylation

- Most of the ATP created from the energy stored in the glucose is produced by oxidative phosphorylation when NADH and $FADH_2$ donate their electrons to a system of electron carriers embedded in the mitochondrial cristae.
- The electron transport chain consists of a series of increasingly electronegative components, starting with a flavoprotein progressing through an iron-sulfur protein, then to ubiquinone and series of cytochrome proteins with iron-containing heme groups, and finally reaching oxygen, which is very electronegative.
- The components of the chain receive electrons from NADH and $FADH_2$ and shift between reduced and oxidized states, passing electrons down an energy gradient to oxygen, which then picks up a pair of hydrogen ions and forms water. Two mobile components, Q and cytochrome c, transfer electrons between the other electron carriers, which are located in three groups of integrated complexes.

- The structural order of the carriers causes electron transfers at three steps along the chain to translocate H^+ from the matrix to the intermembrane space, storing energy in an electrochemical gradient known as the proton-motive force. As hydrogen ions diffuse back the matrix through ATP synthase complexes on the christae, the exergonic passage of H^+ drives the endergonic phosphorylation of ADP.
- The effects of various respiratory poisons provide evidence for the chemiosmotic model of ATP synthesis.

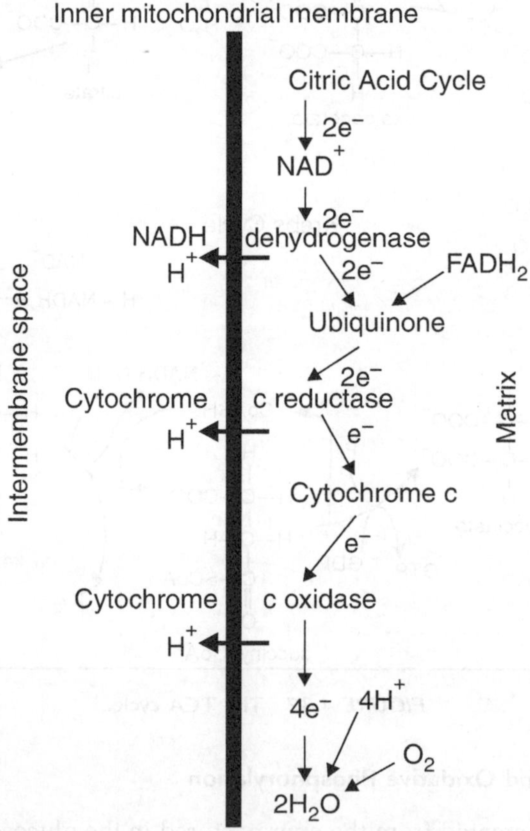

FIGURE 4.33 **Electron transport system (ETS) and oxidative phosphorylation on the inner membrane of mitochondrial cristae.**

- The complete oxidation of glucose to carbon dioxide during aerobic respiration in eukaryotes produces a net yield of about 36 molecules of ATP, compared to only two for incomplete oxidation.
- The actual yield of ATP during respiration varies, owing to differences in the permeability of the christae to H^+ and to partial use of the proton gradient to drive the active transport of certain solutes across the outer mitochondrial membrane.

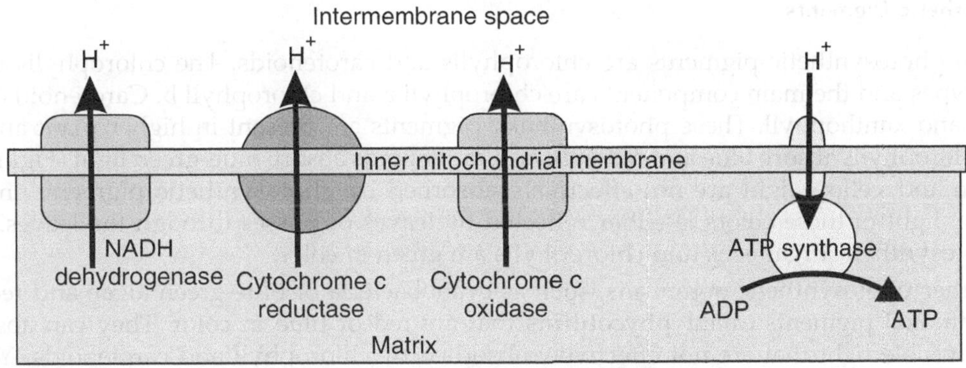

FIGURE 4.34 **The oxidative phosphorylation.**

Photosynthesis

The primary source of energy for nearly all lifeforms on earth is the sun. The energy in sunlight is introduced into the biosphere by a process known as **photosynthesis**, which occurs in plants, algae, and some types of bacteria. All the food we eat and all the fossil fuels we use are products of photosynthesis. Photosynthesis can be defined as the physico-chemical process by which photosynthetic organisms use light energy to drive the synthesis of organic compounds. The photosynthetic process depends on a set of complex pigment and protein molecules that are located in and around a highly organized membrane, which is a part of chloroplast. Through a series of energy transducing reactions, the photosynthetic machinery transforms light energy into a stable form that can last for hundreds of millions of years (for example, fossil fuels).

All photosynthetic organisms convert CO_2 (carbon dioxide) to organic material by reducing this gas to carbohydrates in a rather complex set of reactions. Electrons for this reduction reaction ultimately come from water, by splitting water molecule into oxygen and protons by light energy absorbed by photosynthetic pigments, primarily chlorophylls and carotenoids.

Photosynthetic Organisms

There are two groups of photosynthetic organisms; all of them depend on chlorophyll pigments for conversion of light energy into chemical energy. They are:

- Oxygenic photosynthetic organisms and
- An oxygenic photosynthetic organisms.

Oxygenic Photosynthetic Organisms: The photosynthetic process in all plants and algae as well as in certain types of photosynthetic bacteria involves the reduction of CO_2 to carbohydrate and removal of electrons from H_2O, which results in the release of O_2. In this process, known as oxygenic photosynthesis, water is oxidized by the photosystem II reaction center, a multi-subunit protein located in the photosynthetic membrane.

Anoxygenic Photosynthetic Organisms: Some photosynthetic bacteria can use light energy to extract electrons from molecules other than water for example (H_2S). Therefore, there is no evolution of oxygen. These organisms are of ancient origin, presumed to have evolved before oxygenic photosynthetic organisms. Anoxygenic photosynthetic organisms are certain species of purple bacteria, green sulfur bacteria, green gliding bacteria, and gram positive bacteria.

Photosynthetic Pigments

The main photosynthetic pigments are chlorophylls and carotenoids. The chlorophylls includes various types and the main components are chlorophyll a and chlorophyll b. Carotenoids include carotene and xanthophyll. These photosynthetic pigments are present in higher plant and green algae. Chlorophylls absorb blue and red light and carotenoids absorb blue-green light (Figure 4.36.), but green and yellow light are not effectively absorbed by photosynthetic pigments in plants. Therefore, light of these colors is either reflected by leaves or passes through the leaves. That is why photosynthetic organisms and chlorophylls are green in color.

Other photosynthetic organisms, such as cyanobacteria or blue-green algae and red algae, have additional pigments called **phycobilins** that are red or blue in color. They can absorb the colors of visible light that are not effectively absorbed by chlorophyll and carotenoids. Yet other organisms, such as purple and green bacteria, contain bacteriochlorophyll that absorbs the infrared, in addition to the blue part of the spectrum. These bacteria do not evolve from oxygen, but perform photosynthesis under oxygen-less conditions. These bacteria efficiently use infrared light for photosynthesis. Infrared is light with wavelengths above 700 nm that cannot be seen by the human eye. Some bacterial species can use infrared light with wavelengths of up to 1000 nm.

FIGURE 4.35 **Examples of photosynthetic organisms: leaves from higher plants flanked by colonies of photosynthetic purple bacteria (left) and cyanobacteria (right).**

The Chloroplast

The process of photosynthesis occurs in small organelles known as chloroplasts that are located inside eukaryotic cells. The parts of a chloroplast include an outer membrane and inner membrane with an intermembrane space. The double membrane encloses a matrix called the **stroma.** In the stroma there is a system of double-layered membranes known as **lamellae**. Some of these lamellae are disc-shaped enclosing an inner space separated from the outside by a double membrane. Such disc-double membranes enclosing an inner space are called the **thylakoid membranes**. If a molecule has to go to the outer stroma from the inner space of thylakoid membrane, it has to pass through the double membrane system of thylakoids. Thylakoid membranes are arranged in a stalk to form the structure called **grana.** The chlorophylls and other accessory pigments are built into the membranes of thylakoids. The more primitive photosynthetic organisms such as cyanobacteria, prochlorophytes, and photosynthetic bacteria (anoxygenic organism) do not have chloroplasts or other organelles. In these organisms, the photosynthetic pigments are distributed either in the membrane around the cytoplasm or in the infoldings of cell membrane, or in the thylakoid membrane-like structures as in the case of cyanobacteria.

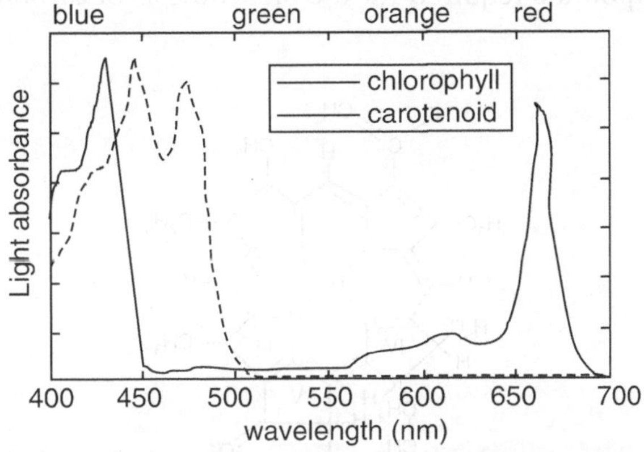

FIGURE 4.36 **Absorption spectrum of isolated chlorophyll and carotenoid species. The color associated with the various wavelengths is indicated above in the graph. The absorption maxima of chlorophyll a are 430 and 662 nm, that of chlorophyll b are at 453 and 642 nm.**

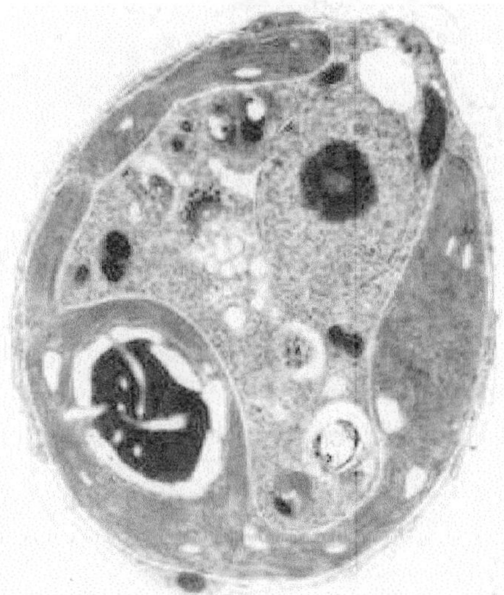

FIGURE 4.37 **Electron micrograph of a thin section of an algal cell. The cup-shaped structure around the edge of the cell (open near the top) is the chloroplast. The structures resembling mostly parallel lines in the chloroplast are the thylakoid membranes.**

The Light Reactions of Photosynthesis

The photosynthetic reactions are traditionally divided into two stages : the 'light reaction', which consists of electron and proton transfer reactions and the 'dark reaction,' which is the biosynthesis of carbohydrates from CO_2. The light reaction occurs in the thylakoid membranes and the dark reaction takes place in the stroma—the matrix of the chloroplast. Stroma is a mixture of all enzymes, substrates, and other compounds required for the biosynthesis of carbohydrates from CO_2.

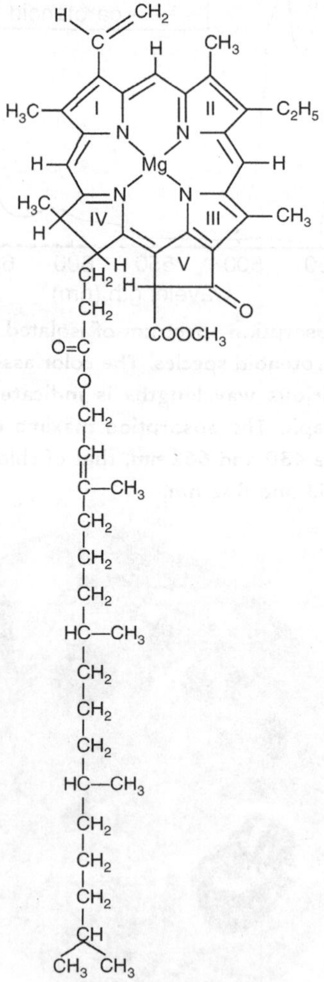

FIGURE 4.38 **Chemical structure of chlorophyll in a molecule.**

The light energy is collected by an array of antenna complexes (chloroplast pigment-protein complexes) and transferred to a special complex called the **reaction center** where the light energy is used to operate a series of electron-transfer reactions (movement of electrons from one molecule to another). These reactions result in a separation of charge across a biological membrane. This charge difference across the membranes (chloroplast membranes) is used for the synthesis of ATP molecules (photophosphorylation) and $NADPH_2^+$, the reducing power. All these reactions including the electron-transport system and the associated photophosphorylation form the light reaction of photosynthesis.

The reducing power comprising of $NADPH_2^+$ and ATP generated in the light reaction of photosynthesis is used for reducing CO_2 to glucose ($C_6H_{12}O_6$). This process of CO_2 fixation known as 'dark reaction' or 'Calvin cycle' is independent of light and takes place through a series of biochemical reactions in the matrix of chloroplast.

Reaction Centers and Antenna

A large number of pigment molecules (100 to 5000) are grouped together to form the antenna, which directly harvests the light energy and transfers the energy to a particular chlorophyll molecule known as the reaction center, where the actual photochemical event leading to the charge separation takes place. The purpose is to maintain a high rate of electron transfer in the reaction center, even at low-light intensities. In many systems, the size of the photosynthetic antenna is flexible, and photosynthetic organisms growing at low light (in the shade, for example) generally will have a larger number of antenna pigments per reaction center than those growing at higher light intensity. However, at high-light intensities if the light intensity exceeds its capacity, part of the photosynthetic electron transport chain may be shut down. This is known as **photoinhibition.**

Photosynthetic Electron-transport System

The light reaction converts energy into several forms. It is diagrammatically represented in Figure 4.39. The antenna pigments absorb the light energy and transfer this energy to the reaction center. When the reaction center obtains the sufficient photons its molecule will be in an excited electronic state and finally an electron is expelled from the molecule. This initial electron-transfer reaction or the charge separation in the photosynthetic reaction center starts a long series of redox (reduction-oxidation) reactions, passing the electron along a chain of cofactors and proteins finally filling up the "electron hole" on the chlorophyll. All photosynthetic organisms that produce oxygen (oxygenic) have two types of reaction centers, named photosystem II and photosystem I (PS II and PS I, for short). Both are pigment-protein complexes that are located in the thylakoid membranes.

Photosystem II (PS II) is the complex where water splitting and oxygen evolution occur. Upon oxidation of the reaction center chlorophyll in PS II, an electron is pulled from a nearby amino acid (tyrosine), which is part of the surrounding protein, which in turn gets an electron from the water splitting complex. From the PS II reaction center, electrons flow to free electron-carrying molecules, plastoquinone in the thylakoid membrane, and from there to another membrane-protein complex, the cytochrome b_6f complex. The other photosystem, PS I, also catalyzes light-induced charge separation in a fashion basically similar to PS II; light is harvested by an antenna, and light energy is transferred to the reaction center chlorophyll, where light-induced charge separation is initiated. However, in PS I electrons are transferred eventually to NADP (nicotinamide adenosine

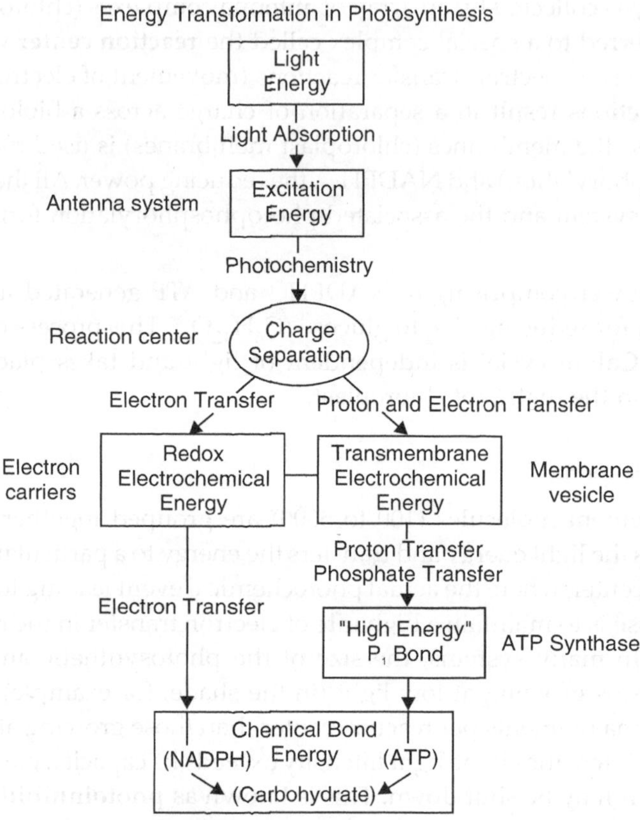

FIGURE 4.39 **Photosynthetic electron-transport system.**

dinucleotide phosphate) to form $NADPH_2$, the reduced form. This can be used for carbon fixation. The oxidized reaction center chlorophyll eventually receives another electron from the cytochrome b_6f complex. Therefore, electron transfer from water through PS II and PS I to NADP results in water oxidation (producing oxygen) and synthesis of $NADPH_2$. The energy for this process is provided by light—two quanta for each electron transported through the whole chain. A schematic overview of these processes is provided in Figure 4.40.

Electron flow from water to NADP requires light and is coupled to the generation of a proton gradient across the thylakoid membrane. This proton gradient is used for synthesis of ATP (adenosine triphosphate), a high-energy molecule. ATP and reduced NADP that result from the light reactions are used for CO_2 fixation in a process that is independent of light.

There are two types of electron transport system operating in higher plants and algae. One is the non-cyclic electron-transport system, which results in the formation of ATP and $NADPH_2$ with the splitting of the water molecule and release of oxygen. This operates under normal conditions at which both PII and PI will work together. But under special circumstances only the PSI will be operating and that results in the production of ATP alone. There is no production of $NADPH_2$ and so there is no evolution of oxygen.

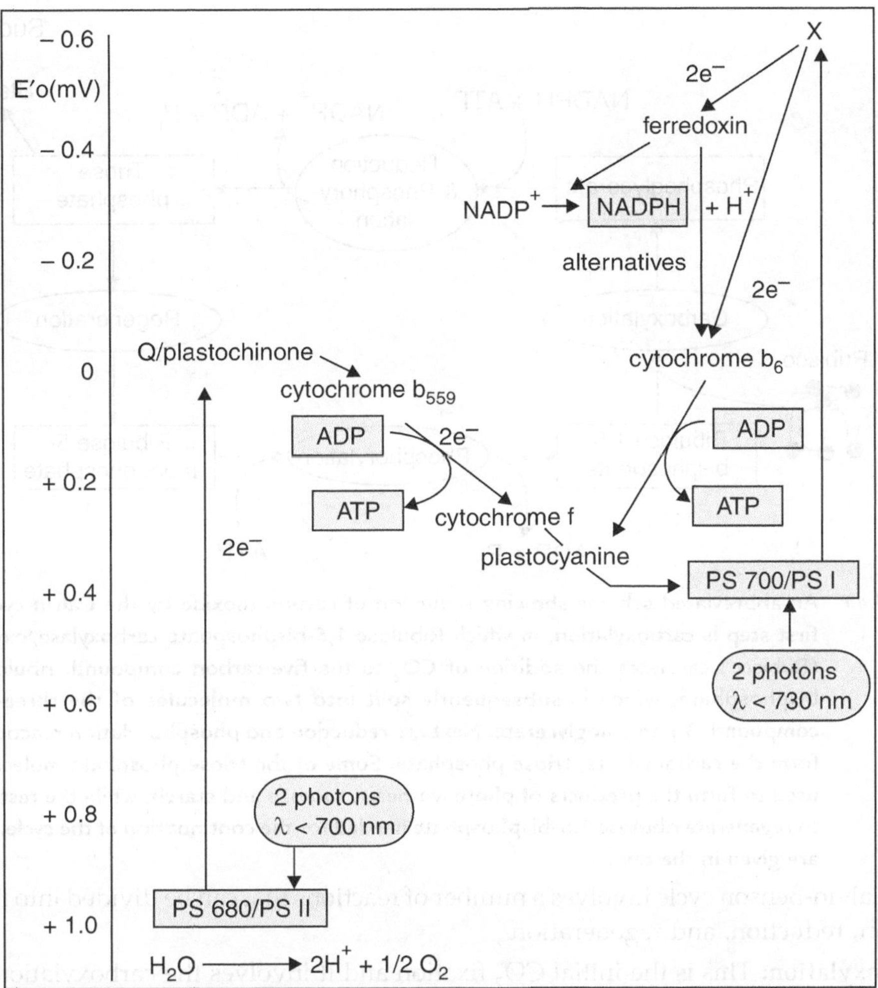

FIGURE 4.40 **The Z scheme of photosynthesis—cyclic andnon-cyclic electron-transport system and photophosphorylation.**

The Dark Reactions of Photosynthesis

The dark reaction or the carbon dioxide fixation takes place in the stroma of chloroplast and is independent of light. During this process, CO_2 is reduced to the level of carbohydrate by the reducing power generated during the light reaction, through a series of enzymatic reactions. The famous scientist M. Calvin and his collaborators, at the University of California, Berkeley, accurately elucidated the sequences of reactions of this process. Therefore, the CO_2 pathway is known as the Calvin-Benson cycle. Due to the use of isotopes, they were able to reveal, completely, the reactions taking place during the incorporation of carbon dioxide into carbohydrates in a relatively short period from 1946–1953. The quick success was based on the use of sensitive methods (two-dimensional paper chromatography, autoradiography), a suitable experimental specimen a single-celled green algae Chlorella pyrenoidosa (Figure. 4.37.), use of $^{12}CO_2$, and the rapid progresses of enzyme biochemistry.

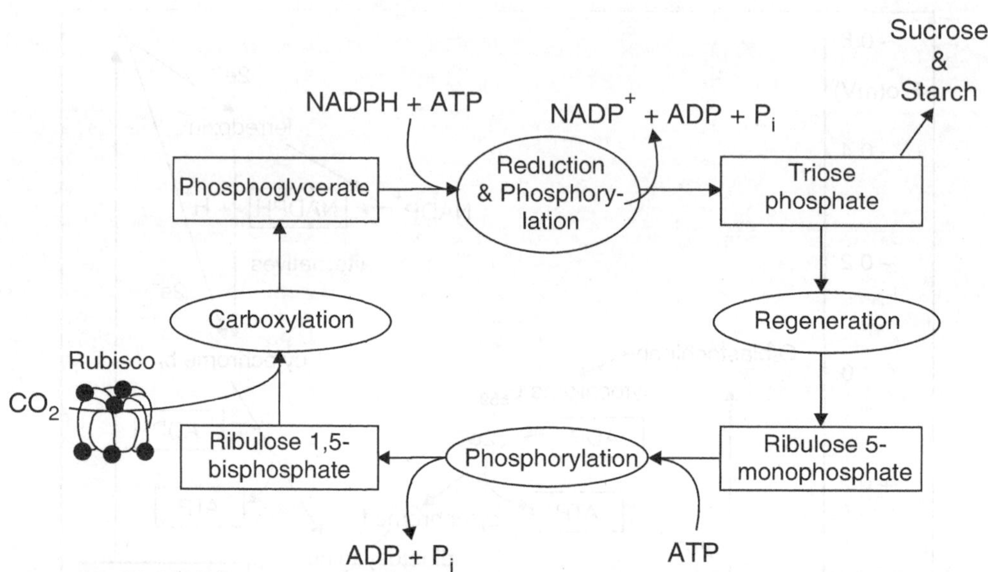

FIGURE 4.41 **An abbreviated scheme showing reduction of carbon dioxide by the Calvin cycle. The first step is carboxylation, in which Ribulose 1,5-bisphosphate carboxylase/oxygenase (Rubisco) catalyzes the addition of CO_2 to the five-carbon compound, ribulose 1,5-bisphosphate, which is subsequently split into two molecules of the three-carbon compound, 3-phosphoglycerate. Next are reduction and phosphorylation reactions that form the carbohydrate, triose phosphate. Some of the triose phosphate molecules are used to form the products of photosynthesis, sucrose and starch, while the rest is used to regenerate ribulose 1,5-bisphosphate needed for the continuation of the cycle. Details are given in the text.**

The Calvin-Benson cycle involves a number of reactions that can be divided into three phases: carboxylation, reduction, and regeneration.

Carboxylation: This is the initial CO_2 fixation and it involves the carboxylation of ribulose bisphosphate. This reaction is catalyzed by the enzyme ribulose-1,5-bisphosphate carboxylase/ oxygenase (RuBisCO), which can react with either oxygen (leading to a process named photorespiration and not resulting in carbon fixation) or with CO_2. The product of carbon-dioxide fixation is 3-phosphoglyceric acid (3-PGA).

Reduction: In the reduction phase 3-PGA is reduced to 3-phospho-glyceraldehyde (3-PGAL) by the reducing power generated through the photochemical reaction. Both ATP and $NADPH_2$ are utilized here to reduce the PGA. Phosphoglyceraldehyde is the basic sugar, the triose, which can be utilized by the cell according to its requirement. It can either be used for the synthesis of starch or it can be exported to the cytoplasm for other metabolic activities such as sucrose synthesis or as an energy source.

In the cytoplasm, there is a modulator compound, fructose-2.6-bisphosphate, which regulates the starch versus sucrose synthesis. The triose phosphate synthesized in the chloroplast is exported to the cytoplasm as dihydroxy acetone phosphate (DHAP) or 3-PGAL in exchange for inorganic phosphate by a specific transporting molecule present in the inner membrane of the chloroplast.

Regeneration Phase: RUBP is the starting compound, which fixes the carbon dioxide initially. So it has to be regenerated for the cyclic reaction to continue. After a number of sugar interconversions, finally the RUBP is regenerated. It again undergoes carboxylation and the cycle continues. The details of the reactions leading to the regeneration of RUBP are represented in Figure 4.42 in detail. Many of the steps in the regeneration reaction are similar to that of the pentose phosphate pathway except that they are operating in the reverse direction. That is why the Calvin cycle is also called the reductive pentose phosphate pathway. The sugar interconversions are mediated fundamentally by transketolase and transaldolase. All the enzymes required for the Calvin cycle are located in the stroma of chloroplast.

For every molecule of triose synthesized from CO_2, 6 $NADPH_2$ and 9 ATP are required. That is, each molecule of CO_2 reduced to a sugar $[CH_2O]_n$ requires two molecules of NADPH and three molecules of ATP.

CAM and C4 Plants

The enzyme RBISCO is a dual enzyme having both carboxylase and oxygenase activity. The probability with which RuBisCO reacts with oxygen versus with CO_2 depends on the relative

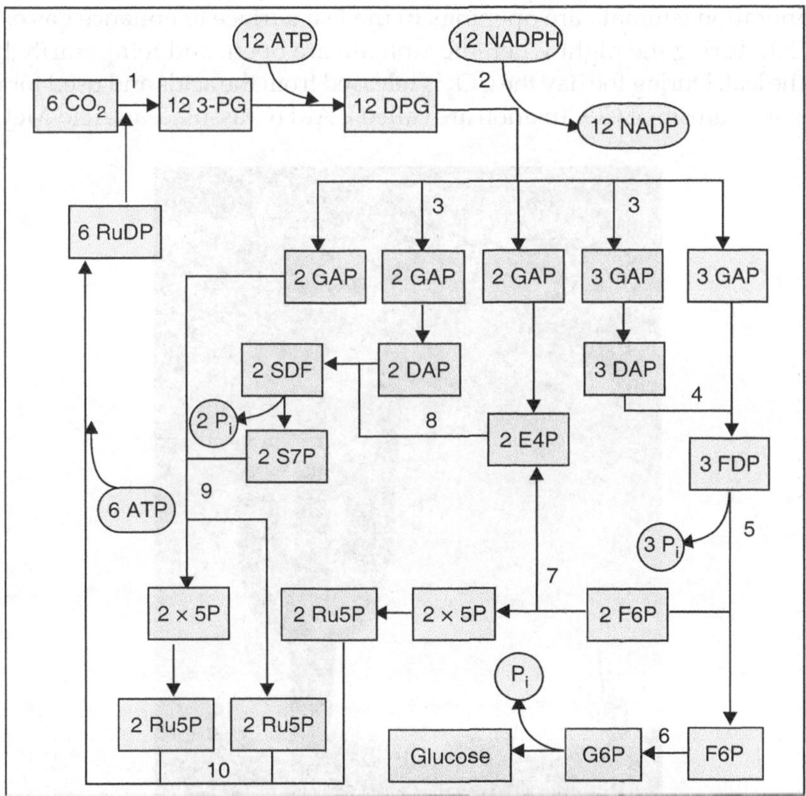

FIGURE 4.42 **Scheme of the Calvin Cycle—scheme of carbon dioxide fixation. Six molecules of CO_2 and six receptor molecules are necessary for the production of one molecule of glucose, consequently six rounds of the cycle are necessary in order to produce one molecule of glucose. For reasons of simplification these six rounds are depicted as one in the scheme above. The colors symbolize the numbers of C-atoms of the molecules.**

concentrations of the two compounds at the site of the reaction. In all organisms, CO_2 is by far the preferred substrate, but as the CO_2 concentration is very much lower than the oxygen concentration, photorespiration does occur at significant levels. To boost the local CO_2 concentration and to minimize the oxygen tension, some plants, referred to as C_4 plants, have set aside some cells within a leaf (named bundle sheath cells) to be involved primarily in CO_2 fixation, and others (named mesophyll cells) to specialize in the light reactions: ATP, CO_2, and reduced NADP in mesophyll cells are used for synthesis of 4-carbon organic acids (such as malate), which are transported to bundle sheath cells. Here, the organic acids are converted releasing CO_2 and reduced NADP, which are used for carbon fixation. The resulting 3-carbon acid is returned to the mesophyll cells. The bundle sheath cells generally do not have PS II activity, in order to minimize the local oxygen concentration. However, they retain PS I, presumably to aid in ATP synthesis. Even though C_4 plants have reduced amounts of photorespiration, the amount of ATP they need per amount of CO_2 fixed is a little higher than in other plants, and therefore their total production rate is similar to that of plants with higher rates of photorespiration.

Some plants living in desert climates, such as cacti, keep their stomata closed during the day to minimize evaporation (stomata are openings in the leaf surface to enhance gas exchange). These plants take up CO_2 during the night when the stomata are open, and temporarily bind the CO_2 to organic acids in the leaf. During the day the CO_2 is released from the acids and used for photosynthesis. Plants using this mechanism of CO_2 fixation are called CAM (Crassulacean Acid Metabolism) plants (Figure 4.43).

FIGURE 4.43 **The light reactions of photosynthesis stop when the sun goes down. However, CO_2 fixation can continue as long as ATP and NADPH are available. In cacti and other succulents CO_2 uptake by the plant occurs primarily at night.**

TABLE 4.7 Difference between C_3 and C_4 plants.

C_3 Plants	C_4 Plants
■ The pathway of carbon assimilation is known as the Calvin cycle.	It is known as Hatch and Slack pathway.
■ It mainly takes place in mesophyll cells.	Occurs in both mesophyll and bundle sheath cell.
■ The first product of CO_2 fixation is a three carbon compound (3-phosphoglyceric acid).	First product is a four carbon compound (Oxalo-acetic acid).
■ Three ATPs are used for the reduction of one CO_2.	5 ATP molecules are used per carbon dioxide.
■ Significant loss by photorespiration.	Photorespiration is almost absent.
■ Comparatively low rate of photosynthesis and biomass production.	High photosynthetic rate.

Ribulose Bisphosphate Carboxylase Oxygenase (RUBISCO)

This is the key enzyme of the carbon dioxide fixation reaction. RUBISCO is a large water-soluble protein complex that has many sub-units. The three-dimensional structure has been determined by *x*-ray crystallography for RUBISCO isolated from different photosynthetic organisms. In nature there are two structurally distinct and functionally similar types of RUBISCO. In photosynthetic bacteria such as Rhodospirillum rubrum, RUBISCO consists of two similar sub-units 50 kD each. Therefore, it is called a homodimer protein. In all other organisms that perform oxygenic photosynthesis, RUBISCO are made up of eight large and eight small sub-units. The large sub-unit is encoded by the chloroplast genome and the small sub-unit is coded by the nuclear genome.

RUBISCO is a bifunctional enzyme that, in addition to binding CO_2 to ribulose bisphosphate, can also bind O_2. This oxygenation reaction produces the 3-phosphoglycerate that is used in the Calvin cycle and a two-carbon compound (2-phosphoglycolate) that is not useful for the plant. In response, a complicated set of reactions (known as photorespiration) is initiated that serve to recover reduced carbon and to remove phosphoglycolate. The RUBISCO oxygenation reaction appears to serve no useful purpose for the plant. Some plants have evolved specialized structures and biochemical pathways that concentrate CO_2 near RUBISCO. These pathways (C4 and CAM) serve to decrease the fraction of oxygenation reactions (see C4 and CAM plants).

Effect of Carbon Dioxide Concentration

The amount of overall CO_2 fixation in plants growing under optimal conditions is limited primarily by the amount of CO_2 available. Therefore, the increase of CO_2 in the atmosphere will lead to somewhat higher rates of plant growth in environments where the CO_2 concentration limits growth rates. This is usually the case in an agricultural setting, where nutrients and water availability are not limiting. However, in natural conditions, where productivity is limited by conditions other than CO_2 concentrations, plant productivity has been found to often increase upon increasing the CO_2 concentration.

Photosynthesis versus Respiration

Virtually all oxygen in the atmosphere is thought to have been generated through the process of photosynthesis. Obviously, all respiring organisms (including plants) utilize this oxygen and produce CO_2. Thus, photosynthesis and respiration are interlinked, with each process depending on the products of the other. The global amount of photosynthesis is on the order of a trillion kg of dry organic matter produced per day, and respiratory processes convert about the same amount of organic matter to CO_2. A large part (probably the majority) of photosynthetic productivity occurs in open oceans, mostly by oxygenic prokaryotes. Without photosynthesis, the oxygen in the atmosphere would be depleted within several thousand years. It should be emphasized that plants respire just like any other higher organism, and that during the day this respiration is masked by a higher rate of photosynthesis.

Nitrogen Fixation and Nitrogen Cycle

The growth of all organisms depends on the availability of mineral nutrients, and none is more important than nitrogen, which is required in large amounts as an essential component of proteins, nucleic acids, and other cellular constituents. There is an abundant supply of nitrogen in the earth's atmosphere—nearly 80% in the form of N_2 gas. However, N_2 is unavailable for use by most organisms because there is a triple bond between the two nitrogen atoms, making the molecule almost inert. In order for nitrogen to be used for growth it must be "fixed" (combined) in the form of ammonium (NH_4) or nitrate (NO_3) ions. So, nitrogen is often the limiting factor for growth and biomass production in all environments where there is suitable climate and availability of water to support life.

The atmospheric nitrogen is converted into a usable form of nitrogen such as ammonium and nitrate ions, which are assimilated by plants and passed to animals in the form of amino acids and proteins. Finally, it is recycled back to the atmosphere. This transformation of nitrogen from atmosphere through organisms and back to atmosphere is called the **nitrogen cycle**. The nitrogen cycle mainly includes three stages: nitrogen fixation, assimilation of inorganic nitrogen (NO_3^-, NH_3^-) to organic nitrogen such as amino acids, and denitrification, which releases molecular nitrogen to atmosphere.

Microorganisms have a central role in almost all aspects of the nitrogen cycle and thus in supporting life on earth.

- Some bacteria can convert N_2 into ammonia by the process of nitrogen fixation. These bacteria are either free-living or form symbiotic associations with plants or other organisms (e.g., termites, protozoa).
- Other bacteria bring about transformations of ammonia to nitrate, and of nitrate to N_2 or other nitrogen gases.
- Many bacteria and fungi degrade organic matter, releasing fixed nitrogen for reuse by other organisms.

All these processes contribute to the nitrogen cycle.

We shall deal first with the process of nitrogen fixation and the nitrogen-fixing organisms, then consider the microbial processes involved in the cycling of nitrogen in the biosphere.

Nitrogen Fixation

Nitrogen fixation is the conversion of molecular nitrogen into reduced forms such as NO_3 and NH_4 ions. The major conversion of N_2 into ammonia, and then into proteins, is achieved by microorganisms by the process of nitrogen fixation (or dinitrogen fixation). A relatively small amount of ammonia is produced by lightning. Some amount of ammonia is also produced industrially by the Haber-Bosch process, using an iron-based catalyst, very high pressures, and fairly high temperature into proteins, which is achieved by microorganisms in the process called nitrogen fixation (or dinitrogen fixation).

TABLE 4.8 Some estimates of the amount of nitrogen fixed on a global scale. The total biological nitrogen fixation is estimated to be twice as much as the total nitrogen fixation by non-biological processes.

Type of fixation	N_2 fixed (10^{12} g per year, or 10^6 metric tons per year)
Non-biological	
Industrial	about 50
Combustion	about 20
Lightning	about 10
Total	about 80
Biological	
Agricultural land	about 90
Forest and non-agricultural land	about 50
Sea	about 35
Total	about 175

(Data from various sources, compiled by DF Bezdicek & AC Kennedy, in Microorganisms in Action (eds. JM Lynch & JE Hobbie). Blackwell Scientific Publications, 1998).

Mechanism of Biological Nitrogen Fixation

Biological nitrogen fixation can be represented by the following equation, in which two moles of ammonia are produced from one mole of nitrogen gas, at the expense of 16 moles of ATP and a supply of 8-electrons and protons (hydrogen ions):

$$N_2 + 8H^+ + 8e^- + 16\ ATP = 2NH_3 + H_2 + 16ADP + 16\ Pi$$

This reaction is performed exclusively by prokaryotes (the bacteria and related organisms), using an enzyme complex termed nitrogenase. This enzyme consists of two proteins—an iron protein and a molybdenum-iron protein, as shown in Figure 4.44.

The reactions occur while N_2 is bound to the nitrogenase enzyme complex. The Fe protein is first reduced by electrons donated by ferredoxin. Then the reduced Fe protein binds ATP and reduces the molybdenum-iron protein, which donates electrons to N_2, producing HN=NH. In two further cycles of this process (each requiring electrons donated by ferredoxin) HN=NH is reduced to H_2N-NH_2, and this in turn is reduced to $2NH_3$.

Depending on the type of microorganism, the reduced ferredoxin, which supplies electrons for this process, is generated by photosynthesis, respiration, or fermentation. There is a remarkable degree of functional conservation between the nitrogenase proteins of all nitrogen-fixing bacteria. The Fe protein and the Mo-Fe protein have been isolated from many of these bacteria, and nitrogen fixation can be shown to occur in cell-free systems in a laboratory when the Fe protein of one species is mixed with the Mo-Fe protein of another bacterium, even if the species are very distantly related.

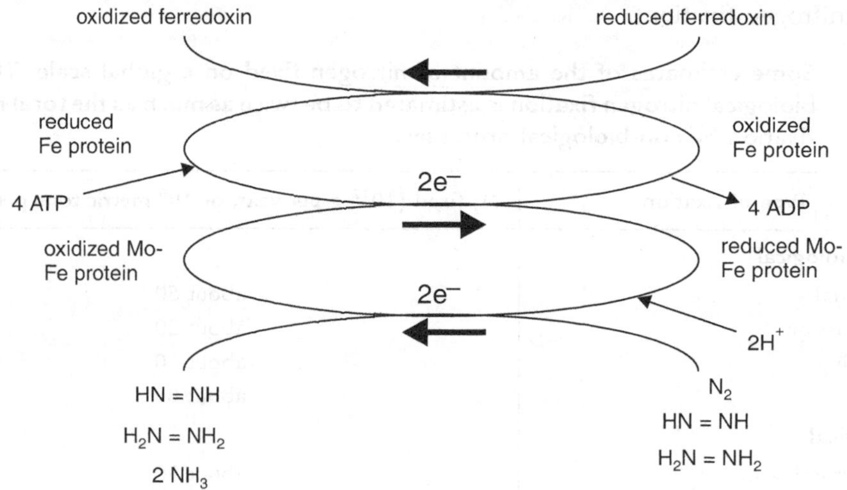

FIGURE 4.44 **Schematic representation of different steps in the process of molecular nitrogen fixation.**

Nitrogen-fixing Organisms

All nitrogen-fixing organisms are prokaryotes (bacteria). Some of them live independently of other organisms—the so-called free-living nitrogen-fixing bacteria. Others live in intimate symbiotic associations with plants or with other organisms (e.g., protozoa). Examples are shown in Table 4.9.

A point of special interest is that the nitrogenase enzyme complex is highly sensitive to oxygen. It is inactivated when exposed to oxygen, because this reacts with the iron component of the proteins. Although this is not a problem for anaerobic bacteria, it could be a major problem for the aerobic species such as cyanobacteria (which generate oxygen during photosynthesis) and the free-living aerobic bacteria of soils, such as azotobacter and beijerinckia. These organisms have various methods to overcome the problem. For example, the azotobacter species have the highest known rate of respiratory metabolism of any organism, so they might protect the enzyme by maintaining a very low level of oxygen in their cells. Azotobacter species also produce copious amounts of extracellular polysaccharide (as do rhizobium species in culture, exopolysaccharides). By maintaining water within the polysaccharide slime layer, these bacteria can limit the diffusion rate of oxygen to the cells. In the symbiotic nitrogen-fixing organisms such as rhizobium, the root nodules can contain oxygen-scavenging molecules such as leghemoglobin, which shows as a pink color when the active nitrogen-fixing nodules of legume roots are cut open. Leghemoglobin may regulate the supply of oxygen to the nodule tissues in the same way hemoglobin regulates the supply of oxygen to mammalian tissues. Some of the cyanobacteria have yet another mechanism for protecting nitrogenase: nitrogen fixation occurs in special cells (heterocysts), which possess only photosystem I (used to generate ATP by light-mediated reactions) whereas the other cells have both photosystem I and photosystem II (which generates oxygen when light energy is used to split water to supply H_2 for synthesis of organic compounds).

TABLE 4.9 Examples of nitrogen-fixing bacteria (*denotes a photosynthetic bacterium).

Free living		Symbiotic with plants	
Aerobic	**Anaerobic**	**Legumes**	**Other plants**
Azotobacter Beijerinckia Klebsiella (some) Cyanobacteria (some)*	Clostridium (some) Desulfovibrio Purple sulfur bacteria* Purple non-sulfur bacteria* Green sulfur bacteria*	Rhizobium	Frankia Axopirillum

Symbiotic Nitrogen Fixation

Legume Symbioses

The most familiar examples of nitrogen-fixing symbioses are the root nodules of legumes (peas, beans, clovers, etc.). In these leguminous associations the bacteria usually are rhizobium

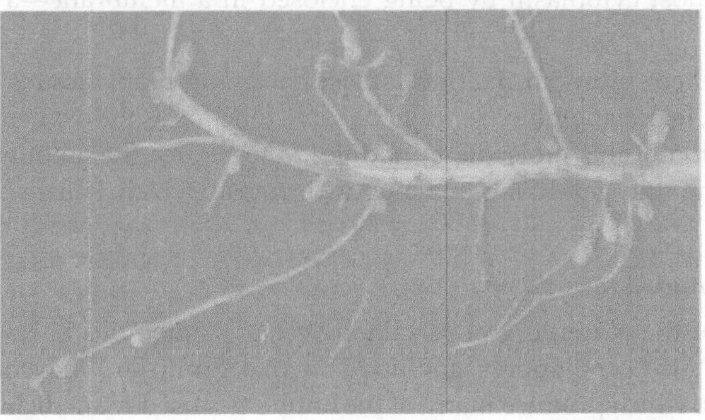

A

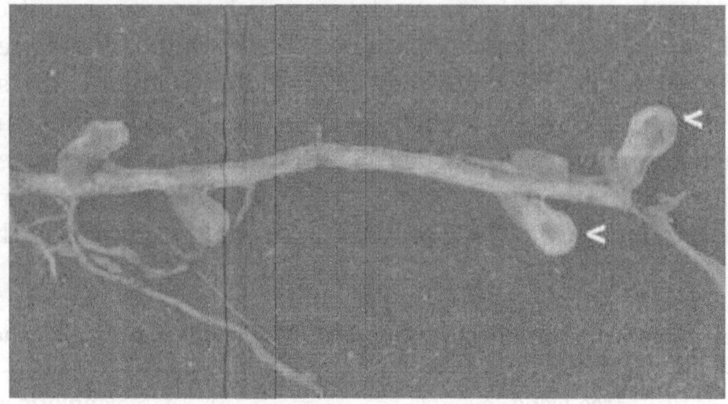

B

FIGURE 4.45 **A Part of a clover root system bearing naturally occurring nodules of rhizobium. Each nodule is about 2 to 3 m long. B-Clover root nodules at higher magnification, showing two partly crushed nodules (arrowheads) with pink-colored contents. This color is caused by the presence of the pigment leghemoglobin—a unique metabolite of this type of symbiosis. Leghemoglobin is found only in the nodules and is not produced by either the bacterium or the plant when grown alone.**

species, but the root nodules of soybeans, chickpeas, and some other legumes are formed by small-celled rhizobia termed bradyrhizobium. Nodules on some tropical leguminous plants are formed by yet other genera. In all cases, the bacteria "invade" the plant and cause the formation of a nodule by inducing localized proliferation of the plant host cells. Yet, the bacteria always remain separated from the host cytoplasm by being enclosed in a membrane—a necessary feature in symbioses.

In nodules where nitrogen-fixation is occurring, the plant tissues contain the oxygen-scavenging molecule, **leghemoglobin** (serving the same function as the oxygen-carrying hemoglobin in blood). The function of this molecule in nodules is to reduce the amount of free oxygen, and thereby to protect the nitrogen-fixing enzyme nitrogenase, which is irreversibly inactivated by oxygen.

Associations with Frankia

Frankia is a genus of the bacterial group actinomycetes; filamentous bacteria that are noted for their production of air-borne spores. Included in this group are the common soil-dwelling streptomyces species, which produce many of the antibiotics used in medicine. Frankia species are slow-growing in culture, and require specialized media, suggesting that they are specialized symbionts. They form nitrogen-fixing root nodules (sometimes called actinorhizae) with several woody plants of different families, such as alder (Alnus species), sea buckthorn (Hippophae rhamnoides), and Casuarina.

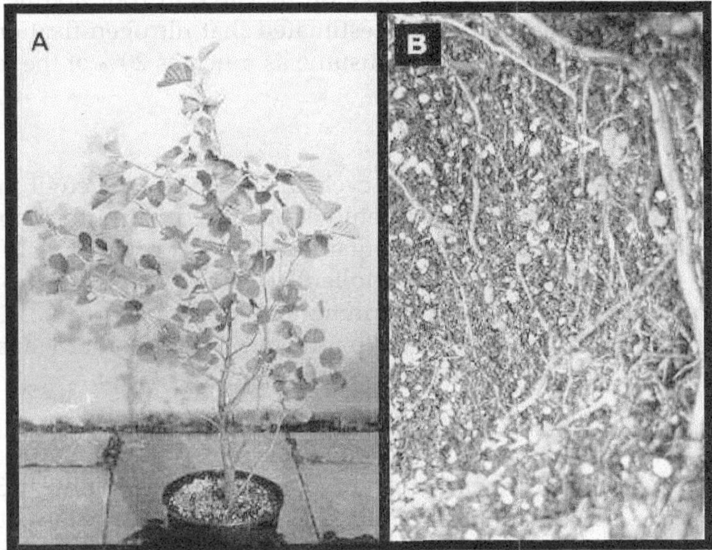

FIGURE 4.46 **A shows a young alder tree (*Alnus glutinosa*) growing in a plant pot. Figure B shows part of the root system of this tree, bearing the orange-yellow colored nodules (arrowheads) containing *Frankia*. Alder and the other woody hosts of *Frankia* are typical pioneer species that invade nutrient-poor soils. These plants probably benefit from the nitrogen-fixing association, while supplying the bacterial symbiont with photosynthetic products.**

Cyanobacterial Associations

Photosynthetic cyanobacteria often live as free-living organisms in pioneer habitats such as desert soils or as symbionts with lichens in other pioneer habitats. They also form symbiotic associations with other organisms such as the water fern Azolla and cycads. The association with Azolla, where cyanobacteria (Anabaena azollae) are harbored in the leaves, has sometimes been shown to be important for nitrogen inputs in rice paddies, especially if the fern is allowed to grow and then ploughed into the soil to release nitrogen before the rice crop is sown. A symbiotic association of cyanobacteria with cycads produces short club-shaped, branching roots that grow into the aerial environment. These aerial roots contain the nitrogen-fixing cyanobacterial symbiont. In addition to these intimate and specialized symbiotic associations, there are several free-living nitrogen-fixing bacteria that grow in close association with plants. For example, azospirillum species have been shown to fix nitrogen when growing in the root zone (rhizosphere) or tropical grasses, and even of maize plants in field conditions. Similarly, azotobacter species can fix nitrogen in the rhizosphere of several plants. In both cases the bacteria grow at the expense of sugars and other nutrients that leak from the roots. However, these bacteria can make only a small contribution to the nitrogen nutrition of the plant, because nitrogen fixation is an energy-expensive process, and large amounts of organic nutrients are not continuously available to microbes in the rhizosphere.

This limitation may not apply to the bacteria that live in root nodules or other intimate symbiotic associations with plants. It has been estimated that nitrogen fixation in the nodules of clover roots or other leguminous plants may consume as much as 20% of the total photosynthate.

The Nitrogen Cycle

Figure 4.47 shows an overview of the nitrogen cycle in soil or aquatic environments. At any one time a large proportion of the total fixed nitrogen will be locked up in the biomass or in the dead remains of organisms (shown collectively as "organic matter"). So, the only nitrogen available to support new growth will be that which is supplied by nitrogen fixation from the atmosphere (pathway 6 in the diagram) or by the release of ammonium or simple organic nitrogen compounds through the decomposition of organic matter (pathway 2). Some other stages in this cycle are mediated by specialized groups of microorganisms.

Nitrification

The term **nitrification** refers to the conversion of ammonium to nitrate (pathway 3-4). This is brought about by the nitrifying bacteria, which are specialized to gain their energy by oxidizing ammonium, while using CO_2 as their source of carbon to synthesize organic compounds. Organisms of this sort are termed chemoautotrophs–they gain their energy by chemical oxidations (chemo-) and they are autotrophs (self-feeders) because they do not depend on pre-formed organic matter. In principle, the oxidation of ammonium by these bacteria is no different from the way in which humans gain energy by oxidizing sugars. Their use of CO_2 to produce organic matter is no different in principle from the behavior of plants.

The nitrifying bacteria are found in most soils and waters of moderate pH, but are not active in highly acidic soils. They almost are always found as mixed-species communities (termed consortia) because some of them (e.g., nitrosomonas species) are specialized to convert ammonium to nitrite (NO_2^-) while others (e.g., Nitrobacter species) convert nitrite to nitrate (NO_3^-). In fact, the accumulation of nitrite inhibits nitrosomonas, so it depends on nitrobacter to convert this to nitrate, whereas nitrobacter depends on nitrosomonas to generate nitrite.

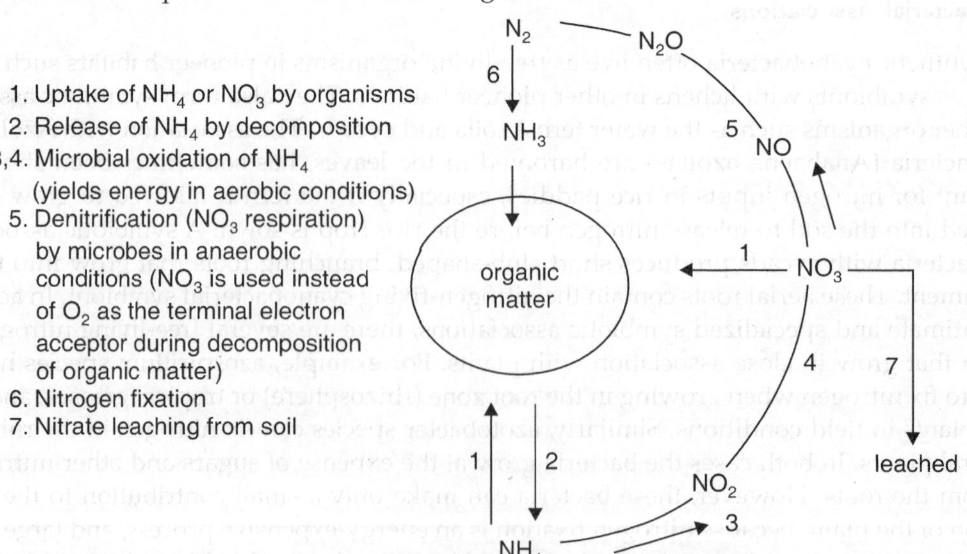

FIGURE 4.47 **Nitrogen Cycle–A schematic representation.**

The nitrifying bacteria have some important environmental consequences, because they are so common that most of the ammonium in oxygenated soil or natural waters is readily converted to nitrate. Most plants and microorganisms can take up either nitrate or ammonium (arrows marked "1" in the diagram). However, the process of nitrification has some undesirable consequences. The ammonium ion (NH_4^+) has a positive charge and so is readily absorbed on to the negatively charged clay colloids and soil organic matter, preventing it from being washed out of the soil by rainfall. In contrast, the negatively charged nitrate ion is not held on soil particles and so can be washed down the soil profile. The process is called leaching (arrow marked 7 in the diagram). In this way, valuable nitrogen can be lost from the soil, reducing soil fertility. The nitrates can then accumulate in groundwater, and ultimately in drinking water. There are strict regulations governing the amount of nitrate that can be present in drinking water, because nitrates can be reduced to highly reactive nitrites by microorganisms in the anaerobic conditions of the gut. Nitrites are absorbed from the gut and bind to hemoglobin, reducing its oxygen-carrying capacity. In young babies this can lead to respiratory distress—the condition known as "blue baby syndrome." Nitrite in the gut can also react with amino compounds, forming highly carcinogenic nitrosamines.

Denitrification

Denitrification refers to the process in which nitrates are converted to gaseous compounds (nitric oxide, nitrous oxide, and N_2) by microorganisms. The sequence usually involves the production of nitrite (NO_2^-) as an intermediate step, is shown as "5" in Figure 4.47. Several types of bacteria perform this conversion when growing on organic matter in anaerobic conditions. Because of the lack of oxygen for normal aerobic respiration, they use nitrate in place of oxygen as the terminal electron acceptor. This is termed anaerobic respiration and can be illustrated as follows:

In aerobic respiration (as in humans), organic molecules are oxidized to obtain energy, while oxygen is reduced to water

$$C_6H_{12}O_6 + 6\,O_2 = 6\,CO_2 + 6\,H_2O + \text{energy}$$

In the absence of oxygen, any reducible substance such as nitrate (NO_3^-) could serve the same role and be reduced to nitrite, nitric oxide, nitrous oxide, or N_2. Thus, the conditions in which we find denitrifying organisms are characterized by (1) a supply of oxidizable organic matter, and (2) absence of oxygen but availability of reducible nitrogen sources. A mixture of gaseous nitrogen products is often produced because of the stepwise use of nitrate, nitrite, nitric oxide, and nitrous oxide as electron acceptors in anaerobic respiration. The common denitrifying bacteria include several species of pseudomonas, alkaligenes and bacillus. Their activities result in substantial losses of nitrogen into the atmosphere, roughly balancing the amount of nitrogen fixation that occurs each year.

4.11 PRECURSOR–PRODUCT RELATIONSHIP

In biological systems there are a large number of biosynthetic pathways. In these bioreactions small molecules are used for the synthesis of large molecules. In these reactions the starting material from which a metabolically active compound is produced is called the precursor. The precursor depends on the type of the metabolically active compound. If we consider the macromolecules as the active product, then its building blocks or the monomers are the precursors. Table 4.10 gives some of the precursors and the products. But if we consider some active specific proteins or some other molecules, its immediate inactive form is called the precursor. For example, the peptide hormone insulin is an active molecule, its inactive form pro insulin is its precursor. Similarly, pro thrombin is the precursor of thrombin, the blood coagulating agent present in the blood.

TABLE 4.10 Some precursors and their products.

Precursor	Products
Glucose	Starch Glycogen Cellulose Fructose Sucrose Lactose Maltose
Adenine	ADP and ATP AMP NADP and NAD Neurotransmitters
Palmitic acid	Phospholipids Fats Waxes
Amino acids	Proteins Peptide hormones Neurotransmitters Alkaloids
Cholesterol	Steroid hormones Vitamin D
Pro insulin	Insulin
Pro thrombin	Thrombin
Carotene	Vitamin A

4.12 SUPRAMOLECULAR ASSEMBLY

The phenomenon of life depends closely on simultaneous and diverse biochemical processes going on inside living cells at every moment of time. All these biochemical processes are possible because of certain unique molecular and cellular structures. For example, the biological membranes and the structures are called **ribosomes**. These are a collection of different types of macromolecules assembled in very specific manner, so that it can carry out their function.

The inorganic molecules of the nature like CO_2, N_2, H_2O are assimilated into complex molecules such as monosaccharides, ammonia, amino acids, fatty acids, etc. These biomolecules form the monomers or the building blocks for synthesizing very complex and diverse types of macromolecules such as proteins, carbohydrates, nucleic acids, lipids, and other types of organic molecules. The macromolecules then interact with each other and produce conjugated molecular assemblies such as lipoproteins, glycoproteins, nucleic acid protein complexes, etc. These macro-molecular complexes assemble and form the specific supramolecular assemblies, which form cellular components such as ribosomes, chromosomes, plasma membranes, membranes of mitochondria, chloroplasts, and other membrane-bound organelles specialized for carrying out different functions. The following are some of the supra-molecular assemblies of cells:

- Proteins + lipids → lipoproteins in association with small percentage of carbohydrates forms the plasma membranes.
- Protein + Nucleic acids → Chromosomes and chromatin materials.
- Protein + carbohydrates → Glycoproteins and peptdoglycans involved in cell to cell communication and hormones, components of bacterial cell walls, etc.
- Carbohydrates and lipids → components of membranes.
- Proteins + Ribonucleic acids → Ribosomes, involved in protein translation.

4.13 BIOINFORMATICS AND BIOMOLECULAR DATABASES

Bioinformatics

Bioinformatics is a recent addition to biotechnology that emerged from the computational data management of biotechnology research such as genomics and proteomics. Powerful data management tools and computational techniques are required now more than ever to store, share, study, and compare the burgeoning library of biological information. A new field of science dealing with issues, challenges, and new possibilities created by the biological databases has emerged: bioinformatics. **Bioinformatics** combines the tools of mathematics, computer science, and biology with the aim of uncovering patterns and associations within and between sets of biological data.

Biological Databases

Biotechnology research in the area of proteomics and genomics generated a huge volume of data and it became necessary to create databases to manage, classify, and access the data systematically.

There are two main functions of biological databases:

- **To make biological data available to scientists.** As much as possible, a particular type of information should be available in one single place (book, site, database). Published data may be difficult to find or access, and collecting it from literature is very time-consuming. And not all data is actually published in an article.

- **To make biological data available in computer-readable form.** Since analysis of biological data almost always involves computers, having the data in computer-readable form (rather than printed on paper) is a necessary first step.

One of the first biological sequence databases was probably the book "Atlas of Protein Sequences and Structures" by Margaret Dayhoff and colleagues, first published in 1965. It contained the protein sequences determined at the time, and new editions of the book were published well into the 1970s. Its data became the foundation for the PIR database.

The computer became the storage medium of choice as soon as it was accessible to ordinary scientists. Databases were distributed on tape, and later on various kinds of disks. When universities and academic institutes were connected to the internet or its precursors (national computer networks), it is easy to understand why it became the medium of choice. And it is even easier to see why the World Wide Web is the standard method of communication and access for nearly all types of biological databases today.

Now databases can be accessed freely from a large number of computer based databases through the World Wide Web. The most widely accessed database of biotechnology and bioinformatics is the http://www. ncbi.nlm.nih.gov. It is the official site of National Institute of Health. It is widely used for various scientific purposes. It has separate databases for genomics, proteomics, and other types of biological databases.

National Center for Biotechnology Information

National Library of Medicine   National Institute of Health

Pubmed	Entrez	BLAST	OMIM	Books	TaxBrowser	Structure
Search	[▾]	for	[ ]	Submit		

SITE MAP

About NCBI

▶ What does NCBI do ?

Established in 1988 as a national resource for molecular biology information, NCBI creates public databases, conducts research in computational biology, develops software tools for analyzing genome data, and disseminates biomedical information—all for the better understanding of molecular processes affecting human health and disease. More ...

GenBank

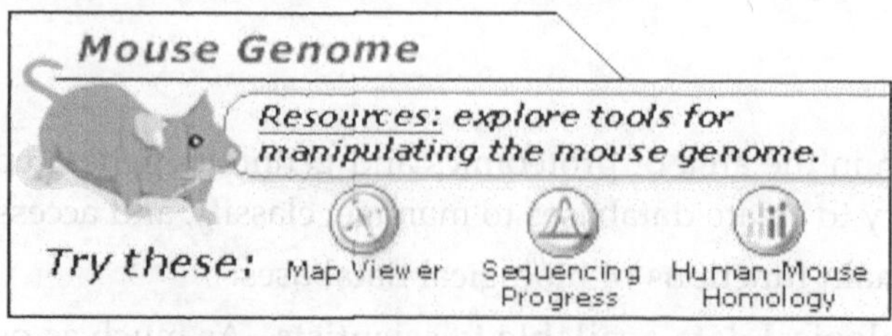

Molecular databases

BLink and get results fast!

Literature databases

Genomic biology

Tools

Research at NCBI

Software engineering

Education

Use BLink to view a graphical alignment of protein sequence similarities, taxonomic trees, 3D structures, and more. BLink provides quick results based on precomputed BLASTp searches against the non-redundant (nr) protein database. More...

CDD joins Entrez

CDD, the Conserved Domain Database, is now available through the Entrez retrieval system. CDD is a collection of sequence alignments and profiles representing protein domains conserved in molecular evolution. The database can be searched by choosing "Domains" from the Entrez drop-down menu. More..

NCBI News

October 2002 marks the 20th anniversary of the creation of GenBank. GenBank has grown from 680,338 base pairs in 1982 to 22 billion base pairs in 2002. In 1984, GenBank was distributed on magnetic tape to 120 institutions and had a daily average of 5 online users. Today, over 30,000 people per day access GenBank online.

FIGURE 4.48 **The Home Page of NCBI.**

REVIEW QUESTIONS

1. Explain optical isomerism.
2. What does anomer mean? What are the possible types of anomers in glucose?
3. Give the total number of possible stereoisomers in aldohexoses.
4. What does the use of L-prefix in the naming of an amino acid signify?
5. Explain the meaning of conformation and configuration.
6. Differentiate between an anomer and epimer.
7. What is a reducing sugar?
8. Is sucrose a reducing sugar? What is the reason?
9. Differentiate between geometrical and optical isomerism.
10. What are the types of reactions responsible for the polymerization of monomers?
11. What is a glycosidic linkage?
12. What is the fate of the pyruvic acid that is produced by glycolysis in muscles cells?
13. Ethanol fermentation is a waste of energy. Why?
14. Differentiate between fermentation and aerobic respiration.
15. Why is the citric acid cycle called an amphibolic pathway?
16. What is photorespiration? Differentiate it from respiration.
17. Compare oxygenic and anoxygenic photosynthesis.

18. What are the types of photosynthetic pigments present in different organisms?
19. What are the various functional groups present in various biomolecules?
20. Differentiate a nucleoside from nucleotide.
21. Differentiate C_4 and CAM plants.
22. Which is the site of oxidative phosphorylation?
23. Explain RUBISCO.
24. Explain the activity of PS I and PS II in operating the cyclic and non-cyclic electron transport systems.
25. What is meant by photophosphorylation?
26. What is nitrogen fixation? What are the different types of nitrogen fixing organisms?
27. What is bioinformatics?
28. What are the uses of biological databases?
29. What does NCBI stand for?
30. What is leghemoglobin?

Chapter 5 STRUCTURE AND FUNCTION OF MACROMOLECULES

In This Chapter

5.1 INTRODUCTION

We have discussed the structure and dynamics of smaller biomolecules, which form the building blocks of the important cellular macromolecules, in earlier chapters. These biomolecules can undergo polymerization or condensation to form specific polymers of high molecular weight known as macromolecules. These macromolecules are of four distinct groups—carbohydrates, proteins, nucleic acids, and lipids. All these macromolecules are specialized for carrying out specific cellular functions, which are very closely related to their functions. So a clear understanding of their structure is required for the proper understanding of their functions in the cell metabolism.

5.2 CARBOHYDRATES

Carbohydrates, as we have discussed in the previous chapter, consist of monosaccharides, disaccharides, oligosaccharides, and polysaccharides. Monosaccharides, or the simple sugars, are the building blocks or the monomers by which other forms are constructed. Disaccharides consist of two monosaccharide residues linked together by **glycosidic bonds**. This bond forms between the OH group of anomeric carbon (carbon No.1) of one sugar and with the OH group of any other carbon atom, preferably of 4th or 6th position of another sugar. The number of monomers varies from three to ten in the case of oligosaccharides, and it is indefinite in the case of polysaccharides. The polysaccharides are defined as the polymers of sugars and their various derivatives, such as glucosamine and galactosamine linked together by **glycosidic bonds.**

There are two types of glycosidic linkages depending on the types of anomers (e.g., α-D glucose or β-D-Glucose) of the sugar. They are **α-glycosidic Linkages** and **β-glycosidic Linkages**. Further polysaccharides are classified on the basis of their function in the biological system into storage polysaccharides and structural polysaccharides. Starch is the main storage form of polysaccharides in plants and glycogen in animals. They form the main energy sources for both plants and animals. There are also polysaccharides that give mechanical support to the cells or cell components or to the organism as a whole. Such polysaccharides are the structural polysaccharides. Cellulose, a structural polysaccharide, is the main component of the plant cell walls and is the most abundant biopolymer in the biosphere. Chitin, the material of the insect exoskeleton and peptidoglycan, the main components of the bacterial cell walls, are other examples of structural polysaccharides.

If the polysaccharides are composed of the same type of monomers or sugars, they are called homopolysaccharides and if more than one type of sugar or sugar derivatives are involved in the formation of the polysaccharides, they are called heteropolysaccharides.

Polysaccharides Containing α-glycosidic Linkages

Starch: Starch is the polymer of glucose units linked together by α-1 > 4 linkages and/or α-1 > 6 linkages. Starch is the main storage form of polysaccharides in plants. In most of the cases, starch is a mixture of two types of glucose chains—**amylose** and **amylopectin**. Amylose molecules are straight chains of glucose units linked by α-1 > 4 linkages, unbranched and water-soluble. They are like most polysaccharides, polydispers, meaning that the length of the molecule is not exactly defined and the number of glucosyl residues is indefinite and can range from several hundred to a few thousand.

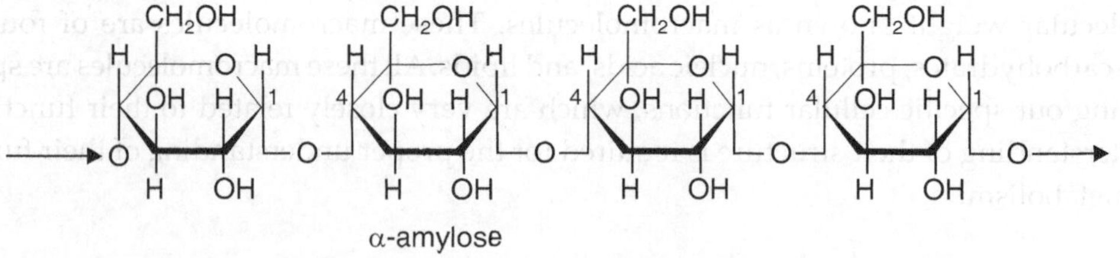

α-amylose

A

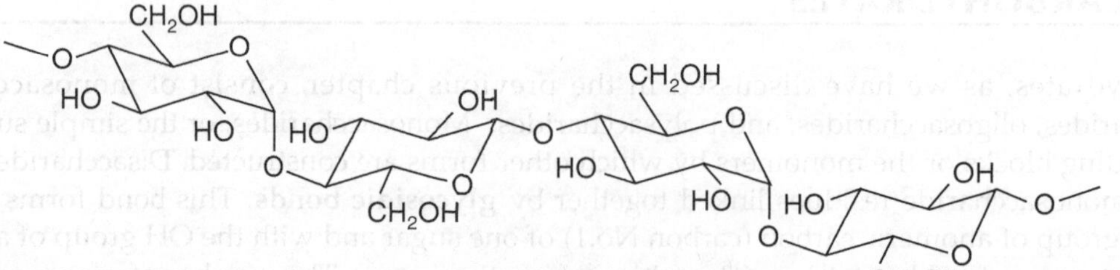

B

FIGURE 5.1 **A. α-1 > 4 Glycosidic bonds in α-amylose.**

B. Glucose units in Chair configuration.

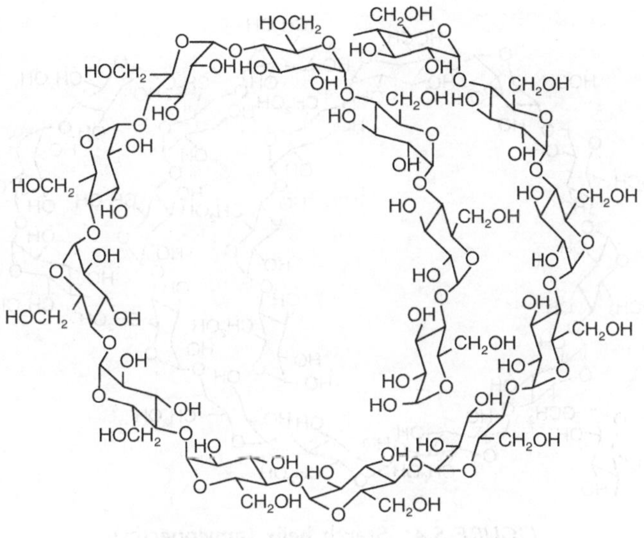

FIGURE 5.2 **Starch helix (amylose).**

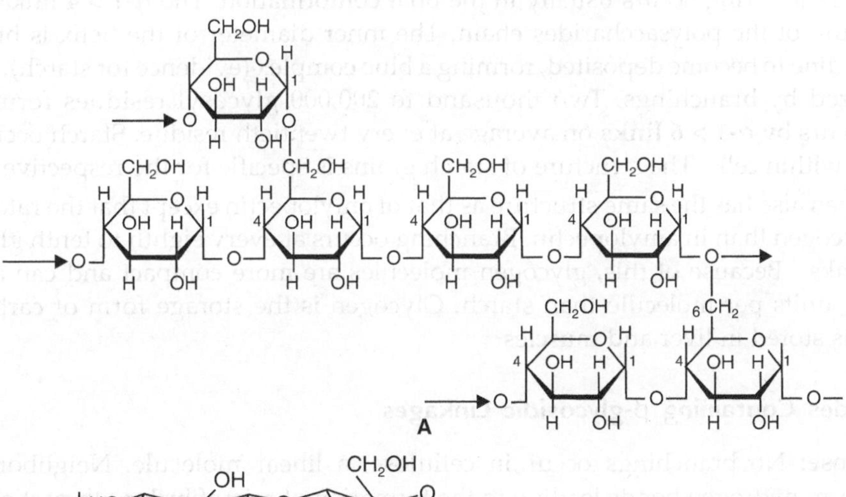

A

B

FIGURE 5.3 **A. Amylopectin with α-1 > 6 links.**
B. Glucose units in chair configuration.

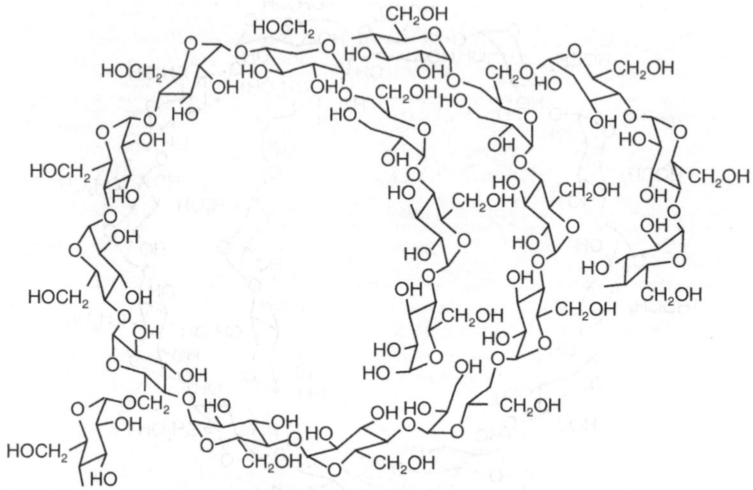

FIGURE 5.4 **Starch helix (amylopectin).**

The pyranose ring occurs usually in the boat conformation. The α-1 > 4 linkages cause the helical structure of the polysaccharides chain. The inner diameter of the helix is big enough for elementary iodine to become deposited, forming a blue complex (evidence for starch). **Amylopectin** is characterized by branchings. Two thousand to 200,000 glycosyl residues form a molecule. Branching occurs by **α-1 > 6 links** on average at every twentieth residue. Starch occurs in layered starch grains within cells. The structure of starch grains is specific for the respective species.

Glycogen also has the same structure as that of amylopectin except that the rate of branching is more in glycogen than in amylopectin. Branching occurs at every eighth to tenth glucose residue by α-1 > 6 links. Because of this, glycogen molecules are more compact and can accommodate more glucose units per molecule than starch. Glycogen is the storage form of carbohydrates in animals and is stored in liver and muscles.

Polysaccharides Containing β-glycosidic Linkages

Cellulose: No branchings occur in cellulose, a linear molecule. Neighboring cellulose chains may form hydrogen bonds leading to the formation of microfibrils with partially crystalline parts (micelles). Cellulose is the most important structural component of nearly all green plants' cell walls. The glucose units are linked by β-1 > **4-glycosidic linkages.**

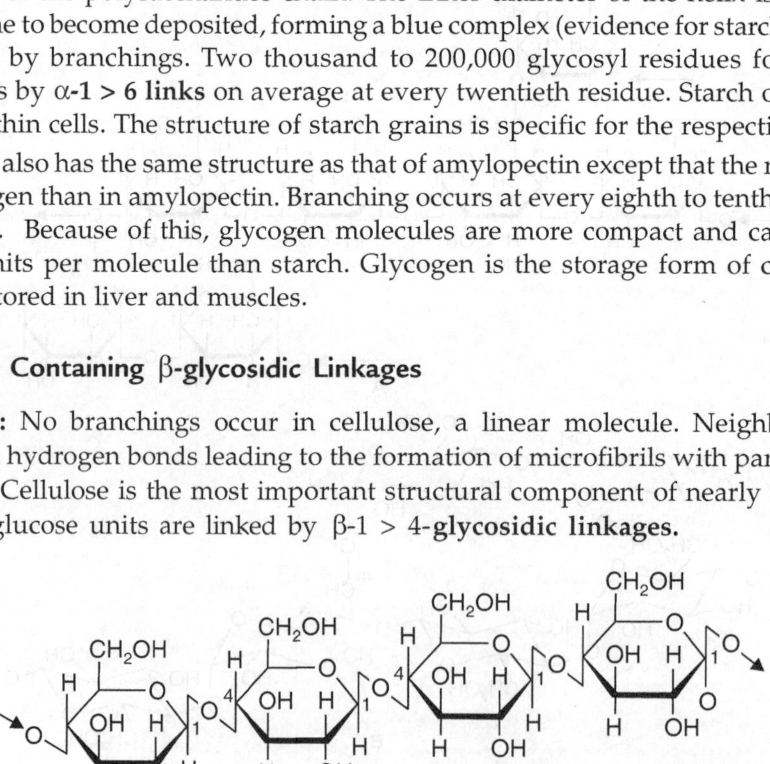

Cellulose

A

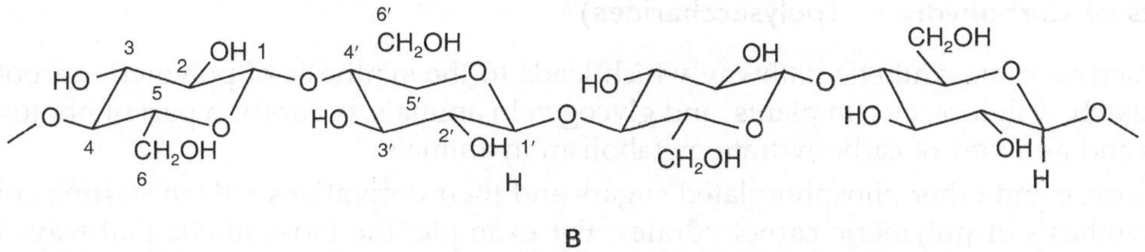

FIGURE 5.5 **A. β-1 > 4-glycosidic linkages. B. Glucose units in chair configuration.**

Callose: This is branched cellulose present in cell walls in very small quantities. In callose, the glucose units are linked via 1, 3 glycosidic linkages. This type of bond causes the polymer to arrange in a helix. The helix binds aniline blue; the resulting complex leads to yellow fluorescence. Callose also contains 1 > 4 and 1 > 6 linkages.

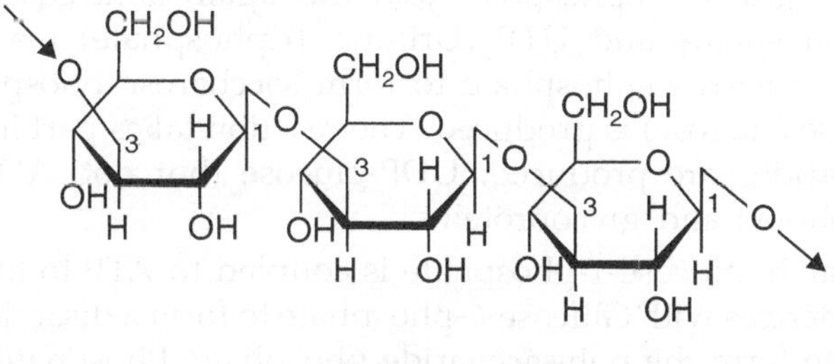

FIGURE 5.6

Chitin is a polymer that consists of glucose derivatives: N-acetyl glucosamine units connected by β-1 > 4 linkages. It is only exceptionally found in plants, in some algae, but it is the main structural component of the cell walls of most fungi.

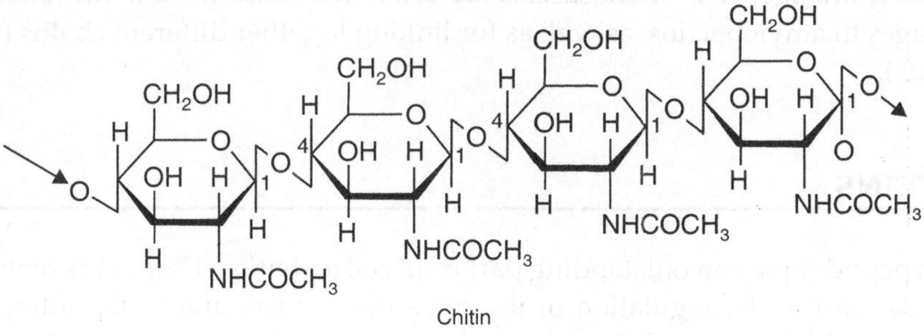

FIGURE 5.7

Peptidoglycan: This is a heteropolysaccharide present in the cell walls of bacteria imparting rigidity and shape to the cells. It is a polymer of alternating units of a sugar or sugar derivatives; e.g., N-acetyl muramic acid and N-acetyl glucosamine. Or it can be defined as the polymer of a disaccharide, consisting of two types of sugars or their derivatives. The monomers are linked by β-1 > 4 linkages as in the case of other structural polysaccharides. Most of the structural polysaccharides are straight chain molecules because of the presence of β-1 > 4 linkages.

Synthesis of Carbohydrates (polysaccharides)

The characteristic biosynthetic pathway which leads to the synthesis of polymeric carbohydrates such as starch, cellulose, etc., in plants, and glycogen in animals, occurs as a part of photosynthesis in plants and as a part of carbohydrate metabolism in animals.

Glucose and other phosphorylated sugars and their derivatives are the starting compound for the synthesis of polymeric carbohydrates. For example, the biosynthetic pathways of many polysaccharides that are the structural elements of the plant cells and the cellulose and storage compounds such as starch, are synthesized using the phosphorylated glucose as the precursors. If the glucose is not already activated, like all other starting compounds of biosynthetic pathways it first has to be activated. This occurs either by phosphorylation or by binding of a sugar residue to a nucleotide (like ATP). Thus, glucose-6-phosphate is formed.

Fructose-6-phosphate is an important intermediate of both photosynthesis and glycolysis. It is at equilibrium with glucose-6-phosphate and this again is at equilibrium with glucose-1-phosphate. Glucose-1-phosphate and UTP (Uridine Triphosphate) react to UDP-glucose that again polymerizes with fructose-6-phosphate to form saccharose phosphate. Upon cleavage of the phosphate, saccharose (sucrose) is produced. The reaction takes part in chloroplasts. Similarly, other types of disaccharides are produced. UDP-glucose (but not ADP-glucose) can also be incorporated into glycolipids and glycoproteins.

In the case of starch, glucose-1-phosphate is coupled to ATP to form ADP-glucose. ADP-glucose that again polymerizes with Glucose-6-phosphate to form a disaccharide, which undergoes further polymerization to form the polysaccharide phosphate. Phosphate is cleaved off from the polysaccharide to generate the starch. This reaction also takes place in part in chloroplasts.

$$\textbf{Glucose-1-phosphate + ATP + } H_2O \rightarrow \textbf{ADP-glucose + 2 } P_i$$

$$\textbf{ADP-glucose + Glucose-1-phosphate} \rightarrow \textbf{ADP-glucose-glucose + } P_i$$

Two enzymes are necessary for starch production: one for the start and the chain elongation (α-1 > 4 glycosidic linkage) with monosaccharide units, the other for the introduction of α-1 > 6 glycosidic linkages in amylopectins, as well as for linking together different chains (via alpha 1 > 4 glycosidic bonds).

5.3 PROTEINS

Proteins or polypeptides play an outstanding part in all cell activities. They act as biological catalysts (= enzymes), take part in the regulation of the cell's metabolism and in the interaction between cells, and are required for the generation of specific structures. They are linear chains consisting of a sequence of 20 amino acids in different combinations linked exclusively by **peptide bonds**. The different combinations and sequences of amino acids are responsible for the diverse nature and functions of proteins (Table 5.1).

$$H_2N-\underset{H}{\overset{R_1}{C}}-COOH + H_2N-\underset{H}{\overset{R_2}{C}}-COOH \xrightarrow{H_2O} H_2N-\underset{H}{\overset{R_1}{C}}-\underset{O}{\overset{}{\underset{\|}{C}}}-\overset{H}{N}-\underset{R_2}{\overset{H}{C}}-COOH$$

Peptide bond

FIGURE 5.8

The peptide bonds are formed by the reaction of the primary amino group of one amino acid with primary carboxyl group of another amino acid with the elimination of a molecule of water (Figure 5.8). Thus, it is a condensation reaction. This type of linkage causes a polarity for the polypeptide chain. One end has the amino group and is called the **N-terminus**, while the other end is terminated by a free carboxyl group and is called the **C-terminus**. Amino-acid sequences are written from N- to C-terminus, the direction in which protein synthesis proceeds. The exact sequence of amino acids (also called the protein's primary structure) is determined by the nucleotide sequence of the gene, the part of the DNA strand, which codes for the protein. The three-dimensional structure and the function of the protein are very closely dependent on the amino-acid sequence of the polypeptide.

TABLE 5.1 Some important proteins and their function.

Protein	Site of location	Function
Collagen	Connective tissue	Gives tensile strength
Thrombin	Blood	Blood coating
Trypsin	Pancreatic juice	Cleaves the polypeptide during protein digestion
Amylase	Salivary secretion	Starch hydrolysis
Insulin	Secreted by islet cells of pancreas into blood stream	Controls the blood sugar level
Immuniglobulins	Blood	Immunity
Endorphins	Brain	Regulates the brain activities
Rhodopsin	Retinal cells of eye	Vision
Cytochrome	Mitochondrial membranes	Cellular respiration

There are three types of proteins based on their complexity. Some proteins are made up of a single polypeptide chain and are known as **simple proteins**. Those proteins having two or more polypeptide chains are called **complex proteins**. In some cases, the protein molecule is associated with a non-protein component known as the **prosthetic group**. Such proteins are known as **conjugated proteins**. The non-protein component may be metallic ions such as Zn^{+} in the case of carbonic anhydrase, hem part of hemoglobin enclosing Fe^{+} ion in it, or organic molecules such as vitamin derivatives such as NAD and NADP, nucleotides such as ATP and GTP, or maybe sugars, oligosaccharides, or various types of lipids.

Amino Acid Composition and Protein-Sequencing

Amino acid Composition

To determine the amino acid composition the peptide is first hydrolyzed into its constituent amino acids by heating in 6 N HCl at 110°C for 24 hours. The amino acids in the hydrolysate can be separated by ion-exchange chromatography and hydrolyzed by reacting them with ninhydrin. Alpha amino acids treated this way give an intense blue color, whereas amino acids, such as proline, give a yellow color. The concentration of amino acids in a solution is proportional to the optical absorbance of the solution after heating it with ninhydrin. This technique can detect a microgram (10 n mol or nanograms) of an amino acid. After getting the information about the amino acid composition and relative quantity of each amino acid, one can proceed to do the sequencing of amino acids for a particular protein or polypeptide.

Amino acid Sequencing

The amino acid sequence of a protein is very important because it is essential to know the structure and function of that protein; and also it can help in identifying and isolating the gene code for the protein. So obtaining at least a partial amino acid sequence is a critical first step in studying many proteins.

The first protein sequenced was the peptide hormone insulin, which controls the glucose level in blood; its deficiency can lead to the metabolic error, diabetes. Nobel Laureate, Frederic Sanger, showed for the first time that proteins have a definite amino acid sequence and a specific three-dimensional structure that is determined by the amino acid sequence. He found out that certain reagents such as fluoro-dinitro-benzene (FDNB) known as **Sanger's reagent** can react specifically with the free NH_2 group of the amino acid at the N-terminal of a polypeptide. Fluoro-DNB reacts with the N-terminal amino acid and on acid hydrolysis it yields a yellow DNP derivative of the amino acid, which can be separated and identified by ion-exchange chromatography. This procedure cannot be used on the same sample of polypeptide because the peptide is totally hydrolyzed on the acid hydrolysis step. But Sanger managed to use this technique with new samples of peptide in each cycle of experiments to sequence insulin. He used more than a gram of insulin for completing this task. Now this method is used only for identifying the N-terminal amino acid of a polypeptide and not for sequencing because of the above-mentioned demerit.

Edman Degradation Reaction

The most popular direct protein-sequencing technique in use today is the **Edman degradation procedure.** The Edman reaction is a series of chemical reactions, which remove one amino acid

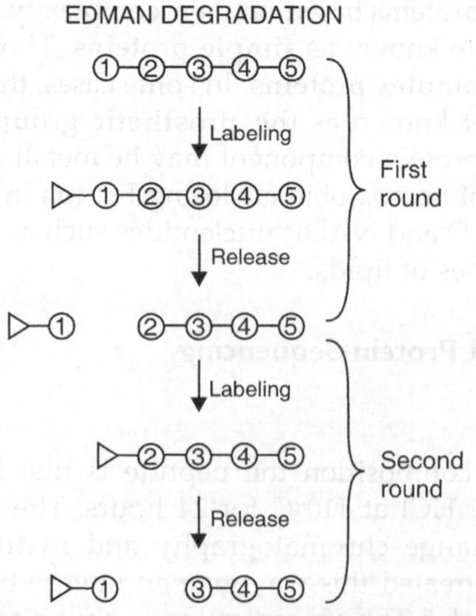

FIGURE 5.9 **Sequential removal of amino acids in the Edman's degradation reaction.**

at a time from amino terminus of a protein, releasing an amino-acid derivative, phenylthiohydantoin (PTH), that may be chromatographically identified (reversed phase). The Edman reagent, Phenylthiocyanate, reacts with the NH_2 group of the terminal amino acid and forms an intermediate, phenylthiocarbomyle derivative. A simplified schematic diagram of the process is shown in Figure 5.9. Release of amino acids from the amino terminus in a sequential manner is possible because the Edman procedure consists of three chemical reactions, which proceed under different pH conditions (Figure 5.10).

The first step is the coupling step (Figure 5.10), which occurs at high pH values and results in formation of phenylthiocarbamoylated (PTC) amino groups on the protein. The second reaction is the cleavage step, which occurs at low pH, resulting in release of an anilionothiazolinone (ATZ) form of the amino acid and regeneration of a free amino terminus on the protein. The ATZ-amino acid is converted to the phenylthiohydantoin (PTH) derivative in a separate reaction, generally exposure to strong acid. The PTH-amino acids are more stable and a bit more amenable to chromatographic resolution. The PTH-amino acid can be separated and identified by HPLC (High Performance Liquid Chromatography). This leaves the intact peptide short of one amino acid. The Edman procedure can be repeated on the shortened peptide obtained in the previous cycle for identifying the second amino acid of the polypeptide from the N-terminal.

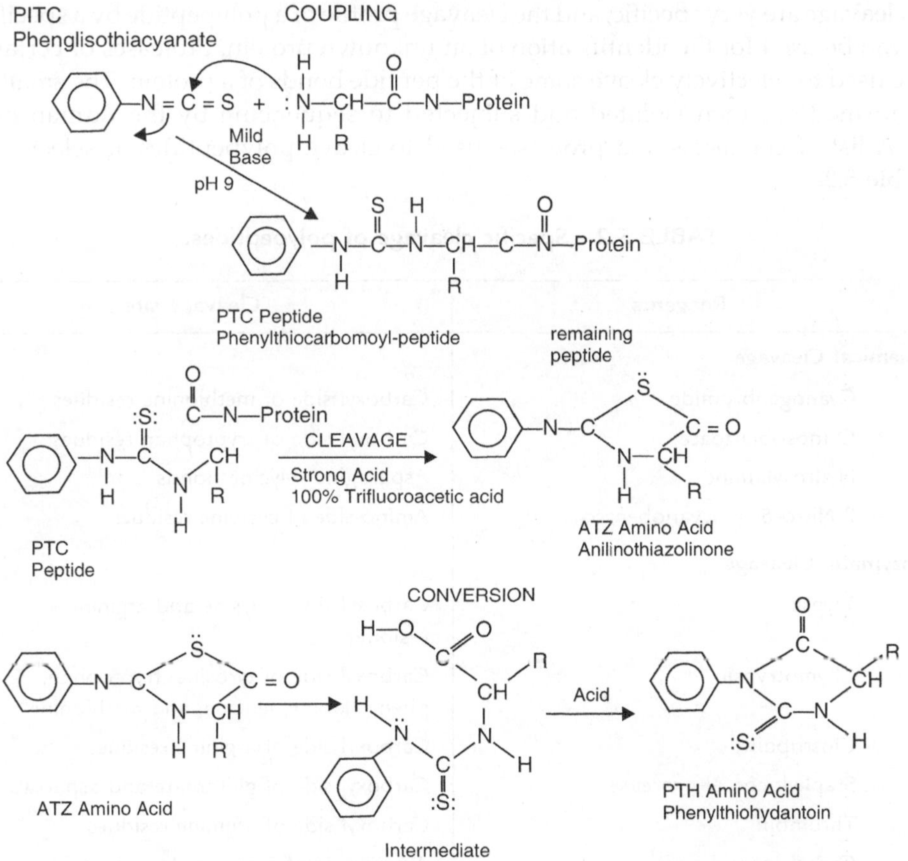

FIGURE 5.10 **Different steps of the Edman degradation reaction as operated in a sequanator.**

The whole method, including the reaction steps and the identification of PTH-amino acid derivative, is automated and the instrument that can carry out these reactions is known as a **sequenator**. Using this automated instrumentation, more than 100 amino acids of a polypeptide can be sequenced efficiently. Polypeptides and proteins of high molecular weight have to be fragmented into smaller polypeptides of 50 to 100 amino acids before carrying out the sequencing by a sequenator. Amino acid sequencing by the Edman degradation reaction can be carried out with very minute quantities of protein to the level of micrograms. There are a number of physical techniques like mass spectrometry, NMR (nuclear magnetic resonance spectrometry), and *x*-ray diffraction techniques used in protein analysis and structural elucidation. There are various types of mass spectrometric techniques for the analysis of proteins. For example, matrix assisted laser desorption ionization (MALDI), mass spectrometry, and tandem mass spectrometry (2D-MS or MS-MS) are efficient techniques developed to identify and sequence proteins rapidly.

Protein-sequencing Strategies

Most proteins are large in size and contain large numbers of amino acid residues. Therefore, it is necessary to fragment these long polypeptide chains into small fragments, which can be sequenced completely by the Edman degradation reaction. Fragmentation of polypeptides into smaller bits can be carried out either by chemical methods or by enzymatic cleavage. Both chemical and enzymatic cleavage are very specific, and the cleavage profile of a polypeptide by a specific chemical or enzyme can be used for the identification of an unknown protein. Proteases or certain chemical reagents are used to selectively cleave some of the peptide bonds of a protein. The smaller peptides fragments formed are then isolated and subjected to sequencing by the Edman degradation procedure. A list of chemicals and proteases used to cleave polypeptides at selected points are given in Table 5.2.

TABLE 5.2 Specific cleavage of polypeptides.

Reagents	Cleavage site
Chemical Cleavage	
Cyanogenbromide	Carboxyl side of methionine residues
O-Idosobenzoate	Carboxyl side of tryptophan residues
Hydroxylamine	Asparagine–glycine bonds
2-Nitro-5-thiocyanobenzoate	Amino side of cysteine residues
Enzymatic Cleavage	
Trypsin	Carboxyl side of lysine and arginine residues
Chymotrypsin	Carboxyl side of tyrosine, tryptophan, phenylalanine, leucine, and methionine
Clostripain	Carboxyl side of arginine residues
Staphylococcal protease	Carboxyl side of glutamate and aspartate
Thrombin	Carboxyl side of arginine residues
Carboxypeptidase A	Amino side of C-terminal amino acids except arginine, lysine, and proline

The chemical reagent, cyanogen bromide (CNBr) reacts specifically with methionine residues to produce peptides with C-terminal homoserine lactone residues and new N-terminal residues. Since most proteins contain very few methionine residues, treatment with CNBr usually produces only a few peptide fragments. Reaction of CNBr with a polypeptide chain containing three internal methionine residues should generate four peptide fragments. Each fragment can then be sequenced from its N-terminus.

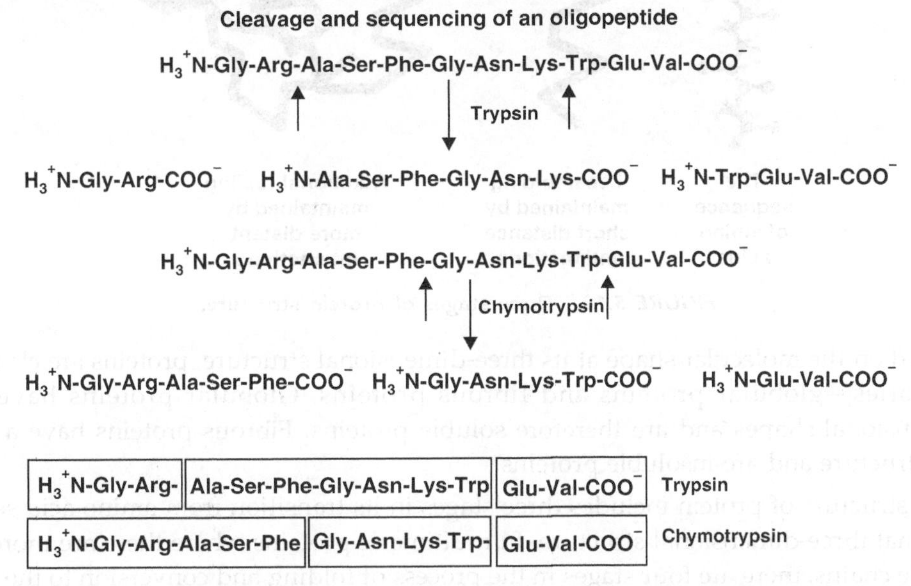

FIGURE 5.11 **Schematic representation of cleavage and sequencing of an oligopeptide.**

In the final stage of sequence determination, the amino acid sequence of the original large polypeptide chain can be deduced by lining up the amino acid sequence of cleaved peptide fragments and matching the overlapping sequences of peptide fragments as illustrated in Figure 5.11.

Three-dimensional Structure of Proteins

All proteins exhibit a specific molecular shape, which is determined by the amino acid sequence or the primary structure of the protein. This was first demonstrated by Christian Anfinsen on enzyme pancreatic Ribonuclease A, which hydrolyzes RNA (ribonucleic acids). He showed that a pure sample of Ribonuclease A lost its three-dimensional structure and most of its properties including catalytic activity when treated with strong chemicals like urea or by heating. The enzyme could regain most of its properties and three-dimensional structure when the denaturing agent (high temperature or the chemicals like urea) was removed. This demonstrated mainly two things. 1) The function of a protein is closely linked to the three-dimensional structure of the protein. 2) All the information necessary for folding the polypeptide into three-dimensional structure is present in the sequence of the amino acid or the primary structure of the protein.

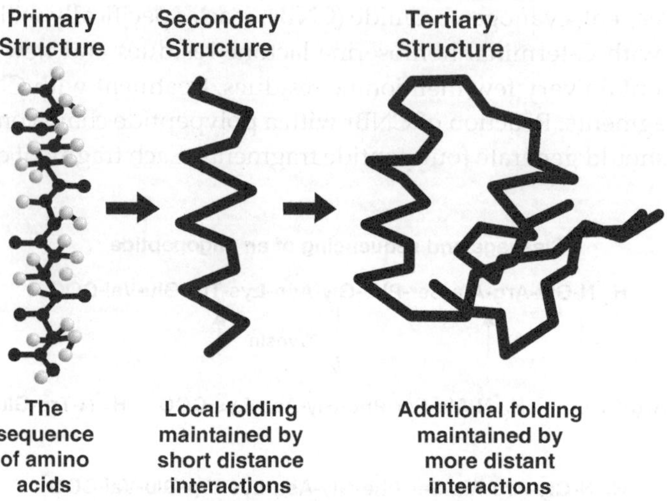

FIGURE 5.12 **Three stages of protein structure.**

Based on the molecular shape at its three-dimensional structure, proteins are classified into two categories—globular proteins and fibrous proteins. Globular proteins have globular three-dimensional shapes and are therefore soluble proteins. Fibrous proteins have a long, thin fiber-like structure and are insoluble proteins.

The structure of protein includes three stages in its transition from amino acid sequence to the functional three-dimensional structure. In multimeric proteins where there are more than one polypeptide chains, there are four stages in the process of folding and conversion to the functional three-dimensional stage. They are primary structure, secondary structure, tertiary structure, and quaternary structure (applicable in multimeric proteins).

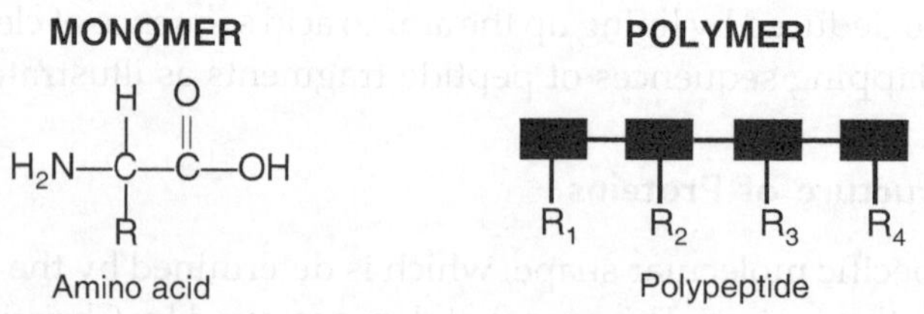

FIGURE 5.13 **Amino acid and primary structure.**

The sequence of R-groups along the chain is called the **primary structure** (Figure 5.13). **Secondary structure** refers to the local folding of the polypeptide chain. **Tertiary structure** is the arrangement of secondary structure elements in three-dimensions and quaternary structure describes the arrangement of a protein's sub-units.

Primary Structure of Proteins

The primary structure of a protein or polypeptide is the sequence of amino acids joined by the peptide bonds formed between the COOH group of one amino acid with the NH_2 group of the other. The primary structure of a protein can be determined by the Edman degradation reaction or it can be obtained from cloning and sequencing the gene responsible for the production of the protein. Size of the protein molecule can be determined by SDS-PAGE (sodium dodecyle sulfate polyacrylamide gel electrophoresis) or electrospray-mass spectrometry (ES-MS).

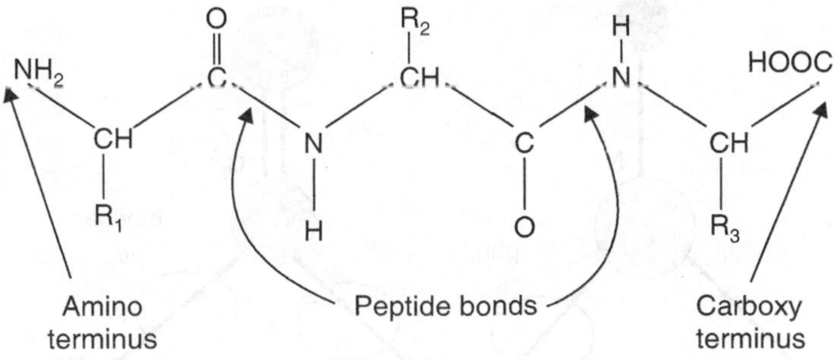

FIGURE 5.14 **Peptide bonds in polypeptide chain.**

The main molecular forces in the primary structure are the covalent linkages—the peptide bonds between amino acids—and the disulphide linkages between the cysteine units of the polypeptides. The peptide linkages are partial double bonds and are comparatively rigid, restricting the free rotation about the bond. All peptide bonds in protein structures are found to be almost planar.

Secondary Structure of Proteins

The peptide bond has some double-bond character (40%) due to resonance. The double bond of C=O alternates with the C-N of the peptide unit. As a consequence of this resonance, all peptide bonds in protein structures are found to be almost planar; i.e., atoms between two Cα atoms (C, O, N and H) are approximately co-planar. This rigidity of the peptide bond reduces the degrees of freedom of the polypeptide during folding. One can visualize the polypeptide chain consisting of rigid planes of peptide units interconnected by Cα atoms of each amino acid residue.

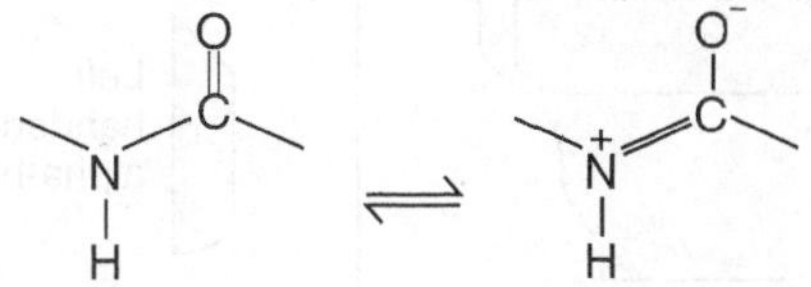

FIGURE 5.15 **Resonance of double bond in peptide linkage.**

The peptide bond nearly always has the *trans* configuration since it is more favorable than *cis*, which is sometimes found to occur with proline residues. Movement in the polypeptide chain is possible only by rotation in the bond angles of Cα-NH- and Cα-CO-. The angles are called the torsion angles. There are three main chain torsion angles of a polypeptide. These are phi (ϕ), psi (ψ), and omega (ω) (Figure 5.16). To make the peptide bond planar, the omega angle is nearly always 180°. It is rarely 0°. The ϕ and ψ angles are present on either side of the Cα atoms. The angle N-Cα bond is the ϕ angle and the Cα-C linkage is the ψ angle. The polypeptide chain is flexible only at these linkages and therefore movements are possible by rotation about ϕ and ψ angles. These movements are also limited by the nature of the side chain (R group) of the amino acids. The

R-groups attached to the Cα- atoms interfere with the values of the torsion angles and only certain values are permitted for the ϕ and ψ angles, to avoid steric hindrance (collision of atoms during bond rotation).

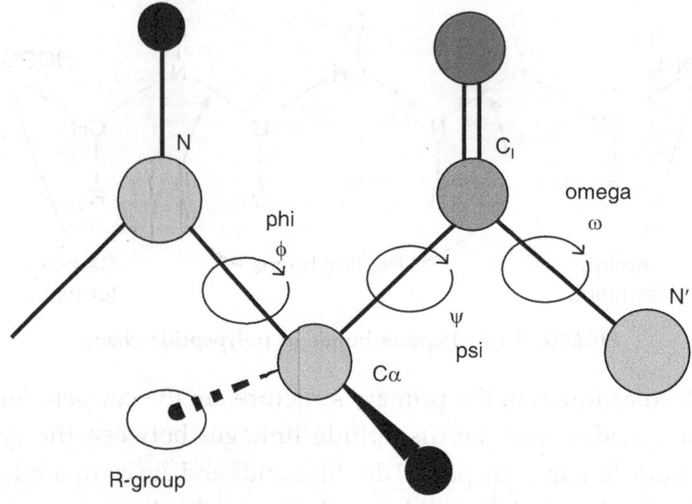

FIGURE 5.16 **Planar nature of peptide bond.**

Noted Indian scientist, Professor G.N. Ramachandran, systematically studied the various possible ϕ and ψ angles via the computer to find stable conformations and predicted the different types of secondary structures in protein molecules. Atoms were treated as hard spheres with dimensions corresponding to their van der Waals radii. Therefore, phi and psi angles, which cause spheres to collide, correspond to sterically disallowed conformations of the polypeptide backbone. The

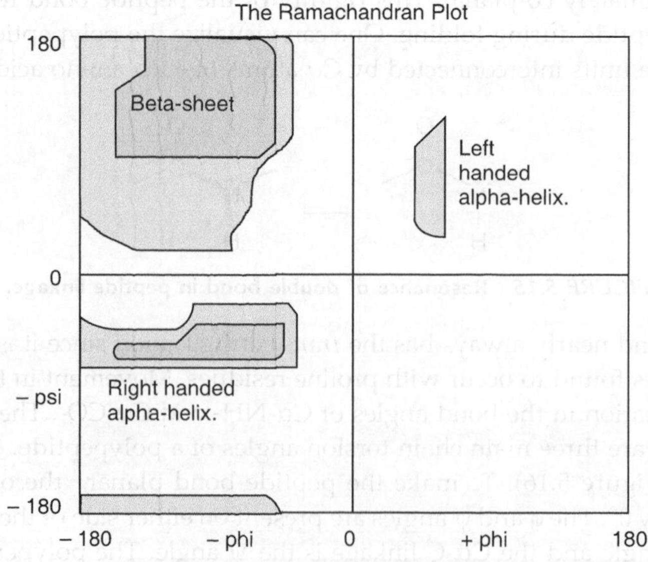

FIGURE 5.17 **Ramachandran plot.**

result which emerged from these calculations in 1962,—now commonly known as the **Ramachandran plot**—was published in the *Journal of Molecular Biology* in 1963 and has become an essential tool in the field of protein conformation (Figure 5.17).

There are mainly four patterns of secondary structures observed in various protein molecules: **alpha helix, beta pleated sheet, random coil**, and **beta turn.**

Alpha Helix

One of the most common secondary structure patterns is called the **alpha helix** or **α-helix** discovered by Pauling and Corey and published in the *Proceedings of the National Academy of Sciences (PNAS)* in 1951. The α-helix can be right-handed or left-handed (see Appendix). But right-handed α-helices are more common. Left-handed helices are observed in certain proteins of connective tissue, such as collagen, and are also known to have unusual amino acid composition. The torsion angles (ϕ and ψ angles) have characteristic values so that the amino acid residues are arranged around an imaginary axis and form a helix—the right-handed α-helix. The distance between each turn in an α-helix is calculated to be 5.4 Å having approximately 3.6 amino acid residues per turn. This is the most frequently observed α-helix among various proteins. This was first observed and described in α-keratin of hair, hence, the name α-helix. The helical conformation is stabilized by hydrogen bonds formed readily between C=O groups and N-H groups in the backbone. The carbonyl oxygen of one amino acid (n) residue makes a hydrogen bond with the hydrogen atom of the amino group of the residue for amino acid residues further ($n + 3$) along the chain. This regular pairing pulls the polypeptide into a helical shape that resembles a coiled ribbon. The α-helix is the main structural component of the protein α-keratin present in hair. You may have noticed that hair strands can

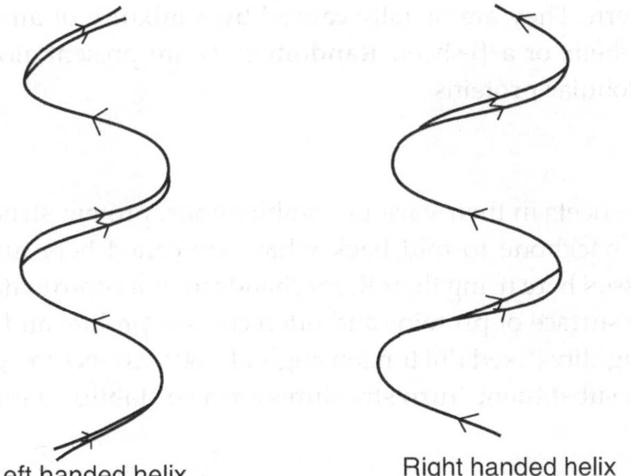

FIGURE 5.18 **Left-handed helix.** ***FIGURE 5.19*** **Right-handed helix.**

be stretched and are more flexible when wet; it is because the intra-strand hydrogen bonding, reduced as water molecules, are available for making hydrogen bonds with the–NH groups making the helix more flexible.

Figures 5.18 and 5.19 show how a right-handed helix differs from a left-handed one. An easy way to remember this is to hold both your hands in front of you with your thumbs pointing up and your fingers curled towards you. For each hand, the thumbs indicate the direction of translation and the fingers indicate the direction of rotation.

Beta-pleated Sheet

This is the second type of secondary structural pattern frequently observed among different classes of proteins. But it is predominant in structural proteins or fibrous proteins. The characteristic torsion angles for a certain combination and sequence of amino acids will allow the polypeptide to form straight chains without any coiling leads to the formation of the structures called **b-pleated sheets.** The intra-strand hydrogen bonding is almost absent in b-pleated sheets. Instead hydrogen bonding is possible between b-strands in an aggregate (i.e., intra-strand hydrogen bonding). The intra-strand hydrogen bonding makes the protein more strong, rigid, and inflexible (see Appendix). The aggregates of b-strands may be running in the same (parallel) or opposite (anti-parallel) directions with respect to the amino terminal and carboxy terminal. Hydrogen bonding between parallel strands is stronger than that of anti-parallel strands. In parallel strands the – CO and the NH groups, between which the hydrogen bond is formed, are closer and are more aligned straight than in anti-parallel strands. The fibroin protein of natural silk and spider webs is fibrous and has b-pleated sheets as the main structural components and has high, tensile strength.

Random Coil

These secondary structure patterns consist of unspecified coils, loops, and sheets, which don't have any constant pattern. They are usually caused by a mixture of amino acid sequence, which cannot form either a α-helix or a β-sheet. **Random coils** are present along with α-helices and β-sheets in most of the globular proteins.

Beta Turns

To combine helices and sheets in their various combinations, protein structures must contain turns that allow the peptide backbone to fold back which are called beta turns or β-turns. Two turn structures will be discussed here using their Ramachandran plot coordinates. These turns can almost always be found on the surface of proteins and often contain proline and/or glycine. Proline gives the backbone a special rigidity (fixed Phi torsion angle at – 60°, Cα -N) and glycine has high flexibility because of its hydrogen substituent. Turn structures are also stabilized through H-bond formation.

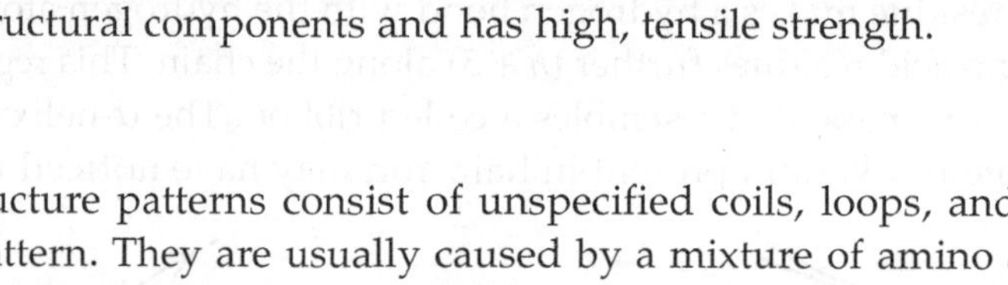

Type I β-bend
$\phi_2 = -60°$ $\phi_2 = -30°$
$\phi_3 = -90°$ $\phi_3 = 0°$

R2 R3 α1 C N H α3 H N C O C O H N α4 α1

FIGURE 5.20 **Type I β-turn.**

Type II β-bend
$\phi_2 = -60°$ $\phi_2 = 120°$
$\phi_3 = 90°$ $\phi_3 = 0°$

FIGURE 5.21 **Type II β-turn.**

Tertiary Structure of Proteins

The regular polypeptide structures or the secondary structures are assembled, or grouped together, by specific folding of the polypeptide units such as helices and sheets grouped together to form a specific shape. It is only with the complete, compact folding into tertiary structure that the proteins attain their "native conformation" and become active proteins (as a result of the creation of active sites). Forces that contribute to tertiary folding include: hydrogen bonds, hydrophobic interactions, and ionic bonds. Sulfhydryl bonds (-S-S- bonds) mentioned earlier, are a force contributing to the tertiary structure, but now they are considered a molecular interaction contributing to the primary structure of polypeptide along with peptide bonds. These are especially important because they are covalent bonds and quite strong compared to H-bonds. It was observed that certain combinations of secondary structures like helix and sheets are conserved in certain classes of proteins. These groups of secondary structures are known as **motifs**. The motifs form the structural elements of proteins since they are conserved. The presence of certain conserved domains and motifs in the tertiary structure of protein can be used for predicting the function of a protein or its identification.

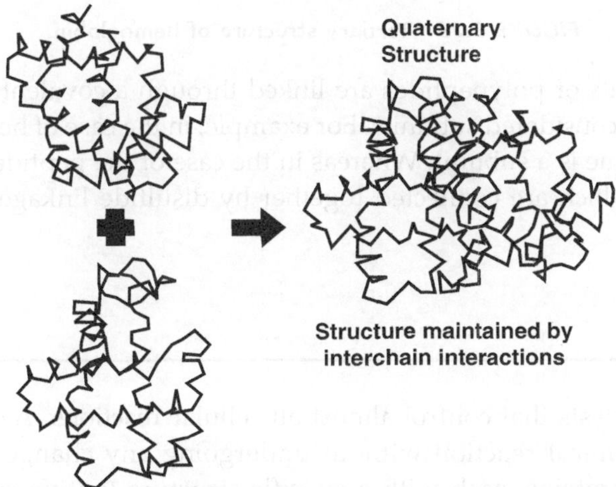

FIGURE 5.22 **Tertiary structure of protein.**

Quaternary Structure of Protein

Some proteins are made of more than one polypeptide chain. These proteins are referred to as multimeric proteins. The polypeptides and their tertiary structures are assembled to form the native conformation and become functional proteins. The individual polypeptides are known as the subunits, which must be assembled together after each individual polypeptide has reached its tertiary structure. Examples include **hemoglobin** (blood protein involved in oxygen transport), which has four subunits (Figure 5.23). **Pyruvate dehydrogenase** (mitochondrial protein involved in energy metabolism) has 72 subunits. There is no involvement of covalent linkages in the assembly of these subunits. The main molecular interactions involved in the assembly of subunits and formation of quaternary structures include hydrophobic and electrostatic attractions in addition to the weak Van der Waal's attractions.

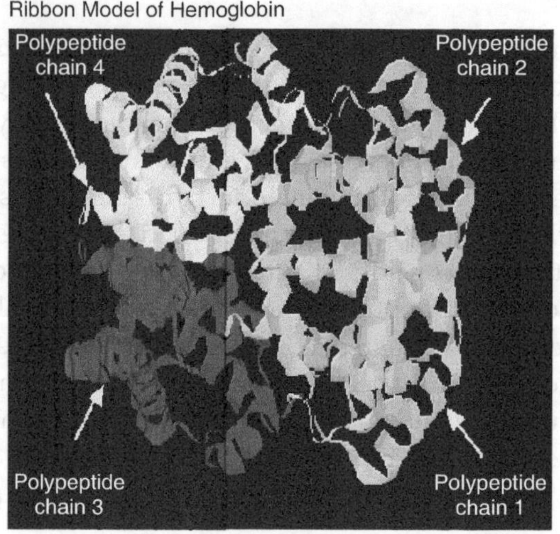

FIGURE 5.23 **Tertiary structure of hemoglobin.**

When two subunits or polypeptides are linked through a covalent bond such as disulfide linkages, they cannot be considered subunits. For example, in the case of hemoglobin there are four polypeptides and each one is a subunit. Whereas in the case of the peptide hormone insulin there are two polypeptides, which are connected together by disulfide linkages and therefore are not subunits.

5.4 ENZYMES

Enzymes are the biocatalysts that control almost all cellular reactions. A catalyst is a compound, which accelerates a chemical reaction without undergoing any change in quality or quantity. Enzymes are globular proteins, each with a specific structure (native conformation), function, distribution of electrical charges, and surface geometry whose specificity depends on their tertiary structure. The tertiary structure determines the three-dimensional shape. They are responsible for control of a single reaction and are thus responsible for control of metabolism. There are about 700 enzymes in a typical prokaryotic cell and there are thousands in a eukaryotic cell.

Enzymes function as **catalysts**, which are substances that facilitate (speed up) reactions without actually entering into the reaction. They are used over and over, and a single enzyme molecule may mediate thousands of reactions in a single second. Even simple reactions like dissolution of carbon dioxide in water will not take place to an appreciable extent by itself. But we can make it dissolve in water in higher concentrations under high pressure. Carbonated drinks have CO_2 under high pressure. On releasing the pressure by removing the cap, lots of CO_2 bubbles will release. But in biological systems the dissolution of CO_2 takes place under normal conditions at a rate more than 10.6 times that of uncatalyzed reactions. This is possible because of an enzyme known as carbonic anhydrase, which mediates the reaction. Similarly, all reactions in biological systems are mediated by one or more enzymes and so reactions take place at higher speeds.

Enzymes operate on reactants, which are known as substrates, and convert them into products. The reaction may require energy or it may release energy. The enzyme is unaffected by the reaction.

A + B + ENZYME ⟶ AB + ENZYME

Enzyme + substrates ⟶ products + enzyme

Chemical reactions depend on the energy of the substrates sufficient to cause the molecules to move and thus collide. This is usually accomplished with heat. Heat increases the motion of molecules and makes their collision more likely. The heat necessary to accomplish this is often inappropriate inside a cell. Enzymes can be used instead of heat to increase the likelihood that reactants will collide in the correct position. Enzymes bring reactants together and hold them in the correct position with respect to each other. Reactions occur between atoms and involve changes in the distribution of electrons. Before going into the details regarding the mechanism of enzyme action, let us discuss the rate of a reaction.

Rate of Enzymatic Reactions

The rate of reaction defines the speed or the velocity of the reaction. The velocity of a reaction can be expressed in two ways. It can be expressed either in terms of the product formed or the substrate consumed. The rate or velocity of a reaction can also be defined as the quantity of the product formed per unit time (per second or per minute) or the quantity of the substrate consumed per minute. The quantity can be expressed in milligrams (mg) or micrograms (μg); and it can also be expressed in moles (M), millimoles (mM), or micromoles (μM).

Rate of a reaction = quantity of product formed per minute

In the case of enzyme-mediated reactions, the rate of reaction is usually represented as **enzyme activity,** which is related to the quantity of enzymes. The quantity of enzymes is expressed as units. It is the international unit (IU) for expressing the quantity of enzymes. One unit of enzymes is defined as the amount of enzymes required to convert one μM of substrate to product or to produce one μM product in unit time under physiological conditions (room temperature, normal pressure, and physiological pH).

Enzymes are Highly Specific

Enzyme molecules, compared to substrate molecules, are very large. The substrate molecules bind to the enzyme molecules at certain specific sites on the surface of the enzyme molecule, known as substrate binding sites or actives sites, to form the enzyme substrate complex (ES complex). The ES complex finally ends up in the product with the release of free and intact enzyme. The active site of the enzyme is responsible for the reactions leading to the formation of the product. The active site in the enzyme molecule is only a very small portion and is complementary to the molecular shape of the substrate. This is the main reason for the substrate specificity of an enzyme. The chemical reactions normally occur because of the molecular collisions between the reactants or the substrates. In enzyme-mediated reactions the substrate molecules are brought in proximity by taking them into the active sites of the enzyme. Thus, the enzyme molecule facilitates the reaction between the substrates without any collision between them. The ability of an enzyme molecule to accelerate a reaction is several orders of magnitude greater than normal inorganic catalysts.

Mechanism of Enzyme Action

Energy of Activation

Reactions involve making and breaking bonds. Some bonds must be broken at the start of a reaction and this requires energy. No matter how exergonic (reactions that proceed with the release of energy) the overall reaction may be, some energy must be added initially to break the necessary bonds and get the reaction started (e.g., match to paper, spark to cylinder). This is the **energy of activation (Ea)**. The activation energy can be defined as the energy needed by a molecule to take part in a reaction. This energy is often supplied as heat but this may not be practical in a cell.

All organic molecules contain energy that could be released by the breakdown of the molecule. Such a reaction would be strongly exergonic, and it does not happen spontaneously because the activation energy must be provided to get the reaction started and this energy is usually not available. To start the reaction, it is necessary to provide the activation energy, or else reduce the amount of activation energy needed.

Enzymes accelerate reactions more likely by reducing the energy required for activation. Enzymes bring down the activation energy by holding the reactants (substrates) in exactly the right orientation to each other so that contact will result in a reaction each time.

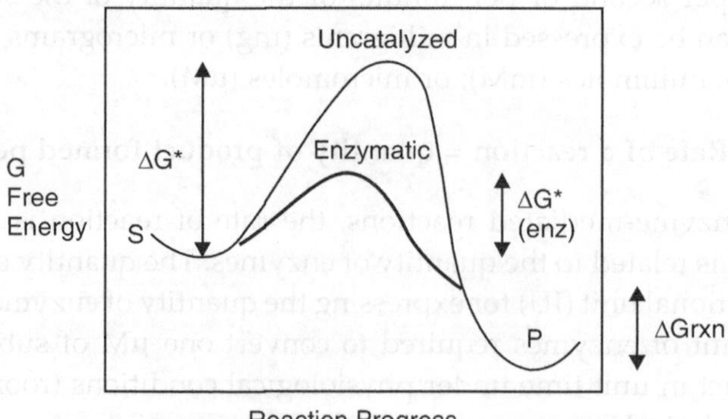

FIGURE 5.24 **Energy of activation.**

Enzymes are thought to operate on a geometric principle. The tertiary and quaternary structures of an enzyme have the substrate binding sites, which have exactly the complementary shape of the substrate molecules. This helps in the binding of the appropriate substrate to the active centers just like a key fits in the keyhole. This hypothesis that explains the mechanism of enzyme substrate interaction is known as the **Lock and Key hypothesis**. Some enzymes change the shape of the active center slightly to accommodate the substrate molecules, a process known as **induced fit**. This theory is known as the **induced fit theory**. The enzyme substrate complex thus formed lowers the energy of activation by either stressing an existing bond or correctly orienting two molecules to favor a reaction. The enzyme holds the substrate molecules in exactly the right position relative to each other to facilitate the reaction due to geometric and electrical configuration. The characteristic functional groups of the amino acid residues present in the active sites of the enzyme molecule can interact with the various bonds of the substrate molecule. The reaction occurs and the new product molecule leaves the enzyme due to diffusion gradients, or to new repulsive electrical forces or the shape changes in either the enzyme or the product. New substrate molecules move into position.

Anabolic reactions are **endergonic**. Endergonic reactions do not occur spontaneously. In them the system moves from low to high potential energy

$$A + B \rightarrow C$$

Catabolic reactions are exergonic

$$C \rightarrow A + B$$

Exergonic reactions usually do not happen spontaneously because the activation energy is not available in the cell. If an enzyme is present to lower the activation energy, there may then be enough energy present for the reaction to occur.

Metabolic Pathways

Enzymes typically work in chains called **metabolic pathways** in which the product of one reaction is the substrate of the next.

substrate → A ⟶ B ⟶ C ⟶ D ⟶ E ⟶ F ⟶ G ⟶ product

Consider the glycolytic pathway, for example, in which glucose is broken down in a stepwise fashion to pyruvic acid, which can be represented as follows. In each step there is an enzyme facilitating the reaction.

Glucose → fructose → phosphoglygeraldehyde → pyruvate

Enzyme Kinetics

The velocity of an enzyme-mediated reaction or the enzyme activity is closely controlled by a number of factors including the environmental (cellular environment) factors such as temperature and pH. The other factors are the substrate and enzyme concentrations.

The rate of a reaction can be measured either by monitoring the formation of the product or by the disappearance of substrate. A specific quantity of enzymes is allowed to interact with an excess amount of substrate for a specific period of time. The product formed can be measured by exploiting any of its physical or chemical properties such as color development, absorbance at a particular wavelength of light radioactivity, etc. For example, if the product formed is colored or can develop any color with some other reagent, the intensity of the color can be measured by an absorption spectrophotometer. The enzyme-mediated reaction has to be conducted under physiological conditions by providing a buffer of appropriate pH and optimum temperature usually between 20°C and 37°C. It is also necessary to provide co-factors and metal ions essential for the specific enzymes. All these physiological factors have to be maintained *in vitro* to keep the protein in its specific three-dimensional conformation and active state. Change in any of the parameters, particularly factors such as temperature, pH, and ionic strength can destroy the enzyme conformation or the three-dimensional structure. This change of conformation in the three-dimensional structure of an enzyme followed by the loss of activity is called denaturation.

Effect of Substrate Concentration

Substrate concentration is one of the factors that affect the rate of an enzyme-catalyzed reaction. Velocity of an enzyme-catalyzed reaction at different substrate concentrations can be measured, and the values can be plotted on a graph with substrate concentration on the x-axis and its corresponding velocity on the y-axis. It can be observed from the graph that as the concentration of substrate increases there is a corresponding increase in the velocity of reaction. However, beyond a

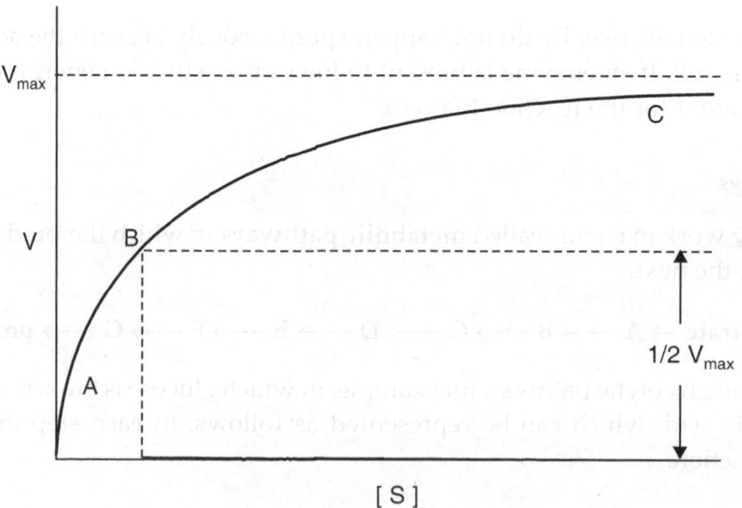

FIGURE 5.25 **Effect of substrate concentration [S] on the velocity of a reaction.**

particular concentration of substrate concentration, the velocity remains constant without any further increase. This maximum velocity of an enzyme-catalyzed reaction under substrate saturation is called the V_{max}. A reaction attains its V_{max} when all the active sites of the enzyme molecules are saturated with its substrate molecules so that the rate of product formation will be at its maximum.There is a substrate concentration at which the reaction attains its V_{max}. Similarly, there is a substrate concentration at which the velocity of the reaction is half of the V_{max}. This substrate concentration is a constant for a particular enzyme and that is known as K_m or Michaelis—Menton constant. K_m value of an enzyme is inversely proportional to affinity of the enzyme toward its substrate. If the K_m value of an enzyme is low it indicates its high affinity toward the substrate. The relationship between the V, K_m, V_{max}, and substrate concentration [S] can be represented by an equation, the **Michaelis–Menton equation:**

$$V = \frac{V_{max}\,[S]}{K_m\,[S]}$$

Turnover Number

The turnover number is the number of substrate molecules converted to product per second by one molecule of enzyme. It can be calculated from the V_{max}. It indirectly indicates the catalytic power of an enzyme. Higher values of turnover number point toward better catalytic power of the enzyme.

Effect of Enzyme Concentration

The effect of enzyme concentration on the rate of reaction is more or less similar to that of substrate concentration. Under excess of substrate concentration, if you increase enzyme concentration there is a corresponding increase in the rate of reaction, but only up to a particular point. After this point there is no further increase in the rate of reaction corresponding to the increase of enzyme concentration.

Effect of Temperature and pH

Enzymes always exhibit typical temperature and pH optima that are characteristics of that enzyme and appropriate for its function. Most of the enzyme works in the physiological pH, which is 7.5, even though their optimal pH is different.

Mammalian enzymes, for example, always show temperature optima around 37°C while those of the ice fish are around 0°C. Most enzymes are denatured at 45°C or so, but not those of thermophilic organisms.

Classification of Enzymes

Enzymes are classified into six major groups or classes based on the reactions they catalyze. They are listed in Table 5.3 along with the reaction they catalyze.

TABLE 5.3 Different classes of enzymes and the reactions they catalyze.

No.	Class	Type of reaction catalyzed
1.	Oxidoreductase	Oxidation reduction reactions (Transfer of electrons, hydrogen, or oxygen)
2.	Transferase	Transfer of functional groups from one compound to another
3.	Hydrolases	Hydrolysis (Breaking of bonds by the addition of water)
4.	Lyases	Addition of groups to double bonds, or formation of double bonds by removing groups
5.	Isomerases	Rearrangements of groups within the molecules resulting in the formation isomers
6.	Ligases	Formation of special covalent linkages like C-C, C-S, C-O and C-N bonds by condensation reaction (with the eliminaion of a water molecule) coupled with ATP cleavage

Enzyme Inhibition

Enzymes may be regulated in several ways: by temperature and pH as mentioned earlier or by control of synthesis or locally (i.e., after its synthesis). We are interested in the latter here. There are certain compounds called enzyme inhibitors, which inactivate the enzyme permanently or temporarily. There are many types of inhibitors. Some are deleterious (poisons) and some are important regulatory mechanisms. Some inhibitions are deliberate control mechanisms; some are detrimental to the organism. Enzymes are often inactivated by some molecule (an inhibitor) that changes their shape or blocks the active site.

Competitive Inhibitors

These are molecules that bind reversibly or irreversibly to the active site. They compete with the substrate for space in the active site. Most naturally occurring competitive inhibitors are irreversible.

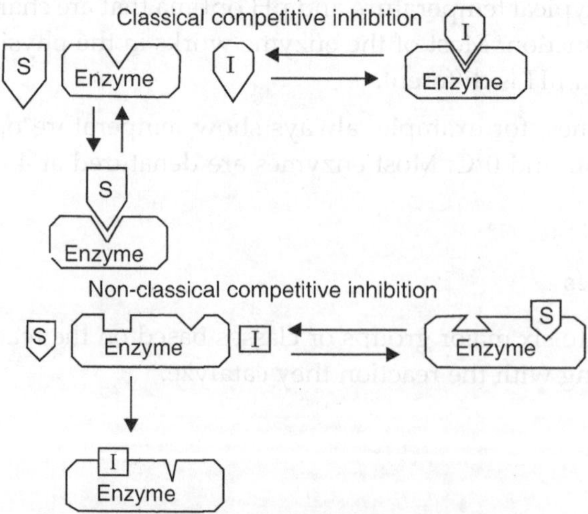

FIGURE 5.26 **Competitive inhibition.**

Reversible Competitor. These competitors are not the substrate molecules but molecules similar to the substrate molecules. Thus, they compete with the substrate to occupy the active center of the enzyme molecule. If they are in high concentration, they may essentially inactivate the enzyme. These are of limited importance in natural systems. The competition is reversible and the competitor can be overwhelmed by high concentration of substrate. For example, fumaric acid is the competitive inhibitor of the enzyme, succinic dehydrogenase. The structure of fumaric acid is similar to that of succinic acid, the actual substrate of the enzyme.

Irreversible Competitor. The competitor and substrate both compete for the active site but the competitor occupies the active site permanently thus deactivating the enzyme. Non-classical competitive inhibition is the binding of S at the active site and prevents the binding of I at a different site. The reverse is also true. Carbon monoxide is an irreversible competitive inhibitor of hemoglobin. Oxygen is the substrate.

Non-competitive Inhibition

In non-competitive inhibition the inhibitor and substrate do not compete for space in the active site. The substrate enters the active site but the inhibitor reacts with some other part of the enzyme molecule. It may be reversible or non-reversible. Reversible non-competitive inhibition is a major metabolic control mechanism.

Reversible Non-competitive. This type of inhibition involves two binding sites on the enzyme molecule: the usual active site and a second **regulatory site** where the inhibitor binds. The inhibitor and the substrate do not compete for the same site. If the regulatory site is occupied by the inhibitor then the shape of the active site is changed so that the substrate molecules cannot fit and react. This type of inhibition, often called **allosteric control**, is very important in regulating metabolic pathways in the cell.

In allosteric control the final product molecule often acts as an inhibitor of one of the enzymes in the pathway, typically the first.

When high concentrations of the final product accumulate, some of them react with the first enzymes at its regulatory site and render it inactive. This eventually results in a decrease in the concentration of the product. When this happens the inhibitor will drop out of the regulatory site and the enzyme will become active again. This is an example of **negative feedback**. This is a type of reaction that we will see over and over again in biology, at all levels from cell to ecosystem. Negative feedback prevents runaway reactions. The thermostat on ACs, furnaces, and incubators is a negative feedback mechanism. When temperature is raised over the control point the heating system of the incubator is turned off.

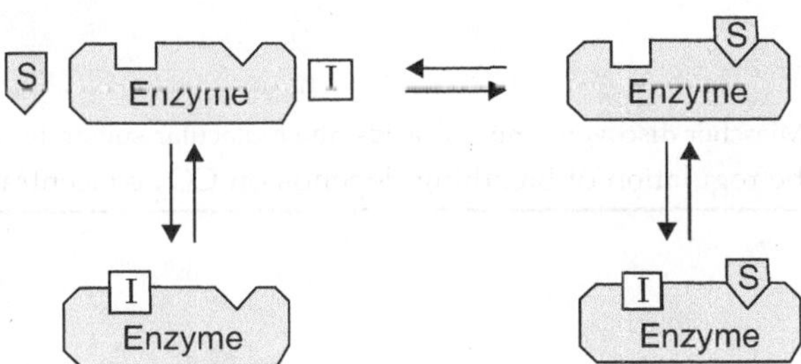

FIGURE 5.27 **Non-competitive inhibition.**

Irreversible Non-competitive. These inhibitors do not compete with the substrate for space in the active site in the sense that they bind irreversible and there is thus no opportunity for competition. They bind irreversible with some other part of the enzyme (or to the active site) and permanently denature or inactivate it. They permanently alter the native conformation of the protein. They include heavy metals, cyanide, nerve gas, and arsenic.

Cofactors

Proteins are often associated with other chemical groups known as **cofactors or prosthetic groups**. Enzymes are no exception.

Often molecules, atoms, or ions other than proteins are necessary for the functioning of enzyme molecules. These may be a permanent part of the molecule, as are most ions (Zn, Mg, Fe) or may be only temporarily associated with the protein. Small, organic prosthetic groups are often called **coenzymes**. Most of the coenzymes are the derivatives of **vitamins**.

In addition to enzymes some non-enzyme proteins also have prosthetic groups or the non-protein parts. They are:

Nucleoproteins: have nucleic acid as the prosthetic group

Glycoproteins: have carbohydrate as the prosthetic group

Lipoproteins: have lipid as the prosthetic group

5.5 NUCLEIC ACIDS

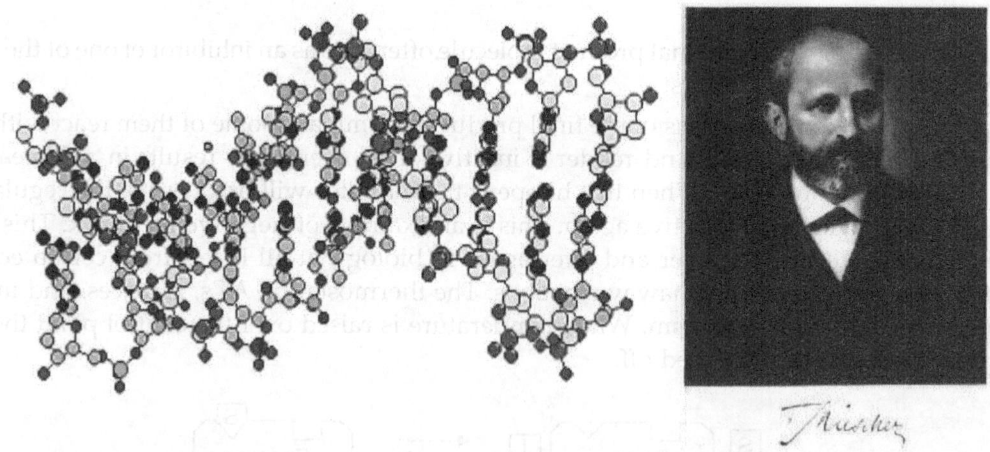

In 1869, Friedrich Miescher discovered nucleic acids, the molecular substrate of the genetic material. He also demonstrated that the regulation of breathing depends on CO_2 concentration in the blood.

Nucleic acids are polymers consisting of nucleotides that are linked by phosphodiester bonds (polynucleotides). Depending on the type of sugar present in the nucleotide (ribose or deoxyribose), they are divided into two classes of nucleic acids: **ribonucleic acids (RNA)** and **deoxyribonucleic acids (DNA)**. DNA is known to be the carrier of genetic information, RNA functions as a messenger (messenger RNA) and takes part in protein synthesis (transferRNA, ribosomalRNA).

As in proteins, the monomers or the nucleotides are linked together to form the polynucleotides, a long chain of nucleotides connected together by **phosphodiester linkages**. To distinguish the C-atoms of the nitrogen bases and those of the sugars from each other, the latter are marked by a ′. It is thus spoken of as 1′, 2′, 3′, 4′, and 5′ positions. Polynucleotides are formed by the condensation of two or more nucleotides. The condensation most commonly occurs between the alcohol of a 5′-phosphate of one nucleotide and the 3′-hydroxyl of a second, with the elimination of H_2O, forming a **phosphodiester bond**. This polymerization reaction involves the breaking of a pyrophosphate residue in each addition of nucleotide. The polynucleotide chain grows (during synthesis) from the 5′ to the 3′ terminus. Since the phosphodiester bond is formed between a 5′ phosphate group of one nucleotide and 3′ hydroxyl group of another nucleotide, it is known as 5′→ 3′ phosphodiester bond. The primary structure of DNA and RNA (the linear arrangement of the nucleotides) proceeds in the 5′ → 3′ direction. The common representation of the primary structure of DNA or RNA molecules is to write the nucleotide sequences from left to right synonymous with the 5′ → 3′ direction as shown:

5′-pGpApTpC-3′ ('p' is the phosphodiester linkage)

Ribonucleic Acids (RNA)

RNA is the non-genetical nucleic acids except in some viruses such as HIV, where it is the genetic material. They are comparatively small and present chiefly as linear, single-stranded molecules. There are four major forms of RNA in cells. They are messenger RNA (mRNA), transfer RNA (tRNA), ribosomal RNA (rRNA), and small nuclear RNA (snRNA). All these RNA molecules are synthesized based on the nucleotide sequence of the genetic material, DNA, for the purpose of protein synthesis.

In contrast to DNA, RNA is normally single-stranded. But it displays also some double-stranded, helical sections that are caused by folding of the single strand. These sequences are mirror images of each other and develop palindromes.

Messenger RNAs are the copies of the coding regions of the genes and have the information to make the protein. It is very unstable and comprises about 5% of the total cellular RNA. It is highly variable in length and has a range of 75 to 3,000 nucleotides.

Transfer RNAs are the different types of small molecules having a length in the range of 90 to 120 nucleotides. They can also fold and form secondary structures and have a clover-leaf-like model and can again fold to form a L-shaped structure as given in the figure below (Figure 5.28). They form about 15% of the total cellular RNA. During protein synthesis they carry the respective amino acids to the ribosomes where the translation of mRNA or the protein synthesis takes place. They contain a number of **rare and modified bases** that stabilize its structure.

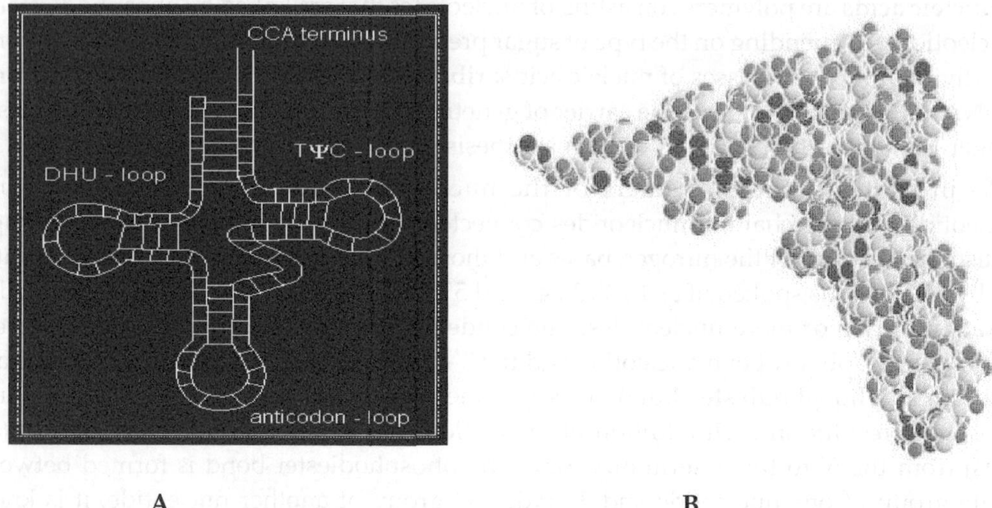

FIGURE 5.28 **The secondary structure (A) and tertiary structure (B) of tRNA molecule. A. The cloverleaf structure of tRNA B. L-shaped 3-D model.**

Ribosomal RNAs are a group of RNA molecules having different sizes. Complex forms of secondary structures can be found in ribosomal RNA (rRNA). They form the structural components of the subcellular particles called ribosomes. Ribosomes are one of the important components needed for protein synthesis. The size of the molecules varies from 120 to 3,000 or more nucleotides. Approximately 80% of the total cellular RNA is rRNA.

There is another class of RNA molecules mostly present in the nucleus of eukaryotic cells known as **small nuclear RNA** or **snRNA**. As the name indicates they are small molecules, but with complicated secondary structures. They form 2% of the total cellular RNA molecules and are involved in the processing or modification, called splicing, of mRNA. They are organized into small nucleoprotein particles called splicisomes. Most of them have catalytic activity and therefore are called **ribozymes.** Ribozymes are the RNA molecules having catalytic (enzyme-like) activity.

Deoxyribonucleic Acid (DNA)

Deoxyribonucleic acid or DNA, forms the genetic material of almost all organisms except some viruses, which have RNA as the genetic material and therefore are called RNA viruses. DNA is a very long polymer and always exists as a double-stranded helical molecule.

DNA contains the nitrogen bases adenine (A), Thymine (T), Cytosine (C), and Guanine (G). Erwin Chargaff isolated DNA samples from different organisms and analyzed nitrogen-based composition and observed that A and T as well as C and G are present in equal amounts. The ratio A + T/C + G is specific for every species. This is known as the **Chargaff rule**. All double-stranded DNA samples irrespective of their source obey the Chargaff rule of base equivalence. There are some DNA viruses, where the genetic material is single-stranded DNA. Single-stranded DNA does not obey the Chargaff rule of base equivalence. Based on these data and *x*-ray diffraction data developed by Rosalind Franklin obtained from crystals of DNA, James Watson and Francis Crick proposed a model, the **Watson-Crick model** for the structure of DNA. They published their findings in 1953 in the journal, *Nature*.

Three-dimensional Structure of DNA

The **Watson-Crick model** (subsequently verified by additional data) explained almost all the properties of DNA as the genetic material including the mechanism of its replication by using both of the DNA strands as matrices. The further characteristics of the model are:

- The DNA molecule exists as a helix of two complementary anti-parallel polynucleotide chains, wound around each other in a rightward direction and stabilized by H-bonding between bases in adjacent strands.
- The purine and pyrimidine bases are directed toward the interior of the helix, the phosphates and the sugar residues are at the outside. The planes of the bases are at a right angle to the axis of the helix.
- The radius of the helix is 10Å (diameter 20 Å), and the distance between neighboring bases is 3.4 Å.
- Both chains are linked by hydrogen bonds that develop between a pyrimidine and a purine base. An adenine is thus always linked with a thymine, a guanine with a cytosine. Between adenine and thymine exist two hydrogen bonds, between cytosine and guanine exist three (Figure 5.29).

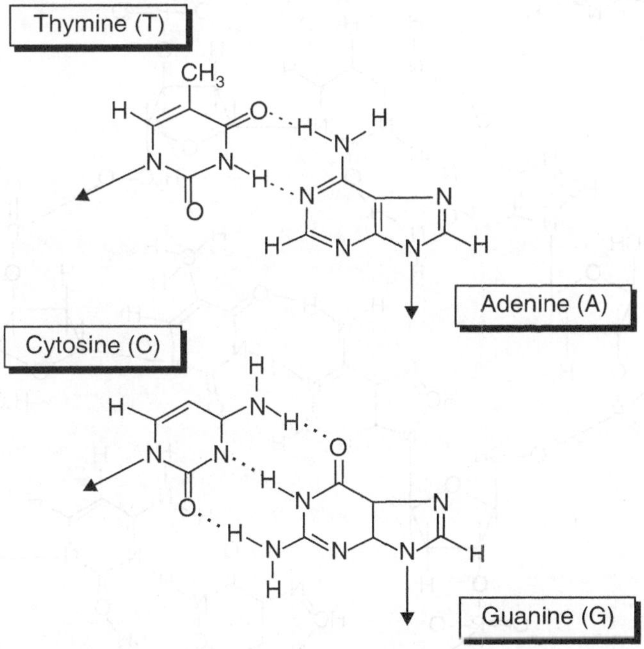

Thymine (T)

Adenine (A)

Cytosine (C)

Guanine (G)

FIGURE 5.29 **Watson-Crick base pairing in DNA strands.**

- The specific base pairing causes complementarity. If the sequence of one strand is known, then that of the other can be predicted.
- The anti-parallel nature of the helix stems from the orientation of the individual strands. From any fixed position in the helix, one strand is oriented in the 5′→ 3′ direction and the other in the 3′ → 5′ direction.

- On its exterior surface, the double helix of DNA contains two deep grooves between the ribose-phosphate chains. These two grooves are of unequal size and termed the **major** and **minor grooves**. The difference in their size is due to the asymmetry of the deoxyribose rings and the structurally distinct nature of the upper surface of a base pair relative to the bottom surface.

The mechanism of replication was elucidated in 1958 by M. Meselson and F. W. Stahl, who experimentally proved a prediction made by Watson-Crick and thus showed that DNA duplicated in a semi-conservative way.

The sequence of nucleotides within a nucleic acid may seem arbitrary at first, but we know that it is this sequence that encodes the genetic information. Though it doesn't mean that every nucleotide sequence contains information. Long, repetitive DNA sequences exist and their function is not known. Since around 1975, methods for the sequencing of nucleotides have been developed.

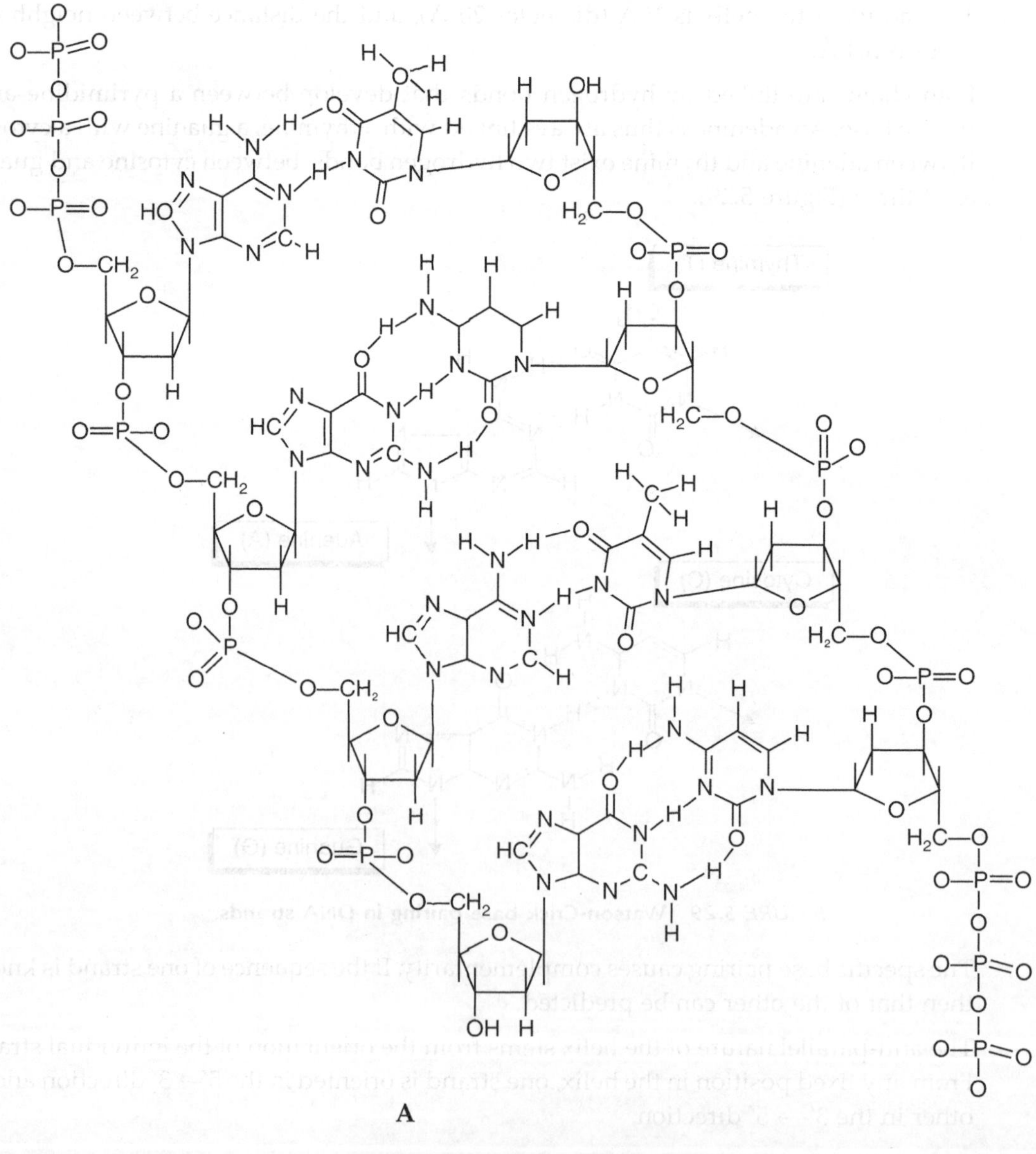

A

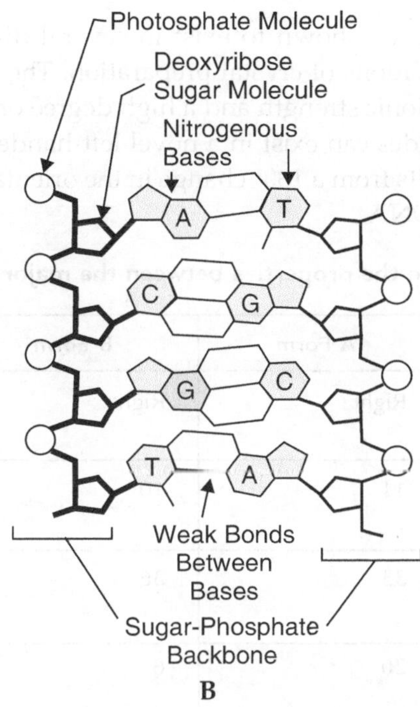

FIGURE 5.30 **A. Molecular structure of DNA.**
B. A slightly more simplified diagram.

Other Forms of DNA

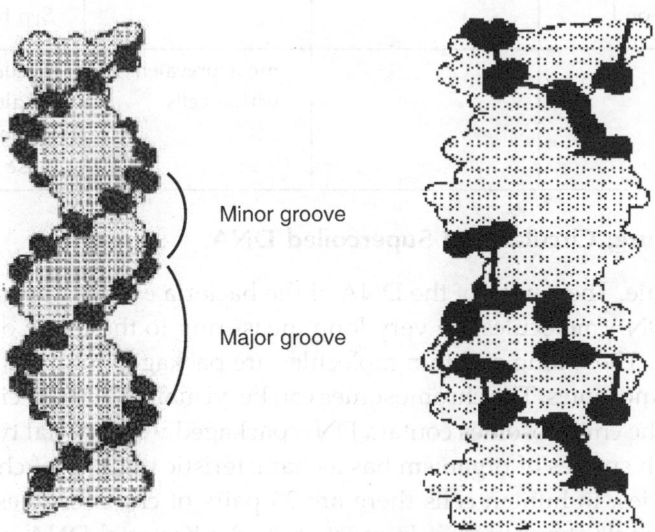

FIGURE 5.31 **Structure of B-DNA structure of Z-DNA.**

The double helix of DNA has been shown to exist in several different forms, depending upon sequence content and ionic conditions of crystal preparation. The **B-form** of DNA prevails under physiological conditions of low ionic strength and a high degree of hydration. Regions of the helix that are rich in pCpG dinucleotides can exist in a novel left-handed, helical conformation termed **Z-DNA**. This conformation results from a 180° change in the orientation of the bases relative to that of the more common A- and B-DNA.

TABLE 5.4 Difference in the properties between the major helical forms of DNA.

Parameters	A Form	B Form	Z-Form
Direction of helical rotation	Right	Right	Left
Residues per turn of helix	11	10	12 base pairs
Rotation of helix per residue (in degrees)	33	36	- 30
Base tilt relative to helix axis (in degrees)	20	6	7
Major groove	narrow and deep	wide and deep	Flat
Minor groove	wide and shallow	narrow and deep	narrow and deep
Orientation of N-glycosidic Bond	Anti	Anti	Anti for Py, Syn for Pu
Comments		most prevalent within cells	occurs in stretches of alternating purine-pyrimidine base pairs

Chromosome Structure; Circular and Supercoiled DNA

DNA is a long molecule. The length of the DNA of the bacteria *e.coli* is approximately 1.44 mm. In higher forms of life DNA molecules are very long measuring to the range of centimeters. But the diameter is only 20 Å. These long and thin molecules are packaged into small, distinct subcellular structures called chromosomes. The chromosomes can be visualized by special staining techniques under microscopes. The chromosomes contain DNA packaged with special types of proteins called **histone proteins**. Each species of organism has a characteristic number of chromosomes, which is constant for that species. In human cells there are 23 pairs of chromosomes, and in bacteria like *e.coli* there is only a single chromosome. In most cases the bacterial DNA is circular without any free end, whereas in higher forms of organisms the DNA molecule of the chromosomes are linear with free ends. These chromosomes are confined to the nucleus in the case of eukaryotic cells. Since the DNA molecules are very long, they are packaged into chromosomes having microns of length in a very specific manner. This packaging involves the process known as supercoiling. Supercoiling in DNA is promoted and maintained by enzymes known as topoisomerases, which keep the DNA in a highly supercoiled state. The supercoiling and packaging with histone proteins produce a beaded state for the DNA molecule, which is an intermediate stage in the formation of chromosomes known as nucleosomes.

It is very interesting to see how cells have managed to package the enormous length of DNA into chromosomes having microns of length. A typical human cell has 20 μm in diameter and the nucleus has about 5 μm. The total length of DNA from all the 23 pairs of chromosomes corresponds to approximately 2 m (total base pairs $= 6 \times 10^9$; length of 1 bp = 3.3 Å). If 2 m have to be packaged into the confines of a 5 μm diameter nucleus the DNA molecules have to be condensed by a factor of 4×10^5. The fragile and long DNA molecules are packaged in chromosomes in such a way that the information stored in the DNA molecules can be retrieved within a very short period. The microchips are very large when compared to this level of miniaturization.

In addition to the circular chromosomal DNA, most bacterial cells contain one or more types of small extra chromosomal circular DNA molecules, which can replicate independently. These circular DNA molecules are known as plasmids. The plasmids are responsible for certain unusual characters of organisms like antibiotic resistance. Plasmids have great importance in gene cloning and recombinant DNA experiments. The vectors commonly used in these experiments are the modified forms of plasmids. Plasmids used for gene cloning and recombinant experiments are known as vectors.

5.6 LIPIDS AND BIOLOGICAL MEMBRANES

Lipids are a class of hydrophobic molecules but very much miscible with organic solvents such as hexane, chloroform, ethyl acetate, etc. They are a group of biologically-important lipids that show amphipathic characteristics. An amphipathic molecule is one in which one end of the molecule exhibits a hydrophobic (water hating) nature and the other end has a hydrophilic (water-loving) nature. They are the phospholipids, the main components of biological membranes. The amphipathic nature of phospholipids makes them ideal candidates for the formation of biomembranes, which are composed of two layers of phospholipids. In aqueous medium phospholipids or any other amphipathic, molecules will arrange themselves in such a way that the hydrophobic ends are kept away from water and the hydrophilic ends are in contact with water molecules. Depending on the molecular weight and quantity of the amphipathic molecules and quantity of water, they can form different structures such as vesicles, bilayers, or single-layered structures.

Phospholipids

Phospholipids are structural molecules forming the major component of all membranes of cells. In addition to lipids there are different types of proteins distributed along the bilayer of membranes depending on the special function of the membranes within the cells. Structure of phospholipid consists of a glycerol molecule with two fatty acids esterified at positions R_1 and R_2 and a phosphoric acid esterified at position X.

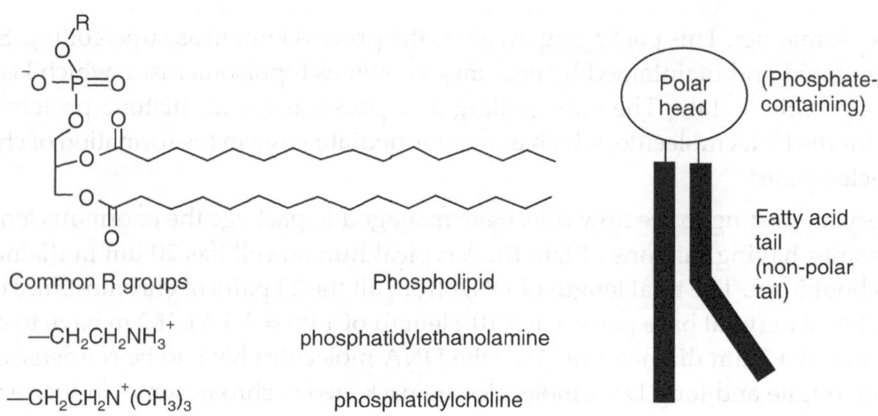

FIGURE 5.32 **Diagram sketch of phospholipid and its derivatives.**

The benefit of the phospholipid structure is that the phosphate region makes the molecule highly amphipathic, ideal for the cell-membrane structure.

- Hydrophilic portion in the phosphate region.
- Hydrophobic portion in the fatty acid tails.

The most common phospholipid is **lecithin**. Phospholipids also make excellent emulsifiers and are used in a number of food and household products. Cholesterol and its derivatives also form another component of the cell membranes, but only in animal cells. It is not found in plant-cell membranes. Different derivatives of phospholipids such as Sphingolipids, sphingomyelin, glycosphingolipids, Cerebrosides, etc., are present in certain specialized cells such as neurons.

All types of cells including both eukaryotes and prokaryotes and the cell organelles present inside the eukaryotic cells are enveloped by lipid bilayer membranes, which are strong and selectively permeable protecting the cytoplasmic content from the osmotic changes and other disturbances. (See Appendix)

Properties of the Lipid Bilayer

As we have already mentioned, the most important property of the lipid bilayer is that it is a highly impermeable structure. Impermeable simply means that it does not allow molecules to freely pass across it. Only water and gases can easily pass through the bilayer. This property means that large molecules and small polar molecules cannot cross the bilayer, or the cell membrane, without the assistance of other structures.

Another important property of the lipid bilayer is its fluidity. The lipid bilayer contains lipid molecules, and, as we will discuss later, it also contains proteins. The bilayer's fluidity allows these structures mobility within the lipid bilayer. This fluidity is biologically important, influencing membrane transport. Fluidity is dependent on both the specific structure of the fatty acid chains and temperature (fluidity increases at lower temperatures).

Structurally, the lipid bilayer is asymmetrical: the lipid and protein composition in each of the two layers is different.

Architecture of Membranes

Subcellular fractionation techniques can partially separate and purify several important biological membranes, including the plasma and mitochondrial membranes, from many kinds of cells. These preparations are often contaminated with membranes from other organelles. However, the plasma membranes from human erythrocytes can be isolated in near purity because these cells contain no internal membranes (see Appendix).

All membranes contain phospholipids and proteins. The protein-lipid ratio varies greatly:

- The inner mitochondria membrane is 76% protein.
- The myelin membrane is only 18% protein.

Because of its high phospholipid content, myelin can electrically insulate the nerve cell from its environment. The lipid composition varies among different membranes. All membranes contain a substantial proportion of phospholipids, predominantly phosphoglycerides, which have a glycerol backbone. All membrane phospholipids are amphipathic, having both hydrophilic and hydrophobic portions. Sphingomyelin, a phospholipid that lacks a glycerol backbone, is also commonly found in plasma membranes. Instead of a glycerol backbone it contains sphingosine, an amino alcohol with a long unsaturated hydrocarbon chain. A fatty acyl side chain is linked to the amino group of sphingosine by an amide bond to form a ceramide. The terminal hydroxyl group of sphingosine is esterified to phosphocholine, thus the hydrophilic head of sphingomyelin is similar to that of phosphatidylcholine.

Functions of Biomembranes

The biomembranes act as a very good barrier, which protects the contents from spilling out or extracellular contents from entering into the cells or cell organelles.

Cell membranes are selectively permeable. Water-soluble compounds such as sugars, organic acids, and salts such as NaCl, KCl cannot freely pass through lipid membranes. There are special protein molecules positioned in the lipid bilayer to facilitate the transport of selected molecules.

The functions of biomembranes are determined by the quality of the proteins, which are distributed along with the lipids in the membranes. This makes a biomembrane distinct from another in its function. For example, the cellular envelop, mitochondrial membranes, and membranes of chloroplasts are specialized for carrying out a specific function that is closely related to the types of protein present in the respective membranes. Certain cell membranes have certain informational molecules such as glycoproteins, which can act as the receptors of signal molecules such as hormones.

REVIEW QUESTIONS

1. What are the differences between starch and cellulose?
2. Differentiate between glycogen and starch.
3. What are the differences between DNA and RNA?
4. What are the special features of DNA that make it the genetic material?
5. What are the various forms of DNA? Explain their structural difference and significance.

6. What is meant by the primary structure of protein?
7. Describe the various types of secondary structures.
8. What is the implication of the presence of tertiary structures and secondary structure in proteins?
9. An insulin molecule has two polypeptides and hemoglobin has four polypeptides in its quaternary structures. But insulin is considered a monomeric protein and hemoglobin is considered a multimeric protein. Why?
10. Enzymes are biological catalysts. What is the significance of catalysts in the biological system?
11. How does an enzyme differ from an inorganic catalyst?
12. What are the factors that influence the velocity of an enzyme-catalyzed reaction?
13. What is one unit of enzyme?
14. Explain the reasons for the high catalytic power and specificity of enzymes.
15. What is the prosthetic group of an enzyme?
16. Give the Michaelis-Menton equation and explain V_{max} and K_m.
17. Name three vitamins and their corresponding co-enzymes.
18. What is the structure of a biological membrane?
19. Even though biological membranes have the same basic lipid bilayer structure, there are great differences in biological functions. Why?
20. What are the functions of plasma membranes?
21. What is a ribozyme?

Chapter 6

BIOCHEMICAL TECHNIQUES

In This Chapter

6.1 INTRODUCTION

Biomolecules can be isolated and purified by applying different techniques, which are based on various chemical and physical properties. The main physical and chemical properties that can be exploited for their separation and characterization of biomolecules are molecular weight and size, interaction with electromagnetic radiations or spectroscopic properties, solubility, molecular charge, and polarity.

The techniques based on size and weight of the molecules are centrifugation, gel filtration, and osmotic pressure.

Techniques that are based on polarity and charge are ion exchange chromatography, electrophoresis, isoelectric focusing, hydrophobic interaction, and partition chromatography.

Techniques based on spectroscopy include colorimetry, UV-visible spectrophotometry, fluorescence spectroscopy, *x*-ray crystallography, and mass spectrometry.

The solubility techniques are the precipitation of molecules with salts and organic solvents.

6.2 TECHNIQUES BASED ON MOLECULAR WEIGHT AND SIZE

Centrifugation

A centrifuge is a device for separating particles from a solution according to their sedimentation rate, which depends on factors like size, shape, density, viscosity of the medium, and centrifugal force (rotor speed). This process of separation of particles based on its sedimentation rate is called centrifugation. In biology, the particles are usually cells, sub-cellular organelles, viruses, and large molecules such as proteins and nucleic acids. The rate of sedimentation will be directly proportional to the molecular weight or size, if all other factors are constant. To simplify mathematical terminology we will refer to all biological material as spherical particles. There are many ways to classify centrifugation.

Analytical and Preparative Centrifugation

The two most common types of centrifugation are analytical and preparative; the distinction between the two is based on the purpose of centrifugation. Analytical centrifugation involves measuring the physical properties of the sedimenting particles such as sedimentation coefficient or molecular weight. Optimal methods are used in analytical ultracentrifugation. Molecules are observed by an optical system during centrifugation, to allow observation of macromolecules in the solution as they move in the gravitational field. The samples are centrifuged in cells having windows that lie parallel to the plane of rotation of the rotor head. As the rotor turns, the images of the cell (proteins) are projected by an optical system onto film or a computer. The concentration of the solution at various points in the cell is determined by absorption of light of the appropriate wavelength (Beer's law is followed). This can be accomplished either by measuring the degree of blackening of a photographic film or by the pen deflection of the recorder of the scanning system, which is fed into a computer.

The other forms of centrifugations are **preparative** and the objective is to isolate specific particles, which can be reused. There are many types of preparative centrifugation such as rate zonal, differential, and isopycnic centrifugation.

Ultracentrifugation *vs* Low-speed Centrifugation

Another system of classification is the rate or speed at which the centrifuge is turning. **Ultracentrifugation** is carried out at speed faster than 30,000 rpm. **High-speed centrifugation** is at speeds between 10,000 and 30,000 rpm. **Low-speed centrifugation** is at speeds below 10,000 rpm (mostly between 3,000 to 9,000 rpm).

Moving Boundary *vs* Zone Centrifugation

A third method of defining centrifugation is by the way the samples are applied to the centrifuge tube. In **moving boundary** or **differential centrifugation**, the entire tube is filled with the sample and centrifuged. Through centrifugation, one obtains a separation of the mixture into two parts—a supernatant and a pellet (Figure 6.1). But any particle in the mixture may end up in the supernatant or in the pellet or it may be distributed in both fractions, depending upon its size, shape, density, and conditions of centrifugation. The pellet is a mixture of all of the sedimented components, and it is contaminated with whatever unsedimented particles were in the bottom of the tube initially. The only component that is purified is the slowest sedimenting one, but its yield is often very low. The two fractions are recovered by decanting the supernatant solution from the pellet. The supernatant can be recentrifuged at higher speeds to obtain further purification with the formation of a new pellet and supernatant.

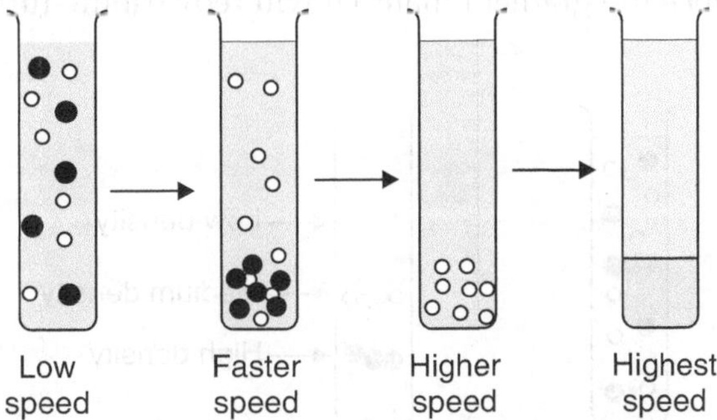

FIGURE 6.1 **Differential centrifugation.**

In **rate zonal centrifugation**, the sample is applied in a thin zone at the top of the centrifuge tube on a density gradient (Figure 6.2). Under centrifugal force, the particles will begin sedimenting through the gradient in separate zones according to their size, shape, and density or the sedimentation coefficient(s). The run must be terminated before any of the separated particles reach the bottom of the tube. S is the sedimentation coefficient and is usually expressed in Svedbergs (S) units.

Rate zonal centrifugation is a type of **density gradient** centrifugation. The other type of density gradient centrifugation is **isopycnic density gradient** centrifugation.

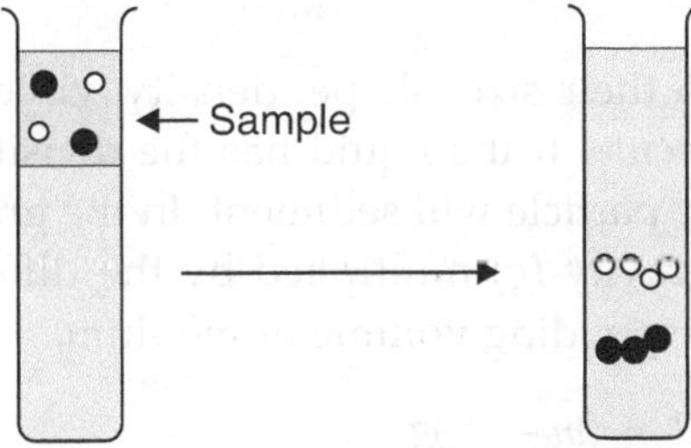

FIGURE 6.2 **Rate zonal centrifugation. Three proteins separate according to size and shape.**

In the **isopycnic technique**, the density gradient column encompasses the whole range of densities of the sample particles. The sample is uniformly mixed with the gradient material. Each of the particles will sediment only to the position in the centrifuge tube at which the gradient density is equal to its own density, and there it will remain (Figure 6.3). The isopycnic technique, therefore, separates particles into zones solely on the basis of their buoyant density differences, independent of time. In many density gradient experiments, particles of both the rate zonal and the isopycnic principles may enter into the final separations. For example, the gradient may be of such a density range that one of the components sediments to its density in the tube and remains there, while another component sediments to the bottom of the tube. The self-generating gradient technique often requires long hours of centrifugation. Isopynically banding DNA, for example, takes 36 to 48 hours in a self-generating cesium chloride gradient. It is important to note that the run time cannot be shortened by increasing the rotor speed; this only results in changing the position of the zones in the tube since the gradient material will redistribute further down the tube under greater centrifugal force.

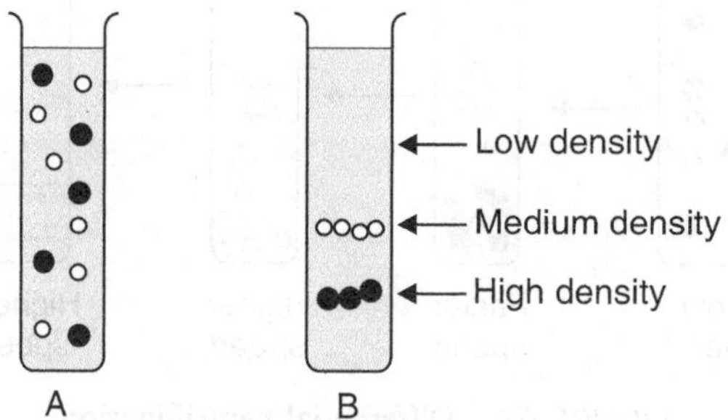

A—Sample is evenly distributed throughout the centrifuge tube.
B—After centrifugation proteins migrate to their isopycnic densities.

FIGURE 6.3 **Isopycnic separation with a self-generating gradient.**

Microcentrifuge

These are special centrifuges designed for handling small volumes like microliters, up to 1,500. These instruments can attain a maximum speed of 12,000 and 10,000 g.

RCF and RPM

Molecules separate according to their size, shape, density, viscosity, and centrifugal force. The simplest case is a spherical molecule. If the liquid has the density of d_o and the molecule has a density of d and if $d > d_o$ then the particle will sediment. In the gravitational field, the motor force (P_g) equals the acceleration of gravity (g) multiplied by the difference between the mass of the molecule and the mass of a corresponding volume of medium.

$$P_g = (m - m_o)g$$

$$P_g = 4/3\ (3.14)\ r^3\ dg\ - 4/3(3.14)\ r^3 d_o g$$

$$P_g = (4/3)r^3\ (3.14)\ (d - d_o)g$$

where, P_g = force due to gravity, g = acceleration of gravity, d_o = density of liquid (or gradient), d = density of molecule, m = mass of the molecule, and m_o = mass of equal to volume of medium.

The acceleration of a centrifuge is usually expressed as a multiple of the acceleration due to gravity – $g = 9.80\ m/s^2$, which is called the **relative centrifugal field (g) or RCF.** RCF depends on the speed of the rotor represented as **Revolution per minute or RPM** (n) and **radius of the rotor** (r). The relation between RPM and RCF is given by the following equation.

$$RCF = 1.118\, r\, (n/1000)^2$$

Using this equation RCF and RPM can be interconverted if the value of r is known.

RPM (n, the speed of rotor) = $945.7\sqrt{(RCF/r)}$. r is the radius of the rotor or the radius of the rotation.

Gel-filtration Chromatography

Gel-filtration chromatography is a separation based on size. It is also called molecular exclusion or gel permeation chromatography. In gel filtration chromatography, the stationary phase consists of porous beads with a well-defined range of pore sizes. The stationary phase for gel filtration is said to have a fractionation range, meaning that molecules within that molecular weight range can be separated.

Proteins that are small enough can fit inside all the pores in the beads and are said to be included. These small proteins have access to the mobile phase inside the beads as well as the mobile phase between beads and elute last in a gel-filtration separation. Proteins that are too large to fit inside any of the pores are said to be excluded. They have access only to the mobile phase between the beads and, therefore, elute first. Proteins of intermediate size are partially included—meaning they can fit inside some but not all of the pores in the beads. These proteins will then elute between the large ("excluded") and small ("totally included") proteins.

Consider the separation of a mixture of glutamate dehydrogenase (molecular weight 290,000), lactate dehydrogenase (molecular weight 140,000), serum albumin (MW 67,000), ovalbumin (MW 43,000), and cytochrome c (MW 12,400) on a gel-filtration column packed with Bio-Gel P-150 (fractionation range 15,000 to 150,000). When the protein mixture is applied to the column, glutamate dehydrogenase would elute first because it is above the upper fractionation limit. Therefore, it is totally excluded from the inside of the porous stationary phase and would elute with the void volume (V_0). Cytochrome c is below the lower fractionation limit and would be completely included, eluting last. The other proteins would be partially included and elute in order of decreasing molecular weight.

These separations can be described with this equation

$$V_r = V_0 + KV_i$$

where V_r is the retention volume of the protein, V_0 is the volume of mobile phase between the beads (outside the beads) of the stationary phase inside the column (sometimes called the void volume), V_i is the volume of mobile phase inside the porous beads (also called the included volume), and K is the partition coefficient (the extent to which the protein can penetrate the pores in the stationary phase, with values ranging between 0 and 1). In the mixture of proteins listed above, the partition coefficient (K) for glutamate dehydrogenase would be 0 (totally excluded), K = 1 for cytochrome c (totally included), and K would be between 0 and 1 for the other proteins, which are within the fractionation range for the column.

In practice, gel-filtration can be used to separate proteins by molecular weight at any point in purification of a protein. It can also be used for buffer exchange; a protein dissolved in a sodium acetate buffer, pH 4.8, can be applied to a gel-filtration column that has been equilibrated with Tris buffer pH 8.0. Using the Tris buffer as the mobile phase, the protein moves into the Tris mobile phase as it travels down the column, while the much smaller sodium acetate buffer molecules are totally included in the porous beads and travel much more slowly than the protein. Similarly, it can be used for the separation of salts and other small molecules from a protein sample.

Osmotic Pressure

Molecules always move from the region of their higher concentration to the region of their lower concentration. It is applicable to both solvent and solute molecules in the case of a solution. When a solution is separated from a pure solvent by a membrane that is permeable to the solvent alone, the molecules of the solvent will move into the solution. The flow of the solvent molecules into the solution can be prevented by applying some amount of pressure that is equal to the pressure exerted by the solvent molecules to enter into the solution compartment through the membrane partition. The pressure that is needed to prevent the entry of the solvent molecule into the solution is called **osmotic pressure.**

The osmotic pressure of a solution depends on the concentration of a solute and the temperature of the solution. It can be used for the calculation of the molecular weight of the solute.

$$\Pi V = nRT \tag{1}$$

where Π is the osmotic pressure, V is the volume of the solution, n is the number of moles of solute, R is gas constant, and T is the absolute temperature.

$$\Pi = n/V \times RT \qquad (n/V = M, \text{ molarity of the solution})$$

Therefore, $\Pi = MRT$

But in equation (1), n, the number of moles = weight of the solute in grams/molecular weight.

i.e., $\Pi V = Wt_g/MW \times RT$

Therefore, $MW = Wt_g/\Pi V \times RT.$

$(Wt_g/\text{Volume} = \text{Concentration, C})$

i.e., $MW = CRT/\Pi.$

Thus, if the osmotic pressure and the concentration of the solution are available the molecular weight of the solute can be determined.

6.3 TECHNIQUES BASED ON CHARGE

Ion-exchange Chromatography (IEC)

Ion-exchange chromatography (IEC) is applicable to the separation of almost any type of charged molecule, from large proteins to small nucleotides and amino acids. It is very frequently used for proteins and peptides, under widely varying conditions. However, for amino acids standardized conditions are used. In protein structural work the consecutive use of gel-permeation chromatography (GPC) and IEC is quite common.

Ion-exchange chromatography is a type of chromatography consisting of a mobile phase and a stationary phase and exchanging ions between the phases depending on their charges. The stationary phase is an insoluble support matrix on which ions are electrostatically linked. The mobile phase contains the ions to be separated. These ions are reversibly exchanged with the ions bound to the immobile phase or the support matrix.

Ion-exchange chromatography separates protein molecules based on their electrical charge, which depends on the distribution of amino acids on the surface and the pH of the medium. The molecules to be separated and binded electrostatically to the stationary matrix are eluted from the column using a gradient solution of increasing ionic strength, since salts tend to disrupt electrostatic interactions. Alternatively, a pH gradient can be used for elution since if pH > pI the average charge is negative and if the pH < pI the average charge is positive. Ion exchangers containing **diethyl aminoethyl (DEAE)** or **carboxymethyl (CM)** groups are most frequently used in biochemistry. These groups are cross-linked to neutral polymers such as cellulose (Example: DEAE cellulose and CM cellulose).

Anion-exchange chromatography is used to separate negatively charged proteins (anions). The column has positively charged groups on the inert phase, which bind negative sites on the protein. The protein may then be exchanged by anions in the solution. Example: DEAE groups.

Cation-exchange chromatography is used to separate positively charged proteins (cations). The column contains a polyanionic matrix, usually sulfonate (—SO_3—) or carboxy (—COO—) groups such as carboxymethyl (CM) groups covalently linked to a cellulose or agarose matrix.

At a pH value below its isoelectric point, a protein (+ surface charge) will adsorb to a cation exchanger (–) such as one containing CM-groups. Above the isoelectric point, protein (– surface charge) will adsorb to an anion exchanger (+) (e.g., one containing DEAE-groups).

Electrophoresis

A great deal of knowledge can be obtained by observing the behavior of charged molecules traversing a uniform electric field. To obtain a uniform electric field with a constant magnitude and direction over a specified volume of space, two flat metal plates are set up parallel to each other as shown in Figure 6.4. When the terminals of a power source with voltage V are connected to these plates, as indicated in the diagram, a uniform electric field E is produced between the plates. Outside of the plates and near the ends, the field is not uniform.

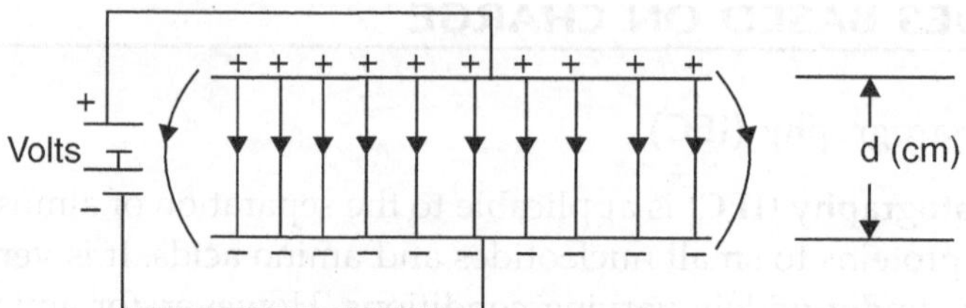

FIGURE 6.4 **Diagram representation of electrophoresis.**

Electrophoresis is the migration of charged molecules such as proteins in an electrical field. The separation of proteins in an electrical field is based on size, shape, and charge. The side chains of the amino acids, which make the protein molecules, contribute to the net electrical charges of the protein molecules. The charge of the protein depends on the isoelectric pH (IpH) of the protein and the pH of the surrounding buffer.

The mobility of the charged particles under the influence of electrical current depends on the net charge of the particles, their size, shape, and the electric field strength. Electrophoresis such as centrifugation is a hydrodynamic technique. A charged particle in an electrical field experiences a force that is proportional to the potential difference (E), or Voltage of the electrical field and inversely proportional to the distance (*d*) between the electrodes. The potential difference divided by the distance (E/*d*) is the field strength of electrophoresis. The force developed is also proportional to the net charge of the molecule (*q*). Therefore, the force experienced by the molecule can be expressed as

$$F = Eq/d$$

The mobility or velocity of a particle in an electrical field is proportional to the electrical field (E/*d*) or the applied voltage, net charge of the molecule, and inversely proportional to the frictional coefficient (i.e., size and shape of the molecule, and the viscosity of the medium) as indicated by the following equation.

Mobility = (applied voltage)Ï (net charge)/(friction coefficient)

Most electrophoretic methods use a supporting media, such as starch, paper, polyacrylamide, or agarose. It should be remembered that the actual environment through which the proteins migrate is composed of a 50% buffer solution. The term zone electrophoresis refers to electrophoresis, which is carried out in a supporting medium, whereas moving-boundary electrophoresis is carried out entirely in a liquid phase. When proteins are visualized on gels and the migration distances are compared to standards the isoelectric pH (Isoelectric focusing) and molecular weights (SDS-PAGE) of the various proteins can be measured. The isoelectric pH and molecular weights are useful in identifying and purifying proteins.

Electrophoresis is a widely used technique to analyze both proteins and nucleic acids. The charges of DNA molecules are always negative and therefore they migrate toward anode at a rate that is dependent on their molecular size. Small and compact DNA molecules move faster than the large and relaxed types of DNA molecules. The relaxed DNA molecules will move slower than the

DNA molecules having compact structure (supercoiling) even though they have the same molecular size because the compact molecules move faster through the pores of the gel matrix than the elongated relaxed molecules.

The net charge of a protein molecule depends on the relative proportion of positively charged and negatively charged amino acid at a particular pH. There are different types of matrices for separating charged particles. For example, there is paper electrophoresis used for the separation of a mixture of amino acids and other similar charged molecules. The matrix used for electrophoresis should be chemically neutral and should be stable. It should eliminate or decrease convection. It should not block the movement of the molecule by binding to the charged molecules. The best and most commonly used matrix is the polyacrylamide, which meets all the requirements.

The polyacrylamide is a polymer matrix synthesized out of monomers—acrylamide and N and N-methylene-bis acrylamide by a polymerization reaction induced by tetramethylenediamine (TEMED) carried out by ammonium persulphate (APS). The polymerization solution is a mixture containing appropriate quantities of acrylamide, N,N-methylene-bis acrylamide and tetramethylenediamine (TEMED), and ammonium persulphate. Once APS is added, it will activate the acrylamide molecules by releasing free radicals. The activated acrylamide molecules then react with successive acrylamide molecules to produce long polymers. These polymer chains of acrylamide are then cross-linked by bis-acrylamide to form a network of acrylamide chains having a specific pore size, which is controlled by the ratio of acrylamide and bis-acrylamide. PAGE is not generally used for the analysis of DNA molecules because of the large size of DNA molecules except for the separation of small nucleic acid molecules such as RNA or oligonucleotides in nucleic acid sequencing. Agarose gels are generally used for the electrophoresis of DNA because of the large pore size of the matrix.

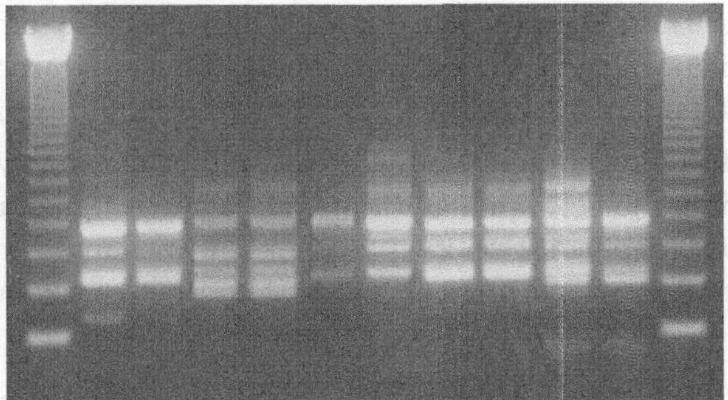

FIGURE 6.5 **Agarose gel electrophoresis of DNA stained with ethidium bromide and illuminated under UV light.**

There are two types of PAGE (polyacrylamide gel electrophoresis):

- Native PAGE
- SDS-PAGE (Denaturing PAGE)

Native PAGE is generally carried out to determine the molecular weight of a protein at its active state in tertiary or quaternary structure. To determine the number of subunits and the nature of the Three-dimensional structure (whether having one or more subunits) and the molecular weight of each subunit, the native gel experiment should be followed by a denaturing gel electrophoresis.

SDS-PAGE or denaturing PAGE is an electrophoretic method for separating protein subunits after they have been denatured by heating under reducing conditions and bound with the non-ionic detergent SDS. During denaturation by boiling, all disulfide bonds in the protein are reduced with 2-mercaptoethanol (sometimes called beta-mercaptoethanol) and the protein subunits (i.e., polypeptide chains) are uniformly bound with the detergent sodium dodecyl sulfate (SDS), which has the structure $CH_3(CH_2)_{11}$–SO_3–Na and the sodium is just a counter ion. The detergent gives the polypeptide a uniform negative charge and binds in proportion to size of the subunit. As a consequence, all the protein subunits in a mixture of proteins have the same charge density and will migrate in an electrical field with the same mobility. However, in an SDS-PAGE system, the pores in the PAGE gel are small enough to cause molecular sieving during electrophoresis so that the polypeptides coated with SDS separate by size. Thus, an SDS-PAGE gel can be calibrated with proteins of known subunit size (or molecular mass, which is called MR) and so the molecular mass of an unknown polypeptide can be determined. This is best done by preparing a plot of the log MR of the protein bands representing the standard proteins versus electrophoretic mobility of the protein bands as compared to the dye front (called relative mobility). The relative mobility of the unknown protein's subunit(s) can be compared to the standard curve and its MR (i.e., molecular size) is estimated.

Isoelectric Focusing

Protein molecules change their charge according to the pH. Such molecules are known as amphoteric molecules. There is a pH at which the net charge of the molecule attains zero. This pH is known as the **isoelectric point** or the **isoelectric pH (IpH).** Molecules differing in their isoelectric pH or Isoelectric point (Ip) can be separated by isoelectric focusing. **Isoelectric focusing (IEF)** is a method of separating proteins according to their isoelectric pH by carrying out electrophoresis in a gel containing a pH gradient. This pH gradient is established by the use of ampholytes. These are low molecular weight amphoteric molecules (a mixture of polyamino polycarboxilic acids) which, when present as a mixture, will migrate in an electric current to their pI, thus producing a pH gradient.

A protein applied to the gel will be either positively or negatively charged depending on the local pH at that point in the gel. Upon application of a current, the protein will move toward either the anode or cathode (depending on its charge) until it encounters that part of the gel, which corresponds to its pI (IpH) at which point it will have no net charge and will, therefore, stop migrating.

An IEF gel can be calibrated with proteins of known pI and a plot of position on the gel against pI allows an estimation of the pI of unknown proteins.

6.4 TECHNIQUES BASED ON POLARITY

Hydrophobic-interaction Chromatography (HIC)

Hydrophobic-interaction chromatography is a type of chromatography in which the molecules are separated according to their hydrophobic interaction between the stationary phase and mobile phase. Proteins and peptides differ from one another in their hydrophobic properties and this difference forms the basis of a **Hydrophobic-interaction Chromatography (HIC)** separation. Salt solutions are often used to mediate the binding of sample molecules to a hydrophilic matrix substituted with a hydrophobic ligand.

Protein molecules differ in their surface hydrophobicity according to the distribution of hydrophobic or non-polar amino acid residues. The hydrophobic residues are scattered over the surface of protein molecules, which control their interaction with other hydrophobic surfaces. In an aqueous medium, a film of water masks these hydrophobic areas of protein molecules and therefore their interaction with other hydrophobic surfaces will be minimum. These hydrophobic surfaces or groups of protein molecules can be exposed by the presence of high salt concentration, which disturbs the water film over the hydrophobic areas facilitating their interaction with other hydrophobic molecules or surfaces. The stationary phase in this chromatography is a matrix attached with suitable hydrophobic groups, which helps in the protein-matrix interaction and binding of proteins to the immobile phase (matrix) according to the strength of hydrophobicity. The most widely used stationary phase in HIC is the supporting matrix such as agarose attached with hydrophobic groups such as alkyls (hexyl , octyl) or phenyls. These bound proteins can be eluted out of the column by mobile phase of varying hydrophobicity or ionic strength, which interfere with the protein-matrix interaction.

Hydrophobic-interaction chromatography has evolved into one of the most powerful methods in preparative biochemistry. Its speed, resolution, and capacity rival ion-exchange chromatography; its selectivity is complementary to ion-exchange chromatography and size exclusion chromatography; and its ability to clear nucleic acids makes it an indispensable tool for the purification of therapeutic proteins.

Partition Chromatography

Chromatography is a technique that exploits differential interactions between the molecules to be separated in a solution, the mobile phase, with molecules immobilized on a solid support, the stationary phase. The mobile phase is normally allowed to pass over the stationary phase in a column. The different interactions will cause different molecules to be eluted at different times (Figure 6.6). Those molecules, which are adsorbed only weakly, are eluted first, whereas those, which interact most strongly with the column matrix, are eluted last. Sometimes it is necessary to change the composition of the mobile phase (buffer) during the elution process to remove the most strongly adsorbed molecules.

In **partition chromatography** the mixture of compounds to be separated is partitioned between the stationary phase and the mobile phase (both are liquids). The stationary phase is a liquid that is supported on a solid matrix either by covalent cross-linking or by hydrogen bonding or other minor molecular interactions. The mobile phase, which is also a solvent, moves over the stationary phase. Compounds that have more affinity for the mobile phase will be removed from the stationary phase and will be eluted out from the chromatography column. So the separation of the components in a mixture mainly depends on the degree of their affinity toward the liquid in the stationary phase or the mobile phase. This affinity is controlled by the polarity of the solvents present in the stationary phase and the mobile phase.

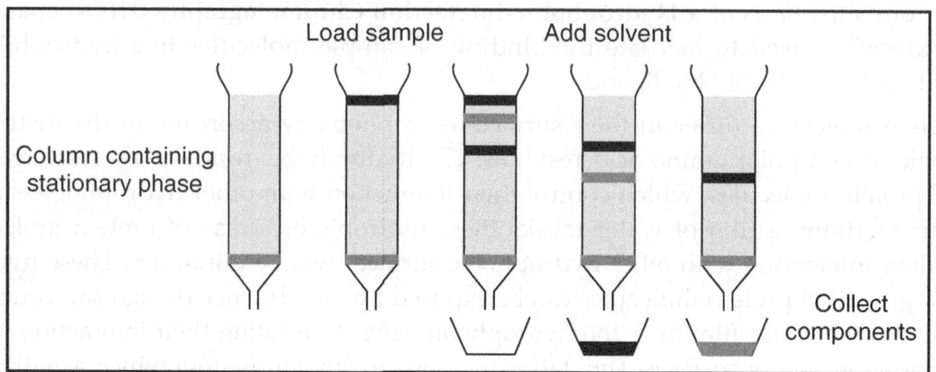

FIGURE 6.6 **A typical chromatography experiment.**

Partition chromatography can be of two types—**normal phase partition chromatography** and **reverse phase partition chromatography.** In normal phase partition chromatography the stationary phase is a polar solvent supported on a solid matrix. For example, in the case of paper chromatography and TLC, water is the stationary phase, which is supported on cellulose (in paper) and silica gel (TLC, thin layer chromatography), respectively. The mobile phase is a non-polar organic solvent or solvents of different polarity.

In the case of reverse phase partition chromatography, the stationary phase is a non-polar solvent, which is chemically attached to a porous support matrix. For example, solvents such as octadecylsilane are chemically bonded to a porous support matrix such as silica. The mobile phase can be a polar solvent starting from water, buffer or alcohol, acetonitrile, etc. Non-polar compounds will interact with the stationary phase effectively and so during elution by the polar solvent, those molecules, which are more polar, will be eluted rapidly and the molecules that are least polar will be eluted at the end. The polarity of the mobile phase or the eluting solvent can be controlled by changing its composition and this in turn controls the retention and separation of the component molecules of the mixture.

All types of liquid chromatography—both adsorption and partition chromatography—can be performed through **HPLC (high-performance liquid chromatography)**. Reverse phase HPLC can be used to separate a wide range of polar, non-polar, and ionic molecules such as proteins, sugars, and oligosaccharides, vitamins, peptides and proteins, amino acids, lipids, etc.

6.5 TECHNIQUES BASED ON SPECTROSCOPY

Spectroscopy is the use of the absorption, emission, or scattering of electromagnetic radiation by matter to qualitatively or quantitatively study the matter or to study physical processes. The matter can be atoms, molecules, atomic or molecular ions, or solids. The interaction of radiation with matter can cause redirection of the radiation and/or transitions between the energy levels of the atoms or molecules.

In the beginning, spectroscopic experiments and techniques were more confined to the use of visible light, particularly for identification of compounds and quantitative estimation. Now with new developments in the instrumentation, the use of spectroscopic techniques has been broadened tremendously. There are various types of spectroscopic techniques available to study the molecules and atoms, using different electromagnetic spectrums such as UV, IR, *x*-rays, microwaves, and even radio waves, and exploiting their specific properties.

Absorption: Molecules absorb a particular electromagnetic radiation of the spectrum which is followed by a transition of electrons from a lower energy level to a higher level with transfer of energy from the radiation to the absorber, atom, molecule, or solid.

Emission: Atoms or molecules absorbing electromagnetic radiation are excited to high energy levels and can decay to lower levels by emitting radiation (emission) in the form of heat or light. If no radiation is emitted, the transition from higher to lower energy levels is called **nonradiative decay.**

Scattering: Redirection of light due to its interaction with matter. Scattering may or may not occur with a transfer of energy (i.e., the scattered radiation might or might not have a slightly different wavelength compared to the light incident on the sample).

Electromagnetic Radiation

Electromagnetic radiation is a transverse energy wave that is composed of an oscillating electric field component, *E*, and an oscillating magnetic field component, *M*. The electric and magnetic fields are orthogonal to each other, and orthogonal to the direction of propagation of the wave (Figure 6.7). A wave is described by the wavelength, λ, which is the physical length of one complete oscillation, and the frequency, ν, which is the number of oscillations per second.

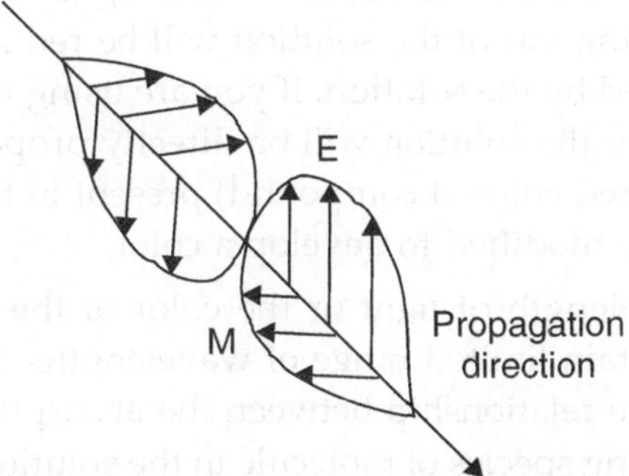

FIGURE 6.7 **Schematic representation of an electromagnetic wave.**

For convenience in talking about radiation, we divide electromagnetic radiation into different spectral regions. The radiation in all of these regions is still electromagnetic waves, but because of their very different energies they interact with matter very differently. For example, the human eye can only detect radiation that is in the visible region of the spectrum (hence the name), which is both transmitted by the lens of the human eye and absorbed by the photoreceptors in the retina. There is no fundamental difference in the nature of electromagnetic radiation of 350 nm versus 400 nm, other than we can see the 400 nm photons directly. Some of the boundaries between regions are not well defined.

Visible Spectrum

The visible region of the electromagnetic radiation is approximately 400 to 750 nm (Figure 6.8). The short wavelength cutoff is due to absorption by the lens of the eye and the long wavelength cutoff is due to the decrease in sensitivity of the photoreceptors in the retina for longer wavelengths. Light at wavelengths longer than 750 nm can be seen if the light source is intense.

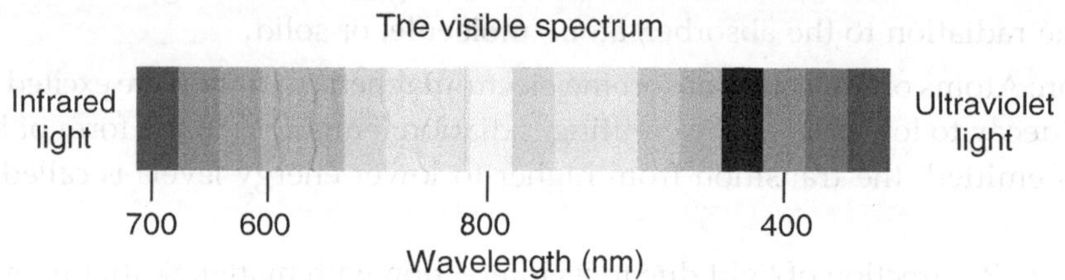

FIGURE 6.8 **The visible spectrum of electromagnetic radiation.**

The spectrophotometric and fluorometric techniques offer rapid and efficient methods of qualitative and quantitative estimation of biomolecules.

Colorimetry

Colorimetry is the interaction of light with colored solutions. The instrument used for this is known as colorimeter, which is the forerunner of the spectrophotometer. As light passes through a colored solution, some of the wavelengths of light will be absorbed by the solution. Wavelengths that are absorbed depend on the color of the solution. If a white light is allowed to pass through a red colored solution, the light coming out of the solution will be red in color. All other wavelengths, except the red color are absorbed by the solution. If you are using the complementary color of red, the amount of light absorbed by the solution will be directly proportional to the concentration of the light-absorbing molecules (red-colored compound) present in the solution. If the compound is not colored it can be chemically modified to develop a color.

In colorimetry the wavelength of light or the color of the light is selected using colored filters, which absorb only a certain limited range of wavelengths. This limited range is known as the bandwidth of the filter. The relationship between the absorption of light by the solution and the concentration of the absorbing species of molecule in the solution is based on the **Beer-Lambert Law**. This law is a combination of two separate laws. One relates the absorbance (intensity of light absorbed) with the concentration of the absorbing molecules and the second law relates the intensity of light absorbed to the path length of light or the thickness of the absorbing medium. The color filter in these experiments should be of the complementary color to that of the test solution.

A colorimeter, the instrument used for colorimetry, consists of a light source (tungsten bulb), the suitable color filter, the cuvette (the specially made transparent light insensitive tubes which hold the sample), and a photosensitive detector to monitor the transmitted light.

Beer-Lambert Law

The **Beer-Lambert Law (or Beer's Law)** is the linear relationship between absorbance and concentration of an absorbing species. The absorbance of a solution at a given wavelength is directly proportional to the path length or the thickness of the absorbing medium and the concentration of the absorbing species of the molecules in the solution (Figure 6.9). These two laws are combined into the Beer-Lambert law, which is usually written as:

$$A = a(\lambda) . l . c$$

where 'A' is the measured absorbance, '$a(\lambda)$' is a wavelength-dependent absorptivity coefficient, 'l' is the path length, and 'c' is the analyte concentration. When working in concentration units of molarity, the Beer-Lambert law is written as:

$$A = \varepsilon\, l\, c$$

where 'ε' is the wavelength-dependent molar absorptivity coefficient or molar extinction coefficient.

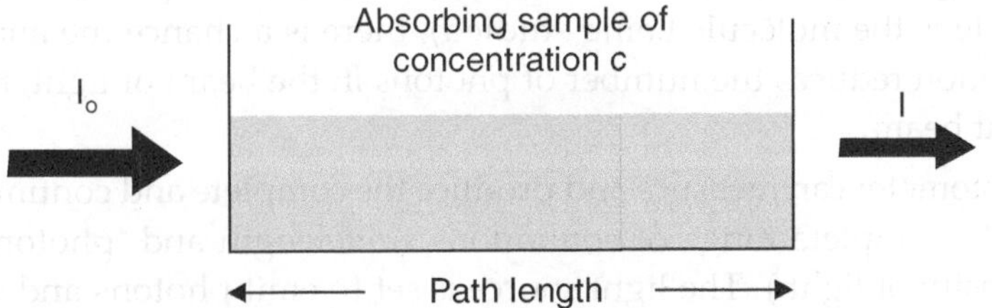

FIGURE 6.9 **Absorption of light by a sample.**

Experimental measurements are usually made in terms of transmittance (T), which is defined as:

$$T = I/I_o$$

where I is the light intensity after it passes through the sample and I_o is the initial light intensity. The relation between A and T is:

$$A = -\log T = -\log (I/I_o).$$

or

$$A = \log (I_o/I).$$

It is assumed that the incident light is parallel and monochromatic (light of single wavelength), and that the solute and solvent molecules in the solution are randomly oriented. The molar extinction constant (ε) varies according to the nature of absorbing compound, solvent, pH (if the light absorbing molecules are in equilibrium with an ionization state), and wavelength of the light. The maximum value for ε is 10^6, and hence at least 10^6 colored molecules are needed to absorb the color.

In Beer-Lambert law, $A = \varepsilon\, l\, c$

'ε' is a constant and if the path length 'l' also can be maintained constant, then the concentration of the solution will be directly proportional to the absorbance. Thus, for practical purposes, the path length is the standard for all spectrophotometric experiments. The Beer-Lambert law is applicable in all spectroscopic techniques, which involve the absorption and emission of electromagnetic radiations.

UV-Visible Spectrophotometry

The technique of spectrophotometry is generally used for the qualitative and quantitative estimation of biomolecules such as proteins, sugars, carbohydrates, amino acids, nucleic acids, vitamins, etc. This technique is also based on the Beer-Lambert law and the instrument is known as a spectrophotometer.

A **spectrophotometer** is employed to measure the amount of light that a sample absorbs. The instrument operates by passing a beam of light through a sample and measuring the intensity of light reaching a detector. The **UV-visible spectrophotometer** is used to measure the absorbance in the UV and visible regions of the spectrum. This instrument is an advanced form of colorimeter in which it can provide a monochromatic light. A prism or a grating will split the light into its component colors and can direct the monochromatic light of our choice to the sample solution to be analyzed.

The beam of light consists of a stream of photons. When a photon encounters an analyte molecule (the analyte is the molecule being studied), there is a chance the analyte will absorb the photon. This absorption reduces the number of photons in the beam of light, thereby reducing the intensity of the light beam.

A spectrophotometer can measure and produce the complete and continuous range of visible spectrum. ('spectro'—complete range of continuous wavelength and 'photometer'—a device for measuring the intensity of light.) The light source is set to emit photons and some of the photons are absorbed (removed) as the beam of light passes through the cell containing the sample solution. The intensity of the light reaching the detector is less than the intensity emitted by the light source. In a colorimeter, the bandwidths of the filters are very broad so it is not possible to get monochromatic light. Therefore, it is difficult to study two compounds of closely related absorption.

Compounds that can absorb ultraviolet light can be studied with a UV-visible spectrophotometer, which can produce the complete range of UV light also. In a UV-visible spectrophotometer, there are two source lamps, one a tungsten filament that provides wavelengths of visible range and a hydrogen or deuterium lamp that forms the source of light in the UV range.

In older instruments, a **single-beam mode** of operation was common. Here, a cell containing solvent alone was first placed in the light beam in order to measure any **background** absorption, then the absorbing solution was placed in the cell and the solute absorbance determined from the difference between the two measurements. To be accurate, this method required a highly stable radiation source to avoid errors due to source instability. Most modern spectrophotometers operate in a **double-beam mode**. Here, the source beam is split (by various means) into two components, one of which is passed through a **reference** cell containing solvent only and the other one is passed through a **sample** cell containing the absorber solution. Electronic comparison of the two beam intensities provides the absorbance.

Using a spectrophotometer it is possible to scan the entire wavelength range starting from the least of UV range to the maximum of visible range and provide the **absorption spectra** of the compound. Absorption spectra is the graphical representation of absorptions against its corresponding wavelength for a particular compound.

For the very common case of absorption measurements, the absorbing solution is held in a container called a **cell or cuvette** with optically flat faces held perpendicular to the radiation beam.

Fluorescence Spectroscopy (Fluorimetry)

Light emission from atoms or molecules can be used to quantitate the amount of the emitting substance in a sample. Atoms or molecules on absorbing electromagnetic radiation are excited to high energy levels and can decay to lower levels by emitting radiation (emission or luminescence). For atoms excited by a high-temperature energy source this light emission is commonly called **atomic** or **optical emission** and for atoms excited with light it is called **atomic fluorescence**. For molecules it is called fluorescence if the transition is between states of the same electron spin and phosphorescence if the transition occurs between electron states of different spin. The emission intensity of an emitting substance is linearly proportional to analyte concentration at low concentrations, and is useful for quantitating the emitting species of molecule.

Fluorescence spectroscopy can be used for the detection and estimation of non-fluorescence compounds also by coupling with a fluorescent probe called fluor. This type of fluorescence is called **extrinsic fluorescence**. In the case of intrinsic fluorescence, the native compound exhibits the phenomenon of fluorescence. Fluorescent probes are now extensively used for labeling biomolecules for their detection in amino acids and DNA sequencings. Detection and estimation of fluorescent-labeled molecules is 1,000 times more sensitive than other methods such as absorption spectroscopy. For example, fluorescent compounds such as dansyl chloride or o-phthalaldehyde can be used to detect amino acids, peptides, or proteins separated by chromatography or electrophoresis.

X-ray Crystallography

X-ray crystallography is a powerful technique used to study the three-dimensional structure of crystals including macromolecules such as protein and nucleic acids. The technique is also known as the **x-ray diffraction technique**. There are a number of methods for studying three-dimensional structure. But *x*-ray crystallography is the most effective of these techniques at present. Certainly the other techniques complement crystallography and have a valued place in the set of tools that we use for studying the structure of molecules.

If you think about how you determine the shape of objects around you, the most obvious is just to look at them. If they're small, you use a microscope. But there's a limit to how small an object can be seen under a light microscope (Figure 6.10). The limit (the "diffraction limit") is that you cannot image things that are much smaller than the wavelength of the light you are using. The wavelength for visible light is measured in hundreds of nanometers, while atoms are separated by distances of the order of 0.1nm, or 1Å. Looking at the electromagnetic spectrum, *x*-rays are most suitable to study and resolve the atomic arrangement in a molecule or crystal because they have the right wavelength range.

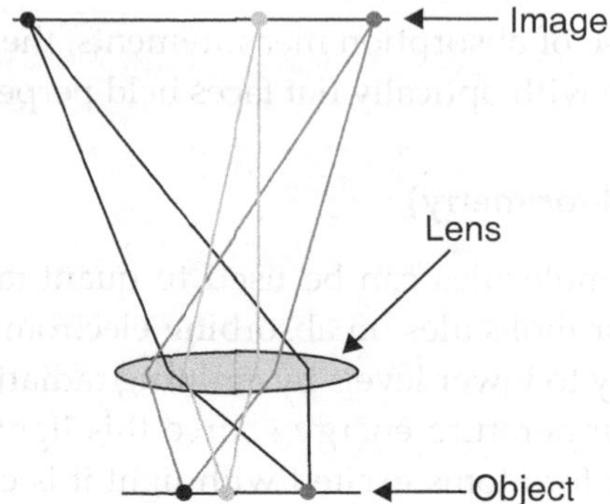

FIGURE 6.10 **Object seen under light microscope.**

But we can't build an *x*-ray microscope to look at molecules because we don't have an *x*-ray lens to converge and diverge the *x*-rays as in the case of an electron microscope and a light microscope. However, we can (in effect) simulate an *x*-ray lens on a computer. You can think of a microscope as working in two stages. First, light strikes the object and is diffracted in various directions. The lens collects the diffracted rays and reassembles them to form an image (Figure. 6.11). With *x*-rays, we can detect diffraction from molecules, but we have to use a computer to reassemble the image. It's not that simple, but that's the essence of the method.

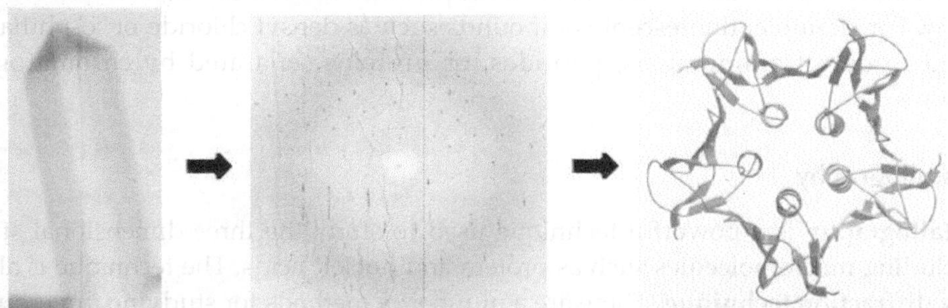

FIGURE 6.11 **Different steps in x-ray crystallography—crystal, x-ray diffraction pattern, and the final three-dimensional structure of protein.**

As noted above, the use of electromagnetic radiation to visualize objects requires the radiation to have a wavelength comparable to the smallest features that you wish to resolve. We often use *x*-rays emitted from copper targets bombarded with accelerated high-energy electrons, which emit at several characteristic wavelengths: the one we use in crystallographic studies has a wavelength of 1.5418 Å. This is very similar to the distance between bonded carbon atoms, so it is well suited to the study of molecular structure.

There are three basic requirements for *x*-ray crystallographic studies. They are an *x*-ray source, the crystal to be studied, and an *x*-ray detector. A narrow beam of *x*-ray of suitable wavelength is allowed to fall on a suitably mounted protein crystal and a diffraction pattern is produced on a photographic plate placed behind the crystal or a radiation counter detects the diffraction pattern. As the *x*-ray passes through the crystal, it is diffracted by the electrons in each atom present in the molecule. The numbers of electrons in the atom determine the intensity of the scattering of *x*-ray.

The intensity of *x*-rays scattered by carbon will be six times greater than that of a hydrogen atom. The scattered *x*-rays from individual atoms can reinforce each other or cancel each other out giving rise to the characteristic pattern for each type of molecule. A series of patterns taken from different angles contains the information needed to determine the three-dimensional structure. The information available in the *x*-ray diffraction pattern is extracted and converted into the image of the three-dimensional structure through a mathematical process known as **Fourier Transform**, which is a computer-aided program. Even though the process is computer-aided it involves a large number of lengthy mathematical calculations.

Finally, what we see as the result of a crystallographic experiment is not really a picture of the atoms, but a map of the distribution of electrons in the molecule (i.e., an electron density map). However, since the electrons are mostly tightly localized around the nuclei, the electron density map gives us a pretty good picture of the molecule.

X-ray scattering from a single molecule would be unimaginably weak and could never be detected above the noise level, which would include scattering from air and water. A crystal arranges huge numbers of molecules in the same orientation, so that scattered waves can add up in phase and raise the signal to a measurable level. In a sense, a crystal acts as an amplifier.

Mass Spectrometry

In **mass spectrometry**, a substance is bombarded with an electron beam having sufficient energy to fragment the molecule. The positive fragments, which are produced (cations and radical cations) are accelerated in a vacuum through a magnetic field and are sorted on the basis of mass-to-charge ratio. Since the bulk of the ions produced in the mass spectrometer carry a unit positive charge, the value m/e is equivalent to the molecular weight of the fragment. The analysis of mass spectroscopy information involves the reassembling of fragments, working backward to generate the original molecule. A schematic representation of a mass spectrometer is shown in (Figure 6.12).

A very low concentration of sample molecules is allowed to leak into the ionization chamber (which is under a very high vacuum) where they are bombarded by a high-energy electron beam. The molecules fragment and the positive ions produced are accelerated through a charged array

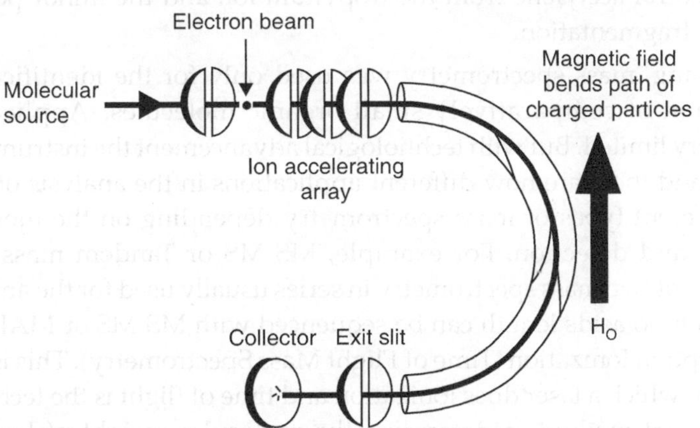

FIGURE 6.12 **Schematic representation of a mass spectrometer.**

into an analyzing tube. The path of the charged molecules is bent by an applied magnetic field. Ions having low mass (low momentum) will be deflected most by this field and will collide with the walls of the analyzer. Likewise, high momentum ions will not be deflected enough and will also collide with the analyzer wall. Ions having the proper mass-to-charge ratio, however, will follow the path of the analyzer and exit through the slit and collide with the collector. This generates an electrical current, which is then amplified and detected. By varying the strength of the magnetic field, the mass-to-charge ratio which is analyzed, can be continuously varied.

The output of the mass spectrometer shows a plot of relative intensity versus the mass-to-charge ratio (m/e). The most intense peak in the spectrum is termed the **base peak** and all others are reported relative to its intensity. The peaks themselves are typically very sharp, and are often simply represented as vertical lines.

The process of fragmentation follows simple and predictable chemical pathways and the ions, which are formed, will reflect the most stable cations and radical cations, which that molecule can form. The highest molecular weight peak observed in a spectrum will typically represent the parent molecule, minus an electron, and is termed the **molecular ion (M+)**. Generally, small peaks are also observed above the calculated molecular weight due to the natural isotopic abundance of ^{13}C, ^{2}H, etc. Many molecules with especially labile protons do not display molecular ions; an example of this is alcohol, where the highest molecular weight peak occurs at m/e one less than the molecular ion (m-1). Fragments can be identified by their mass-to-charge ratio, but it is often more informative to identify them by the mass, which has been lost. That is, loss of a methyl group will generate a peak at m-15; loss of an ethyl, m-29, etc.

As an example the mass spectrum of toluene (methyl benzene) is shown in Figure 6.13. The spectrum displays a strong molecular ion at $m/e = 92$, small m+1 and m+2 peaks, a base peak at $m/e = 91$, and an assortment of minor peaks $m/e = 65$ and below. The molecular ion, again, represents loss of an electron and the peaks above the molecular ion are due to isotopic abundance. The base peak in toluene is due to loss of a hydrogen atom to form the relatively stable benzyl cation. This is thought to undergo rearrangement to form the very stable tropylium cation, and this strong peak at $m/e = 91$ is a hallmark of compounds containing a benzyl unit. The minor peak at $m/e = 65$ represents loss of neutral acetylene from the tropylium ion and the minor peaks below this arise from more complex fragmentation.

In the beginning, mass spectrometry was used only for the identification and molecular weight determination of comparatively small organic molecules. Application in the case of biomolecules was very limited. But with technological advancement the instrumentation has become more sophisticated and there are now different applications in the analysis of protein and nucleic acids. There are different types of mass spectrometry depending on the mode of ionization, ion sorting, or analysis and detection. For example, MS-MS or Tandem mass spectrometry is an instrument consisting of two mass spectrometry in series usually used for the amino acid sequencing. Polypeptides of 50 amino acids length can be sequenced with MS-MS or MALDI-TOF MS (Matrix assisted Laser Desorption Ionization–Time of Flight Mass Spectrometry). This is another specialized mass spectrometry in which a laser does ionization and time of flight is the technique of ion sorting. This is another efficient method to determine the molecular weight of larger proteins and its identification by generating the **molecular fingerprint**. The molecular fingerprint is the pattern of ionization (MS spectrum) of a protein molecule, which is very specific for that molecule. The protein molecule can be identified by comparing its molecular fingerprint with that of known samples. **Electron spray ionization** or **ESI** is the common method of ionization for protein studies. Protein studies become more simple and rapid when separation techniques such as capillary electrophoresis and chromatographic techniques are combined with mass spectrometry. In such systems, the outlet of the chromatography column or capillary electrophoresis forms the inlet of mass spectrometry and the protein sample purified or fractionated can be directly delivered to the ionization chamber of mass spectrometry and identified or sequenced. This type of instrumentation and protein identification is known as **Multidimensional Protein Identification Technology** or **MuD-PIT.**

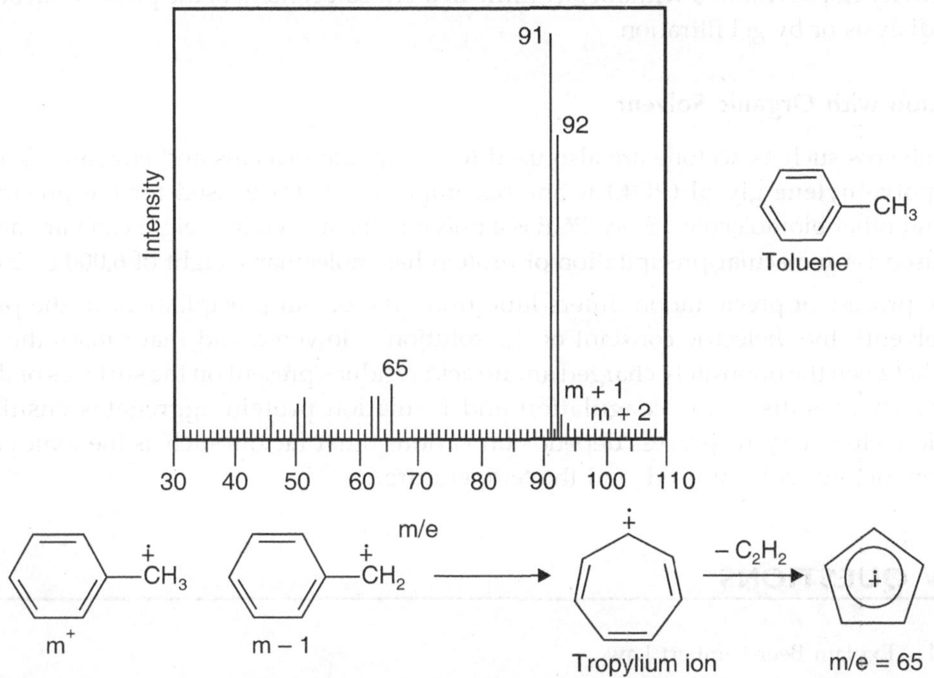

FIGURE 6.13 **Mass spectrum of toluene.**

6.6 TECHNIQUES BASED ON SOLUBILITY

Salt Precipitation

This is a very simple and widely used technique to isolate proteins and enzymes from other macromolecules such as carbohydrates and nucleic acids. The solubility of protein in an aqueous medium is dependent on [1] ionic strength (number of dissolved chemical species (i.e., salts ~6M $(NH_4)_2SO_4$)), [2] pH of the solution (a protein's minimum solubility is at the pI), [3] nature of the solvent, [4] other dissolved compounds (i.e., urea and [5] temperature). A change in one (or all) of these factors can cause a protein to "drop out" of solution. This is known as **denaturing.** Under gentle denaturing conditions, one can usually return the protein to full biological activity ("the native conformation") by restoring approximate *in vivo* conditions (via dialysis). Salt precipitation is one of the mild methods to denature protein leading to its precipitation without damaging the protein viability.

When salt is added to the extract, it dissolves to give ions that become hydrated by hydrogen bonding and it disturbs the interaction of water molecules with the protein molecules. This will expose the hydrophobic patches of protein and results in the hydrophobic interactions between protein molecules. This finally leads to the aggregation of protein molecules and their precipitation. The salt-precipitated proteins are not permanently denatured. They can be redissolved in a buffer and the activity can be restored without any difficulty. The salt content of the protein can be removed either by dialysis or by gel filtration.

Precipitation with Organic Solvent

Organic solvents such as acetone are also used to precipitate proteins and enzymes. The **organic polymer polyethylene glycol (PGE)** is another important reagent used for the precipitation of proteins and other biomolecules. Since PGE is a polymer its molecular weight can vary and the one which is used for molecular precipitation of protein has molecular weight of 6,000 or 20,000.

The process of precipitation differs little from that of salt precipitation. In the presence of organic solvents, the dielectric constant of the solution is lowered and that causes the increased attraction between the oppositely charged amino acid residues present on the surfaces of the protein molecules. This results in the coagulation and formation protein aggregates ensuing in the precipitation. However, the process depends on various other factors such as the ionic strength of the solution, organic solvent used, and the temperature.

REVIEW QUESTIONS

1. Explain Beer-Lambert Law.
2. Define spectroscopy. What are the different types of spectroscopy?
3. Name the techniques that can be used for separating molecules of different sizes.
4. What is the driving force in electrophoresis?
5. Addition of salts in an aqueous solution of protein causes the precipitation of protein. How?
6. What is the basis of gel permeation chromatography?
7. What is the use of high salt concentration in hydrophobic interaction chromato-graphy?
8. Distinguish between absorption chromatography and partition chromatography.
9. A sample of protein mixture having a pH 7.5 is given along with CM cellulose and DEAE cellulose. You are asked to separate a protein of isoelectric point 5.4 from the mixture by ion exchange chromatography. Which ion exchanger you will select, and why?
10. Describe the principle and use of *x*-ray crystallography.
11. What is meant by absorption spectroscopy? What are the applications in biochemistry?
12. Explain the principle of mass spectroscopy and its application in protein science.

13. Define the following:
    (*a*) Zonal centrifugation
    (*b*) Amphoteric molecules
    (*c*) Extrinsic fluorescence
    (*d*) Isoelectric pH
    (*e*) Reverse phase chromatography
    (*f*) Isoelectric focusing
    (*g*) Osmotic pressure
    (*h*) Isopycnic centrifugation
    (*i*) Ampholytes
14. Explain how osmotic pressure can be used for the determination of molecular weight of a solute.
15. Differentiate between a colorimeter and spectrophotometer.

Part 3 THE CELL AND DEVELOPMENT

A cell is the basic unit of life, the unit of structure and function for all organisms. Part 3 discusses the various aspects of the cell. There are three chapters in this part. In the first chapter we shall discuss the cell and its organization at different levels. We shall study the structure and function of the cell and its various organelles, organization of cell into tissues, and tissues into organs. We further discuss various organs and organ systems and how these organ systems are coordinated to make up a functional organism. A group of organisms forms a population, which is affected by a number of genetical and environmental factors. It is also discussed how various other factors such as adaptation and natural selection contribute to the evolution of population and biodiversity. The last part of the chapter is for discussing the interaction of organisms among themselves and with the ecosystem and how this can affect environmental and climatic conditions, and finally the biosphere.

The second chapter basically discusses various activities of cells. How cells multiply, cell division and cell cycles, how cells communicate with their environment and with other neighboring cells, cell-cell interaction, etc. Various physiological activities of cells such as their role in the movements, nutrition, internal transport, gaseous exchange, defense mechanisms, etc., are also discussed in detail. The last part of the chapter is mainly devoted to discussing the reproduction and development of plants and animals and the natural mechanism by which cells die.

The third chapter mainly deals with various techniques, traditional and modern, which are used for the detailed studies of the structure and functions of cells even at the molecular level. Techniques such as microscopy—both light microscopy and various other types such as electron microscopy, fluorescent microscopy, flow cytometry, cell fractionation techniques, measurement of cell growth etc.—are discussed in detail.

Part  THE CELL AND DEVELOPMENT

A cell is the basic unit of life, the unit of structure and function for all organisms. Part 3 discusses the various aspects of the cell. There are three chapters in this part. In the first chapter we shall discuss the cell and its organization at different levels. We shall study the structure and function of the cell and its various organelles, organization of cell into tissues, and tissues into organs. We further discuss various organs and organ systems and how these organ systems are coordinated to make up a functional organism. A group of organisms forms a population, which is affected by a number of genetical and environmental factors. It is also discussed how various other factors such as adaptation and natural selection contribute to the evolution of population and biodiversity. The last part of the chapter is for discussing the interaction of organisms among themselves and with the ecosystem and how this can affect environmental and climatic conditions, and finally the biosphere.

The second chapter basically discusses various activities of cells. How cells multiply, cell division and cell cycles, how cells communicate with their environment and with other neighboring cells, cell-cell interaction, etc. Various physiological activities of cells such as their role in the movements, nutrition, internal transport, gaseous exchange, defense mechanisms, etc., are also discussed in detail. The last part of the chapter is mainly devoted to discussing the reproduction and development of plants and animals and the natural mechanism by which cells die.

The third chapter mainly deals with various techniques, traditional and modern, which are used for the detailed studies of the structure and functions of cells even at the molecular level. Techniques such as microscopy—both light microscopy and various other types such as electron microscopy, fluorescent microscopy, flow cytometry, cell fractionation techniques, measurement of cell growth etc.—are discussed in detail.

Chapter 7 THE BASIC UNIT OF LIFE

In This Chapter

7.1 CELL STRUCTURE AND COMPONENTS

Cells are structural and functional units of life. According to the cell theory all living things are composed of one or more cells. One-celled organisms are called **unicellular** organisms and those with more than one cell are called **multicellular** organisms. Virus particles do not have any cells and therefore, are termed as **acellular**. No matter what type of cell we are considering, all cells have certain features in common: **cell membrane, nucleic acids, cytoplasm, and ribosomes.** Cells are small 'sacks' composed mostly of water. The 'sacks' are made from a phospholipid bilayer. The membrane is semi-permeable, allowing some things to pass in or out of the cell and blocking others. Microscopes make it possible to magnify small objects such as cells in order to see the details of their structure. Both light and electron microscopes are used to study cells. Study of cells with a microscope is called **cytology**. There are some fundamental activities which are common for most of cell types from bacteria to the nerve cells in humans. The study of these basic cellular processes is called **cell biology**.

Cells are 90% fluid (cytoplasm), which consists of free amino acids proteins, glucose, and numerous other molecules. The cell environment (i.e., the contents of the cytoplasm, and the nucleus, as well as the way the DNA is packed) affects the gene expression/regulations, and thus is very important part of inheritance.

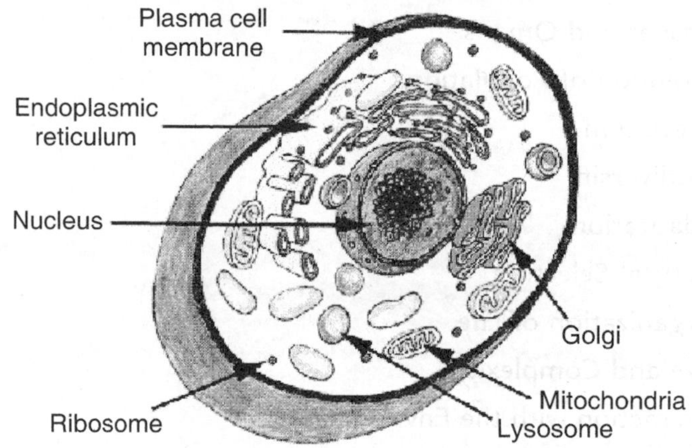

FIGURE 7.1 **Structure of a typical animal cell.**

Cells basically fall into two groups: **prokaryotic cells** and **eukaryotic cells.**

Prokaryotes, Eukaryotes, and Viruses

Cells are classified as prokaryotes or eukaryotes based on their basic structure and the way by which they obtain energy. Cells are also classified according to their need for energy. Autotrophs are "self feeders" that use light or chemical energy to make food. Plants are an example of autotrophs. In contrast, heterotrophs ("other feeders") obtain energy from other autotrophs or heterotrophs. Many bacteria and animals are heterotrophs.

Characteristics of Prokaryotic Cells

Prokaryotes include bacteria and blue-green algae (cyanobacteria). Simply stated, prokaryotes are molecules surrounded by a membrane and cell wall. Prokaryotic cells lack characteristic eukaryotic subcellular membrane-enclosed "organelles," but may contain membrane systems inside a cell wall as an extension or infoldings of the cell membrane. The nucleus is not well-organized and is without any membrane.

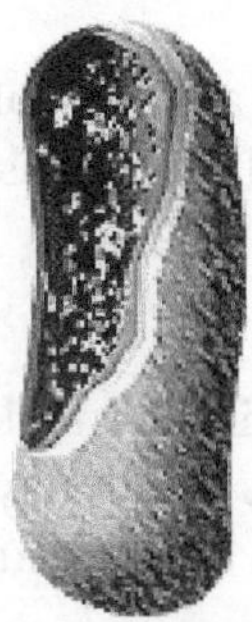

FIGURE 7.2 **A bacteria cell.**

Prokaryotic cells may have photosynthetic pigments, such as is found in cyanobacteria ("blue-green algae"). Some prokaryotic cells have external whip-like flagella for locomotion or hair-like pili for adhesion. Prokaryotic cells come in multiple shapes: cocci (round), baccilli (rods), and spirilla or spirochetes (helical cells). All prokaryotes are unicellular organisms and eukaryotes include both unicellular and multicellular organisms.

Multicellular organisms are created from a complex organization of cooperating cells. There must be new mechanisms for cell-to-cell communication and regulation. There also must be unique mechanisms for a single fertilized egg to develop into all the different kinds of tissues of the body. In humans, there are 10^{14} cells comprising 200 kinds of tissues!

Eukaryotes

Basic Structure

The basic eukaryotic cell contains the following:

- Plasma membrane
- Nucleus
- Cytoplasm (semifluid)
- Cytoskeleton—microfilaments and microtubules that suspend organelles, give shape, and allow motion
- Presence of characteristic membrane enclosed subcellular organelles.

Characteristic Biomembranes and Organelles

Plasma Membrane

A lipid-protein-carbohydrate complex, providing a barrier and containing transport and signaling systems.

Nucleus

Double membrane surrounding the chromosomes and the nucleolus. Pores allow specific communication with the cytoplasm. The nucleolus is a site for synthesis of RNA making up the ribosome.

Mitochondria

Surrounded by a double membrane with a series of folds called cristae. Functions in energy production through cellular respiration and metabolism. Contains its own DNA, and is believed to have originated as a captured bacterium.

Chloroplasts (plastids)

Surrounded by a double membrane, containing stacked thylakoid membranes. Responsible for photosynthesis, the trapping of light energy for the synthesis of sugars. Contains DNA, and like mitochondria, is believed to have originated as a captured blue-green algae. They are present only in plants.

Rough endoplasmic reticulum (RER)

A network of interconnected membranes forming channels within the cell. Covered with ribosomes (causing the "rough" appearance), which are in the process of synthesizing proteins for secretion or localization in membranes.

Smooth endoplasmic reticulum (SER)

A network of interconnected membranes forming channels within the cell. A site for synthesis and metabolism of lipids. Also contains enzymes for detoxifying chemicals including drugs and pesticides.

Ribosomes

Protein and RNA complex responsible for protein synthesis.

Golgi apparatus

A series of stacked membranes. Vesicles (small membrane surrounded by bags) carry materials from the RER to the Golgi apparatus. Vesicles move between the stacks while the proteins are "processed" to a mature form. Vesicles then carry newly formed membrane and secreted proteins to their final destinations including secretion or membrane localization.

Lysosymes

A membrane-bound organelle that is responsible for degrading proteins and membranes in the cell, and also helps degrade materials ingested by the cell.

Vacuoles

Membrane surrounded "bags" that contain water and storage materials in plants.

Peroxisomes or Microbodies

Produce and degrade hydrogen peroxide, a toxic compound that can be produced during metabolism.

Cell wall

Plants have a rigid cell wall in addition to their cell membranes. It is absent in animals.

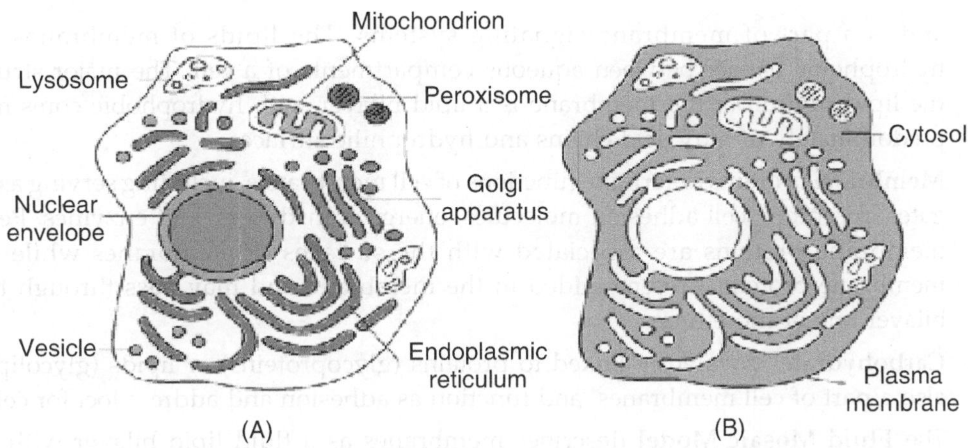

FIGURE 7.3 **(A) membrane-bounded compartments that exist within eukaryotic cells, each specialized to perform a different function. (B) The rest of the cell, excluding all these organelles, is the cytosol, and is the site of many vital cellular activities.**

Viruses

Basic Characteristics of Viruses

Simply stated, viruses are merely genetic information surrounded by a protein coat. They may contain external structures and a membrane. Viruses are obligate intracellular parasites meaning that they require host cells to reproduce. In the viral life cycle, a virus infects a cell, allowing the viral genetic information to direct the synthesis of new virus particles *by the cell*. There are many kinds of viruses. Those infecting humans include polio, influenza, herpes, smallpox, chickenpox, and human immunodeficiency virus (HIV) causing AIDS.

Cell Membrane

1. Cell membranes are selective barriers that separate individual cells and cellular compartments.
2. Membranes are assemblies of **carbohydrates**, **proteins**, and **lipids** held together by non-covalent forces. They regulate the transport of molecules, control information flow between cells, generate signals to alter cell behavior, contain molecules responsible for cell adhesion in the formation of tissues, and can separate charged molecules for cell signaling and energy generation.
3. Cell membranes are dynamic, constantly being formed and degraded. Membrane vesicles move between cell organelles and the cell surface. Inability to degrade membrane components can lead to **lysosomal storage diseases.**
4. **Lipids** of cell membranes include phospholipids composed of glycerol, fatty acids, phosphates, and a hydrophobic organic derivative such as choline or phosphoinositol. Cholesterol is a lipid component of cell membranes that regulates membrane fluidity

and is a part of membrane signaling systems. The lipids of membranes create a hydrophobic barrier between aqueous compartments of a cell. The major structure of the lipid portion of the membrane is a lipid bilayer with hydrophobic cores made up predominately of fatty acid chains and hydrophilic surfaces.

5. **Membrane proteins** determine functions of cell membranes, including serving as pumps, gates, receptors, cell adhesion molecules, energy transducers, and enzymes. Peripheral membrane proteins are associated with the surfaces of membranes while integral membrane proteins are embedded in the membrane and may pass through the lipid bilayer one or more times.
6. **Carbohydrates** covalently linked to proteins (glycoproteins) or lipids (glycolipids) are also a part of cell membranes, and function as adhesion and address loci for cells.
7. The **Fluid Mosaic Model** describes membranes as a fluid lipid bilayer with floating proteins and carbohydrates (see Appendix).
8. **Cell junctions** are a special set of proteins that anchor cells together (desmosomes), occlude water passing between cells (tight junctions), and allow cell-to-cell direct communication (gap junctions).

Cell Organelles

Cells contain a variety of subcellular particles specialized for carrying out a specific function. Most of them are membrane-enclosed structures. The material that is seen within the cell enclosed by the cell membrane can be divided into a central concentrated structure known as the nucleus and the surrounding semifluid, the cytoplasm. Both the cytoplasm and nucleus together form the protoplasm. The protoplast includes the cell membrane also.

Nucleus

The **nucleus** is the hallmark of eukaryotic cells; the very term eukaryotic means having a "true nucleus." It is the largest of the organelles, and is the location of the majority of different types of nucleic acids. It has got an important role in controlling the shape and features of the cell. Deoxyribonucleic acid (DNA) is the physical carrier of inheritance and with the exception of plastid DNA and mitochondrial DNA (cpDNA and mDNA), all DNA is restricted to the nucleus. Ribonucleic acid (RNA) is formed in the nucleus by coding off of the DNA bases. RNA moves out into the cytoplasm. The nucleolus is an area of the nucleus (usually two nucleoli per nucleus) where ribosomes are constructed.

The nucleus is enveloped in a pair of membranes enclosing a lumen that is continuous with the endoplasmic reticulum. However, the nuclear envelope is perforated by thousands of **nuclear pore complex (NPCs)** that control the passage of molecules in and out of the nucleus. The nuclear pore is formed by the fusion of outer and inner membrane. The outer membrane is the continuation of endoplasmic reticulum. The term nucleoplasm is still used to describe the contents of the nucleus.

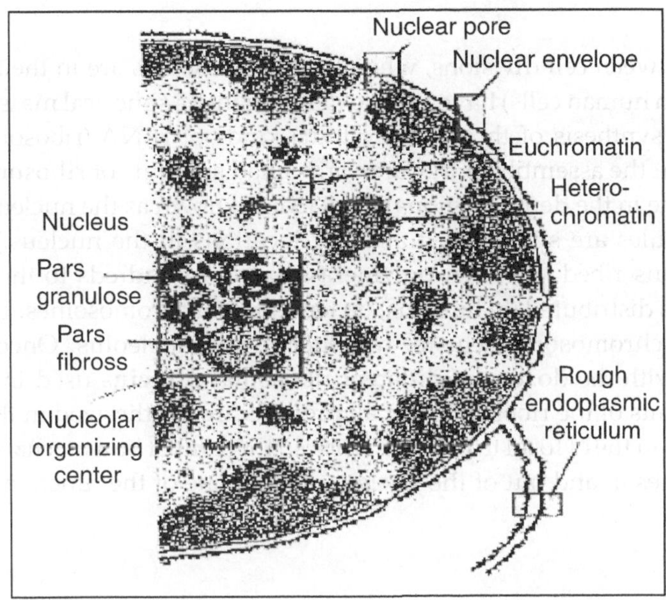

FIGURE 7.4 **Electron micrograph showing the structure of the nucleus.**

Chromatin

The nucleus contains the chromosomes of the cell. Each chromosome consists of a single molecule of DNA complexed with an equal mass of proteins. Collectively, the DNA of the nucleus with its associated proteins is called **chromatin**. Most of the protein consists of multiple copies of five kinds of **histones**. These are basic proteins, bristling with positively charged arginine and lysine residues. (Both Arg and Lys have a free amino group on their R group, which attracts protons (H^+) giving them a positive charge.) Chromatin also contains small amounts of a wide variety of **non-histone proteins**. Most of these are transcription factors. (e.g., the steroid receptors) and their association with the DNA is more transient.

During cell division chromatin becomes more condensed and becomes **chromosomes** of specific shapes. During the interphase, the chromatin becomes dispersed and becomes very light and cannot be visible very easily (except for special cases like the polytene chromosomes of drosophila and some other flies). However, the density of the chromatin (that is, how tightly it is packed) varies throughout the nucleus:

Dense regions are called **heterochromatin.**

Less dense regions are called **euchromatin.**

Heterochromatin is found in parts of the chromosome where there are few or no genes, such as centromeres, telomeres, transposons, and other "junk." DNA is densely packed. Those genes present in heterochromatin are generally inactive; that is, not transcribed.

Euchromatin is found in parts of the chromosome that are active in gene transcription; it is loosely packed in loops of 30-nm fibers. These are separated from adjacent heterochromatin by insulators. The loops are often found near the nuclear pore complexes (which makes sense for the gene transcripts to get to the cytosol).

Nucleolus

During the period between cell divisions, when the chromosomes are in their extended state, one or more of them (ten in human cells) have loops extending into a spherical mass called the nucleolus. They are the sites of synthesis of three (of the four) kinds of rRNA (ribosomal RNA) molecules (28S, 18S, 5.8S) used in the assembly of the large and small subunits of **ribosomes**. The synthesis of these rRNA takes place in the denser regions of nucleolus known as the nucleolar organizer region. (The 5S rRNA molecules are synthesized at other locations in the nucleus.). 28S, 18S, and 5.8S ribosomal RNA is transcribed (by RNA polymerase I) from hundreds to thousands of tandemly arranged **rDNA genes** distributed (in humans) on ten different chromosomes. The rDNA-containing regions of these ten chromosomes cluster together in the nucleolus. Once formed, the rRNA molecules associate with the dozens of different ribosomal **proteins** used in the assembly of the large and small subunits of the ribosome. But all proteins are synthesized in the cytosol and all the ribosomes are needed in the cytosol to do their work so there must be a mechanism for the transport of these large structures in and out of the nucleus. This is one of the functions of the nuclear pore complexes.

Nucleosomes

Two copies of each of four kinds of histones—H2A, H2B, H3, and H4 form a core of protein, the **nucleosome core**. Around this is wrapped 147 base pairs of DNA to form the structures called **Nucleosomes on DNA** chain just like beads of a chain.

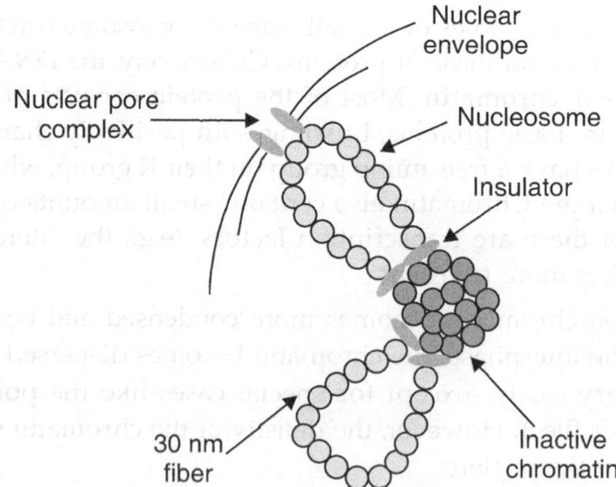

FIGURE 7.5 **Nucleosomes in heterochromatin and euchromatin regions.**

Cytoplasm

The cytoplasm was defined earlier as the material between the plasma membrane (cell membrane) and the nuclear envelope. Fibrous proteins that occur in the cytoplasm, referred to as the cytoskeleton, maintain the shape of the cell as well as anchoring organelles, moving the cell and controlling internal movement of structures. Microtubules function in cell division and serve as a "temporary scaffolding" for other organelles. Actin filaments are thin threads that function in cell division and cell motility. Intermediate filaments are between the size of the microtubules and the actin filaments.

Endoplasmic Reticulum

Endoplasmic reticulum is a network of tubules, vesicles, and sacs that are interconnected. They may serve specialized functions in the cell including protein synthesis, sequestration of calcium, production of steroids, storage and production of glycogen, and insertion of membrane proteins. There are two types of **endoplasmic reticulum (ER): rough endoplasmic reticulum,** which gets its name from the presence of ribosomes on its surface during protein synthesis and **smooth ER** without any ribosomes. Rough endoplasmic reticulum may either be vesicular or tubular. Or it may consist of stacks of flattened cisternae (like sheets) that may have bridging areas connecting the individual sheets. The ribosomes sit on the outer surfaces of the sacs (or cisternae). They resemble small beads sitting in rosettes or in a linear pattern.

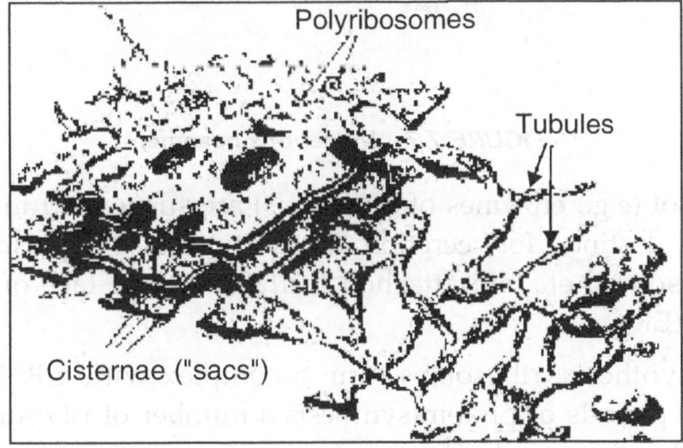

FIGURE 7.6 **Rough endoplasmic reticulum.**

Vacuoles and Vesicles

Vacuoles are single-membrane organelles that are essentially part of the outside that is located within the cell. The single membrane in plant cells is known as a tonoplast. Many organisms will use vacuoles as storage areas. Vesicles are much smaller than vacuoles and function in transport within and to the outside of the cell.

Ribosomes

Ribosomes are protein-synthesizing machines of the cell. They translate the information encoded in **messenger RNA (mRNA)** into a protein. They are not membrane-bound and thus occur in both prokaryotes and eukaryotes. Eukaryotic ribosomes are slightly larger than prokaryotic ones. The ribosomes of the prokaryotes are 80S and those found in bacteria and eukaryotic organelles like mitochondria and chloroplasts are 70S. The presence of bacterial (prokaryotic) ribosomes (70S) in chloroplasts and mitochondria shows the prokaryotic origin of these organelles.

Despite these differences, the basic structure and operations of bacterial and eukaryotic ribosomes are very similar. Structurally, the ribosome consists of a small and larger subunit. These two units join together when the ribosome attaches to messenger RNA to produce a protein in the cytoplasm. Biochemically the ribosomes are ribonucleoprotein complexes consisting of **ribosomal RNA (rRNA)** and some 50 structural proteins. In eukaryotes, ribosomes that synthesize proteins

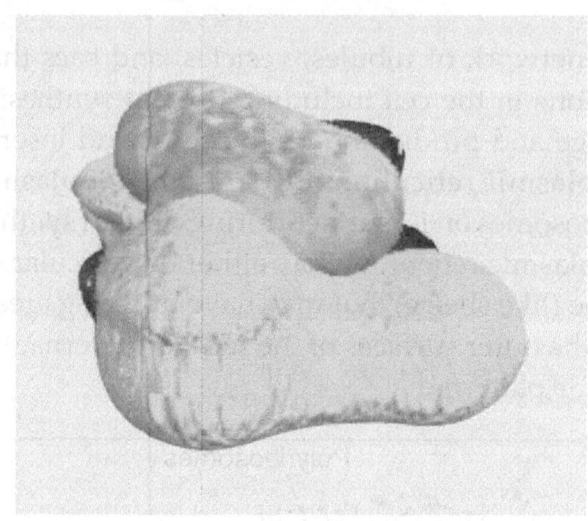

FIGURE 7.7 **Model of ribosome.**

for use within the cytosol (e.g., enzymes of glycolysis) are suspended in the cytosol. Ribosomes that synthesize proteins destined for secretion (by exocytosis), the plasma membrane (e.g., cell surface receptors), lysosomes, etc. are attached to the cytosolic face of the membranes of the endoplasmic reticulum (ER).

During protein synthesis, ribosomes form a complex with mRNA and move along the molecule. During active periods of protein synthesis a number of ribosomes attach to the rRNA and such structures are known as **polyribosomes**.

Golgi Complex

Golgi Complex is another important membrane-bound organelle present in eukaryotic cells, both in plants and animals. It is usually present in association with or near ER. The Golgi complex consists of a pile of flattened membrane-bound sacs called **cisternae**, which are formed by the fusion of vesicles. The vesicles are formed from the ER. The Golgi complex has an outer convex surface and an inner concave surface. The vesicles from the ER fuse together and form cisternae on the outer or convex surface and vesicles bud off constantly from the lower or concave side. Thus, an entire stack consists of a number of cisternae, which are moving from the outer convex to the inner concave face of the Golgi complex.

The nascent protein transported from ER in the form of vesicles undergoes post-synthetic modifications inside the cisternae of the Golgi complex. One of the important modifications on protein molecules is the addition of carbohydrate moiety to the proteins, a process known as **glycozilation**. Glycoproteins function as the receptors on the cell surfaces along with phospholipids of the membranes. Another important function of the Golgi complex is the production of **lysosomes**. Lysosomes are the vesicles containing hydrolytic enzymes.

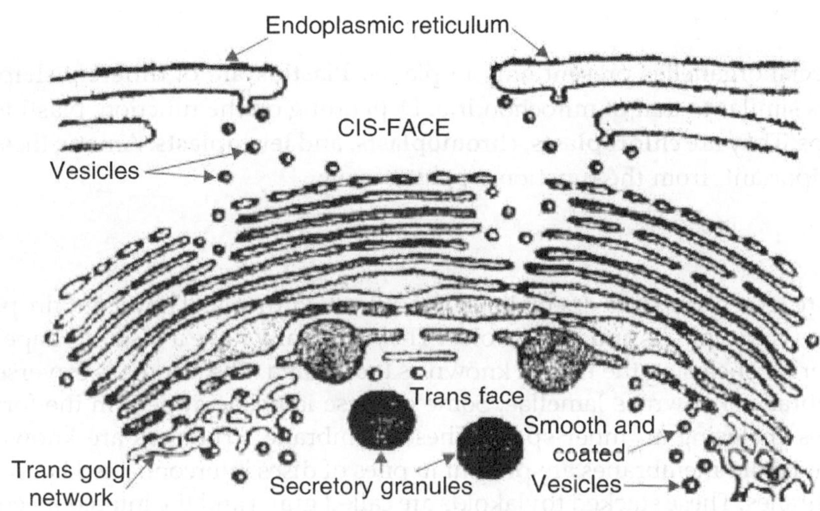

FIGURE 7.8 **Golgi complex showing the interphase with ER.**

Lysosomes

These are single, membrane-bound vesicles present in eukaryotic cells containing hydrolytic enzymes such as nucleases, proteases, and lipases. These enzymes are synthesized in ER and are transported to the Golgi complex through their cisternae and vesicles. From the Golgi complex these are removed as vesicles known as **primary lysosomes**. The lysosomes are the defense units of cells and in unicellular eukaryotic organisms such as protozoans they form the digestive enzymes for digesting the food materials engulfed by the organisms.

Lysosomes fuse with the phagocytic vesicle or the phagosomes (the material engulfed by the cells) to form phagolysosomes. The enzymes of the lysosomes will digest the food material or the invading organisms present within the phagocytic vesicle. Therefore, lysosomes are abundant in those cells which are specialized for phagocytosis.

Mitochondria

Mitochondria are another very important membrane-bound organelle present in both animal and plant cells. They are short cylindrical structures. There are two membranes covering this organelle, an outer membrane and an inner membrane with a space in between. The inner membrane has a number of infoldings into the matrix known as cristae. The matrix of mitochondria contains 70S ribosomes and a circular DNA in addition to a large number of enzymes. On the surfaces of the cristae there are a large number of granular structures known as **oxysomes.**

The main function of mitochondria is that they are the centers of power generation inside the cells. The final stage of respiration, the cellular respiration, takes place in mitochondria. Cellular respiration or the oxidative breakdown of glucose occurs in three steps – glycolysis, Kreb's cycle and electron transport system and oxidative phosphorylation (synthesis of ATP). The site of glycolysis is outside the mitochondria in the cytoplasm. The other two steps take place inside the mitochondria. The site of Krebs cycle is in the matrix and that of ETS is on the inner side of the cristae.

Plastids

Plastids are special organelles present only in plants. **Plastids** are of different shapes covered by two membranes similar to that of mitochondria. Depending on the function, plastids are classified into three groups. They are **chloroplasts, chromoplasts,** and **leucoplasts**. Among these, chloroplasts are the most important from the functional point of view.

Chloroplasts

Chloroplasts are green plastids specialized for photosynthesis. They contain photosynthetic pigments such as chlorophylls and carotenoids. They generally have a discoid shape with a double membrane covering enclosing the matrix known as the stroma. The stroma is traversed by a system of double membranes known as lamellae. Some of these lamellae appear in the form of flattened and discoid sacs enclosing an inner space. These membrane structures are known as thylakoid membranes. Thylakoid membranes are present in piles of discs interconnected by the normal types of double membranes. These stacked thylakoids are called grana and the interconnecting membrane is called grana lamellae. There are membranes without any grana traversing the stroma and they are known as stroma lamellae. The photosynthetic pigments are present on the thylakoid membranes as pigment protein complexes. They are organized into light-harvesting systems known as pigment systems I and II (PSI and PSII). Thylakoid membranes are the centers of light reaction, where the light energy is converted into chemical energy. The components required for the light harvesting and its conversion into the reducing powers of $NADPH_2$ and ATP are distributed on the thylakoid membrane in a very organized manner.

The second reaction of the photosynthesis—the dark reaction or the Calvin cycle—occurs in the matrix or the stroma of the chloroplast. The **stroma** is a mixture containing all the required enzymes and other compounds including the intermediates of the Calvin cycle essential for operating the absorption and reduction of carbon dioxide to carbohydrate. The stroma also contains a circular DNA molecule and 70S ribosomes needed for protein synthesis. The presence of these things indicates that chloroplasts like mitochondria are semi-autonomous organelles.

The presence of a circular DNA molecule and 70S ribosomes and the absence of membrane-bound organelles and the presence of membrane infoldings postulate the prokaryotic origin of chloroplasts and mitochondria. It is believed that chloroplasts evolved from a symbiotic type of blue-green algae (cyanobacteria) and mitochondria might have evolved from a symbiotic type of bacteria.

Chromoplasts

Chromoplasts are non-photosynthetic colored plastids. They do not have non-photosynthetic pigments. They contain red, orange, or yellow pigments and are present in the colored parts of the plant such as flowers, fruits, etc.

Leucoplasts

These plastids do not contain any pigments and therefore are colorless. These plastids sometimes form the precursors of chloroplasts or change into structures for food storage. Depending on the type of stored food they are further classified into amyloplasts, lipidoplasts, and proteoplasts.

Amyloplasts contain starch and are also known as starch grains, which are present in the storage regions of the plants such as tubers and grains. Lipidoplasts store lipids and proteoplasts or aleuron grains store proteins. These are also present in the storage parts of plants such as endosperm of seeds and pulses.

Peroxisomes

These are small, single membrane-bound granular structures seen in eukaryotic cells. They are small lysosomes containing the enzyme catalase or peroxidase. These structures are involved in a number of metabolic reactions, which involve oxidation. Peroxidase enzymes are needed for the oxidation of peroxides formed in the cells as byproducts of other metabolic reactions.

Cytoskeleton

This is a network of fibers made up of fibrous proteins present inside the cells, which gives mechanical support and shape to the cells. In addition to this, there are some other specific functions for which it is modified. The fibers of cytoskeleton are of two types: microtubules and microfilaments. The other functions of these fibers of cytoskeleton include movements of cells, as the components of cilia and flagella, and the movements of organelles and components within the cells. For example, movements of chromosomes by spindle fibers during cell division.

Microtubules

Microtubules are present in spindle fibers, cilia, and flagella. These fibers are hollow cylindrical fibers consisting of proteins tubulin. In flagella, cilia, and spindle the microtubules undergo sliding movements. They are responsible for the movement of chromosomes attached to the spindle fibers during cell division and for the rhythmic movements of cilia and flagella. They are also responsible for the movements of other organelles such as lysosomes, Golgi complex, etc. In addition to all these they give mechanical support to the cells, particularly in the case of animal cells, which do not have a rigid cell wall.

Microfilaments

These filaments are present in bundles or sheets just below the cell surface membranes. These are very thin filaments and are composed of fibrous protein actin. They are supposed to be involved in the phagocytosis or endocytosis and exocytosis. In addition to this they are also involved in the movements of cells and cellular parts.

7.2 TISSUES AND ORGANS

In multicellular organisms cells cooperate for the well-being of the whole body. Groups of cells differentiate and emphasize certain functions while losing others. Basic categories of differentiated cells include **secretory** cells in both plants and animals with highly developed powers of secretion, for example, of enzymes, hormones (in animals), honey, and mucilage (in plants); **fat** cells for storage of fat; **muscle** cells which are highly contractile; **nerve** cells with high irritability; and **reproductive** cells that produce gametes. Each kind of cell has reduced or given up some of the functions of the other kinds.

The cells with the same function and structure are arranged together to form tissues. Each tissue carries out a specific function for which the cells are specialized. Different types of tissues are organized together and form specific structures called organs, which cooperate with other similar organs to carry out specific functions of the body. In animals, different organs cooperate to form a system which carries out a specific function of the body. For example, the circulatory system, digestive system, nervous system, excretory system, etc.

Animal Tissues

In animals, there are four basic types of tissues: **epithelial** or linings, **connective** or supporting, **muscular**, and **nervous**. An organ of the body may have all the four types of tissues. For example, the stomach, an organ of the digestive system has all the four types of tissues. (See Appendix)

Epithelial Tissue

The cells are arranged in single or multilayered sheets. They basically form the covering on the external and internal surfaces of the organs and body parts. Epithelial cells are not supplied with blood vessels. They protect the internal tissues from physical injury and infection. The free surface of the epithelial tissue may be of different types depending on its special function such as secretory, absorption, or excretory functions. Epithelial cells are basically classified according to their shapes.

There are three basic cell shapes in epithelial tissues: **columnar**, **cubical**, and **squamous** (scale-like). The deep columnar cells often have a secretory function, and the nucleus is pushed to the bottom by the made and stored secretions near the surface from which they will exit (e.g., the cells lining the stomach, which secrete mucus). Cubical cells form the walls of small ducts as from salivary glands. Squamous cells are very flat, and the nucleus may form a bulge; they look something like a fried egg. The thinness permits diffusion of molecules across membranes (e.g., alveolar walls in lungs). Thick layers of cells (e.g., skin) prevent diffusion. In addition to the above three basic types there are some modified forms of these tissues. They are the following. **Ciliated epithelium**, the columnar cells with numerous cilia on their free surface, which lines the respiratory passage. **Psudostratified epithelium**, which forms a single layer of cells but on sectional view appears to be multilayered. The last one is the **stratified epithelium**, which are multilayered and form a very tough and impervious barrier.

The secretory or glandular cells may be present individually as in the case of goblet cells or in groups forming multicellular glands. An epithelial tissue having many goblet cells that secrete mucus is called **mucus membrane**. If the glandular cells or glands discharge their secretion on the surface of the cells or through a duct, they are called **exocrine glands**. But there are glands that discharge their secretion directly into the bloodstream and do not have any ducts. They are called the **ductless glands**.

Connective Tissue

This includes the various types of supporting tissues in the body. Connective tissues are cells in a matrix. The matrix may be a fluid, semi-fluid, or a composite structure made up of secretory products of cells such as fibrous proteins. **Blood** is a connective tissue in which cells are embedded in a fluid matrix. In **fibrous connective tissue** cells are scattered among the **collagen fibers** (fibrous protein) they secrete. In **bone and cartilage**, cells are scattered throughout the hard or pliable matrix. In cartilage, the cells known as chondroblasts deposit in the matrix. The cell, along with the matrix, forms the chondrocytes. The cartilage is hard but flexible because the matrix is compressible and elastic. Bone is a calcified connective tissue. The cells are embedded in a hard matrix. The cells in the bone tissue are called osteoblasts, which are present in **lacunae**. Lacunae are present throughout the tissue. The main inorganic component of bone is **hydroxyapatite**.

Muscle Tissues

Muscle tissues are made up of highly differentiated contractile cells or fibers held together by connective tissues. Muscle tissues are of three types. **Striated** muscle cells are large, multinucleate, and column-shaped cells; they are chiefly attached to the skeleton and are known as skeletal muscles or voluntary muscles. Voluntary muscles are under the control of the voluntary nervous system. They show powerful rapid contractions. They are attached to the bones in the trunk, limbs, and head. **Smooth** muscle cells are small and mononucleate; they are found in the walls of tubes such as blood vessels, glandular ducts, and the digestive system. They are also known as unstriated or involuntary muscles. The involuntary muscles are under the control of the autonomic nervous system and show sustained rhythmical contraction and relaxation movements. **Cardiac** muscle cells of the heart are small, striated, and branched. They are present only in the heart. They show rapid rhythmical contractions and relaxation movements with long refractory periods and do not show any fatigue.

Nervous Tissues

Nervous tissues consist of nerve cells, the **neurons** and associated neuroglial cells. Neurons are capable of generating and transmitting electrical impulses. These cells also act as supporting connective tissue in the brain and spinal cord. The neurons transmit the stimuli from receptors such as skin to the effectors such as muscles and glands that then react to the stimuli.

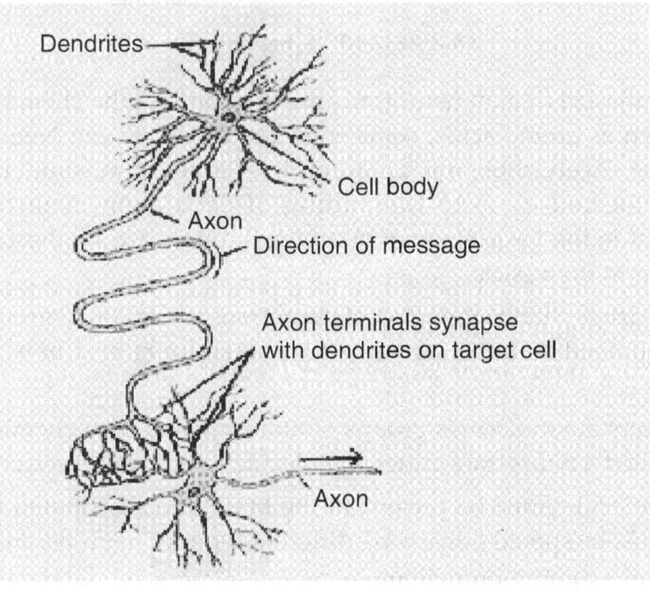

FIGURE 7.9 **Neurons showing structure and synapse.**

Each neuron may have thousands of branches that connect it to other neurons. The branches are called **dendrites** or **axons.** The nerve fibers or the axons are completely ensheathed by a myelin sheath formed by **Schwann cells**. Such nerve fibers are called **myelinated**. When bundles of nerve fibers are ensheathed in connective tissue, they form the nerves. Dendrites carry messages toward the cell body; axons carry messages away from the cell body to another neuron. Axons extend for as long as four feet in humans. In some animals, axons are even longer.

In the beginning, we thought that axons and dendrites simply ran through the body continuously, like wires. Then we discovered a space between each axon and dendrite. We call this space a **synaptic gap**, or **synapse** (Figure 7.10). The synapse is the space between the axon of one neuron and the dendrites of the next neuron in a nerve pathway. That gap is extremely small—about one-millionth of an inch. Researchers originally thought that electrical impulses jumped these gaps, like electricity jumps across the gap in a spark plug. Now we know this is not true. Chemicals, not electrical impulses, travel across the gaps. These chemicals are **neurotransmitters**. Today, we know of about 50 neurotransmitters. Undoubtedly there are more things waiting to be discovered.

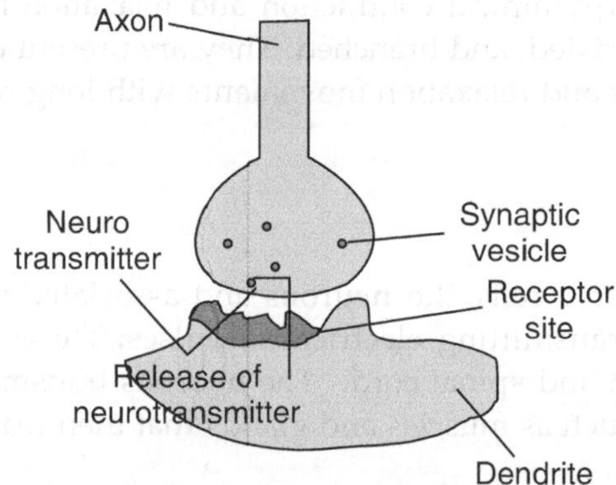

FIGURE 7.10 **Synaptic gap.**

Our bodies synthesize (make) neurotransmitters. Some of the chemical building blocks for neurotransmitters, such as amino acids, come from the foods we eat. Neurons include places to store neurotransmitters. Acetylcholine is an example of a neurotransmitter. These storage areas, called **vesicles**, are located close to the ending of each axon. Neurons synthesize some neurotransmitters right in the vesicle. Other neurotransmitters are synthesized in the body of the cell and shipped down to the vesicle.

Most addictive drugs change the effect of neurotransmitters on neurons. To understand how these drugs work, we need to know about neurotransmitters and how they act as chemical "messengers."

Neurotransmitters are molecules-groups of atoms, joined by a chemical bond, that act as a unit. In order to be called a neurotransmitter, a molecule must meet three criteria:

- First, the molecule must be present in the brain and distributed unevenly. That is, the molecule must be spread out among different types of neurons, and across regions of the brain that have different functions.

- Second is a chemical criterion. The enzymes that help to create the neurotransmitter must be present in the brain. Also, these enzymes must be present in areas where the neurotransmitter is found.
- Third is the criterion of mimicry. Suppose that we directly inject a neurotransmitter into a part of the brain known to contain certain neurons. This injection should mimic (imitate) the effects of electrically stimulating the same neurons.

Combinations of these tissues make up the organs in the human body. Organs are united into systems: digestive, circulatory, respiratory, excretory, endocrine, nervous, locomotory, and reproductive systems.

Plant Tissues

Vascular plants have distinctive cell types, all of which are surrounded by a cell wall of cellulose fibers and other molecules secreted by the cells. Just as in animals, cells are organized into tissues that perform different functions, but plants do not have organ systems like those of animals. The tissues of plants are grouped into three basic kinds: ground, vascular, and dermal. Meristem is a special embryonic tissue.

Plants differ from animals in that the tips of roots and stems, called **apical meristem**, remain **embryonic** and retain the ability to form new structures (e.g., leaves, stems, flowers, and roots). Hormones secreted by meristem cells are transported elsewhere in the plant; meristem is in part analogous to the endocrine system in animals.

Ground Tissues (Simple tissues)

Ground tissues or the simple tissues include parenchyma, collenchyma, and sclerenchyma. Thin-walled **parenchyma** cells (Figure 7.11) have a variety of functions such as photosynthesis, starch storage, and secretion; they retain the capacity to divide and are important in repair of damage. They form the large part of the bulk of various organs such as stem, root, etc. In some parts they are modified to perform some special functions. For example:

Epidermis: This is a single-layered tissue that covers the whole plant body. It protects the internal part from infection and loss of water. This layer of cells has a waxy coating on the surface, which is secreted by the cells. This waxy layer is called cuticle, which helps to reduce the water loss.

Mesophylls: These types of parenchyma cells are found in the leaves between the two epidermal layers. These are specialized for carrying out photosynthesis. Parenchyma cells containing chlorophyll are also known as chlorenchyma. If the cells of the mesophyll tissue are tightly packed without air space, they are known as palisade parenchyma or mesophyll; if a lot of air space is present it is called spongy parenchyma. **Endodermis, pericycle, and companion cells,** etc. are also an example of modified parenchyma cells.

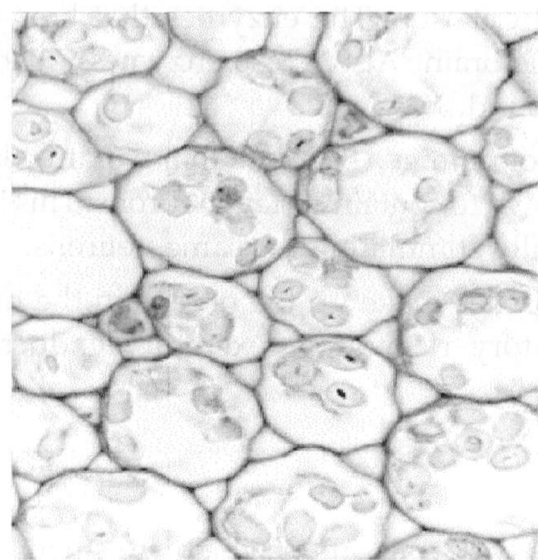

FIGURE 7.11 **Parenchyma cells.**

Collenchyma: These cells resemble parenchyma cells but are characterized by the presence of extra cellulose at the corners of the cells. Their walls are thickened and made strong with cellulose and pectin. **Collenchyma cells** help strengthen the plant parts in which they occur. Celery strings are an example. (See Appendix)

Sclerenchyma: **Sclerenchyma cells** have very thick secondary walls that are commonly impregnated with lignin, which makes them quite rigid. The lignified sclerenchyma of flax plants is made into linen threads for weaving, sewing, and paper making. Wood is made of lignified xylem cells. The hardness of a coconut shell or a peach pit is caused by lignified cells. Ground tissues are analogous to the supporting connective tissue and skeletal elements in animals. Sclerenchyma cells act as supporting elements in plants. Mature sclerenchyma cells can't elongate. The two types of sclerenchyma cells are fibers and sclereids. Fibers are long, slender, and tapered cells that occur in bundles. Sclereids are shorter than fibers and shaped irregularly. Nutshells and seed coats are composed of sclereids. Sclereids scattered among the soft parenchyma tissue of the pear give it a gritty texture. (See Appendix)

Complex Tissues

Complex tissues consist of more than one type of cells. **Vascular tissues** of plants include xylem and phloem; this is the plant's circulatory system. Xylem and phloem are the complex tissues.

Xylem consists of four types of cells—**trachieds, vessel elements, parenchyma**, and **fibers**. Trachieds are single cells that are elongated and lignified. At maturity, trachieds cells are dead and form interconnected tubes throughout the plant. Vessels are long, tubular structures formed by the fusion of several cells end to end in a row. They conduct water and dissolve nutrients that the plant absorbs from the soil; their thick, sclerified walls allow them to give mechanical support to the plant. Wood is made of xylem cells. Xylem parenchyma has thin cellulose cell walls and living contents similar to the typical parenchyma cells. Xylem fibers are shorter and thinner than trachieds and have much thicker walls.

Phloem consists of tubular cells modified for translocation. These tubular cells have interconnected cytoplasm and they conduct other solutes, chiefly nutrients (e.g., carbohydrates) from areas of food production such as leaves to areas of food storage such as tubers. There are five types of cells in phloem. They are sieve tube elements, companion cells, parenchyma, fibers, and schlerids. Sieve tubes are long tube-like structures formed by the end to end fusion of sieve tube elements. Adjacent to the sieve tube elements lie the companion cells with dense cytoplasm. Phloem parenchyma is similar to the ordinary parenchyma cells and the phloem fibers are like the sclerenchyma fibers.

Dermal tissues include epidermis and cuticle. The **epidermis** is a continuous layer of tightly packed cells. It is usually coated with a **cuticle** of waxes embedded in a fatty substance; this is analogous to keratinized outer layer of skin, including your own, in animals that live on land. Leaf epidermis is perforated by **stomata** for gas exchange between the photosynthetic mesophyll (parenchyma) and the surrounding atmosphere. Thus, leaves function in part like lungs. All these tissue types—both simple and complex tissues—are distributed all over the plant parts, but their position and orientations are different in different organs like stem, roots, leaves flowers, fruits, etc.

Organs

Various types of tissues are associated together to carry out a specific function of the body and such structures are known as organs. In animals, stomach, heart, brain, etc., are specific organs carrying out specific functions due to the interaction of various tissues. The heart is involved in the pumping of blood, which in turn is circulated to other organs and tissues by a network of arteries. Blood from various organs and tissues is brought back to the heart by another network of tubes called veins. Thus, forms an important system called the circulatory system. Kidneys are another example of organs. They are involved in the excretion of metabolic waste and other toxins produced in the body. Similarly, there are a large number of organs such as stomach, intestine, liver, pancreas muscles, reproductive organs such as testes, ovaries, external genitalia, etc., that carry out specific functions in association with other organs. There are varieties of glandular organs, both ductless glands such as pituitary, thymus, adrenal, etc., and glands with ducts such as the salivary glands that execute their function primarily by secreting specific enzymes and hormones to carry out various metabolic activities.

7.3 EVOLUTION OF POPULATION

Evolution is a change in the gene pool of a population over time. A gene is a **hereditary unit** that can be passed on unaltered for many generations. The **gene pool** is the set of all genes in a species or population. A **population** is a group of organisms of the same species usually found in a clearly defined geographical area.

The English moth or the peppered moth, ***biston betularia***, is a frequently cited example of observed evolution. In this moth there are two color morphs, light and dark. Dr. Henry Bernard Davis Kettlewell, a British lepidopterist and medical doctor, is notable for his experiments on the peppered moth, most of which were done in Manchester, England. He found that dark moths constituted less than 2% of the population prior to 1848. The frequency of the dark morph increased in the years following. By 1898, 95% of the moths in Manchester and other highly industrialized areas were of the dark type. Their frequency was less in rural areas. The moth population changed from mostly light colored moths to mostly dark colored moths. The moths' color was primarily determined by a single gene. So, the change in frequency of dark colored moths represented a change in the gene pool. This change was, by definition, evolution.

The increase in relative abundance of the dark type was due to **natural selection**. The late eighteenth century was the time of England's industrial revolution. Soot from factories darkened the birch trees the moths landed on. Against a sooty background, birds could see the lighter colored moths better and ate more of them. As a result, more dark moths survived until reproductive age and left offspring. The greater number of offspring left by dark moths is what caused their increase in frequency. This is an example of natural selection.

Populations evolve. In order to understand evolution, it is necessary to view populations as a collection of individuals, each harboring a different set of traits. A single organism is never typical of an entire population unless there is no variation within that population. Individual organisms do not evolve; they retain the same genes throughout their life. When a population is evolving, the ratio of different genetic types is changing—each individual organism within a population does not change. For example, in the previous example, the frequency of black moths increased; the moths did not turn from light gray to dark in concert. The process of evolution can be summarized in three sentences: Genes mutate. Individuals are selected. Populations evolve.

The word evolution has a variety of meanings. The fact that all organisms are linked via descent to a common ancestor is often called evolution. The theory of how the first living organisms appeared is often called evolution. This should be called abiogenesis. And frequently, people use the word evolution when they really mean natural selection—one of the many mechanisms of evolution. **Phenotype** is the morphological, physiological, biochemical, behavioral, and other properties exhibited by a living organism. **Genotype** is the genetic make up of an organism.

Evolution can occur without morphological change; and morphological change can occur without evolution. Humans are larger now than in the recent past, a result of better diet and medicine. Phenotypic changes like this, induced solely by changes in environment, do not count as evolution because they are not heritable; in other words, the change is not passed on to the organism's offspring. Most changes due to environment are fairly subtle, for example, size differences. Large-scale phenotypic changes are obviously due to genetic changes, and therefore are evolution.

Genetic Variation

Evolution requires genetic variation. If there were no dark moths, the population could not have evolved from mostly light to mostly dark. In order for continuing evolution there must be mechanisms to increase or create genetic variation and mechanisms to decrease it. Mutation is a change in a gene. These changes are the source of new genetic variation. Natural selection operates on this variation.

Genetic variation has two components: **allelic diversity** and **non-random associations of alleles.** Alleles are different versions of the same gene. For example, humans can have A, B, or O alleles that determine one aspect of their blood type. Most animals, including humans, are diploid—they contain two alleles for every gene at every locus, one inherited from their mother and one inherited from their father. Locus is the location of a gene on a chromosome. Humans can be AA, AB, AO, BB, BO, or OO at the blood group locus. If the two alleles at a locus are of the same type (for instance two A alleles) the individual would be called homozygous. An individual with two different alleles at a locus (for example, an AB individual) is called heterozygous. At any locus there can be many different alleles in a population, more alleles than any single organism can possess. For example, no single human can have an A, B, and an O allele.

Allele frequency: The number of organisms in a population carrying a particular allele of gene determines the allele frequency. In population genetics the allele frequency is usually expressed as decimals. Thus, a frequency of 99% is represented as 0.99 and the 1% frequency would be 0.01, because the total population represents 100% or 1.0. In population genetics these are represented as:

$$p + q = 1,$$

where p–frequency of dominant allele

q–frequency of recessive allele

Thus, in the above example total frequency is 0.99 + 0.01 = 1.0.

If we know the frequency of one allele (gene), the frequency of the other allele can be determined.

Non-random breeding: In most of the natural population, mating is non-random. But there are many structural and behavioral mechanisms that prevent the random mating. In populations where there is no random mating, fewer heterozygotes (an organism that has two different alleles at a locus) are found than would be predicted under random mating. A decrease in heterozygotes can be the result of mate choice, or simply the result of population sub-division. Most organisms have a limited dispersal capability, so their mate will be chosen from the local population.

Genetic Drift: The variation in allele frequencies can occur only by chance. This is called genetic drift. Drift is a binomial sampling error of the gene pool. What this means is, the alleles that form the next generation's gene pool are a sample of the alleles from the current generation. When sampled from a population, the frequency of alleles differs slightly due to chance alone. Alleles can increase or decrease in frequency due to drift.

Gene Flow: New organisms may enter a population by migration from another population. If they mate within the population, they can bring new alleles to the local gene pool. This is called gene flow. In some closely related species, fertile hybrids can result from interspecific matings. These hybrids can vector genes from species to species. Gene flow between more distantly related species occurs infrequently. This is called horizontal gene transfer.

Mutation: The cellular machinery that copies DNA sometimes makes mistakes. These mistakes alter the sequence of a gene. This is called a mutation. There are many kinds of mutations. A point mutation is a mutation in which one "letter" of the genetic code is changed to another. Lengths of DNA can also be deleted or inserted in a gene; these are also mutations. Finally, genes or parts of genes can become inverted or duplicated. Typical rates of mutation are between 10^{-10} and 10^{-12} mutations per base pair of DNA per generation. Most mutations are thought to be neutral with regards to fitness. Only a small portion of the genome of eukaryotes contains coding segments. And, although some non-coding DNA is involved in gene regulation or other cellular functions, it is probable that most base changes would have no fitness consequence.

Most mutations that have any phenotypic effect are deleterious. Mutations that result in amino acid substitutions can change the shape of a protein, potentially changing or eliminating its function. This can lead to inadequacies in biochemical pathways or interfere with the process of development. Organisms are sufficiently integrated that most random changes will not produce a fitness benefit. Only a very small percentage of mutations are beneficial. The ratio of neutral to deleterious to beneficial mutations is unknown and probably varies with respect to details of the locus in question and environment.

Genetic Load: The existence of disadvantageous alleles in heterozygous genotypes within the population is known as genetic load. The disadvantageous alleles when come as homozygous will affect the organism negatively for their phenotype and their existence. Such organisms may be eliminated from the population when these alleles (if they are deleterious in nature) occur in homozygous condition. If the allele is recessive, its effect won't be seen in any individual until a homozygote is formed. The eventual fate of the allele depends on whether it is neutral, deleterious, or beneficial.

7.4 SPECIATION

Speciation is the process of a single species becoming two or more species. Many biologists think speciation is key to understanding evolution. According to biological species concept, species are groups of actually or potentially interbreeding natural populations which are reproductively isolated from other such groups. Speciation is thus seen in terms of the evolution of isolating mechanisms and is said to be complete when reproductive barriers are sufficient to prevent gene flow between the two new species. The problem is that the capacity to interbreed cannot always be tested neither the potential for interbreeding. For asexually reproducing organisms and fossils, this concept does not apply.

Modes of Speciation

Biologists recognize two types of speciation: **allopatric** and **sympatric** speciation. The two differ in geographical distribution of the populations in question.

Allopatric speciation is thought to be the most common form of speciation. It occurs when a population is split into two (or more) geographically isolated subdivisions that organisms cannot bridge. Eventually, the two populations gene pools change independently until they cannot interbreed even if they were brought back together. In other words, they have speciated.

Sympatric speciation occurs when two sub-populations become reproductively isolated without first becoming geographically isolated. Insects that live on a single host plant provide a model for sympatric speciation. If a group of insects switched host plants they would not breed with other members of their species still living on their former host plant. The two sub-populations could diverge and speciate.

7.5 BIODIVERSITY

Biodiversity is the occurrence of all lifeforms in the biosphere. The phenomenon of speciation increases biodiversity. Biodiversity can be for a specific region or geographical area and similarly can be within a species. Within a species there can be varieties or sub-species, strains, and types. This variation within a species constitutes the biodiversity within a species. It is directly linked to the stability of the ecosystem. The magnitude of the biodiversity is not completely studied. The total number of species collected, named, and classified in taxonomic groups is around 1.5 million. This number is only a small fraction, about 10% of all living organisms, in this biosphere. The remaining, more than 90%, remains to be identified and classified.

Out of this 1.5 million known species, 750,000 are insects. The remaining part includes 280,000 animal species and 250,000 numbers of plant species. There are approximately 69,000 fungi, 27,000 algae, 3,000 protozoans, and about 3,000 prokaryotes including eubacteria and archaebacteria. Among these known groups, some have been studied extensively and others have been studied very poorly.

Biodiversity, which is created by speciation and evolution, has a direct impact on the stability of the ecosystem and the biosphere. Due to many man-made changes in the environment through deforestation and construction of big dams, there is disturbances in the habitat of the species, slowly leading to their mass extinction and destabilization. This loss of biodiversity is non-reversible unless we take special precautions. The phenomenon of extinction is opposite to that of speciation.

Extinction is the ultimate fate of all species. The reasons for extinction are numerous. A species can be competitively excluded by a closely related species, the habitat a species lives in can disappear, and/or the organisms that the species exploits could come up with an unbeatable defense. Some species enjoy a long tenure on the planet while others are short-lived. Some biologists believe species are programed to go extinct in a manner analogous to organisms being destined to die. This is **ordinary extinction**. The majority, however, believe that if the environment stays fairly constant, a well-adapted species could continue to survive indefinitely.

Mass extinctions shape the overall pattern of macroevolution. If you view evolution as a branching tree, it's best to picture it as one that has been severely pruned a few times in its life. The history of life on this earth includes many episodes of mass extinction in which many groups of organisms were wiped off the face of the planet. Mass extinctions are followed by periods of radiation where new species evolve to fill the empty niches left behind. It is probable that surviving a mass extinction is largely a function of luck. Thus, contingency plays a large role in patterns of macroevolution.

7.6 ADAPTATION

The existence of an organism in its environment or habitat is closely related to the special features of that organism or the adaptations. Adaptation is the special feature of an organism's morphology, anatomy, and physiology, which improves its interaction with its environment. Adaptations usually have the following characteristics.

- Special features are specially suited to a specific habitat.
- These special features are often complex.
- These special features help organisms to live in their environment and capture food, regulate the body's physiology, reproduce, disperse, and defend against enemies.

Adaptation is one of the important factors that drives the process of evolution. Adaptations are created through mutation and natural selection. Evolution requires genetic variation. In order for continuing evolution there must be mechanisms to increase or create genetic variation and mechanisms to decrease it. Mutation is a change in a gene. These changes are the source of new genetic variation. Natural selection operates on this variation. If these variations are suited to the changed environment that organism will outperform the others, which leads to the evolution of the population. If these new changes created through mutation are not suitable for existence in that environment, they will lead to extinction.

7.7 NATURAL SELECTION

Some types of organisms within a population leave more offspring than others. Over time, the frequency of the more prolific type will increase. The difference in reproductive capability is called natural selection. **Natural selection** is the only mechanism of adaptive evolution; it is defined as differential reproductive success of pre-existing classes of genetic variants in the gene pool.

The most common action of natural selection is to remove unfit variants as they arise via mutation. In other words, natural selection usually prevents new alleles from increasing in frequency. This led a famous evolutionist, George Williams, to say "Evolution proceeds in spite of natural selection."

Natural selection can maintain or deplete genetic variation depending on how it acts. When selection acts to weed out deleterious alleles, or causes an allele to sweep to fixation, it depletes genetic variation. When heterozygotes are more fit than either of the homozygotes, however, selection causes genetic variation to be maintained. (A heterozygote is an organism that has two different alleles at a locus; a homozygote is an organism that has two identical alleles at a locus.) This is called **balancing selection**. An example of this is the maintenance of sickle cell alleles in human populations subject to malaria. Variation at a single locus determines whether red blood cells are shaped normally or sickled. If a human has two alleles for sickle cell, he/she develops anemia—the shape of sickle cells precludes them from carrying normal levels of oxygen. However, heterozygotes who have one copy of the sickle cell allele coupled with one normal allele enjoy some resistance to malaria—the shape of sickle cells make it harder for the plasmodia (malaria-causing agents) to enter the cell. Thus, individual homozygous for the normal allele suffer more malaria than heterozygotes. Individual homozygous for the sickle cell are anemic. Heterozygotes have the highest fitness of these three types. Heterozygotes pass on both sickle cell and normal alleles to the next generation. Thus, neither allele can be eliminated from the gene pool. The sickle cell allele is at its highest frequency in regions of Africa where malaria is most pervasive. Balancing selection involves opposing selection forces. An equilibrium results when two alleles selected in the homozygous state are retained because of the superiority of heterozygotes. Balancing selection is rare in natural populations.

Types of Selection

Following are some important types of selection:

1. **Stabilizing Selection:** Most adaptive character is preserved as long as the adaptive peak does not change. H. Bumpus presented a typical example of stabilizing selection in 1899. He measured the wings of house sparrows (*passer domesticus*) killed in a storm in New York. He found that those with markedly long or short wings were more frequently killed. Stabilizing selection does not allow new variations to emerge. Once a

character is optimized, natural selection keeps it as it is, like keeping the number of fingers, egg size and number, mate choice adaptations, seasonal timing of migrations, birth weight in humans, etc., stable. Both extremes in variation are selected against and eliminated. Aristotle's description of wild animals and plants, written 2500 years ago, is still accurate as natural selection must have been preventing their further evolution (from a stable state). Thus, natural selection cannot be equated to evolution as sometimes it prevents evolution, but it is a major mechanism involved in evolution.

2. **Directional Selection:** Strong selection favors one of the extreme phenotypes. This type of selection decreases variation (sexual selection of male characters). Antibiotic resistance by bacteria and insecticide resistance by insects are other examples. By favoring those who are resistant, the variation in the population decreases, and the population eventually consists of only resistant individuals. A common form of directional selection causes character displacement when two species compete. When there are two species of finches on an island, each will evolve to have a different size beak (small and large) whereas either of the species alone can lead to an intermediate beak size elsewhere.
3. **Disruptive Selection:** Both extreme phenotypes are preferred over the intermediate one. This selection occurs as a result of the heterozygous being at a disadvantage, thus the two homozygotes are selected (under dominance). In a bi-allelic polymorphism, if one of the alleles is remarkably more rare than the other, the rare allele may be lost. Selection of two different colors in North American lacewings but not of intermediate color is an example. This happens because the two extreme color patterns provide the best camouflage in the two different niches, but the intermediate one does not offer any protection. Similarly, the African swallowtail butterfly produces two distinct morphs, both of which mimic distasteful butterflies of other species with aposematic coloration (Batesian mimicry). This type of selection tends to increase or maintain the diversity in the population. It might even cause one species to evolve into two.

7.8 ORGANIZATION OF LIFE

Life on Earth is incredibly extensive and to make it easier to study, biologists have broken living systems up into generalized hierarchical levels as follows:

- Molecules
- Organelles
- Cells
- Tissues
- Organs
- Organisms
- Populations
- Communities
- Ecosystems
- Biosphere

The lowest level of the biological hierarchy begins with molecules. Examples include proteins, DNA, lipids, etc. Many such specialized molecules are organized into cells, the basic unit of life. There are single-celled organisms such as bacteria, amoeba, yeast, etc., in which the body consists of a single cell. When the body consists of more than one cell it is called multicellular. Multicellular organisms are collection of various types of specialized cells. A group of specialized cells carrying out a specific function is called a tissue. For example, muscle tissue, nervous tissue, connective tissue, etc. When different types of tissues are organized together to perform a common function it is called an organ. Examples include, liver, stomach, heart, etc. When a number of organs function together to accomplish a specific function of the body, it forms an organ system. For example, the stomach, liver, intestine, pancreas, salivary glands, etc. work together to form the digestive system. In an organism there are a number of organ systems that work in an associated way to form the organism and its life activities. Each individual organism is a member of a large population, which exists in a habitat.

A population is a group of organisms belonging to a species. A group of different species that live and interact in a particular area or environment is known as a community. The communities, along with the environment in which they exist, are known as ecosystems. An ecosystem consists of biomes, which are large geographical areas of the world. Each biome is a part of the biosphere, which includes the entire living population on the Earth along with its physical environment.

7.9 SIZE AND COMPLEXITY

Living organisms greatly differ in size and complexity of their body. They range from minute unicellular bacteria to very big multicellular organisms such as blue whales and redwood trees. Primitive cell forms such as bacteria and blue-green algae are very simple in organization and function. The cells and organisms are very small and cannot be seen with the naked eye.A microscope is needed for observing these microorganisms. The multicellular organisms and their cells are very complex in organization and function. The body of higher plants and animals consists of billions of various types of specialized, structurally and functionally complex cells. Therefore, their body and its function is highly complicated.

Variations in the body size affect various other body measurements differently. This is because the volume of cells and so the volume of the entire organism increases much faster than the surface area. Entire single-celled organisms and most primitive multicellular organisms use their cell surfaces to acquire nutrients and dispose of wastes. But the amount of nutrients needed, and the quantity of wastes produced, is related to cell volume. Since the surface area to volume ratio of a cell decreases as its size increases, cells have an upper limit on how much volume they can sustain with a given surface area. Large organisms have less surface area relative to mass than do small organisms. This relationship affects the efficient exchange of material between the body and the environment.

Allometric relationships describe the effect of body size on biological features. These relationships can reveal general patterns of how organisms function; for example, how much they sleep, their food requirements, and their brain size. Allometric relationships also have practical applications, as in the proper determination of drug dosages for animals of differing body sizes. For multicellular organisms, increases in overall body size are mostly due to increases in cell number, not cell size. This is also because of surface area to volume ratio limitations on cells.

The evolution of complexity in multicellular organisms is driven by the specialization of cells. Multicellular complexity requires coordination among body cells. Internal communication mechanisms such as hormones and the nervous system help make this possible. Complexity also requires many body cells to give up reproduction in support of a relatively few cells that do reproduce.

7.10 INTERACTION WITH THE ENVIRONMENT

Living organisms and the physical environment have a close relationship. They interact with each other. The biosphere is the parts of Earth inhabited by living organisms. The biosphere consists of specific geographical areas known as biomes. A biome is a collection of different types of ecosystems. The ecosystems include grasslands, rain forests, streams, lakes, sea, deserts etc., with various types of organisms starting from bacteria, fungi, algae, and various other types of plants and animals. There are millions of known species of organisms and there are many millions to be discovered. Each organism lives in a specialized regional environment within the ecosystem known as the habitat. An organism can live in a specific habitat because it is adapted to live in that habitat. Deep-sea vent, bottom of sea, arctic rivers, and river banks, etc. are examples of habitat.

Organisms living in a specific environment interact with the environment and also with themselves in very different ways. There are big trees growing along the bank of the stream. Since the trees are very big, they make half of the stream a shady area, and this may make the temperature of the water a bit lower than that in the middle region. It is because the water at the bank side is not directly heated by the sun. Similarly, there are many algae floating in the water freely, and this may reduce the penetration of sunlight. So the light intensity under the water may be decreased. All the organisms living in an environment along with that physical environment form an ecosystem. The organisms living in a fresh water pond, along with the pond, form an aquatic ecosystem. All organisms on land along with their environment form the terrestrial ecosystem.

The energy flow in an ecosystem obeys the laws of thermodynamics. It is an open system. An open system allows the free flow or exchange of energy and matter such as water, carbon dioxide, nitrogen, food materials, and even the movement of organisms from one ecosystem to another. There are producers, consumers, and decomposers in ecosystems. The producers are the photosynthetic organisms or the autotrophs. The producers of the ecosystem take energy from sunlight and convert it into chemical energy. This energy is passed on to consumers and then to decomposers, which cycles back the materials to the environment. But the energy flows only in one direction and is not cycled back. Herbivorous animals consume the organic food synthesized by the producers, which form the primary consumers. These herbivores form the food of carnivores, which are the secondary consumers. And finally the decomposers act on the dead remains of all these organisms including the producers, decompose the organic materials into inorganic materials, and thus cycle back the materials to the environment. This forms the food chain in the ecosystem. In each step of the food chain energy is also transferred. In each step a portion of the energy is lost in the form of heat. Thus, heat is flowing in one direction and is not cycled back. The energy enters the ecosystem from the sun through producers and leaves the ecosystem in all steps of the food chain in the form of metabolic heat.

The materials in the form of nutrients required for life are cycled between organisms and the environment. The materials are absorbed by the producers for synthesizing the nutrients and are cycled among the consumers and finally returned to the environment by the activity of saprophytes and other decomposers such as fungi and bacteria. Considering the flow of energy and nutrients in the ecosystem and in the biosphere, it can be considered a single living organism.

REVIEW QUESTIONS

1. Plasma membrane has the basic fluid mosaic structure, mainly consisting of phospholipids and proteins. Even then, plasma membrane at different regions of the cells such as mitochondria, chloroplast, and cell surface carries out different types of functions. Explain.
2. What are the functions of endoplasmic reticulum?
3. What are the similarities between chloroplasts and mitochondria?
4. Give the structure of a neuron.
5. What is genetic drift?
6. Biosphere can be considered a single living organism. Explain.
7. What is speciation? Give the difference between allopatric and sympatric speciation.
8. What are the factors that control population evolution?
9. Describe the complex tissues in plants.
10. What are the differences between prokaryotic and eukaryotic cells?
11. What is the importance of adaptation in evolution?
12. What are the different types of natural selection? How are they important in the process of evolution?
13. What are the various types of tissues present in animals?
14. What is the special feature of ribosomes when compared to the other organelles of eukaryotes?
15. What is the type of tissue to which blood belongs?

Chapter 8

CELL GROWTH AND DEVELOPMENT

In This Chapter

8.1 CELL DIVISION

Cell division is an integral part of the process of reproduction, growth, and repair of the body. During cell division, genetic material is distributed among the daughter cells. This is the important phenomenon taking place during all types of cell divisions. There are two types of cell division observed in eukaryotic cells. They are mitosis and meiosis. The distribution of genetic materials among the daughter cells is so important because their future development and generation of cells developing from that will be entirely dependent on the genetic material that they receive from the parent cell. Mitosis is the cell division that occurs in the somatic cells (body cells) and meiosis takes place in the sex organs for the production of gametes. The main features and the process of mitosis and meiosis are discussed below.

Mitosis

Mitosis is the process that facilitates the equal partitioning of replicated chromosomes into two identical groups. Two new daughter cells arise from one original cell. All the cells created through mitosis are genetically identical to one another and to the cell from which they came. The main purpose of mitosis in eukaryotic cells is:

- Growth of the individual,
- To repair tissue, and
- To reproduce asexually.

Mitosis is a nuclear division in which the daughter cells receive the same number of chromosomes as that of the parent cell. The nuclear division is sometimes referred to as karyokinesis, which is followed by the cytoplasmic division known as cytokinesis. The daughter cells resulting from mitosis are identical to each other and also to the parent cell in the quantity and quality of genetic material. The genetic information, which the cell is copying and distributing during mitosis, is contained in the form of chromosomes.

The period between two successive cell divisions is referred to as the **interphase**. It is not a part of mitosis, but forms the preparatory stage for cell division. The main metabolic activities during this stage of cell are respiration, protein synthesis, and the duplication of genetic material (DNA replication). Cells usually are small with no large vacuoles.

The interphase stage is subdivided into three parts: **G1, S, and G2**. In the G1 stage the cell carries out its "housekeeping" functions while it collects the materials it will need to divide. S stands for DNA synthesis and this is the stage in which the DNA makes a copy of itself. An enzyme (DNA polymerase) assists each DNA double helix in making a copy of it in a process known as **DNA replication**. The two identical DNA double helices are represented in the chromosomes as the **sister chromatids**, and they are held together by the **centromere**, which is visible as a constricted area somewhere along the length of the chromosome. The G2 stage of mitosis comes next, and in this stage the cell checks to see that everything is ready to begin mitosis. When the cell has duplicated its DNA and checked it to make sure that there are no errors, it is ready to begin the distribution of the DNA to two separate cells.

Mitosis is divided into four stages—**prophase, metaphase, anaphase, and telophase.** The important changes that the cells undergo in each phase are described below.

Prophase

During prophase the nuclear membrane and nucleolus disappear. The chromosomes themselves condense from long, thin filaments into compact rods so that they can be more easily moved. The apparatus needed to move the chromosomes around is set up. In animal cells this is done by the two centrioles, which move toward opposite poles of the cell and begin to form the **mitotic spindle**.

Plant cells do not have centrioles; they use a different mechanism to form the spindle fibers. Each chromosome has a spindle fiber attached to it at the centromere and extending off to either side of the cell where the centrioles are located. These spindle fibers are exerting tension on the chromosomes and when the centromere splits later on, this will allow the chromosomes to be pulled to either end of the cell.

Metaphase

Metaphase is an easy stage to recognize because all of the chromosomes are lined up at the equator of the cell. Depending on where the viewer is "standing," it can look as though the chromosomes are lined up east to west or it can look as though the chromosomes are aligned north to south.

Anaphase

In anaphase the centromere on each chromosome splits. Because the spindle fibers are exerting tension on the chromosomes, when the centromere splits each chromatid is pulled toward the spindle pole that it faces. Once a chromatid has its own centromere it can be called a chromosome.

Telophase

In telophase the chromosomes reach the opposite poles of the cell and their attached spindle fibers disappear. A new nuclear envelope forms, the nucleolus reappears, and the condensed chromosomes expand once more. The cytoplasm divides in a process called **cytokinesis**, and this forms the two daughter cells. In animal cells the cell membrane pinches in from either side by a constriction and separates into two. In a plant, a new cell wall begins to form in the middle of the cell and gradually grows longer from each end as it works its way toward the edges of the cell.

Mitosis is finished and there now are two cells, which are identical to each other, and identical to the cell from which they came (see Appendix).

Cytokinesis

Mitosis ends in the formation of two daughter nuclei on either poles of the cell. In the next stage the cytoplasm divides into two resulting in the formation of two daughter cells, each with a single nuclei. The process of cytoplasmic division is called **cytokinesis**. The process of cytokinesis is different in animal cells and plant cells. In animal cells, following the separation of chromatids and the reappearance of nuclei at the end of anaphase, a cleavage furrow cuts between the two daughter nuclei or the separating chromatids resulting in two daughter cells. Thus, the two cells receive the identical genetic materials. In higher plants after the nuclear division, a cell plate is formed between the two nuclei by the deposition of actin filaments of the microtubules. The cell wall materials are deposited on either side of the cell plate resulting in the formation of two separate cells.

Meiosis

Meiosis is the reduction division. It involves a replication of the genetic material followed by two successive nuclear divisions resulting in four daughter nuclei with a haploid set of genome or chromosomes. This process of cell division occurs in reproductive organs and is essential for the production of gametes. The formation of haploid gametes and the fusion of male and female gametes resulting in the diploid zygote are involved in the process of sexual reproduction. The absence of meiotic cell division could result in diploid gametes, which in turn would result in the doubling of genomes with each new generation. This reduction division can occur at different stages of the life cycle depending on the type of organism. For convenience meiosis is divided into two parts: meiosis I and meiosis II.

Meiosis I

The first meiotic division is the reduction because at the end of the first division the chromosome number is reduced to half because of the segregation of homologous chromosomes. It is divided into prophase I, metaphase I, anaphase I, and telophase I, and the important events in each stage are summarized as follows:

Prophase I

The prophase of Meiosis I is the longest of all stages and is further divided into five stages for convenience. They are leptotene, zygotene, pachytene, diplotene, and diakinesis. During leptotene, chromosomes start condensing. Chromosomes are visible as long thread-like structures. During zygotene the homologous chromosomes start pairing. The pairing of homologous chromosomes is called **synapsis** and the paired chromosomes are known as **bivalent chromosomes** or **tetrads of chromatids**. The stage pachytene starts when the process of synapsis ends. Pachytene is very important because the phenomenon of crossing over between the homologous chromosomes occurs during this stage. The exhange of parts between homologous chromosomes is called **crossing over**. Crossing over has important genetic significance because it is responsible for generating genetic variation among the progenies. The nuclear membrane starts disappearing. The next stage, diplotene, is marked by the tendency of the bivalent chromosomes to separate from each other. As the homologous chromosomes try to pull away from each other, they seem to remain attached to certain specific points known as **chiasmata**. Chiasmata are the points on the bivalent chromosomes at which cross over has taken place. During these processes the chromosomes are undergoing progressive condensation and become shorter. The last stage of prophase I is diakinesis. During this phase, the meiotic spindle is assembled and the chromosomes become very short. Nuclear membrane and nucleolus disappear completely. The movement of chromosomes to the equitorial plate marks the end of prophase I and diakinesis. (See Appendix)

Metaphase I

The tetrads line up along the equatorial plate of the cell by their kinetochore. The chromosomes are arranged in such a way that the sister chromatids of each chromosome face the same pole. Because of this the spindle fiber from the same pole is connected to both the chromatids of a single chromosome.

Anaphase I

Homologous chromosomes separate and move toward opposite poles of the cell. It is important to note that the chromatids are still attached to each other and they do not separate at this time.

Telophase I

The homologous chromosomes are segregated and are dragged to the poles of the cells during this phase. Nuclear envelope reappears around the chromosomes and nucleolus also appears in the nucleus. The cytoplasm splits into two producing two daughter cells. In some cases the cytoplasmic division takes place after the meiosis I and II.

Prophase II

Prior to prophase II cells may enter a kind of interphase, however, there is no S-phase here and the DNA amount remains the same. Chromosomes condense. Note that there is no pairing up of homologous chromosomes because there are only single chromosomes comprised of two chromatids.

Metaphase II

The chromosome line up along the equator of the cell. Note that unlike metaphase I, each chromosome is held in place by spindle fibers, from each pole of the cell (as in mitosis).

Anaphase II

As in mitosis, anaphase II results in sister chromatids being separated and moving toward opposite poles of the cell. However, here there is half the number of chromosomes because DNA was not replicated following telophase I.

Telophase II/Haploid Cell Formation

The nuclei form at the opposite poles of the cells. Following cytokinesis, four cells are formed, each with half the DNA of the original cell at the start of meiosis (i.e., haploid cells if the original cell was diploid). These cells then differenetiate into gametes or spores.

TABLE 8.1 Mitosis versus meiosis. (See Appendix)

	Mitosis	Meiosis
Purpose	Produces somatic cells (Body, growth)	Produces reproductive cells
Process	Cell duplication (Diploid → diploid)	Reduction division (diploid → haploid)
Number of Divisions	1 cell division	2 cell division
Product	1 → 2 daughter cells Daughter cells identical (To each other and mother cell)	1 → 4 cells (gametes) daughter cells different

TABLE 8.2 Differences in cell division between animal cells and plant cells.

Animals cells	Plant cells
Late in anaphase, the plasma membrane pulls inward forming a cleavage furrow, which divides the cell in two during the telophase.	A centriole is found at each pole during mitosis.
After anaphase, a new cell wall forms between the two new nuclei to create two cells.	No centrioles are found in plant cells.

8.2 CELL CYCLE

A eukaryotic cell undergoes alternate phases of division and non-division. The sequences of events occurring from the completion of one division to the next division form one cell cycle. The stage, or the interval, between two consecutive mitotic divisions is the interphase. Therefore, a cell cycle can be considered as interphase + mitosis. The interphase is metabolically very dynamic and is further divided into S-phase, G1, and G2 phase. The period between M-phase and S-phase is called G_1; that between S and M is $\mathbf{G_2}$. A cell cannot divide into two, the two into four, etc., unless two processes alternate: **doubling** of its genome (**DNA**) in **S-phase** (synthesis phase) of the cell cycle and **halving** of that genome during mitosis (**M-phase**).

Stages of Cell Growth

1. **G1 phase:** primary growth phase. Cell does its housekeeping activities.
2. **S-phase:** DNA replication.
3. **G2 phase:** chromosome condensation, cell organelle replication.
4. **M-phase:** mitosis (nuclear division) (prophase, metaphase, anaphase, and telophase).
5. **C phase:** cytokinesis (cytoplasmic division), daughter cells form.

Interphase is the longest stage of the cell cycle. Human cells contain 46 chromosomes during the G1 stage of interphase. This is doubled to 92 during the S stage of interphase.

A typical cell takes about 16 hours to complete the cell cycle. The actual process of cell division or mitosis occupies only a small part of this cycle, approximately one hour. The lengths of S and G_2 are almost equal in all cell types, but the length of the G1 phase varies considerably between cells.

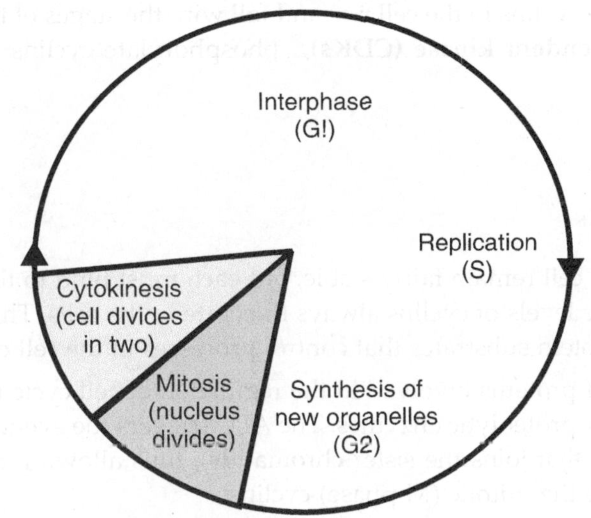

FIGURE 8.1 **Cell Life Cycle. The repeating sequence of growth and division through which eukaryotic cells pass each generation.**

Normally the cells at G_1 phase can follow either of the two paths. The cell after the cell division may withdraw from the cell cycle and enter into a resting phase called the **G_0 phase**, or it can enter into the G_1 phase of the cell cycle. Cells in the G_0 phase are viable and metabolically active but can be stimulated to enter into the G1 phase at any time and start the cell cycle again. Often, G_0 cells are terminally differentiated: they will never re-enter the cell cycle but instead will carry out their function in the organism until they die.

For other cells, G_0 can be followed by re-entry into the cell cycle. Most of the lymphocytes in human blood are in G_0. However, with proper stimulation, such as encountering the appropriate antigen, they can be stimulated to re-enter the cell cycle (at G_1) and proceed onto new rounds of alternating **S phases** and **mitosis.** G_0 represents not simply the absence of signals for mitosis but an active repression of the genes needed for mitosis. Cancer cells cannot enter G_0 and are destined to repeat the cell cycle indefinitely.

Regulation of the Cell Cycle

The cell cycles in almost all eukaryotic cells are essentially the same with minor variations in the duration of each phase. There are some proteins in the cytoplasm that control and coordinate the passage of a cell in the correct order through the cell cycle. The major proteins involved in the regulation are called **cyclins.** There are three groups:

- **G_1 cyclins**
- **S-phase cyclins**
- **M-phase cyclins**

The levels of these cyclins in the cell rise and fall with the stages of the cell cycle. The group of enzymes, **Cyclin-dependent kinase (CDKs),** phosphorylate cyclins. Again, there are three groups of CDKs.

- **G_1 CDKs**
- **S-phase CDKs**
- **M-phase CDKs**

Their levels in the cell remain fairly stable, but each must bind to the appropriate cyclin in order to be activated. The levels of cyclins always fluctuate in the cells. The **CDKs** add phosphate groups to a variety of protein substrates that control processes in the cell cycle.

The third group of proteins involved in the regulation of cell cycle is **anaphase-promoting complex (APC)** and other proteolytic enzymes. The APC triggers the events leading to destruction of the cohesin (a protein that joins the sister chromatids), thus allowing the sister chromatids to separate. It also degrades the mitotic (M-phase) cyclins.

The Important Steps in the Cell Cycle

- A rising level of **G_1 cyclins** signals the cell to prepare the chromosomes for replication.
- A rising level of **S-phase promoting factor (SPF)** prepares the cell to enter S-phase and duplicate its DNA (and its centrioles).
- As DNA replication continues, one of the cyclins shared by G_1 and S-phase CDKs (cyclin E) is destroyed and the level of mitotic cyclins begins to rise (in G_2).
- **M-phase promoting factor** (the complex of mitotic cyclins with M-phase CDK) initiates the assembly of the mitotic spindle, breakdown of the nuclear envelope, and condensation of the chromosomes.
- These events take the cell to the **metaphase** of mitosis.
- At this point, the M-phase promoting factor activates the **anaphase promoting complex (APC)**, which allows the sister chromatids at the metaphase plate to separate and move to the poles (= anaphase), completing mitosis. It destroys the M-phase cyclins. It also turns on synthesis of G_1 cyclins for the next turn of the cycle and degrades **geminin**, a protein that has kept the freshly-synthesized DNA in S-phase from being re-replicated before mitosis.

Checkpoints: Quality Control of the Cell Cycle

The cell has several systems for interrupting the cell cycle if something goes wrong. There are some points in the cell cycle where the cell can check the sequence of activities related to its replication. These points are known as **checkpoints.** The following are the important checkpoints in the cell cycle, which monitor the healthy progress of cell division:

- **DNA damage checkpoints.** These sense DNA damage before the cell enters S-phase (a G_1 checkpoint), during S-phase, and after DNA replication (a G_2 checkpoint). The cell seems to monitor the presence of okazaki fragments on the lagging strand during DNA replication. The cell is not permitted to proceed in the cell cycle until these have disappeared.

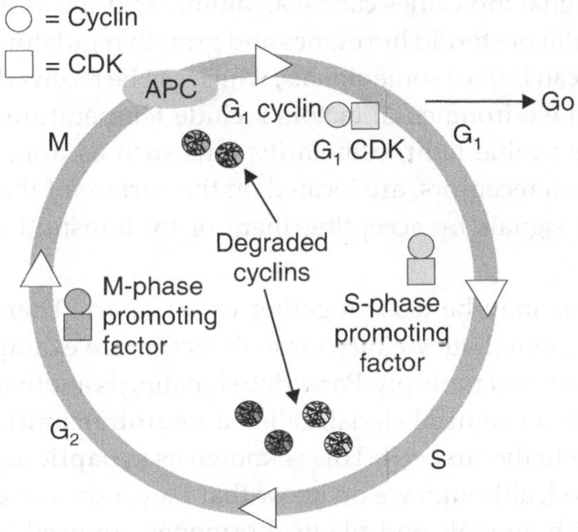

FIGURE 8.2 **Regulation of the cell cycle.**

- **Spindle checkpoints.** These are the final checkpoints also known as M-phase checkpoints. They detect any failure of spindle fibers to attach to **kinetochores** and arrest the cell in **metaphase** (M checkpoint). They monitor the alignment and detect improper alignment of the spindle itself and block cytokinesis. If the damage caused to the DNA is irreparable, it triggers the process of apoptosis or cell death.

All the checkpoints examined require the services of a complex of proteins. Mutations in the genes encoding have been associated with cancer; that is, they are oncogenes. This should not be surprising since checkpoint failures allow the cell to continue dividing despite damage to its integrity.

8.3 CELL COMMUNICATION AND SIGNAL TRANSDUCTION PATHWAYS

The cells, unicellular or multicellular, respond to changes in the environment. In multicellular organisms, cells communicate with the neighboring cell in order to coordinate the functions of the system and body. Cell-to-cell communication is essential for normal functioning of multicellular organisms, and this is because they carry special abilities to maintain a well-defined communication network. In recent past, "cell signaling" has become a fascinating field. It explains a lot of critical issues regarding embryological development, hormone action, and the development of diseases such as cancer.

Cell Communication

Cell-to-cell communication is essential for multicellular organisms for coordinating various metabolic activities including activation of immune systems and gene expression. Interestingly, the same fundamental cell communication strategies are evident in many different types of cells. The cell-to-cell communication usually occurs through certain chemicals called as **signal molecules.** Signal molecules are small organic molecules, which can interact with certain specific proteins known as receptors. The signal molecules can be an amino-acid-like tyrosine and its derivatives or small peptides such as insulin or steroid hormones and growth regulators such as cytokines. Certain environmental factors also can impart some signals, which can be received by some receptor proteins. Signals that originate from environmental factors include temperature, electromagnetic waves of different spectrum such as visible light, osmolarity, ions such as iron, etc. Recipient cells, which receive these signals through receptors, are located on the surface of the cell membranes. Then the cells can respond to these signals by accepting them or by transmitting them to the next target molecule or cells.

Communicating cells may be close together or far apart. There are local regulators that influence cells in the more immediate vicinity. Growth factors are examples of local regulators that stimulate nearby cells to grow and multiply. Paracrine signaling is another local signaling in animals. When a nerve cell produces a chemical signal, called a **neurotransmitter**, that diffuses to a single target cell that is very close to the first cell. This is known as **synaptic signaling.** Local signaling in plants is not well understood, although we do know that they must use different mechanisms since they have cell walls. In both animals and plants, hormones are used for cell signaling at greater distances. In **endocrine (hormone) signaling,** the specialized cells release hormone molecules into the circulatory system that then carry the hormones to the target cells in other parts of the body.

Cells may also communicate by direct contact. Cell junctions provide cytoplasmic continuity between cells and signaling substances in one cell can therefore diffuse into the cytoplasm of the adjacent cell.

Signal Transduction

There are three stages in the process of cell signaling or communication:

- Reception—a protein at the cell surface detects chemical signals.
- Transduction—a change in protein stimulates other changes including signal-transduction pathways.
- Response—almost any cellular activity.

Once the target cell receives the signal molecule it converts the signal to a form that can bring about a specific cellular response. This often occurs in a series of steps called a **signal transduction pathway**. Signal molecules bind to receptor protein, in cell membranes, and generally cause a conformational change in the proteins. This change in conformation is transmitted to the cytoplasmic domain or part of the receptor molecule. The transformed molecule interacts with the **information-relaying molecules** in the cytoplasm. These molecules are small molecules present in the cytoplasm known as **secondary messengers**. Calcium ions and Cyclic AMP (cAMP) are examples. This further starts a series of chain reactions, which ultimately reaches the target gene and causes its expression or repression. A single cell may have several types of receptors each binding to a specific signal molecule. A cell can receive a number of different types of signal molecules simultaneously. Once the signals are relayed into the cells, they are selectively routed through various signal pathways to the target, which may be a gene or a protein. Usually the cellular response for a signal molecule may be a change in gene expression, change in ion permeability, or a change in the enzyme activity or protein three-dimensional structure, which ultimately affects the metabolism of the cell or organism.

Some types of signal molecules pass through the cell membrane and directly activate the gene or proteins, without the involvement of any secondary messengers. Lipid-soluble molecules such as steroids (steroid hormones) or small molecules (such as nitric oxide) are examples. Molecules such as interferon and interleukin can also do the same type of direct activation of a gene. Through cell-to-cell signaling and signal transduction, the information that the cells acquire from the environment and from other neighboring cells is effectively monitored and responded into the appropriate manner.

8.4 MOVEMENT

Movement is an important characteristic of living things. Animals are able to move from one place to another. But plants are restricted to some types of movements induced by water currents, turbidity, osmosis, etc. Animals are able to move because of the presence of contractile muscle fibers. Prokaryotes and other single-celled eukaryotic organisms move by hair-like structures called **flagella** and **cilia** or parts of protoplasts (pseudopodia) as in the case of amoeba.

Amoeboid Movement

This type of movement is observed in some single-celled eukaryotes such as amoebas, slime molds, and eukaryotic cells such as lymphocytes of humans. Amoebas move by means of pseudopods (a foot-like extension at the front of a cell) attaching and detaching from the substratum. Locomotion may be correlated with the forward flow of fluid cytoplasm (endoplasm) into advancing pseudopodia through a surrounding gel-like ectoplasmic tube. The ectoplasm forms at the pseudopodia tip in a region called the **fountain zone**. As the amoeba advances, the ectoplasmic tube "liquefies" at the posterior end to form endoplasm. As the cell moves, new pseudopodia are formed in the direction of movement while the earlier ones are withdrawn. Amoeboid movement has a striking similarity with cytoplasmic movement known as **cyclosis**. The cytoplasmic streaming movement is a common characteristic of all plant and animal cells. This has an important role in intracellular transport. The cytoplasmic streaming movement and amoeboid movement by pseudopodia may be dependent on the contractile activity of protein fibers such as actin and myosin present in the cytoplasm along with the cytoskeleton.

Movement by Cilia and Flagella

Bacteria move using a rotating flagellum. Non-animal eukaryotes achieve motion in many ways, including moving by cilia and flagella, growing roots and hyphae, or dispersing gametes, spores, and seeds with the help of wind, water, and animals.

Cilia and flagella are the main locomotory organs (organelle) of prokaryotes such as bacteria and other single-celled organisms. Both flagella and cilia have identical internal structures. Microtubules form the internal structures of both cilia and flagella. They are also involved in other cellular movements such as movement of chromosomes and streaming movement of protoplasts, etc. The combined stroking movement of cilia propels certain unicellular organisms, such as paramecia, through their environment and moves fluid and particles over the surface of ciliated cells of higher eukaryotes. Flagella provide the propulsion for motile cells ranging from bacteria to sperm of higher eukaryotes. The main difference between cilia and flagella is in their stroking pattern and also in their length and number per cell. Flagella beat with a symmetrical undulation that is propagated as a wave along entire length of flagella. Flagella are very long and less in number. Flagellated cells usually carry one or few flagella. Ciliated cells usually have thousands of cilia all over the cell surfaces and are very short. A cilium beats symmetrically with a fast stroke in one direction followed by slower recovery stroke. Cilia and flagella are also present in many multicellular organisms. There are some types of tissues in higher organisms where cilia are present. For example, in mammals the ciliated epithelium helps in the transport of certain materials on the internal surfaces as in the case of the movement of mucus in the respiratory tract. The sperm of most animals move with help of single flagella.

Muscle and Movement

Even though all organisms can move directly or indirectly, the animals have an especially impressive range of movement. The presence of muscles contributes a special ability for animals in movement. The contraction of muscle tissue powers animal movement. Muscle contains thin filaments of the proteins **actin** and **myosin**; myosin filaments lie between the actin filaments within a **sarcomere,** the basic contractile unit of muscle. A series of sacromeres attached end-to-end makes up a single fiber called **myofibrils**. A number of myofibrils bundled together form a muscle fibril. Myofibrils are bundled to form a muscle fiber, which is further bundled to constitute a functional body muscle (e.g., the human biceps). Since muscles only produce force when contracting, they must work as opposing pairs to give the full range of possible motion.

Sarcomeres appear as bands under a microscope. Each sarcomere is bounded on either side by dark lines, the **Z discs**. The cells contain the contractile proteins, actin and myosin, which are arranged in a very specific manner. Z discs form the anchor points for the actin filaments. From this point actin filaments extend toward the center of the sarcomere. Myosin filaments are present in between the actin filaments toward the middle.

Muscles contract as the actin and myosin filaments slide past one another. The filaments do not decrease in length during the process of contraction and relaxation. ATP binds to the myosin head at a specific binding site and releases myosin from the actin filament. Myosin hydrolyzes the ATP into ADP and Pi. (Note that myosin is an ATPase.) Some of the energy is used to change the position of the myosin from a "bent" low-energy state, to an "open" high-energy state. Energized myosin binds to a site on the actin filament. ADP and Pi are released from the myosin head, and this causes the head to "spring back" to its original bent configuration moving the actin filament toward the sarcomere center, pulling the Z discs inward causing contraction of the muscle. The muscles are anchored to the skeleton for exerting the force caused by the contraction. Following contraction, the muscle fibers return to their original position.

Muscle strength is increased somewhat by stretching existing muscle cells through exercise, but mostly by increasing the number of muscle cells. In contrast, muscle speed is mainly influenced by muscle length and the percentage of fast or slow muscle fibers.

Muscles and skeleton work together to control the strength and speed of animal movement. The physical model of a lever system explains how vertebrate skeleton and muscles interact to achieve adaptive movement.

Animals move by swimming, running, or flying, with all three requiring the production of thrust. Environmental drag, caused by friction with the surrounding molecules and the pressure of those molecules sticking to the body, resists thrust, and thereby restricts animal's motion. Natural selection has favored organisms that minimize drag.

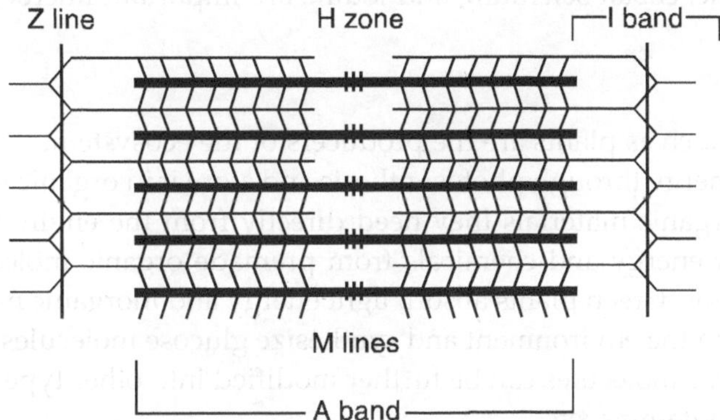

FIGURE 8.3 **Sarcomere, the 'stripes' or 'striations' across each skeletal muscle cell.**

A band. Appears dark under microscopy

I band. Appears light under microscopy

Z line. Lies in the middle of each I band

H zone. Lies in the middle of each A band

M line. Lies in the middle of each H zone.

8.5 NUTRITION

The energy that powers life comes from the sun. Producers are organisms that capture sunlight energy through photosynthesis and store it in organic molecules. They obtain all the energy and inorganic materials they need directly from the environment. Consumers are organisms that harvest energy and chemicals from premade organic molecules contained in the bodies of other organisms. Nutrition is the method by which organisms obtain materials and energy needed for sustaining life. Animals, plants, and microbes have different methods to obtain energy from sources and nutrients from the environment.

Elements of Nutrition

Organisms depend on only a small number of elements to sustain life. These include carbon, hydrogen, oxygen, nitrogen, phosphorus, and sulfur. Of these, carbon, hydrogen, and oxygen form the framework of organic molecules such as carbohydrates, proteins, fats, and nucleic acids. The other elements are restricted in distribution. Sulfur and nitrogen are present in proteins; phosphorous and nitrogen are present in nucleic acids. Many other elements also have equally important and essential roles similar to that of the biomolecules such as proteins carbohydrates, lipids, and nucleic acids. They may not be present as a component of macromolecules, but may be present as ions or as a part of other organic molecules such as vitamins and cofactors. There are a group of elements known as microelements or trace elements, which are required only in very small quantities but play an important role in metabolism. So as nutrients they are very important. Elements such as manganese, copper, zinc, cobalt selenium, and iodine are important microelements.

Plant Nutrition

Autotropic organisms such as plants are the producers of the ecosystem. Producers are organisms that capture sunlight energy through photosynthesis and store it in organic molecules. They obtain all the energy and inorganic materials they need directly from the environment. Consumers are organisms that harvest energy and chemicals from premade organic molecules contained in the bodies of other organisms. Green plants absorb light energy and inorganic molecules such as water and carbon dioxide from the environment and synthesize glucose molecules with a process known as photosynthesis. These molecules can be further modified into other types of biomolecules such as amino acids, lipids, vitamins, etc.

The chlorophyll-containing parts of the plants such as leaves are the sites of photosynthesis. The leaves are the organs specially designed to trap the maximum amount of sunlight energy and transform it into a usable form of chemical energy in organic molecules such as lipids, carbohydrates, and proteins. Leaves are flat exposing a maximum surface area against sunlight with its photosynthetic tissue, the palisade parenchyma on the upper surface. The spongy parenchyma specialized in the absorption of carbon dioxide are present on the lower half of the leaf tissue.

The root hairs are the organs adapted for the absorption of water and minerals from the soil. The root hairs are the thin-walled extension of the epidermal cells of the root and have a big vacuole in the middle. Root hairs provide increased surface area for the absorption of water and minerals actively (using energy) or passively (by osmosis). Higher plants are also able to convert inorganic nitrogen (nitrates, nitrites, and ammonia) into organic nitrogens such as amino acids and proteins but they cannot utilize elemental forms of nitrogen even though it is abundant in the atmosphere. They are also able to synthesize all types of fatty acids and other vitamins needed. But plants have symbiotic association with nitrogen-fixing bacteria. These bacteria living in the root nodules can absorb elemental forms of nitrogen and can be converted to nitrates and ammonia, and that will be further transformed into organic forms by host plants. Plants of the legume family have the symbiotic association of nitrogen-fixing bacteria in their root nodules.

Animal Nutrition

Animals and all other organisms that depend on organic food materials synthesized by producers are called **heterotrophs** and form the **consumers** of the ecosystem. The biomolecules synthesized by the green plants are the sources of energy and raw materials for the synthesis of new biomolecules needed by their systems. For example, animals consume the protein of plants and in the stomach it is converted into amino acids. These amino acids are absorbed into the cells and are used as the starting materials for the synthesis of new protein molecules. But animals are unable to synthesize amino acids. Organisms must first digest the complex macromolecules in their food before absorbing simpler chemicals into their bodies. In animals, food is first mechanically broken down and then chemically digested before being absorbed. The amount of absorptive surface area that organisms devote to nutrient acquisition is related to the type of nutrients they need. Most organisms use active transport to absorb nutrients, and large absorptive surface areas increase the rate of transport.

Some bacteria in the digestive track help the animals with their nutrition by providing some essential vitamins. Vitamins are nutrients needed in the diet in very small quantities but are very essential for normal functions of the system. Animals cannot synthesize most vitamins and should be supplemented along with the diet. There are two types of vitamins—fat-soluble vitamins (A, D, and E) and water-soluble vitamins (all B complex vitamins, C, and folic acid). Deficiency of any of the nutrients, micronutrients, or macronutrients can cause serious diseases or disorders or even deformities in the system. For example, lack of proteins in the diet can cause stunted growth in children and deficiency of certain vitamins and iron can cause anemia. Deficiency of vitamin B complex can lead to the problems related to the nervous system. The absence of vitamin C can cause scurvy (bleeding gums), deficiency of vitamin A can cause weak eyesight, and deficiency of vitamin D causes bone deformities.

The heterotrophs are classified according to the mode of nutrition. Those organisms which directly depend on the producers, are the **herbivores** and those animals that eat other animals are the **carnivores**. There are some organisms that depend on the dead organic matter of both plants animals and they are the **saprophytes** or (**detritivores**). The digestive systems of animals are diverse, but show common structural features. Most have one-way flow that allows specialized activities such as chewing, storage, digestion, absorption, and the elimination of wastes. Unlike plants, animals can move and forage for their food. Herbivores and highly mobile predators use active foraging, while less mobile predators employ sit-and-wait foraging, often with the aid of a trapping mechanism. In animals, the more difficult the diet is to digest, the more digestive specializations are present to increase the rate of nutrient acquisition. Large herbivores, in particular, have many specializations to deal with the daunting task of trying to digest enough cellulose to meet energy and nutrient needs. There are some bacteria living in the rumen of cattle and other similar herbivores, which can produce cellulase to break the cellulose present in the fodder into glucose molecules.

Nutrition in Microbes

Microorganisms include both autotrophic and heterotrophic organisms with respect to the mode of nutrition. The autotrophs include photosynthetic and chemosynthetic bacteria, and algae including cyanophyceae. The chemosynthetic bacteria use chemical energy for the synthesis of organic materials. Thus, they also belong to the producers of the ecosystem. The other bacteria, which form a greater part of the microbial population and all fungi, form the heterotrophs, which consume premade organic materials produced by plants. They live on the dead organic materials of plants and animals and are known as saprophytes. They secrete a number of enzymes to the substrate, the surface on which they grow. These enzymes digest the organic materials present on the substrate into simpler forms. These simple molecules are absorbed into the cells and are used for synthesizing new organic molecules.

8.6 GASEOUS EXCHANGE

All living organisms exchange carbon dioxide and oxygen with their environment for carrying out different functions. Organisms must exchange gases with their environment in order to carry out photosynthesis, cellular respiration, and other essential processes. This gas exchange is accomplished exclusively through passive diffusion.

Oxygen and carbon dioxide are the two most important gases that organisms exchange. Certain microorganisms are also capable of taking up nitrogen gas for the purpose of nitrogen fixation. Three primary factors govern gas exchange: the concentration gradient of the gas, the amount of surface area available for gas exchange, and the distance over which the diffusion takes place. Gas concentrations differ between air and water. The concentration of oxygen is much higher in air than in water, whereas carbon dioxide concentration is similar in both. Density differences between water and air influence the uptake of gases. Diffusion is slower in water because water is denser than air. Aquatic organisms must therefore move large quantities of water over their gas exchange surfaces to get the same amount of gas found in a much smaller volume of air. Oxygen has a higher concentration in the atmosphere than does carbon dioxide. Oxygen therefore diffuses into animals more rapidly than carbon dioxide diffuses into plants.

Organisms use different methods and organs for the purpose of gas exchange, which depends on their habitat.

Body surface. Small organisms and plants accomplish gas exchange by simple diffusion across their cell or body surfaces. Multicellular organisms, which have a large surface area relative to their body volume, exchange gases across their body surface. Earthworms, algae, etc. are examples. Body surfaces specialized for gas exchange are moist, which allows easy exchange of gases.

Gills. Larger, more complex organisms use specialized gas exchange structures such as gills and lungs. Gills provide large surface area for gas exchange. Gills are convoluted growths covered by thin epithelial layers, containing a rich supply of blood vessels. Since gases cannot reach individual body cells directly from these structures, large organisms use transport fluids to move gases around their bodies. The surface area devoted to gas exchange in an organism is directly proportional to the organism's size and metabolic needs. Gas exchange on land is a problem because of the water loss associated with exposing the exchange surfaces to dry air. Terrestrial gas exchange is thus a compromise between obtaining necessary gases and avoiding water loss.

Tracheal system. This is a network of tubes present in insects. These tracheal systems open outside through openings called spiracles, through which oxygen enters into the network of tubes. Thus, there is direct exchange of gases between the cells and the tracheal tubes.

Alveoli. These are the structures present in the lungs. Lungs are the well-developed organ system for the effective exchange of gases between atmosphere and the blood present in the terrestrial organisms. Lungs are elastic sacs that allow the animal to pump air in and out of the body. The trachea of the lungs divide into two branches, which in turn divide into many sub-branches known as bronchioles. The bronchioles end in small, thin-walled sacs known as alveoli. The presence of alveoli increases the effective surface area for gaseous exchange. Alveoli have a rich supply of fine blood vessels with very thin walls through which gas can easily diffuse into and out.

Gas Exchange in Plants

In plants gas exchange usually takes place through stomata, small openings present on the epidermis of the leaves. The stomata open into spongy parenchyma and the gas exchange takes place between the cells and the gas filled in the air space. Gas exchange is needed for both respiration and photosynthesis.

8.7 INTERNAL TRANSPORT

Living things must be capable of transporting nutrients, wastes, and gases to and from cells. Single-celled organisms use their cell surface as a point of exchange with the outside environment. Multicellular organisms have developed transport and circulatory systems to deliver oxygen and food to cells and remove carbon dioxide and metabolic wastes. Simple multicellular organisms such as sponges, multicellular fungi, and algae have a transport system. Sea water is the medium of transport and is propelled in and out of the sponge by ciliary action. Simple animals, such as hydra and planaria, lack specialized organs such as hearts and blood vessels, and instead use their skin as an exchange point for materials. This, however, limits the size an animal can attain. To become larger, they need specialized organs and organ systems. In lower plants such as algae and fungi, transport of material takes place through the body surface and cytoplasmic streaming movements.

Any system of moving fluids which reduces the functional diffusion distance that nutrients, wastes, and gases must traverse may be referred to as an internal transport or circulatory system.

Internal Transport in Animals

All animals must maintain a homeostatic balance in their bodies. This requires the circulation of nutrients, metabolic wastes, and respiratory gases through the animal's body. Multicellular animals do not have most of their cells in contact with the external environment and so have developed circulatory systems to transport nutrients, oxygen, carbon dioxide, and metabolic wastes. Components of the circulatory system include:

- Blood: a connective tissue of liquid plasma and cells.
- Heart: a muscular pump to move the blood.
- Blood vessels: arteries, capillaries and veins that deliver blood to all tissues.

There are several types of circulatory systems. Organisms, such as hydra, have a fluid-filled, internal gastrovascular cavity. This cavity supplies nutrients for all body cells lining the cavity, obtains oxygen from the water in the cavity, and releases carbon dioxide and other wastes into it. The gastrovascular cavity of a flatworm, such as the planarian, is more complex than that of the hydra. The open circulatory system is common to molluscs and arthropods. Open circulatory systems (evolved in insects, mollusks, and other invertebrates) pump blood into a hemocoel with the blood diffusing back to the circulatory system between cells. Blood is pumped by a heart into the body cavities, where tissues are surrounded by the blood. The resulting blood flow is sluggish.

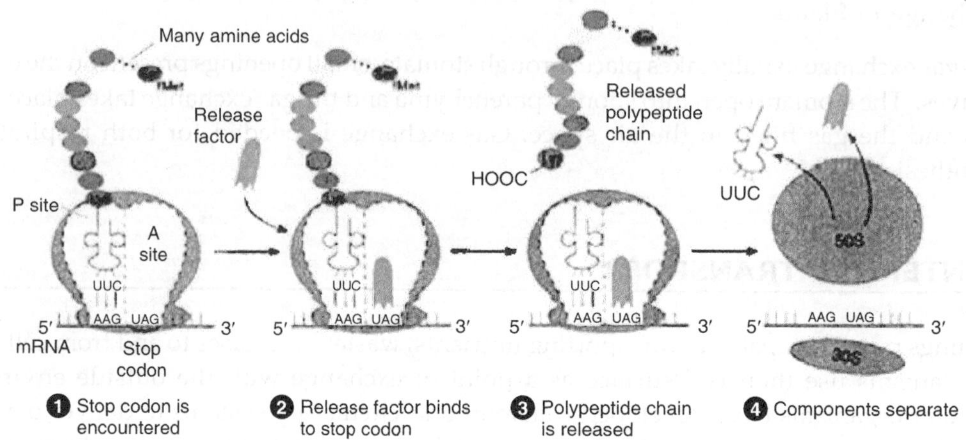

FIGURE 8.4 **Circulatory system of an insect.**

Higher animals such as vertebrates, and a few invertebrates, have a closed circulatory system. Closed circulatory systems have the blood enclosed within blood vessels of different sizes and wall thicknesses. It is not released in between the cells. In this type of system, blood is pumped by a heart through vessels, and does not normally fill body cavities. Blood flow is not sluggish. Hemoglobin causes vertebrate blood to turn red in the presence of oxygen; but more importantly, hemoglobin molecules in blood cells transport oxygen. The closed circulatory system present in humans is called the **cardiovascular system**. A circulatory or cardiovascular system is a specialized system that moves the fluid medium, hemolymph or blood, in a specific direction determined by the presence of unidirectional blood vessels.

The vertebrate cardiovascular system includes a heart, which is a muscular pump that contracts to propel blood out to the body through arteries, and a series of blood vessels. The upper chamber of the heart, the atrium, is where the blood enters the heart. Passing through a valve, blood enters the lower chamber, the ventricle. Contraction of the ventricle forces blood from the heart through an artery. The heart muscle is composed of cardiac muscle cells. Arteries are blood vessels that carry blood away from heart. Arterial walls are able to expand and contract. The aorta is the main artery leaving the heart. The pulmonary artery carries deoxygenated blood to the lungs. In the lungs, gas exchange occurs—carbon dioxide diffuses out and oxygen diffuses in. **Arterioles** are small arteries that branch into collections of capillaries known as capillary beds. **Capillaries** are thin-walled blood vessels in which gas exchange occurs. Capillaries are concentrated into capillary beds. Nutrients, wastes, and hormones are exchanged across the thin walls of capillaries.

The circulatory system functions in the delivery of oxygen, nutrient molecules, and hormones and the removal of carbon dioxide, ammonia, and other metabolic wastes. Capillaries are the points of exchange between the blood and surrounding tissues. Materials cross in and out of the capillaries by passing through or between the cells that line the capillary. (See Appendix)

Blood leaving the capillary beds is collected into a progressively larger series of **venules** (venules are smaller veins that gather blood from capillary beds into veins), which in turn join to form veins. **Veins** carry blood from capillaries to the heart. With the exception of the pulmonary veins, blood in veins is oxygen-poor. The pulmonary veins carry oxygenated blood from lungs back to the heart.

Internal Transport in Plants

Land plants require a transport system because unlike their aquatic ancestors, photosynthetic plant organs have no direct access to water and minerals. The internal transport of plant system involves the lifting of water and minerals to great heights to the tips of all branches and leaves, against the gravitational pull.

Three levels of transport occur in plants:

- Uptake of water and solutes by individual cells.
- Short-distance, cell-to-cell transport at the level of tissues and organs.
- Long-distance transport of sap in xylem and phloem at the whole-plant level.

Transport at the cellular level depends on the selective permeabilities of membranes.

Active and Passive Transport of Solutes

The plasma membrane's selective permeability controls the movement of solutes between a plant cell and the extracellular fluids. Solutes may move by passive or active transport.

Passive transport occurs when a solute molecule diffuses across a membrane down a concentration gradient with no direct expenditure of energy by the cell. Transport proteins embedded in the cell membrane may increase the speed at which solutes cross. **Transport proteins** may facilitate diffusion by serving as carrier proteins or forming selective channels. **Carrier proteins** bind selectively to a solute molecule on one side of the membrane, undergo a conformational change, and release the solute molecule on the opposite side of the membrane. **Selective channels** are passageways by which selective molecules may enter and leave a cell; some gated selective channels are stimulated to open or close by environmental conditions.

Active transport occurs when a solute molecule is moved across a membrane against a concentration gradient. It is an energy-requiring process. The **proton pump** is an active transporter important to plants.

Water Potential and Osmosis

Osmosis results in the net uptake or loss of water by the cell and depends on which component, the cell or extracellular fluids, has the highest water potential. **Water potential** is the free energy of water that is a consequence of solute concentration and applied pressure. Water potential is the physical property predicting the direction of water flow. Water will always move across the membrane from the solution with the higher water potential to the one with lower water potential. Pure water has a water potential of zero, and addition of solutes lowers water potential into the negative range. Increased pressure raises the water potential into the positive range. A negative pressure may also move water across a membrane; this bulk flow (movement of water due to pressure differences) is usually faster than movement caused by different solute concentrations. Plant cells will gain or lose water to intercellular fluids depending upon their water potential. A flaccid cell placed in a **hyperosmotic** solution will lose water by osmosis; the cell will **plasmolyze** (protoplast moves away from cell wall). A flaccid cell placed in a **hypoosmotic** solution will gain water by osmosis; the cell will swell and a turgor pressure develops; when pressure from the cell wall is equal to the osmotic pressure, equilibrium is reached and no net movement of water occurs.

Transport Within Tissues and Organs

The symplast and apoplast both function in transport within tissues and organs. Lateral transport is usually along the radial axis of plant organs, and can occur by three routes in plant tissues and organs. These are:

1. **Across the plasma membrane and cell walls.** Solutes move from one cell to the next by repeatedly crossing plasma membranes and cell walls.
2. **The symplast route.** A **symplast** is the continuum of cytoplasm within a plant tissue formed by the **plasmodesmata**, which passes through pores in the cell walls. Once water or a solute enters a cell by crossing a plasma membrane, the molecules can enter other cells by traveling through plasmodesmata.
3. **The apoplast route.** An **apoplast** is the continuum between plant cells, which is formed between the continuous matrix of cell walls. Water and solute molecules can move from one area of a root or other organ via the apoplast without entering a cell.

Water and solute molecules can move laterally in a plant organ by any one of these routes or by switching from one to another.

Bulk Flow Functions in Long-distance Transport

This type of transport is usually along the vertical axis of the plant from root to leaves and vice versa. Vascular tissues are involved in this type of transport, as diffusion would be too slow. Bulk flow (movement due to pressure differences) moves water and solutes through **xylem vessels** and **sieve tubes**. Transpiration reduces pressure in the leaf xylem; this creates a tension, which pulls sap up through the xylem from the roots. Hydrostatic pressure develops at one end of the sieve tubes in the phloem; this forces the sap to the other end of the tube.

Absorption of water and minerals by roots: Water and minerals enter through the root epidermis, cross the cortex, pass into the stele, and are carried upward in the xylem.

Active accumulation of mineral ions: The cells cannot get enough mineral ions from the soil by diffusion alone. The soils solution is too dilute. **Active transport** of these ions must occur. Specific carrier proteins in the plasma membrane attract and carry their specific mineral into the cell. H^+ is pumped out of the cell causing a change in pH and a voltage across the membrane. This helps drive the anions and cations into the cell. Water and minerals cross the cortex either by symplast, which is the living continuum of cytoplasm, connected by plasmodesmata or by apoplast, which is nonliving matrix of cell walls. At the endodermis the casparian strip blocks the apoplastic route. This is a ring of suberin around each endodermal cell. Here, water and minerals must enter the stele through the cells of the endodermis. Water and minerals enter the stele via symplast, but xylem is part of the apoplast. Transfer cells selectively pump ions out of the symplast into the apoplast so they may enter the xylem. This action requires energy.

The ascent of xylem sap depends mainly on transpiration and the physical properties of water. The shoot depends on the efficient delivery of its water supply. Xylem carries sap containing dissolved mineral and nutrients from the roots to the leaves. Water is pulled up through the xylem by the force of transpiration. Water transported up from roots must replace that lost through transpiration. The connection between the xylem gives a continuous column of water to form between roots and leaves. This column acts like a thread that moves upward by the pull of transpiration. Transpiration pulls the xylem sap upward, and cohesion of water transmits the upward pull along the entire length of xylem. The forces responsible for the ascent of sap through xylem are Transpiration, Adhesion, Cohesion, and Tension **(TACT)**.

Translocation

The transport of sap containing dissolved products of photosynthesis through phloem is called translocation. **Translocation** is the transport of the products of photosynthesis by phloem to the rest of the plant. In angiosperms, sieve-tube members are the specialized cells of phloem that function in translocation. Sieve-tube members are arranged end-to-end forming long sieve tubes. Porous cross walls called **sieve plates** in between the members allow phloem to move the solution freely along the sieve tubes. Phloem sap contains primarily sucrose, but also minerals, amino acids, and hormones. Phloem sap movement is not unidirectional; it moves through the sieve tubes from source to the sink. **Source** is the organ where sugar is produced by photosynthesis or by the breakdown of starch (usually leaves) and **sink** is the organ that consumes or stores sugar (growing parts of plant, fruits, non-green stems and trunks, and others). Sugar flows from source to sink. Source and sink depend on season. A tuber is a sink when stockpiling in the summer, but it is a source in the spring. The sink is usually supplied by the closest source. Direction of flow in phloem can change, depending on locations of source and sink.

8.8 MAINTAINING INTERNAL ENVIRONMENT

The extremes of environmental conditions upset the internal environment of cells and organisms, which can severely damage their system and performance or even life. There is only a very narrow range of environmental conditions under which organisms and cells can maintain their life activities. A stable internal environment must be maintained, even in the face of sudden changes in external conditions. For example, when you feel "cold," that perception sets in motion a sequence of events that involve three different hormones, and three different types of tissues: nerve cells, muscle cells, and endocrine cells. The result is the production of body heat, to maintain a constant body temperature. Maintaining a constant internal environment in spite of changes in the external environment is known as **homeostasis.** Homeostasis involves both **sensors** (in this example, heat-sensing nerve cells) and **effectors** (which initiates a response, such as shivering muscle cells).

There are two factors, which play a role in maintaing the internal environment. They are regulation of water and regulation of temperature.

Regulation of Water Balance

Water is a very important element of life. There is a constant and steady loss of water from the body in different ways. For example, transpiration and excretion can reduce the water content of the body substantially and therefore should be replaced. Water is the medium for biochemical reactions and is needed in sufficient quantities for the easy diffusion and flow of dissolved and suspended materials within the body. A disturbance in the water balance in the body can disturb entire metabolic activities.

In animals, the site of homeostasis is the kidney. Kidneys are concerned with the isolation and excretion of metabolic waste products. Kidneys filter a large volume of water, solutes, and wastes everyday. But major amounts of water and solutes are reabsorbed. Only a small amount of water, solutes, and all wastes are excreted as urine. This is enough to create a shift in the water balance of the body. But we take in a lot of water along with food and otherwise. In the case of fresh water fishes, since the solute concentration in the cells is higher than the surrounding water excess water gets into their cells. The excess water is removed by excreting very dilute urine. Whereas, fish living in salt water have a different mechanism for water regulation. They take in a lot of salt water and the excess salt is pumped out through their gills. This is an active mechanism and needs lot of energy. In addition to this they produce highly concentrated urine in their specialized kidneys. Thus, they can remove a lot of salt by conserving water. This is the mechanism of homeostasis in marine organisms.

In plants, different mechanisms and modifications regulate water loss. Terrestrial plants regulate water balance through roots and leaves. For example, during dry conditions certain plants drop their leaves to check water loss. In certain soil conditions where there is high salt concentration, certain plants actively pump minerals through their roots, which increases the solute concentration of the cell sap in the root hair. As the salt concentration is higher in the cell sap, water diffuses into the cell by osmosis. Plants have several mechanisms to conserve water. In desert plants, there is a very thick waxy coating all over the body to prevent water loss by evaporation. In most cases, stomata are located on the lower side of the leaves and also there is mechanisms to close the stomata during the day in those plants that are growing in dry weather.

Regulation of Temperature

Regulation of temperature is known as **thermoregulation.** It is almost as important as osmoregulation. Biochemical reactions of the cells are taking place under an optimal temperature. Therefore, the stability of the optimal body temperature is very essential for the normal functioning of all biological reactions. The variations in the body temperature can affect the functioning of the organisms in a very lethal way. Depending on the influence of the environmental temperature on the body temperature the animals can be of two types: **ectotherms** and **endotherms** organisms.

Ectotherms use the external environment and behavioral mechanisms to maintain a thermal balance as much as possible. But their metabolic machinery must be generalized. The enzymes and metabolic reactions have to be functional in a wide temperature ranges. Their enzymes have the ability to work in a wide range of temperatures. For example, insects tend to be maxitherms when given the choice (fish, to digest, move down to cooler temperatures to digest food, while feeding in warm surface waters). Many ectotherms show behavioral and structural adaptations, which help them to adjust to temperature changes. For example, amphibians undergo hibernation during the hot summer. Another example, marine iguanas, feed in cold ocean waters, where they lose heat quickly. They try to gain heat by orienting their black-colored body parts directly to the radiant heat of sun. They also try to get heat by pressing their belly against the rocks that have been warmed by the sun.

Endotherms are organisms that have a constant body temperature and use a variety of physiological mechanisms to maintain a constant internal temperature. The basic mechanism is to have a constantly active metabolic machine. Endotherms such as mammals and birds have very effective temperature control methods. They have a very high respiratory rate, which generates metabolic heat. They also conserve heat by minimizing the heat loss. The heat loss can be minimized by decreasing conductivity or by increasing insulation, or by vasoconstriction or vasodilatation, or by changing the color of insulation. For example, feathers. Air trapped in fur or feathers acts as an insulator, which prevents the loss of heat.

Panting by birds and mammals and sweating in mammals are effective methods for cooling the system. In certain organisms minimized respiration prevents the loss of excess heat from the body. In mammals there is the mechanism of thermogenesis by shivering, a type of active thermoregulation controlled by the hypothalamus of the brain through a negative feedback thermostat.

There are some leaves with a shiny leaf surface to cut down the radiant heat by reflection. Certain desert plants are provided with folded ridges on the trunks, so that sunlight falls only at shallow angles resulting in reduced absorption of radiant heat.

All living organisms are provided with different structural and physiological modifications and adaptations to maintain homeostasis.

8.9 REPRODUCTION

Reproduction is one of the most important characteristics of living organisms. It can be defined as the ability of microorganisms to produce individuals of the same species. This process transfers the genetic material from the parental generation to the next generation. Reproduction is necessary to maintain the genetic continuity among a species and it allows the increase in the total number of members in a species or variety. The process of reproduction involves the participation of special types of cells and processes of cell division involving genetic recombination. Therefore, reproduction can act as means to generate genetic variations and diversity among species.

There are different methods of reproduction among various types of living organisms. It can be broadly classified into **asexual** reproduction, **sexual** reproduction, and **vegetative** reproduction.

Reproduction in Microbes

Microorganisms such as bacteria and other single-celled organisms mainly reproduce by asexual methods. In bacteria, a single organism divides into two identical cells, the progenies. This type of asexual reproduction is called **binary fission**. By this reproduction, the number of bacteria increases exponentially and this phase of growth is called the **exponential growth phase** or **logarithmic growth**. In the asexual method of reproduction there is no mixing of genetic materials of two cells; therefore, all progenies are of the same genetic makeup. If any variation is observed in the population of an asexually reproducing organisms, it may be due to random variations caused by mutation that occurs during DNA replication.

There is a primitive type of sexual reproduction in bacteria known as **conjugation**. It does not involve the fusion of any gametes or fertilization. But the essential feature of sexual reproduction, namely genetic recombination (mixing of genetic material), occurs. During the process of conjugation a part of genetic material (DNA) is transferred to the recipient cell through a temporarily formed conjugation tube established between the cells.

There are three processes by which genetic recombination can occur in bacteria. They are **transformation, conjugation,** and **transduction.**

Transformation: There is no contact between the donor cell and the recipient cell. It is the random picking of DNA fragments released by some other cells. The recipient cell actively takes the DNA fragment and inserts it into the genome of the recipient bacteria.

Conjugation: This is the DNA transfer between two bacteria through a long protoplasmic connection established between the cells. Conjugation occurs between two different types of bacteria known as positive and negative strains, equivalent to male and female cells. They are called **F+** and **F–**. The donor or the F+ strains of bacteria harbor a circular molecule of DNA known as **F factor** (F for fertility) or the sex factor. This factor is absent in the F–strains. The presence of F factor is responsible for the presence of a special type of pilus, known as **F pilus** or **sex pilus**. The sex pilus is responsible for the formation of the cell-to-cell contact and the formation of the protoplasmic connection, known as the **conjugation tube**, between the cells. After the establishment of the conjugation tube, one of the strands of the F–factor DNA passes through the sex pilus or the conjugation tube from the donor (F+) to the recipient (F–). In some cases, the F factor plasmid gets integrated with the genomic DNA and mobilizes the transfer of the genomic DNA to the recipient. Since such strains show high-frequency recombination, they are called **Hfr strains**.

Transduction: This is another important type of genetic recombination in bacteria which takes place through bacteriophages. A fragment of DNA is transferred to a recipient through a bacteriophage. A **bacteriophage** is a virus, which infects bacteria. When a bacteriophage infects a bacterium, it injects its DNA into the bacterial cell and it gets integrated with the bacterial DNA. This phage DNA undergoes multiplication along with bacterial DNA. After some multiplications, the phage DNA comes out of the bacterial genome and will be encapsidated in protein coats to daughter phages. In this process a small bit of bacterial DNA will also be taken along with phage DNA. When this phage infects another bacteria, the DNA bit of the old bacterial host will be transferred to the new bacterial genome and causes a genetic recombination. The old bacterial host is the genome and the new bacterial host of the phage is the recipient.

Reproduction in Plants

In higher plants there are mainly two methods of reproduction—sexual reproduction that involves the formation of gametes and asexual or vegetative reproduction, in which there the vegetative parts are used for propagation.

Sexual Reproduction

Typically, plant life history involves alternation of generations, during which a diploid sporophyte gives rise to a haploid gametophyte. The gametophyte generation produces gametes that, through syngamy (fusion of gametes during fertilization), provide for another generation of diploid sporophytes, to continue the cycle. The sporophyte does not produce gametes but rather, meiosis occurs in spore mother cells and produces haploid spores. These spores divide mitotically to form gametophytes, which subsequently produce gametes via mitosis. Within the plant kingdom the dominance of phases varies. In nonvascular plants such as mosses and liverworts, the gametophyte phase is dominant. Vascular plants show a progression of increasing sporophyte dominance from the ferns and "fern allies" to angiosperms.

In angiosperms, flowers are the organs of reproduction. A typical flower has four parts arranged in circles: **sepals, petals, androecium,** and **gynoecium** from periphery to center. Androecium and gynoecium are the male and female reproductive organs.

Androecium: The male reproductive organ of the flower is composed of units called **stamens**. Each stamen consists of a **filament** and an **anther.** The anther produces the male spores called **pollen grains**.

Gynoecium: The female reproductive organ consists of units called **pistils**. Each pistil consists of terminal filament called **style** with **stigma** at its terminal part and an **ovary**, from which the style starts. The ovary contains **ovules**. The flowers may be bisexual with both male and female organs in the same flower, or it may be unisexual with any one type of sex organs. The male flowers are the **staminate flowers** and the female flowers are the **pistillate flowers**.

Development of Spores and Pollination

Gametophyte formation occurs within the flowers of plants. Development of the male gametophyte (pollen grain) occurs within four pollen sacs (microsporangia) in the anther; development of the female gametophyte (embryo sac) occurs within the ovule. Meiosis of the micro and megasporocyte occurs, resulting in four haploid spores. However, only one of the megaspores survives. Each microspore undergoes mitosis to yield a pollen grain, which contains a generative cell and a pollen tube nucleus. Subsequently, the generative cell divides to form two sperm nuclei inside the pollen tube. The megaspore undergoes three mitotic divisions, producing eight nuclei within the mature embryo sac. Pollination is the phenomenon whereby the pollen grain reaches the stigma. There are two mechanisms whereby pollination occurs:

Animal pollination or insect pollination. Food or another reward is provided and "advertised" via color or fragrance. This mechanism can be effective in promoting pollination even in sparse populations.

Wind pollination requires dense populations.

Self-pollination (which would result in inbreeding) is frequently prevented in plants, even though they might have perfect flowers. Flowers may be imperfect, plants may be dioecious, or compatibility genes might be required for successful pollination to occur.

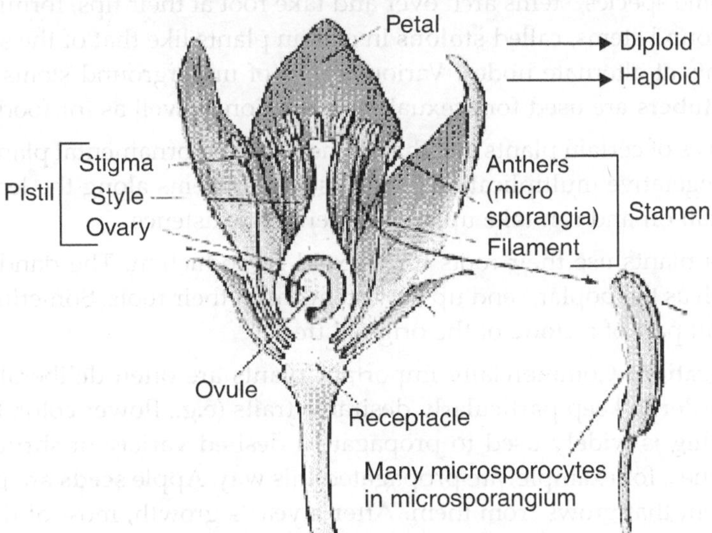

FIGURE 8.5 **Parts of a flower.**

Once pollination occurs, the pollen coat ruptures and the pollen tube grows downward through the style to the ovary. Once it enters the embryo sac via the micropyle, the pollen nuclei migrate downward through the tube and double fertilization occurs.

Double Fertilization

During **double fertilization**, the two sperm nuclei enter the embryo sac. One fuses with the egg nucleus (this syngamy leads to the diploid nucleus of the zygote) while the other fuses with the two polar nuclei in the center (leading to a triploid endosperm nucleus). Later, mitosis of the endosperm and fertilized egg lead to formation of the cotyledons and embryo, respectively.

Sexual reproduction causes a lot of genetic variations and thereby new genetic combinations or genotypes by the mendalian segregations and recombination.

Asexual Reproduction

Asexual reproduction in plants does not involve any process of fertilization or syngamy. New plants are produced from somatic cells or from unfertilized gametes. Asexual reproduction in plants is of two types: vegetative reproduction and apomixis.

Vegetative Reproduction

It is reproduction by mitosis allowing a new, genetically identical individual to be produced. When a very desirable combination of traits is found, sexual reproduction risks losing them in the randomness of the process. Asexual reproduction does not allow genetic variation, but guarantees reproduction (no dependence on others). It rapidly increases the numbers of an organism and keeps its desired combination of traits. Many plants use a combination of sexual and asexual reproduction to get the benefits of both methods.

Most plant organs have been used for **asexual** reproduction, but stems are the most common.

Stems: In some species, stems arch over and take root at their tips, forming new plants. The horizontal aboveground stems, called **stolons** in certain plants like that of the strawberry, produce new daughter plants at alternate nodes. Various types of underground stems such as **rhizomes**, **bulbs**, **corms**, and **tubers** are used for asexual reproduction as well as for food storage.

Leaves: Leaves of certain plants like that of the common ornamental plant bryophyllum acts as the organs for vegetative multiplication. Mitosis at meristems along the leaf margins produce tiny plantlets that fall off and can take up an independent existence.

Roots: Some plants use their roots for asexual reproduction. The dandelion is a common example. Trees, such as the poplar, send up new stems from their roots. Sometimes, an entire grove of trees may form all part of a **clone** of the original tree.

Plant Propagation: Commercially important plants are often deliberately propagated by asexual means in order to keep particularly desirable traits (e.g., flower color, flavor, resistance to disease etc.). **Grafting** is widely used to propagate a desired variety of shrub or tree. All apple varieties, rose varieties, for example, are propagated this way. Apple **seeds** are planted only for the root and stem system that grows from them. After a year's growth, most of the stem is removed and a twig (**scion**) taken from a mature plant of the desired variety is inserted in a notch in the cut stump (the **stock**). So long as the cambiums of scion and stock are united and precautions are taken to prevent infection and drying out, the scion will grow. It will get all its water and minerals from the root system of the stock. However, the fruit that it will eventually produce will be identical (assuming that it is raised under similar environmental conditions) to the fruit of the tree from which the scion was taken. Cuttings may be taken from the parent and rooted. The same method of grafting is applicable in the case of rubber plantations.

Apomixis

Citrus trees and many other species of angiosperms use their seeds as a method of asexual reproduction; a process called **apomixis**. There are **obligate apomixis** and **facultative apomixis**. The first one is present in those plants in which sexual reproduction is completely absent. The second one is present in angiosperms. The facultative apomixis includes many types. In one form of apomixis, the egg is formed with **2n** chromosomes and develops without ever being fertilized. In another version, the cells of the ovule (**2n**) develop into an embryo instead of—or in addition to—the fertilized egg. Hybridization between different species often yields infertile offspring. But in plants, this does not necessarily trouble the offspring. Many such hybrids use apomixis to propagate themselves.

In a rare cypress, the pollen grains are diploid, not haploid, and can develop into an embryo when they happen to fall on either the female cones (female reproductive organ) of their own species (rare) or those of a much more common species of cypress. This can be considered **paternal apomixis** in a surrogate mother, a desperate attempt to avoid extinction.

Reproduction in Animals

Asexual Reproduction

In animals the asexual reproduction is limited to unicellular and lower forms organisms. There are different methods of asexual reproductions.

Binary fission

The parent cell divides into two or more daughter cells by a cleavage of the protoplast after the nuclear division. Normally two identical cells are formed. Therefore, it is known as **binary fission**. It is usually observed in organisms such as plasmodium, paramecium, amoeba, etc. When the parent cell is divided into a number of progenies it is called **multiple fission**. First, the nucleus undergoes repeated division, which is followed by cytoplasmic division into a number of daughter cells. The cell undergoing multiple fission is called **schizont** and the process is known as **schizogamy**.

Budding

Here, offspring develops as a growth on the body of the parent. In some species (e.g., jellyfishes) the buds break away and take up an independent existence. In others (e.g., corals) the buds remain attached to the parent and the process results in **colonies** of animals. Budding is also common among parasitic animals such as tapeworms.

Fragmentation

In certain tiny worms, as they grow to full size, they spontaneously break up into eight or nine pieces. Each of these fragments develops into a mature worm and the process is repeated.

Parthenogenesis

In **parthenogenesis** ("virgin birth"), the female produces eggs, but these develop into young without ever being fertilized. Parthenogenesis occurs in some fishes, several kinds of insects, and a few species of lizards. In a few species it is the only method of reproduction, but more commonly animals turn to parthenogenesis only at certain times. For example, aphids use parthenogenesis in the spring when they find themselves with ample food. Reproduction by parthenogenesis is more rapid than sexual reproduction, and the use of this mode of asexual reproduction permits the animals to quickly exploit the available resources.

Parthenogenesis is forced on some species of wasps when they become infected with bacteria such as the genus **Wolbachia**. In these wasps (as in honeybees), fertilized eggs (diploid) become females; unfertilized (haploid) eggs become males. However, in Wolbachia-infected females, all their eggs undergo endoreplication producing diploid eggs that develop into females without fertilization; that is, by parthenogenesis. Treating the wasps with an antibiotic kills off the bacteria and "cures" the parthenogenesis.

Sexual Reproduction

Sexual reproduction is the most common method of reproduction in higher forms of animals. Sexual reproduction involves the formation of male and female gametes and their fusion resulting in the formation of a zygote. The gametes are produced after the meiosis and are **haploid** and the fusion product of gametes, the zygote, is **diploid**. The process of syngamy or fertilization causes the mixing of genetic materials of the male and female parent.

Animal development commonly proceeds through several stages.

Gametogenesis, during which the egg and sperm mature within the reproductive organs of the parents. **Fertilization**, which begins when a sperm penetrates an egg and is completed when the sperm and egg nuclei fuse forming a **zygote**. The zygote undergoes mitotic cell divisions that form the early multicellular embryo, which develops into an organism gradually through different stages. Sexual reproduction thus shows the formation diploid and haploid stages in the life cycle alternately.

The two types of gametes—the male gametes and female gametes—may be produced in the same body or separately by the male and female parent. When separate male and female individuals are present in a species, they are called **unisexual organisms**. All higher forms of animals and humans are examples. Some species are capable of producing both male and female gametes in the same organisms (body) and such species are known as **bisexual** or **hermaphrodites**. Some lower forms of organisms such as earthworms, tapeworms, snails, and some fish belong to this class.

Human Reproduction

Reproduction in humans occurs only by sexual methods. The gametes in human reproduction are the sperm and egg. Sperms are the male gametes and egg is the female gamete. Both are produced after meiosis and therefore are haploid. Humans are unisexual and thus male and female gametes are produced in separate individuals. Humans have a pair of primary reproductive organs; sperm-producing testes in males and egg-producing ovaries in females, along with accessory ducts and glands. Testes and ovaries also produce hormones that influence reproductive functions and secondary sexual traits. The hormones testosterone, LH (luteinizing hormone), and FSH, (follicle stimulating hormone) control sperm production. The hormones estrogen, progesterone, FSH, and LH control egg maturation and release, as well as changes in the lining of the uterus, the endometrium. The **testis** is an ovoid-shaped gland consisting of coiled tubules called **seminiferous tubules**. The **testes** are placed in a scrotal sac outside the abdominal cavity. The male gametes or sperms are produced in seminiferous tubules by a complex process called **spermatogenesis**. Sperm produced in the testes are carried to the copulating organ, the **penis** via **epididymis** (stores sperm until they have matured) and vas deference. **Vas deference** is a straight tube, which along with spermatic artery and vein, forms the spermatic cord. Seminal vesicle, prostate gland, Cowper's gland, etc. are the accessory parts present along with male genital organs.

The female reproductive organs consist of **ovaries** and **fallopian tubes.** Ovaries produce the female gamete **ovum** (egg) and fallopian tubes are a pair of ducts, one from each ovary, that catches the ovum released at **ovulation** each month. This tube carries the egg and houses the fertilized egg through its embryonic development and is lined with smooth muscle, which contracts, moving the ovum toward the uterus. The fertilization of ovum with sperm usually occurs in this tube and the initial development of resulting **zygote** into an *embryo* also occurs here. The **uterus** is lined with **endometrium**; this is where the embryo implants and completes its development the **cervix** is the muscular ring at the mouth of the uterus and the **vagina** is a thin-walled chamber into which sperm are directly deposited during intercourse. The **urethra** is part of the female urinary tract, but not part of the female reproductive tract, unlike males. The production of sperm is a continuous process starting from puberty and lasting throughout life in males. But in females the production of female gamete or the egg is a cyclic process with a periodicity of about 28 days. During these periods there is a great change in the structure and function of the entire reproductive system. At birth, the female has about 2 million primary oocytes, which give rise to what will become a mature egg, or ovum. Unlike the male, no more primary oocytes or cells that will give rise to a mature gamete are produced after a female is born. At birth the primary oocytes are in a resting state and will not develop any further until they are triggered by the hormone FSH released from the pituitary, at which point a few at a time will resume meiosis. Only 400 of the original 2 million primary oocytes actually develop into mature eggs.

During copulation sperm is deposited near the cervix in the vagina. These sperms are motile and active for some times—at least for three days. They move toward the fallopian tubes where they may come in contact with the egg cell. During fertilization, the sperm cell injects its nucleus into the cytoplasm of the oocyte and fertilization takes place. Only one sperm can fertilize an egg and further fusion with sperm is prevented. By fertilization the diploid number of chromosomes is restored in the fertilized egg, which is known as the zygote. The zygote starts its development in the fallopian tube and continues its development in the uterus.

8.10 ANIMAL AND PLANT DEVELOPMENT

Animal Development

The developmental pattern of a fertilized egg into an embryo is almost identical in all animal forms at least in the initial stages. Following fertilization, the zygote undergoes a series of divisions that leads from a single cell to a collection of cells in the form of a hollow sphere, known as **blastula.** The cells increase in number very rapidly at the same time the size of the cells decreases. During this period some of the blastula cells begin to differentiate into **endoderm, mesoderms** and **ectoderm.** The endoderm generally gives rise to epithelial lining of the gut, the mesoderm forms the muscles, internal skeleton; and the ectoderm develops into nerves and the outer covering of the animal. During this stage there is the rearrangement in the position of the cellular layers by a process known as **gastrulation.** After gastrulation the endoderm becomes the innermost layer of cells and the mesoderm surrounds the endoderm and the outermost layer is the ectoderm. Endoderm can further specialize into liver, pancreas, lung or many other cell types, but cannot reverse course and become ectoderm or mesoderm. This stage of embryo is known as **gastrula.** There is no significant growth in size between zygote and gastrula. All these stages are so important that all vertebrates, in spite of their great anatomical and physiological differences, follow this developmental pattern.

Once the gastrulation has taken place with the rearrangement of different layers, the cell differentiation starts rapidly. The three layers of cells differentiate and develop into various organs needed to make a functional individual. Just three weeks after fertilization, human embryos will develop a heart and by the eighth week of development the head is completely identifiable in an embryo with a 2.5 cm length. Almost all types of tissues and organs start developing by this time. By the twelfth week of development it develops external recognizable parts such as sex organs, fingers, nails, and toes. A gut also develops from the endoderm during this period.

How does an animal develop from a single cell? The answer to this fundamental question of embryology and developmental biology is based on asymmetry in the egg cell and instructions in the DNA of the developing animal. Structures form in the developing embryo under the guidance of the DNA instructions that are the same in each cell, and external cues that let the cell know where it is and what type of cell it should become. Signals include information from neighbor-neighbor contact and from gradients of protein or small-molecule morphogens.

Plant Development

The process of double fertilization results in the formation of zygote and an endosperm cell inside the embryo sac of ovule. The ovule is within the ovary. The zygote follows series of mitotic cell divisions and differentiates into a small plant later known as the embryo. The endosperm cell develops into the endosperm or the **cotyledons**. Ovules develop into the seed and the ovary forms the **fruit**.

The zygote within the embryo sac undergoes a number of repeated mitotic divisions to form a group of cells surrounded by the endosperm tissue, which is also under development. This structure is known as the **proembryo**. In the proembryo the cells are arranged in three layers:

- Protoderm, which forms the surface tissues such as the epidermis.
- Procambium, which forms the vascular tissues.
- Ground meristem, which gives rise to ground tissues.

At this stage, the embryo takes on the shape of an axis with meristems at both ends. These meristems are the apical shoot meristem and the apical root meristem, from which structures of the shoot system and root system will ultimately develop. In addition, two bumps appear near the anterior; these are the two cotyledons, characteristic of dicot embryos. The cotyledons rapidly elongate, and the embryo is divided into regions, with respect to the cotyledons. The region above the attachment of the cotyledons is the **epicotyl**, which contains the apical shoot meristem. The region below the attachment of the cotyledons is the **hypocotyl**, which ends with the **radicle**, containing the apical root meristem. Typically, the embryonic axis will have to fold, to fit within the embryo sac. Endosperm may or may not be absorbed into the cotyledons. It may be consumed completely in the maturation of the embryo, or some may remain for germination. One of the main differences in the growth and development of plant systems from that of animal tissues is that in plants the growing ends or the meristems are very small but repeated many times above the ground as the terminal parts of shoot systems. These meristems are always active and never stop their embryonic nature. Because of this they continue to produce new tissues and cells throughout their life.

8.11 IMMUNE RESPONSE IN HUMANS AND ANIMALS

All living organisms whether plants, animals, or microbes are always prone to the attack of pathogenic organisms such as bacteria, fungi, viruses, and other types of parasitic protozoans. There is a well-developed defense mechanism in humans and animals to fight against these parasitic and pathogenic organisms. This defense mechanism in animals and human is known as the **immune system** and the protective response of the body against the invading organism is called the **immune response**. The immune system always guards the body against the various types of microbes and parasites present in the environment. If the immune system is not responding properly, even a minor infection can become fatal.

The Immune System

The immune system is well developed and is very complex in mammals and higher forms of vertebrates. The complexity of the immune system decreases as we go down the evolutionary scale. Organisms such as birds, reptiles, amphibians, fish, etc. have comparatively simple types of immune systems. There is no immune system in invertebrates such as starfish, hydras, earthworms, insects, etc. The immune system consists of certain specialized cells known as **immune cells** and certain specialized organs called **lymphoid organs.** The lymphoid organs in which the immune cells originate and mature are called the primary lymphoid organs, which include **bone marrow** and **thymus**. After maturation they migrate to other organs, the secondary lymphoid organs, where they settle down and function. These organs include the **lymph nodes** and the **spleen.**

The immune cells are distributed all over the body. Some of them reside in tissues and others circulate in the body through body fluids such as blood and lymph. The cells that carry out the immune response include **phagocytic cells** and **natural killer cells** (NK). The phagocytic cells include the **white blood cells** or **lymphocytes** and the **macrophages**. The macrophages engulf the invading organisms. The lymphocytes are the main immune cells and are further divided into different types. Morphologically, all lymphocytes are identical and cannot be distinguished. They can be classified based on the presence of certain specific molecules on the surface of the cell membrane and the function they perform. The most important lymphocyte groups are **B-lymphocytes** or **B-cells** and **T-lymphocytes** or **T-cells**.

Macrophages consist of different types of phagocytic cells. They are neutrophils, eosinophyls, and basophyls. These phagocytic cells are also called **granulocytes** because of the presence of granules in the cytoplasm and because they have a multi-lobed nucleus. There is another cell without any granules in the cytoplasm and without any lobes in the nucleus. These cells are called **monocytes**. Monocytes are the precursors of macrophages present in the tissues. The monocytes migrate into the tissues from blood and change into macrophages. Macrophages are large cells with extensive cytoplasm and have many vacuoles. The macrophages of tissues are generally called **histiocytes**. Those macrophages present in the liver tissues are called **kupfer cells**, those present in linings is known as **alveolar**, and those present in the **peritoneal cavity** are called **peritoneal macrophages**.

The Immune Response

The immune response is the protective response against the invading microorganism shown by the immune cells. When a foreign cell enters the body, the body fluid, lymph, takes these cells to the lymph nodes. The lymph nodes are the filter-like organs scattered in several parts of the body. The lymph nodes contain all the cell components of the immune system—B-cells, T-cells, and macrophages. The macrophages engulf the pathogen and are digested by the enzymes present in the lysosomes of the macrophages. These digested components of the pathogens are presented to the lymphocytes. Then a number of cellular mechanisms occur and a number of substances are secreted as a result of the immune response, which is based on the nature of the antigen (the degraded product of the pathogen). Basically, there are two types of immune response, which is dependent on two cellular systems: the **humoral** or **circulating antibody system, (B-cells immunity)**, and **cell-mediated immunity, (T-cells immunity).**

Both immune responses work by identifying antigens (foreign proteins or polysaccharides) either as part of a virus or bacterium or as a partially degraded byproduct. Both systems also recognize human antigens not made by the individual resulting in graft rejection.

The humoral antibody system (B-cell response) produces secreted antibodies (proteins), which bind to antigens and identify the antigen complex for destruction. Antibodies act on antigens in the serum and lymph. B-cell-produced antibodies may either be attached to B-cell membranes or free in the serum and lymph. The cell-mediated system acts on antigens appearing on the surface of individual cells. T-cells produce T-cell receptors, which recognize specific antigens, bound to the antigen presenting structures on the surface of the presenting cell (see Appendix).

Humoral-Antibody System: B-cells

Each B-lymphocyte, or B-cells produces a distinct antibody molecule (immunoglobulin or Ig). Over a million different B-lymphocytes are produced in each individual. Thus, each individual can recognize over a million different antigens. The antibody molecules are glycoproteins in nature and each one is composed of two copies of two different proteins. There are two copies of a heavy chain, over 400 amino acids long, and there are two copies of a light chain, over 200 amino acids long. There are five different kinds of antibodies. They are **IgG, IgM, IgA, IgD,** and **IgE.** Each antibody molecule can bind two antigens at one time, thus, a single antibody molecule can bind to two viruses, which leads to clumping. When a new antigen comes into the body, it binds to the B-cell, which is already making an antibody that matches the antigen. The antigen-antibody complex is engulfed into the B-cell and partially digested. The antigen is displayed on the cell surface by a special receptor protein **(MHC II)** for recognition by helper T-cells. The B-cell is activated by the helper T-cell to divide and secrete antibodies, which circulate in the serum and lymph. Some B-cells become memory cells to produce antibody at a low rate for a long time (long-term immunity) and to respond quickly when the antigen is encountered again. The response is regulated by a class of T-cells called suppressor T-cells. (See Appendix)

Cell-Mediated System: T-cells

T-cells mature in the thymus, which is why it is called a T-cell. A large number of different kinds of T-cells, each producing a different receptor in the cell membrane, are present in the system. Each receptor is composed of one molecule each of two different proteins. Each receptor binds a specific antigen but has only one binding site. Receptors recognize only those antigens, which are "presented" to it by another membrane protein of the MHC type (major histocompatibility complex). T-cell receptors recognize antigens presented by B-cells, macrophages, or any other cell type. T-cells, B-cells, and macrophages use MHC-II receptors for presentation; all other cells use MCH-I (responsible for most of tissue graft rejection). When a T-cell is presented with an antigen, its receptor binds to the antigen and it is stimulated to divide and produce helper T-cells to activate B-cells with bound antigen, suppressor T-cells to regulate the overall response, and cytotoxic "killer" T-cells to kill cells with antigen bound in MHC-I.

8.12 APOPTOSIS

Apoptosis is defined as programmed cell death, which occurs very systematically. Normally cell death can occur in two ways. One is by this apoptosis and the other is by **necrosis**, which occurs under pathogenic conditions or deficiencies. Apoptosis is a highly ordered process. During apoptosis the cells are disassembled very systematically. They detach from the neighboring cells of the tissue and its protoplasm condenses. The membrane-bound organelles such as mitochondria disintegrate by releasing its contents into the cytoplasm. The enzymes, **endonucleases**, act on the chromatin materials and break the DNA into fragments. At the final stage the cell membrane starts forming blebs and the cell fragments into apoptosis bodies. This type of cell death is a process of normal physiology and always occurs during organ development. Compared to apoptosis necrosis occurs in a disordered manner and occurs due to the action of toxins produced by the pathogens on the cell.

8.13 DEFENSE MECHANISMS IN PLANTS

Like animals, plants are also exposed to a wide variety of enemy organisms, which can damage the plants. These organisms include insect pests, nematodes, pathogenic fungi, bacterium, viruses and many other organisms. Plants are also exposed to many types of environmental stress called **abiotic stresses**. The stress of living organisms is known as **biotic stress**. Biotic stress can cause a severe reduction in the quantity as well as quality of the crops. In spite of the attack by pathogenic organisms and other animals, they remain healthy. This is because plants also have a defense mechanism to fight against the invading organisms. Studies about plant-defense mechanisms are very important because the identification and isolation of any genes related to the defense response can be used for genetically engineering other crop plants if needed. The defense system can be classified into two categories based on the defense response: **passive** or **constitutive** if it is a pre-existing method of response and **active** or **inducible** if the method of response is a new type developed after the infection or attack by the pathogen.

Passive Defense

This type of defense response is due to the presence of some structural components or some type of metabolites present in the body of the plant. The outer covering of the plant surface may be a special type such as cuticle or wax, which cannot be attacked or digested by the infecting fungus or bacteria. The presence of strong material such as lignin, tough bark, cuticle, etc. can effectively prevent the organisms from penetrating the plant surface. There are a large number of secondary metabolites such as alkaloids, tannins, phenols, resins, etc., which are toxic to pests and pathogens. Some of these compounds may have antimicrobial, antibacterial, or insecticidal properties. In addition to the secondary metabolites, there are certain proteins or peptides that have antimicrobial properties. For example, the antifungal peptides present in the seeds, which help in preventing the seeds from fungal infection; hydrolytic enzymes, which can lysing the bacteria and fungus; and proteins that inactivates the viral particle by digesting its coat protein and nucleic acids.

Active Defense

The defense response, which is produced newly and is not present previously in the cell or body, is called the active defense. The plant-cell wall is one of the sites where the change due to the defense response can be observed. All changes that happen in the cell wall due to an infection are collectively known as **wall apposition**. When a microorganism such as a fungus or bacteria starts infecting the plant body through the surface, immediately cell-wall thickness at that part is increased to make the penetration impossible. The change in thickness is due to the addition of new wall materials to the cell wall, specifically to the area of infection. Another interesting mechanism or response is called **hypersensitive response (HR)**. In this response, the cells around the site of infection become necrotic. The metabolic activities of these cells also change. Their respiration becomes very slow or completely stopped. They begin to accumulate toxic compounds. Thus, an inhibitory effect or an unfavorable condition is created for the further growth and spread of the pathogen around the site of infection. The plant system or those cells (cells around the site of infection) also produce certain new chemicals in response to the infection known as **phytoalexins**. Phytoalexins are small molecular weight compounds produced when there is microbial attack or under conditions of stress, which are completely absent in healthy tissues.

It has been experimentally observed that if the phytoalexins production by an infected tissue is blocked or inhibited using some selective inhibitors, the resistance of the plant against the infection has reduced substantially. Similarly, it has been demonstrated that those pathogens, which can produce the enzyme for degrading the phytoalexins, had a pathogenisity that was very high compared to those that cannot produce such enzymes.

Apart from biotic stress, there are a large number of adverse environmental factors known as **abiotic stress** acting on the plant body. Plants respond to environmental factors such as salinity, high and low temperatures, heavy metal toxicity, drought, mineral deficiency, etc. in a very specific way. Plants have developed a number of physical as well as physiological adaptations to manage such environmental situations. These may be the modifications in the concerned plant parts such as leaf, root, or stem; or it may be in the form of a physiological or biochemical change by which the effect of change of environment on the plant system is neutralized. For example, abiotic stress such as drought, high temperature, and high salt concentration in the soil or salinity, ultimately cause loss of water from the cells or cellular dehydration. The plant system effectively counters it by maintaining water content inside the cells by accumulating various types of metabolites and solutes (osmolytes), which are non-toxic and inert (known as compatible solutes). These compatible solutes include various types of sugars such as sucrose, fructose, trehalose, etc. and other sugar alcohols such as mannitol and inositol (*myo* inositol). It has been further observed that under certain abiotic stress conditions a set of new genes are activated and expressed. These gene products (proteins) may have an important role in creating tolerance against the stress. For example, **LEA proteins (late embryogenesis abundant protein)** expressed in seeds kept under desiccation and by the vegetative parts underwater, dehydrins expressed during dehydrated conditions, **heat shock proteins (HSP)** in response to high temperatures, etc.

Detailed studies on plant-pathogen interactions and abiotic stresses have revealed the participation of multiple signal transduction pathways in the expression of stress proteins. Extensive research has been carried out to identify the genes associated with abiotic stresses, so that the relevant genes can be used for genetic engineering of crop plants for various stress tolerance. Transgenic rice and other crops have already been produced for increased salt tolerance, drought resistance, and other abiotic stresses by transferring the respective gene or genes.

8.14 PLANT-PATHOGEN INTERACTION

When an insect or microbe attacks a plant it triggers a chain of reactions leading to the production of some compounds that can act against the invading organisms. This process is called **elicitation** and the compounds produced in responds to the insect or microbial attack are called **phytoalexins**. Studies on plant-pathogen interactions are very important to understand the molecular mechanism of insect resistance or disease resistance in plants so that adequate precautions can be taken to prevent crop loss in agriculture. It was observed that plants can synthesize certain polypeptides in response to a microbial attack as was demonstrated in the case of the tobacco plant when attacked with the tobacco mosaic virus. These proteins are called **pathogen-related proteins** or **PR proteins**. In addition to the formation of PR proteins, an array of other defense-related responses such as production of enzymes responsible for the expression of genes related to the phytoalexins synthesis, wall-bound phenolics, hydrolyzing enzymes, and hydroxy-proline rich glycoproteins were detected. These metabolites were synthesized and accumulated around the site of infection for preventing the entry of the invading organism. These information will help in devising strategies, including genetic engineering, to protect crop plants from disease and attack of insects, thus to reduce the crop loss.

8.15 SECONDARY METABOLISM

Biomolecules such as amino acids, lipids and carbohydrates such as glucose, fatty acids, etc. are involved in the synthesis of various macromolecules such as proteins, nucleic acids, starch and lipids such as triglycerides, steroids, etc. These types of molecules and other related activities such as generation of energy, immunological activities, etc., are essential for the existence of life activities. Such reactions are known as **primary metabolism** and their products are called **primary metabolites**. But organisms, when they become mature or when the cells come to the lag phase of growth, operate additional metabolic pathways to synthesize certain compounds, which are not essential for carrying out normal life activities. Such compounds are known as **secondary metabolites** and their biosynthetic pathway is known as **secondary metabolism**. Their biosynthesis starts from some primary metabolite or from intermediates of the primary metabolism. These secondary metabolites are produced in small quantities and are believed not to have any function in the body. But it has been observed that in some cases they have some role in the defense against microorganisms or insects and pests. Even some secondary metabolites are produced in response to the attack of some microorganisms or insects. Some compounds impart special odors to body parts such as flowers or leaves, which attracts or keeps away insects and predators. In plants, there are various types of secondary metabolites such as alkaloids, steroids, terpenes, latex, tannins, resins etc., which are produced and stored in specialized cells, in most cases.

In the case of bacteria and fungus these secondary metabolites are produced at the stationary phase of the growth. Compounds such as antibiotics are the secondary metabolites accumulated by the bacterial cultures and fungus at their stationary phase of growth. The biosynthesis of secondary metabolism is dependent on the state of growth and growth conditions. It is possible to alter the sequence of secondary metabolism by altering growth conditions and media compositions, in the case of bacteria and fungi. Even a change in the pH can alter the route of secondary metabolism in microbial cultures.

Secondary metabolites such as alkaloids, latex, and antibiotics are of great economic value for man. Therefore, these compounds can be synthesized in large quantities by manipulating culture conditions, or by adding the necessary precursors in the media of the cultures.

8.16 DEFENSE STRATEGIES IN MICROBES AND INSECTS

It was once believed that lower organisms do not have defense mechanisms. But this is not true. Even though a well-developed defense mechanism is present only in vertebrates, simple forms of defense mechanisms and immune systems are also available in lower forms of organisms. In vertebrates, as we move down from mammals to birds, to reptiles, to amphibians and to fish, the immune system becomes simpler. Lower organisms do not have a well-established defense mechanism or immune system as we have discussed in the case of vertebrates, but these organisms can also protect themselves from enemies and competitors by other types of defense mechanisms, some of which are specific for a particular organism.

Defense in Microbes

Bacteria and other microorganisms have various types of structural and physiological or biochemical means for defense against their enemies and adverse situations. A large number of bacteria and fungi produce digestive enzymes and toxic chemicals such as antibiotics and antimicrobial compounds. These compounds can kill or damage the cells of organisms in which they are in contact.

Many bacteria and other microorganisms are capable of producing capsules. The phagocytic cells cannot engulf and destroy the capsulated bacterial cells. Phagocytes such as macrophages and neutrophils can easily engulf the non-capsulated bacterial cells and destroy them easily.

During unfavorable conditions, a large number of bacteria can be changed into spores, known as **endospores**. The cytoplasm of the cell is detached from the cell wall and changed into a spore in the middle of the cell. The spore has a very thick covering, and since it is formed within the cell, it is called an endospore. Endospores are highly resistant to heat, UV-radiations, and chemicals and antibiotics. Even though the formations of endospores are a mechanism to overcome the unfavorable conditions, the formation of such endospores can be considered a defense mechanism.

Defense in Insects

There are different types of defense mechanisms in insects, which depends on the species of the insect. One of the main mechanisms is the production of antibacterial peptides. These toxic compounds are capable of killing bacteria. A large number of other insects also produce these types of toxic antibacterial peptides and are called cecropins. Bees produce a type of toxin called melittin, which is present in their venom and are able to do the hemolytic activity (lysing the RBC). In insects like the fruit fly (*drosophila*) the reproductive organs produce antibacterial peptides, the andropins. In certain other insects antibacterial peptides are present in the hemolymph. Some of these antibacterial peptides are produced in response to the bacterial infection in insects. In addition to these defense mechanisms, they also possess phagocytic cells. These cells can attack and destroy the invading microorganisms and pathogens.

REVIEW QUESTIONS

1. Explain and compare mitosis and meiosis.
2. Explain alternation of generation. Explain it with respect to angiosperms.
3. Explain the methods of genetic recombination in bacteria.
4. Name the components of the immune system.
5. How are macrophages involved in immune responses?
6. Name the two proteins responsible for the contraction and relaxation of muscle fibers.
7. What are the types of defense mechanisms in plants?
8. Compare B-cell mediated and T-cell mediated immune responses.
9. Describe a muscle contraction.
10. Define apomixes. Give an example.
11. What are phytoalexins? What is their relevance in crop improvement?
12. What is double fertilization and triple fusion?
13. Explain the hypersensitive response in plants.
14. Explain the method of asexual reproduction in bacteria.
15. What are the structural stages of gametophytic generations in angiosperms?
16. Define blastula and gastrula. Name the cell layers from which neurons develop.
17. What is parthenogenesis? Compare it with apomixes.
18. Distinguish between primary metabolism and secondary metabolism.
19. Explain signal transduction pathway.
20. What are the forces that help in the ascension of sap through Xylem vessels?
21. Explain the process of fertilization in plants.
22. What is homeostasis?
23. Explain thermoregulation in animals.
24. Differentiate between apoptosis and necrosis.
25. Explain the structure of an immuniglobulins with the help of a diagram.

Chapter 9

CELLULAR TECHNIQUES

In This Chapter

9.1 INTRODUCTION

The microscope is an invaluable tool in today's research and education. It is used in a wide range of scientific fields, where major discoveries in biology, medicine, and materials research are dependent on advances in microscopy. Cell biology mainly deals with the fine structure and function of cells, including the molecular basis of cell activities. Since the size of the cell is very small, it is very difficult to examine its complex structure and to watch how the individual components work. Technological improvement in the instrumentation and techniques to study the cells and its metabolic activities at a molecular level has contributed much to the understanding of cells biology.

From the simple light microscope different techniques have evolved, aimed at making it possible to see certain objects or processes. Scientists use electron microscopes in order to get extraordinary resolution, microscopes that give three-dimensional images of surfaces or biological molecules, and microscopes that mark out specific substances. This chapter outlines the various cellular and histological techniques used to study cell structure and function.

9.2 MICROSCOPY

The discovery of the cell and further developments in the structural details of it was possible solely because of the discovery and developments of different types of microscopes. In 1667, Robert Hooke studied various objects with his microscope and published the results in 'Micrographia.' Among his works was a description of a cork and its ability to float in water. In 1675, Anton van Leeuwenhoek used a simple microscope with only one lens to look at blood, insects, and other objects. He was the first to describe cells and bacteria, seen through his very small microscope. In the eighteenth century, several technical innovations made microscopes better and easier to handle, which led to microscopy becoming more popular among scientists.

Later M. J. Schleiden and T. Schwann formulated cell theory with the help of microscopes. The developments in the field of cytology and cell biology followed closely behind developments in microscopy. The detailed structural analysis of a cell, with the help of a microscope, is called **cytology**. The detailed study of all aspects of the structure and function, metabolism, etc. of cells at a molecular level using a variety of cellular techniques including microscopy is called **cell biology**.

The principle of the microscope is the ability of a lens or a group of lenses to converge or diverge light so that the image of the object can be magnified. The light microscope makes it possible to see structure, that are beyond the resolution of the naked eye. Basically, there are two types of light microscopes: **simple microscopes** and **compound microscopes**. A simple microscope is relatively simple with a single lens and is used for studying the morphological details of small objects such as the internal structure of a flower. The compound light microscope contains a number of lenses arranged in a definite order, which can be adjusted to get the correct magnification. All these lenses are organized into two parts: the **objective** and the **eyepiece**. There are different types of objectives with different combinations of lenses with different magnifications. The objectives and eyepiece are arranged in such a way that the distance between them can be adjusted. The distance between the object and objective can also be varied to get optimum magnification with different objects. In addition to the objective and eyepiece, there is another lens or a group of lenses known as the **condenser**, which is fixed below the plateform for keeping the object. The object or the specimen is placed between the condenser and objective on a specimen platform. The specimen platform is also moveable and adjustable just like the set of lenses.

Resolving Power

The human eye is capable of distinguishing objects down to a fraction of a millimeter. With the use of light and electron microscopes it is possible to see down to an angstrom and study everything from different cells and bacteria to single molecules or even atoms.

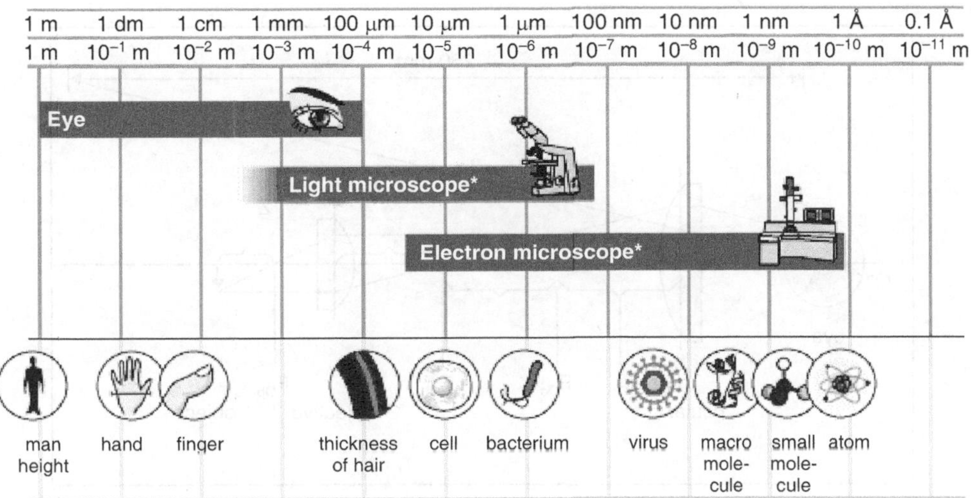

FIGURE 9.1 **Resolving power of eye and different types of microscopes.**

The resolving power of a microscope is the ability to see a small object clearly and distinctly from another similar-sized object separated by a very small distance. The resolving power of object *d* is given by the equation:

$$\text{Resolution for the object } d = \frac{0.61\lambda}{\eta \sin\theta}$$

where λ is the wavelength of the light, η is the refractive index of the medium (it is usually air in ordinary light microscopes and oil in oil immersion microscopes), and θ is half angular aperture of the beam entering the objective lens from condenser through the object.

Refractive index is the ratio of the angle of incidence to the angle of refraction of the light. **Numerical aperture** is the ability of the lenses to collect the converged lights.

The specimens or tissues studied by light microscopes are usually processed by staining with proper stains, after cutting to very thin slices of appropriate thickness to facilitate the transmission of light through it. There is an instrument known as a **microtome,** which is used to make thin slices of tissues to the level of micrometers thickness.

Almost all cell components, except some organelles such as chloroplast, are colorless and therefore absorb more or less the same wavelength of light. So it is difficult to differentiate the cellular parts. To observe all cellular parts under a light microscope the tissue or cells should be stained with a suitable stain or with a number of stains. The staining of tissues and cells imparts different colors to different parts and cellular organelles. Histological and cytological stains are the colored chemicals that selectively bind to the cell components depending on the chemical nature of the constituents. There are a number of stains used in cytology such as safranin, acetocarmine, hematoxylin, eosin, etc.

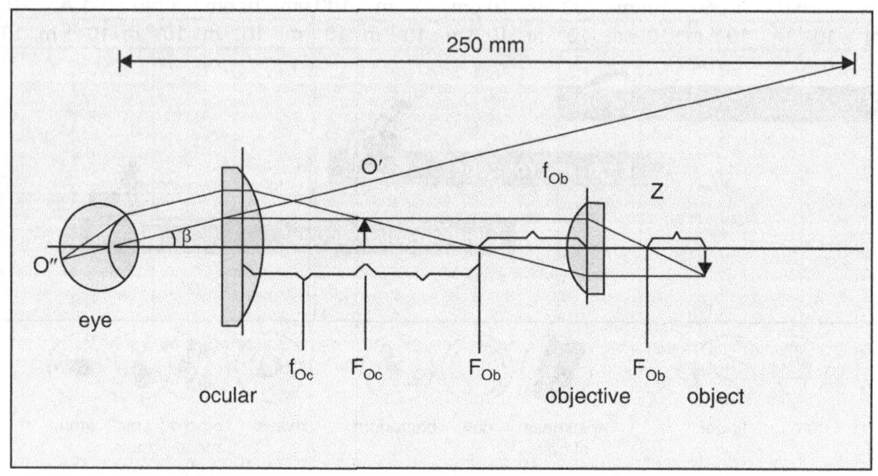

FIGURE 9.2 **Light path in the microscope: F = Focal plane, O = Object (Specimen), Ob = Objective, and Oc = ocular (eyepiece).**

Safranin binds to cell walls of the plant cell and microbes because it interacts with the cellulose and materials such as pectin. Acetocarmine is used for staining chromosomes specifically because it can selectively bind to the basic proteins (histones) of the chromosomes. Similarly, hematoxylin binds to the basic amino acids and proteins and eosin binds to acidic molecules. Because of the different binding properties of these stains, different cellular components are colored distinctly and so are visible clearly under a light microscope.

Fluorescence Microscopy

Fluorescence microscopy is a powerful microscopic technique used to locate and quantify specific molecules or organelles in a cell. It can be applied to study the various activities of living cells. In fluorescence microscopy, the sample you want to study is itself the light source. The technique is used to study specimens, which can be made to fluoresce. The fluorescence microscope is based on the phenomenon that certain material emits energy detectable as visible light when irradiated with the light of a specific wavelength. The sample can either be fluorescing in its natural form such as chlorophyll and some minerals, or treated with fluorescing chemicals.

Fluorescence microscopy is a rapidly expanding technique, both in the medical and biological sciences. The technique has made it possible to identify cells and cellular components with a high degree of specificity. For example, certain antibodies and disease conditions or impurities in organic material can be studied with the fluorescence microscope.

The basic operation of the fluorescence microscope includes the excitation of the specimen by illuminating it with radiation and then sorting it out of the much weaker radiation emitted by it at the visible range to make up the image. First, the microscope has a filter that only lets through radiation with the desired wavelength that matches the fluorescing material. The radiation collides

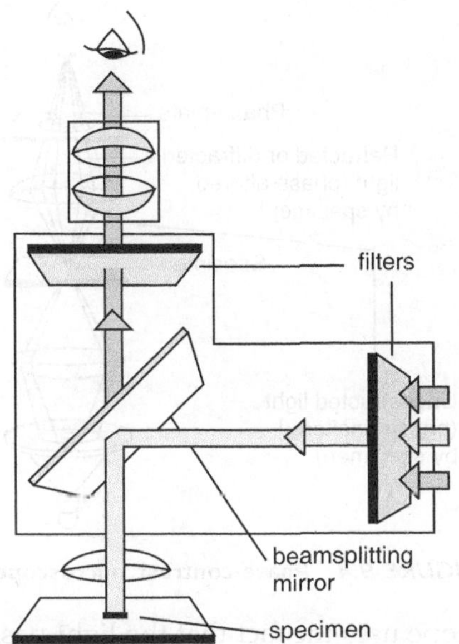

FIGURE 9.3 **Diagram representation of the fluorescence microscope.**

with the atoms in the specimen and electrons are excited to a higher energy level. When they relax to a lower level, they emit light. To become visible, the emitted light is separated from the much brighter excitation light in a second filter. The fluorescing areas can be observed in the microscope and shine out against a dark background with high contrast.

The principle behind excitation and fluorescence is that the atom absorbs energy and becomes excited. The electron jumps to a higher energy level. Soon, the electron drops back to the ground state, emitting a photon (or a packet of light), which forms the fluorescence of that molecule or atom. A chemical is said to be fluorescent if it absorbs light at one wavelength and emits light at another specific and longer wavelength. There are a large number of fluorescent staining dyes such as rhodamine, texas red, which emits red light, Cy 3, which emits orange light, and fluorescein, which emits green light. By staining with different dyes (staining at different wavelengths), multiple proteins can be visualized in a cell. Fluorescent dyes such as fura-2 are used for the measurement of the concentration of free calcium ions in cytoplasm to study the signal changes in cell metabolism. There are certain fluorescent proteins such as **green fluorescent protein (GFP)**, which can be used as a marker in genetic engineering experiments. GFP is a naturally-occurring fluorescent protein present in the jellyfish, *aequorea victoria*.

Phase-Contrast Microscopy

The phase-contrast microscope is widely used for examining such specimens as biological tissues without staining. It is a type of light microscopy that enhances contrasts of transparent and colorless objects by influencing the optical path of light. A phase-contrast microscope is able to show components in a cell or bacteria, which would be very difficult to see in an ordinary light microscope.

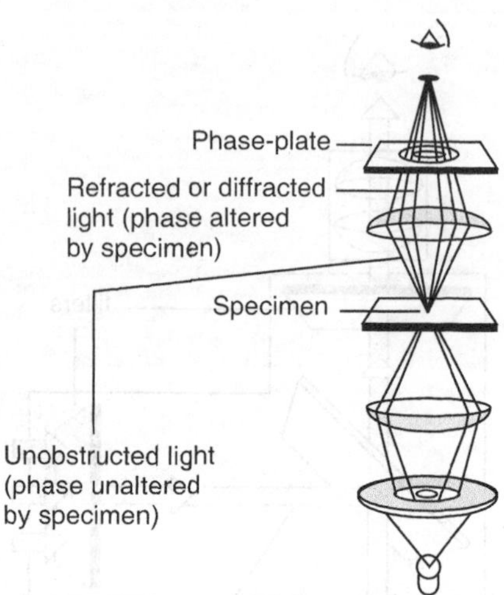

FIGURE 9.4 **Phase-contrast microscope.**

A phase-contrast microscope uses the fact that the light passing through a transparent part of the specimen travels slower and, due to this its phase, is shifted compared to the uninfluenced light. This difference in phase is not visible to the human eye. However, the change in phase can be increased to half a wavelength by a transparent phase-plate in the microscope and thereby causing a difference in brightness. This makes the transparent object shine out in contrast to its surroundings.

The phase-contrast microscope is a vital instrument in biological and medical research. When dealing with transparent and colorless components in a cell, dyeing is an alternative but at the same time stops all processes in it. The phase-contrast microscope has made it possible to study living cells, and cell division is an example of a process that has been examined in detail with it. The phase-contrast microscope was awarded with the Nobel Prize in Physics, 1953.

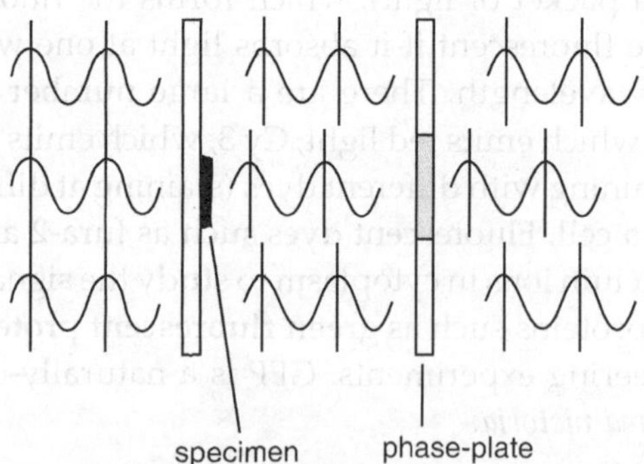

FIGURE 9.5 **Phase difference in a phase-contrast microscope.**

The phase-plate increases the phase difference to half a wavelength. Destructive interference between the two sorts of light when the image is projected results in the specimen appearing as a dark object.

Dark-Field Microscopes

These microscopes have a dark-field condenser instead of the normal one. The dark-field condenser focuses light obliquely on the object, so that only the deflected light enters the objective. The entire background field appears black and only the object is bright against the dark background. The contrast between specimen and the background permits the visualization of even the smallest of objects such as bacteria and virus particles.

Electron Microscope

In microscopy there is limitation in the resolving power of a light microscope. It is not possible to observe a particle that has a size lesser than the wavelength of visible light. The discovery of the electron microscope has solved that problem, because electron beams have a wavelength lesser than lightwaves. Detailed studies about viral particle subcellular structures and organelles, etc. can be carried out with an electron microscope. There are two basic types of electron microscopes: **Transmission Electron Microscope (TEM)** and **Scanning Electron Microscope (SEM).**

Transmission Electron Microscope

The **transmission electron microscope (TEM)** operates on the same basic principles as the light microscope but uses electrons instead of light. What you can see with a light microscope is limited by the wavelength of light. TEMs use electrons as the "light source," and their much lower wavelength makes it possible to get a resolution a thousand times better than with a light microscope. You can see objects to the order of a few angstrom (10^{-10} m). For example, you can study small details in the cell or different materials down to near atomic levels. The possibility of high magnifications has made the TEM a valuable tool in medical, biological, and materials research.

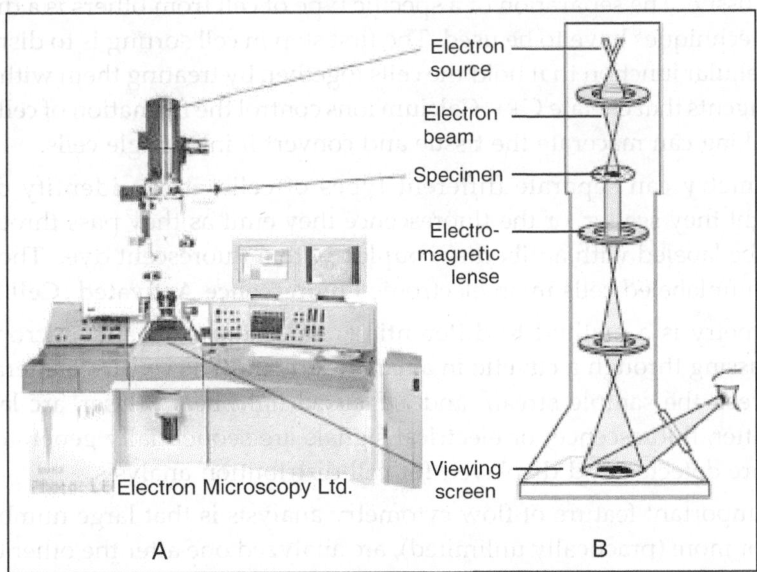

FIGURE 9.6 **A. Transmission electron microscope.**
B. Diagram representation of electron path.

A "light source" (electron source) at the top of the microscope emits the electrons that travel through the vacuum in the column of the microscope. Instead of glass lenses focusing the light in the light microscope, the TEM uses **electromagnetic lenses** to focus the electrons into a very thin beam. The electron beam then travels through the specimen studied. Depending on the density of the material present, some of the electrons are scattered and disappear from the beam. At the bottom of the microscope, the unscattered electrons hit a fluorescent screen, which gives rise to a "shadow image" of the specimen with its different parts displayed in varied darkness according to their density. The image can be studied directly by the operator or photographed with a camera. Unstained and fixed specimens can be observed by the electron microscope if they are frozen in hydrated form. The technique is known as **cryoelectron microscopy**.

Scanning Electron Microscopy

The **scanning electron microscope (SEM)** is a type of electron microscope that shows the surface images of a sample. In the SEM, the structure of a surface is studied using a stylus that scans the surface at a fixed distance from it. The study of surfaces is an important part of physics, with particular applications in semiconductor physics and microelectronics. In chemistry, surface reactions also play an important part, for example, in catalysis. The SEM works best with conducting materials, but it is also possible to fix organic molecules on a surface and study their structures. For example, this technique has been used in the study of DNA molecules.

9.3 CELL SORTING

Cell sorting is a method by which a particular population of cells can be separated from other types of cells in a tissue. The separation of a specific type of cell from others is a difficult technique, and a number of techniques have to be used. The first step in cell sorting is to disrupt extra cellular matrix and intercellular junction that hold the cells together, by treating them with some proteolytic enzymes or with agents that chelate C++. Calcium ions control the formation of cell-to-cell adhesion. Now a gentle shaking can macerate the tissue and convert it into single cells.

Flow cytometry can separate different types of cells, it can identify different cells by measuring the light they scatter, or the fluorescence they emit as they pass through a laser beam. Specific cells can be labeled with antibodies coupled with a fluorescent dye. The labeled cells can be separated from unlabeled cells in an electronic **Fluorescence Activated Cell Sorter (FACS)**.

Flow cytometry is a method to differentiate and count cells and microparticles. Cells in suspension are passing through a cuvette in a narrow stream. The sensitive interaction zone is the interception between the sample stream and a focused light beam (laser, arc lamp) or electrical field. Optical (scatter, fluorescence) or electrical signals are sequentially generated by each single particle. Signals are detected and displayed for cell-distribution analysis.

The most important feature of flow cytometry analysis is that large numbers, for example, 100,000 particles or more (practically unlimited), are analyzed one after the other typically in about one minute. The detection limit is as low as 100 fluorescent molecules per cell. Sub-micron particles (small bacteria, phytoplankton) are well resolved.

In contrast, microscopic analysis is based on a very limited number of cells seen on a slide (1 to 100). The visualization of sub-micron particles and structures in a microscope is not easy either. Therefore, a flow cytometer provides useful information about a broad range of cells and their functions.

Fluorescence analysis makes it possible to quantify fluorescence from single cells up to millions of cells after a single sample run out of one test tube. Statistical data such as mean fluorescent intensity (and their shifts in time or their dependence on cell function) have proven to be very reliable.

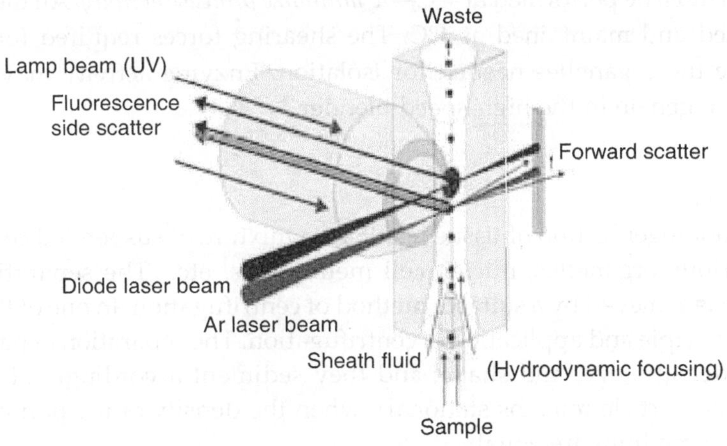

FIGURE 9.7 **Cell sorting by fluorescence flow cytometery.**

9.4 CELL FRACTIONATION

Cell biology involves the disruption of cells and tissues into its components by fractionation for detailed biochemical and physiological studies of the cell and its organelles.

The first step for isolating any subcellular particle is to break the cell wall (if it is present) and cell membrane to release the cellular components. Techniques of cell fractionation proceed in two consecutive stages:

1. **Homogenization** (disrupts the tissue and releases cellular components).
2. **Centrifugation** (separates the individual components according to density, size, and shape).

A tissue homogenizer can carry out homogenization of rigid cells and tissues mechanically. A solid abrasive such as sand, silica, alumina, etc. can be included to generate high shearing forces in the case of plant tissues. Liquid shear methods generate mild shearing forces generated between the tissue and liquid medium (homogenization buffer), which can be used for the homogenization of soft tissues. Microhomogenizers are designed to disrupt small amounts of tissue/cells by generating low shearing forces. Soft tissues and cells like that of animals can be ruptured by giving an osmotic shock by keeping the cells in a hypotonic solution. Another simple method for cell rupturing is by a freezing and thawing technique, which uses ice crystals to rapture the cells. Single-celled organisms such as bacteria and yeast can be effectively homogenized by ultrasound waves in a sonicator. Non-physical methods such as enzymatic methods also can be adopted for cell fractionation. For example, the enzymatic digestion of the cell wall in the case of plant tissue results in protoplasts, which are subsequently subjected to one of the previous methods of cell/ protoplast disruption. A mixture of enzymes—chitinases, pectinases, lipases, proteases, and cellulases—is used for the production of protoplasts.

During the process of cell fractionation the following things need to be performed. All stages of cell fractionation must be performed at 4°C—*to minimize protease activity.* All media and apparatus should be precooled and maintained at 4°C. The shearing forces required for breaking the cell walls may damage the organelles needed for isolation. Enzyme activity may be lost if there is foaming of the homogenate in the high-speed blender.

Centrifugation

Cell disruption or homogenization of tissue results in a mixture of suspended cellular components, which include various organelles, nuclei, cell membranes, etc. The separation of the desired subcellular fraction is achieved by a suitable method of centrifugation. In one of the earlier chapters, we discussed the principle and application of centrifugation. The separation of particles (organelles) depends on their density, size, and shape; and they sediment accordingly at different rate in a centrifugal field. The particle remains stationary when the density of the particle and the density of the centrifugation medium are equal.

Types of Centrifugation

1. Differential centrifugation
2. Rate-zonal centrifugation
3. Isopycnic centrifugation

Differential centrifugation separates particles as a function of size and density. In differential centrifugation at relatively low speeds larger cell components such as nuclei and unbroken cells sediment to form a pellet at the bottom of the centrifuge tube. At slightly higher speeds a pellet of chloroplasts and mitochondria is formed. Even at very high speeds of centrifugation and longer periods, closed vesicles and finally ribosomes will form a pellet.

With rate-zonal centrifugation or density gradient centrifugation, specific components can be separated as layers. Centrifugation medium is characterized by a positive increment in density and therefore known as density gradient centrifugation. The sample is applied on the top of the density gradient as a narrow band. Particles separate into a series of bands in accordance to centrifugal field, size, and shape of the particle and difference in density between the particle, and the suspending medium. The rate at which each of the components sediment can be expressed, in terms of its sedimentation coefficient, as **S value.**

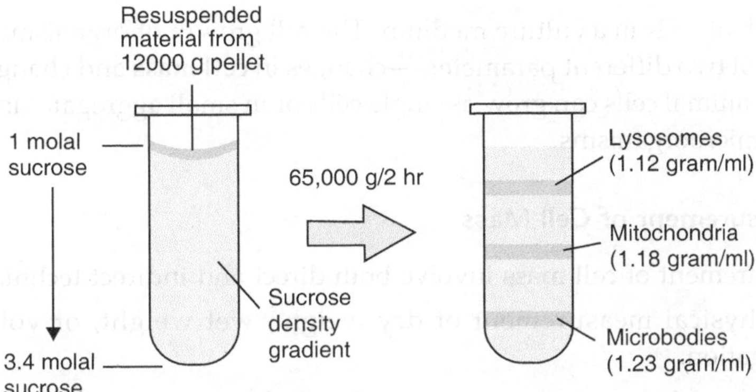

FIGURE 9.8 **Isopycnic centrifugation.**

Cellular components can also be separated by **isopycnic centrifugation** in which there is no sedimentation of particles but separation into layers in a density gradient solution. Isopycnic centrifugation uses the density gradient of the separating medium, and the separation is solely dependent on the density of the particles and separation medium, the density-gradient medium. The process is unaffected by the size or the shape of the particles. This type of advanced ultra-centrifugation technique is used for the separation of different types of biomolecule complexes such as nucleic acids, ribosomes, etc. This centrifugation technique is so sensitive that it can be used to separate radio-labeled compounds from normal types of molecules.

The density-gradient medium used for centrifugation should have the following characteristics:

- Inert toward the biological material, the centrifuge tubes and rotor.
- No interference with the monitoring of the sample material.
- Easy separation from the fraction after centrifugation.
- Easy to monitor the concentration of the gradient medium.
- Stable in solution and available in a pure and analytical form.
- Exert minimal osmotic pressure.

9.5 CELL-GROWTH DETERMINATION

Growth is an orderly increase in the quantity of cellular constituents. It depends upon the ability of the cell to form new protoplasm from nutrients available in the environment. In most organisms, including bacteria, growth involves increase in cell mass and number of cell organelles (ribosomes in bacteria), duplication of the genetic material or chromosome, synthesis of new cell wall and plasma membrane, partitioning of the doubled chromosomes, septum formation, and finally, the cytoplasmic division. This is the method of growth in eukaryotic cells, but this cell division is the method of asexual process of reproduction in unicellular organisms such as bacteria. This process of asexual reproduction is known as binary fission. Plant and animal cells can also be grown as suspension cultures in liquid media just like unicellular organisms. There are a number of techniques to assess the growth of cells in a culture medium. The cell growth of organisms in a culture can be measured in terms of two different parameters—changes in **cell mass** and changes in **cell numbers**. Since the plant and animal cells can grow as single cells or in small aggregates in culture, its growth is also like that of microorganisms.

Methods for Measurement of Cell Mass

Methods for measurement of cell mass involve both direct and indirect techniques.

1. Direct physical measurement of dry weight, wet weight, or volume of cells after centrifugation.
2. Direct chemical measurement of some chemical component of the cells such as total N, total protein, or total DNA content.
3. Indirect measurement of chemical activity such as rate of O_2 production or consumption, CO_2 production, or consumption, etc.
4. Turbidity measurements employ a variety of instruments to determine the amount of light scattered by a suspension of cells. Particulate objects such as bacteria scatter light in proportion to their numbers. The turbidity or optical density of a suspension of cells is directly related to cell mass or cell number, after construction and calibration of a standard curve. The method is simple and nondestructive, but the sensitivity is limited to about 10^7 cells per ml for most bacteria.

Methods for Measurement of Cell Numbers

1. **Direct microscopic counts:** These measuring techniques involve direct counts, visually or instrumentally, and indirect viable cell counts. Direct microscopic counts are possible using special slides known as **counting chambers**. Dead cells cannot be distinguished from living ones. Only dense suspensions can be counted (>10^7 cells per ml), but samples can be concentrated by centrifugation or filtration to increase sensitivity. A variation of the direct microscopic count has been used to observe and measure growth of bacteria in natural environments. In order to detect and prove that thermophilic bacteria were growing in boiling hot springs, T. D. Brock[1] immersed microscope slides in the springs and withdrew them periodically for microscopic observation. The bacteria in the boiling water attached to the glass slides naturally and grew as microcolonies on the surface.
2. **Electronic counting chambers:** Electronic counting chambers count numbers and measure size distribution of cells. For cell size of bacteria the suspening medium must be very clean. Such electronic devices are more often used to count eukaryotic cells such as blood cells. Indirect viable cell counts, also called plate counts, involve plating out (spreading) a sample of a culture on a nutrient agar surface. The sample or cell suspension can be diluted in a nontoxic diluent (e.g., water or saline) before plating. If plated on a suitable medium, each viable unit grows and forms a **colony**. Each colony that can be counted is called a **colony-forming unit (cfu)** and the number of cfus is related to the viable number of bacteria in the sample.

---

[1] "Life at High Temperatures", *Science, 11 Oct., 1985, Vol. 230, no. 4722, pp. 132–138.*

Advantages of the technique are its sensitivity (theoretically, a single cell can be detected), and it allows for inspection and positive identification of the organism counted. Disadvantages are (1) only living cells develop colonies that are counted; (2) clumps or chains of cells develop into a single colony; and (3) colonies develop only from those organisms for which the cultural conditions are suitable for growth. The latter makes the technique virtually useless to characterize or count the **total number of bacteria** in complex microbial ecosystems such as soil or the animal rumen or gastrointestinal tract. Genetic probes can be used to demonstrate the diversity and relative abundance of prokaryotes in such an environment, but many species identified by genetic techniques have so far proven uncultivable.

Culture-based Growth Determination

A viable cell under suitable environmental and nutritional conditions is able to multiply and give a visible change in the media where it is growing. This may be by the formation of a colony on an agar medium or formation of turbidity in a liquid medium.

In the **spread or pour plate method,** a small volume (about 50 to 100ml) is placed on an agar petriplate and spread uniformly with an 'L' shaped glass rode. Then it is incubated under appropriate conditions. The number of bacterial colonies formed in the agar plates can be related to the number of organisms present in the volume of liquid spread on the agar plates.

Determination of Growth by Turbidity

This is another effective technique for the determination of growth not only for bacterial cultures but also for almost all types of cells that can grow in liquid cultures. This method can be used for the determination of growth pattern of cells growing in cultures by generating a growth curve. The growth can be determined by the increase in the turbidity of the medium, which can be monitored by a spectrophotometer at a suitable wavelength.

Bacterial Growth Curve

In the laboratory, under favorable conditions, a growing bacterial population doubles at regular intervals. Growth is by geometric progression: 1, 2, 4, 8, etc., or 2^0, 2^1, 2^2, 2^3.........2^n (where n = the number of generations). This is called **exponential growth**. In reality, exponential growth is only part of the bacterial life cycle, and not representative of the normal pattern of growth of bacteria in nature.

When a fresh medium is inoculated with a given number of cells, and the population growth is monitored over a period of time, plotting the data will yield a typical bacterial growth curve (Figure 9.9).

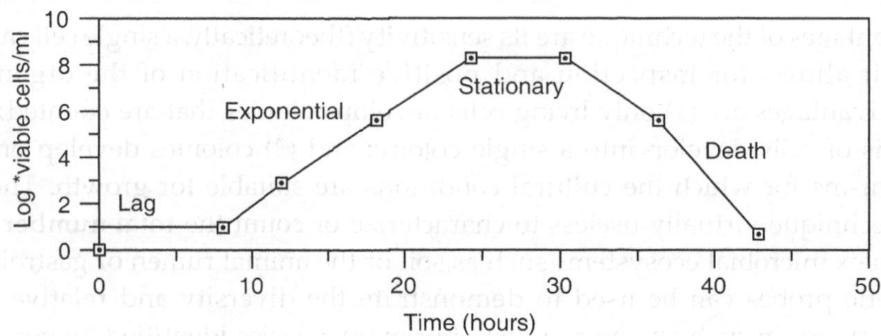

FIGURE 9.9 **Typical bacterial growth curve.**

Four characteristic phases of the growth cycle are recognized.

1. **Lag Phase.** Immediately after inoculation of the cells into a fresh medium, the population remains temporarily unchanged. Although there is no apparent cell division occurring, the cells may be growing in volume or mass, synthesizing enzymes, proteins, RNA, etc., and increasing in metabolic activity. The length of the lag phase is apparently dependent on a wide variety of factors including the size of the inoculum; time necessary to recover from physical damage or shock in the transfer; time required for synthesis of essential coenzymes or division factors; and time required for the synthesis of new (inducible) enzymes that are necessary to metabolize the substrates present in the medium.
2. **Exponential (log) Phase.** The exponential phase of growth is a pattern of balanced growth wherein all the cells are dividing regularly by binary fission, and are growing by geometric progression. The cells divide at a constant rate depending upon the composition of the growth medium and the conditions of incubation.
3. **Stationary Phase.** Exponential growth cannot be continued forever in a batch culture (e.g., a closed system such as a test tube or flask). Population growth is limited by one of three factors: 1. Exhaustion of available nutrients; 2. Accumulation of inhibitory metabolites or end products; 3. Exhaustion of space, in this case called a lack of "biological space." During the stationary phase, if viable cells are being counted, it cannot be determined whether some cells are dying and an equal number of cells are dividing, or the population of cells has simply stopped growing and dividing. The stationary phase, such as the lag phase, is not necessarily a period of quiescence. Bacteria that produce secondary metabolites, such as antibiotics, do so during the stationary phase of the growth cycle; (Secondary metabolites are defined as metabolites produced after the active stage of growth). It is during the stationary phase that spore-forming bacteria have to induce or unmask the activity of dozens of genes that may be involved in sporulation process.
4. **Death Phase.** If incubation continues after the population reaches the stationary phase, a death phase follows, in which the viable cell population declines. (**Note:** if counting by turbidimetric measurements or microscopic counts, the death phase cannot be observed.) During the death phase, the number of viable cells decreases geometrically (exponentially), essentially the reverse of growth during the log phase.

TABLE 9.1 Some methods used to measure bacterial growth

Method	Application	Comments
Direct microscopic count	Enumeration of bacteria in milk or cellular vaccines	Cannot distinguish living from nonliving cells
Viable cell count (colony counts)	Enumeration of bacteria in milk, foods, soil, water, laboratory cultures, etc.	Very sensitive if plating conditions are optimal
Turbidity measurement	Estimations of large numbers of bacteria in clear liquid media and broths	Fast and nondestructive, but cannot detect cell densities less than 10^7 cells per ml
Measurement of total N or protein	Measurement of total cell yield from very dense cultures	Only practical application is in the research laboratory
Measurement of Biochemical activity; e.g., O_2 uptake, CO_2 production, ATP production, etc.	Microbiological assays	Requires a fixed standard to relate chemical activity to cell mass and/or cell numbers
Measurement of dry weight or wet weight of cells or volume of cells after centrifugation	Measurement of total cell yield in cultures	Probably more sensitive than total N or total protein measurements

REVIEW QUESTIONS

1. Explain the principle of fluorescence microscopy. How it is different from ordinary light microscope?
2. Name the different types of light microscopes and its special purpose.
3. Explain the difference between fluorescence and phosphorescence.
4. What is a phase-contrast microscope?
5. Which light microscope is suitable for viewing tiny particle such as viruses and bacteria?
6. Explain the principle and function of FACS.
7. Define the resolving power of a microscope.
8. Explain the principle and the function of a transmission electron microscope.
9. What is the use of SEM? How is it different from TEM?
10. Explain the methods used for cell fractionation. What precautions should be taken during the process?
11. What is the principle of flow cytometry? How is it similar to fluorescence microscopy?
12. Differentiate between isopycnic centrifugation and rate-zonal centrifugation.
13. What are the different methods used for the determination of growth in a liquid culture?
14. What is meant by a growth curve? Explain the nature of a typical growth curve.
15. Explain the pour-plate method in growth determination.

Part 4

GENETICS AND MOLECULAR BIOLOGY

The science of genetics originated when Gregor Johann Mendel published his pioneering work on inheritance in pea plants in 1866 in *Proceedings of the Natural History Society of Brünn*. But even before that people were aware of inheritance. Farmers used to select seeds of crop plants from those having good traits and even used to adopt breeding techniques to improve the agricultural traits of crop plants. But it was not always a science. Mendel explained his experimental results in the form of laws of heredity. He predicted that there are factors that control each trait, which is transmitted from generation to generation; and that is the subject of the first chapter of this part. It discusses the principles of genetics, the nature of genes and their interaction with environment, genetic recombination, and mutations, their role in variations and evolution, and gene frequencies in a population.

The second chapter mainly deals with the molecular basis of inheritance and the chemical nature and mode of action of genes. It explains the replication of genes, the expression of genes as proteins, and molecular mechanisms of genes regulation. It also discusses the possible involvement of errors in gene regulation and the basis of uncontrolled cell division resulting in the development of cancer.

The third chapter is about the various biomolecular techniques used for studying the genetic materials, such as chromosomes, DNA, recombination in bacteria, breeding methods in plants, isolation and analysis of DNA, etc.

The last chapter of this part briefly discusses recent methods of gene manipulations, recombinant DNA techniques, and transgenic organisms and their application in biotechnology.

Chapter 10 THE PRINCIPLES OF GENETICS

In This Chapter

10.1 HISTORICAL PERSPECTIVES

Knowledge of the principles of heredity is so basic to our fundamental understanding of the biological sciences that it is hard to believe that these principles were discovered only in the 1860s (and their importance was realized less than a century ago). However, a practical knowledge of the hereditary process came long before its mechanism was understood. Archeologists have discovered that as long as 7,000 years ago farmers in China and south Asia were improving crops by planting hybrid seeds that had developed preferred characteristics. Over 6,000 years ago, the Chinese learned how to develop superior strains of rice. An ancient Babylonian tablet shows a pedigree of a family of horses through five generations, with detailed information about height, length of the mane, and other traits, revealing that they had some knowledge that these traits were transmitted. Farmers and gardeners have continued to practice this type of selective breeding in both plants and animals. Each time an individual plant or animal appeared with a desired characteristic, it was bred again to produce more with similar traits. For example, at harvest time farmers would select heads of wheat that had the most or largest kernels and save them to use as seed the next year.

For thousands of years farmers and breeders have been selectively breeding their plants and animals in order to produce more productive hybrids. It was somewhat of a hit or miss process since the actual mechanisms governing inheritance were unknown. Knowledge of these genetic mechanisms finally came as a result of careful laboratory breeding experiments carried out over the last century and a half.

By the 1890s, the invention of better microscopes allowed biologists to discover the basic facts of cell division and sexual reproduction. The focus of genetics research then shifted to understanding what really happens in the transmission of hereditary traits from parents to children. A number of hypotheses were suggested to explain heredity, but Gregor Mendel, a little known Central European monk, was the only one who got it more or less right. His ideas had been published in the 1860s but largely went unrecognized until after his death. His life was spent in relative obscurity teaching high-school biology and physics in Brno (now in the Czech Republic). In his later years, he became the abbot of his monastery and put aside his scientific work. While Mendel's research was with plants, the basic underlying principles of heredity that he discovered also apply to humans and other animals because the mechanisms of heredity are essentially the same for all complex lifeforms. The modern science of genetics started from the rediscovery of the laws of inheritance in 1900, which were originally postulated by Gregor Johann Mendel. Three scientists—Hugo de Vries, Tshermark, and Carl Correns—rediscovered the principles of genetics independently. Sutton and Boveri observed the parallelism between the Mendelian 'factors' and the chromosomes during meiosis. They established the chromosomal basis of inheritance. The term 'genetics' was proposed by Bateson in 1906 to denote all the subjects concerned with heredity and variation. He defined it as the science of heredity and variation. Bateson along with Punnet observed the variations of Mendelian inheritance and explained the phenomenon of linkage. In 1910, another scientist W. Johannsen, introduced the term gene mainly to represent those factors or units, which are responsible for the inheritance of a trait of an organism. Now it is genetically and biochemically proven fact that a gene is a stretch of DNA that codes for a specific polypeptide.

Mendelian Genetics

Through the selective growing of common pea plants (*pisum sativum*) over many generations, Mendel discovered that certain traits show up in offspring plants without any blending of parent characteristics. For instance, pea flowers are either purple or white; intermediate colors do not appear in the offspring of cross-pollinated pea plants. Mendel observed seven traits that are easily recognized and apparently only occur in one of two forms:

1. Flower color is purple or white
2. Flower position is axil or terminal
3. Stem length is long or short
4. Seed shape is round or wrinkled
5. Seed color is yellow or green
6. Pod shape is inflated or constricted
7. Pod color is yellow or green

FIGURE 10.1 **Gregor Johann Mendel, the "Father of Genetics."**

This observation that there are traits that do not show up in offspring plants with intermediate forms was critically important because the leading theory in biology at the time was that inherited traits blend from generation to generation. Most of the leading scientists in the 19th century accepted this "blending theory." Charles Darwin proposed another equally wrong theory known as "pangenesis." This held that hereditary "particles" in our bodies are affected by the things we do during our lifetime. These modified particles were thought to migrate via blood to the reproductive cells and subsequently could be inherited by the next generation. This was essentially a variation of Lamarck's incorrect idea of the "inheritance of acquired characteristics."

Mendel picked common garden pea plants for the focus of his research because they can be grown easily in large numbers and their reproduction can be manipulated. Pea plants have both male and female reproductive organs. As a result, they can either self-pollinate themselves or cross-pollinate with another plant. In his experiments, Mendel was able to selectively cross-pollinate purebred plants with particular traits and observe the outcome over many generations. This was the basis for his conclusions about the nature of genetic inheritance.

In cross-pollinating plants that either produce yellow or green peas exclusively, Mendel found that the first offspring generation (f1) always has yellow peas. However, the following generation (f2) consistently has a 3:1 ratio of yellow to green.

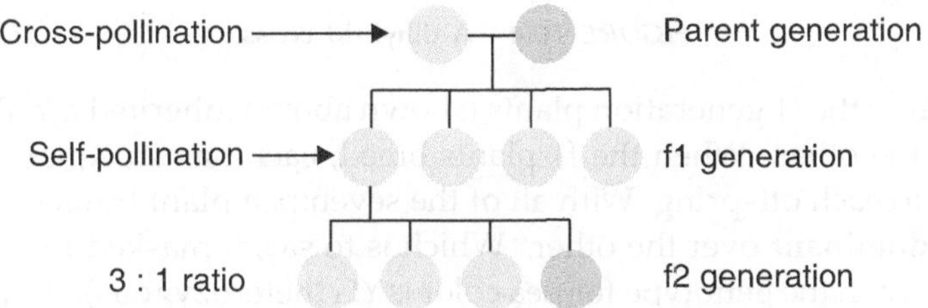

FIGURE 10.2 **Monohybrid cross and monohybrid ratio.**

This 3:1 ratio occurs in later generations as well. Mendel realized that this is the key to understanding the basic mechanisms of inheritance.

Mendel came to three important conclusions from these experimental results:

- That the inheritance of each trait is determined by "units" or "factors" (now called genes) that are passed on to descendents unchanged.
- That an individual inherits one such unit from each parent for each trait.
- That a trait may not show up in an individual but can still be passed on to the next generation.

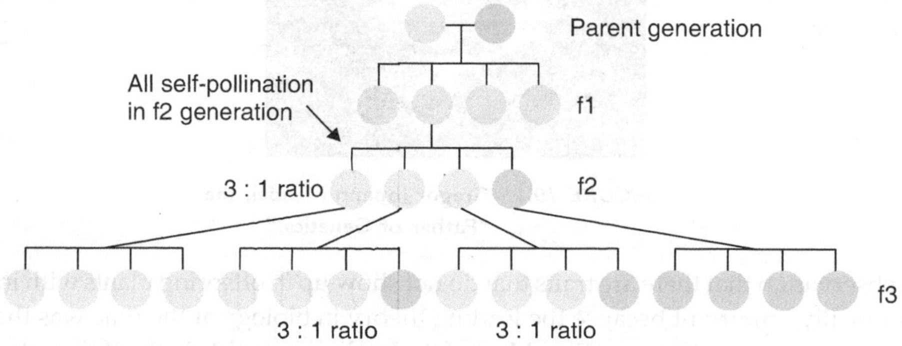

FIGURE 10.3 **Dihybrid cross and monohybrid ratio.**

It is important to realize that in this experiment the starting parent plants were homozygous for pea color. That is to say, they each had two identical forms (or alleles) of the gene for this trait—two yellows or two greens. The plants in the f1 generation were all heterozygous. In other words, they each had inherited two different alleles—one from each parent plant. It becomes clearer when we look at the actual genetic makeup, or genotype, of the pea plants instead of only the phenotype, or observable physical characteristics.

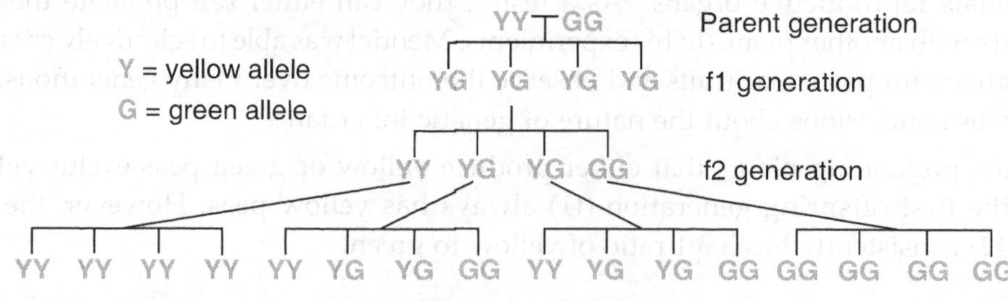

FIGURE 10.4 **A dihybrid cross.**

Note that each of the f1 generation plants (shown above) inherited a Y allele from one parent and a G allele from the other. When the f1 plants breed, each has an equal chance of passing on either Y or G alleles to each offspring. With all of the seven pea plant traits that Mendel examined, one form appeared **dominant** over the other. Which is to say, it masked the presence of the other allele. For example, when the genotype for pea color is YG (heterozygous), the phenotype is yellow. However, the dominant yellow allele does not alter the recessive green one in any way. Both alleles can be passed on to the next generation unchanged.

Mendel's observations from these experiments can be summarized in three principles:

1. Law of Dominance
2. Law of Segregation
3. Law of Independent Assortment

The **law of dominance** states that there are at least two alleles for a gene and one of the alleles can be dominant and the other is recessive. The dominant allele, if it is present, will always determine the trait. For example, in pea plant tall is dominant over dwarf.

According to the **law of segregation**, for any particular trait, the pair of alleles of each parent separate and only one allele passes from each parent on to an offspring. Which allele in a parent's pair of alleles is inherited is a matter of chance. We now know that this segregation of alleles occurs during the process of sex-cell formation (i.e., meiosis).

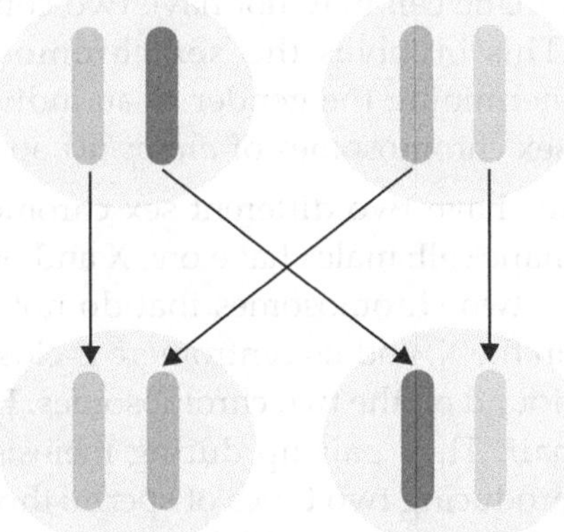

FIGURE 10.5 **Segregation of alleles in the production of sex cells.**

According to the **law of independent assortment**, different pairs of alleles are passed to offspring independently of each other. The result is that new combinations of genes present in either parent are possible. For example, a pea plant's inheritance of the ability to produce purple flowers instead of white ones does not make it more likely that it would also inherit the ability to produce yellow peas in contrast to green ones. Likewise, the principle of independent assortment explains why the human inheritance of a particular eye color does not increase or decrease the likelihood of having six fingers on each hand. Today, we know this is due to the fact that the genes for independently assorted traits are located on different chromosomes.

These two principles of inheritance, along with the understanding about unit of inheritance and dominance, were the beginnings of our modern science of genetics. However, Mendel did not realize that there are exceptions to these rules.

Role of Chromosomes in Inheritance

As described elsewhere, chromosomes are structures made of chromatin, which is a mixture of DNA and a specific family of proteins called histones. During cell division, chromosomes are highly condensed, becoming visible by light microscopy. During interphase, chromosomes are relatively decondensed, and are not visible under the microscope.

However, chromosomes are not uniform. Some regions of the chromosome consist of chromatin that is always highly condensed, even during interphase. These regions are called heterochromatin ('different' chromatin). The other regions, which are uncoiled during interphase, and highly condensed during cell division, are called euchromatin ('good' or 'true' chromatin). Each chromosome of a cell consists of regions of euchromatin interspersed with regions of heterochromatin. These various regions appear as light and dark bands when mitotic chromosomes are stained with various dyes. Each chromosome has a different banding pattern, so chromosomes can easily be identified when stained in this way. Photographs of stained mitotic chromosomes for the purpose of chromosome identification (such as that shown in Figure 10.6) are known as **karyotypes**.

Sex Chromosomes

There is one example where a somatic cell may not have two copies of a particular chromosome under normal circumstances. This involves the **sex chromosomes**. As we shall see, these chromosomes are involved in determining the gender of an individual. Our consideration of sex chromosomes will focus on the sex chromosomes of *drosophila* and humans.

Both fruit flies and humans have two different sex chromosomes: X and Y. Females have two X chromosomes in every somatic cell; males have one X and one Y. The X and Y chromosomes are quite different, so males have two chromosomes that do not appear as a natural pair. The Y chromosome is much smaller than the X, and its centromere is closer to one end. There is also little similarity in the DNA sequences found on the two chromosomes. **However, the X and Y do behave as members of a homologous pair**. They pair up during meiosis, just as other homologues do. They separate during meiosis I, producing two types of sperm: those containing an X chromosome (female-producing sperm) and those containing a Y chromosome (male-producing sperm). Because the sex chromosomes are somewhat unique in their properties, they are classified separately from the other chromosomes. This is indicated by their name: the sex chromosomes. The remaining chromosomes are called **autosomes**.

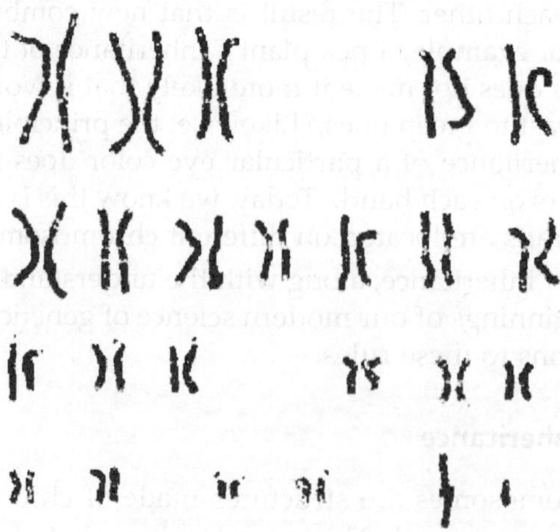

FIGURE 10.6 **Human karyotype.**

Chromosome Theory of Heredity

We've seen how chromosomes are segregated during cell division, but what exactly does this have to do with the inheritance of physical traits (i.e., genes)? Fairly soon after the rediscovery of Mendel's work, many biologists believed that genes were situated on the chromosomes, but this idea required proof. Evidence came in from an experiment conducted by T. H. Morgan on an eye color mutation (white eyes) in *drosophila*. By comparing karyotypes of specific flies with their phenotype, Morgan demonstrated that the white eye mutation was inherited along with the X chromosome. Further work led to the identification of other genes that were transmitted along with the X chromosome, and genes that were transmitted along with particular autosomes. Each chromosome appeared to contain a particular set of genes. Mapping studies showed that each gene resided at a particular locus on a chromosome, indicating that chromosomes are linear arrays of genes.

The realization that chromosomes are linear arrays of genes made it possible to explain Mendel's postulates in terms of chromosome behavior during meiosis.

Law of Segregation

To explain Mendel's law of segregation, let's consider a premeiotic cell from an individual who is heterozygous for a particular gene (A*a*). Each member of the homologous pair of chromosomes will have a different allele of that gene.

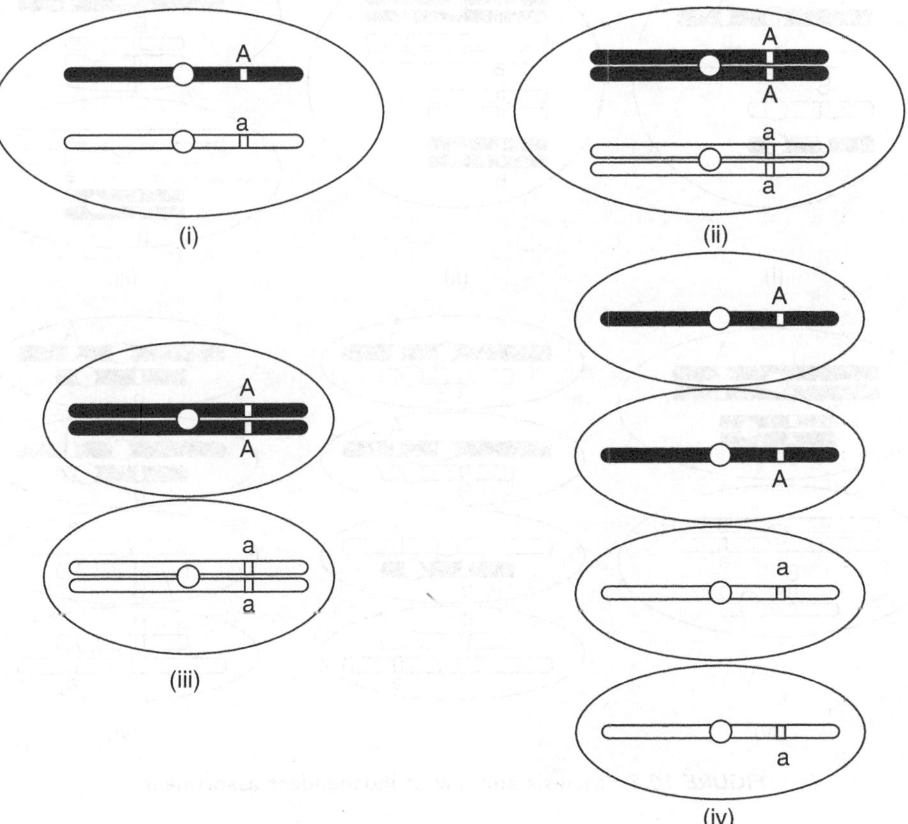

FIGURE 10.7 **Meiosis and law of segregation.**

After replication, each chromosome of the homologous pair consists of two sibling chromatids. During meiosis I, the two members of the homologous pair segregate into separate cells.

After meiosis II, two of the gametes contain a chromosome with the dominant allele, and two gametes contain a chromosome with the recessive allele. Each of these gametes has an equal chance of participating in fertilization. Therefore, if two heterozygotes mate, the offspring will have an equal chance of inheriting a dominant or recessive allele from the male parent, and an equal chance of inheriting a dominant or recessive allele from the female parent. Therefore, there is an equal chance of having each of the following genotypes: AA, A*a*, *a*A, and *aa*.

Law of Independent Assortment

To illustrate Mendel's law of independent assortment, we'll consider a premeiotic cell from an individual heterozygous at two gene loci (A*a*, B*b*). Each of the larger chromosomes has a different allele for the A gene, and each of the smaller chromosomes has a different allele for the B gene.

After replication, each of the chromosomes consists of two sibling chromatids joined at the centromere.

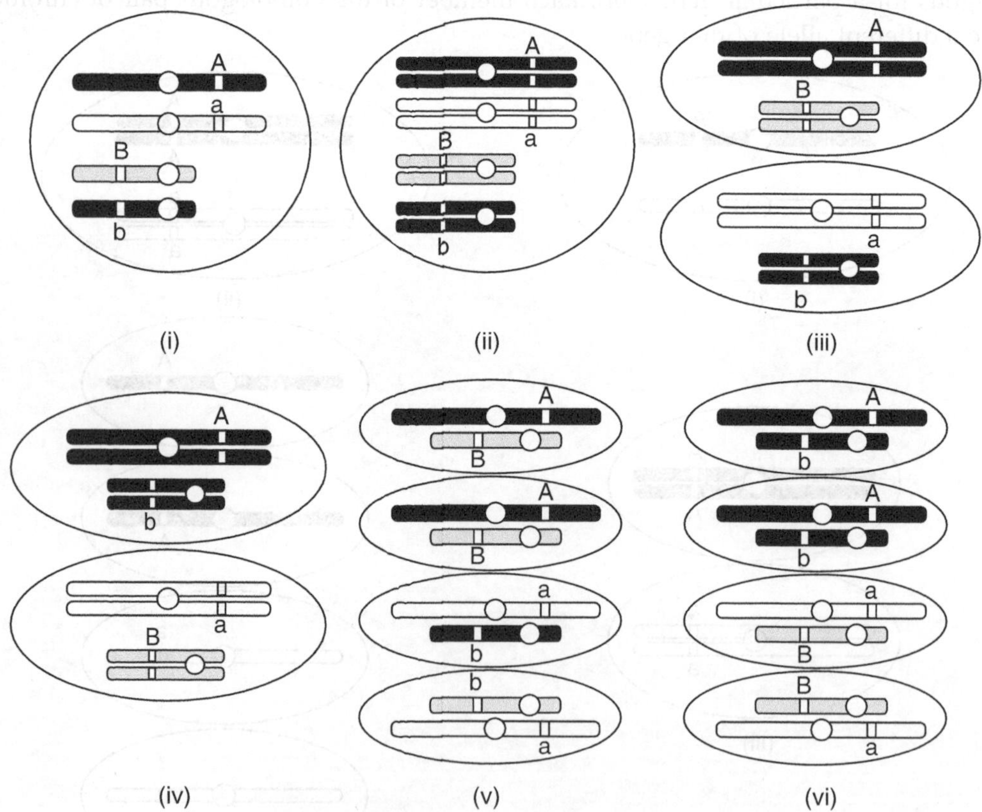

FIGURE 10.8 **Meiosis and law of independent assortment.**

The first meiotic division produces two possible combinations of chromosomes in the daughter cells, because the large and small chromosomes assort independently. One of the possible combinations (let's call it 'combination 1') is shown to the right (iii), where the large chromosome with the dominant allele of the A gene has segregated along with the small chromosome that has the dominant allele of the B gene. Likewise, the chromosomes containing the recessive allele of each gene have segregated together.

Here is the other possible combination of chromosomes (we'll call it 'combination 2') after meiosis I. In this case, the chromosome with the dominant allele of the A gene has segregated with the chromsome that has the recessive allele of the B gene, and vice versa (iv). Because the larger chromosome pair assorts independently of the small chromosome pair (i.e., segregation of the homologues of the large pair does not affect segregation of the homologues of the small pair), both of the possible combinations will occur in a large group of cells undergoing meiosis.

The chromosome complement of each gamete after meiosis II resulting from combination 1 is shown in Figure 10.8 (v).

The chromosomal complement of the four gametes after meiosis II, resulting from combination 2 is also shown in Figure 10.8 (vi). You can see that the allele combinations in the gametes are different between combination 1 and combination 2, but that between the two possible chromosome combinations, all possible combinations of alleles are obtained. In an individual, many cells would be undergoing meiosis simultaneously, and statistical probability dictates that all of the chromosome/allele combinations represented here would be present in the gamete population.

But it was shown in some insects, that male insects have one X chromosomes (XO) per cell and the female cells have two X-chromosomes (XX) . Therefore, male insects produced two types of gametes, those with X and without X in equal proportions. An example is the insect *Protenor,* which produced two types of spermatids (male gametes) cells, with X chromosomes (X) and without X chromosomes (O) in equal numbers.

10.2 MULTIPLE ALLELES

In a diploid (2n) organism every gene must exist in pairs, and each member of that pair is called an allele or allelic pair. One of the alleles is supplied by the male gamete (n) and the other one is given by the female gamete (n) during the process of fertilization. These alleles are situated in the homologous pairs of chromosomes at specific positions called **loci**. Over a period of time random mutations can create changes in these genes resulting in altered forms of the gene. When there are two altered forms of a gene present in a diploid cell or in an allelic pair, it is known as heterozygous and when the members of an allelic pair are of the same type of genes, they are known as homozygous. But when we consider a population of a species there are innumerable copies of a single gene at any given time. By random mutation a gene can be mutated to more than one form in a population. Thus, more than one alternate form of a gene are created naturally. When a gene is present in more than one form it is called multiple alleles. However, in a diploid organism only two of them are represented in the homologous chromosomes as heterozygous alleles. For example, the gene for flower color may be present as that of red color, yellow color, or white color. But in an individual only two of them are present in alleles, the heterozygous allele.

For the sake of simplicity, we usually use examples with only two possible alleles (A and *a*). But a single gene can actually have many possible alleles (A, *a*, A1, A2, A', etc.). For example, hair color in mice is determined by a single gene with a series of alleles, each resulting in different coloration. There are alleles for black, brown, agouti, gray, albino, and others. The twist here is that the same allele can be dominant or recessive depending on context. Allelic series are often written as agouti > black > albino. This means that agouti is dominant to black, and black is dominant to albino. (And agouti is necessarily also dominant to albino.) If the black allele is in the presence of an agouti allele, the mouse will be agouti because black is recessive to agouti. If that same black allele is paired with an albino allele, the mouse will be black since black is dominant to albino. Similarly, human blood type (A, B, and O) characteristics are due to the interactions between three alleles (designated IA, IB, and i), which are present at the same locus of homologous chromosomes. Cellular antigens result from the presence of IA and/or IB, but not i. Four blood types result from the various allele combinations as follows:

- **Type A blood: IAIAIAi**
- **Type B blood: IBIBIBi**
- **Type AB blood: IAIB**
- **Type O blood: ii**

Note that the alleles A and B are co-dominant and both are expressed in individuals with type AB blood.

10.3 LINKAGE AND CROSSING OVER

Sometimes, the predictions (i.e., for dihybrid crosses) based on our understanding of Mendel's principles are seemingly violated; it is often found that the number of offspring obtained for each phenotype is significantly different from a 9:3:3:1 ratio, the dihybrid ratio. Numbers within the recombinant phenotypes are higher than predicted and numbers of the two parental phenotypes are fewer than predicted. Typically, when this occurs, it is because the genes of the trait that we considered for the study are present on the same set of chromosomes (linkage) and crossing over during prophase I affects the outcome of meiosis. When genes distribute after meiosis varies depending on whether or not crossing over has occurred, the genes are said to be linked.

Up until now we have assumed that all genes were inherited independently. However, we have also said that genes are arranged on chromosomes, which are essentially long strands of DNA residing in the nucleus of the cell. This certainly opens the possibility that two otherwise unrelated genes could reside on the same chromosome. Does independent inheritance hold for these genes?

To start with, we need to consider the rather complex process that forms gametes (egg and sperm cells, each with only one copy of each chromosome or haploid) from normal cells or diploid with two copies of each chromosome, one derived from each parent. At one stage of this process (please see the details of meiosis), the maternally derived chromosome lines up with the corresponding paternally derived chromosome, and only one of the two goes to a specific gamete. A chromosome is not a single gene, in fact it is a group of genes arranged in lines, which migrate as single unit. A dog with 39 chromosomes does not only have 39 genes, but rather there are a large number of genes in a chromosome and they move together as a single unit. Thus, Mendel's law of independent assortment is based on the independent orientation of genes that are located on different chromosomes. Luckily, all the seven characteristics that Mendel studied in the pea plant were located on different chromosomes and that is why he was able to get the expected 9:3:3:1 ratio, the dihybrid ratio in the F_2 generation. But it was observed that all the genes do not obey the law of independent assortment and result in the modified forms of dihybrid ratios. Because those genes, which are located in the same chromosomes, cannot move independently but sometimes behave like a single gene, it gives monohybrid ratios for dihybrid crosses. But in fact, in dihybrid crosses in addition to the formation of monohybrid ratios, there is the formation of variations of dihybrid ratios, too. This shows that things are a little more complicated yet, because while the paternal and maternal chromosomes are lined up, they can and do exchange segments, so that at the time they actually separate, each of the two chromosomes will most likely contain material from both parents.

At this point we need to define a couple of terms. Two genes that are located in the same chromosomes are linked if they are close together and thus tend to be inherited together. Such genes are called **linked genes** or **linkage groups**. But if those genes located on the same chromosomes are present far apart, there is the chance of exchanging their positions between the homologous chromosomes. This interchange of genes between the homologous chromosomes is called **crossing over**. The process of crossing over gives some degree or percentage of independent assortment, which is closely related to the physical distance between the linked genes in a chromosome. In true linkage, there is always the possibility that linked genes can cross over. Crossing over involves the breaking and rejoining of DNA segments (exchange) of genes between the homologous chromosomes. Thus, the process of crossing over breaks the original linkage between genes present in a chromosome. This happens often enough that genes far apart on long chromosomes appear to be inherited independently, but if genes are close together, a break is much less likely to form between them than at some other part of the paired chromosomes. The process of crossing over results in the genetic recombination of paternal and maternal traits and the amount

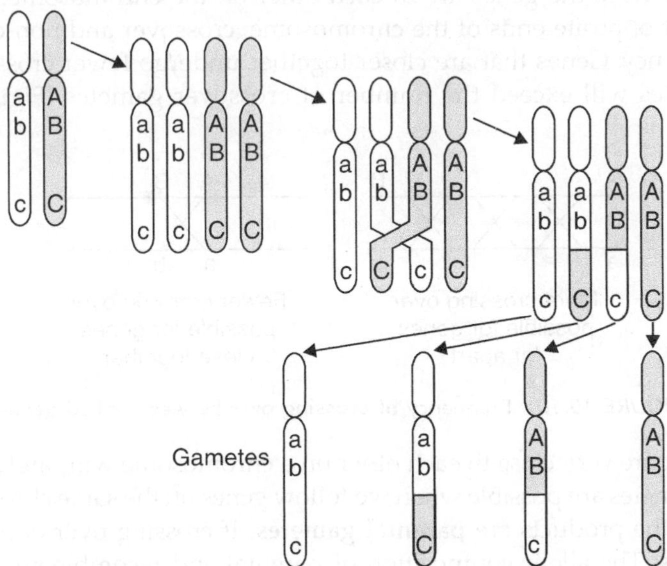

FIGURE 10.9 **Crossing-over and recombination during meiosis.**

of such crossing over depends on the relative distance between the genes. Modifications of dihybrid ratios depend on this physical distance, and the frequency of recombination between two genes can be used to estimate the relative distance between them on the chromosomes. The recombination frequencies can be calculated from the phenotypes of the parental and new recombinants formed in the offspring of heterozygous individuals.

Genetic recombination is any exchange of genetic materials between two chromosomes. Normally, during meiosis there is the chance of genetic recombination due to the exchange of segments of DNA between the pairs of homologous chromosomes. This creates new combinations of genes (Figure 10.9). This process takes place during meiosis by the crossing over. It allows the mixing of paternal and maternal traits located on the same chromosomes. The process of genetic recombination is very important for the generation of variation among the progenies and in the population of sexually reproducing organisms. Mutation also produces variations, but there is a difference. Recombination gives rise to different combinations of the existing genes and its alleles. The process of sexual reproduction can create innumerable combinations of genes on which the natural selection has operated for millions of years to produce adaptable phenotypes and genotypes. Genetic recombination can take place only during the process of meiosis and it is absent in mitosis. In mitosis, cell division is an equal division with the equal distribution of genetic materials in diploid conditions. There is no pairing of chromosomes (synapsis) and related crossing over during mitosis. This method of cell division is essential for the growth and repair of the body. Thus, mitosis occurs in somatic cells and meiosis occurs in specialized cells of reproductive organs during the formation of gametes. The cells formed by meiosis are quantitatively and qualitatively different from the mother cells and dissimilar in quality among the daughter cells.

10.4 GENETIC MAPPING

Recombination can occur between any two genes on a chromosome. The amount of crossing over is dependent on how close the genes are to each other on the chromosome. If two genes are far apart, for example, at opposite ends of the chromosome, crossover and non-crossover events will occur in equal frequency. Genes that are closer together undergo fewer crossing over events and non-crossover gametes will exceed the number of crossover gametes. Figure 10.10 shows this concept.

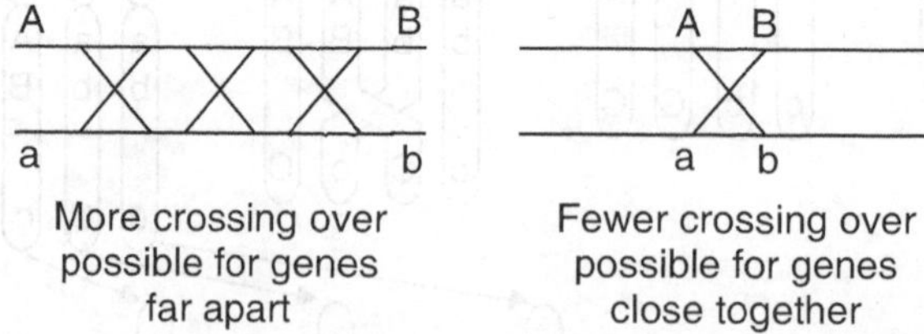

FIGURE 10.10 **Frequency of crossing over between linked genes.**

Two genes that are very close to each other on a chromosome will rarely experience crossing over. Two types of gametes are possible when we follow genes on the same chromosomes. If crossing over does not occur, the products are **parental gametes**. If crossing over occurs, the products are **recombinant gametes**. The allelic composition of parental and recombinant gametes depends on whether the original cross-involved genes are in coupling or repulsion phase. Figure 10.11 depicts the gamete composition for linked genes from coupling and repulsion crosses.

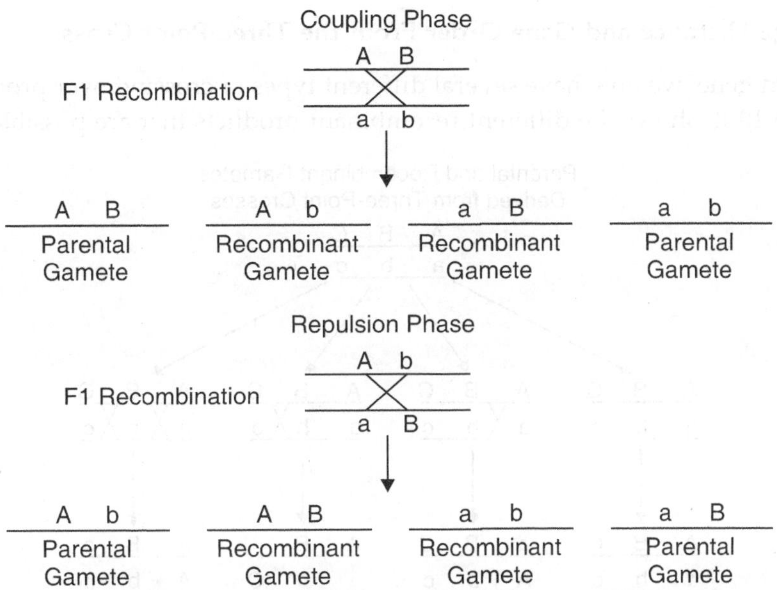

FIGURE 10.11 **Coupling or repulsion in gamete formation.**

It is usually a simple matter to determine which gametes are recombinants. These are the gametes that are found in the lowest frequency. This is the direct result of the reduced recombination that occurs between two genes that are located close to each other on the same chromosome. Also, by looking at the gametes that are most abundant, you will be able to determine if the original cross was a coupling or repulsion phase cross. For a coupling phase cross, the most prevalent gametes will be those with two dominant alleles or those with two recessive alleles. For repulsion phase crosses, gametes containing one dominant and one recessive allele will be most abundant (see Figure 10.11). Understanding this fact will be important when you actually calculate a linkage distance estimate from your data.

The important question is how many recombinant chromosomes will be produced. If the genes are far apart on the chromosome a crossover will occur every time that pairing occurs and an equal number of parental and recombinant chromosomes will be produced. Test cross data will then generate a 1:1:1:1 ratio. (A test cross is the crossing of the F1 hybrid back with the recessive parent and the test cross ratio is 1:1:1:1 for a dihybrid cross.) But as two genes are closer and closer on the chromosome, fewer crossover events will occur between them and thus fewer recombinant chromosomes will be derived. We then see a deviation from the expected 1:1:1:1 test cross ratio.

How can we decide how close two genes are on a chromosome? Because fewer crossover events are seen between two genes physically close togehter on a chromosome, the lower the percentage of recombinant phenotypes will be seen in the testcross data. By definition, one map unit (m.u.) is equal to one percent recombinant phenotypes, and one m.u. is called one centimorgan (cM). To determine the linkage distances, simply divide the number of recombinant gametes by the total gametes produced. The frequency of recombination between distantly placed genes and the order genes in a chromosome can be determined by different mapping techniques. One such method is the **three-point cross.**

Deriving Linkage Distance and Gene Order From the Three-Point Cross

By adding a third gene, we now have several different types of crossing over products that can be obtained. Figure 10.12 shows the different recombinant products that are possible.

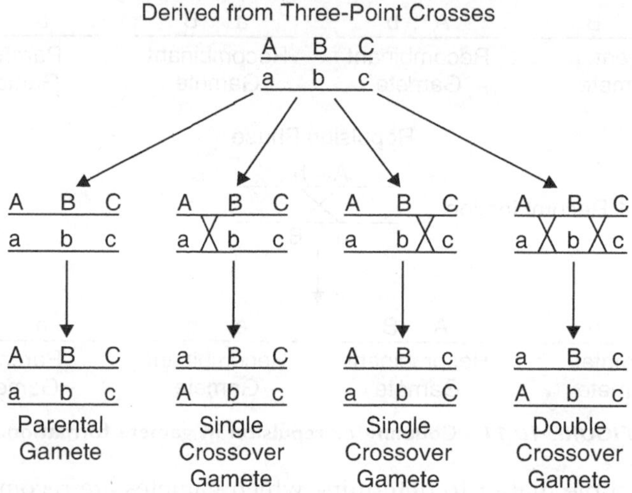

FIGURE 10.12 **A three-point crossing over (trihybrid cross).**

Now if we were to perform a testcross with F1, we would expect a 1:1:1:1:1:1:1:1 ratio. As with the two-point analysis described in Figure 10.12 shows, up ratios deviation from this expected ratio indicates the occurrence of a linkage. By analyzing gamete formation in a single crossover, double crossover, and then a triple crossover systematically it is possible to get the respective recombination frequencies for the individual genes, and from that the relative distance between them, known as **genetic distance**, can be calculated.

It was Thomas Hunt Morgan who first demonstrated the occurrence of linkage between genes in *drosophila* in 1911 and related the frequency of recombination with the linear distance between the genes along the chromosomes. He observed that a large number of hereditary characteristics associated together in groups called **linkage groups**. He showed that the white eyes, miniature wings, yellow body color, single bristles, etc. are a group of traits linked to the sex chromosomes. Thus, they are sex-linked traits associated with male members. Thus, each chromosome is a linkage group. So that the haploid number of chromosomes of an organism are its linkage groups. There are four linkage groups in *drosophila,* 23 in man, 7 linkage groups in peas, and 10 in maize.

By knowing the genetic distance and the order of genes in a chromosome it is possible to make the genetic map of that chromosome.

10.5 GENE INTERACTION OR POLYGENES

In a monohybrid cross, the F2 generation will show two phenotypes in the ratio 3:1 known as the **monohybrid ratio**. This happens only when there is complete dominance. In dihybrid crosses the

F2 generation will have four phenotypes in the ratio 9:3:3:1. This is applicable when genes are independent (absence of linkage) and show complete dominance. In the case of incomplete dominance the heterozygous ratio of a monohybrid cross will modify into 1:2:1, which is actually the genotypic ratio. Similarly, the number of phenotypes in the F2 generation of a dihybrid cross also undergoes modification.

Multiple alleles are different forms of the same gene. That means the sequence of the bases is slightly different in the genes located on the same place of the chromosome. In polygenic characteristics, etc., there is more than one gene involved, and there may be multiple alleles for each of these genes affecting a specific character. This is far more complex, involving potentially a number of chromosomes. Phenotypic characteristics such as high blood pressure (hypertension) are not the result of a single "blood pressure" gene with many alleles (a 120/80 allele, a 100/70 allele, a 170/95 allele, etc.). This phenotype is the result of interactions between a person's weight (one or more obesity genes), cholesterol level (one or more genes controlling metabolism), kidney function (salt transporter genes), smoking (a tendency to addiction), and probably lots of others, too. Each of the contributing genes can also have multiple alleles. Similarly, eye color in man is due to complex interactions of two genes each having pairs of incompletely dominant alleles. Skin color is determined by the interactions between at least three independent genes having two or more alleles.

These multiple gene interactions are divided into two types: **additive gene interactions** and **non-additive gene interactions** or **epistasis**. When a phenotype is influenced by two or more independent genes and each one of them contributes equally to the formation of the trait, it is called an **additive gene interaction**. For example, two independent genes having their own alleles determine the seed coat color in wheat.

Epistasis is a **non-additive gene interaction** in which a gene influences another gene in different ways and the ratio of the phenotypes obtained indicate the type of gene interaction or epistasis. Any type of suppression of the effect of a gene by another, non-allelic gene can be considered epistasis. The following are some commonly occurring ratios for epistasis.

- Both genes are complementary to give phenotype 9:7.
- One gene inhibits the expression of another 13:3.
- One gene supplementing the effect of other 9:3:4.
- One gene hides the effect of other 12:3:1.

10.6 SEX-LINKED INHERITANCE

Thomas Hunt Morgan, a famous geneticist, very clearly demonstrated that a specific trait namely, white eye color, was linked to the sex chromosome. Since the Y chromosome in males did not possess any gene for eye color it gave a distorted segregation in the progeny. Therefore, a white-eyed female will produce white-eyed male progenies, but in the female progenies one may expect 50% white-eyed and 50% red-eyed individuals. Now it is very well known that the distorted segregation is because genes of eye color and sex determination are tightly linked on the X chromosome. This was the first to prove the existence of linked genes.

Sex determination in humans results from the action of a pair of chromosomes, the **sex chromosomes** (the other 22 pairs are called autosomes). Those chromosomes, which are responsible for the determination of the sex of an organism, are termed as sex chromosomes or X chromosomes. All other chromosomes of the cell are **autosomes**. There are two types of sex chromosomes: X chromosomes and Y chromosomes. X chromosomes in homozygous conditions determine femaleness and association of X chromosomes and Y chromosomes produce maleness. We designate the male XY and the female XX. An understanding of Mendelian principles allows us to predict that males and females are likely to be conceived with equal probability.

The X chromosome contains a significant number of genes. In contrast, the Y chromosome contains very few. Therefore, genes on the X chromosome (often called 'X-linked' genes) are in a unique situation. Females have two copies of each gene, just like the normal situation, with autosomal (in humans, there are 22 pairs of autosomes) genes. Males, on the other hand, since they have only one X chromosome, have only one copy of each X-linked gene. Because of this, males cannot be homozygous or heterozygous; they are referred to as hemizygous (The condition of having only one allele of a pair.) Therefore, alleles that are recessive in a female are automatically expressed in a male (because there is no second allele to overshadow the recessive one). The normal rules of dominance do not apply to males in this case. For this reason, problems associated with X-linked recessive alleles, such as hemophilia (inability of blood to clot) and color blindness are more common in males than they are in females.

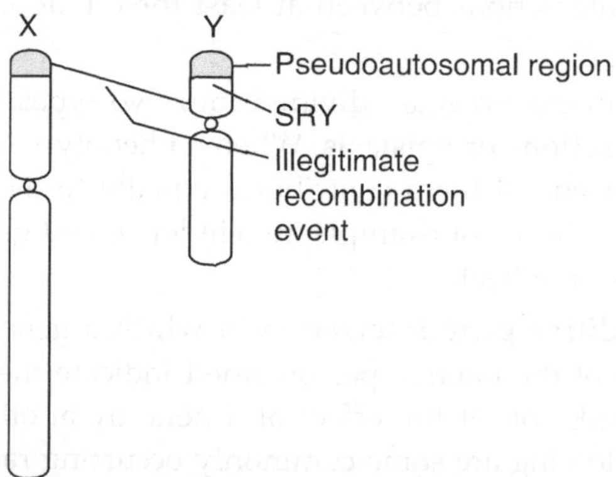

FIGURE 10.13 **Human sex chromosomes.**

Let's consider hemophilia as an example. For a female to be affected by hemophilia, she has to be homozygous for the hemophilia allele (which we will designate 'h'). If an affected female mates with an unaffected male:

P1 : ♀hh X H– ♂

↓

F1 : ♀Hh, ♂h–

All of the male offspring would be affected by hemophilia. All of the female offspring would be unaffected, but would be carriers. This criss-cross pattern of inheritance of a phenotype (from mother to sons) is one of the features of X-linked inheritance. If the female parent is a carrier, and the male parent is a hemophiliac, the following offspring are obtained:

P1 : ♀Hh X h– ♂

↓

F1 : ♀1/2 Hh, 1/2 hh

♂1/2 H–, 1/2 h–

Half of the offspring of each gender will, on an average, be hemophiliac.

The origin of the hemophilia that plagued the crowned families of Europe is actually of English/German origin. Queen Victoria was the original carrier. Her son the Prince of Wales did not inherit it, and so the British royal family has been free of the disease. The royal families of France, Spain, and Russia, whose daughters were carriers of the hemophilia gene were not so lucky.

Sex-influenced traits are related to sex but not coded by genes on the sex chromosomes (baldness, gout, allergies, and cleft palate are all generally more prevalent in men than women). Typically, they are influenced by the hormone, testosterone (in both men and women).

Chromosomal non-disjunction is the improper segregation of chromosomes during cell division. During meiosis this phenomenon can result in a gamete (and hence, possibly an offspring) having an abnormal number of chromosomes. Often, it occurs in sex chromosomes of males or females but it may occur in an autogenous model as well. For example, in a female cell undergoing meiosis, if non-disjunction occurs once during the first meiotic division, the result with respect to sex chromosomes could be as shown below.

sex chromosomes in the original cell: XX
after interphase: XXXX
after meiosis I: XXXX + O
defective eggs: XX + XX + O + O

Note that it would be possible for a nucleus to be XXXX, if non-disjunction were to occur during both meiosis I and II.

Defective sperm could result from non-disjunction in a male: O, XX, YY, XY.

Unfortunately, such defective gametes sometimes participate in fertilization, yielding zygotes. Resulting offspring could have **Trisomy X** (fertile females with XXX), **Klinefelter's Syndrome** (sterile males with XXY), **Turner's Syndrome** (sterile females with X), or be XYY males. Down's Syndrome results from non-disjunction involving autosomal chromosomes. All of these genetic abnormalities and others can be diagnosed using **amniocentesis**, a procedure in which fetal cells obtained from the amniotic fluid of a pregnant woman are karyotyped. Another technique, **chorionic villi sampling,** may also be used to determine the genetic characteristics of a fetus.

10.7 EXTRA NUCLEAR INHERITANCE

According to traditional ideas the genome is located in the nucleus of a eukaryotic cell.

All genetic loci discussed to date have been located on the chromosomes in the cell's nucleus. But there are some genes present in the cytoplasm, inside the organelles, such as mitochondria and chloroplasts. These organelles have a circular DNA molecule similar to that of a eukaryotic system. The characteristics inherited by them are known as **extra chromosomal inheritance** or **cytoplasmic inheritance**. We have studied the process of fertilization. During fertilization only the nucleus of the male gamete enters the egg, leaving the cytoplasm outside. Thus, the cytoplasm or the cytoplasmic genes of the zygote are contributed only by the egg and not by the male gamete. Therefore, the extra chromosomal inheritance is also known as **maternal inheritance**. There are three types of cytoplasmic inheritance known today:

A. Maternal influence (egg cytoplasm influences the phenotype of the offspring)
B. Organelle heredity (mitochondria and chloroplasts)
C. Infectious heredity (an infectious particle is transmitted during conjugation)

Cytoplasmic inheritance shows certain special features. (1) Lack of Mendelian segregation and typical Mendelian ratios. (2) Persistence of characteristics for many generations. (3) Controlled by mitochondrial and chloroplast DNA. (4) Shows maternal inheritance as these characters are transmitted only by female gametes.

Variegation in Four o'clock Plant and Maternal Inheritance

The classic study of maternal inheritance was performed by Carl Correns on the four o'clock plant. This plant can have green, variegated (white and green), or white leaves. Flower structures can develop at different locations on the plant and the flower color corresponds to the leaf color. When Correns crossed the different colored flowers from different locations on the female plant with pollen obtained from flowers of the three different colors, the progeny that resulted from the cross always exhibited the color of the leaf of the female. That is, regardless of whether the pollen was from a leaf that was green, variegated, or white. If the female flower came from a region where the leaves were green, all the progeny were green. Similar results were seen when the female was from a region on the plant where the leaves were either variegated or white. In comparison to traits controlled by maternal effects, those traits controlled by maternal inheritance, the female phenotypes are always expressed in offspring.

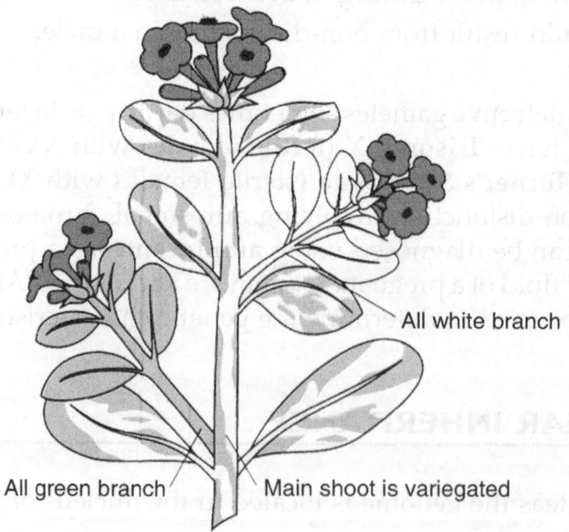

FIGURE 10.14 **Four o' clock plant (*mirabils jalapa*).**

Female	Male	Progeny Phenotype
Green	Green, Variegated, or White	Green
Variegated	Green, Variegated, or White	Variegated
White	Green, Variegated, or White	White

The results can be explained in the following manner. **All of the organelle DNA that is found in an embryo is from the female.** The egg cell is many times larger than the pollen cells and contains both mitochondria and chloroplasts. Pollen is small and is essentially devoid of organelles, and thus organelle DNA. So any trait that is encoded by the organelle DNA will be contributed by the female. In the case of the four o'clock plant, the different colors of the leaves are a result of the presence or absence of chlorophyll in the chloroplast, a trait that can be controlled by the chloroplast DNA. Thus, green shoots contain chloroplasts that have chlorophyll, the chloroplasts in the white shoots contain no chlorophyll, and the variegated shoots contain some chloroplasts with chlorophyll and some without chlorophyll. Thus, depending on the location in the plant where the flower comes from, the egg can have chloroplast with chlorophyll, without chlorophyll, or a mixture of the two types of chloroplasts. This is the biological basis of maternal inheritance.

Snail Shell Coiling and Maternal Effects

The embryo is formed when a female gamete unites with a male gamete. In the vast majority of species, the female gamete is physically larger than the male gamete and provides the cytoplasm for the developing embryo. Within this cytoplasm are factors that were released by the nuclear genes of the female. Those factors may have specific effects upon the developing embryo. The female cytoplasm also contributes the mitochondria for all species as well as the chloroplast for plant species. These two organelles contain DNA and control certain traits in the offspring. Those phenotypes that are controlled by nuclear factors found in the cytoplasm of the female are said to express a **maternal effect**. Those phenotypes controlled by organelle genes exhibit **maternal inheritance.**

The classic phenotype, which exhibits maternal effects, is the coiling direction of snail shells. The **coiling phenotype** that is seen in the offspring is **controlled by the genotype of the mother.** The following crosses were made between pure line snails, and the following results were seen. By convention, the female is always given first.

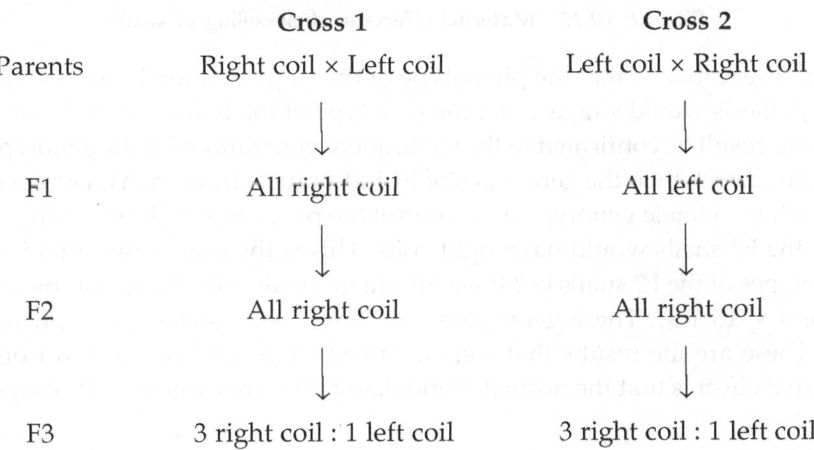

These results at first glance appear to be at odds with Mendel's laws. First, the F1 phenotype is not the same for both crosses. With other experiments, the results of reciprocal crosses (complementary crosses were the phenotypes of females and males are reversed in the initial parental cross) were equivalent, but with this experiment it appears that the female controls the phenotype. Yet, the F2 appears to contradict this hypothesis because the left- and right-coiled F1 individuals produced all right progeny. Furthermore, the 3:1 Mendelian ratio is not seen in the F2, but rather appears in the F3 generation.

How can this result be explained? First, let's look for results that are familiar. The F3 ratio of 3 rights : 1 left for both crosses suggests that right-coiled shells are dominant to left-coiled shells. If this is the case, then we can assign the following genotypes to the pure lines:

- *Right-coiled shell:* s^+s^+
- *Left-coiled shell: ss*

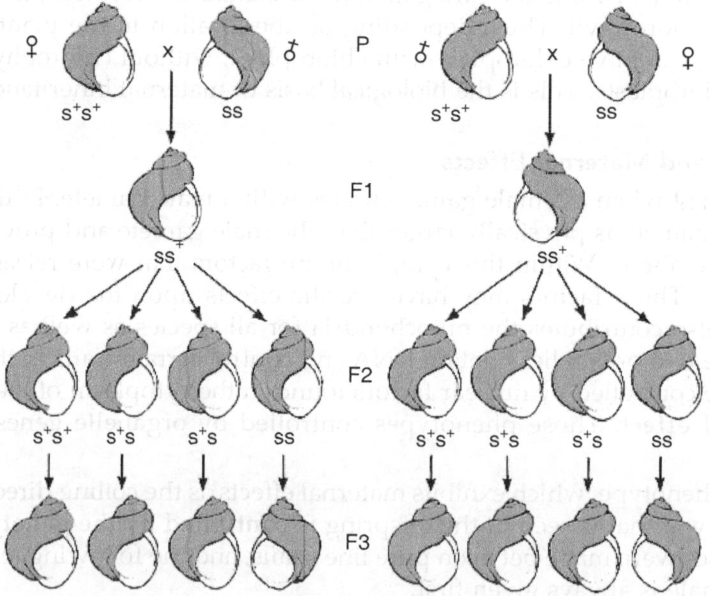

FIGURE 10.15 **Maternal effects in shell coiling of snails.**

The next observation is that the phenotype of the F1 generation is always that of the female parent. One hypothesis would suggest that the genotype of the female controls the genotype of its offspring. Can this result be confirmed in the subsequent generations? If the genotypes we assigned to the parents are correct, then the genotypes of F1 individuals from each cross are s^+s (from s^+s^+ x ss and ss x s^+s^+). If the female genotype does control the phenotype of its offspring, then we would predict that all the F2 snails would have right coils. This is the exact result that is seen. But what would the genotypes of the F2 snails be? If we intermate snails with the genotype s^+s the genotypic ratio should be 3 s^+ to 1 ss. These genotypes would not be expressed as a phenotype until the F3 generation. These are the results that were obtained. A general conclusion from all traits that express a maternal effect is that the normal Mendelian ratios are expressed in one generation more than expected.

Inheritance of Pyrenoids in Chloroplasts of *Spirogyra Triformus*

Spirogyra contains two chloroplasts in each cell. In 1920, VanWisselingh found a cell in which one chloroplast had pyrenoids and the other chloroplast did not have pyrenoids. When this cell divided, both daughter cells were like the mother cell. Genes in the nucleus couldn't have caused this because both chloroplasts would have been the same. Genes in the cytoplasm outside of the chloroplast couldn't have caused this because both chloroplasts are in the same cytoplasm and would be the same. This could only be caused by genes in the chloroplast itself. This work proved that genes are present in the chloroplasts themselves, which control the phenotype of chloroplasts. Genes in the nucleus also control the phenotype of chloroplasts. There are hundreds of known genes located on chromosomes that control the phenotype of chloroplasts. For example, albino gene in maize.

Petite Mutations in *Saccharomyces Cerevisiae*

Yeast cells usually form large colonies called grande on nutrient agar plates. Rarely small or minute colonies known as petite also appear among them. Petite yeasts produce small colonies because they are unable to carry out aerobic respiration as they have defective mitochondria (with defective electron transport proteins; cytochromes a, b, and c). Segregational petites are due to genes in the nucleus. Neutral and suppressive petites are due to genes in the cytoplasm. Thus, there are genes in the nucleus and also in the cytoplasm that control the phenotype of the mitochondria. There are three types of petite mutants that differ in their mode of inheritance. The inheritance of a nuclear gene of haploid yeast shows Mendelian segregation of asosopres into a 2:2 ratio. Such a ratio is found when the first category of petite mutants is crossed with the grande wild type and thus a nuclear gene controls this phenotype, the segregational petites. In the second type the cross between the grande and petite phenotypes results in progeny with all normal phenotypes known as neutral petites. In the third category the cross between grande and petite phenotypes produces all petite progeny, and they are known as suppressive petites. (Figure 10.16).

The inheritance of poky phenotype (similar to petite yeasts) in *Neurospora crassa* and cytoplasmic male sterility in maize are also governed by the mitochondrial gene. **Cytoplasmic male sterility (CMS)**, a cytoplasmic mutation, is extremely important for agriculture. CMS plants are male-sterile and female-fertile. These mutations are in the mitochondria and are maternally inherited.

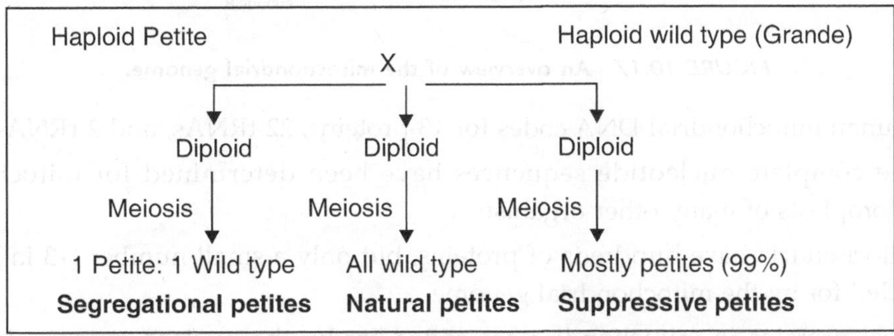

FIGURE 10.16 **Three categories of petite yeasts based on segregation patterns.**

Infectious heredity, an invading microorganism, causes a mutant phenotype that is then transmitted by maternal inheritance.

Genetic Information is Present in Chloroplasts and Mitochondria

DNA was shown to be present in chloroplasts (1963) and in mitochondria (1964). Chloroplasts and mitochondria have complete transcriptional and translational apparatuses. The complete nucleotide sequence of human mitochondrial DNA has been determined (16,569 nucleotides).

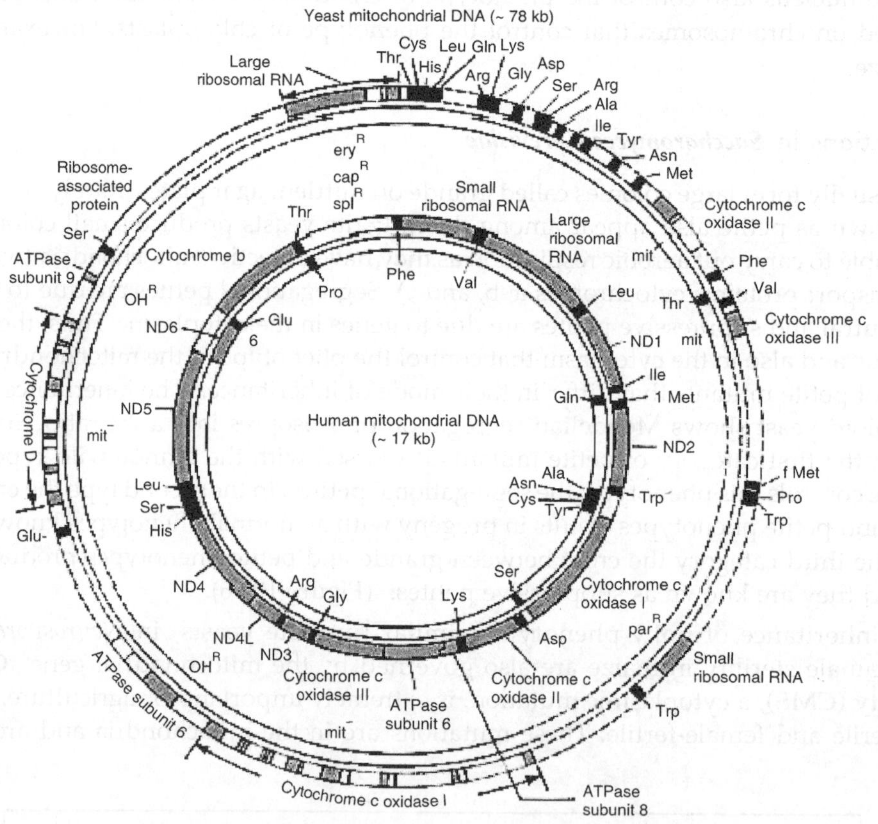

FIGURE 10.17 An overview of the mitochondrial genome.

1. Human mitochondrial DNA codes for 13 proteins, 22 tRNAs, and 2 rRNAs.
2. The complete nucleotide sequences have been determined for mitochondria and chloroplasts of many other organisms.
3. Mitochondria have hundreds of proteins, but only a small number (13 in humans) are coded for by the mitochondrial genome.
4. Most proteins in mitochondria are coded for by the nuclear genome, translated on cytoplasmic ribosomes, and imported into the mitochondria (Figure 10.17).

Evolution of Eukaryotic Cells

Endosymbiont hypothesis, the hypothesis that is most widely accepted and the possible steps involved in the evolution of eukaryotic cells are as follows:

- An endosymbiont is a symbiont that is present inside a cell.
- Anaerobic prokaryotes were thought to be the first living things on earth.
- Anaerobic prokaryotes evolved into aerobic prokaryotes.
- The oxygen released by aerobic prokaryotes was toxic to the anaerobic prokaryotes, and this presumably was the driving force that caused anaerobic prokaryotes to evolve into aerobic prokaryotes.
- Anaerobic prokaryotes evolved into anaerobic eukaryotes.
- The anaerobic eukaryotes took up aerobic prokaryotes as an endosymbiont, and the aerobic prokaryotes became mitochondria.
- These cells then took up photosynthetic prokaryotes as an endosymbiont, and the photosynthetic prokaryotes became chloroplasts.
- Most of the genes of the original endosymbionts were lost because the same types of genes were present in the nucleus of the host eukaryotic cell. For this reason, the genomes of mitochondria and chloroplasts are much smaller than prokaryotic genomes.

10.8 QUANTITATIVE INHERITANCE

Many genes control the inheritances of certain single traits that are measured in a quantitative manner. Some of the examples for quantitative inheritance include:

A. Height

B. Weight

C. Yield in crops

D. Growth rate in farm animals

E. IQ, etc.

In addition to quantitative inheritance, inheritance of these traits is often referred to as "cumulative gene action" or "polygenic-inheritance".

Genetic analysis of contributing genes can best be examined using simplified models. As an example, let's consider plant height in grain sorghum (often called milo). Wild sorghums brought from Africa are often 10' tall; those grown from grain are nearly 2' tall, which greatly simplifies harvest. The "combine" types are sometimes referred to as "4-dwarf" sorghums.

The model for sorghum height is as follows:

There are four genes involved in height determination. Each gene has two alleles, one of which adds about one foot of height; the other is "null." The alleles that contribute to height are called "contributing alleles," and are designated by primes, A′, B′, C′, etc. Non-contributing (null) alleles are designated with A, B, C etc. (Sometimes a, b, c may be used.) A plant with no contributing alleles is 2′ tall (AA, BB, CC, DD). Each gene is equivalent and shows incomplete dominance; that is, A′A′ adds 2 feet, A′A adds 1 foot, and AA adds 0 feet to the base height. The same is true for the B, C, and D gene loci. A plant with the genotypes A′A′, BB, CC, DD, or AA, BB C′C′, DD will be 4′ tall.

A′A′, BB, CC, DD (4′) X AA, BB, CC, DD (2′)

↓

A′A, BB, CC, DD (3′)

If we allow the 3′ tall F1 progeny to self pollinate, which is normal in sorghum, we expect to see a 1:2:1 ratio of F2 plants that are 4′:3′:2′ tall, respectively. This is true because only one gene with incomplete dominance is "segregating" (heterozygous) in the F1 hybrid. (A′A X A′A gives 1/4th A′A′: 2/4th A′A:1/4th AA). The same would be true if AA, B′B′, CC, DD, or any other 4′ tall true breeding plant (homozygous at all 4 gene loci) was crossed to the AA, BB, CC, DD parent. What if we cross A′A′ BB, CC, DD to AA, B′B′, CC, DD? Both parents are 4′ tall as is the F1, but the F1 is heterozygous for 2 genes (A′A, B′B, CC, DD). Now in the F2, we can predict that, 1/4th of the progeny will be A′A′ and that 1/4th of these will also be B′B′; thus, 1/16th of the progeny should be 6' tall! Likewise, 1/16th should be AA, BB, CC, DD, or only 2' tall having no primes. When the range in the F2 progeny goes beyond the original parents, we see **transgressive segregation**. Transgressive segregation implies that the parents donate contributing alleles from different genes to the hybrid.

There should also be plants in the F2 generation that have one contributing allele (A′A, BB, CC, DD, or AA, B′B, CC, DD). These should represent 1/4th (1/8th + 1/8th) of the progeny, and they will be 3' tall. The same is true for plants with three contributing alleles (A′A′, B′B, CC, DD, and A′A, B′B′, CC, DD).

Genes that contribute to quantitative traits are referred to as polygenes, or **QTLs,** which stands for **quantitative trait loci.** When this model applies, the hybrid will always show a phenotype that is the average of the parents, and there will be more variation among the F2 progeny than in either P1 or F1 progeny. It would not be difficult to imagine cases where some genes made larger or smaller contributions than others, or where one or more genes may be dominant.

Environmental Effects

Most or all quantitative traits are also influenced by environmental factors. In the case of sorghum height, plants will not reach their genetic potential without water, nutrients (fertilizer), and sunlight. Environmental effects means that some differences can be seen within a purebred parent or in an F1 population, where all of the plants have the same exact genotype. Environmental effects will also "smooth out" the stair-step effect expected for genotypes in the F2 that differ in the number of contributing alleles. Since it is not possible to "count" the number of classes in an F2 population when environmental effects smooth away the genotypic differences, or to identify individuals in the extremes, the number of genes that contribute to the trait cannot be simply estimated.

Normal statistical methods can be used for studying the inheritance of quantitative traits and to analyze the contributions of genes and environment to a trait. The concept is to partition the sources of variation that lead to differences among individuals in the sample, and to identify the portion of variation that results from segregating genes. Total variation (Vt), which is often called phenotypic variation (Vp), arises from differences in genotype (Vg), the environment (Ve), and may also result from interactions (Vgxe) where some genotypes do better in one environment and others in another.

$$\mathbf{Vt = Vp = (Vg + Ve + Vgxe).}$$

In many cases, the interaction component cannot be measured, so it is ignored or handled by working within a specific environment or only working with a specific "breed" or "cross".

Heritability (H^2 or broad-sense heritability) is the fraction of variation due to genetic differences (i.e., Vg. $H^2 = Vg/Vt$). It is relatively simple to make H^2 estimates in plants, since pure-breeding completely homozygous parents can be maintained. Any variation within a pure-breeding homozygous parent, or in the F1 progeny of a cross between two pure-breeding parents must result from Ve, since all plants within each of the populations have the exact same genotype. Thus, these plants can be used to estimate Ve (variation) among F2 plants, which arises both from differences in genotype and from local environments. So the variation in the F2 is a measure of Vt. In the example below, two true breeding corn parents, one with an average row number of six, is crossed to another with an average of fourteen rows. The F1, as expected, is right between the parents, having an average of ten rows. Variance (V) for row number in the parents and F1 is low (arbitrarily measured as 1) Variance in the F2 is eight.

Vt(8) = Vg + Ve (1); ignoring Vgxe since all plants were grown in the same environment, Vg must be 7. Therefore, $H^2 = 7/8$.

It is critical to realize that:

1. Heritability measures are only valid for the population that was measured.
2. Genetic differences will not be measured unless the parents have different alleles.
3. In the environment, gene interactions may be important, but are generally ignored.

Plant and animal breeders are interested in heritability, because it allows them to predict if selective breeding can be used to "improve" a trait. For example, if rate of weight gain in nursing Hampshire pigs is highly heritable, saving those that grow the fastest for breeding purposes will lead to improved weight gain in future generations. If H^2 is low, changes in the diet may be more important. Selection will make progress as long as genetic differences can be combined to make an improved genotype. Selection for one trait may be balanced by loss in another; for example, selection for increased egg size in leghorns is successful, but the hens lay fewer eggs.

10.9 GENE AT THE POPULATION LEVEL

Most species live in groups or populations and each individual of the population has a specific genome mix up or allele combination at the individual level. In individual cases we have seen various types of allelic combinations and their interaction, which lead to a specific type of phenotype. In a diploid organism every gene exists in a pair, the allelic pair or allele. When both the members of the allele are identical the allelic pair is homozygous and when they are different forms, the allele or the allelic pair is heterozygous. Then the phenotype of the organism depends on the nature of the genes, whether they have complete dominance, incomplete dominance, or co-dominance.

Various types of interactions between the members of multiple alleles and the nature of dominance may generate several alternate phenotypes. There may be the effect of polygenes in which a trait may be governed by more than one gene and its alleles. Similarly, there can be the effect of a cumulative gene or interactions of various types similar to that of epistasis. But all these are at the individual level.

Population genetics is concerned with changes in the **gene frequencies** of sexually reproducing populations over time. Major changes in the **gene pool** create a potential for genetic differences, which may ultimately result in speciation and adaptive variation.

Gene Pool

The term gene pool refers to the total number and variety of genes and alleles present in a sexually reproducing population, which are available to be passed on to the next generation.

Suppose we have 100 homozygous wild type (++) female drosophila and 100 homozygous vestigial (vv) male drosophila in a population. In the gene pool, half the genes for this character would be '+' and the other half would be 'v'. To put it another way, we could say that the **gene frequency** of both + and v is 0.5 (50%). In the next generation, the phenotypes of the flies produced by crossing ++ females with vv males will be wild and their genotypes will be +v. The proportions of the genes present have not changed and therefore their gene frequencies will remain at 0.5. By using colored counters, we can make a model to represent this situation and then trace the relative frequencies of genes in the population from generation to generation.

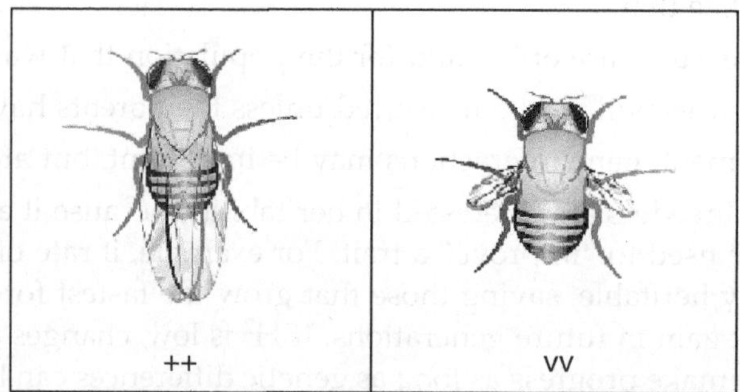

FIGURE 10.18 **(++) Female drosophila and (vv) male drosophila.**

Now consider a mathematical model of a gene in a population (gene pool): Consider a population of drosophila, which started with a mixture of wild and vestigial winged flies (Figure 10.18). Assume that during the period of the investigation there has been:

- A large population (say 250+).
- Random mating between all genotypes.
- No mutation (or mutation is in equilibrium).
- No selective advantage for any genotype (all genotypes survive equally well).
- No gene flow between this population and others (no immigration/emigration or balanced migration).
- Normal meiosis (i.e., chance is the only factor affecting gamete formation).

In this population there will be three genotypes, ++, +v, and vv.

Let the frequency of allele + in the gene pool = **p**.

Let the frequency of allele v in the gene pool = **q**.

Then **p + q = 1** (100% of the gene pool).

Flies of genotype ++ can only be produced by the union of two gametes each carrying a + allele. The probability of one + bearing gamete uniting with another + bearing gamete is the product of their frequency in the population. The frequency of genotype ++ is therefore,

$$\mathbf{p \times p = p^2}$$

Similarly, flies of genotype vv can only be produced by the union of gametes each carrying a v allele. The probability of this occurring is the product of their frequencies in the population.

The frequency of genotype vv is therefore,

$$\mathbf{q \times q = q^2}$$

Flies of genotype +v can be produced by the union of a + carrying sperm and a v carrying egg. The probability of the genotype being formed in this way is therefore,

$$\mathbf{p \times q = pq}$$

Flies of genotype +v can also be formed by the union of a v sperm with a + egg. The probability of this occurring is therefore,

$$\mathbf{q \times p = qp}$$

The total frequency of genotype +v is therefore,

$$\mathbf{pq + qp = 2pq}$$

Since the frequencies of the three genotypes together **must add up to 1** (i.e., 100% of the population), we can write

$$\mathbf{p^2 + 2pq + q^2 = 1}$$

This basic equation of population genetics was worked out in 1908 by the English mathematician G. H. Hardy and a German physician W. Weinberg and is usually referred to as the **Hardy-Weinberg Equation.** They realized that provided certain conditions were met (the assumptions given above) the gene frequencies in a population will not change from one generation to the next. In the same way we can make a mathematical model for the distribution of a gene in humans with its two alleles T and *t*. The allele T is for tasting a chemical PTC and *t* for the absence of the ability to taste the chemical. T is dominant over the recessive *t*. In exactly the same way we can find out the frequency (p and q) of distribution of these genes in the population.

According to the Hardy-Weinberg law, at any point of time the total frequency of these two alleles will be equal to one, provided the above mentioned criteria are fulfilled. This is called the **Hardy-Weinberg Law.** A population whose gene frequencies are not changing is said to be in **Hardy-Weinberg Equilibrium.** It is not evolving. If we know the value of p or q or q^2 in any population, we can then calculate the frequencies of the other genotypes. This may enable us to detect long-term changes in gene frequencies resulting from immigration, emigration, mutation, and selection. But for a normal population the Hardy-Weinberg law is not applicable because it cannot satisfy some or all of the conditions for Hardy-Weinberg equilibrium, and therefore the population will not be static and will be under the process of evolution.

Hardy and Weinberg established the conditions under which populations remain in genetic equilibrium. (i.e., their gene frequencies remain constant from generation to generation).

It follows that the opposites of those conditions are the basic reasons for changes in a population's gene frequencies, and are therefore the basic cause of evolutionary change.

- The population is small (e.g., < 100).
- Mating is not random (e.g., how do wingless *drosophila* catch a mate?)
- Mutation is not in equilibrium.
- Some genotypes have a selective advantage (e.g., wingless *drosophila* on a windy island).
- Gene flow is not balanced.
- Meiosis is not normal.

Note that the Hardy-Weinberg Law does not apply to non-sexually reproducing populations or monoploid populations.

10.10 DISCOVERY OF DNA AS GENETIC MATERIAL

Mendel helped to establish that heredity was controlled by "factors" and chromosomes were soon suspected of carrying the factors or genes. This followed the discovery of the scientific proof that DNA is the genetic material or the material of the genes and chromosomes. DNA was first identified in 1868 by Friedrich Miescher, a Swiss biologist, in the nuclei of pus cells obtained from discarded surgical bandages. He called the substance **nuclein,** and in 1914 Feulgen perfected a specific DNA stain (the Feulgen stain); however, the connection between DNA and heredity was not made until many years later. It was only in the middle of the twentieth century that the role of DNA as genetic material was established. The following are two of the classic experiments that helped to demonstrate the role of DNA as the carrier of genetic information.

In 1928, the British scientist Frederick Griffith was working with the bacterium *streptococcus pneumoniane,* a causing agent of pneumonia. He worked with two types of bacteria, the strain *S*, which is highly infectious and grows in smooth and shiny colonies, and the strain *R*, which is harmless and grows in rough colonies. The *S* strain is surrounded by a polysaccharide coat, which is responsible for its virulence. The *R* strain lacks that coat. Griffith injected mice with different strains of the bacterium. Initially, he used *R* strains and the mice were unaffected. Then, he used *S* strains and the mice died of pneumonia. Living bacteria could be isolated from the dead mice. Next, heat-killed *S* bacteria were injected into the mice, and the mice survived, indicating that the bacteria has to be alive and has to have the coat to be infective. Finally, Griffith injected mice with a mix of living *R* and heat-killed *S* strains of the pathogen. To his surprise, the mice died and living *S* bacteria was recovered from the mice's blood. Griffith concluded that some *R* bacteria had been transformed into *S* by interaction with the dead *S* bacteria. He thought that there was a **transforming principle** responsible for this change in genetic material, and that it could be a protein (Figure 10.19).

Avery's Transforming Principle

Oswald Avery, together with his co-workers Colin MacLeod and Maclyn McCarty, continued Griffith's experiments in the 1930s and 1940s to identify the transforming principle. They cultured *S* cells and broke them open, and separated the different cellular components of that cell extract (i.e., lipids, polysaccharides, proteins, and nucleic acids). Then, they tested each of these macromolecular components to see which of them could transform *R* cells into *S* bacteria. They found that the only components capable of transformation were the nucleic acids. At this point RNA was not separated from DNA so they conceived another experiment to identify which of these two nucleic acids was the transforming principle. They took the nucleic acids and treated them with the enzyme **ribonuclease**, which degrades RNA and not DNA, but the transforming principle was still present. Then, they treated the nucleic acids with **deoxyribonuclease,** which degrades DNA but not RNA, and no transformation resulted. So, they suggested that the genetic material was DNA.

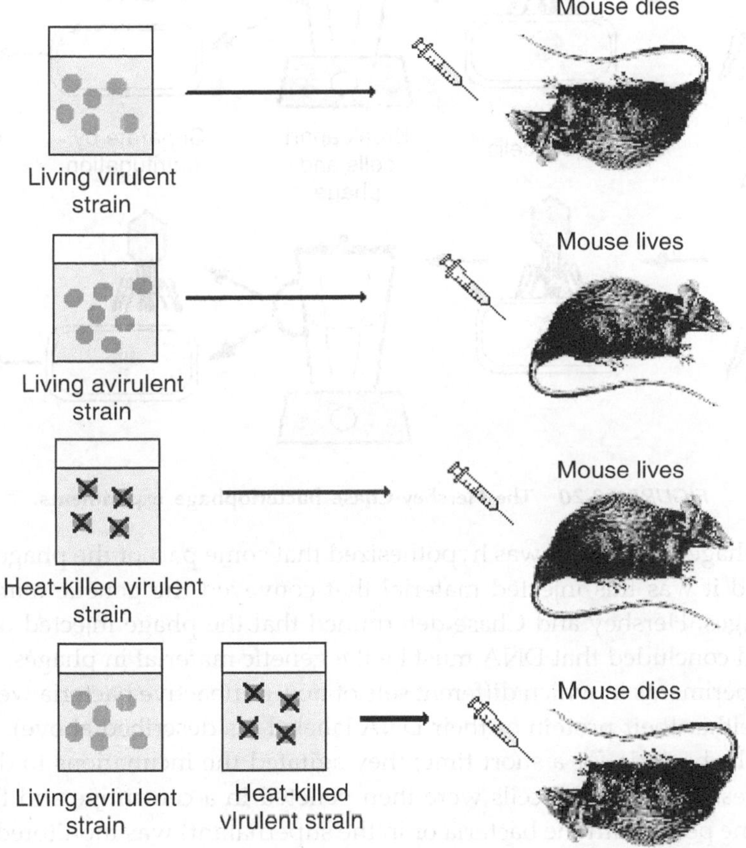

FIGURE 10.19 **Griffith's transformation experiment.**

Hershey-Chase Bacteriophage Experiments

In 1952, the research team of Hershey and Chase published a report that concluded that DNA is the genetic material of the bacteriophage (a virus that infects bacteria) T2. They knew that when a bacteria was infected with this phage, the bacteria soon became a machine that produced new phages. Phages have a very simple structure; they are composed of a strand of DNA surrounded by a protein coat. Hershey and Chase set out to determine which component (the DNA or the protein) was responsible for the ability of the phage to 'take over' and control the metabolic activity of the bacteria to produce new phages. In their experiments, they used two radioactive markers to label the proteins and the DNA of the phages. The proteins were labeled with ^{35}S (a radioactive form of sulfur) and the DNA was labeled with ^{32}P, a radioactive form of phosphorus. This allowed the researchers to easily differentiate between a sample that contained protein (thus, would have ^{35}S present) and a sample that only contained DNA (thus, would have ^{32}P present).

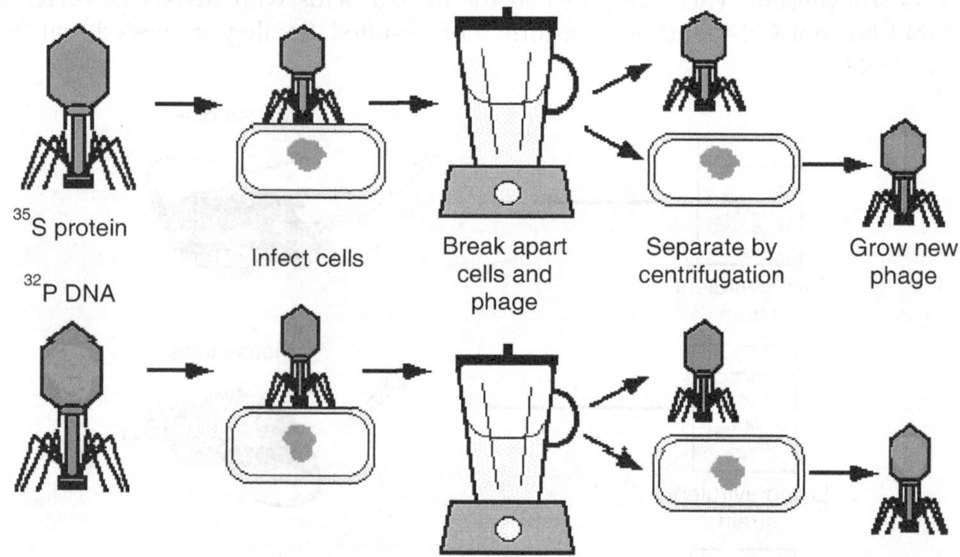

FIGURE 10.20 **The Hershey-Chase bacteriophage experiments.**

During a phage infection, it was hypothesized that some part of the phage was injected into the bacterium and it was this injected material that conveyed the genetic material necessary to produce new phages. Hershey and Chase determined that the phage injected only the DNA into the bacterium and concluded that DNA must be the genetic material in phages. This was done by a series of two experiments in which different sets of non-radioactive bacteria were incubated with phages that had either their protein or their DNA labeled (as described above). They allowed the phages to infect the bacteria for a short time; they agitated the incubations to dislodge any loose parts of the phages. The bacterial cells were then pelleted in a centrifuge and the location of the radioactivity (in the pellet with the bacteria or in the supernatant) was monitored. They found that the radioactive DNA was always found with the bacterial cells and that the radioactive protein was always in the supernatant. This suggested that the DNA was injected into the bacteria but the protein coat was not. Thus, all of the information needed to produce new viruses was contained in the DNA and not the protein.

10.11 MUTAGENESIS

Genetic material, DNA, has high fidelity and so genetic information is transferred from generation to generation without changing its integrity. But changes may occur in the quality of the genetic material, which is transferable to the next generation. Such heritable changes occurring in the DNA is known as **mutation**. Mutation may be due to a chemical change in the DNA or by recombination. Mutations form an important cause of diversity for evolution to act upon. In a population, high genetic variability of the forces of natural selection can select for advantageous genetic changes and result in speciation. The mutations that are happening in the germ line cells responsible for the production of gametes are passed onto the next generation. Generally, mutations are random and recessive in nature. It can be both harmful and beneficial to the organisms, but is particularly very useful for the genetic improvement of crops and microbial strains. Some mutations cause genetic disorders and even result in the development of cancers. It was Hugo de Vries who proposed the theory of mutation in 1901 based on his observations on evening primrose. Similarly, in 1928, H. J. Muller used x-rays for inducing mutations in fruit fly, for the first time.

Types of Mutations

Mutations may be spontaneous or induced. Spontaneous mutations are natural, which occur naturally by changes in the DNA sequences during replication or by recombination of genetic materials or by some environmental factors–chemical or physical factors. The following are the major types of mutations observed in a living system.

Spontaneous Mutations and Induced Mutations

They can be subdivided into mutations occurring during replication or post- replication periods. They are caused by errors occurring during DNA replication by incorporating the incorrect nucleotide in the growing DNA chain; this, however, is a very infrequent event because of the very effective DNA repair systems, which evolved in all organisms. During the post-replication periods mutations are mainly caused by spontaneous physicochemical alterations of a base. There are two ways in which this can be brought about:

- By isomerization of bases such as adenine (amino into imino form) or thymine (keto into enol form). In both cases a AT/GC transition will occur. This event is not recognized by the cellular repair functions and can therefore escape their activity.
- By spontaneous deamination of bases such as cytosine to uracil or 5-methylcytosine to thymine (note: 5% of cytosines in bacteria and bacteriophages are methylated). Although cells have evolved a repair mechanism to remove uracil from DNA, it will of course not remove thymine because it is a normal component of native DNA. Therefore, a GC pair is converted into an AT pair.

Induced mutations are those that happen in the genome as a whole in the DNA as described above, but induced by some environmental factors called mutagens. These can be physical mutagens such as UV light, x-rays, gamma rays, etc., or chemical mutagens such as alkylating agents such as nitrosoguanidine and ethylmrthane sulphonate or acridine compounds such as acriflavine.

Genetic recombination also can cause large-scale genetic changes, which includes exchange of polynucleotide fragments between chromosomes or insertion or transpositions of fragments of DNA from one position to another position in a chromosome.

Somatic and Germline Mutations

Eukaryotic organisms have two primary cell types: germ and somatic. Mutations can occur in either cell type. If a gene is altered in a germ cell, the mutation is termed a germinal mutation. Because germ cells give rise to gametes, some gametes will carry the mutation and it will be passed onto the next generation when the individual successfully mates. Typically, germinal mutations are not expressed in the individual containing the mutation. The only instance in which it would be expressed is if it negatively (or positively) affected gamete production.

Somatic cells give rise to all non-germline tissues. Mutations in somatic cells are called somatic mutations. Because they do not occur in cells that give rise to gametes, the mutation is not passed along to the next generation by sexual means. To maintain this mutation, the individual containing the mutation must be cloned. Two examples of somatic clones are navel oranges and red apples. Horticulturists first observed the mutants. They then grafted mutant branches onto the stocks of "normal" trees. After the graft was established, cuttings from that original graft were grafted on to tree stocks. In this way the mutation was maintained and proliferated. Most tissues are derived from a cell or a few progenitor cells. If a mutation occurs in one of the progenitor cells, all of its daughter cells will also express the mutation. For this reason, somatic mutations generally appear as a sector on the mutated individual.

Cancer tumors are a unique class of somatic mutations. The tumor arises when a gene involved in cell division, a protooncogene, is mutated. All of the daughter cells contain this mutation. The phenotype of all cells containing the mutation is uncontrolled cell division. This results in a tumor that is a collection of undifferentiated cells called **tumor cells**.

Mutations Based on Functions

Morphological Mutants or Phenotypic Mutations

This type of mutation generates a visible phenotypic or morphological alteration. Plant height mutations could change a tall plant to a short one, or from having smooth to round seeds.

The molecular explanations of these mutations are threefold: **Frameshift mutation** (addition or deletion of one or two nucleotides); ethidium bromide or other intercalating agents are effective in producing this type of mutation. **Chain-termination mutations** (nonsense mutations) by mutation of an amino acid encoding nucleotide triplett into a stop-codon leading to a premature dissociation of the ribosomal translation machinery from the mRNA. **Missence mutation** leads to the substitution of one nucleotide by another one and result in the replacement of one often crucial amino acid in the polypeptide by another one with different pysicochemical properties. An agent causing missense mutation is nitrous acid.

Biochemical Mutations

These types of mutations produce one or more lesions in one specific step of an enzymatic pathway. This is because the specific enzyme, which carries out the reaction, has changed. For bacteria, biochemical mutants need to be grown on a media supplemented with a specific nutrient. Such mutants are called **auxotrophs**. Often though, morphological mutants are the direct result of a mutation in a biochemical pathway. In humans, albinism is the result of a mutation in the pathway that converts the amino acid tyrosine to the skin pigment melanin. Similarly, cretinism results when the tyrosine to thyroxine pathway is mutated. Therefore, in a strict genetic sense, if appropriate experiments are performed, a morphological mutation can be explained at the biochemical level.

Conditional Mutations

The expression of these types of mutations is dependent on the individual's specific environmental conditions. If these mutated genes' expression is restricted to a particular condition and not in any other conditions, it is called the **restrictive condition**. But if the individual grows in any other environment (**permissive condition**), the wild-type phenotype is expressed. These are called **conditional mutations.** Mutations that are only expressed at a specific temperature (temperature-sensitive mutants), usually elevated, can be considered **conditional mutations**.

Lethal Mutations

As the term implies, the mutations lead to the death of the individual. Death does not have to occur immediately, it may take several months or even years. But if the expected longevity of an individual is significantly reduced, the mutation is considered a lethal mutation.

Loss-of-function Mutations

Wild-type alleles typically encode a product necessary for a specific biological function. If a mutation occurs in that allele, the function for which it encodes is also lost. The general term for these mutations is **loss-of-function mutations**. The degree to which the function is lost can vary. If the function is entirely lost, the mutation is called a **null mutation**. It is also possible that some function may remain, but not at the level of the wild-type allele. These are called **leaky mutations**. Loss-of-function mutations are typically recessive. When a heterozygote consists of the wild-type allele and the loss-of-function allele, the level of expression of the wild-type allele is often sufficient to produce the wild-type phenotype. Genetically, this would define the loss-of-function mutation as recessive. Alternatively, the wild-type allele may not compensate for the loss-of-function allele. In those cases, the phenotype of the heterozygote will be equal to that of the loss-of-function mutant, and the mutant allele will act as a dominant.

Gain-of-function Mutations

Although it would be expected that most mutations would lead to a loss of function, it is possible that a new and important function could result from the mutation. In these cases, the mutation creates a new allele that is associated with a new function. Any heterozygote containing the new allele along with the original wild-type allele will express the new allele. Genetically, this will define the mutation as a dominant. This class of mutations is called **gain-of-function mutations**.

Resistant Mutations

These mutations are mainly seen in microbial populations. Certain microbes of the population become resistant to certain antibiotics such as *ampicillin* and *streptomycin* resistance.

Regulatory Mutations

These mutations occur in the regulatory regions of a gene and they loose their ability to control the expression of gene or genes.

Genome, Chromosome, and Gene Mutations

Genome and chromosome mutation include the change in the number, rearrangements in the chromosomes, or change in a single gene. The first two involve changes in the entire genome or in the whole chromosome. But the third one occurs in an individual gene. Both chromosomal mutations and gene mutations can occur spontaneously or can be induced by physical or chemical mutagens.

Down's Syndrome is one well-known example of **non-disjunction** mutation. Non-disjunction is when the spindle fibers fail to separate during meiosis, resulting in gametes with one extra chromosome and other gametes lacking a chromosome.

If this non-disjunction occurs in chromosome 21 (Figure 10.21) of a human egg cell, a condition called Down's Syndrome occurs. This is because the mutant possesses 47 chromosomes as opposed to the normal chromosome compliment in humans of 46.

Genome Mutation or Polyploidy

Polyploidy is a change in the quantity of total genomes. This type of mutation affects chromosome content of an organism. Humans are diploid creatures; that is, for every chromosome in our body, there is another one to match it. If an organism possesses multiples of the haploid number of chromosomes, it is called **euploid**. Human and other eukaryotes are diploid (2*n*). The following are the various types of euploids:

- Haploid creatures have one of each chromosome.
- Diploid creatures have two of each chromosome.
- Triploid creatures have three of each chromosome.
- Polyploid creatures have three or more of each chromosome.

They can be represented by '*n*' where *n* equals haploid, 2*n* equals diploid, and so on. It is possible for a species, particularly a plant species, to produce offspring that contains more chromosomes than its parent. This can be a result of non-disjunction, where normally a diploid parent would produce diploid offspring, but in the case of non-disjunction, one of the parents produces a polyploid. In the case of triploid, although the creation of particular triploids in species is possible, they cannot reproduce themselves because of the inability to pair homologous chromosomes at meiosis, therefore preventing the formation of gametes. Polyploids have important applications in plant breeding and agriculture and are responsible for the creation of thousands of species in today's planet, and will continue to do so. They are also responsible for increasing genetic diversity and producing species showing an increase in size, vigor, and increased resistance to disease. **Autopolyploids** arise due to doubling of 2*n* genome (diploids) to a 4*n* genome (autotetraploids). These are also very important in crop breeding as they increase the size of plant parts such as seeds, fruits, leaves, etc. The haploids can be converted into homozygous diploids with a process called **diploidization,** which uses chemical mutagenic agents such as colchicine. These homozygous diploids can be used as parents in hybridization.

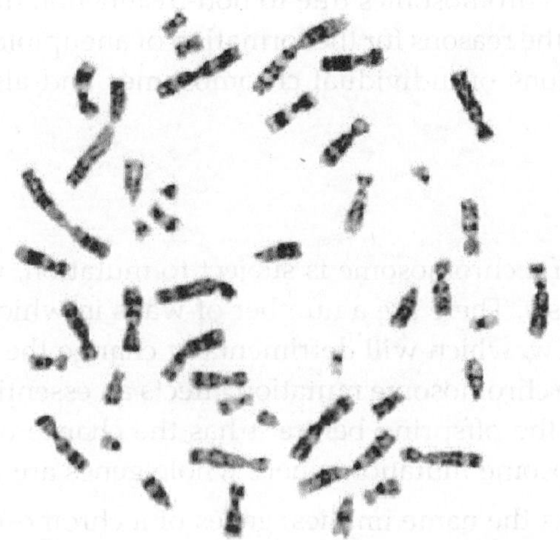

FIGURE 10.21 **Normal karyotype of a human.**

There are some types of chromosome multiplication, which do not involve the complete set. They are known as **aneuploids.** In the case of aneuploids the chromosome number is abnormal and differs from the wild type by only the addition or deletion of one or a small number of chromosomes. The following are some of the common types aneuploids:

- Monosomic—One less than diploid number ($2n - 1$).
- Trisomic—One more than diploid number ($2n + 1$).
- Nullisomic—Two less than diploid number ($2n - 2$).

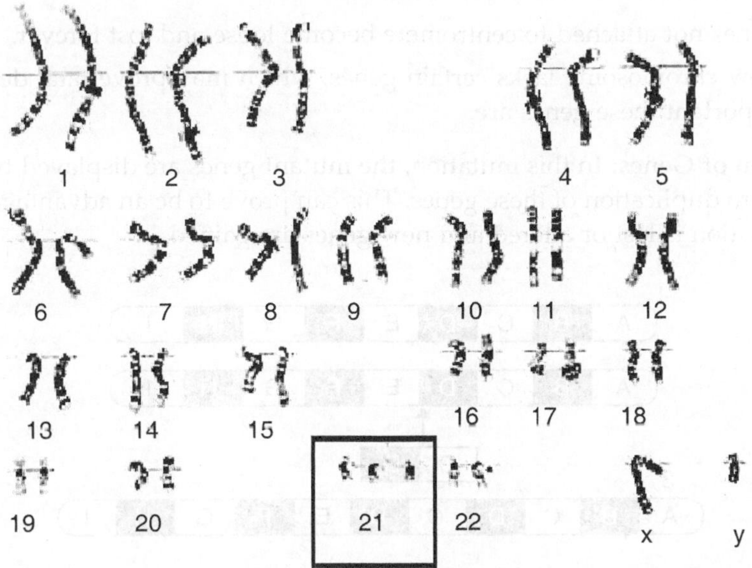

FIGURE 10.22 **In the Down's Syndrome, the chromosomes have been arranged in pairs. Note the extra chromosome.**

Unequal separation of chromosomes due to non-disjunction during mitosis and meiosis or chromosome duplication are the reasons for the formation of aneuploidy. Aneuploids are important for understanding the functions of individual chromosomes and also important in engineering specific crop genotypes.

Chromosome Mutations

The fundamental structure of a chromosome is subject to mutation, which will most likely occur during crossing over at meiosis. There are a number of ways in which the chromosome structure can change, as indicated below, which will detrimentally change the genotype and phenotype of the organism. However, if the chromosome mutation affects an essential part of DNA, it is possible that the mutation will abort the offspring before it has the chance of being born. The following indicates the types of chromosome mutation where whole genes are moved.

Deletion of a Gene: As the name implies, genes of a chromosome are permanently lost as they become unattached to the centromere.

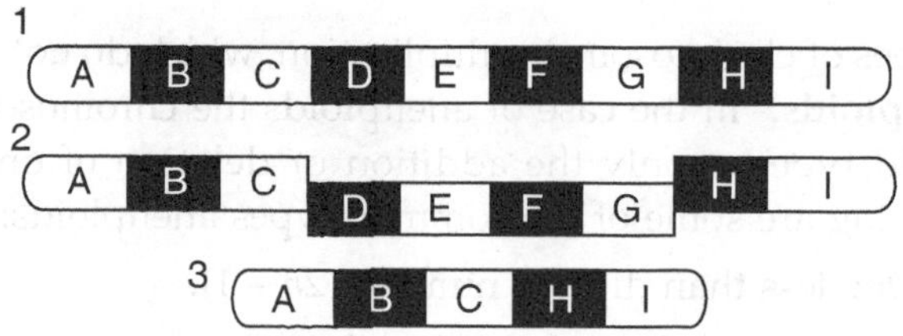

FIGURE 10.23 **Deletions of genes.**

1. Normal chromosome before mutation.
2. Genes not attached to centromere become loose and lost forever.
3. New chromosome lacks certain genes, which may prove fatal depending on how important these genes are.

Duplication of Genes: In this mutation, the mutant genes are displayed twice on the same chromosome due to duplication of these genes. This can prove to be an advantageous mutation as no genetic information is lost or altered and new genes are gained.

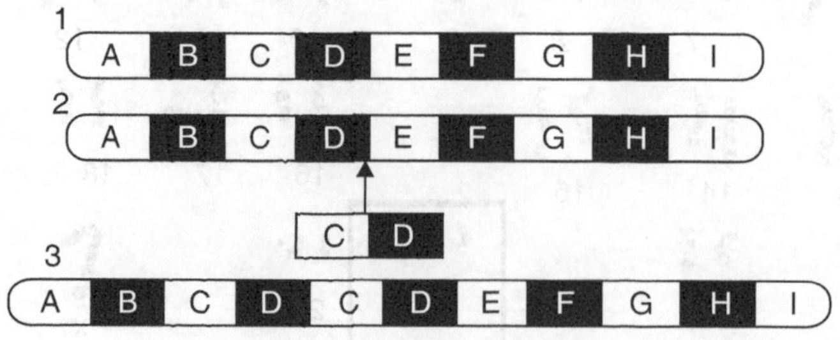

FIGURE 10.24 **Duplication of genes.**

1. Normal chromosome before mutation.
2. Genes from the homologous chromosome are copied and inserted into the genetic sequence.
3. New chromosome possesses all its initial genes plus a duplicated one, which is usually harmless.

Inversion of Genes: This is where particular order of genes is reversed as seen below:

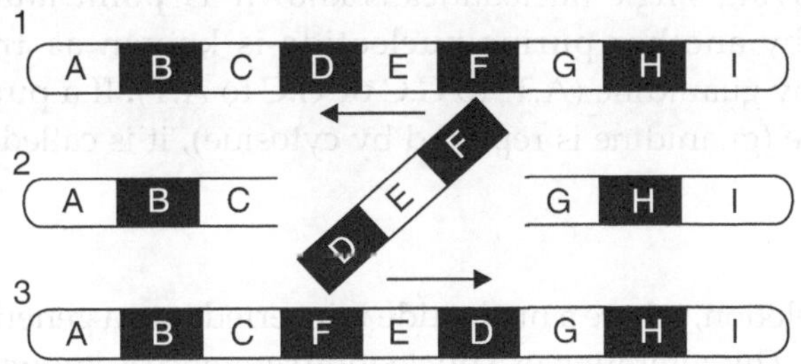

FIGURE 10.25 **Inversion of genes.**

1. Normal chromosome unaltered.
2. The connection between genes break and the sequence of these genes are reversed.
3. The new sequence may not be viable enough to produce an organism, depending on which genes are reversed. Advantageous characteristics from this mutation are also possible.

Translocation of Genes: This is where information from one of two homologous chromosomes breaks and binds to the other. Usually this sort of mutation is lethal.

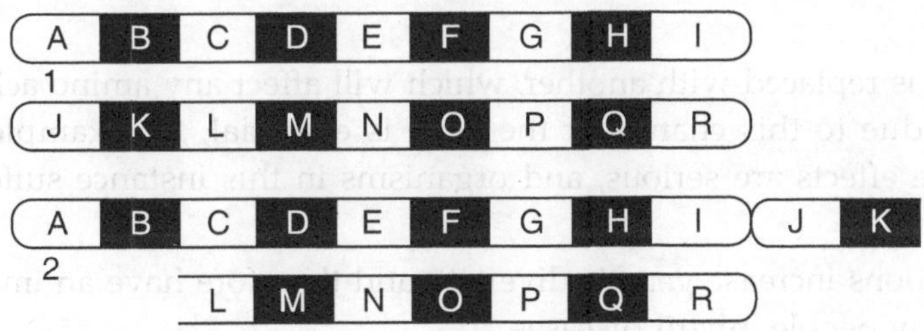

FIGURE 10.26 **Translocation of genes.**

1. An unaltered pair of homologous chromosomes.
2. Translocation of genes has resulted in some genes from one of the chromosomes attaching to the opposing chromosome.

Gene Mutation (Alteration of a DNA Sequence)

The previous examples of mutation have investigated changes at the chromosome level. The sequences of nucleotides on a DNA strand are also susceptible to mutation.

Deletion

Here, certain nucleotides are deleted, which affects the coding of proteins that use this DNA sequence. If, for example, a gene coded for alanine, with a genetic sequence of C-G-G, and the cytosine nucleotide was deleted, then the alanine amino acid would not be created, and some other amino acids that are supposed to be coded from this DNA sequence will also be unable to be produced because each successive nucleotide after the deleted nucleotide will be out of place.

The changes involving single nucleotides is known as **point mutation**. A replacement of one purine nucleotide by another purine nucleotide is known as **transition**. For example, replacement of adenine by guanidine (A.T to G.C or G.C to A.T). If a purine molecule is replaced by a pyrimidine molecule (guanidine is replaced by cytosine), it is called **transversion**.

Insertion

Similar to the effects of deletion, where a nucleotide is inserted into a genetic sequence and therefore alters the chain thereafter. This alteration of a nucleotide sequence is known as **frameshift mutation**. The effect of frameshift mutation is the change of the reading frame of codons downstream from the mutation. This results in a completely changed gene product (protein) and finally the phenotype or physiology.

Inversion

Where a particular nucleotide sequence is reversed (is not as serious as the above mutations). This is because the nucleotides that have been reversed in order only affect a small portion of the sequence at large.

Substitution

A certain nucleotide is replaced with another, which will affect any amino acid to be synthesized from this sequence due to this change. If the gene is essential, for example, for the coding of hemoglobin then the effects are serious, and organisms in this instance suffer from a condition called sickle cell anemia.

Genetic mutations increase genetic diversity and therefore have an important role to play. This is also why many people inherit diseases.

Silent mutations are when a point mutation occurs that change a codon to another codon for the same amino acid so no change results in the gene product (protein).

Nonsense mutations change a codon to a stop, leading to premature chain termination and a piece of random peptide or a truncated protein will be translated.

(*a*) Point mutations and small deletions

Wild-type sequences

Amino acid	N-Phe	Arg	Trp	Ile	Ala	Asn-C
mRNA	5′-UUU	CGA	UGG	AUA	GCC	AAU-3′
DNA	3′-AAA	GCT	ACC	TAT	CGG	TTA 5′
	5′-TTT	CGA	TGG	ATA	GCC	AAT 3′

Missense

	3-AA T	GCT	ACC	TAT	CGG	TTA-5
	5-TT A	CGA	TAG	ATA	GCC	AAT-3
	N- Leu	Arg	Trp	Ile	Ala	Asn-C

Nonsense

	3-AAA	GCT	A T C	TAT	CGG	TTa-5
	5-TTT	CGA	T A G	ATA	GCC	AAT-3
	N-Phe	Arg	Stop			

Frameshift by addition

	3-AAA	GCT	ACC	A TA	TCG	GTT A-5
	5-TTT	CGA	TGG	T AT	AGC	CAA T-3
	N-Phe	Arg	Trp	Tyr	Ser	Gln

Frameshift by deletion

GCTA
CGAT

	3-AAA	CCT	ATC	GGT	TA-5
	5-TTT	GGA	TAG	CCA	AT-3
	N-Phe	Gly	Stop		

FIGURE 10.27 **Types of gene mutations.**

Molecular Mechanism of Mutations

Causes of Mutations

Spontaneous DNA mutations arise from a variety of sources including the errors during DNA replication. It can be caused by radiation (either UV or ionizing), alkylation, deamination, introduction of base analogs, and tautomeric shifts. In **tautomeric shifts** the normal keto form of each base changes to enol or amino, which results in transitions by mispairing. Mutation results in the tautomeric shift between the keto and enol forms of thymine since one form pairs with A and one with G. If it happens to be in the form that pairs with G at the moment of replication, then G will be transcribed in A's place making the mutation. **Depurination** is the removal of purine base by breaking glycosidic bond between sugar and the base to form a gap known as the **apurinic site.** The apurinic site is filled with another base resulting in transition, transversion, or no change. Another mechanism of point mutation is **deamination**. In this mechanism, cytosine is deaminated to uracil and if it is not repaired, it pairs with adenine in the next cycle of DNA replication.

Radiations are two types of ionizing radiations and non-ionizing radiations. Ionizing radiations such as x-rays, gamma rays, and thermal neutrons cause mutations by ionization of DNA molecules. Non-ionizing radiations such as UV light cause dimerization of adjacent bases in the DNA strand, especially the thymine dimmers known as photodimers.

Chemical mutagens include alkylating agents, base analogs, intercalating agents, etc., and can cause chemical modification of the nitrogen bases leading to the mutations. Alkylating agents add alkyl group, methyl or ethyl group, to the appropriate nitrogen bases (mainly guanidine). The alkylated base undergoes mispairing and results in mutation during DNA replication, if not corrected.

A base analog is a compound sufficiently similar to one of the four DNA bases but with different pairing properties. For example, 5-bromouracil is the analog of thymine but sometimes pairs with guanine and 2-aminopurine is the analog of adenine but sometimes pairs with cytosine. The incorporation of a base analog will result in mispairing and ultimately in a complete change of a base pair during DNA replication.

Certain DNA-binding chemicals such as ethidium bromide can intercalate and slip into the space between the sequence of nitrogen bases and increase the distance between them. This can lead to additions and deletions in the DNA strand. Most chemical mutagens have mutational specificity, as each mutation is due to a specific chemical reaction with DNA.

Nitrous acid is another chemical that affects DNA complementation. The acid randomly modifies the base adenine so that it will pair with cytosine instead of thymine. This change is made evident during DNA replication when a new base pair appears in daughter cells in a later generation.

Even though mutations are randomly happening in large numbers, most of them can be reversed before they get stabilized and transferred to the next generation. Mutations can also be reversed spontaneously or can be induced by mutagens. For example, if an alkylating agent like EMS is creating a transition mutation such as GC to AT, it can be reversed by 5-bromouracil as it causes both types of transition, GC to AT and AT to GC. Because of this same reason, mutations caused by base analogs 5-bromouracil can be reversed by the same mutagens.

There are a number of methods developed for the screening of suspected mutagenic compounds. They mainly include cytological and cytochemical techniques, which are time consuming. In 1970, Bruce Ames developed a very simple but effective assay technique to check genotoxicity. This test is known as the **Ames test**. The test consists of two auxotrophic strains of *salmonella typhimurium*. Ames devised a method to simulate the human metabolism in the bacterial cultures by adding the rat liver enzymes to the bacterial suspension containing the suspected compound. The liver enzymes are involved in the detoxification of harmful compounds. The mutagencity of the unknown compound can be determined by reversion of histidine mutations. The reversion of the histidine mutation can be detected by the growth and formation of colonies on the nutrient medium lacking histidine. Bacterial growth cannot happen unless histidine mutation is reversed.

10.12 DNA REPAIR

As a major defense against environmental damage to cells, DNA repair is present in all organisms including bacteria, yeast, *drosophila*, fish, amphibians, rodents, and humans. DNA repair is involved in processes that minimize cell killing, mutations, replication errors, persistence of DNA damage, and genomic instability. Abnormalities in these processes have been implicated in cancer and aging (Figure 10.28). The genetic material has to be guarded against the various types of damage to keep the high fidelity. There are various types of enzymatic machineries to check and repair the possible errors in the total DNA molecules at different stages of cell cycle. The error checking and repair occurs just before the starting the DNA replication and operates during the replication and continues checking even after the replication. The DNA repair enzyme can recognize the damaged sequences and repair the damage very promptly. There are some mechanisms, which even neutralize the potentially damaging compounds before they even react with the DNA. One example is the enzymes that detoxify the superoxide radicals, which can do oxidative damage to DNA.

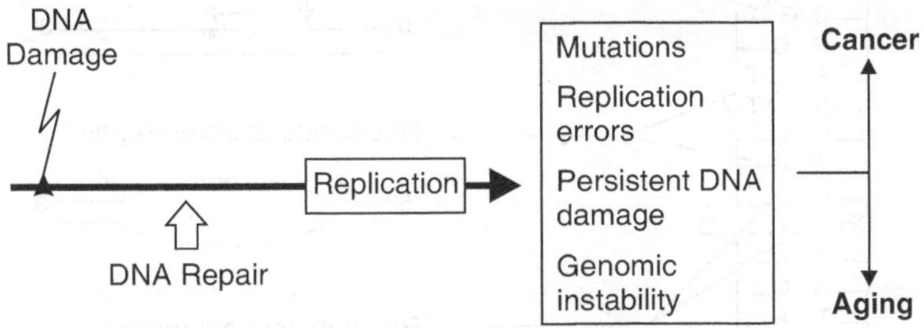

FIGURE 10.28 **The functions of DNA repair.**

The repair mechanisms for DNA damage can be broadly classified into two categories: **light repair system** and **dark repair system**. The repair mechanism that operates in the presence of light is the light repair mechanism and light is not necessary for the dark repair mechanism to operate. It can function at all times.

The light repair system is called a **photorectivation mechanism**. It can remove photodimers or thymine dimmers induced by UV radiations of sunlight. The photoreactivating enzyme known as photolyase can bind to the thymine dimmers and split it in the presence of light and folic acid, which is the co-enzyme of the photolyase. Folic acid takes the light energy and helps the enzyme to split the photodimers. Photodimers can also be removed by dark repair systems.

The dark repair system consists of enzymes that are able to recognize damaged DNA strands and correct it. These repair mechanisms exist to correct most of the chemical changes that come to the bases. The following are five types of dark repair:

1. **Base Excision Repair (BER)** is the removal of specific damaged bases. The enzymes, DNA glycosylases, scan the DNA to recognize the wrong bases and repair them. For example, uracil, if it is present in DNA, can be efficiently removed by uracil-DNA glycosylase. Now the AP endonucleases can recognize the apurinated or apyrimidinated (AP) sites and repair them.
2. **Nucleotide Excision Repair (NER)** is the removal of an entire nucleotide or a fragment of the single DNA strand along with the modified base. This is used over base excision repair when the mutation results in a change of more than just the base but a chemical change to the sugar or phosphate as well. This is a more general form of DNA repair. It can recognize several kinds of DNA damage including the photodimers. In the first step an enzyme recognizes the damaged nucleotides and makes single strand nicks a few base pairs away on either side of the damaged part. In the second step the segment of

DNA containing the damaged nucleotide is removed by leaving a gap. A DNA polymerase, using the opposite strand as the template, fills this gap and finally the DNA ligase joints the nick. DNA polymerase III is the main enzyme involved in the cutting, removal, and synthesis of the damaged DNA strand.

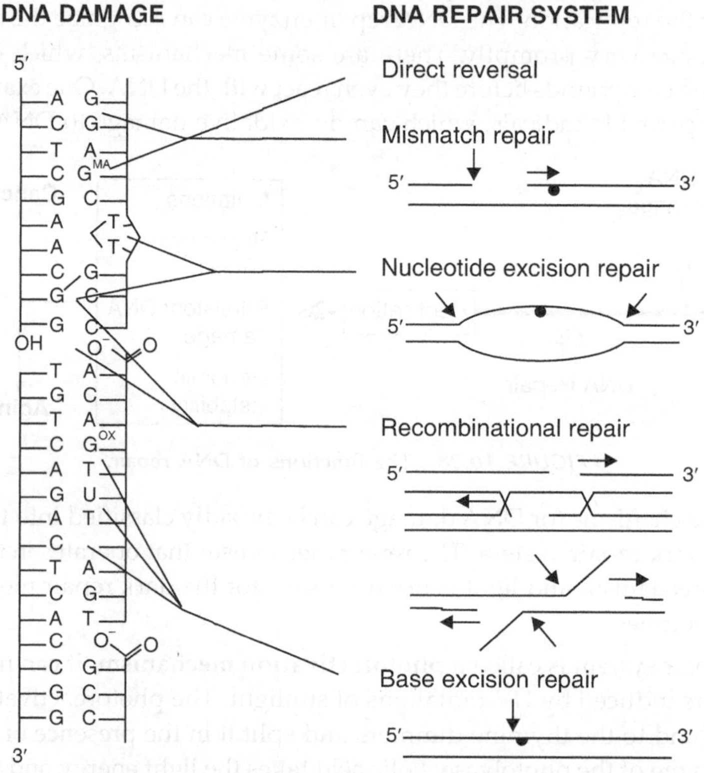

FIGURE 10.29 **Various types of DNA damage and the DNA repair mechanisms.**

3. **Mismatch Repair (MMR)** is the post-replication mismatch repair, which is the normal checking and removal of mismatched nitrogen bases or nucleotides by DNA polymerase during DNA replication or just after that.
4. **Recombination Repair (RR)** is when the DNA damage is very extensive, which includes a large part of gene or DNA and has to be corrected by the recombination method. If a major part of the gene is damaged or has lost its correct sequence in both the strands, it cannot be repaired by normal mechanisms. Because damage has occurred in both strands of the DNA, the template is not available for correcting the error. In this case, the damage correcting system will check the entire genome to see the similar sequences so that it can be copied and exchanged with the damaged part of the DNA. This damage correction mechanism is called the **retrieval system.**
5. **SOS Repair** is when there is very extensive DNA damage, which inactivates the DNA, and the repair mechanisms are not able to repair it, then the DNA cannot replicate, which leads to the death of the cell. The DNA cannot replicate because the modified bases

formed by mutation will block the functioning of the DNA polymerases. Under such situation **SOS induction** occurs. This is the last resort to minimize the degree of mutations and allow for cell survival. Therefore, this repair pathway is called **error prone repair**. SOS also modifies the DNA replication complex so that the DNA replication proceeds even though the DNA contains some error. Since it is the last resort for the survival of the cell, it is called the SOS (Save Our Soul) mechanism.

10.13 GENETIC DISORDERS

In all organisms, including humans, genes play an important role in the development of a disease. Whatever the disease, ultimately it ends in the expression of or lack of a gene. But the term genetic disease or genetic disorder indicates that the mutation of one or more genes is primarily responsible for that particular disorder or diseased state. Genetic diseases such as hemophilia, sickle cell anemia, and cystic fibrosis, etc. are caused by mutations. These changes are inherited and passed on from affected parents to their children. All these disease-causing mutations are recessive in nature and happen for a single gene. But translocation also creates some types of genetic disorders including cancer. Some are due to aneuploidy (loss or gain of one or more chromosomes). Klinefelters Syndrome is a genetic disorder in males due to the Trisomic condition, XXY, and shows feminine characteristics. Turner's Syndrome affects females and is due to the presence of a monosomic condition, XO, and shows male features. Down's Syndrome affects both males and females and is due to a trisomy of the 21st chromosome and shows mental retardation and short stature.

In addition to these any damage or mutation to the genes responsible for DNA repair will cause a variety of genetic disorders and cancer susceptibilities. Xeroderma pigmentosum is one such disorder. Xeroderma pigmentosum develops when there is a failure of the DNA polymerases to carry out their repair functions. The patient will have all forms of skin pigmentation problems. Because of the lack of ability to repair mutations they will usually develop skin cancer in either the first or second decade even if they are very careful to stay out of the sun. A recessive mutation results in the absence or reduced activity of the excision repair mechanism, and therefore they are susceptible to UV light.

Even though genetic disorders are due to gene mutations, all of them may not be hereditary. If a genetic disorder is transmitted to the next generation then it is a hereditary disease. Mutations related to the development of cancers occur in somatic cells and are often associated with genes that control the cell cycle. Cancerous cells show uncontrolled cell growth resulting in tumor formation.

Sickle Cell Anemia (SCA) is characterized by episodes of pain, chronic hemolytic anemia, and severe infections, usually beginning in early childhood. SCA is an autosomal recessive disease caused by a point mutation in the hemoglobin beta gene (HBB) found on chromosome 11. Carrier frequency of HBB varies significantly around the world, with high rates associated in zones of high malaria incidence, since carriers are somewhat protected against malaria. A mutation in HBB results in the production of structurally abnormal hemoglobin (Hb), called HbS. Hb is an oxygen carrying protein that gives red blood cells (RBC) their characteristic color. Under certain conditions, such as low oxygen levels or high hemoglobin concentration, in individuals who are homozygous for HbS the abnormal HbS crystallize and cluster together, distorting the RBCs into sickled shapes. These deformed and rigid RBCs become trapped within small blood vessels and block them, producing pain and eventually damaging organs. Though, as of yet, there is no cure for SCA, a combination of fluids, painkillers, antibiotics, and transfusions are used to treat symptoms and complications. Hydroxyurea, an antitumor drug, has been shown to be effective in preventing painful crises. Hydroxyurea induces the formation of fetal Hb (HbF), an Hb normally found in the fetus or newborn, which, when present in individuals with SCA, prevents sickling. A mouse model of SCA has been developed and is being used to evaluate the effectiveness of potential new therapies for SCA.

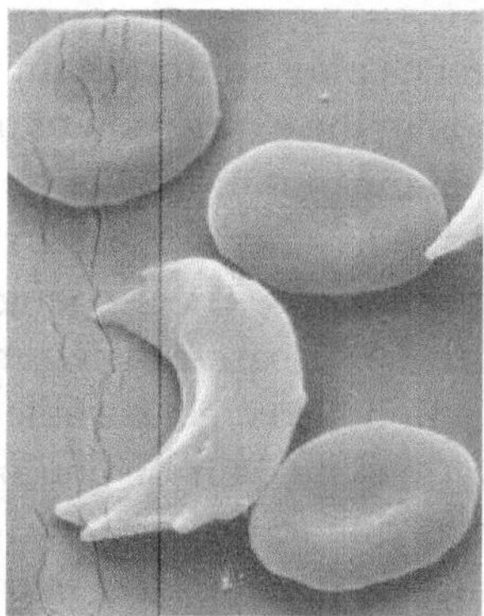

FIGURE 10.30 **Normal and sickled RBC.**

Thalassemia is another genetic disorder related to hemoglobin. This diseased condition occurs when one or more copies of one of the Hb genes have been recombined to give an unrecognizable sequence. The lack of the gene will lead to insufficient production of Hb and obvious problems in transporting O_2 in the bloodstream.

10.14 TRANSPOSONS

Genes are supposed to be static, having a fixed position in the chromosomes. They will not change their position unless some chromosome mutation occurs. But in 1940, Barbara Mc Clintock, demonstrated the presence of moving genes or jumping genes in the maize. These genes move from one location to another position within the genome and are called **transposable elements** or **jumping genes**. The transposons isolated from the maize were named Ac and Ds elements. Later such mobile genes or transposable elements (transposons) were discovered from a large number of organisms including microbes. These moving fragments of DNA are responsible for a variety of genetic instabilities including chromosome breakages and other mutations. Transposons can be used as marker genes in genetic mapping and genetic transformation experiments.

A transposable DNA fragment has unique nucleotide sequences at each of its ends, which enable its transfer to other locations. The transposable elements are always a part of chromosomes, plasmids, or a viral DNA. Their ends are characterized by short sequences of inverse repeats or palindromic sequences. An enzyme known as **transposase** encoded by the same element or another element of the same cell carries out the process of **transposition**. This enzyme makes use of the repeat DNA sequences at the ends to transfer all the DNA of the element to another location within the chromosome or to another chromosome (Figure 10.31). Many transposable elements during their transfer leave a copy behind by replication. The simplest of the transposable elements has an insertional sequence with a single transposase gene. Certain other types of transposable elements, **known as T_n elements,** have antibiotic resistant genes, in addition to transposase genes and repeat sequences. The transposable elements present in plants are found to be more complex than those present in microbes.

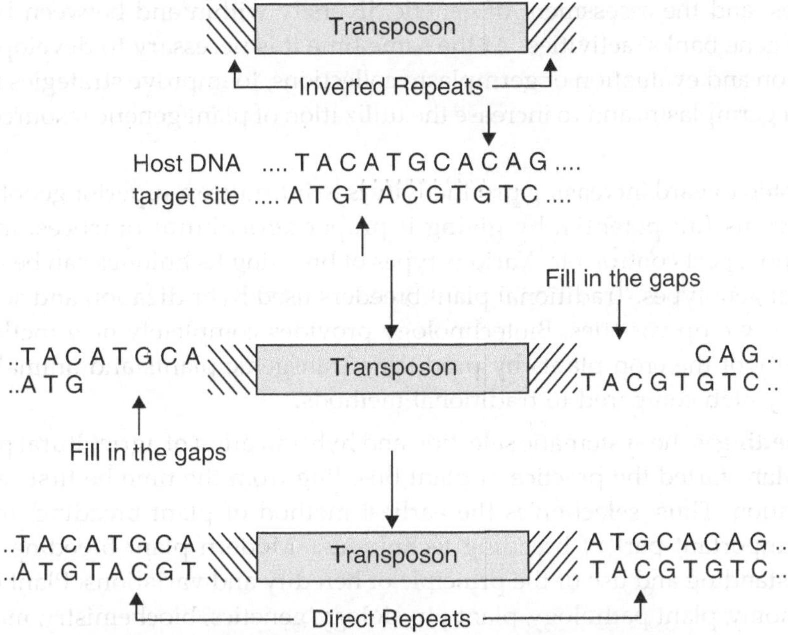

FIGURE 10.31 **Cutting and pasting of transposable elements.**

10.15 ANIMAL AND PLANT BREEDING

The most pressing problems facing the human species can be summarized in a few words—population, food security, and the environment. The world's population is forecast to rise from 6.0 to 8.7 billion by the year 2030. At the same time, the area of arable land per person is decreasing because of loss of land due to human activity (currently, one million hectares of arable land is lost per year), as well as the increasing number of people. Feeding more people from less land will require massive increases in agricultural production (amount of product) and productivity (output per unit input), while at the same time the sustainability of production systems must be improved.

The total number of plant species, which are cultivated as agricultural or horticultural crops, can be estimated to be close to 7,000 botanical species. Nevertheless, it is often stated that only 30 species "feed the world," because the major crops are made up by a very limited number of species. The latter is also the major reason that six million accessions collected and conserved in gene banks belong to a very limited number of species compared to the total number of species, which contribute to food security. About half of all accessions maintained in collections, *ex situ*, are advanced cultivars or breeders' lines, while just over a third of them are made up of land races, or old cultivars, and about 15% are wild relatives of crop species, weedy plants, or wild plants. Only a third of all accessions are characterized. There is obviously a gap in the collections of minor crops and underutilized species, in particular landraces and wild relatives of crops, which are under-represented in gene banks.

Therefore, the further exploration of minor and underutilized species, the collection of these genetic resources, and the assessment of genetic diversity within and between landraces should have priority in gene banks' activities. At the same time it is necessary to develop better methods of characterization and evaluation of germplasm collections, to improve strategies for conservation and collection of germplasm, and to increase the utilization of plant genetic resources for increasing productivity.

The first step toward increasing productivity is to introduce a superior genotype. The second step is to achieve its full potential by giving it proper agricultural practices, including proper irrigation, nutrition, pest control, etc. Various types of breeding techniques can be used to generate superior forms of genotypes. Traditional plant breeders used hybridization and selection methods for developing new crop varieties. Biotechnology provides completely new methods to improve the genetic quality of the crop plants by producing transgenic plants and animals. The accuracy and speed is very high compared to traditional methods.

Plant breeding is the systematic selection and hybridization of agricultural plants for genetic improvement. Man started the practice of plant breeding from the time he first selected a seed of plant for cultivation. Thus, selection is the earliest method of plant breeding. In present times, selection is an important part of breeding techniques. Modern plant breeding is based on the thorough understanding and use of the principle of heredity and variations. Plant breeders should also know agronomy, plant pathology, plant physiology, genetics, biochemistry, molecular biology, and statistics. The goal of a plant breeder is to develop a new genotype of plant having elite qualities such as productivity, greater biotic and abiotic stress tolerance, and improved qualities of agricultural and food products.

Plant-breeding procedures are closely related to the type of reproduction of the plant (i.e., self-pollinated or cross-pollinated). Accordingly, the approach will vary. The first step in a breeding procedure is the selection of the correct varieties of parents having the desired characteristics. Depending on the purpose of the breeding the plants can be homozygous or heterozygous. In a self-fertilizing population entire individuals will be homozygous for a particular trait. For example, consider flower color. Both white-colored and red-colored varieties are in the population. Since the plant is naturally self-pollinated the genotype will be 'RR' for red color and 'rr' for white color and these varieties can produce only one type of gamete with either 'R' or 'r'. If the plant is a natural cross-breeding type, then the population of that variety will be a mixture of heterozygous and homozygous plants for flower color, with a smaller percentage of homozygous individuals. The heterozygous condition of a red flower-colored plant will be 'Rr' (hybrid) in which the trait 'R' is dominant over 'r'. These plants will produce two types of gametes: 50% with the 'R' gene and the rest with the 'r' gene. You can make homozygous plants for a particular trait in naturally cross-pollinating plants by adopting artificial self-pollination procedures for a number of generations. But it is not possible to produce a plant that is homozygous for all characteristics of a plant. A practical state of homozygosity can be attained after five to six generations of self-pollination. A mixed population of self-pollinated plants is really a mixture of homozygous individuals for different genes. Such populations are known as heterogeneous but homozygous for a particular gene. Some crops such as sorghum and cotton undergo both self-pollination and cross-pollination to different degrees and their population will be a mixture of both depending on the level of self-pollination. High percentages of self-pollination will keep more homozygosity in the population for that gene. Once the parents are selected, the hybridization can be conducted by self-pollinating the plants. The hybrids obtained can be evaluated and further selection and hybridization techniques can be adopted.

Another important breeding technique is the introduction of new varieties to a new area of cultivation. Some of the important wheat varieties that caused the famous Green Revolution were actually introduced varieties such as Sonsora 64 and Lerma Rojo from Mexico. They were actually developed in Mexico and introduced in India and hybridized with our local varieties. The first step for introduction of a variety is to select and collect a number of prospective varieties called **collections**. On the basis of limited field trials, evaluation of their performance along with locally adapted varieties and selection, they can be introduced as a new variety.

Selection is one of the oldest breeding techniques and is the basis of all crop improvements. Selection is a process in which individual plants or group of plants are sorted out from a mixed population. There are methods of **mass selection** and **pure line selections**, which accompany the processes of **hybridization** and **evaluation of hybrids**. Many of the varieties released are through artificial hybridization followed by selection for high yield and quality, disease resistance, etc. A well-known method is the **pedigree method** where a detailed record of parents and selection criteria adopted in each generation is maintained. The most widely adopted breeding methods used for cross-pollinated crops are **mass selection, recurrent selection**, and methods for creating hybrid varieties and composite varieties. These methods exploit the generally high level of heterozygosity present in the cross-pollinated crops. A number of hybrid varieties have been released; the two most important cross-pollinated crops are sorghum and maize.

The availability of plant genetic resources is an important factor in designing new hybridization and selection procedures. The most effective methods for conservation of diversity for the respective plant groups considered (i.e., crops, their wild relatives, weeds, and wild plants) are different. In some cases, a combination of different strategies is the most effective way. The strategies include *ex situ* conservation (management of gene banks), conservation and management on farms (monitoring and protection of agro-ecosystems), and *in situ* conservation (monitoring and protection of natural ecosystems). The most effective way of assessing genetic diversity within a given taxon is also a combination of different methods, combining morphological, agronomic, and molecular characterization of genetic diversity.

There is an increase in demand for animal products, particularly for export, and is projected to outstrip that for plant products. The demand for milk and milk products is always on the rise, so there will be even more pressure on the animal sector to increase productivity. Satisfying these demands will require effective and integrated action in many areas (e.g., political, sociological, economic, and trade), in addition to directly increasing animal production.

The present **global animal genetic resources (AnGR)** are what we have to meet the animal product needs of the current human population. We need to utilize them as efficiently as possible. Any loss of these resources will restrict our options for livestock improvement, both now and in the future.

The breeding of agricultural livestock is aimed at efficient food production and thus makes a major contribution toward feeding the world's population. For many years biotechnological procedures such as artificial insemination and embryo transfer have been an integral part of modern animal husbandry, and they have resulted in the well-known and recognized improvements in the performance of agricultural animals. But certain disadvantages could not be countered by these techniques: the relatively slow annual rate of genetic progress (1 to 3%), the lack of a way to separate desirable from undesirable traits by breeding, and the impossibility of transferring genetic information between species.

New biotechnology and novel molecular-genetic tools already available and others under development indicate that it will be possible to overcome these limitations to breeding. Today "biotechnology in farm animals" basically includes techniques in reproductive and molecular biology intended to enhance performance, efficiency, and health for sustainable animal production. In the very near future the complete sequencing of the genomes of important domestic animals will make it possible to distinguish molecular phenotypes and thus improve the use of genetic resources. In view of the world's limited resources and increasing population, biotechnology and novel genetic-molecular tools will provide important resources for making animal production more efficient, environmentally appropriate, and economically viable.

Cloning and transgene technology will open new horizons both for biomedicine and for many agricultural applications, particularly in the area of product diversification. The development and application of biotechnology and genetic technology in animal breeding must be accompanied by interdisciplinary research leading to more rational and factual social and ethical discourse.

REVIEW QUESTIONS

1. Describe the basic laws of heredity.
2. What is meant by gene interaction?
3. Describe the molecular basis of sickle cell anemia.
4. What is a DNA repair mechanism?
5. Differentiate between multiple alleles and multiple genes.
6. What are photodimers? Describe two mechanisms by which they can be repaired.
7. If a point mutation occurs in the coding region of a gene, what will be the nature of its protein?
8. What is a frame shift mutation?
9. Explain xeroderma pigmentosum.
10. Give definitions for the following:
    (*a*) Mutation
    (*b*) CentiMorgan
    (*c*) Missense mutation
    (*d*) Aneuploids
    (*e*) Photolyase
    (*f*) Conditional mutations

(*g*) Hardy-Weinberg law

(*h*) Dominant and recessive allele

(*i*) Incomplete dominance

(*j*) Transposase

11. What are cytoplasmic genes? How do they affect the heredity of organisms?
12. Why is the inheritance of cytoplasmic genes called maternal inheritance?
13. What is a linkage group?
14. Describe quantitative inheritance.
15. How can Mendel's Law of Inheritance be explained by chromosomal movements during meiosis?
16. What are phenotypes and genotypes?
17. What is the SOS response? When does it become functional?
18. Describe the Ames test.
19. Define the following terms:

    (*a*) Autotetraploids

    (*b*) Monohybrid cross and monohybrid ratio

    (*c*) Lethal mutation

    (*d*) Reverse mutation

    (*e*) Auxotrophs

    (*f*) Trisomics

    (*g*) Hemophilia

    (*h*) Genetic mapping

    (*i*) Alkylating agents

    (*j*) Petite yeast

    (*k*) AP site

    (*l*) Tautomeric shifts

    (*m*) Base excision repair

    (*n*) Down's Syndrome

    (*o*) Sex determination

    (*p*) Sex-linked inheritance
20. Describe the inheritance of hemophilia.
21. What is the contribution of linkage and crossing over toward evolution?
22. What is spontaneous mutation and induced mutation?
23. What are the various types of gene mutations?
24. Explain biochemical mutations.
25. Differentiate between euploids and aneuploids.
26. What are the various types of chromosome mutations?
27. Explain the cytoplasmic inheritance in *mirabilis jalapa*.
28. 5-bromouracil can reverse the mutation effect created by EMS. Explain.
29. Explain the photorectivation mechanism of DNA repair.

Chapter **11** *GENOME FUNCTION*

In This Chapter

11.1 GENOME ORGANIZATION

The complete genetic or DNA complement of an organism is called the **genome.** It includes the genetic material of the nucleus and cytoplasm. All organisms have a genome made up of DNA, containing genes. Genomes may vary in their size, number of genes, number of chromosomes, and how genes are organized within chromosome(s), and the DNA may be circular or linear. Generally, genome size increases with the complexity of the organism. There is considerable variation in the size of the genome among organisms. As the size of the genome increases the number of genes also increases correspondingly. But there are exceptions. Among the various groups of organisms, viral genomes are usually relatively small and come in sizes ranging from 5 kb (SV40) to about 250 kb (vaccinia-virus and cytomegalovirus). (1 kb = 1,000 base pairs). Bacterial genomes, on the other hand, range from 600 kb (mycoplasm) to more than 7,000 kb (streptomycetes), but smaller than eukaryotic genomes. Genome size and organization vary greatly even among eukaryotic organisms.

Viral Genomes

Viruses are non-living particles that must infect and subvert cellular machinery of a host to reproduce. Viruses consist of a protein coat and a core that contains the genome. Even though the genomes of viruses are the most simplest and smallest, they are the most diverse due to diverse evolutionary histories as well as diversity in strategies for packing genome in a coat. It can be made of single-stranded DNA (ssDNA), double-stranded DNA (dsDNA), single-stranded RNA (ssRNA), or even double-stranded RNA (dsRNA). These molecules of nucleic acids may be linear or circular. The complete nucleotide sequences of many viral genomes have been determined. Viral genomes are very small, have little intergenic space, and few genes (herpes virus contains ~200 genes, and many others have far fewer genes). There are various types of viruses. They can be morphologically variable and even complex. Their morphology is one way of classifying viruses. They contain DNA or RNA but never both. Although they have a protein coat around their hereditary material, they lack properties of cells such as membranes, ribosomes, enzymes, and ATP synthesis ability. Thus, they can only proliferate using a host cell's translational machinery. In other words, they are obligate intracellular parasites. Their replication cycle consists of attachment and entry into the cell; replication of viral nucleic acid; synthesis of viral proteins; and finally, assembly of viral components and escape from host cells. Bacteriophages have received the greatest attention from molecular geneticists. A **bacteriophage** is a virus that infects bacteria only.

Bacteriophages have a simple structure, which consists of their double-stranded DNA (in some cases single stranded, e.g., M13) surrounded by a protein coat. Only the DNA enters into the bacteria. Some phages supply their own enzyme components (own factors) to instruct the bacteria to transcribe phage genes preferentially. Bacteriophages usually infect only one species of bacteria, but there are some who can infect several species even in different genera. Their life cycle may be either lytic (virulent phage) or lysogenic (temperate phage). Their DNA may be integrated into the host chromosome and remain as a prophage. Integration is achieved by recombination between a 15 bp sequence called **att** (for attachment) in the host chromosome and an identical sequence in the phage chromosome. This recombination requires an integrase (Int) enzyme encoded by the phage. Bacteriophages are used for gene cloning in molecular biology. DNA fragments can be inserted into a phage and following transfection of competent bacteria, many copies of the desired DNA fragment can be obtained.

The size of the bacteriophage genome also varies in different strains. For example, M13 and T7 contain genomes that are 6.4 kb and 39 kb, respectively (kb = Kilo base). The complete nucleotide sequences of many phage genomes have been determined and reveal that they contain relatively few genes. For example, M13 phage have only ten genes, that include genes of coat protein and enzymes needed for phage DNA replication. These genomes, in most cases, have overlapping genes (i.e., genes within genes, coding for different proteins). The eukaryotic viral genome is more or less similar to that of bacteriophages, but have introns (non-coding regions between a gene), which is a unique feature of eukaryotic genomes.

Prokaryotic Genomes

In prokaryotes, genome size is relatively small, but bigger than that of a virus. There is considerable variation in the size among prokaryotes and in some cases the genome is bigger than certain eukaryotic cells. For example, the genome size of *bacillus megasterium* is 30 Mb and that of yeast is

12.1 Mb, which is the same as a eukaryote. The genome sizes in other bacteria are smaller than this. For example, *mycoplasma genitalium* has a genome size of 580 kb, *methanococcous jannaschii* has a genome of 1739 kb, and the genome size in *e.coli* is 4639 kb. In most cases, the genome consists of a single circular chromosome, but there are exceptions. They also possess small independently replicating, circular, double-stranded DNA molecules known as **plasmids**. Plasmids can be 1 to 300 kb long and may exist as multiple, free copies. They also carry genes, which are absent in bacterial chromosome. *Vibrio Cholera* has two different circular chromosomes and *Borrelia burgdoferi* (Lyme disease) has a linear chromosome and 17 plasmids.

Generally, the bacterial genome has a high gene density (i.e., genes are close together) with very small intergenic regions. Unlike eukaryotic systems, the major part of the genome is functional. About 85 to 90% of total DNA is involved in protein coding, about 20-10% is the intergenic DNA, and less than 2% of DNA is made up of transposable elements. Functionally related genes located together as a group called **operons** and **introns** are extremely rare. Overlapping genes are also common in bacterium as in the case of viruses. Bacterial genomes have associated proteins, used to pack it into a **nucleoid**, but the proteins are not similar to the histones of eukaryotes.

In *e. coli*, 4,639 kb DNA contains about 2,400 genes, which make up about 80% of the total DNA. The remaining 20% is made up of intergenic regions. There are three cases of overlapping genes and about 27% of the transcriptional units are organized into 75 polycistronic groups or operons. The genes in an operon are functionally related. (In *Aquifex aeolicus*, operon genes are not functionaly related as per our ideas about operons.)

Another example, *methanococcus jannaschii*, is an extreme thermophilic bacteria that has three circular chromosomes; one 166 Mb, plus two smaller ones of 58.4 and 16.5 kb. The entire genome was sequenced in 1996; 58% of its genes do not match any known genes. The majority of genes involved in energy production, cell division, and general metabolism, more closely resemble eubacterial genes. They have some similarities with that of eukaryotic genes involved in RNA synthesis, protein synthesis, and DNA synthesis. They contain histone genes and DNA organized into chromatin, and introns in tRNA genes.

Eukaryotic Genomes

Eukaryotic genomes are much more complex than bacterial genomes. They cover an enormous size range (more than three orders of magnitude) and display different levels of structural complexity. Initially, it is important to obtain an idea of the relative sizes of various eukaryotic genomes and put them into perspective with bacterial and viral genomes. In principle, the rule holds that an increase in structural complexity of an organism goes hand in hand with a more complex genome (since it takes obviously more DNA to encode more proteins!), but there are some exceptions to this rule such as the extraordinary large sizes of some amphibian and plant genomes. There are two types of genomes in a eukaryotic system. They are nuclear genomes and cytoplasmic genomes. Cytoplasmic genomes include mitochondrial genomes in the case of animals and both mitochondrial and chloroplast genomes in the case of plants. For example, the human genome, the term used to describe the total genetic information (DNA content) in human cells, consists of two genomes: a complex nuclear genome, which accounts for 99.9995% of the total genetic information, and a simple mitochondrial genome, which accounts for the remaining 0.0005%.

Nuclear Genomes

The size of genomes in eukaryotes varies significantly. The genomes of the structurally simplest unicellular or filamentous eukaryotes such as fungi, *saccharomyces cerevisiae*, or *schizosaccharomyces pombe* are around 12 and 18 Mb, respectively (i.e., only about four times the size of the average bacterial genome). Multicellular eukaryotes with differentiated cells (tissues) again contain more DNA in their somatic/germ line cells: nematodes (*caenorhabditis elegans*) and small dicot plants (such as *arabidopsis thaliana*) are around 100 Mb; insects (dipters) have genomes nearly twice as large again (165 Mb for *drosophila melanogaster*). But then there is a large jump in genome size when one looks at the animal phylum of vertebrates: bird genomes have around 1,100 Mb to 2,000 Mb, reptiles between 1,600 and 5,100 Mb and mammals between 2,300 and 5,600 Mb; fish genomes range from 2,600 to 7,000 Mb. The genomes of many monocot plants (cereals) also lie in the lower Mb range but the maize genome has 15 Mb. Few genomes tend to be even more complex (salamander genome: 90 Gbp, some flowering plants 120 Gbp), but this increased size is mostly not linked to more structural complexity.

In prokaryotic genomes, a large portion of the total DNA is functional, with codes for different proteins and RNAs. In eukaryotes a great portion of the genome is non-functional and the coding region is relatively small. According to the draft human genome project the total number of genes present in a human genome is only about 33,000, around 2 to 5% of the total genome.

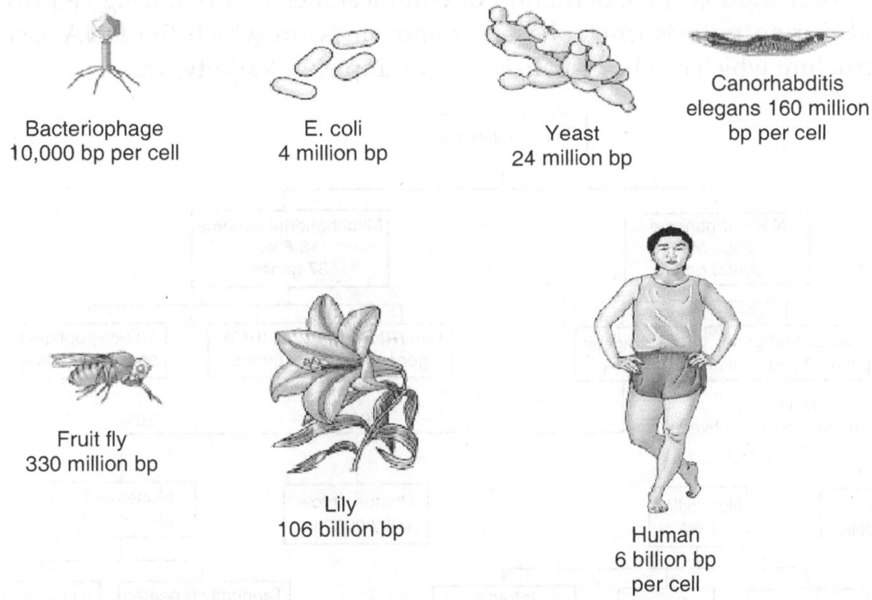

FIGURE 11.1 **Amount of eukaryotic genomic DNA.**

Most genomic DNA are in an intergenic region and are non-coding. It includes the highly repetitive sequences and intones. The repetitive sequences are individual short or small sequences, which repeat either in tandem arrays or interspersed throughout the genome. **Simple sequence repeats (SSR)** or **microsatellites** are the regions of DNA where one to few bases are tandemly repeated for few to hundreds of times. Some of the repetitive DNA sequences may be present several million times in the genome and others only as ten to several hundred copies.

The length of DNA in the nucleus is far greater than the size of the compartment in which it is contained. To fit into this compartment the DNA has to be condensed in some manner. The degree to which DNA is condensed is expressed as its packing ratio. It may be worth recapitulating that DNA comes as a double-stranded, right-handed helix with three isoforms, two with right-handed turns (A- and B-conformation) with a major and minor grooves, and one left-handed one (Z-DNA) with only one groove. The B-conformation with 10.5 bp per helical turn is the *in vivo* conformation. In order to accommodate the immense length of eukaryotic DNA inside the minute size of the cell, the DNA strand is wound up into so-called **superstructures.**

The first level of such a superstructure is the nucleosome in which the DNA is wound in two full turns as a left-handed superhelix around a protein disc consisting of histone proteins forming the so-called **histone octamer**. These nucleosomes are separated by short linker DNA regions stabilized by histone H1 giving the structure some resemblance to "beads on a string." The next higher structural order is the 30 nm chromatin structure, which is clearly visible under an electron microscope. This structure arises by a further supercoiling of the string of nucleosomes around a fixed axis with about six-seven nucleosomes per turn. The next more complex structural feature is called the 100 nm chromsomal domain in which 20 to 100 kb of DNA are held together by so-called matrix proteins; this structure is visible under a light micrsocope in **lampbrush chromosomes** of *Chironomus tentans oocytes*, for example.

Further (but transient) condensation of chromosomes occurs during cell division in late G2 phase and during mitosis (metaphase chromosomes) in which the DNA assumes a clear cytological structure which can be visualized, for example, in **karyotyping**.

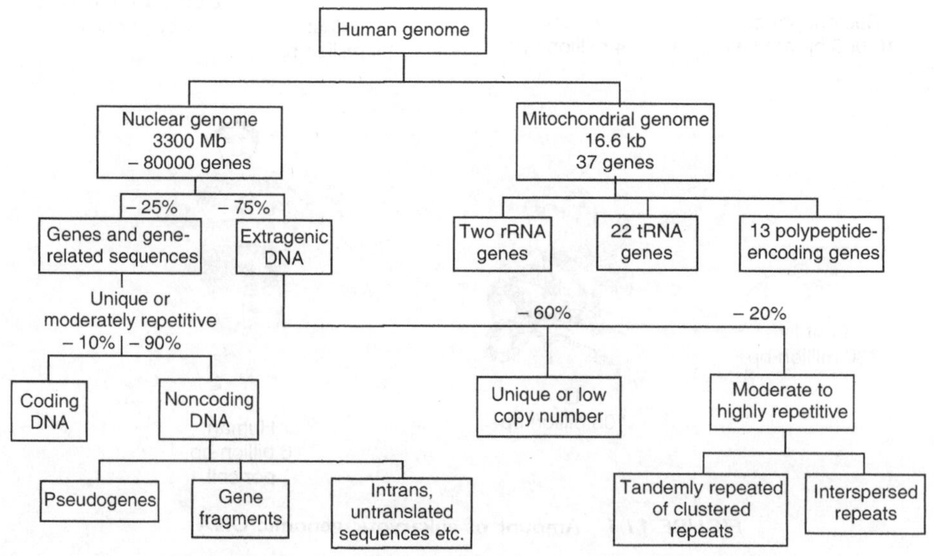

FIGURE 11.2 **Organization of a human genome.**

Organellar Genomes

The eukaryotic organelles, mitochondria and chloroplast, also contain their own genome in the form of circular, double-stranded DNA that carry a limited number of genes, which are essential for the function and maintenance of the organelles. They have the small ribosomes similar to that of the prokaryotes, bacteria and cynobacteria. There is a great variation in the size of mitochondrial genomes from organism to organism, but in the case of chloroplast, genome variation is very low among different types of plants. Genome organization, the genes and their expression, share several similarities with that of the prokaryotic system. This supports the theory of prokaryotic origin of chloroplasts and mitochondria.

11.2 GENOME-SEQUENCING PROJECTS

The genomes of a large number of organisms have been sequenced and a number of them are in the final stages of different genome project groups. The genome projects are having a profound impact on health-care discoveries. The main purpose of genome projects is to access the entire genome sequences that can be used to find out the probable genes and their functions in various organisms. Some important genomes that are partially or completely sequenced include that of *e.coli*, mycobacterium tuberculosis, yeast, arabidopsis, drosophila, mice, humans, rice, wheat, etc. Complete sequences are available for arabidopsis, yeast, humans, mycobacterium, etc. The availability of these databases for researchers will help them to better understand the gene and its expression, and thereby better understand fundamental biological processes. For example, the human genome project will greatly help in the development of gene therapy. The final and the complete sequence of the human genome will unravel the mystery of heredity and reveal the sources of many diseases such as diabetes, heart disease, cancer, hemophilia, Alzheimer's, and cystic fibrosis. It can also help unravel the intricacies of development and differentiation. The genome sequencing of a number of important pathogenic organisms such as the malarial parasite will help to design methodologies to control the disease.

11.3 DNA REPLICATION

Reproduction is one of the basic characteristics of life and it involves the faithful transmission of genetic material from the parents to the next generation. **Genes** are small heredity units on chromosomes and are made of DNA molecules. Before a cell divides, the DNA must replicate so that each of the two new cells will have the organism's genetic code. If foreign materials invade the cell while the DNA is duplicating, they may be incorporated into the molecule. If these miscoded cells are not killed by the body's own defense systems, they will multiply and could take over, disrupting the cell's normal activities or dividing rapidly and erratically, crowding out the normal cells. It is important to understand the construction of the DNA molecules; how and why they divide; and how good nutrition and personal habits can help maintain the genetic code.

The Principle of DNA Replication

The structure of DNA suggested to Watson and Crick that the mechanism by which DNA—genes—could be copied faithfully. They proposed that when the time came for DNA to be replicated, the two strands of the molecule would be separated from each other but would remain intact as each served as the template for the synthesis of a complementary strand. When the replication process is completed, two DNA molecules—identical to each other and identical to the original—have been produced. This mode of replication is described as **semi-conservative replication;** one-half of each new molecule of DNA is old and the other half new.

In DNA replication, the two chains separate from each other as the hydrogen bonds, which link the bases of one chain with the bases of the other, are not very strong. Any free nucleotides then come along and form hydrogen bonds with each of the two chains. These nucleotides then join together through their sugar and phosphate groups and two DNA molecules result. The complementary relationship between the bases ensures that each of the DNA molecules is identical to the original one. Because the sequence of the bases in the two daughter molecules is exactly the same as that of the parent molecule (i.e., A to T and C to G), accurate replication occurs. The enzyme that carries out this reaction is called **DNA polymerase**. DNA replication occurs only at a specific step in the cell cycle. The following table describes the cell cycle for a hypothetical cell with a 24-hour cycle.

TABLE 11.1 Different phases of the cell cycle.

Stage	Activity	Duration
G1	Growth and increase in cell size	10 hr
S	DNA synthesis	8 hr
G2	Post-DNA synthesis	5 hr
M	Mitosis	1 hr

Meselson and Stahl Experiment

When Watson and Crick published their famous short paper on the structure of DNA, they suggested a mechanism by which a molecule would replicate: that each strand would be copied into a new strand with a complementary sequence. However, it fell to Matthew Meselson and Franklin Stahl at Caltech to demonstrate that this was true.

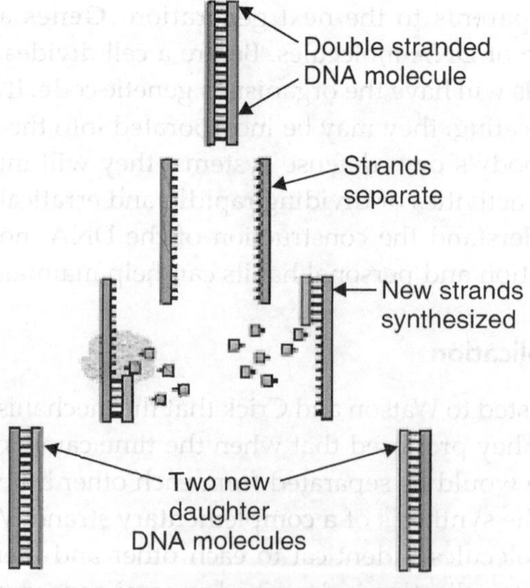

FIGURE 11.3 **DNA replication.**

Meselson and Stahl, in 1957, presented experimental evidence suggesting each DNA strand serves as a template for new synthesis in a process called semi-conservative replication.

They grew *e. coli* in a medium using ammonium ions (NH_4^+) as the source of nitrogen for DNA (as well as protein) synthesis. ^{14}N is the common isotope of nitrogen, but they could have also used ammonium ions that were enriched for a rare heavy isotope of nitrogen, ^{15}N. After growing *e.coli* for several generations in a medium containing $^{15}NH_4^+$, they found that the DNA of the cells was heavier than normal because of the ^{15}N atoms in it. The difference could be detected by extracting DNA from the *e. coli* cells and spinning it in an ultracentrifuge (Figure 11.4). The density of the DNA determines where it accumulates in the tube. They then transferred more living cells that had been growing in $^{15}NH_4^+$ to a medium containing ordinary ammonium ions ($^{14}NH_4^+$) and allowed them to divide just *once*. The DNA in this new generation of cells was exactly intermediate in density between that of the previous generation and the normal. It tells us that half of the nitrogen atoms in the new DNA are ^{14}N and half are ^{15}N. But it tells us nothing about their arrangement in the molecules. However, when the bacteria were allowed to divide *again* in normal ammonium ions ($^{14}NH_4^+$), two distinct densities of DNA were formed: half the DNA was normal and half was intermediate.

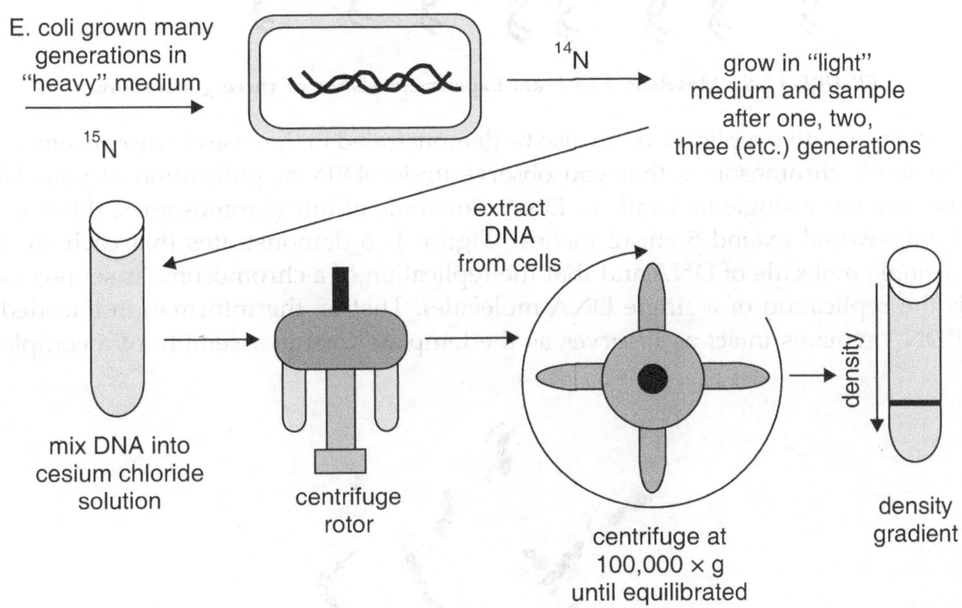

FIGURE 11.4 **Meselson and Stahl experiment.**

DNA of heavy, light, and intermediate densities can be separated by centrifugation.

As this interpretative figure indicates, their results show that DNA molecules are not degraded and reformed from free nucleotides between cell divisions, but instead, each original strand remains intact as it builds a complementary strand from the nucleotides available to it. This is called semi-conservative replication because each daughter DNA molecule is one-half "old" and one-half "new."

E. coli is a prokaryote, but semi-conservative replication of DNA also occurs in eukaryotes. And because each DNA molecule in a eukaryote is incorporated in one chromosome, the replication of entire chromosomes is semi-conservative as well.

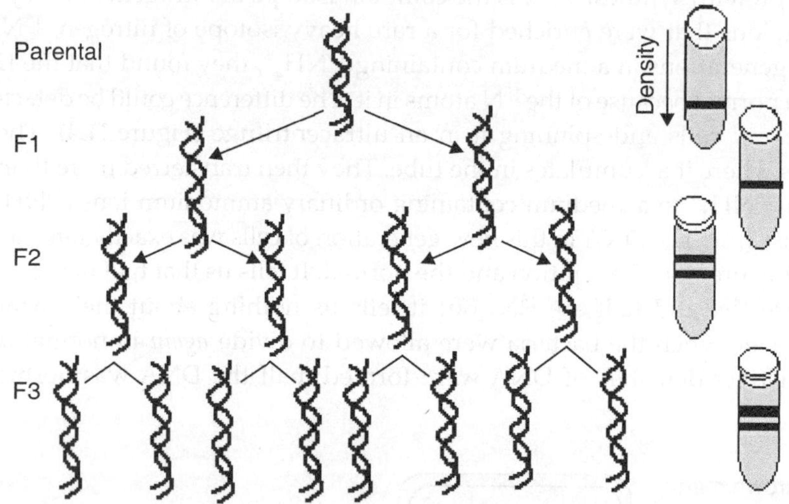

FIGURE 11.5 **Meselson and Stahl Experiment—DNA of three generations.**

Semi-conservative replication can also be demonstrated in the case of chromosomes. Actually each of the single chromosomes that you observe under 440X magnification of your laboratory microscope contains a single molecule of DNA. For some of our chromosomes, this molecule—if stretched out—would extend 5 cm (2 inches). Figure 11.6 demonstrates that each chromosome contains a single molecule of DNA and that the replication of a chromosome is semi-conservative (just as is the replication of a single DNA molecule). That is, the information encoded in each strand of DNA remains intact as it serves as the template for the assembly of a complementary strand.

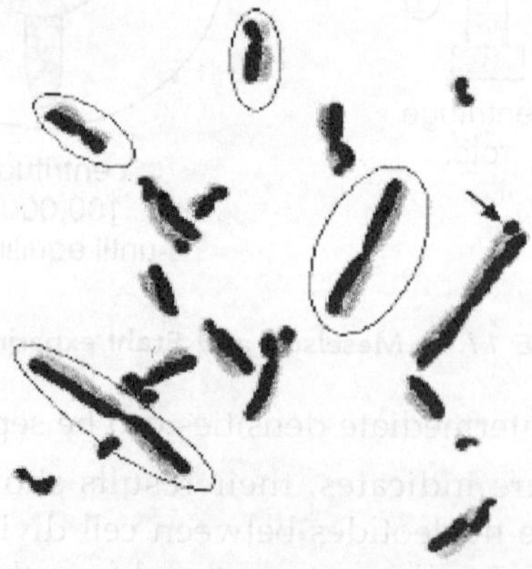

FIGURE 11.6 **Semi-conservative replication of chromosomal DNA.**

If cells dividing in culture are treated with the pyramidine analog **bromodeoxyuridine** (**BrdU**), during S phase, the cells are fooled into incorporating it—instead of thymidine—into their

DNA. One of the properties of the resulting DNA is that it fails to take up stain in a normal way. When cells are allowed to duplicate their chromosomes *once* in BrdU, the chromosomes that appear at the next metaphase stain normally. However, when the cells duplicate their chromosomes a second time in BrdU, one of the sister chromatids that appears at the next metaphase stains normally, while its sister chromatid does not. Each chromatid is normally completely stained or not (circled in Figure 11.6). If chromosomes were made up of numbers of different DNA molecules, we would expect to find, at best, a variegated pattern of stained and unstained regions and, more likely, no clear pattern at all.

But there can be exceptions, as shown with the arrow in the figure, that are due to the spontaneous exchange of segments between the sister chromatids and the exchange is reciprocal.

Biochemical Mechanism of DNA Replication

It is very important to know that DNA replication is not a passive and spontaneous process. Many enzymes are required to unwind the double helix and to synthesize a new strand of DNA. DNA replication has two important requirements in addition to the enzymes and the nucleotides: DNA template and free 3' -OH group. The parent strands provide the template and the free 3' -OH group is provided by a primer, a small oligo nucleotide. We will approach the study of the molecular mechanism of DNA replication from the point of view of the machinery that is required to accomplish it. The unwound helix, with each strand being synthesized into a new double helix, is called the **replication fork**.

The Enzymes of DNA Replication

1. **Topoisomerase** is responsible for the initiation of the unwinding of the DNA. The tension holding the helix in its coiled and supercoiled structure can be broken by nicking a single strand of DNA. Try this with string. Twist two strings together, holding both the top and the bottom. If you cut only one of the two strings, the tension of the twisting is released and the strings untwist.
2. **Helicase** accomplishes the unwinding of the original double strand, once supercoiling has been eliminated by the topoisomerase. The two strands very much want to bind together because of their hydrogen-bonding affinity for each other, so the helicase activity requires energy (in the form of ATP) to break the strands apart.
3. **DNA polymerase** proceeds along a single-stranded molecule of DNA, recruiting free dNTPs (deoxy-nucleotide-triphosphates) to hydrogen bond with their appropriate complementary dNTP on the single strand (A with T and G with C), and to form a covalent phosphodiester bond with the previous nucleotide of the same strand. The energy stored in the triphosphate is used to covalently bind each new nucleotide to the growing second strand. There are different forms of DNA polymerase. DNA polymerase I (pol I) was the first enzyme discovered with polymerase activity, and it is the best characterized enzyme. Although this was the first enzyme to be discovered that had the required polymerase activities, it is not the primary enzyme involved with bacterial DNA replication. That enzyme is DNA polymerase III (pol III). Three activities are associated with DNA polymerase I:

- 5′ to 3′ elongation (polymerase activity)
- 3′ to 5′ exonuclease (proofreading activity)
- 5′ to 3′ exonuclease (repair activity)

The second two activities of DNA pol I are important for replication, but DNA polymerase III (pol III) is the enzyme that performs the 5'-3' polymerase function. DNA polymerase cannot start synthesizing *de novo* on a bare single strand. It needs a **primer** with a **3′ OH group** on to which it can attach a dNTP. DNA polymerase is actually an aggregate of several different protein subunits, so it is often called a **holoenzyme**. The holoenzyme also has proofreading activities, so that it can make sure that it inserted the right base, and nuclease (excision of nucleotides) activities so that it can cut away any mistakes it might have made.

4. **Primase** is actually part of an aggregate of proteins called **primeosomes**. This enzyme attaches a small RNA primer to the single-stranded DNA to act as a substitute 3′ OH for DNA polymerase to begin synthesizing from. This RNA primer is eventually removed by **RNase H** and the gap is filled in by DNA polymerase I.
5. **Ligase** can catalyze the formation of a phosphodiester bond given an unattached but adjacent 3′ OH and 5′ phosphate. This can fill in the unattached gap left when the RNA primer is removed and filled in. The DNA polymerase can organize the bond on the 5′ end of the primer, but ligase is needed to make the bond on the 3′ end.
6. **Single-stranded binding proteins** are important to maintain the stability of the replication fork. Single-stranded DNA is very labile, or unstable, so these proteins bind to it while it remains single stranded and keep it from being degraded.

Semi-discontinuous Replication

The two strands in a DNA molecule are antiparallel. One strand runs in the 3-5' direction and the other strand runs in the opposite direction (i.e., the 5'-3' direction). This suggests that during DNA replication one of the newly synthesized strands is polymerized in the 5'-3' direction while the other strand is polymerized in the 3'-5' direction. But it is known that all known DNA polymerases can carry out the polymerization only in the 5'-3' direction. That is, it can add fresh nucleotides only to the 3′-OH groups of sugar and not to the 5' terminal of the growing nucleic acid chain. This presents a problem: How does the synthesis of the 3'-5' direction take place? The answer is that the strand which grows in the 3'-5' direction (copying of the 5'-3' strand), known as the **lagging strand,** is synthesized discontinuously as a series of short DNA fragments known as **Okazaki fragments** (after its discoverer Reiji Okazaki). Each of these fragments is synthesized by DNA polymerase III, with usual 5'-3' polarity, as shown in Figure 11.9.

In bacterial systems the Okazaki fragments are 1,000 to 2,000 nucleotides in length, but in eukaryotes these fragments are relatively small and, in most cases, less than 200 nucleotides long. Finally, as the replication proceeds, these Okazaki fragments are joined together to produce an intact daughter strand. Thus, during DNA replication one strand is synthesized continuously and the other strand is synthesized discontinuously; therefore, DNA replication is called semi-discontinuous.

We already have observed that DNA polymerase does not have the *de novo* synthesis of DNA strands. It can only carry out the addition of new nucleotides to a growing nucleic acid strand. Thus, DNA synthesis is primed by short segments of RNA made by the enzyme primase or DNA polymerase (in the case of eukaryotes). The leading strand requires only single priming, which is in the beginning or initiation step and does not need any more priming. But for the lagging strand, the priming is a routine process and it repeats every time a new Okazaki fragment is initiated. Finally, the RNA primer fragments are excised by the 5' to 3' exonuclease action of polymerase I, and the gap so created will be filled by the DNA polymerase I with its 5'-3' polymerase activity. The adjacent Okazaki fragments are finally joined by a DNA ligase enzyme resulting in an intact continuous lagging strand. The movement of the replication fork (Figure 11.7) may be unidirectional or bi-directional. The final product is a pair of identical DNA molecules or chromosomes conserving half of the strand of the parent molecule, and it does not have any RNA stretches in it.

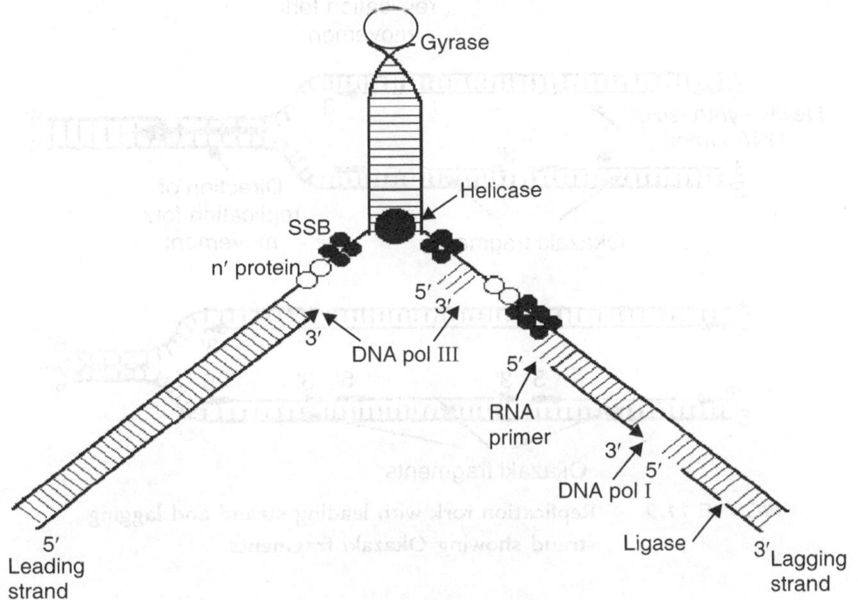

FIGURE 11.7 DNA replication fork.

FIGURE 11.8 An addition of a dNTP to a growing oligo nucleotide.

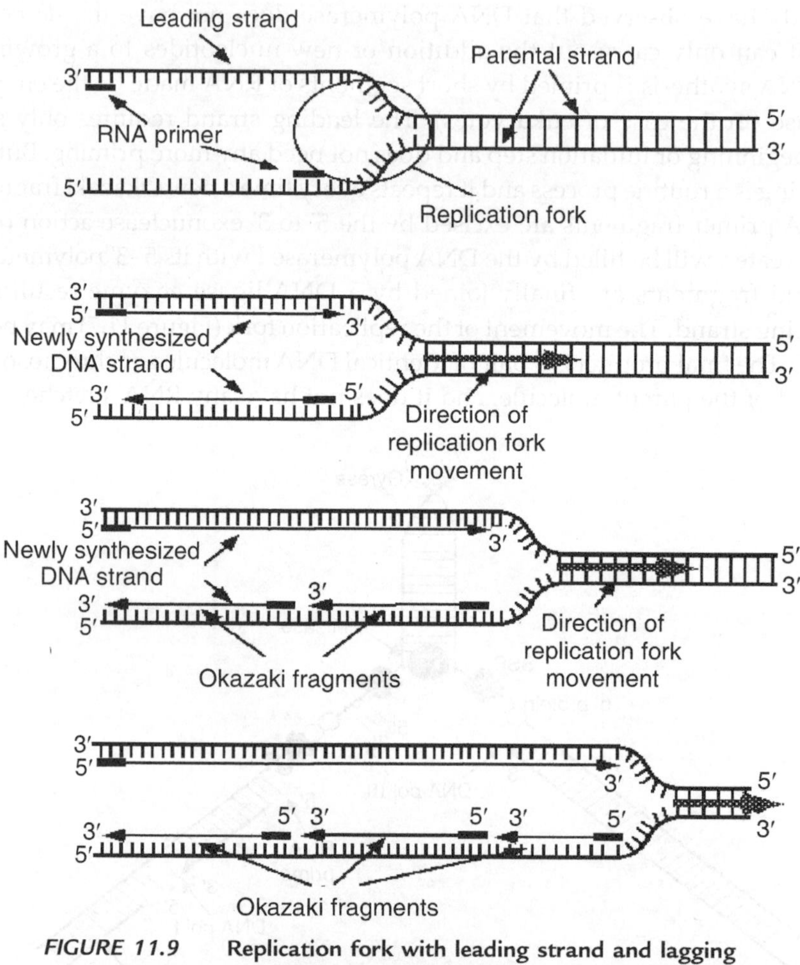

FIGURE 11.9 **Replication fork with leading strand and lagging strand showing Okazaki fragments.**

REVIEW QUESTIONS

1. How does the gene number influence genome size and organization?
2. Explain the possible benefits of human-genome projects in the medical science and drug industries.
3. Differentiate between eukaryotic and prokaryotic RNA polymerase.
4. What are Okazaki fragments?
5. Differentiate between leading strand and lagging strand.
6. List the enzymes involved in the process of DNA replication.
7. What are the various types of DNA polymerases present in prokaryotes?
8. Which is the main polymerase involved in the process of polymerase?
9. Why is the replication of DNA called semi-conservative replication?
10. What is meant by origin of replication?
11. Explain the gene organization in prokaryotes and eukaryotes.

Chapter **12** *GENE EXPRESSION*

In This Chapter

12.1 FINE STRUCTURE OF A GENE

A gene is the unit of heredity—it is the unit of recombination, mutation, and cystron. Based on the data obtained from the experiments on pea plants, Gregor Johann Mendel very correctly predicted the presence and nature of genes (which he interpreted as particulate factors) including the alternate forms of genes called alleles. Today, we know both its biochemical as well as genetical aspects. A gene is defined as a segment of DNA molecule (A DNA molecule = a chromosome), which carries all the biochemical information of a polypeptide or RNA molecule, along with its regulatory elements. A complete gene should have the coding region of the protein or RNA along with its control elements. The size of genes varies greatly. The minimum number or the average number of nucleotides in a gene is 1,000 bp. The part of the gene that holds the sequences for the polypeptide chain or the RNA molecule (in the case of non-protein genes such as rRNA and tRNA, etc.) is called the **cistron** or the **coding region**. Some genes, as in the case of prokaryotes, have more than one cistron in a gene controlled by a single regulatory element and such genes are known as **polycistronic genes**. The controlling elements are known as **promoters** and are located upstream to the coding region of a gene. The end of the cistron, or the coding region, is marked by specific signals called **termination codons**. In some cases, the regulatory elements also include certain sequences in addition to the promoters known as **enhancers** near or away from the promoters.

The biochemical information present in genes is in the form of specific nucleotide sequences or instructions known as **codons** for the synthesis of RNA molecules that ultimately direct the synthesis of the corresponding polypeptides or proteins. These processes (i.e., DNA-directed RNA synthesis and RNA directed protein synthesis) are known as **gene expressions** and form the central dogma of life. According to what has been called the central dogma of molecular genetics, the function of DNA is to store information and pass it on to RNA, while the function of RNA is to read, decode, and use the information received from DNA to make proteins.

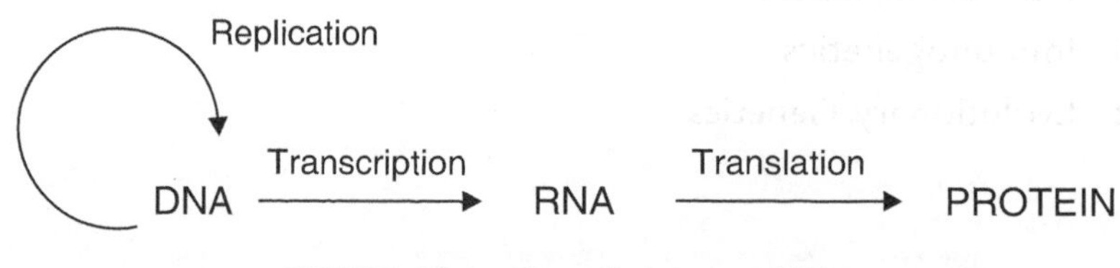

FIGURE 12.1 **Central dogma of life.**

The three fundamental processes that take place in the transfer and use of genetic information are the following:

1. **Replication** is the process by which a replica, or an identical copy, of DNA is made. Replication occurs every time a cell divides so that information can be preserved and handed down to offspring. This is similar to making a copy of a file on to a disk so you can take that file to a different computer.
2. **Transcription** is the process by which the genetic messages contained in DNA are 'read' or transcribed. The product of transcription, known as **messenger RNA (mRNA),** leaves the cell nucleus and carries the message to the site of protein synthesis.
3. **Translation** is the process by which the genetic messages carried by mRNA are decoded and used to build proteins.

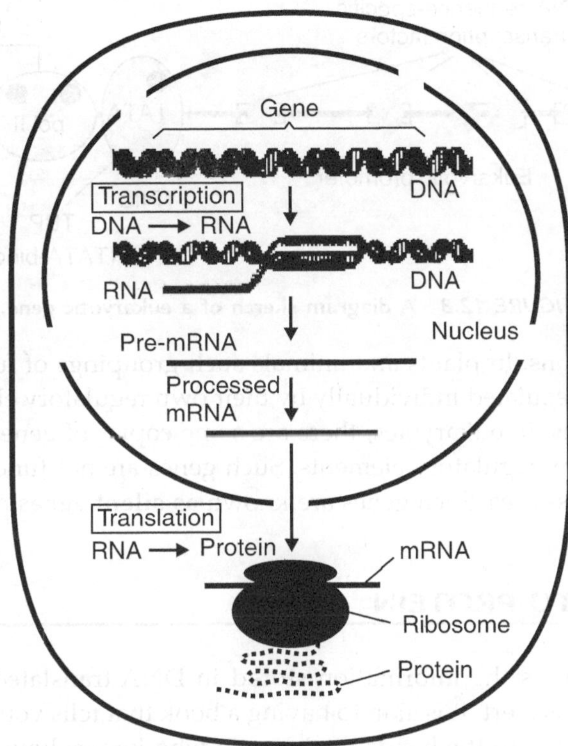

FIGURE 12.2 **Pathway of gene expression.**

Intergenic DNA separates the genes from one another. In the case of a virus genome the genes are packed very closely and therefore these intergenic regions are very low or almost nil. In higher organisms the genes are more spread out and are usually separated by very long intergenic regions. Intergenic regions are comparatively less in the case of prokaryotes. The number of genes varies greatly from organism to organisms. *E. coli* carries approximately 2,400 genes, yeasts with 16 chromosomes have about 6,600 genes. It is estimated that the human genome contains about 30,000 to 35,000 genes. But it forms only about 2 to 3% of the total DNA. The remaining major parts of the DNA are non-coding and non-functional, consisting of various types of repeated intergenic regions. In addition to these intergenic repeated sequences there are non-coding regions within the coding regions of the genes known as **introns**. The coding regions of the genes, which are intervened by introns, are called **exons**. Thus, the coding regions of the genes in eukaryotes are split apart by the introns, and such genes are called the **split genes**. They are also known as **mosaic genes** or **discontinuous genes**. Split genes and introns are very rare in the case of prokaryotes. In higher organisms where the introns are very common, there is a mechanism known as RNA splicing to remove the introns from the transcribed RNA molecule. Since prokaryotes do not have the introns, the splicing mechanism is absent in them.

Genes may occur individually or in clusters. Clusters of genes are very common in bacteria and viruses, where the DNA gene ratio is low. In bacteria, the genes which are functionally related, are grouped together under a single regulatory element, and such gene clusters or polycistronic

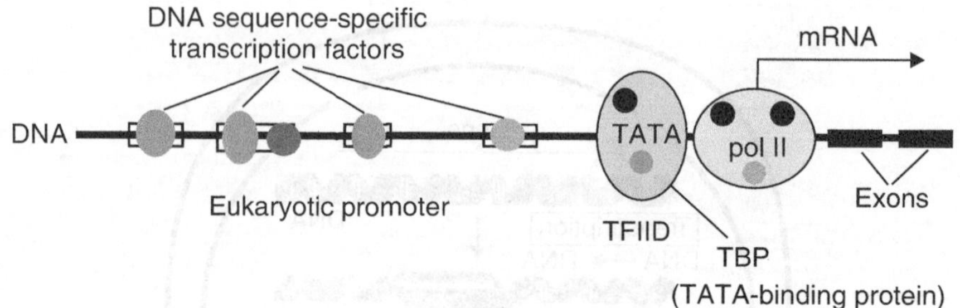

FIGURE 12.3 **A diagram sketch of a eukaryotic gene.**

genes are known as **operons**. In plants and animals such groupings of functionally related genes are present, but they are regulated individually by their own regulatory elements. Such groups are called **multi-gene families**. In eukaryotes, there are some copies of genes with coding regions or cistrons alone, without any regulatory elements. Such genes are not functional due to the lack of promoters and other accessories. Such genes are known as **silent genes** or **pseudogenes**.

12.2 FROM GENE TO PROTEIN

Now the question is, how is the information stored in DNA translated into biological action? Remember, DNA by itself is inert. It is akin to having a book that tells you the recipe for dynamite. By itself, the book is useless. In the hands of someone who knows how to read the instructions, collect the raw materials, and has the requisite technical skill, though, the information is potentially quite explosive.

One Gene–One Enzyme

The action of enzymes on their substrates is the basis of most of the metabolic and biochemical reactions that occur in biological systems. Each metabolic reaction is carried out by an enzyme that has been specified by a unique gene. The synthesis of proteins and enzymes directed by the corresponding genes is known as process of **gene expression**. Now it is almost clear that a gene specifies an enzyme or a polypeptide and that it determines the fate of metabolic reaction.

12.3 GENE EXPRESSION

Think of genetic information as a recipe. If you had only one copy of your recipe, and any damage would be irreparable, you would want to safeguard that information. If someone approached you for that recipe, you would not let him walk off with it. You might instead give him a copy that you did not need back. In the same way, the first step in gene expression is to copy the genetic information into mRNA (transcription). Then, the protein synthetic machinery (ribosomes, tRNA, amino acyl tRNA synthetases, amino acids, biosynthetic pathways) can use it to construct the appropriate protein. The mRNA serves as the template for the synthesis of proteins in the cytoplasm along with other components, and this process is known as **translation**.

It is easy to see how base pairing can specify the sequence of RNA on a DNA template and how mRNA can specify the protein. With only four RNA monomers (A, C, G, U), the genetic word must be at least three characters long in order to specify 20 amino acids. These codons are read one at a time by tRNA adapters that match the correct amino acid to its codon. The whole process is carried out on the ribosome, and demands a constant supply of charged tRNAs.

12.4 TRANSCRIPTION

Transcription is the synthesis of a single-stranded RNA directed by the DNA using one of its strands as the template. During this process the genetic information present in the DNA or the gene is transferred to the RNA molecule. A gene that is meant for a protein or enzyme will be transcribed into RNA called messenger RNA or mRNA, because it carries the message of genetic information for protein synthesis to the site of protein synthesis in the cytoplasm. The two strands of the DNA molecule are separated from each other, exposing the nitrogenous bases. Only one strand is actively used as a template in the transcription process; this is known as the **sense strand**, or **template strand**. The complementary DNA strand, the one that is not used, is called the **nonsense** or **antisense strand**.

The two strands of the DNA molecule are separated from each other, exposing the nitrogenous bases. Only one strand is actively used as a template in the transcription process, this is known as the **template strand** or the **antisense strand**. The complementary DNA strand, the one that is not used, is called the **sense** or the **coding strand**. The RNA molecule transcribed is a direct **complementary copy** of the nitrogenous bases in the template or antisense strand and is exactly identical to the non-template strand of the DNA except for the presence of uracil instead of thymine. In short, the RNA transcribed will be an exact copy of the sense strand or the coding strand and complementary to the template strand or the antisense strand.

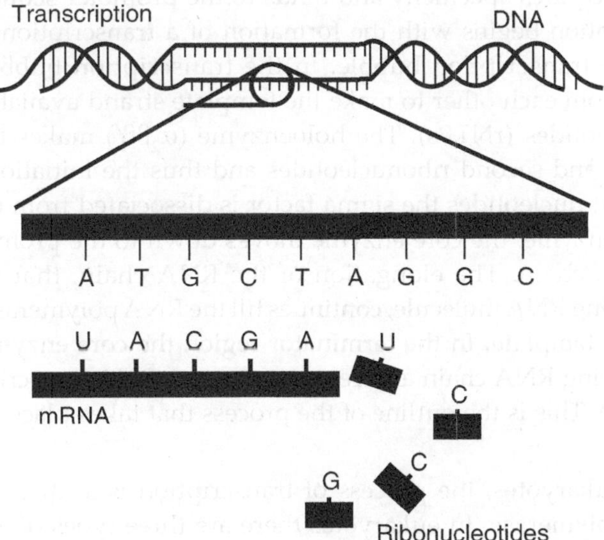

FIGURE 12.4 **The process of transcription.**

A multi-subunit enzyme called RNA polymerase or more specifically DNA- dependant RNA polymerase catalyzes the process of transcription. It needs the ribonucleoside triphosphates (rNTPs) ATP, UTP, CTP, and GTP, double-stranded DNA for the template. One at a time, this enzyme adds ribonucleotides to a growing RNA strand by joining incoming ribonucleoside triphosphates to the ribose sugar molecule of the last nucleotide of the growing RNA strand. Two of the phosphate groups are removed from the triphosphate and a covalent bond is formed between the remaining phosphate and the third carbon atom of the ribose sugar at the end of the RNA strand. There is an important difference between RNA polymerase and DNA polymerase. RNA polymerase does not need a primer and can start the synthesis of a new chain, whereas DNA polymerase cannot synthesize a new strand of DNA, but can only add new nucleotides to an existing strand. That is, DNA polymerase is primer dependent.

Initiation of the transcription process begins with the binding of the RNA polymerase enzyme to the DNA molecule at a region known as the **promoter site**. This site is right in front of the gene where transcription will begin. In bacteria, this region contains two short sequences of bases that appear to be the same or similar in all, or most, promoters. The RNA polymerase enzyme does not copy the promoter into the RNA, but begins the synthesis of the RNA at a specific nucleotide sequence called the **start signal** or **initiation site**, which is often the base GTA on the DNA (which then becomes the basic CAU on the RNA molecule). The first nucleotide corresponding to the position at which transcription begins is marked as +1 and is the start site.

RNA polymerase of *e. coli* has been studied extensively and is considered a model to explain the activity of the enzyme in other bacteria. In bacteria, a single enzyme is responsible for the transcription of all types of RNA molecules including mRNAs, rRNAs, and tRNAs. This enzyme is a multimeric molecule consisting of five subunits. The enzyme molecule can be divided into two components—a core enzyme and a sigma factor. The core enzyme consists of four subunits—two α subunits, one copy of β, and β' subunits. The core enzyme ($\alpha_2\beta\beta'$) has a great affinity for DNA molecules and its binding is random. But when it is joined by the σ factor ($\alpha_2\beta\beta'$) it loses its affinity for random DNA but acquires specificity and binds to the promoter sequences of the DNA. Now the process of transcription begins with the formation of a transcription bubble. RNA synthesis takes place within the transcription bubble. In the transcription bubble, the two strands are transiently separated from each other to make the template strand available for base pairing with the incoming ribonucleotides (rNTPs). The holoenzyme ($\alpha_2\beta\beta'$) makes the first phosphodiester bond between the first and second ribonucleotides and thus the initiation is achieved. After the addition of four to eight nucleotides the sigma factor is dissociated from the holoenzyme and the remaining part of the enzyme, the core enzyme moves down to the promoter and carries out the elongation of the RNA chain. The elongation of the RNA chain, that is, the addition of new nucleotides to the growing RNA molecule, continues till the RNA polymerase reaches the terminator sequences on the DNA template. In the terminator region the core enzyme stops the addition of nucleotides to the growing RNA chain and releases the new RNA transcript. Finally, it dissociates from the DNA template. This is the outline of the process that takes place during the transcription in prokaryotes.

In the case of eukaryotes, the process of transcription is a little complex. The difference starts from the RNA polymerase. In eukaryotes there are three types of RNA polymerase—RNA polymerase I, II, and III, each type transcribing different class of RNA molecules. RNA polymerase I transcribes rRNAs, RNA polymerase II transcribes mRNA, and RNA polymerase III transcribes tRNA and other RNAs such as small nuclear RNAs (snRNAs). Each of these RNA polymerases contain more than eight to fourteen subunits. There is no difference in the basic mechanism of transcription between eukaryotic and prokaryotic RNA polymerases. In eukaryotes the RNA is also synthesized in the 5′–3′ direction using the four rNTPs under the direction of the DNA template. Another important difference between the two types of RNA polymerases is that none of the eukaryotic RNA polymerases can directly recognize the promoters of the eukaryotic genes. Therefore, the recognition of the promoter and initiation of RNA synthesis results in a large number of small protein molecules known as **transcription factors (TF)**, which are responsible for the recognition of promoters by the RNA polymerase. TF molecules and their interaction with different types of promoters will differ from tissue to tissue or cell to cell. They are controlled by many other factors. They are the basis of differential expression of genes or tissue-specific or tome-bound expression of certain specific genes in a specific cell or tissue.

In eukaryotes all the types of RNAs—mRNA, rRNA, and tRNA—are transcribed in the form of a precursor RNA molecule known as the **primary transcript**. The primary transcript is processed to create the actual functional form of RNA. This mechanism of maturation of the primary transcript, leading to the formation of functional form RNA molecules, is called the **post-transcriptional modification**. For example, the primary transcript of the mRNA molecule is called the **heterogenous RNA** or **hnRNA**. They are processed in the nucleus before they are transported to the cytoplasm as mRNA for translation.

The following are the major post-transcriptional modifications for the formation of a mature mRNA from hnRNA:

- **Capping at the 5′ end:** An extra guanosine residue is first added to the 5′ terminal nucleotide of the primary transcript. This guanosine is further modified by the addition of methyl groups. This is the 5′ cap, which prevents the 5′ end of the mRNA from exonuclease enzymes present in the cytoplasm. It also helps in the transport of this mRNA out of the nucleus and helps in the initiation of the translation process.
- **Addition of Poly A tail at 3′ end:** At the 3′ end of the mRNA, a string of adenine residues is added to form a poly A tail. This also saves the mRNA from the attack of exonuclease.
- **Splicing:** Removal of intervening non-coding sequences or introns, which are also transcribed along with the coding sequences. This process of cutting and removing the introns and joining the parts of exons is known as **mRNA splicing**. This is very important for the formation of a functional enzymes or proteins.

The mature RNA thus formed leaves the nucleus and gets transported to the cytoplasm. The mRNA in prokaryotes does not need the process of splicing because the genes are without any introns, and therefore the RNA produced can be used for the process of translation.

12.5 GENETIC CODE

DNA is a two-stranded molecule. Each strand is a polynucleotide composed of **A** (adenosine), **T** (thymidine), **C** (cytidine), and **G** (guanosine) residues polymerized by 'dehydration' synthesis in linear chains with specific sequences. Each strand has polarity, such that the 5′-hydroxyl (or 5′-phospho) group of the first nucleotide begins the strand and the 3′-hydroxyl group of the final nucleotide ends the strand; accordingly, we say that this strand runs 5′ to 3′ (*"Five prime to three prime"*). It is also essential to know that the two strands of DNA run **antiparallel** such that one strand runs 5′ → 3′ while the other one runs 3′ → 5′. At each nucleotide residue along the double-stranded DNA molecule, the nucleotides are complementary. That is, **A** forms two hydrogen bonds with **T** and **C** forms three hydrogen bonds with **G.** In most cases the two-stranded, antiparallel, complementary DNA molecule folds to form a helical structure, which resembles a spiral staircase. This is the reason DNA has been referred to as the "double helix."

One strand of DNA holds the information that codes for various genes; this strand is often called the template strand or antisense strand (containing anticodons). The other, and complementary, strand is called the coding strand or sense strand (containing codons). Since mRNA is made from the template strand, it has the same information as the coding strand. Figure 12.5 refers to triplet nucleotide codons along the sequence of the coding or sense strand of DNA as it runs 5′ → 3′ ; the code for the mRNA would be identical but for the fact that RNA contains **U** (uridine) rather than **T.**

An example of two complementary strands of DNA would be:

(5′ → 3′) **ATGGAATTCTCGCTC** (Coding, sense strand)

(3′ ← 5′) **TACCTTAAGAGCGAG** (Template, antisense strand)

(5′ → 3′) **AUGGAAUUCUCGCUC** (mRNA made from template strand)

Since amino acid residues of proteins are specified as triplet codons, the protein sequence made from the above example would be Met-Glu-Phe-Ser-Leu... (MEFSL...).

Practically, codons are 'decoded' by transfer RNAs (tRNA), which interact with a ribosome-bound messenger RNA (mRNA) containing the coding sequence. There are 64 different tRNAs, each of which has an anticodon loop (used to recognize codons in the mRNA). Sixty one of these have a bound amino acyl residue; the appropriate 'charged' tRNA binds to the respective next codon in the mRNA and the ribosome catalyzes the transfer of the amino acid from the tRNA to the growing (nascent) protein/polypeptide chain. The remaining three codons are used for 'punctuation'; that is, they signal the termination (the end) of the growing polypeptide chain.

Lastly, the genetic code in Figure 12.5 has also been called "the universal genetic code." It is known as 'universal,' because all known organisms use it as a code for DNA, mRNA, and tRNA. The universality of genetic code encompasses animals (including humans), plants, fungi, archaea, bacteria, and viruses. However, all rules have their exceptions, and such is the case with the genetic code; small variations in the code exist in mitochondria and certain microbes. Nonetheless, it should be emphasized that these variances represent only a small fraction of known cases, and that the genetic code applies quite broadly, certainly to all known nuclear genes.

First Position		Second Position of Codon: T	C	A	G		Third Position
	T	TTT Phe [F]	TCT Ser [S]	TAT Tyr [Y]	TGT Cys [C]	T	
	T	TTC Phe [F]	TCC Ser [S]	TAC Tyr [Y]	TGC Cys [C]	C	
	T	TTA Leu [L]	TCA Ser [S]	TAA Ter [end]	TGA Ter [end]	A	
	T	TTG Leu [L]	TCG Ser [S]	TAG Ter [end]	TGG Trp [W]	G	
	C	CTT Leu [L]	CCT Pro [P]	CAT His [H]	CGT Arg [R]	T	
	C	CTC Leu [L]	CCC Pro [P]	CAC His [H]	CGC Arg [R]	C	
	C	CTA Leu [L]	CCA Pro [P]	CAA Gln [Q]	CGA Arg [R]	A	
	C	CTG Leu [L]	CCG Pro [P]	CAG Gln [Q]	CGG Arg [R]	G	
	A	ATT Ile [I]	ACT Thr [T]	AAT Asn [N]	AGT Ser [S]	T	
	A	ATC Ile [I]	ACC Thr [T]	AAC Asn [N]	AGC Ser [S]	C	
	A	ATA Ile [I]	ACA Thr [T]	AAA Lys [K]	AGA Arg [R]	A	
	A	ATG Mot [M]	ACG Thr [T]	AAG Lys [K]	AGG Arg [R]	G	

		GTT Val [V]	GCT Ala [A]	GAT Asp [D]	GGT Gly [G]	T
		GTC Val [V]	GCC Ala [A]	GAC Asp [D]	GGC Gly [G]	C
	G	GTA Val [V]	GCA Ala [A]	GAA Glu [E]	GGA Gly [G]	A
		GTG Val [V]	GCG Ala [A]	GAG Glu [E]	GGG Gly [G]	G

FIGURE 12.5 **The universal genetic code—triplet codons and their corresponding amino acids.**

12.6 TRANSLATION

The process of protein synthesis is called **translation** because there is a change in the language. The languages of DNA and RNA are the same, the alphabets being the four letters of nucleotides A, T (U), G, and C. The language of protein is different. It consists of the alphabets of 20 amino acids. The story written in the language of nucleotides (four letters) is copied into another paper (mRNA) and taken to the cytoplasm, where it is translated into another language—the language of amino acids (20 letters). That is why the process of protein synthesis is called translation and synthesis of RNA is called transcription.

The tool needed for translation is a single strand of RNA copied from a DNA molecule. This is the **messenger RNA (mRNA)** molecule that carries the codes; ribosomes, which are the sites of assembly of the polypeptide; all 20 amino acids, which must be present at the same time; small highly specific RNA molecules called **transfer (tRNA)** to which amino acids are covalently bound and which play a part in the decoding of the genetic code; ATP, which is required as a source of energy for this highly non-spontaneous (energy-consuming) process; and various proteins and enzymes (for example, the set of enzymes, which specifically join the amino acids to the tRNA molecules).

Transfer RNA (tRNA): The Adapter Molecules

Transfer RNA (tRNA) molecules are the information adapter molecules. They are the direct interface between the amino acid sequence of a protein and the information in DNA. Therefore, they decode the information in DNA. There are 20 or more different types of tRNA molecules. Most tRNA molecules are about 76 nucleotides in length, but they range from 60 to 95. All tRNAs from all organisms have similar structures, indeed a human tRNA can function in yeast cells.

There are four arms and three loops. The **acceptor**—it is so named because amino acids become attached to this arm. The **TψC arm** is named for the presence of this triplet sequence and ψ stands for pseudouridine. The **anticodon arm** always contains anticodon triplet in the center of the loop. And the **DHU arm** contains the modified uridine, the dihydrouridine. Sometimes tRNA molecules have an extra or variable loop present between the DHU loop and anticodon loop.

Transfer RNA is synthesized in two parts. The body of the tRNA is transcribed from a tRNA gene. The acceptor stem is the same for all tRNA molecules and is added after the body is synthesized. It is replaced often during the lifetime of a tRNA molecule. Nine hydrogen bonds hold the molecule into an 'L' shaped tertiary structure (shown in Figure 12.6). The anticodon is on a loop at the opposite end of the molecule to the amino acid binding region.

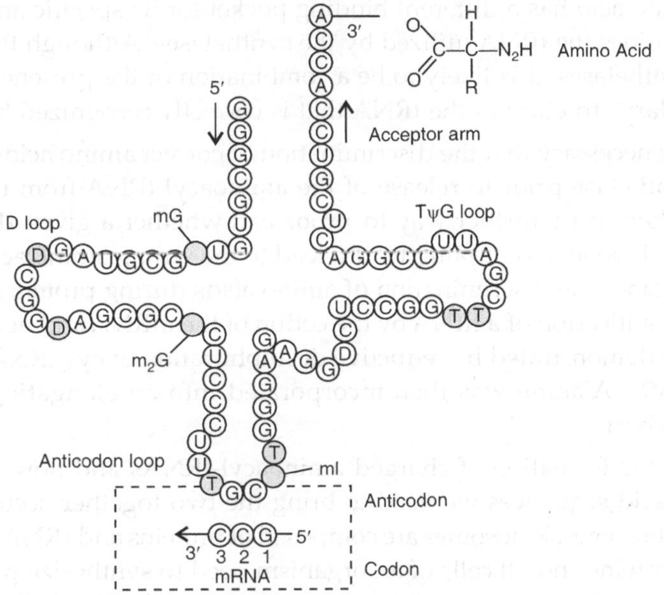

FIGURE 12.6 **Structure of a typical tRNA: yeast alanine tRNA (tRNA^{Ala}).**

The acceptor stem is the site at which a specific amino acid is attached by an amino-acyl-tRNA synthase. Amino acids are covalently joined to the acceptor arm of the molecule that is not looped. tRNA molecules become 'charged' or **aminoacylated** in a two-step reaction in which an enzyme first attaches most of an **ATP molecule (AMP)** to the amino acid to create a temporary intermediate. In the second step, the tRNA molecule displaces the AMP from the amino acid and joins with it, forming an amino acid-tRNA complex. The enzymes recognize unique features on every different tRNA molecule and will only join the correct amino acid to the correct tRNA. The anticodon reads the information in an mRNA sequence by base pairing. The anticodon of the tRNA pairs with the codon of mRNA.

Charging the tRNA: Attachment of Amino Acid to tRNA

Activation of amino acids is carried out by a two-step process catalyzed by **aminoacyl-tRNA synthetases**. Each tRNA, and the corresponding amino acid, is recognized by individual aminoacyl-tRNA synthetases. This means there exists at least 20 different aminoacyl-tRNA synthetases (there are actually at least 21 since the initiator met-tRNA of both prokaryotes and eukaryotes is distinct from non-initiator met-tRNAs).

The two-step process begins, first when the enzyme attaches the amino acid to the α-phosphate of ATP with the concomitant release of pyrophosphate. This is called an **aminoacyl-adenylate intermediate.** In the second step, the enzyme catalyzes the transfer of the amino acid to either the 2′- or 3′-OH of the ribose portion of the 3′-terminal adenosine residue of the tRNA generating the **activated aminoacyl-tRNA.** Although these reactions are freely reversible, the forward reaction is favored by the coupled hydrolysis of PP_i.

Accurate recognition of the correct amino acid as well as the correct tRNA is different for each aminoacyl-tRNA synthetase. Since the different amino acids have different R groups, the enzyme for each amino acid has a different binding pocket for its specific amino acid. It is not the anticodon that determines the tRNA utilized by the synthetases. Although the exact mechanism is not known for all synthetases, it is likely to be a combination of the presence of specific modified bases and the secondary structure of the tRNA that is correctly recognized by the synthetases.

It is absolutely necessary that the discrimination of correct amino acids and correct tRNA be made by a given synthetase prior to release of the aminoacyl-tRNA from the enzyme. Once the product is released there is no further way to **proofread** whether a given tRNA is coupled to its corresponding tRNA. Erroneous coupling would lead to the wrong amino acid being incorporated into the polypeptide since the discrimination of amino acids during protein synthesis comes from the recognition of the anticodon of a tRNA by the codon of the mRNA and not by recognition of the amino acid. This was demonstrated by reductive desulphuration of cys-$tRNA^{cys}$ with raney nickel generating ala-$tRNA^{cys}$. Alanine was then incorporated into an elongating polypeptide where cysteine should have been.

We have seen the formation of charged aminoacyl-tRNAs and now to convert nucleotide sequences to amino acid sequences we need to bring the two together accurately and efficiently. This is the job of the ribosomes. Ribosomes are composed of proteins and rRNAs. All living organisms need to synthesize proteins and all cells of an organism need to synthesize proteins, therefore, it is not hard to imagine that ribosomes are a major constituent of all cells of all organisms. The make-up of the ribosomes, both rRNA and associated proteins, is slightly different between prokaryotes and eukaryotes.

Ribosome Structure and Function

Ribosomes are cytoplasmic organelles found in prokaryotes and eukaryotes. They are large complexes of proteins and three (prokaryotes) or four (eukaryotes) rRNA (ribosomal ribonucleic acid) molecules called subunits made in the nucleus. The main function of ribosomes is to serve as the site of mRNA translation (protein synthesis, the assembly of amino acids into proteins); once the two (large and small) subunits are joined by the mRNA from the nucleus, the ribosome translates the mRNA into a specific sequence of amino acids, or a polypeptide chain. Ribosomes are found in cells in two ways: free and bound. In an electron micrograph as shown in Figure 12.7, ribosomes appear as dark granules. Ribosomes exist at various locations within the cell; however, the location of ribosomes depends on the function of the cell:

Free Ribosomes

- Are found in the cytosol.
- May occur as a single ribosome or in groups known as **polyribosomes** or **polysomes**.
- Occur in greater number than bound ribosomes in cells that retain most of their manufactured protein.

- Responsible for proteins that go into solution in the cytoplasm or form important cytoplasmic structural or motile elements.

Bound Ribosomes

- Are found bound to the exterior of the endoplasmic reticulum (ER) constituting rough ER.
- Occur in greater number than free ribosomes in cells that secrete their manufactured proteins (e.g., pancreatic cells, producers of digestive enzymes).
- Responsible for proteins that become a part of membranes or packaged into vesicles for storage in the cytoplasm or export to the cell exterior.

Ribosomes are also located in mitochondria and chloroplasts of eukaryotic cells; they are always smaller than cytoplasmic ribosomes and are comparable to prokaryotic ribosomes in both size and sensitivity to antibiotics; however, the sedimentation values *s* (*s* = Svedberg units: a measure of the rate of sedimentation of a component in a centrifuge, related both to the molecular weight and the three-dimensional shape of the component) vary somewhat in different phyla. Prokaryotic and eukaryotic ribosomes perform the same functions by the same set of chemical reactions; however, eukaryotic ribosomes are much larger than prokaryotic ones and most of their proteins are different. Mitochondrial and chloroplast ribosomes resemble bacterial ribosomes.

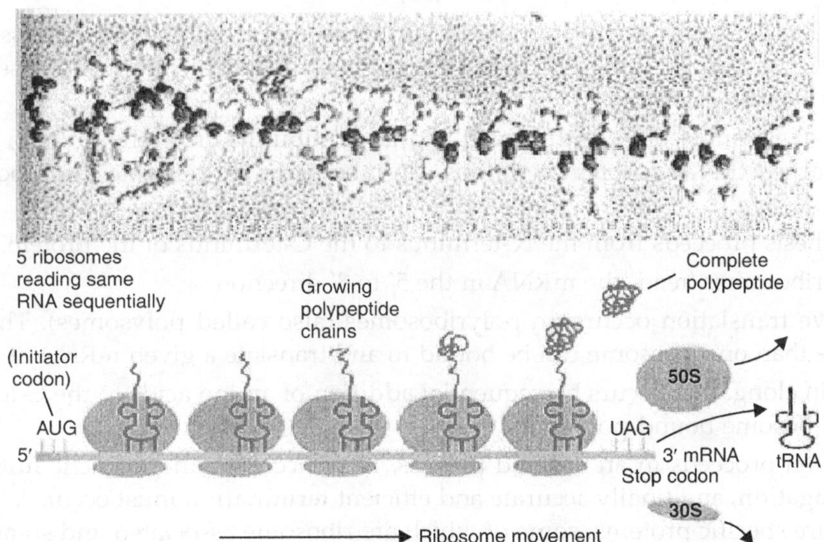

FIGURE 12.7 **Electron micrograph showing translation: polyribosome and its diagram sketch.**

There are three sites on a ribosome—the A site or the acceptor site, P site or the peptide site, and E site or exit site. A site is the entry site for an aminoacyl-tRNA, the P site is the binding site for the peptidyl-tRNA carrying the nascent polypeptide under synthesis, and the E site is the exit for the leaving tRNA.

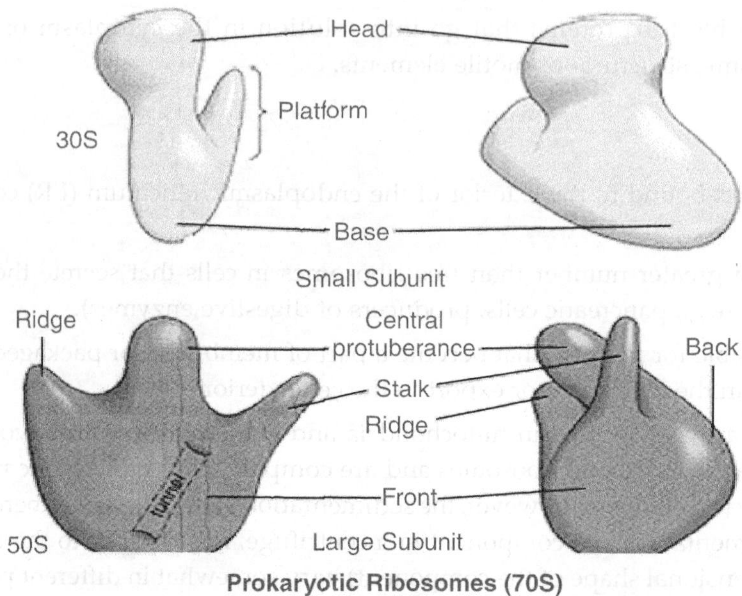

FIGURE 12.8 **Model of 70S Ribosomes—small and large subunit.**

Order of Events in Translation

The ability to begin to identify the roles of the various ribosomal proteins in the processes of ribosome assembly and translation was aided by the discovery that ribosomal subunits will self-assemble *in vitro* from their constituent parts.

Following assembly of both the small and large subunits on to the mRNA, and given the presence of charged tRNAs, protein synthesis can take place. To reiterate the process of protein synthesis:

1. Synthesis proceeds from the N-terminus to the C-terminus of the protein.
2. The ribosomes 'read' the mRNA in the 5′ to 3′ direction.
3. Active translation occurs on polyribosomes (also called polysomes). This means that more than one ribosome can be bound to and translate a given mRNA at any one time.
4. Chain elongation occurs by sequential addition of amino acids to the C-terminal end of the ribosome bound polypeptide.

Translation proceeds in an ordered process. First, accurate and efficient **initiation** occurs, then chain **elongation**, and finally accurate and efficient **termination** must occur. All three of these processes require specific proteins, some of which are ribosome associated and some of which are separate from the ribosome, but may be temporarily associated with it.

Initiation

Initiation of translation in both prokaryotes and eukaryotes requires a specific initiator tRNA, $tRNA^{met}{}_i$, that is used to incorporate the initial methionine residue into all proteins. In *e. coli* a specific version of $tRNA^{met}{}_i$ is required to initiate translation, ($tRNA^{fmet}{}_i$). The methionine attached to this initiator tRNA is formylated. Formylation requires N^{10}-formy-THF and is carried out after the methionine is attached to the tRNA. The fmet-tRNA$^{fmet}_i$ still recognizes the same codon, AUG, as regular tRNA^{met}. Although tRNA$^{met}_i$ is specific for initiation, in eukaryotes it is not a formylated tRNA^{met}.

The initiation of translation requires recognition of an AUG codon. In the polycistronic prokaryotic RNAs this AUG codon is located adjacent to a Shine-Delgarno element in the mRNA. The Shine-Delgarno element is recognized by complimentary sequences in the small subunit rRNA (16*S* in *e. coli*). In eukaryotes, initiator AUGs are generally, but not always, the first encountered by the ribosome. A specific sequence context surrounding the initiator AUG aids ribosomal discrimination. This context is $^A/_G$CC$^A/_G$CCAUG$^A/_G$ in most mRNAs.

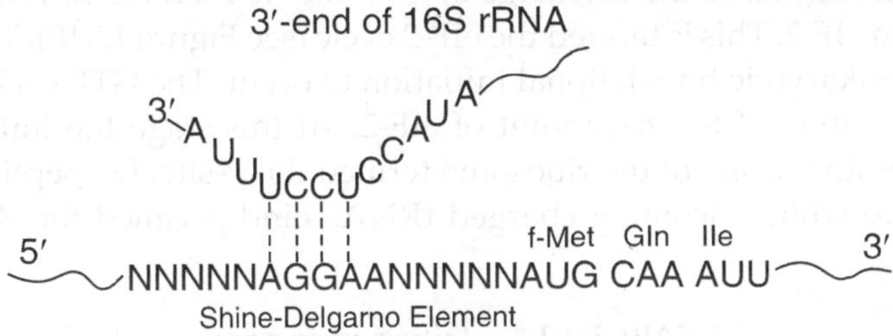

FIGURE 12.9 **Shine-Delgarno sequence and its complement in 16S rRNA.**

Eukaryotic Initiation Factors and their Functions

The specific non-ribosomally associated proteins required for accurate translational initiation are called **initiation factors**. In *e. coli,* they are IFs, in eukaryotes, they are eIFs. Numerous eIFs have been identified.

Specific Steps in Translational Initiation

Initiation of translation requires four specific steps:

1. A ribosome must dissociate into its 40S and 60S subunits.
2. A ternary complex called the **pre-initiation complex** is formed consisting of the initiator, GTP, eIF-2, and the 40S subunit.
3. The mRNA is bound to the pre-initiation complex.
4. The 60S subunit associates with the pre-initiation complex to form the 80S initiation complex.

The initiation factors eIF-1 and eIF-3 bind to the 40S ribosomal subunit favoring anti-association to the 60S subunit. The prevention of subunit re-association allows the pre-initiation complex to form.

The first step in the formation of the pre-initiation complex is the binding of GTP to eIF-2 to form a binary complex. eIF-2 is composed of three subunits, a, b, and g. The binary complex then binds to the activated initiator tRNA, met-tRNA^{met}, forming a ternary complex that then binds to the 40*S* subunit forming the 43S pre-initiation complex. The pre-initiation complex is stabilized by the earlier association of eIF-3 and eIF-1 to the 40S subunit.

The cap structure of eukaryotic mRNAs is bound by specific eIFs prior to association with the preinitiation complex. Cap binding is accomplished by the initiation factor eIF-4F. This factor is actually a complex of three proteins; eIF-4E, A, and G. The protein eIF-4E is a 24 kDa protein, which physically recognizes and binds to the cap structure. eIF-4A is a 46 kDa protein, which binds and hydrolyzes ATP and exhibits RNA helicase activity. Unwinding of mRNA secondary structure is necessary to allow access of the ribosomal subunits. eIF-4G aids in binding of the mRNA to the 43S preinitiation complex.

Once the mRNA is properly aligned on to the preinitiation complex and the initiator met-$tRNA_{met}$ is bound to the initiator AUG codon (a process facilitated by eIF-1) the 60S subunit associates with the complex. The association of the 60S subunit requires the activity of eIF-5 which has first bound to the preinitiation complex. The energy needed to stimulate the formation of the 80S initiation complex comes from the hydrolysis of the GTP bound to eIF-2. The GDP bound form of eIF-2 then binds to eIF-2B, which stimulates the exchange of GTP for GDP on eIF-2. When GTP is exchanged eIF-2B dissociates from eIF-2. This is termed the eIF-2 cycle (see Figure 12.10). This cycle is absolutely required in order for eukaryotic translational initiation to occur. The GTP exchange reaction can be affected by phosphorylation of the α-subunit of eIF-2. At this stage the initiator met-$tRNA^{met}$ is bound to the mRNA within a site of the ribosome termed the P-site, for peptide site. The other site within the ribosome to which incoming charged tRNAs bind is called the A-site, for amino acid site.

TABLE 12.1 Initiation factors.

Initiation Factor	Activity
eIF-1	repositioning of met-tRNA to facilitate mRNA binding
eIF-2	ternary complex formation
eIF-2A	AUG-dependent met-$tRNA_{met}^{i}$ binding to 40S ribosome
eIF-2B (also called GEF)	GTP/GDP exchange during eIF-2 recycling
eIF-3 composed of ~10 subunits	ribosome subunit antiassociation, binding to 40S subunit
eIF-4F composed of 3 subunits: eIF-4E, eIF-4A, eIF-4G	mRNA binding to 40S subunit, ATPase-dependent RNA helicase activity
eIF-4A	ATPase-dependent RNA helicase
eIF-4E	5′ cap recognition
eIF-4G	acts as a scaffold for the assembly of eIF-4E and -4A in the eIF-4F complex
eIF-4B	stimulates helicase, binds simultaneously with eIF-4F
eIF-5	release of eIF-2 and eIF-3, ribosome-dependent GTPase
eIF-6	ribosome subunit antiassociation

Elongation

The process of elongation, like that of initiation, requires specific non-ribosomal proteins. In *e. coli* these are **EFs** and in eukaryotes **eEFs.** Elongation of polypeptides occurs in a cyclic manner such that at the end of one complete round of amino acid additions, the A site will be empty and ready to accept the incoming aminoacyl-tRNA dictated by the next codon of the mRNA. This means that not only does the incoming amino acid need to be attached to the peptide chain but the ribosome must move down the mRNA to the next codon. Each incoming aminoacyl-tRNA is brought to the ribosome by an **eEF-1a-GTP complex.** When the correct tRNA is deposited into the A site the GTP is hydrolyzed and the eEF-1a-GDP complex dissociates. In order for additional translocation events the GDP must be exchanged for GTP. This is carried out by **eEF-1bg,** which is similar to the GTP exchange that occurs with eIF-2 catalyzed by eIF-2B.

The peptide attached to the tRNA in the P site is transferred to the amino group at the aminoacyl-tRNA in the A site. This reaction is catalyzed by ***peptidyltransferase.*** This process is called **transpeptidation.** The elongated peptide now resides on a tRNA in the A site. The A site needs to be freed in order to accept the next aminoacyl-tRNA. The process of moving the peptidyl-tRNA from the A site to the P site is called **translocation.** Translocation is catalyzed by **eEF-2** coupled to GTP hydrolysis. In the process of translocation the ribosome is moved along the mRNA such that the next codon of the mRNA resides under the A site. Following translocation, eEF-2 is released from the ribosome. The cycle can now begin again.

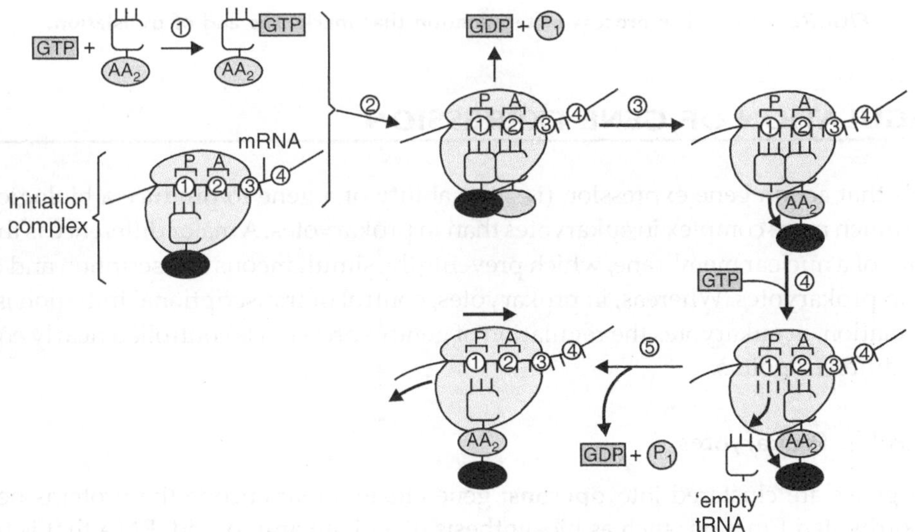

FIGURE 12.10 **Translation—elongation of polypeptide.**

Termination

The signals for termination are the same in both prokaryotes and eukaryotes. These signals are termination codons present in the mRNA. There are three termination codons, UAG, UAA, and UGA.

In *e. coli*, the termination codons UAA and UAG are recognized by RF-1, whereas RF-2 recognizes the termination codons UAA and UGA. The eRF binds to the A site of the ribosome in conjunction with GTP. The binding of eRF to the ribosome stimulates the peptidyl transferase activity to transfer the peptidyl group to water instead of an aminoacyl-tRNA. The resulting uncharged tRNA left in the P site is expelled with concomitant hydrolysis of GTP. The inactive ribosome then releases its mRNA and the 80S complex dissociates into the 40S and 60S subunits are ready for another round of translation.

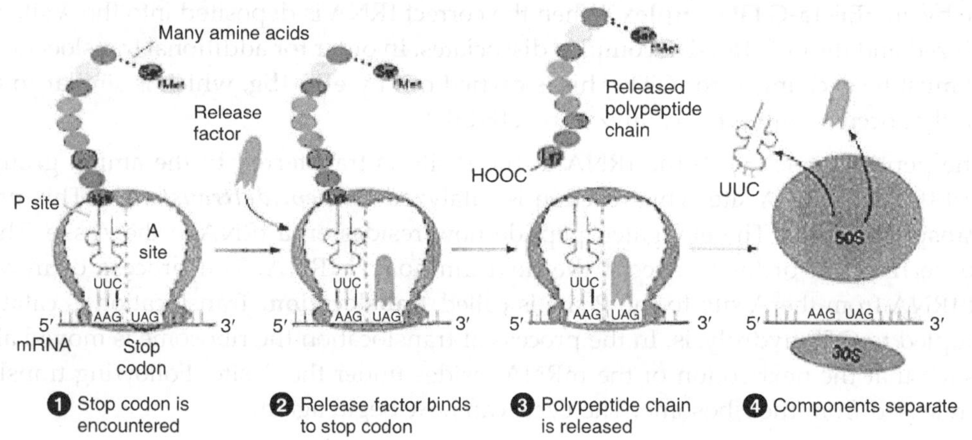

FIGURE 12.11 **The process of termination that marks the end of translation.**

12.7 REGULATION OF GENE EXPRESSION

The controls that act on gene expression (i.e., the ability of a gene to produce a biologically active protein) are much more complex in eukaryotes than in prokaryotes. A major difference is the presence in eukaryotes of a nuclear membrane, which prevents the simultaneous transcription and translation that occurs in prokaryotes. Whereas, in prokaryotes, control of transcriptional initiation is the major point of regulation, in eukaryotes the regulation of gene expression is controlled nearly equivalently from many different points.

Gene Control in Prokaryotes

In bacteria, genes are clustered into **operons:** gene clusters that encode the proteins necessary to perform coordinated function, such as biosynthesis of a given amino acid. RNA that is transcribed from prokaryotic operons is **polycistronic**, a term implying that multiple proteins are encoded in a single transcript.

In bacteria, control of the rate of transcriptional initiation is the predominant site for control of gene expression. As with the majority of prokaryotic genes, initiation is controlled by two DNA-sequence elements that are approximately 35 bases and 10 bases, respectively, upstream of the site of transcriptional initiation and as such are identified as the -35 and -10 positions. These two sequence elements are called **promoter sequences,** because they *promote* recognition of transcriptional start sites by ***RNA polymerase.*** The consensus sequence for the -35 position is TTGACA, and for the -10 position, TATAAT. (The -10 position is also known as the **Pribnow-box.**) These promoter sequences are recognized and contacted by **RNA polymerase.**

The activity of RNA polymerase at a given promoter is in turn regulated by interaction with accessory proteins, which affect its ability to recognize start sites. These regulatory proteins can act both positively (activators) and negatively (repressors). The accessibility of promoter regions of prokaryotic DNA is in many cases regulated by the interaction of proteins with sequences termed operators. The operator region is adjacent to the promoter elements in most operons and in most cases the sequences of the operator bind a repressor protein. However, there are several operons in *e. coli* that contain overlapping sequence elements, one that binds a repressor and one that binds an activator.

As indicated above, prokaryotic genes that encode the proteins necessary to perform coordinated function are clustered into operons. Two major modes of transcriptional regulation function in bacteria (*e. coli*) to control the expression of operons. Both mechanisms involve repressor proteins. One mode of regulation is exerted upon operons that produce gene products necessary for the utilization of energy; these are catabolite-regulated operons. The other mode regulates operons that produce gene products necessary for the synthesis of small biomolecules such as amino acids. Expression from the latter class of operons is attenuated by sequences within the transcribed RNA.

A classic example of a catabolite-regulated operon is the ***lac* operon,** responsible for obtaining energy from β-galactosides such as lactose. A classic example of an attenuated operon is the ***trp* operon,** responsible for the biosynthesis of tryptophan.

The *lac* Operon

The *lac* operon (see Figure 12.12) consists of one regulatory gene (the *i* gene) and three structural genes (*z, y,* and *a*). The *i* gene codes for the repressor of the *lac* operon. The *z* gene codes for **β-galactosides** (β-gal), which is primarily responsible for the hydrolysis of the disaccharide, lactose into its monomeric units, galactose and glucose. The gene ***'y'*** codes for **permease,** which is responsible for the high permeability of bacterial membranes for the β-galactosides. The *'a'* gene encodes a **transacetylase.**

During normal growth on a glucose-based medium, the *lac* repressor is bound to the operator region of the *lac* operon, preventing transcription. However, in the presence of an inducer of the *lac* operon, the repressor protein binds the inducer and is rendered incapable of interacting with the operator region of the operon. RNA polymerase is thus able to bind at the promoter region, and transcription of the operon ensues.

The *lac* operon is repressed, even in the presence of lactose, if glucose is also present. This repression is maintained until the glucose supply is exhausted. The repression of the *lac* operon under these conditions is called **catabolite repression** and is a result of the low levels of cAMP that result from an adequate glucose supply. The repression of the *lac* operon is relieved in the presence of glucose if excess cAMP is added. As the level of glucose in the medium falls, the level of cAMP increases. Simultaneously there is an increase in inducer binding to the *lac* repressor. The net result is an increase in transcription from the operon.

The ability of cAMP to activate expression from the *lac* operon results from an interaction of cAMP with a protein called **CRP** (for **cAMP receptor protein**). The protein is also called **CAP** (for **catabolite activator protein**). The cAMP-CRP complex binds to a region of the *lac* operon just upstream of the region bound by RNA polymerase and that somewhat overlaps the repressor-binding site of the operator region. The binding of the cAMP-CRP complex to the *lac* operon stimulates RNA polymerase activity 20 to 50-fold.

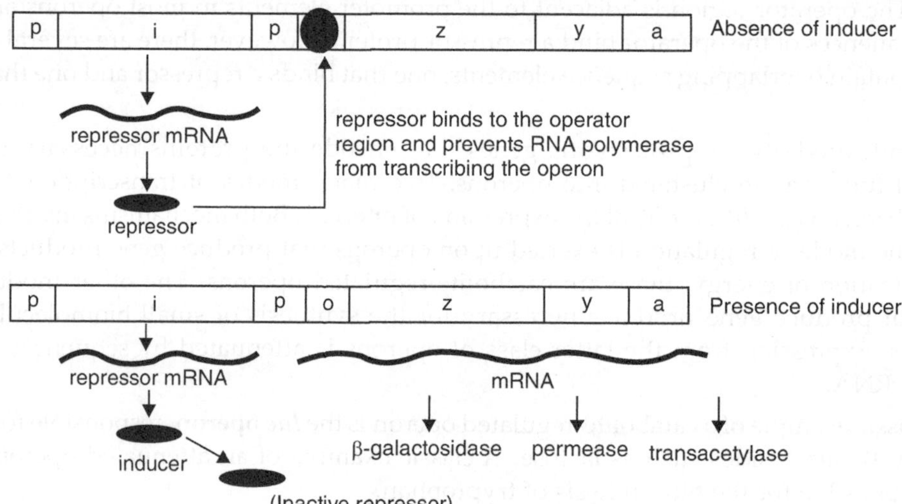

FIGURE 12.12 **Regulation of the *lac* operon in *e. coli*. The repressor of the operon is synthesized from the *i* gene. The repressor protein binds to the operator region of the operon and prevents RNA polymerase from transcribing the operon. In the presence of an inducer (such as the natural inducer, allolactose) the repressor is inactivated by interaction with the inducer. This allows RNA polymerase access to the operon and transcription proceeds. The resultant mRNA encodes the β-galactosidase, permease, and transacetylase activities necessary for utilization of β-galactosides (such as lactose) as an energy source. The *lac* operon is additionally regulated through binding of the cAMP-receptor protein, CRP (also termed the catabolite activator protein, CAP) to sequences near the promoter domain of the operon. The result is a 50-fold enhancement of polymerase activity.**

The *trp* Operon

The *trp* operon (see Figure 12.13) encodes the genes for the synthesis of tryptophan. This cluster of genes, like the *lac* operon, is regulated by a repressor that binds to the operator sequences. The activity of the *trp* repressor for binding the operator region is enhanced when it binds tryptophan; in this capacity, tryptophan is known as a **co-repressor.** Since the activity of the *trp* repressor is enhanced in the presence of tryptophan, the rate of expression of the *trp* operon is graded in response to the level of tryptophan in the cell.

Expression of the *trp* operon is also regulated by **attenuation.** The attenuator region, which is composed of sequences found within the transcribed RNA, is involved in controlling transcription from the operon after RNA polymerase has initiated synthesis. The attenuator of sequences of the RNA is found near the 5′ end of the RNA called the **leader region** of the RNA. The leader regions are located prior to the start of the coding region for the first gene of the operon (the *trp E* gene). The attenuator region contains codons for a small leader polypeptide, which contains tandem tryptophan codons. This region of the RNA is also capable of forming several different stable stem-loop structures.

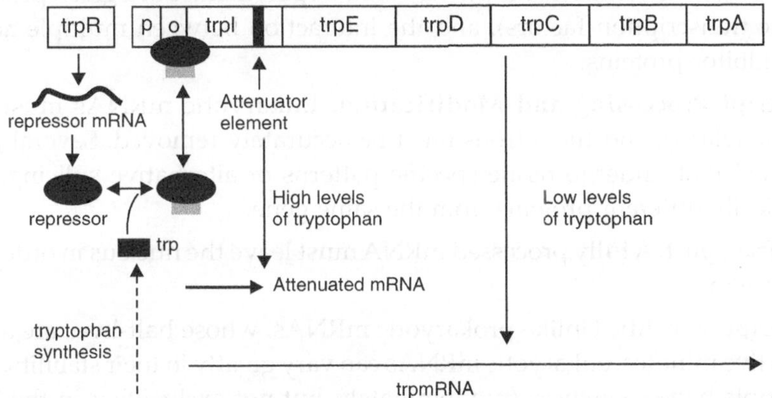

FIGURE 12.13 ***trp* Operon. Regulation of the *trp* operon in *e. coli*. The *trp* operon is controlled by both a repressor protein binding to the operator region as well as by translation-induced transcriptional attenuation. The *trp* repressor binds the operator region of the *trp* operon only when bound to tryptophan. This makes tryptophan a co-repressor of the operon. The *trpL* gene encodes a non-functional leader peptide which contains several adjacent *trp* codons. The structural genes of the operon responsible for tryptophan biosynthesis are *trpE, D, C, B,* and *A*. When tryptophan levels are high some binds to the repressor, which then binds to the operator region and inhibits transcription.**

Depending on the level of tryptophan in the cell—and hence the level of charged trp-tRNAs—the position of ribosomes on the leader polypeptide and the rate at which they are translating allow different stem-loops to form. If tryptophan is abundant, the ribosome prevents stem-loop 1–2 from forming and thereby favours stem-loop 3–4. The latter is found near a region rich in uracil and acts as the transcriptional terminator loop as in the RNA synthesis. Consequently, RNA polymerase is dislodged from the template. The operons coding for genes necessary for the synthesis of a number of other amino acids are also regulated by this attenuation mechanism. It should be clear, however, that this type of transcriptional regulation is not feasible for eukaryotic cells.

Gene Control in Eukaryotes

In eukaryotic cells, the ability to express biologically-active proteins comes under regulation at several points:

1. **Chromatin Structure.** The physical structure of the DNA, as it exists compacted into chromatin, can affect the ability of transcriptional regulatory proteins (termed transcription factors) and RNA polymerases to find access to specific genes and to activate transcription from them. The presence of the histones and CpG methylation most affect accessibility of the chromatin to RNA polymerases and transcription factors.
2. **Transcriptional Initiation.** This is the most important mode for control of eukaryotic gene expression. Specific factors that exert control include the strength of promoter elements within the DNA sequences of a given gene, the presence or absence of enhancer sequences (which enhance the activity of RNA polymerase at a given promoter by binding specific transcription factors), and the interaction between multiple activator proteins and inhibitor proteins.
3. **Transcript Processing and Modification.** Eukaryotic mRNAs must be capped and polyadenylated, and the introns must be accurately removed. Several genes have been identified that undergo tissue-specific patterns of alternative splicing, which generate biologically different proteins from the same gene.
4. **RNA Transport.** A fully processed mRNA must leave the nucleus in order to be translated into protein.
5. **Transcript Stability.** Unlike prokaryotic mRNAs, whose half-lives are all in the range of One to five minutes, eukaryotic mRNAs can vary greatly in their stability. Certain unstable transcripts have sequences (predominately, but not exclusively, in the 3-non-translated regions) that are signals of rapid degradation.
6. **Translational Initiation.** Since many mRNAs have multiple methionine codons, the ability of ribosomes to recognize and initiate synthesis from the correct AUG codon can affect the expression of a gene product. Several examples have emerged demonstrating that some eukaryotic proteins initiate at non-AUG codons. This phenomenon has been known to occur in *e. coli* for quite some time, but only recently has it been observed in eukaryotic mRNAs.
7. **Post-translational Modification.** Common modifications include gycosilation, acetylation, fatty acylation, disulphide bond formations, etc.
8. **Protein Transport.** In order for proteins to be biologically active following translation and processing, they must be transported to their site of action.
9. **Control of Protein Stability.** Many proteins are rapidly degraded, whereas others are highly stable. Specific amino acid sequences in some proteins have been shown to bring about rapid degradation.

Control of Eukaryotic Transcription Initiation

Transcription of different classes of RNAs in eukaryotes is carried out by three different polymerases. RNA pol I synthesizes the rRNAs, except for the 5S species. RNA pol II synthesizes the mRNAs and some small nuclear RNAs (snRNAs) involved in RNA splicing. RNA pol III synthesizes the 5S rRNA and the tRNAs. The vast majority of eukaryotic RNAs are subjected to post-transcriptional processing.

The most complex controls observed in eukaryotic genes are those that regulate the expression of RNA pol II-transcribed genes, the mRNA genes. Almost all eukaryotic mRNA genes contain a basic structure consisting of coding exons and non-coding introns and basal promoters of two types and any number of different transcriptional regulatory domains (see Figure 12.14). The basal promoter elements are termed **CCAAT-boxes** (pronounced *cat*) and **TATA-boxes** because of their sequence motifs. The TATA-box resides 20 to 30 bases upstream of the transcriptional start site and is similar in sequence to the prokaryotic Pribnow-box (consensus TATA$^{T}/_{A}$A$^{T}/_{A}$, where $^{T}/_{A}$ indicates that either base may be found at that position).

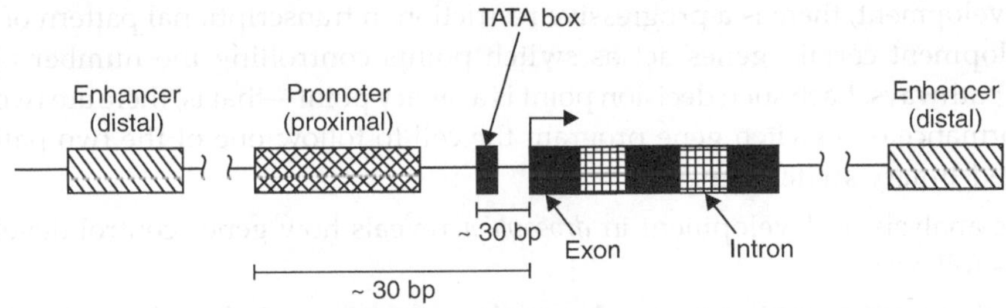

FIGURE 12.14 **Structure of a typical eukaryotic mRNA gene.**

Numerous proteins identified as **TFIIA, B, C,** etc. (for transcription factors regulating RNA pol II), have been observed to interact with the TATA-box. The CCAAT-box (consensus GG$^{T}/_{C}$CAATCT) resides 50 to 130 bases upstream of the transcriptional start site. The protein identified as **C/EBP** (for CCAAT-box/Enhancer Binding Protein) binds to the CCAAT-box element.

There are many other regulatory sequences in mRNA genes as well, that bind various transcription factors. These regulatory sequences are predominantly located upstream (5′) of the transcription initiation site, although some elements occur downstream (3′) or even within the genes themselves. The number and type of regulatory elements to be found vary with each mRNA gene. Different combinations of transcription factors also can exert differential regulatory effects upon transcriptional initiation. Each of various cell types express characteristic combinations of transcription factors; this is the major mechanism for cell-type specificity in the regulation of mRNA gene expression.

12.8 GENETIC BASIS OF DIFFERENTIATION AND DEVELOPMENT

All multicellular organisms developed from a single cell, the fertilization product of two gametes—the zygote. Embryonic development involves a series of cell division cells and development events, morphogenesis, and cell differentiation, which ultimately give rise to a multicellular adult organism. A fully developed adult organism is a collection of millions of cells organised into a cohesive and coordinated unit. The development of such an organism from a single cell, like any other life activities, is governed by a large number of genes, which express in a very organized and time-bound manner.

Researchers study development in model organisms to identify general principles: *science as a process.* Development is the attainment of differentiated state of various cells in multicellular organisms. At differentiated state of an organism, the cells show the differential gene expression.

Differential Gene Expression

All different types of cells in an organism have the same DNA. But different cell types make different proteins, usually as a result of transcriptional regulation. Maternal molecules in the cytoplasm (maternal influence or cytoplasmic influence) and signals from other cells direct transcriptional regulation the cell—cell interactions.

Genetic and Cellular Mechanisms of Pattern Formation

In embryo development, there is a progressive restriction in transcriptional pattern of the genome. During development certain genes act as switch points controlling the number of alternative development pathways. Each such decision point is a binary point—that is, there are two alternatives and the performance of a switch gene program the cell to follow one of the two pathways. Such genes are called **binary switch genes.**

Genetic analysis of development in *drosophila* reveals how genes control development. An overview is as follows:

1. Starting with a fertilized egg, *drosophila* passes through five distinct phases before it becomes an adult fly. These phases are the embryo, three larval stages, and a pupal stage, which give rise to the adult. The adult fly emerges after ten days of fertilization. During the divisions of the fertilized egg, the cytoplasm divides into a series of maternally divided molecular gradients. These gradients have an important role in deciding the developmental fate of the nuclei that migrates into specific regions of the embryo. Gradients of maternal molecules in the early embryo control axis formation. The first group of genes that are required for development of *drosophila* is transcribed by maternal genome during oogenesis. The mRNA or those proteins are transported into the oocyte and are stored there. These proteins and RNA are used in the early phases of development and convert the egg into an embryo with anterior-posterior and dorsal-ventral axes. These are the developments influenced by the maternal genes.
2. The maternal genes through their products trigger a cascade of zygotic gene activations, which sets up the segmentation pattern in *drosophila.*
   - Homeotic genes direct the identity of body parts. As segmentation boundaries are being established by the action of segmentation genes, a group of selector genes is activated. The expression of these genes determine the structures to be formed by each segment such as antennae, mouthparts, legs, thorax, wings, etc. These genes are known as **homeotic genes.**
   - Molecular biology studies of these developments have shown that each of the homeotic genes encodes a transcription factor having a DNA-binding domain. This domain is encoded by a 180-bp sequence called homeobox. **Homeobox genes** have been highly conserved in evolution.

The work in the development of *drosophila* has inspired others to study the molecular basis of development and differentiation in other organisms also. In nematodes such as *coenorabiditis*, it was observed that neighboring cells instruct other cells to form particular structures or to follow a particular pattern of development. This is an example of developmental control by cell signalling and induction. Similarly, in plant development, process depends on cell signalling and transcriptional regulation. A lot of research studies of *arabidopsis* have contributed to the understanding of organ identity in plants.

12.9 HOUSEKEEPING GENES

It was observed that the level of gene expression qualitatively and quantitatively is highly variable from cell to cell and also from tissue to tissue. Some genes are tightly regulated at different levels of their expression depending on the stages of development, age, and environmental factors. Expression of these genes is often regulated at the level of transcription. Some of these genes are induced at certain specific conditions, by altered environmental factors, temperature, or certain chemicals or by certain biological factors such as insect or pathogen attack. Such factors or chemicals responsible for the induction or expression of certain genes or gene groups are known as **inducers**.

But there are some genes, which express irrespective of the factors or inducers. Such genes code for proteins which are essential for maintaining normal metabolic activities such as respiration and energy transactions, protein synthesis and metabolism, and essential cellular structural components such as cell membranes and cell wall. These genes are expressed constitutively at relatively low levels in all cells of the body system. The genes that code for the basic metabolic enzymes and cellular components that are essential for maintaining the life activities of a living system are known as the **housekeeping genes.**

Genes that code for the formation of enzymes of glucose metabolism, enzymes of DNA replication, enzymes of metabolic pathways that lead to the production of amino acids, nucleotides, ATPase, etc., are examples of housekeeping genes.

12.10 GENETICS OF CANCER

Most, if not all, cancer cells contain genetic damage that appears to be the responsible event leading to tumorigenesis. The genetic damage present in a parental tumorigenic cell is maintained (i.e., not correctable) such that it is a heritable trait of all cells of subsequent generations. Genetic damage found in cancer cells is of two types:

1. Dominant and the genes have been termed **proto-oncogenes.** The distinction between the terms proto-oncogene and oncogene relates to the activity of the protein product of the gene. A proto-oncogene is a gene whose protein product has the capacity to induce cellular transformation given that it sustains some genetic insult. An oncogene is a gene that has sustained some genetic damage and, therefore, produces a protein capable of cellular transformation.

   The process of activation of proto-oncogenes to oncogenes can include retroviral transduction or retroviral integration, point mutations, insertion mutations, gene amplification, chromosomal translocation and/or protein-protein interactions.

Proto-oncogenes can be classified into many different groups based upon their normal function within cells or based on sequence homology to other known proteins. As predicted, proto-oncogenes have been identified at all levels of the various signal-transduction cascades that control cell growth, proliferation, and differentiation. The list of proto-oncogenes identified to date is too lengthy to include here. Proto-oncogenes that were originally identified as a resident in transforming retroviruses are designated as **c-indicative** of the cellular origin as opposed to **v-** to signify original identification in retroviruses.

2. Recessive, and the genes are variously termed **tumor suppressors, growth suppressors, recessive oncogenes,** or **anti-oncogenes**.

   Given the complexity of inducing and regulating cellular growth, proliferation, and differentiation, it was suspected for many years that genetic damage to genes encoding growth factors, growth factor receptors, and/or the proteins of the various signal-transduction cascades would lead to cellular transformation. This suspicion has proven true with the identification of numerous genes, whose products function in cellular signaling, that are involved in some way in the genesis of the tumorigenic state. The majority of these proto-oncogenes were identified by either of two means: as the transforming genes (oncogenes) of transforming retroviruses or through transfection of DNA from tumor cell lines into non-transformed cell lines and screening for resultant tumorigenesis.

Viruses and Cancer

Tumor cells can also arise by non-genetic means through the actions of specific tumor viruses. Tumor viruses are of two distinct types. There are viruses with DNA genomes (e.g., papilloma and adenoviruses) and those with RNA genomes (termed **retroviruses**).

RNA tumor viruses are common in chickens, mice, and cats but rare in humans. The only currently known human retroviruses are the **human T-cell leukemia viruses (HTLVs)** and the related retrovirus, **human immunodeficiency virus (HIV).** Retroviruses can induce the transformed state within the cells they infect with two mechanisms. Both of these mechanisms are related to the life cycle of these viruses. When a retrovirus infects a cell its RNA genome is converted into DNA by the viral encoded RNA-dependent DNA polymerase **(reverse transcriptase).** The DNA then integrates into the genome of the host cell where it can continue being copied as the host genome is duplicated during the process of cellular division. Contained within the sequences at the ends of the retroviral genome are powerful transcriptional promoter sequences called **long-terminal repeats (LTRs).** The LTRs promote the transcription of the viral DNA leading to the production of new virus particles.

At some frequency, the integration process leads to rearrangement of the viral genome and the consequent incorporation of a portion of the host genome into the viral genome. This process is called transduction. Occasionally, this transduction process leads to the virus acquiring a gene from the host that is normally involved in cellular growth control. Because of the alteration of the host gene during the transduction process as well as the gene being transcribed at a higher rate due to its association with the retroviral LTRs, the transduced gene confers a growth advantage to the infected cell. The end result of this process is unrestricted cellular proliferation leading to tumorigenesis. The transduced genes are called oncogenes. The normal cellular gene in its unmodified, non-transduced form is called a proto-oncogene since it has the capacity to transform cells if altered in some way or expressed in an uncontrolled manner. Numerous oncogenes have been discovered in the genomes of transforming retroviruses.

The second mechanism by which retroviruses can transform cells relates to the powerful transcription promoting effect of the LTRs. When a retrovirus genome integrates into a host genome it does so randomly. At some frequency, this integration process leads to the placement of the LTRs close to a gene that encodes a growth-regulating protein. If the protein is expressed at an abnormally elevated level it can result in cellular transformation. This is called **retroviral-integration-induced transformation**. It has recently been shown that HIV induces certain forms of cancers in infected individuals by this integration-induced transformation process.

Cellular transformation by DNA tumor viruses, in most cases, has been shown to be the result of protein-protein interaction. Proteins encoded by the DNA tumor viruses, called **tumor antigens** or T antigens, can interact with cellular proteins. This interaction effectively sequesters the cellular proteins away from their normal functional locations within the cell. The predominant types of proteins that are sequestered by viral T antigens have been shown to be of the tumor-suppressor type. It is the loss of their normal suppressor functions that results in cellular transformation.

Tumor-suppressor Genes

Tumor-suppressor genes were first identified by making cell hybrids between tumor and normal cells. On some occasions, a chromosome from the normal cell reverted the transformed phenotype. Several familial cancers have been shown to be associated with the loss of function of a tumor-suppressor gene. Some of the tumor-suppressor genes identified from different organisms include the **retinoblastoma susceptibility gene (RB), Wilms' tumors (WT1), neurofibromatosis type-1 (NF1), familial adenomatosis polyposis coli (APC or FAP),** and those identified through loss of heterozygosity such as in colorectal carcinomas (called **DCC** for **deleted in colon carcinoma**) and **p53** which was originally thought to be a proto-oncogene. However, the wild-type p53 protein suppresses the activity of mutant alleles of p53, which are the oncogenic forms of p53.

Translocation and Chromosomes

Chromosomal mutations are associated with various forms of cancers. One of the best-studied examples is translocation between chromosome 9 and chromosome 22. This aberration is associated with **chronic myelogenous leukemia (CML)**. Originally, this translocation was described as abnormal chromosome 21 and named as Philadelphia chromosome. Now it is clearly known that Philadelphia chromosome results from the exchange of genetic material between chromosome 9 and 22 and is never seen in normal chromosomes, but only in white-blood cells having this leukemia.

Role of Environmental Agents in Cancer

Environmental factors play an important role in the incidence of cancer developments in addition to personal behavior. Environmental factors such as UV light and other types of ionizing and non-ionizing radiations, mutagenic chemicals such as asbestos, and personal behaviors such as use of tobacco, certain diets, drugs, etc., also trigger the development of cancers. Certain types of viruses also play an important role in cancer development.

12.11 IMMUNOGENETICS

The vertebrates, particularly the mammals, have a well-developed immune system. It consists of lymphoid organs, various types of cells such as B and T lymphocytes, macrophages, and NK cells. These immune cells interact with each other and also with other cells through the receptors present on their membrane surface. The development of antibodies or the humoral immune response is the outcome of these interactions. This is also known as cell-mediated immune response or CMI. The main purpose of immunogenetics is to study the genetic basis of receptors present on the cells of the immune system in addition to the genetic basis of autoimmune disorders. The autoimmune disorder or disease is characterized by the damage of the immune system by its own organs and cells by generating an immune response. This is a highly disabling state and no cure is available for most of these diseases.

The important molecules involved in the process of immune response are immunoglobulins, T-cell receptors, and major histocompatibility antigens. *Tge* genes for histocompatibility antigens are grouped together and are present in a chromosome and form the **major histocompatibility complex or MHC**. It has been identified in different organisms. In mouse it is located in the 6th chromosome and is named as H2 complex. In humans, it is located in the 17th chromosome and is known as the **human leukocyte antigen** or **HLA complex**. The genes of these complexes are very closely linked, both in the case of mice and humans. For every locus, there are alleles in the homozygous chromosomes. The number of alleles may be many in most of the cases. The complete set of alleles for all the loci situated in H2 or HLA are haplotype and are inherited as a single unit. Thus, a child inherits a haplotype from the father and another from the mother.

MHC plays an important role in the successful transplantation of organs such as kidneys, heart, bone marrow, etc., between two individuals. The MHC or the HLA of the donor and recipient should closely match each other for successful tissue or organ transplantation. In the cases of mismatches, the transplanted organ is liable to be rejected by the immune system of the recipient. Therefore, organ transplantation is usually preferred between blood relatives or close relatives after ensuring adequate MHC matching.

Certain autoimmune disorders are associated with certain genes of MHC. The presence of a particular allele in the MHC is found to be more prone to certain autoimmune disorders. For example, the development of ankylosing spondylosis is linked to the presence of the allele HLA-B27. In this state the vertebrate tend to fuse and this results in the inflammation and discomfort. Some alleles of this complex (HLA) make the person more susceptible to diabetes. All this information clearly indicates the relevance of immunogenetics in medical applications.

12.12 EVOLUTIONARY GENETICS

In the previous chapters we discussed the involvement of genetics and mutation in the process of development of new traits and their combinations. These form an important driving force in the process of evolution. According to the law of speciation proposed by Charles Darwin, new species of organisms are developed from pre-existing ones by the natural selection. For the force of natural selection to play, there should be enough number of variations created by mutations on a particular loci or gene. Thus, mutations and the gene frequency of different alleles in the population is an important factor for natural selection to drive speciation. If a new mutation has a selective advantage over its allele, in that particular environment, its frequency will increase rapidly. And the other may be eliminated from the population. This may cause a complete change in the appearance and behavior of the organism.

Sexual barriers can also create new species. Mutations can directly cause sexual isolation or create sexual barriers because of reasons such as incompatibility or sterility with certain genotypes. These types of sexual barriers check the free mixing of new forms genes created by mutations in the population. Geographical isolation can also create a sexual barrier and prevent the mixing of the variation among the population and gradually evolve into a different species. The process of natural selection changes the gene and gene frequencies in the population. New variations are produced by rare genetic recombinations or various types of spontaneous mutations and these variations are a key factor in the process of speciation. If the variation is not genetic, that is, not the result of gene change, the selection will not have any consequences. Such variations may be epigenetic-induced by environmental factors, such as the acquired characteristics such as muscle strength features obtained by regular exercise and other characteristics are not transmitted to the next generation and, therefore, do not have any impact on *evolution*.

REVIEW QUESTIONS

1. What is a polycistronic gene?
2. Define the following terms:
   (*a*) Operon (*b*) Introns
   (*c*) Translation (*d*) Pseudogenes
   (*e*) Oncogenes (*f*) MHC
3. Differentiate between a multigene family and an operon.
4. What is the relevance of MHC in medical science?
5. What are the tumor suppress genes?
6. Explain the relationship between cancer and a virus.
7. What is the special feature of a retrovirus?
8. Differentiate between oncogenes and proto-oncogenes.
9. Distinguish between a promoter and an operator.
10. What are the various steps involved in the process of translation?
11. Differentiate between a *lac* operon and a *trp* operon.
12. Explain the genetic basis of development using drosophila as an example.
13. What are transcriptional factors?
14. Differentiate between introns and exons.
15. Explain how genetics is linked to evolution.
16. Distinguish between sense and antisense strands.
17. Give the structure of ribosomes and its role in translation.
18. Explain the major mechanism of gene regulation in eukaryotes.

Chapter 13 GENETIC TECHNIQUES

In This Chapter

13.1 INTRODUCTION

This chapter mainly discusses the various techniques commonly used in molecular biology and genetics. There are a lot of methodologies developed to study the behavior of chromosomes during mitosis and meiosis. When it was found that chromosomes are the seats of hereditary factors, there was a very active effort to understand more about them. This has led to the development of lot of histological and cytological techniques to view and study the movements and structure of chromosomes. In addition to this, microscopy has also facilitated the cytological studies of chromosomes and their function. Chromosome techniques such as chromosome banding and painting have been developed because of the various staining methods and histological and cytological techniques. Some of these techniques are so well developed that they are now routinely used in various laboratories for diagnosis of certain diseases and identifications. Clinical laboratories and hospitals are offering karyotyping and pedigree analysis in humans for the identification of various syndromes and chromosome abnormalities. Creation of various mutant lines helped in identifying a number of important genes within a very short span of time.

In bacteria, the process of recombination has helped to understand the basic mechanism and it has been used in other organisms also. Gene therapy and site-directed mutagenesis is the indirect outcome of these processes. The techniques of plant breeding and new methodologies in that area have helped in bringing about the Green Revolution. Several of the basic techniques used to study chromosomes and their mutations, bacterial recombination, some aspects of plant breeding, human pedigree analysis, and isolation and purification of DNA are discussed in this chapter.

13.2 CHROMOSOMAL TECHNIQUES

Cytological studies of chromosomes actually started in 1956 when Joe Hin Tjio and Albert Levan used a hypotonic solution to break open the cell and release the chromosomes. They showed for the first time that the human chromosome number is 46 ($2n = 46$). Following this, chromosomal techniques were developed to study the chromosomal basis of inheritance and the relationship between genetic syndromes and chromosomal aberrations. In 1959, it was shown that Down's Syndrome is due to an additional chromosome ($2n + 1 = 47$). Later, it was identified as chromosome number 21. New methods and staining techniques have been developed to produce well-spread preparations of human chromosomes.

Human mitotic chromosomes can be prepared from lymphocytes by arresting the cell division at the metaphase using the chemical colchicine. Then the chromosomes can be released by treating the cells with a hypotonic solution. Various staining techniques can be used to view the chromosomes. In the beginning, identification of chromosomes was based on the position of the centromere and length of the chromosomes. But at present time there are a number of other selective staining and banding techniques used to identify certain specific regions of the chromosome, which include labeling of the chromosomes with fluorescent dyes or radiolabeled compounds. Another powerful technique of modern times is the *in situ* hybridization of specific chromosomes with radiolabeled or fluorescently labeled nucleic acid probes to locate the position of specific genes. The following are some of the routinely used chromosomal techniques. The same types of techniques were conducted with plant cells also, with root tips and flower buds mainly to study about the behavior of chromosomes during mitosis and meiosis; and also to study about polyploids and other types of chromosomal aberrations.

Staining Techniques for Nucleic Acids

Since nucleic acids are an essential component of chromosomes, certain nucleic acid-staining techniques can be used for the detection of chromosomes and their specific areas. There are mainly three staining techniques that can be used for the visualization of chromosomes.

(*a*) **Histochemical stains:** These stains selectively bind to certain cellular parts or components depending on their chemical nature. Some of the important dyes and their applications are given in Table 13.1.

TABLE 13.1 Different types of histological stains.

Name of Dye	Applications
Acridine orange	RNA/DNA identification, flow cytometry, cell-cycle studies, permeant
Acridine homodimer	Impermeant, A-T selective, high affinity, DNA bunding
Ethidium bromide	Impermeant, DsDNA intercalator, electrophoresis, flow cytometry, chromosome counterstain
Propidium iodide	Chromosome and nuclear counterstain, impermeant dead cell stain
Aceto-carmine	Permeant, chromosome counterstain, cell-cycle analysis
Aceto-orcein stain	Drosophila, salivary gland chromosome, permeant

(*b*) **Stains based on antibodies:** Stains based on antibodies, which are highly selective in their binding. They are able to bind specific gene sequences very specifically. The nucleic acid parts or chromosome parts can be visualized if the antibodies are fluorescently labeled. This type of staining is known as immuno-staining.

(*c*) **Radiolabeled stains:** This staining technique can be used for visualizing the nucleic acids within the nucleus. Here, we use radiolabeled nucleotides (labeled with 3H; for example, 3H labeled uridine, which may be used specifically to detect and quantify RNA content). This is another technique of *in vivo* labeling. Radiolabeling has to be coupled with autoradiography for visualization or detection.

Chromosomal Banding Patterns

Most of chromosomes, at the prophase and the metaphase, are characterized by a banding pattern. But this banding pattern is more evident and clear in the case of larger chromosomes such as the polytene chromosomes of drosophila melanogaster, or the fruit fly. The banding patterns are the regions rich in heterochromatin, where the histone-DNA interaction is more. These complexes can be stained very easily by the conventional nuclear dyes or chromosomal dyes such as orcein. The regions between the bands are actually the active regions of the chromatin where more genes are present, but the quantity of DNA is very low and therefore the histone proteins. That is why they appear unstained or colored lightly. These interband regions can be detected with immuno-staining by using fluorescently labeled antibodies against the DNA-dependent RNA polymerase, which is usually seen with euchromatin regions required for the process of transcription.

Specialized staining techniques are now available, which enable one to differentiate or precisely identify individual chromosome homologes, chromosome regions, and/or chromosome bands. A renewed interest in the chromosomal or cytogenetic status of various species has been generated by the advancements of genetic mapping techniques utilizing fluorescence *in situ* hybridization or FISH. Depending upon the type of dye or fluorochrome or the chromosome pretreatment, there can be different types of banding patterns. They include banding patterns such as G-banding, Q-banding, C-banding, and R-banding. The data generated by multiple chromosome banding techniques can be used for karyotypic analysis.

Q-banding: This banding pattern is obtained by treating with a fluorochrome or the fluorescent dye quinacrin. They can be identified by a yellow fluorescence of different intensity as shown in the figure in the Appendix. Most parts of the stained DNA are heterochromatin. Quinacrin binds those regions which are rich in A-T and G-C, but fluorescences only A-T-quinacrin regions. A-T regions are seen more in heterochromatin than in euchromatin. Therefore, by this banding method heterochromatin regions are labeled preferentially. The characters of the banding regions and the specificity of the fluorochrome are not exclusively dependent on their affinity to regions rich in A-T, but it depends on the distribution of A-T and its association with other molecules such as histone proteins.

G-banding: This technique is not a fluorochrome-based pretreatment. It is well suited for animal cells. It resembles the C-banding technique without pretreatment. During mitosis, the 23 pairs of human chromosomes condense and are visible with a light microscope. A karyotype analysis usually involves blocking cells in mitosis and staining the condensed chromosomes with Giemsa dye. The dye stains regions of chromosomes that are rich in the base pairs Adenine (A) and Thymine (T) producing a dark band (see Appendix). A common misconception is that bands represent single genes, but in fact the thinnest bands contain over a million base pairs and potentially hundreds of genes. For example, the size of one small band is about equal to the entire genetic information for one bacterium.

C-banding: The name C-banding originated from centromeric or constitutive heterochromatin. The centromere appears as a stained band compared to other regions (Figure 13.1). The technique involves a pretreatment with alkali before staining. The alkaline pretreatment leads to the complete depurination of the DNA. The remaining DNA is again renatured and stained with Giemsa solution consisting of methylene azure, methylene violet, methylene blue, and eosin. In this staining the heterochromatin take a lot of stain but the rest of the chromosomes stain only a little. This banding technique is well suited for the characterization of plant chromosomes.

R-banding: This is known as a reverse banding technique. This technique results in the staining of areas rich in G-C that is typical for euchromatins. G-, Q-, and R-bandings are not observed with plant chromosomes.

Hy-banding: This is a common technique used with plant cells. The technique involves a pretreatment of the cells in which the cells are warmed in the presence of HCl and then stained with acetocarmine. The pattern of Hy-band is different from that of C-bands. The binding of histone protein to DNA and its complete extraction has an impact on the binding ability of acetocarmine and formation of bands.

Further variations in the procedure of the pre-treatment choice of dyes and fluorochromes further enhanced the resolution of the banding techniques. Many of the techniques are well suited for animal chromosomes, but face many difficulties with plant chromosomes. The reason for this is not well understood. The banding pattern of plant chromosomes with any of these techniques never comes to the same degree as that of animal chromosome banding patterns. The consistent banding patterns of the constitutive heterochromatin and the remaining chromatin are exactly constant in many species with an intraspecific variable karyotype.

Karyotyping

Karyotyping is a valuable research tool used to determine the chromosome complement within somatic or cultured cells. It is important to keep in mind that karyotypes evolve with organisms. Because of this evolution, it is important for the interpretation of biochemical or other data, that the karyotype of a specific subline be determined. Many morphological and physiological problems can be traced to the change in the karyotype. Numerous technical procedures have been reported that produce banding patterns on metaphase chromosomes. A band is defined as that part of a chromosome, which is clearly distinguishable from its adjacent segments by appearing darker or lighter. The chromosomes are visualized as consisting of a continuous series of light and dark bands. A G-staining method resulting in G-bands uses a Giemsa dye mixture or Leishman dye mixture as the staining agent. What follows is a brief description of the steps involved in assembling a karyotype.

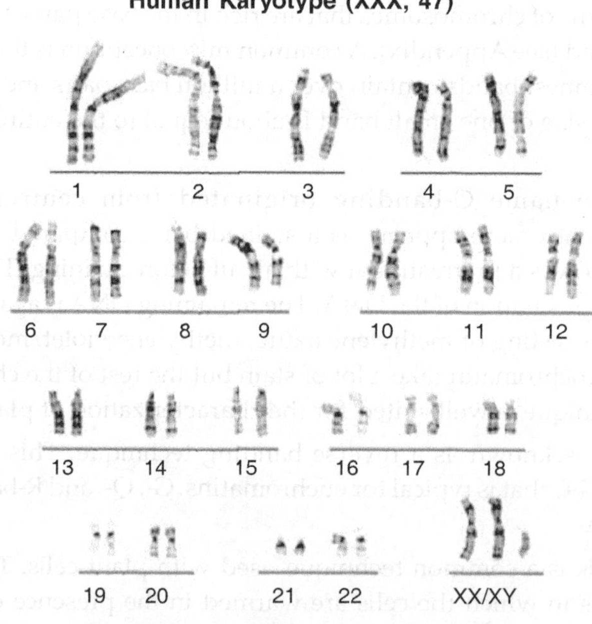

FIGURE 13.1 **Human karyotypes.**

Karyotypes are usually prepared from cells in which chromosomes can be readily distinguished, counted, and measured. Chromosomes at the mitotic metaphase, meiotic metaphase II, and pachytene of meiosis are best suited to make and evaluate the karyotypes. After taking the microphotograph of the complete chromosomes, a photograph karyotype may be prepared by cutting out the chromosomes from the microphotograph and arranging them in ordered pairs. A diagrammatic representation of karyotype is called an idiogram. It can be prepared by taking measurements and drawing the chromosomes with all their relative differences. An idiogram represents the diploid complement of chromosomes. It shows the number, size, and shape and allows easy comparison of chromosomes within the karyotype and also with other organisms.

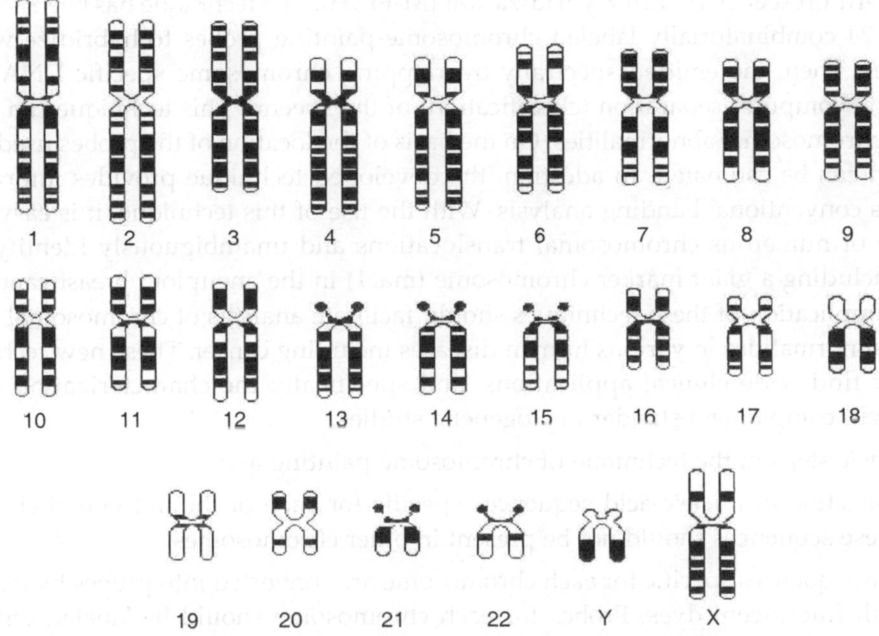

FIGURE 13.2 **An idiogram of human chromosomes.**

The position of the centromere with respect to the length of arms is called **arm ratio of the chromosomes**.

$$\text{Arm ratio} = \frac{\text{Length of Long arm}}{\text{Length of Short arm}}$$

Karyotyping is often used for the parental diagnosis and detection of variations in the chromosome number and structure, aberrations, and anomalies, which are the common cause of many congenital defects and spontaneous abortions.

Chromosome Painting

Chromosome painting is a powerful tool for chromosomal analysis. The technique of labeling chromosomes with different colored dyes is known as **chromosome painting**. This is carried out by the technique known as **FISH** or Fluorescent In Situ Hybridization.

FISH has been used to detect the location of specific genomic targets using probes that are labeled with specific fluorochromes. The technique allows detection of simple and complex chromosomal rearrangements. In addition, complex chromosomal abnormalities can be identified that could not be detected by the conventional cytogenetic banding techniques (see Appendix).

In the July 26, 1996 issue of *Science*, Schrock, et al. reported the development of a similar technique that allows the multi-color detection of human chromosomes[2]. The technique is known

[2] Schrock, et al., *Science, vol. 273, pp. 494-497, Jul. 26, 1996* "Multicolor Spectral Karyotyping of Human Chromosomes".

as **Multiplex-fluorescence In Situ Hybridization (M-FISH)**. The technique has been accomplished by allowing 24 combinatorially labeled chromosome-painting probes to hybridize with human chromosomes. Then, the emitted spectrally overlapping chromosome specific DNA probes are resolved using computer separation (classification) of the spectra. This technique can be used for detection of chromosomal abnormalities. On the basis of the location of the probes used, the size of the alteration can be estimated. In addition, the developed technique provides information that complements conventional banding analysis. With the use of this technique, it is easy to identify the presence of numerous chromosomal translocations and unambiguously identify structural alterations including a giant marker chromosome (mar1) in the aneuploid breast cancer cell line, SKBR3. The application of these techniques should facilitate analysis of chromosomal aberrations and genetic abnormalities in various human diseases including cancer. These new techniques will undoubtedly find wide clinical applications, and specifically the characterization of complex karyotypes will complement standard cytogenetic studies.

The basic steps in the technique of chromosome painting are:

1. Collection of nucleic-acid sequences specific for each of the individual chromosomes. These sequences should not be present in other chromosomes.
2. The sequences specific for each chromosome are converted into probes by labeling them with fluorescent dyes. Probes for each chromosome should be labeled with different (colors) fluorescent dyes.
3. *In situ* hybridization of each probe with the target chromosomes within the cells. Simultaneous hybridization with all probe set results in a chromosome spread preparation, in which each of the set of homologous chromosomes appears a different color when viewed with a fluorescent microscope.

Applications of Chromosome Painting

To find out the location of a specific gene located on a specific chromosome: FISH hybridization is done with the appropriate gene-specific probe labeled with the fluorescent dye. The test will give the binding of the gene-specific probe labeled with the fluorescent dye to the respective chromosome at the specific position, where the gene is located.

Detection of translocation: Chromosomes that have undergone translocation will have two segments. When it is subjected to the technique of chromosome painting it will take different probes and appear in two colors or multicolored depending on the number of translocations.

Detecting chromosome abnormalities: FISH has improved the efficiency of screening cells for chromosome abnormalities in mutagenic studies and for testing the mutagenic ability of chemicals and other potent mutagens in the environment. It has also improved the detection of chromosome aberrations and rearrangements associated with tumour and cancer.

To find out the chromosomal similarity between divergent species: Use of the same chromosome paint for chromosomes of different species reveals the extent of chromosome rearrangements since divergence of the species. Such studies reveal extensive synteny between fairly divergent species.

Clinical applications: With the use of this technique, it is possible to easily identify the presence of numerous chromosomal translocations and unambiguously identify structural alterations in cancer cell lines (for example, a giant marker chromosome (mar1) in the aneuploid breast cancer cell line, SKBR3). The application of these techniques should facilitate analysis of chromosomal aberrations and genetic abnormalities in various human diseases including cancer.

13.3 MUTAGENIC TECHNIQUES

Mutations are heritable changes that occur in the genome or DNA. Normally, these changes are harmful to the organism, a small percentage of mutations are useful in the process of evolution. Mutations can be induced with the help of different mutagenic agents and chemical or physical methods to create improved qualities in microbial systems and in plants for producing improved crop varieties. For example, improved agronomic characteristics such as disease resistance, salt tolerance, early flowering, pest resistance, early maturing, grain size, etc.

Bacterial Mutagenesis

The best-suited method for inducing mutation in microbes such as bacteria is radiation, particularly UV-radiation. The following is an experiment to induce mutation in bacteria with UV-radiation:

1. Make a bacterial culture by inoculating a bacterial colony (s strain of *e.coli*) to a small volume of broth culture (liquid medium; LB medium or the minimal medium) and grow the culture overnight in an incubator shaker at 37°C.
2. The next day, spread 0.1 ml of overnight culture on LB agar plates using a bent 'L' shaped glass rod (spreader) that has been flamed after being dipped in alcohol. Each plate should be labeled properly to avoid confusion before starting the experiment.
3. Now keep the agar plate containing the bacteria to be mutagenized under the UV lamp in a laminar flow chamber or hood. Remove the cover from the plate and close the door of the hood. Turn the UV lamp on and note the time of exposure of the bacteria to the UV light. (Exposure should be timed in seconds. You can find out the optimal time of exposure by repeating the experiment between 5 and 240 seconds to determine what is optimal.)

   The UV lamp is very strong. Do not expose your skin to the UV light. Do not operate the UV lamp with the hood open. UV light can cause serious and permanent damage to your eyes. The glass in the hood door will absorb the UV light. Never look at the UV lamp when it is on without wearing eye protection. At the end of the time turn the lamp off.
4. Replace the cover on the plate, remove it from the UV box, and place it in a 37°C incubator. Plates should be incubated upside down. This is important to prevent accumulation of condensation of moisture on the surface of the agar.
5. Check plates after 24 hours and count or estimate the number of colonies on the plate, and check for the type of mutant that you are looking for.

Depending on the type of mutant that you are looking for, prepare another set of the agar media plates to grow and select the mutants. If your aim is to get an auxotrophic mutant for Arginine (Arg-), you prepare agar media plates with minimal media, which contains only the essential components in the form of salts and elements and no organic components except the carbon source in the form of glucose. Now transfer the colonies from the master plates to the selection plate by replica plating. Take a circular filter paper that fits inside the petriplate and gently keep it over the colonies on the master plate. Slowly take the filter paper and put it in the selection media and allow growing overnight inside the incubator. By comparing the colonies in the selection media with that of the master plate, you can find out the colonies on the master plate, which cannot grow on the minimal media. This can be confirmed by monitoring the growth of the colonies again on the minimal media, which is supplemented with arginine (arg + plates).

Seed Mutagenesis

In the case of agricultural plants seeds are the parts that can be used for inducing mutations. We can use both chemical as well as physical mutagens for creating mutations. For this experiment, we can use a mutagenic chemical–EMS (ethyl methanesulphonate) for inducing mutations in wheat or any other experimental plants such as arabidopsis. Take a specific number of healthy seeds and soak them in water overnight. The next day, the water is blotted off with tissue paper or filter paper. Incubate the seeds in an aqueous solution of EMS of suitable concentration for about 2 hours at room temperature. Incubate some seeds in water under similar physical conditions and use them as the control for the experiment. Note the concentration and the time of treatment of the experiment. After the exposure, take the seeds out and wash them thoroughly with water to remove the traces of the mutagen. The treated seeds have to be sown in the controlled environment along with the control separately. After germination, the seedlings of the treated seeds can be compared with that of the control to evaluate the desired mutations. The experiment can be repeated with different strengths (concentration) of the mutagens and time of exposure.

13.4 RECOMBINATION IN BACTERIA

Genetic recombination is the transfer of DNA from one organism to another. The transferred donor DNA may then be integrated into the recipient's nucleoid by various mechanisms.

Mechanisms of genetic recombination include:

1. Transformation: DNA fragments (usually about 20 genes long) from a dead degraded bacterium bind to DNA-binding proteins on the surface of a competent recipient bacterium. Nuclease enzymes then cut the bound DNA into fragments. One strand is destroyed and the other penetrates the recipient bacterium. This DNA fragment from the donor is then exchanged for a piece of the recipient's DNA by means of Rec A proteins.
2. Transduction: Transfer of fragments of DNA from one bacterium to another bacterium by a bacteriophage.
   (*a*) Generalized transduction: During the replication of a lytic phage, the capsid sometimes assembles around a small fragment of bacterial DNA. When this phage infects another bacterium, it injects the fragment of donor bacterial DNA into the recipient where it can be exchanged for a piece of the recipient's DNA. Plasmids, such as the penicillinase plasmid of *Staphylococcus aureus*, may also be carried in a similar manner.

(*b*) Specialized transduction: This may occur occasionally during the lysogenic life cycle of a temperate bacteriophage. During spontaneous induction, a small piece of bacterial DNA may sometimes be exchanged for a piece of phage genome (that remains in the nucleoid). This piece of bacterial DNA replicates as a part of the phage genome and is put into each phage capsid. The phages are released, adsorbed into recipient bacteria, and injected into the donor bacterium DNA/phage DNA complex and into the recipient bacterium where it inserts into its nucleoid.

3. Bacterial conjugation: Transfer of DNA from a living donor bacterium to a recipient bacterium. In gram-negative bacteria, a sex pilus produced by the donor bacterium binds to the recipient. The sex pilus then retracts, bringing the two bacteria into contact. In gram-positive bacteria sticky surface molecules are produced that bring the two bacteria into contact. DNA is then transferred from the donor to the recipient.

(*a*) F^+ conjugation: This results in the transfer of an F^+ plasmid (coding only for a sex pilus) but not chromosomal DNA from a male donor bacterium to a female recipient bacterium. One plasmid strand enters the recipient bacterium while one strand remains in the donor. Each strand then makes a complementary copy. The recipient then becomes an F^+ male and can make a sex pilus. Other plasmids present in the cytoplasm of the bacterium, such as those coding for antibiotic resistance, may also be transferred during this process.

(*b*) Hfr (high-frequency recombinant) conjugation: An F^+ plasmid inserts or integrates into the nucleoid to form an Hfr male. The nucleoid then breaks in the middle of the inserted F^+ plasmid and one DNA strand begins to enter the recipient bacterium. The bacterial connection usually breaks before the transfer of the entire chromosome is completed so the remainder of the F^+ plasmid seldom enters the recipient. As a result, there is a transfer of some chromosomal DNA, that may be exchanged for a piece of the recipient's DNA, but not maleness.

(*c*) Resistance plasmid conjugation: This results in the transfer of a resistance plasmid (R-plasmid) from a donor bacterium to a recipient. One plasmid strand enters the recipient bacterium while one strand remains in the donor. Each strand then makes a complementary copy. The R-plasmid has genes coded for multiple antibiotic resistance and sex-pilus formation. The recipient becomes antibiotic resistant and male and is now able to transfer R-plasmids to other bacteria.

This is a big problem in treating opportunistic gram-negative infections (urinary tract infections, wound infections, pneumonia, septicemia) with such organisms as *e. coli, proteus, klebsiella, enterobacter, serratia,* and *pseudomonas,* as well as with intestinal infections by organisms such as *salmonella* and *shigella.*

Genes in bacteria can also be altered artificially through recombinant DNA technology. In recombinant DNA technology, endonucleases and ligase enzymes are routinely employed. Restriction endonuclease enzymes are naturally occurring enzymes in bacteria that help protect bacteria from viral attacks by cutting up the foreign viral DNA while not harming the bacterium's own DNA. Restriction endonuclease enzymes recognize specific palindromic deoxyribonucleotide base sequences (base sequences that read the same forward and backward on the complementary DNA strands), and then split each DNA strand at a specific site within that sequence. For example, *escherichia coli* makes a restriction endonuclease called *eco* R1 that recognizes the deoxyribonucleotide base sequence G-A-A-T-T-C and cuts the DNA strand between the G and the A. Since the complementary strand has the sequence CTTAAG, it is also cut between the G and the A. This leaves short, complementary, single-stranded sticky ends capable of hydrogen bonding with the complementary sticky ends of DNA fragments cut by the same enzyme.

Experiment to Carry Out Transformation

Bacterial transformation is routine work in all molecular biology laboratories as part of recombinant DNA experiment or gene cloning. In rDNA experiments or gene cloning, we prepare recombinant DNA or the gene or plasmid to be cloned, which has to be transferred to a host cell so that the DNA will multiply inside the bacterial cell. Transfer of the plasmid or the rDNA is carried out by bacterial transformation.

The first step is to select a suitable host cell such as a suitable strain of *e.coli* like DH5 α, a common strain available in all molecular biology laboratories, which can take foreign DNA easily. For this we have to treat the grown bacterial cultures at its log phase of growth, with $CaCl_2$. Centrifuge the cells growing at the log phase under low rpm (3,000–5,000 for 10 minutes) at 4°C and collect the cells. Suspend the cells in chilled $CaCl_2$ of 0.1 M. The cells in calcium chloride are able to accept the small DNA molecules. These cells in $CaCl_2$ can be stored for a long time under low temperatures such as –20 or –70°C.

Sudden exposure of this cell to the room temperature or higher can force the cell to take the DNA from outside. Take the stored competent cells, which are in the frozen condition and add the DNA sample to these cells and expose them to a higher temperature, at 42°C for two to three minutes. Some of these cells take the DNA from outside and will be transformed by intercalating with its genome. These cultures can be plated on a selection agar plate and the transformed colonies can be selected against the untransformed ones.

This transformation is extensively used in genetic engineering experiments. Any gene or DNA, before transferring into an organism, can be tested in a selected host by this transformation method. New promoters can be checked for their strength of expression. Commercially-useful enzymes and therapeutic proteins can be prepared in industrial scales. In short, any genetic engineering or gene cloning cannot be accomplished without bacterial transformation.

An Experiment of Conjugation

Take two strains of bacterial cultures of *e.coli*. One is the male strain or the F+, which is auxotrophic for biotin and methionine (Bio^-, Met^-). This bacterium can grow in the minimal medium, only if these two components are supplemented. Similarly, the female bacteria or the F^- strain is able to produce both biotin and methionine, but are auxotrophic for threonine and leucine (Thr^-, Leu^-). These bacteria cannot grow in the minimal medium unless the respective nutrients are supplemented.

But when these two populations are mixed and grown in media with only the salts and the carbon source (minimal media) some of the cells could grow without the supplementation of the additional amino acids and vitamin, biotin. This indicates that when grown together some female cells receive the functional genes of Thr and Leu from the male strains by conjugation. Similarly, some of the male strains receive the functional genes for biotin and methionine from the female strain. These new genetically transformed strains can grow in the minimal medium without any additional nutrient supplementation.

This is a common method of natural recombination in bacteria resulting in the formations of variants. In this context it is very important because conjugation can produce drug resistance among pathogenic bacteria. Therefore, the mechanism of conjugation has to be clearly understood and should be aware that in heterogonous cultures there is the chance of bacterial conjugations and genetic recombinations resulting in new strains with new weapons. Contaminated laboratory cultures, organic factory effluents, and sewage water are good media for bacterial conjugations.

Significance of Genetic Recombination in Bacteria

All the recombination methods in bacteria were discovered in laboratories accidentally. Since microbial populations are in close interaction with humans, these genetic transformations have great significance in human life and environment, particularly in the era of genetic modifications. The occurrence of a high degree of variability seen among some pathogenic organisms such as mycobacterium and developments of drug and antibiotic resistance are due to these mechanisms. There is the presence of restriction endonuclease, the enzymes specialized for destroying the foreign DNA entering to its cytoplasm. There wouldn't be any need of these enzyme systems if these genetic recombination mechanisms were absent. The genome sequencing projects have revealed that the gene sequences of unrelated organisms have a great degree of homology, which suggests that in the past, genes were moved from one species to another through any of the natural mechanisms, the genetic engineering methods of nature, explained above.

13.5 BREEDING METHODS IN PLANTS

Breeding plants to create new varieties and improve upon old ones is a hobby that nearly everyone can engage in. The crossing techniques are easy to learn and breeders can experiment with many kinds of plants. Generally, amateur plant breeders work with traits that are fairly easy to change—for example, flower color, fruit shape, or plant size. Nevertheless, although experiments may be simple, it is possible to produce unusual or beautiful plants. In order to breed plants successfully it is important to understand the principles of plant reproduction, which have been discussed in earlier chapters. The purpose of this is to explain the simple techniques that can be used to produce new varieties or strains of plants.

Plant Selection

The first step in the plant hybridization procedure is the selection of parent plants with the desired characteristics. Plant characteristics can be changed after many generations by a process of selection. There are two types of selection—natural and artificial. **Natural selection** is the process that occurs in nature whereby strong and well-adapted plants survive while weak and poorly adapted plants eventually die out. This process has taken place since the beginning of life on earth and it is still occurring in nature. **Artificial selection** is the process that humans use to obtain more desirable types of plants. Thousands of years ago people learned that saving seed from the kind of plant they wanted to continue growing would increase the chances of getting a plant similar to the original. But our ancestors didn't know what their chances of success were nor did they understand the processes by which traits were changed or maintained. It wasn't until the eighteenth and nineteenth centuries that humans began to understand the laws of heredity and the processes of plant reproduction. Even today, these fundamentals aren't completely understood. But enough is known so that we can select plants for breeding with considerably more assurance of success than our ancestors did.

In our experiment we have to select the parents for the process of hybridization. The plants selected for breeding should be sturdy and healthy. It is usually easier to tell which ones are healthy after a few flowers on the plant have bloomed. Some plants have natural barriers to cross-or self-pollination. It is advisable to check for this before breeding, for although barriers can be overcome, some plants cannot be artificially pollinated. An example of a barrier that cannot be overcome is the self-pollination prohibitor of some orchids; the stigmas of certain orchids produce a substance, which kills the pollen of flowers of the same plant. The mechanism that performs this cannot be removed without destroying the pistil. In choosing a pollen parent (male parent), select one that has a heavy yellow powder on the anther. This powder is the pollen. In choosing a seed parent, examine the stigma. It should have either a glistening substance on it that is sticky to the touch or a "hairy" surface. It is this substance or surface that retains the pollen, thus making fertilization possible. Once the seed parents and pollen have been selected, you are ready to begin pollination.

Hybridization

The equipment required for plant breeding is relatively inexpensive and easy to use. Here are some useful items:

- Magnifying glass (10 or 15 power)
- Tweezers
- Small sharp-pointed scissors
- Camel-hair brush
- Small containers or vials
- Alcohol
- Rubber bands or soft wire
- Paper or cellophane bags
- Paper clips
- Tags
- Notebook

The first step is to mark those flowers that are to serve as pollen parents and those that are to serve as seed parents. This can be done with colored thread, one color for the male and another color for the female. Or paper labels can be used, covered with varnish to protect them from the weather.

Emasculation

The next step is to protect the plant from unwanted pollen. If the plant is to be cross-pollinated, the stamens will have to be removed to prevent the possibility of self-pollination or selfing. The removal of the stamen is called **emasculation**. It should be performed before the anthers split open to release pollen. This may require opening the flower by hand before it is ready to bloom. Emasculation can be accomplished by: (1) pinching off the stamens or anthers with tweezers, or (2) snipping off the stamens or anthers with sharp-pointed scissors, or (3) removing the petals to which the stamens are sometimes attached. A magnifying glass is particularly useful in emasculation.

Protecting the Plant from External Influence

Both the seed and pollen parents should be protected from contamination by foreign pollen. This can be done by one of the following methods:

Closing the flower

In many flowers, such as morning-glories, petunias, and lilies, the petals can be closed around the floral organs with a piece of soft wire, string, or rubber band. Care should be taken not to tear the petals.

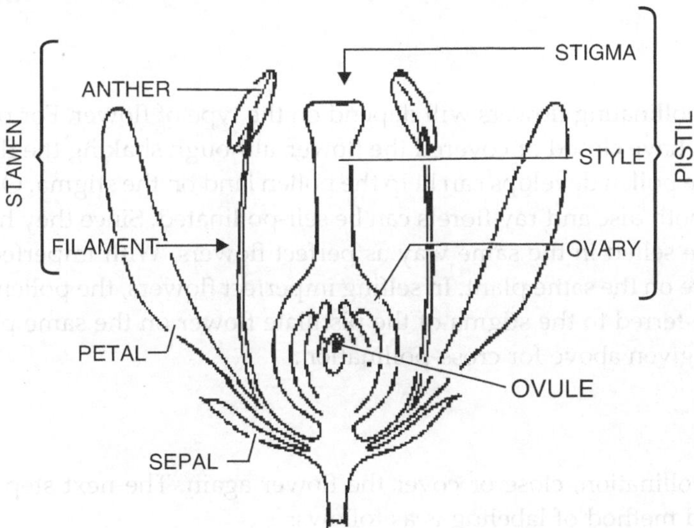

FIGURE 13.3 **Parts of a flower. The flower shown here is a perfect flower; that is, it has male and female reproductive organs. The stamen is the male organ and the pistil is the female organ.**

Covering the flower

Some flowers, such as composite flowers, cannot be closed. To protect them from unwanted pollen, you can cover the flower with a paper bag. Or, if you wish to observe the flower at all times, you can cover it with a cellophane bag. The bag should be held securely in place with a paperclip or string.

Pollination

Crossing

There are several methods that can be used for cross pollinating flowers. Here are the most common methods.

Place the stamens in a small container. Remove the protector from the seed parent. Holding a stamen with the tweezers, gently brush the anther across the stigma. Replace the protector.

If the stamens are too small or too difficult to grasp, you can cross-pollinate by transferring the pollen from the anthers with a brush into a container and then brushing the pollen on to the stigma.

Shake the bagged pollen parent so that the pollen is collected in the bag that is covering it for protection. Remove the bag from the pollen parent, being careful not to spill the pollen. Remove the protector from the seed parent and place the bag containing the pollen over the seed parent and shake the bag so that pollen falls on the stigmatic surfaces. This is usually done on corn.

Each time you use different pollen, be sure to first wash with alcohol the camel-hair brush, tweezers, and any other item that might have touched some pollen. This step is very important to prevent pollination of the seed parent with unwanted pollen that has adhered to the equipment. After washing the instruments be sure that they are dry before using them again.

Selfing

Procedures for self-pollinating flowers will depend on the type of flower. For perfect flowers, your job is done once you have closed or covered the flower, although shaking the flower once a day for several days after the pollen develops can help the pollen land on the stigma. Only those composite flowers containing both disc and ray florets can be self-pollinated. Since they have both pistils and stamens, they can be selfed in the same way as perfect flowers. With imperfect flowers, self only those flowers that are on the same plant. In selfing imperfect flowers, the pollen from the staminate flower must be transferred to the stigma of the pistillate flower on the same plant. To do this, use any of the methods given above for cross-pollination.

Labeling

Immediately after pollination, close or cover the flower again. The next step is to label the seed parent. The standard method of labeling is as follows:

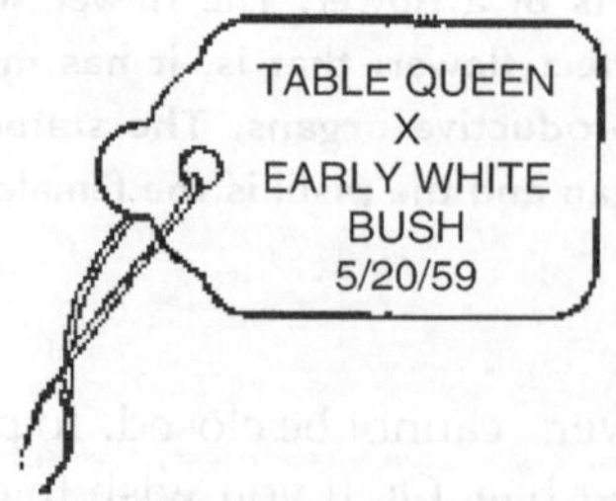

FIGURE 13.4 **A metal tag commonly used in plant-breeding experiments.**

Write on the label in the following order: (a) the number that you have assigned or the variety name of the seed parent; (b) the letter X; (c) the number or variety name of the pollen parent; and (d) the date of the cross was made (Figure 13.4). Attach the label to the stem just below the flower that has been pollinated.

Once the seed parent is labeled, the next step is to record the cross or self in a notebook. Keeping complete and accurate records of breeding operations is very important. The recorded information should contain all essential facts regarding the cross or self so that you can refer to it at a later time and even do the job again from the beginning, if necessary. A separate form or page should be used for each cross or self. An easy way to keep track of the offspring is to assign consecutive numbers to each generation resulting from each crossing or selfing.

Collection of Seeds

The protector or the bag cover used to avoid unwanted pollens, can be used to collect the seeds if the seeds are very small.

Screening

Collected seeds are screened for the desired characters by generating seedlings. These plants can be used for the next steps of experiments.

Back Crossing

It is very important step. The new progeny crossed with the parental varieties until the desired characteristic that is present in one of the parents is achieved.

Uniformity Checking and Establishment

This is the final step in which the hybrid plant is verified for that specific characteristic by suitable methods such as selfing or establishing the stability of the characteristic by self-pollination or by vegetative propagation to safeguard the characteristic segregation during gamete formation. The genetical, morphological, and biochemical methods can be adopted to evaluate the plant for stabilization of the desired characteristic.

Hybrid Vigor

The progenies produced by crossing two known, but different varieties of parents will perform better than both the parents. This phenomenon shown by the hybrid organisms is known as heterosis or hybrid vigor. Lots of crop plants and horticultural plants including the garden plants are exploiting this phenomenon of hybrid vigor.

13.6 PEDIGREE ANALYSIS IN HUMANS

Humans are unique among organisms in many ways. But one way, which is near and dear to a geneticist's heart, is that humans are not susceptible to genetic experimentation. In practice, we humans actually share this characteristic with many long-lived organisms who delay first births. In short, it is not terribly convenient to perform experimental crosses if one has to wait 15 years between generations. However, for humans, one also has to add that our system of morality uniquely does not allow such experimentation on humans. This is an unfortunate state of affairs since there is no other organism for which practical knowledge of their genetics would be more useful, especially in the case of the genetics of heritable diseases. It has been found that human genetics may readily be inferred so long as good records have been kept within large families. This formal mechanism of inference is called **pedigree analysis**. In this section we will discuss many aspects of human genetics, in particular considering strategies of **pedigree analysis** whereby we will attempt to infer the genetics of human conditions based on knowledge of marriage (mating) and affliction in large extended families.

Symbols Used in Pedigree Charts

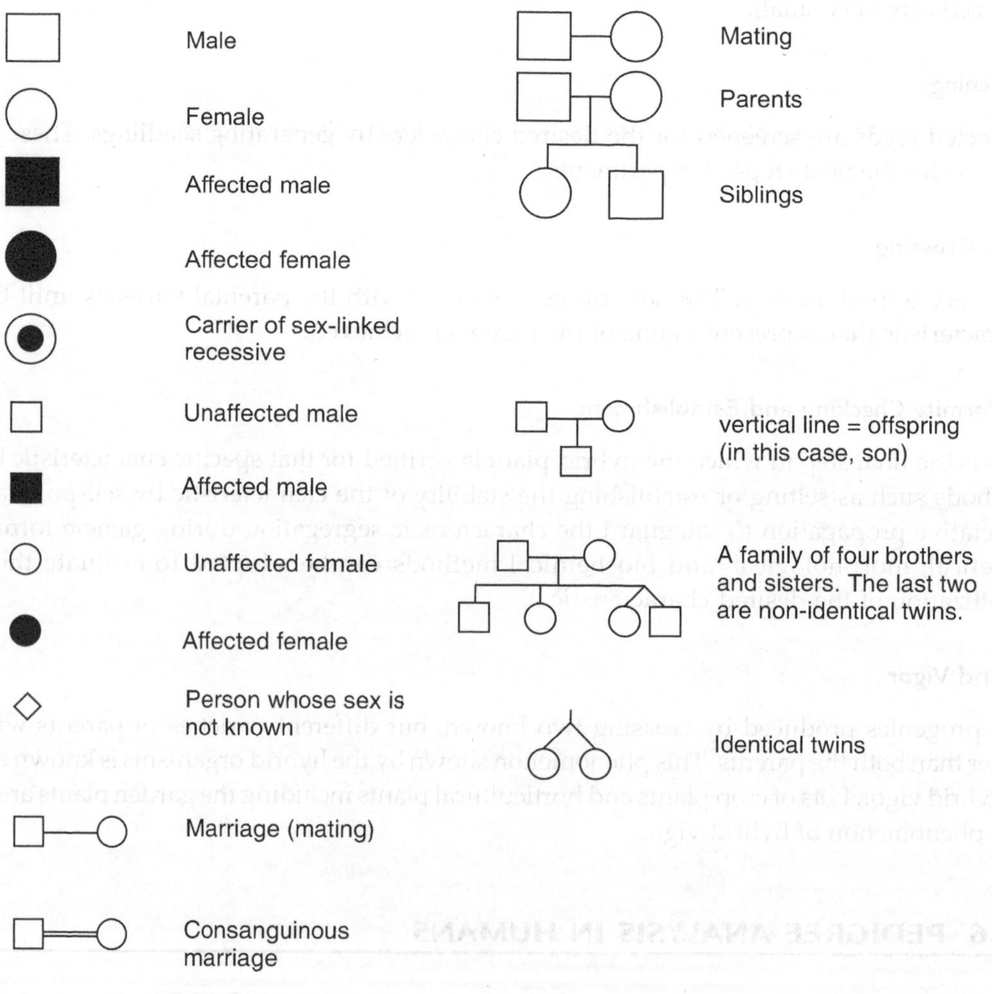

FIGURE 13.5 **Various symbols used in pedigree charts.**

Pedigree Analysis Procedures

Pedigree analysis is one of the central tasks of the human geneticist. It involves the construction of family trees. Family history information is often collected at major family gatherings. A pedigree is used to trace inheritance of a trait over several generations.

Three primary patterns of inheritance in man are the following:

- Autosomal recessive
- Autosomal dominant
- Sex-linked (X-chromosomal)

Autosomal Dominant Inheritance

A dominant condition is transmitted in unbroken descent from each generation to the next. Most matings will be of the form M/m × m/m (i.e., heterozygote to homozygous recessive). We would therefore expect every child of such a mating to have a 50% chance of receiving the mutant gene and thus of being affected. A typical pedigree might look like this:

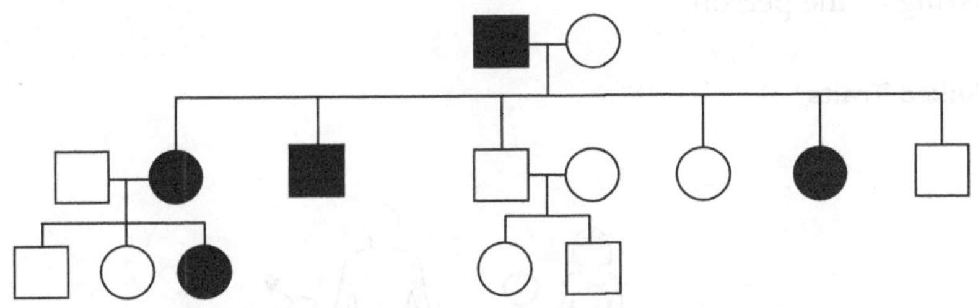

FIGURE 13.6 **Inheritance of an autosomal dominant trait (i.e., Huntington's disease).**

In this pedigree, an affected father passes the trait to half of his six children, including two daughters and a son. One of the daughters passes the same trait to one of her three children.

Autosomal Recessive Pedigree

A recessive trait will only manifest itself when homozygous. If neither parent has the characteristic phenotype (disease) displayed by the child, the trait is recessive. If it is a severe condition it is unlikely that homozygotes will live to reproduce and thus most occurrences of the condition will be in matings between two heterozygotes (or **carriers**). An autosomal recessive condition may be transmitted through a long line of carriers before, by the ill chance, two carriers mate. Then there will be a one fourth chance that any child will be affected.

A typical pedigree of recessive autosomal inheritance is given in Figure 13.7:

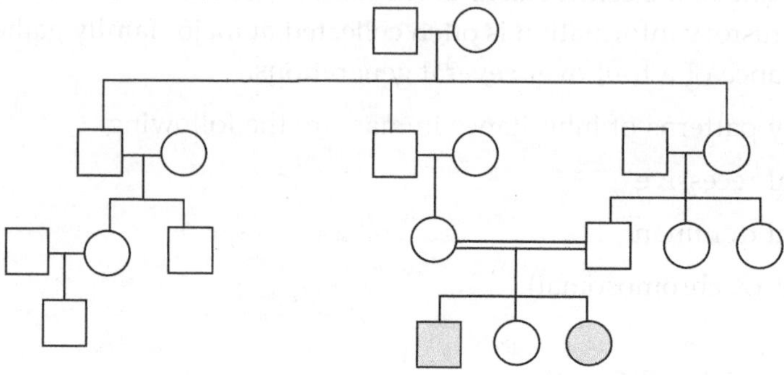

FIGURE 13.7 **Inheritance of an autosomal recessive trait.**

If the parents are related to each other, perhaps by being cousins, there is an increased risk that any gene present in a child may have two alleles **identical by descent**. The degree of risk that both alleles of a pair in a person are descended from the same recent common ancestor is the degree of **inbreeding** of the person.

Pedigree of Sex-linked Traits

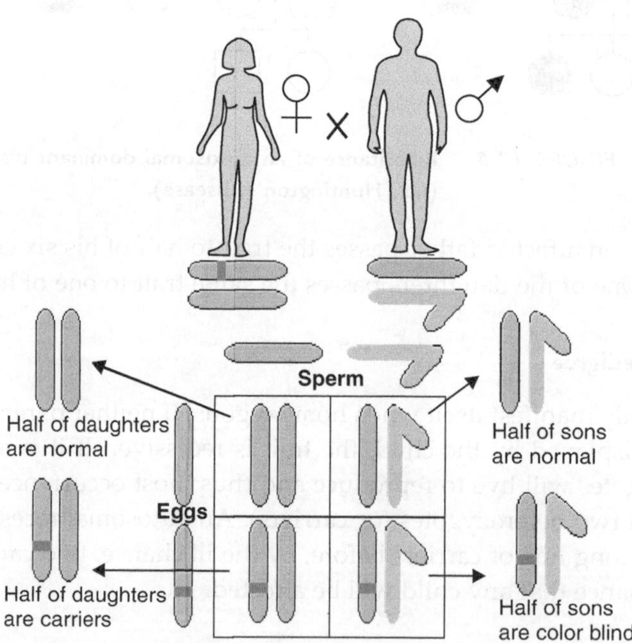

FIGURE 13.8 **Inheritance of red-green color blindness.**

The transmission of X-linked traits is in a zigzag manner. Females transmit X chromosomes to both sons and daughters. Males transmit the X chromosomes only to daughters and Y chromosomes to sons. The X-linked traits, which are recessive are preferentially seen in males, who are always hemizygous for the X chromosomes. Females are heterozygous and form the "carriers" of that trait. Most X-linked traits are recessive. Inheritance of red-green color blindness, an X-linked, recessive trait is represented in Figure 13.8.

A color-blind man is the father of "carrier" daughters and normal sons. Carrier daughters have a 50% chance to have color-blind sons and a cross between a color-blind male and carrier female can produce color-blind daughters. Hemophilia also has the same type of inheritance. Duchenne muscular dystrophy is another example of X-linked inheritance.

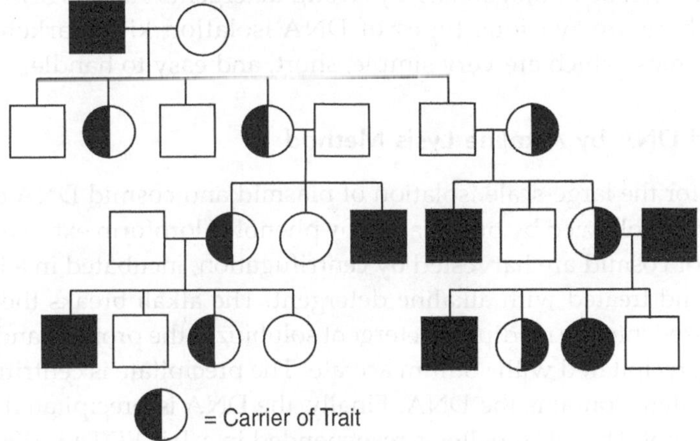

FIGURE 13.9 **Inheritance of X-linked traits. (Example: red-green color blindness).**

Transmission of Y-linked Genes

Men are homozygous for Y-linked genes present on the non-homologous parts. All these genes will be expressed in all conditions. These genes are always transmitted from father to sons and never to daughters. There are no essential genes in the Y-chromosomes except the locus for the maleness and fertility.

Pseudoautosomal Inheritance

There are some homologous regions in the X chromosomes and Y chromosomes. These homologous parts pair during meiosis and may undergo crossing over. Therefore, genes in these homologous regions show inheritance similar to autosomal genes and are called **pseudoautosomal inheritance**. Such genes or characters are very rare.

13.7 DNA ISOLATION

DNA isolation and purification is a routine technique used in all laboratories engaged in molecular biology experiments. There are a number of standardized techniques and variations, which can be adopted according to the type of cells or tissues. The isolation and purification methods used in earlier times were lengthy and tiresome with the use of ultracentrifugation. But now with the advancement of separation techniques, the procedure is very simple and short. In any method of extraction and purification, there are three main steps:

1. Breaking of the cells
2. Extraction of DNA
3. Purification

Cells can be broken in different ways. One common method for lysis of bacterial cultures is alkaline lysis. In the case of animals, cells can be lysed by simple detergents or by hypotonic solutions. Plant tissues can be homogenized by strong detergents such as SDS and heating at high temperatures. But there are various types of DNA isolation kits marketed by a number of biotechnology companies, which are very simple, short, and easy to handle.

Isolation of Plasmid DNA by Alkaline Lysis Method

The method is used for the large-scale isolation of plasmid and cosmid DNA by a modification of alkaline lysis procedure, followed by purification by phenol chloroform extraction. Cells containing the desired plasmid or cosmid are harvested by centrifugation, incubated in a lysozyme buffer (re-suspension buffer), and treated with alkaline detergent. The alkali breaks the cells and the DNA and proteins are released into the medium. Detergent solubilizes the proteins and DNA. The proteins and membranes are precipitated with sodium acetate. The precipitate is centrifuged out at a higher RPM and the supernatant contains the DNA. Finally, the DNA is precipitated out by adding 95% ethyl alcohol or propanol. The DNA pellet is resuspended in a Tris EDTA buffer. This DNA sample contains some DNA-binding proteins, which have to be removed. This can be carried out by phenol-chloroform extraction. There are several variations to this protocol, which is suited to the situations and type of bacterial cultures.

Phenol Chloroform Purification

In this method, the DNA pellet is resuspended in RNase buffer or RNase A is added to the diluted solution of DNA. The RNase should be heated to remove any traces of DNase present. Nuclease treatment is necessary to remove the RNA by digestion since RNA will be precipitated along with DNA. After RNase treatment, the proteins present in the sample can be precipitated by adding and mixing with an equal volume of phenol equilibrated with buffer. Again, the precipitate can be separated by centrifugation at a higher RPM (10,000 to 15,000). This step can be repeated two or three times, and finally with a mixture of chloroform and isobutyl alcohol. Finally, the DNA can be precipitated with ethyl alcohol. The DNA precipitate can be dried and dissolved in the minimum quantity of TE buffer. The DNA recovery and its purity can be measured by taking absorbance readings at 260 and 280 nanometers (nm). The ratio of absorbance at 260 by absorbance at 280 gives the value of its purity. The ratio should be two or below two for a pure sample of DNA without any protein contamination. After concentration, the DNA is assayed by restriction digestion and can also be used for other experiments such as cloning or rDNA experiments.

Phenol Extraction of DNA Samples

Phenol extraction is a common technique used to purify a DNA sample (1). Typically, an equal volume of TE-saturated phenol is added to an aqueous DNA sample in a microcentrifuge tube. The mixture is vigorously vortexed, and then centrifuged to enact phase separation. The upper, aqueous layer is carefully removed to a new tube, avoiding the phenol interface, and then is subjected to two ether extractions to remove residual phenol. An equal volume of water-saturated ether is added to the tube, the mixture is vortexed, and the tube is centrifuged to allow phase separation. The upper, ether layer is removed and discarded, including phenol droplets at the interface. After this extraction is repeated, the DNA is concentrated by ethanol precipitation.

Protocol

1. Add an equal volume of TE-saturated phenol to the DNA sample contained in a 1.5 ml microcentrifuge tube and vortex for 15 to 30 seconds.
2. Centrifuge the sample for five minutes at room temperature to separate the phases.
3. Remove about 90% of the upper, aqueous layer to a clean tube, carefully avoiding proteins at the aqueous-phenol interface. At this stage the aqueous phase can be extracted a second time with an equal volume of 1:1 TE-saturated phenol-chloroform, centrifuged and removed to a clean tube as above. But this additional extraction is usually not necessary if care is taken during the first phenol extraction.
4. Add an equal volume of water-saturated ether, vortex briefly, and centrifuge for three minutes at room temperature. Remove and discard the upper, ether layer, taking care to remove phenol droplets at the ether:aqueous interface. Repeat the ether extraction.
5. Ethanol precipitate the DNA by adding 2.5 to 3 volumes of ethanol-acetate, as discussed as follows.

Concentration of DNA by Ethanol Precipitation

Typically, 2.5 to 3 volumes of an ethanol/acetate solution is added to the DNA sample in a microcentrifuge tube, which is placed in an ice-water bath for at least ten minutes. Frequently, this precipitation is performed by incubation at –20°C overnight. To recover the precipitated DNA, the tube is centrifuged, the supernatant discarded, and the DNA pellet is rinsed with a more dilute ethanol solution. After a second centrifugation, the supernatant again is discarded, and the DNA pellet is dried in a vacuum.

Protocol

1. Add 2.5 to 3 volumes of 95% ethanol/0.12 M sodium acetate to the DNA sample contained in a 1.5 ml microcentrifuge tube, invert to mix, and incubate in an ice-water bath for at least ten minutes. It is possible to place the sample at –20°C overnight at this stage.
2. Centrifuge at 12,000 rpm in a microcentrifuge for 15 minutes at 4°C, decant the supernatant, and drain inverted on a paper towel.

3. Add 80% ethanol (corresponding to about two volume of the original sample), incubate at room temperature for five to ten minutes and centrifuge again for five minutes, and decant and drain the tube, as above.
4. Place the tube in a vacuum and dry the DNA pellet for about five to ten minutes, or until dry.
5. Always dissolve dried DNA in 10 mM Tris-HCl, pH 7.6-8.0, 0.1 mM EDTA (termed 10:0.1 TE buffer).
6. It is advisable to aliquot the DNA purified in large-scale isolations (i.e., 100 μg or more) into several small (0.5 ml) microcentrifuge tubes for frozen storage because repeated freezing and thawing is not advisable.

Elution of DNA Fragments from Agarose

DNA fragments are eluted from low-melting temperature agarose gels. Here, the band of interest is excised with a sterile razor blade, placed in a microcentrifuge tube, frozen at –70°C, and then melted. Then, TE-saturated phenol is added to the melted gel slice, and the mixture is again frozen and then thawed. After this second thawing, the tube is centrifuged and the aqueous layer removed to a new tube. Residual phenol is removed with two ether extractions, and the DNA is concentrated by ethanol precipitation.

Protocol

1. Place excised DNA-containing agarose gel slice in a 1.5 ml microcentrifuge tube and freeze at –70°C for at least 15 minutes, or until frozen. It is possible to pause at this stage in the elution procedure and leave the gel slice frozen at –70°C.
2. Melt the slice by incubating the tube at 65°C.
3. Add one-volume of TE-saturated phenol, vortex for 30 seconds, and freeze the sample at –70°C for 15 minutes.
4. Thaw the sample, and centrifuge in a microcentrifuge at 12,000 rpm for five minutes at room temperature to separate the phases. The aqueous phase then is removed to a clean tube, extracted twice with equal volume ether, ethanol precipitated, and the DNA pellet is rinsed and dried.

Genomic DNA Isolation from Blood

Genomic DNA isolation is performed according to the standard protocol suggested by Federal Bureau of Investigation, USA. After the blood samples (stored at –70°C in EDTA vacutainer tubes) are thawed, a standard citrate buffer is added, mixed, and the tubes are centrifuged. The top portion of the supernatant is discarded and additional buffer is added, mixed, and again the tube is centrifuged. After the supernatant is discarded, the pellet is resuspended in a solution of SDS detergent and proteinase K, and the mixture is incubated at 55°C for one hour. The sample then is phenol-extracted once with a phenol/chloroform/isoamyl alcohol solution, and after centrifugation the aqueous layer is removed to a fresh microcentrifuge tube. The DNA is ethanol-precipitated, resuspended in buffer, and then ethanol-precipitated a second time. After the pellet is dried, buffer is added and the DNA is resuspended by incubation at 55°C overnight, and the genomic DNA solution is assayed by the polymerase chain reaction.

DNA Isolation from Plant Tissues

Plant tissues bring up several problems during DNA isolation. Plant cells have a rigid cell wall and the tissue contains a number of toxic metabolites, which can interact with the DNA and change its nature and make it useless for other experimental purposes. Metabolites such as mucilage and other carbohydrates can very easily form complexes with DNA and it can be damaged. Therefore, the extraction buffer should be supplemented with some compounds that can protect the DNA against these metabolites.

Many DNA-isolation techniques widely employed by plant molecular biologists use a **CTAB (Cetyltrimethylammonium bromide) extraction buffer.** This compound forms a complex with DNA and thus protects it from other toxic metabolites such as mucilage and phenolics. The procedure involves the following essential steps:

1. The plant tissue should be very thin and soft. The plant tissue can be powdered in liquid nitrogen with mortar and pestle. The fine powder is then transferred to a centrifuge tube.
2. Add the extraction buffer, which contains SDS and CTAB, and mix very well, incubate at 60°C for one hour in a water bath. Now centrifuge the solution and remove the precipitate, and transfer the supernatant to another centrifuge tube.
3. Extract this solution with chloroform-isoamyl alcohol. All the organic and lipids will separate to the chloroform layer and DNA will be in the aqueous layer. Take the aqueous layer and again extract it with chloroform and repeat the step.
4. Finally, the aqueous layer containing DNA is precipitated with ice-cold isopropanol or ethyl alcohol. The DNA should appear as white or creamy strands or fibers and can be separated or taken out with a glass rod. Or centrifuge it and collect the pellet, air-dry the pellet to remove the excess alcohol and dissolve in a minimum volume of Tris EDTA buffer.

The DNA, isolated and purified by any of these methods, can be used for a variety of experimental purposes. It can be used for restriction digestion analysis, cloning, ligation, transformation experiments, *in vitro* transcription, PCR amplification, RFLP (restriction length polymorphism), fingerprinting, RAPD (random amplification polymorphic DNA), sequencing, nick translation and radiolabeling, preparation of genomic DNA library and cDNA library, etc.

REVIEW QUESTIONS

1. What are the various kinds of recombinations in bacteria?
2. What are the various steps involved in the hybridization of two different genotypes?
3. What is transduction? Explain various types of transduction.
4. Describe the various steps involved in bacterial conjugation.
5. What is a karyotype? How it is different from an idiogram?
6. Explain the technique of chromosome banding.
7. Differentiate between Q-banding and C-banding.

8. Differentiate between FISH and chromosome painting.
9. What are the various uses of chromosome painting?
10. Explain the practical uses of bacterial transformation.
11. What is the significance of bacterial genetic recombination?
12. Define the following:
    (*a*) Heterosis
    (*b*) *In situ* hybridization
    (*c*) R-banding
    (*d*) Giemsa stain
    (*e*) Emasculation
    (*f*) Pedigree analysis
    (*g*) Plasmid DNA
    (*h*) G-banding
    (*i*) Sex-linked inheritance
    (*j*) Holandric genes or inheritance
    (*k*) Hemizygous
13. Explain the zigzag inheritance with respect to red-green color blindness.
14. Explain the inheritance of sickle cell anemia.
15. Explain a suitable method for the isolation and purification of plasmid from bacterial cultures.
16. Expand the following:
    (*a*) EDTA
    (*b*) CTAB
    (*c*) SDS
    (*d*) FISH and M-FISH
    (*e*) RAPD
    (*f*) RFLP
17. Explain the clinical application of M-FISH or chromosome painting.

Part 5

Protein and Gene Manipulations

This part is all about proteins, genetic engineering, genomics and bioinformatics, and accordingly this part is divided into four chapters. The first chapter gives more information about proteins, their structure-function relationships, and the role of proteins in the biological systems and their commercial applications. It discusses the primary, secondary, tertiary, and quaternary structures of proteins, the molecular forces involved in the structural stabilization of the protein molecules and their involvement in the functional properties of proteins. The altered-protein structure and its relationship with certain diseased states is also described. The commercial application of proteins in different areas such as medical and pharmaceutical, food and agriculture, and other industries, its commercial extraction and purification, etc., is also discussed. The last part of this chapter deals with the proteomics, an upcoming and very promising field, its applications, and various techniques involved in its studies and applications.

The second chapter is about recombinant DNA technology (rDNA Technology). The tools of recombinant-DNA techniques such as restriction enzymes or molecular scissors, various types of plasmids and vectors, and their use in the technique of gene cloning genetic engineering are introduced. Construction of DNA libraries, transformation of host cells by recombinant DNA, screening and identification of recombinants or genetically-transformed cells are also introduced in this chapter. The other important techniques such as Polymerase Chain Reaction (PCR), DNA hybridization reactions such as Southern hybridization, DNA sequencing, and site-directed mutagenesis are discussed in the last part of the chapter.

The third and the fourth chapters discuss genomics and bioinformatics and their impact on modern biotechnology. The various genome-sequencing projects and data generated from those have enriched genomics and bioinformatics and become an important part of modern biotechnology. These chapters also deal with functional genomics, comparative genomics, and SNPs in addition to the general bioinformatics and the tools commonly used in genomics.

Chapter 14 PROTEIN STRUCTURE AND ENGINEERING

In This Chapter

14.1 INTRODUCTION TO THE WORLD OF PROTEINS

Proteins are essential to maintain the structure and function of all lifeforms. The word 'protein' itself is derived from the Greek word *protos*, meaning "primary" or "first." Proteins are vital for the growth, repair, and maintenance of muscles, blood, internal organs, skin, hair, and nails, and their functions are endless. Each and every property that characterizes a living organism is affected by proteins, whether it is a bacteria or a human body. Nucleic acids, another major biological macromolecule, are also essential for life; they encode genetic information—mostly specific for the structure of proteins—and the expression of that information depends almost entirely on proteins. The fertilization of an egg with a sperm and the development and differentiation of the resultant zygote into a fully developed organism and its growth and maintenance of life activities up to its death is controlled and programed by a large number of proteins.

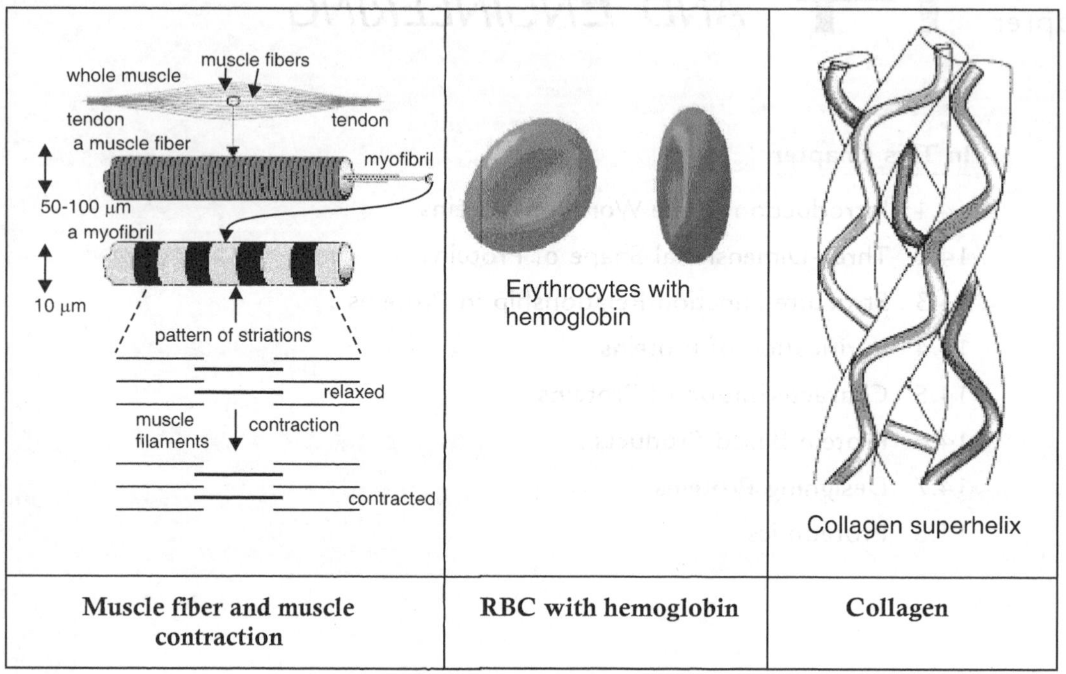

FIGURE 14.1 **Various roles of proteins.**

In our body, when we breath, oxygen present in lungs will be taken by the hemoglobin present in the RBC of blood to the various cells of the system for the process of cellular respiration. Movements and activities of body parts and systems including lungs, heart, stomach, etc. are happening due to the contractions and relaxations of various types of muscles. Myosin, actin, and collagen are the protein molecules involved in body structure, protection, and muscular contraction and relaxation. The structure of cells, and the extracellular matrix in which they are embedded, is largely made of protein. Plants and many microbes depend on carbohydrates such as cellulose for support. All biological activities of cells are mediated and regulated by a large number of catalytic proteins called enzymes. The function of the human brain and the speed at which the electric impulses are generated and transmitted to coordinate various activities of the systems are meticulously done by a large number of proteins that act as enzymes and receptors. The receptors and hormones are another class of proteins, which act as signal molecules that are involved in the coordination of different metabolic functions of the system. There are the proteins called transcription factors, which turn the genes on and off, to guide the differentiation of the cells and development, and there are many more activities in which proteins are involved. Thus, proteins are diverse in their functions and are truly the physical basis of life. To understand the diversity in the biological function of proteins, their molecular structure and shape has to be studied in detail, since the function is closely related to the structure.

The absence or malfunctioning of one or more proteins in the system can cause serious life-threatening diseases. The malfunctioning of proteins can be traced to some type of structural abnormality due to variations in the chemical composition. For example, the absence of one of the subunit, beta chain of the oxygen-carrier protein hemoglobin of RBC, can cause thalassaemia. This metabolic error due to abnormal hemoglobin affects many children who can only survive on repeated blood transfusion. Another type of abnormal hemoglobin is where the beta chain is mutated and the glutamic acid at position six is replaced with valine and results in deformed RBC and a condition known as sickle cell anemia. The absence of an enzyme adenosine-deaminase, an important enzyme in nucleotide metabolism, can cause the disease known as SCID (severe combined immunodeficiency) in children. These children cannot survive infancy. There are some types of infectious protein particles known as 'prions', which can turn normal proteins to rogue proteins or incorrectly shaped proteins and can cause diseases such as mad cow disease. To understand more about the relationship between the disease and the structural abnormality of the protein we should know more about the structure and its relationship with biological activity.

In spite of these diverse biological functions, proteins have relatively homogeneous compositions. All proteins are linear polymers of the same 20 types of amino acids in different combinations. The major difference between proteins is in the sequence in which the amino acids are assembled into polymeric chains. The secret to their functional diversity lies partly in the chemical diversity of the 20 amino acids, but primarily in the diversity of the three-dimensional structures that these amino acid building blocks can form by linking in different sequences. The amazing functional properties of proteins can be understood only in terms of their relationship to the three-dimensional structures of proteins.

Now we know that the amino acid sequence of a protein and thereby its three-dimensional structure is specified by a gene. But, this is not completely true. Even though the gene sequence specifies the amino acid sequence of the protein, the three-dimensional structure is also influenced by a number of other factors. The number of proteins produced in a system always exceeds the number of genes. The *Human Genome Project* has announced the presence of about 35,000 genes. But the actual number of proteins encoded by these genes exceeds the number of genes. This is mainly because of the various types of molecular modifications such as deletion of amino acids, chemical modifications of certain amino acids, addition of other macromolecules and groups such as phosphate groups (phosphorylation), acetyl groups (acetylations), sugar and other types of carbohydrates (glycosylations), lipids, etc. All these chemical modifications of proteins just after their formation are collectively called **post-translational modifications**. Actually, these post-translational modifications are responsible for the diversity in the three-dimensional structures and functions along with the amino acid sequence prescribed by the respective genes. A number of proteins are expressed in all cells irrespective of their functional specialization. Such proteins are called **housekeeping proteins**, required for the basic life activities of all cells. But there are certain proteins, which are unique to certain cells. Hemoglobin in erythrocytes (RBC), collagen, myosin, etc., in muscle cells are some examples. This is called **cell-specific or organ-specific or tissue-specific gene expression**. The expressions of these genes are under the control of very specific regulatory proteins or other types of small molecules called **transcription factors**.

The proteins, which are produced, have various lifespans or half-life ranging from a few seconds to many months or years. Each protein can be identified by its unique amino acid sequence, the three-dimensional shape of the protein (that mainly depends on amino acid sequence and other environmental factors), function of the protein, and cellular location. The proteins expressed in a cell may be intracellular (present within the cell) or extracellular (secreted outside the cells), or circulating in the blood like hormones, immunoglobulins, etc., constantly interacting with other molecules such as proteins, lipids, sugars, DNA, RNA, metal ions, vitamins, etc., or with other cells. The functions of these circulating proteins are also influenced by their interaction with other molecules that are present nearby. Even though the total number of genes estimated is about 35,000, the actual number of proteins is much higher than this. About 17, 000 proteins were identified at the gene level but information regarding their function and biological role is still being investigated in detail. The type and total number of genes in an organism will be stable (static) and are identical in all somatic cells of an organism. But, the total number of proteins expressed by a cell (protein profile) of an organism is always variable (dynamic). The protein profile of a cell depends on its metabolic state, stage of development, and other micro- and macro-environmental factors, which influence the expression of a set of genes at a particular time. Therefore, the challenge for the future is to determine the actual total number of proteins expressed in each cell type and find out the functions and biological role of these proteins in metabolism, health, and disease. Through these topics you will be exposed to the three-dimensional structure of proteins and their relationship to proteomics, recombinant DNA technology, genomics, and finally bioinformatics.

14.2 THREE-DIMENSIONAL SHAPE OF PROTEINS

A single cell develops into a multicellular embryo through a large number of complicated biochemical reactions mediated and controlled by various types of proteins expressed during the course of its development. Every function in the living cell depends on proteins. They make us who we are and make our cells operate properly. A cell cannot function without proteins. The shape of a protein determines its biological activity. A single protein may have a varying structure and more than one function. Proteins have many different biological functions. Proteins are even classified according to their biological roles. The key to appreciating how different proteins function in these different ways lies in an understanding of protein structure and their three-dimensional shape. Proteins interact with other molecules such as small molecules, other proteins, nucleic acids, lipids, etc., and these interactions form the basis of their biological roles. Structural complementarity is the means of molecular recognition that allows molecules to interact. The structure of one molecule is complementary to that of its partner(s) in the interaction, like pieces in a puzzle, or a lock and its key. Proteins, by virtue of their architectural diversity, are ideal for such complementary interactions. In short, the structure or the molecular shape of the protein determines its function. Therefore, to understand the function and biological role of a protein it is essential to understand the structure and three-dimensional shape of the protein in detail.

The detailed study of the structure of proteins requires protein extraction and purification to its homogeneity. The purified protein has to be analyzed by various biochemical and instrumental methods to get the details about its chemical composition. The pure protein obtained has to be crystallized to study its three-dimensional shape by x-ray crystallography (an x-ray diffraction technique). Another equally powerful technique to elucidate the three-dimensional structure of protein is NMR spectroscopy. The structural study of protein, thus has two parts—the first part is determination of amino acid sequence and the second part is the elucidation of the three-dimensional shape of the protein formed by the specific folding of the polypeptide chain controlled by a number of molecular forces. The extraction and purification and its crystallization are the preconditions for the detailed structural and functional studies and are explained in later parts of this chapter.

The ability to sequence polypeptides was a major step forward in the understanding of the relationship between protein structure and function. It was Dr. Frederick Sanger who developed the basic chemical method for sequencing proteins during the 1940s. He showed for the first time, that proteins are a linear polymer of amino acids, linked in a continuous sequence by peptide bonds. He received the Nobel Prize in 1958 for determining the sequence of the peptide hormone insulin. The peptide bond is formed between the alpha-amino and alpha-carboxyl groups of two adjacent amino acids. Pehr Edman modified this process of amino acid sequencing by introducing a new reagent—phenyl-isothio-cyanate for the sequential removal of amino acids and their identification in a protein. This method of sequencing is now automated and called the **Edman Degradation Reaction** and the instrument is called the **Sequenator.**

In the beginning, the process of protein sequencing was very difficult and time consuming and also required large quantities of pure protein samples. For example, in the 1940s, Sanger used about 300 mg of pure insulin and more than 10 years to sequence its amino acids and elucidate its structure. Whereas today, amino acid sequencing of proteins can be carried out with about one to two mg of pure proteins using the Edman Degradation Reaction in a sequenator within a very short period of time. The advancement of analytical techniques has contributed greatly to the elucidation of three-dimensional structures of proteins and their relationship to functions. Scientists such as Linus Pauling, G.N. Ramachandran, Max Perutz, and John Kendrew started studying the three-dimensional structures of proteins by x-ray crystallography and computational methods such as modeling. They studied protein shapes, protein folding, and the basic forces and laws that govern the folding of the polypeptide chain to form the secondary and tertiary structures and stabilize them. They elucidated much about the planar nature of the peptide bond, the torsion angles, various types of secondary structures such as the alpha helix, beta-pleated sheets, beta turns, and random coils, studied the concepts, domain and motifs, in the secondary and tertiary structures. Some important points of protein structures are revived here briefly:

Proteins and peptides are biopolymers composed of amino acid residues interlinked by amide bonds. Their structures can be discussed in terms of four levels of complexity and are as follows:

- Primary structure
- Secondary structure
- Tertiary structure
- Quaternary structure

The linear unbranched chain of amino acids linked together by covalent bonds known as peptide bonds is called the **primary structure** of a protein. In addition to peptide bonds, other types of covalent linkages such as disulphide bonds, if present, are also included in the primary structures. The types and the sequence of amino acids present in the polypeptide chain determine the nature of the **secondary structures** at different regions of the chain. The secondary structure of a segment of a polypeptide chain is the local spatial arrangement of its main-chain atoms without regard to the conformation of its side chains or to its relationship with other segments. The major types of secondary structures observed in protein molecules are **alpha (α) helices** and **beta (β)**

pleated sheets in addition to **random coils** and **beta turns.** All these secondary structures may be present independently or may be together in the secondary or tertiary structures of a single polypeptide chain. The secondary structure undergoes further folding and reorganization within the molecule resulting in higher order compact structures or the **tertiary structure.** There

Primary structure	Secondary structure	Tertiary structure
amino acid; peptide; protein made from two polypeptide chains, linked by covalent -S-S- bonds	Random coils; α helix; β sheet; β turn	

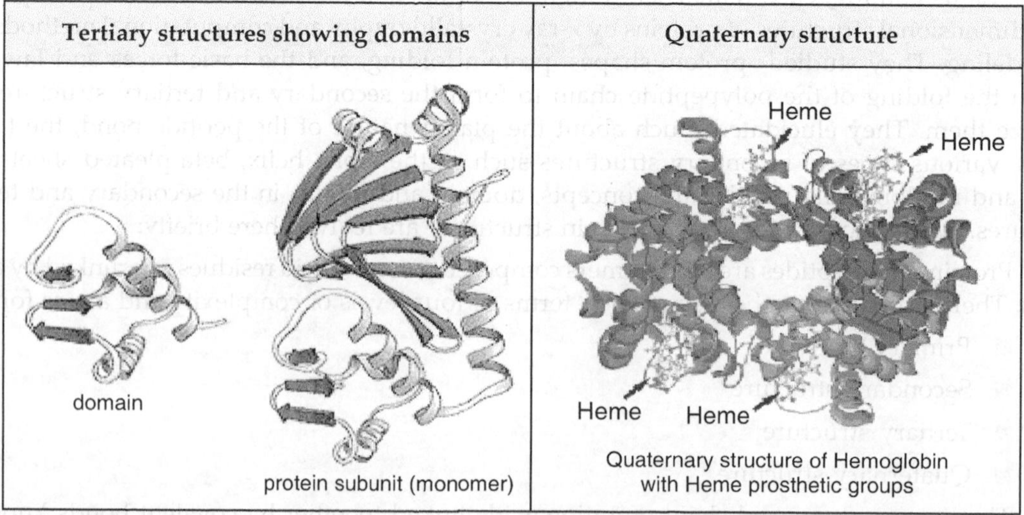

FIGURE 14.2 **Different levels of hierarchical organization of protein three-dimensional structure.**

are structural components comprising a few alpha-helices or beta-strands, which are frequently repeated within structures, called **supersecondary structures** (being intermediate to secondary and tertiary structures). These compact structurally distinct elements are known as **motifs.** When these structurally distinct regions of protein molecules are associated with a specific function, those structurally and functionally distinct units are called a **domain.** Structurally-related domains are found in different proteins, which perform similar functions. The molecular forces, which are responsible for the secondary and tertiary structures, are the non-covalent interactions between the various amino acid side chains within the molecule and with the water molecule surrounding it. The main molecular force responsible for the various secondary structures are the **hydrogen bonds** and the molecular forces behind the tertiary structures are the **ionic bonds, hydrogen bonds, hydrophobic** and **hydrophilic interactions,** and **van der Waals force.** Secondary and tertiary structures represent the most thermodynamically stable conformations or shapes for the molecule in a solution. The **quaternary structure** is the assembly of two or more independent polypeptides or proteins at their tertiary stage to form a **multimeric protein.** The individual component peptides of the multimeric proteins are known as **subunits** and are held together via non-covalent forces. The subunits of a multimeric protein may be similar or dissimilar. For example, hemoglobin contains four polypeptide chains (2α chains and 2β chains) held together non-covalently in a specific conformation as required for its function.

The major molecular forces that cause the linear polypeptide chain to undergo a specific type of coiling and folding in space to a characteristic three-dimensional shape are the non-covalent forces. These forces, to a greater extent, lie in the chemical and structural properties of the constituent amino acid residues of the polypeptide chain. There are 20 types of amino acids by which the entire protein of the living system is composed. These amino acids can be broadly classified into three categories—**hydrophobic** (tryptophan, phenylalanine, leucine, etc.), **polar** (glutamine, serine, etc.), and **charged** (aspartic acid, lysine, etc.) amino acids. Therefore, these amino acids are capable of interacting with each other within the protein molecules via various non-covalent interactions leading to a very characteristic shape and biological property for the protein molecule.

Non-covalent Molecular Interactions

The three-dimensional structure of macromolecules are the heart of life. If the three-dimensional structure is destroyed by heat, life ceases to exist. Weak, non-covalent forces play a key role in replication of DNA, the folding of proteins for the specific recognition of substrates, and the detection of signal molecules. All biological structures and processes depend on non-covalent interactions as well as covalent bonds. The four fundamental non-covalent interactions are:

- Electrostatic bonds (or ionic bonds)
- Hydrogen bonds
- Van der Waals forces (bonds)
- Hydrophobic interactions

These molecular forces differ in their geometry, strength, and specificity, and are also affected by the presence of water. Non-covalent forces are weak forces of bonding strength of one to seven kcal mol^{-1} (4–29 kJ mol^{-1}) as compared to the strength of covalent bonds, which have a bonding strength of about 50 kcal mol^{-1}.

Electrostatic Bonds (or Ionic Bonds)

A charged group on an amino acid residue can attract an oppositely charged group on another amino acid residue. For example, the positively charged amino residues of lysine and arginine attract the positively charged carboxyl groups of glutamic acid and aspartic acid. The attraction is strongest in vacuum and is weakest in water because of its dielectric strength. The electrostatic

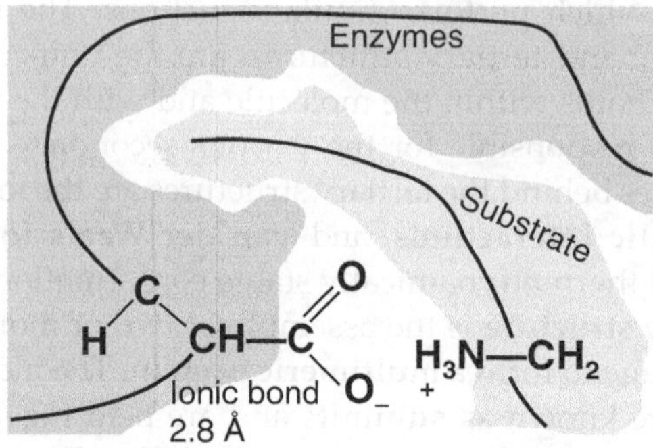

FIGURE 14.3 **The electrostatic interaction between protein and its substrate.**

interactions are more prone to changes in pH and ionic strength of the medium. This kind of attraction is also called an ionic bond, salt linkage, salt bridge, or ion pairs. The distance between oppositely charged groups in an optimal electrostatic attraction is 2.8 Å. Figure 14.3 is an example of a charged group on a substrate, which is attracted to an oppositely charged group on an enzyme.

Hydrogen Bonds

Hydrogen bonds can be formed between uncharged molecules as well as charged ones. In a hydrogen bond, a hydrogen atom is shared by two other electronegative atoms. The atom to which the hydrogen is more tightly linked is called the hydrogen donor and is a polar covalent bond, whereas the other atom is the hydrogen acceptor. The acceptor has a partial negative charge that attracts the hydrogen atom. In fact, a hydrogen bond can be considered an intermediate in the transfer of a proton from an acid to a base.

Hydrogen Donor Hydrogen Acceptor

—O—H ... N—

2.88 Å

Hydrogen Donor Hydrogen Acceptor

—N—H .... O—

3.04 Å

The donor in a hydrogen bond in biological systems is an oxygen or nitrogen atom that has a covalently attached hydrogen atom. The acceptor is either oxygen or nitrogen, which has a partial negative charge. The hydrogen bond is stronger than van der Waals bonds but much weaker than covalent bonds. The very small size of these atoms allow them to approach each other very close resulting in very strong hydrogen bonds. Strong hydrogen bonds are highly directional. This means that the donor, the hydrogen, and the acceptor atoms all lie in a straight line and attain a partial covalent character. Non-linear hydrogen bonds are weaker. The strength of hydrogen bond is only one to two kcal/mol; however, the hydrogen bond for water is five kcal/mol.

The helix, a recurring motif in a protein structure, is stabilized by hydrogen bonds between amid (—NH) and carbonyl (—co) group. The specific binding of substrate to enzyme is in part due to hydrogen bonding. Here are two amino acids (Thry-45 and Ser-123) of ribonuclease, which bind to uracil through three hydrogen bonds. Substrate without the two C=0 and one NH groups would be unable to bind or would bind less tightly to the enzyme (Figure 14.4).

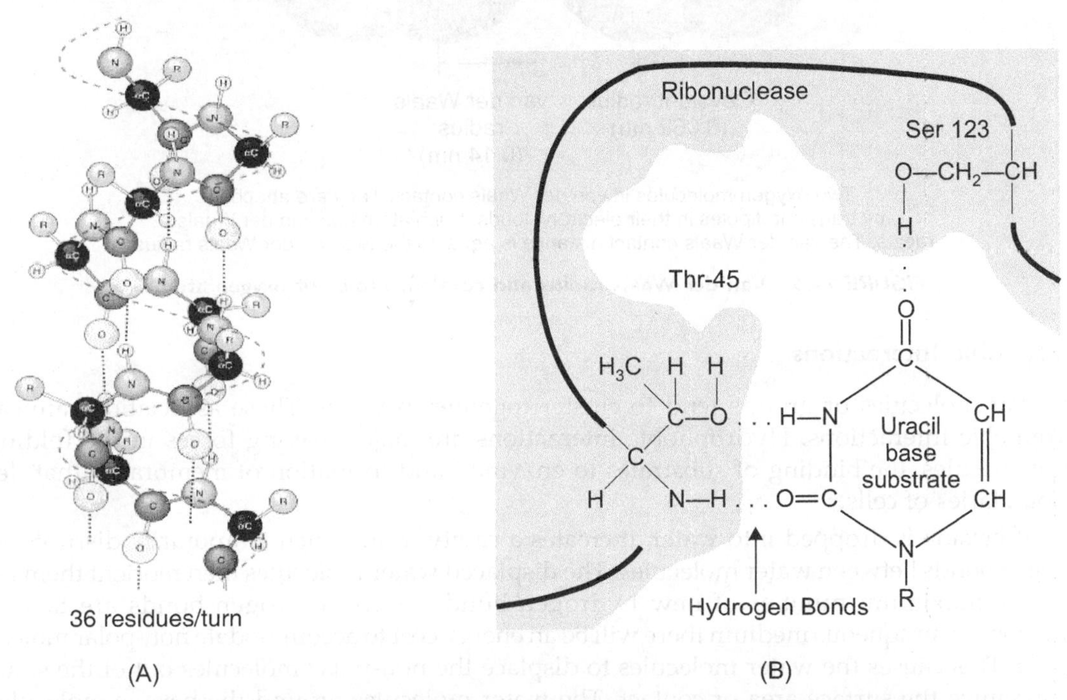

FIGURE 14.4 **Hydrogen bonds (A) in α-Helix and (B) in the interaction between ribonuclease (enzyme) and RNA (substrate).**

Van der Waals Forces (Bonds)

Van der Waals bonds are non-specific and very weak, attractive, or repulsive forces between any two atoms, when they are three to four Å apart. These forces are caused by transient dipoles in all atoms. Momentary random fluctuations in the distribution of the electrons of any atom give rise to a transient unequal distribution of electrons, which create dipoles. If two non-covalently bonded atoms are close enough together, the transient dipole in one atom will perturb the electron cloud of the other and the two molecules will be attracted. The van der Waals forces are weak but are important in biological systems. The basis of a van der Waals bond is that the distribution of electronic charge around an atom changes with time. At any instant, the charge distribution is not perfectly symmetric. This transient asymmetry in the electronic charge around an atom encourages similar asymmetry in the electron distribution around its neighboring atoms. The resulting attraction between a pair of atoms increases as they come closer, until they are separated by the van der Waals contact distance or van der Waals radius (Figure 14.5). At a shorter distance, very strong repulsive forces become dominant because the outer electron clouds overlap. The contact distance between an oxygen and carbon atom, for example, is 3.4 Å, which is obtained by adding 1.4 and 2.0 Å, the contact radii of the O and C atoms.

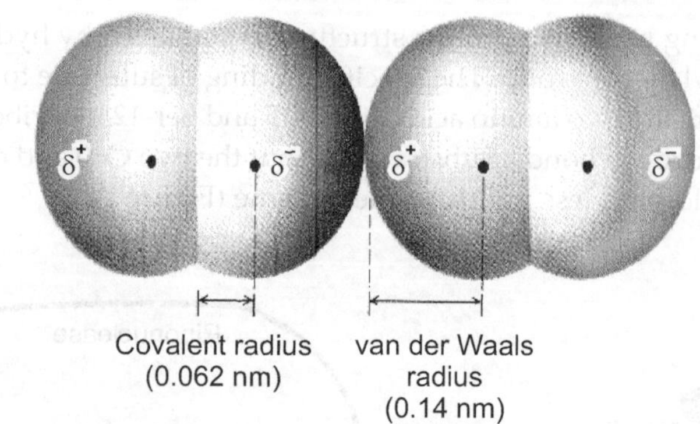

Two oxygen molecules in van der Waals contact. They are attached by transient dipoles in their electron clouds. Each atom has van der Waals radius. The van der Waals contact distance is equal to the two van der Waals radius.

FIGURE 14.5 **Van der Waals radius and covalent radius of oxygen atoms.**

Hydrophobic Interactions

Non-polar molecules or groups tend to cluster together in water. These associations are called hydrophobic interactions. Hydrophobic interactions are major driving forces in the folding of macromolecules, the binding of substrates to enzymes, and formation of membranes that define the boundaries of cells.

If hexane is dropped into water, it creates a cavity in it, which temporarily disrupts some hydrogen bonds between water molecules. The displaced water molecules then reorient themselves to form a maximum number of new hydrogen bonds. Since hydrogen bonds are favorable interactions in an aqueous medium there will be an energy cost to accommodate non-polar molecules in water. This causes the water molecules to displace the non-polar molecules out of the solution and minimize the surface area of contact. The water molecules around the hexane molecule are much more ordered than elsewhere in the solution. If two hexane molecules are put in water they will come together. Non-polar solute molecules are driven together in water not primarily because they have a high affinity for each other but because water bonds strongly force them together. The hydrophobic interaction is thus a manifestation a hydrogen-bonding network in water. This hydrophobic interaction is one of the most important molecular forces that shape the protein molecules in water and hold proteins together to make complex protein structures such as muscles, biological membranes, form the DNA double helix, cause DNA protein interactions, etc. In protein molecules, the hydrophobic amino acids usually keep away from the surface of the molecule which is in contact with water and forms the core of the protein molecule.

14.3 STRUCTURE-FUNCTION RELATIONSHIP IN PROTEINS

Proteins are the workhorses of cells and in every activity there is the involvement of one or more proteins in different ways. Now, it is very clear that each protein has a specific three-dimensional shape determined by the amino acid sequence and various other intermolecular interactions. This three-dimensional shape has a great influence on the biological function that it performs in the cells. We consider two proteins, as an example, to understand the importance of a three-dimensional structure on its specific function: a proteolytic enzyme, chymotrypsin, and the oxygen carrying protein, hemoglobin.

Chymotrypsin—A Protein-digesting Enzyme

Chymotrypsin is a member of a family of enzymes, all of which cleave peptide bonds through the action of an active site serine (the serine proteases). This family includes the pancreatic enzymes chymotrypsin, trypsin, and elastase as well as a variety of other proteases (e.g., cocoonase, thrombin, acrosomal protease, etc.). Chymotrypsin, trypsin, and elastase show a high degree of similarity in their overall tertiary structure, but have different substrate specificities determined by a specific substrate-binding site on each enzyme. Chymotrypsin is one of the protein-hydrolyzing enzymes produced by the digestive gland pancreas. The protein present in the food that we eat is digested mainly by two proteases—trypsin and chymotrypsin in the beginning of the small intestine (the duodenum). These two digestive enzymes are produced by pancrease and are released into the duodenum through the pancreatic duct. Thus, the site of production of these enzymes is the pancreas and their site action is the duodenum. Trypsin and chymotrypsin cut the linear polypeptide chains into short peptides by cutting at specific sites. These short peptides thus produced are acted upon by other peptidases releasing amino acids. But the pancreas is made up of many proteins. How are these proteins protected from the hydrolytic activity of chymotrypsin? These types of hydrolytic enzymes, particularly proteases are produced in an inactive form called **zymogen** and are transported to the site of action, the duodenum—where it is converted into an active enzyme by a process known as ***in situ* activation**. Because of this process, the protein undergoes a major change in its three-dimensional shape, which is now suitable for its interaction with its substrates. The active chymotrypsin enzyme is known as **alpha chymotrypsin** and its inactive form from which it is produced is called **chymotrypsinogen**.

Chymotrypsinogen, the precursor (zymogen) of active chymotrypsin, consists of 245 amino acid residues. Activation of chymotrypsinogen involves proteolytic cleavage at two sites along the chain and removal of two amino acids at each cleavage site. The resultant three peptide chains are A, B, and C. These three chains are held together by five disulfide bonds and fold into a globular structure. This process of folding brings three distantly placed amino acid residues his 57, asp 102, and ser 195 close together in a particular order to form the active center or the reaction center of the enzyme. The overall chymotrypsin molecule is folded into two domains, each containing six beta strands arranged as antiparallel sheets that form a circular structure known as a **beta barrel**. The active site residues (ser 195, his 57, and asp 102) are far apart in the primary sequence but are brought together in a crevice formed between the two protein domains.

The active site of chymotrypsin consists of asp 102 positioned close to his 57 and ser 195. The precise mechanism of action is still debated, but it appears that a hydrogen on the his imidazole ring is transferred to the asp 102 carboxylate either via a "charge relay system" or via a "low barrier H-bond." This shift results in the histidine ring being able to accept the serine195 hydroxyl hydrogen, forming a very nucleophilic serine alkoxide ion. The images in Figure 14.6 describe a proposed mechanism for the hydrolysis of peptide bonds by chymotrypsin. Each arrow in the figure represents the movement of two electrons and the consequent making or breaking of a bond.

Stage 1. A polypeptide substrate moves into the active site of the enzyme. The shape, size, and amino acid sequence of chymotrypsin's active site allow that part of the enzyme to bind a portion of a polypeptide that has non-polar side chains, like those found in phenylalanine. Once the polypeptide is in the active site, an H+

ion moves from the serine amino acid at position 195 (Ser-195) of the enzyme's amino acid sequence to the histidine amino acid at position 57 (His-57). The oxygen atom in serine's hydroxyl group then forms a covalent bond to the carbon of one of the substrate's peptide bonds shifting the two electrons from one of the double bonds up to form a lone pair.

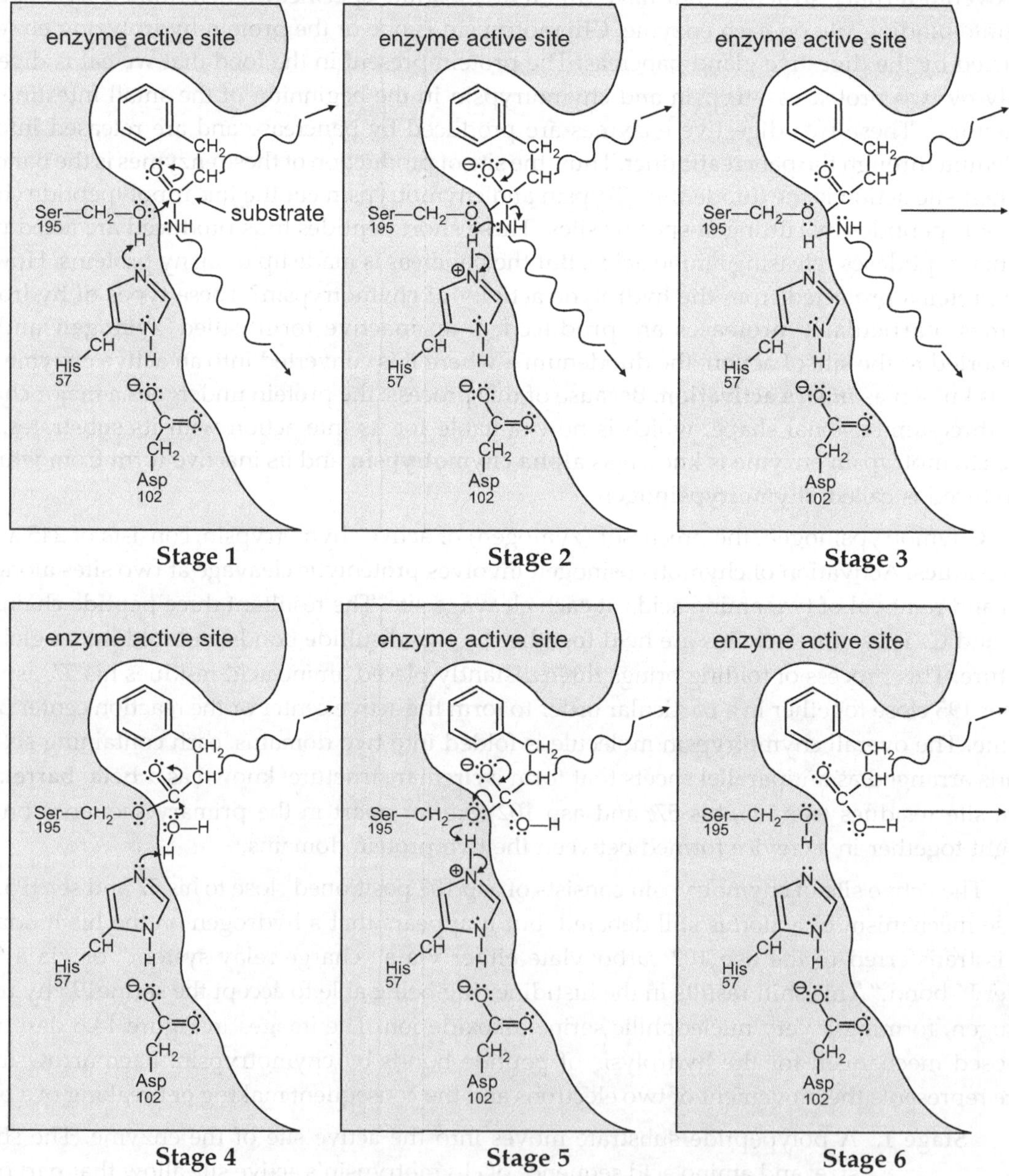

FIGURE 14.6 **The catalytic triad and the mechanism of the charge relay system in chymotrypsin activity resulting in the hydrolysis of peptide bonds.**

Stage 2. The positive charge formed on his 57 is stabilized by the negative charge on the aspartic acid at position 102 (asp 102). When the double bond is reformed between carbon and oxygen in the peptide bond, the bond between the carbon and the nitrogen in the peptide bond is broken. The nitrogen-containing group is stabilized by the formation of a bond with a hydrogen atom from his 57.

Stage 3. The portion of the polypeptide that contains the nitrogen atom from the broken peptide bond moves out of the active site.

Stage 4. A water molecule moves into the active site. The oxygen atom in the water molecule loses an H+ ion to a nitrogen atom on his 57. This allows water's oxygen atom to form a bond with the carbon atom of the remaining portion of the substrate. Like in Stage one, one of the bonds in the double bond shifts up to form a lone pair.

Stage 5. When the double bond is reformed, the bond between carbon and the oxygen of ser 195 is broken. The –OH group on ser 195 is restored with a transfer of an H+ ion from his 57. With this step, the ser 195 and his 57 are both returned to their original forms.

Stage 6. The remaining portion of the substrate moves out of the active site, leaving the active site in its original form, ready to repeat stages one to five with another polypeptide molecule.

The active site of chymotrypsin is hydrophobic and is a large space, which can accommodate only bulk hydrophobic and aromatic amino acid residues such as phenylalanine. This binding brings susceptible peptide bonds close to the catalytic triad serine 195, his 57, and asp 102 residues.

Chymotrypsinogen, the precursor of chymotrypsin, contains certain additional peptides known as **signal peptides**, which block the active center (substrate binding site) of the enzyme and hence it is inactive. These signal peptides also have another function. It directs the protein out of the cell from where it is synthesized to the site of function, the duodenum. Another proteolytic enzyme, trypsin, is involved in the *in situ* activation of chymotrypsinogen to chymotrypsin. The process involves the proteolytic cleavage of the signal peptides by trypsin, which results in a change in the conformation of the protein molecule by correctly folding the polypeptides, bringing the catalytic triads—his 57, asp 102, and ser 195 together in the correct orientation for the proper accommodation and interaction with the substrate. The same pattern of folding and orientation of the catalytic triad is also present in other proteolytic enzymes such as trypsin, subtilisin (an enzyme from *bacillus subtillis*), thrombin (a blood-clotting factor), and acetylcholine esterase (an enzyme involved in the release of the neurotransmitter, acetyl choline in the neurons). All these enzymes have the serine residue in the active center.

There are some compounds, that react with the active center of the enzyme and make it inactive permanently or temporarily, which are known as **inhibitors**. Certain organophosphate compounds can react with the acidic serine residue of the active center and irreversibly inhibit enzyme activity. Derivatives organophosphates such as parathion and malathion have been used as insect repellents and are present in mosquito repellents. These insect repellents are not toxic to humans. Nerve gas is one of the important irreversible inhibitors of the enzyme acetylcholine esterase. It also reacts with acidic serine and prevents the transmission of nerve impulses across the junction of neuron joints in the nervous system causing death. This was used as chemical warfare in many of the wars in the past.

Molecular Disease–Sickle Cell Anemia

Red blood cells (RBCs, for short, or technically, erythrocytes) are membrane-bound cells produced by stem cells in bone marrow. They contain mostly the protein hemoglobin as (Hb), whose function is to bind oxygen molecules (O_2) taken in through the lungs and to carry them to tissues throughout the body. Molecules of Hb are composed of four protein subunits, each of which has an iron atom at the center of a special ring structure, the **porphyrin ring**. The iron atom has a special affinity for molecular oxygen (O_2), such that each subunit picks up an oxygen molecule. Since there are four subunits per Hb molecule, when fully saturated any hemoglobin therefore carries four oxygen molecules (or eight oxygen atoms). Saturation occurs in the lungs, and desaturation in the body tissues, as follows:

In lungs : $Hb + 4O_2 \longrightarrow Hb.O_8$

In tissues : $Hb.O_8 \longrightarrow Hb + 4O_2$

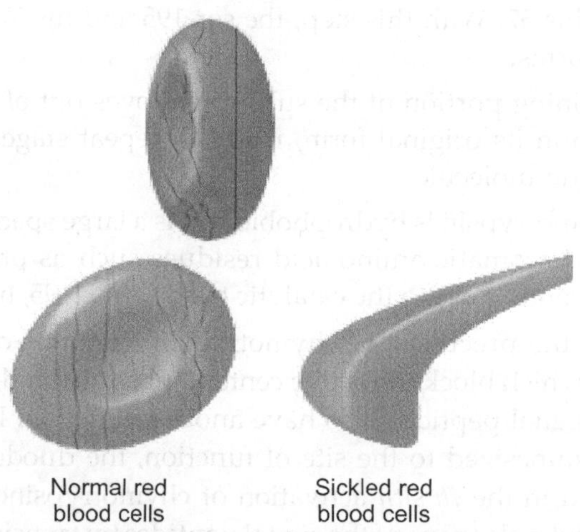

FIGURE 14.7 **Normal and sickled RBC.**

Normally, oxygenated Hb is written simply as HbO_2, where it is understood that all four subunits are occupied by oxygen. The ability of hemoglobin to reversibly bind and release oxygen is one of its most important characteristics, since if either property was predominant, it could not serve as an effective carrier molecule. Hb binds oxygen when the oxygen pressure is high, and releases it when the oxygen pressure is low. In the tissues, hemoglobin picks up a small percentage (25%) of the carbon dioxide released there and transports it back to the lungs where it is released. The remainder of the carbon dioxide is transported to the lungs as H_2CO_3, or carbonic acid, most of which dissociates into hydrogen ions and bicarbonate:

$$H_2CO_3 \rightleftharpoons H^+ + HCO_3^{-2}$$

Any significant decrease in the amount of functional Hb is known as **anemia**. All forms of anemia have serious physiological effects because of reduced oxygen delivery to and reduced carbon dioxide removal from the tissues.

Sickle cell anemia, in particular, creates serious depletion of oxygen through two mechanisms:

1. Because of molecular changes within the sickled cell, oxygen-carrying capacity of the blood is greatly reduced.
2. Because of their peculiar shape, greater rigidity, and tendency to stick together, sickle cells clog smaller vessels in the circulatory system—the arterioles and capillaries, in particular—preventing the blood from delivering oxygen and nutrients, and removing carbon dioxide and wastes from tissues.

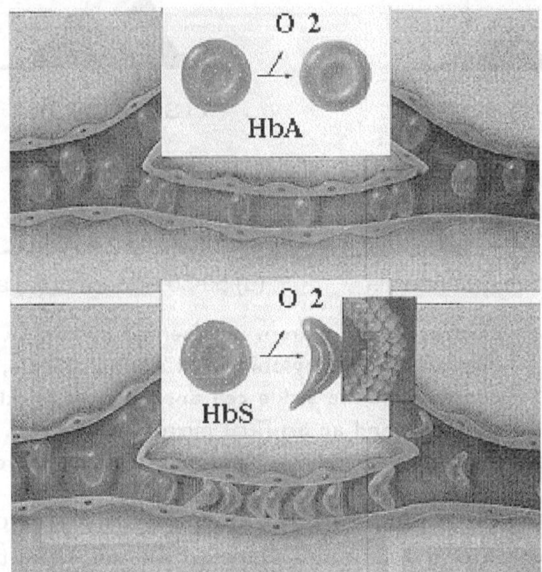

FIGURE 14.8 **This schematic diagram shows the changes that occur as sickle or normal red cells release oxygen in microcirculation. The upper panel shows that normal red cells retain their biconcave shape and move through the microcirculation (capillaries) without a problem. In contrast, the hemoglobin polymerizes in sickled red cells when they release oxygen, as shown in the lower panel. (*Joint Center for Sickle Cell and Thalassemic Disorders, Brigham and Women's Hospital, Boston, MA*). Source: *http://www.nslc.wustl.edu/sicklecell/part1/background.html***

This disease was reported for the first time by a Chicago-based cardiologist James B. Herrick[3] in 1910. It was recognized to be the result of a genetic mutation, inherited according to the Mendelian principle of incomplete dominance. Initially, it was not clear what the actual defect was, that caused the sickling. Various experiments indirectly narrowed down the site of the defect to the hemoglobin molecule. The most direct evidence that mutation affected the hemoglobin molecule came from electrophoretic analysis, a method to separate complex mixtures of large molecules by means of an

[3] Herrick JB. *Peculiar elongated and sickle-shaped red blood corpuscles in a case of severe anemia.* Arch Intern Med 1910; 6:517-21.

electric current on a gel. When hemoglobin from people with severe sickle cell anemia, the sickle cell trait, and normal red blood cells was subjected to electrophoresis, the following interesting results were obtained:

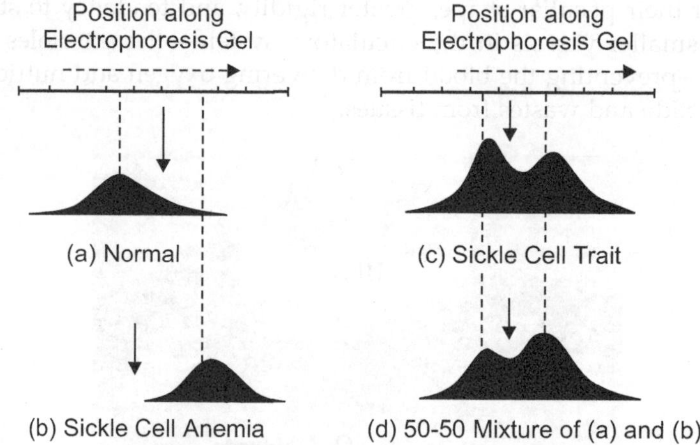

FIGURE 14.9 **Electrophoretic pattern represented as Longsworth scanning diagrams of hemoglobin from normal people, compared to people with the sickle cell anemia trait, sickle cell anemia (disease), and an artificial mixture of the two. Each peak of the curve represents a band on the electrophoretic gel.**

It was clear that hemoglobin molecules of people with sickle cell anemia migrated at a different rate, and thus ended up at a different place on the gel, from the hemoglobin of normal people (diagram, parts *a* and *b*). What was even more interesting was the observation that individuals with the sickle cell trait had about half normal and half sickle-cell hemoglobin, each type making up 50% of the contents of any red blood cell (diagram, part *c*). To confirm this latter conclusion, the electrophoretic profile of people with the sickle cell trait could be duplicated simply by mixing sickle cell and normal hemoglobin together and running them independently on an electrophoretic gel (diagram, part *d*). These results fit perfectly with an interpretation of the disease as inherited in a simple Mendelian fashion showing incomplete dominance. Here, then, was the first verified case of a genetic disease that could be localized to a *defect in the structure of a specific protein molecule*. Sickle cell anemia thus became the first in a long line of what have come to be called **molecular diseases**. Thousands of such diseases (most of them quite rare), including over 150 mutants of hemoglobin alone, are now known.

But what was the actual defect in the sickle cell hemoglobin? What is the molecular defect that makes the sickle cell hemoglobin different from normal hemoglobin? It is the story of one of the first identifications of the molecular basis of a disease. From the electrophoretic mobility of the two hemoglobins, Linus Pauling predicted that sickle cell Hb differed in charge from the normal Hb molecule. He and his co-workers turned their attention to determining the actual difference between normal and sickle cell hemoglobin molecules. Breaking the protein molecules down into shorter fragments called **peptides**, Pauling and co-workers subjected these fragments to another technique called **protein fingerprinting** by paper chromatography.

The technique of protein fingerprinting involves the following steps:

1. Extract and purify hemoglobin from sickle cell RBC and normal RBC separately in a clean test tube.
2. Digest these proteins with a commercial sample of trypsin separately under standard conditions. Trypsin is another type of serine protease that cleaves the peptide bond adjacent to a lysine or arginine residue in a protein molecule.
3. The cleaved peptides are subjected to paper electrophoresis under pH (pH 2.5) and dry the paper.
4. After electrophoresis they are subjected to paper chromatography perpendicular to the direction of electrophoresis using the solvent system water: butanol: acetic acid in the ratio 5:4:1. The peptides will separate depending on their partition coefficient, which further depends on their degree of hydrophobicity. The more hydrophobic peptide will move fast and the less hydrophobic will move slowly.
5. Remove the chromatographic paper and stain with ninhydrin.
6. Examine the peptide spots and compare with the standards.

When this procedure is applied to samples of normal and mutant (sickle) hemoglobin molecules (alpha and beta chains) that had been broken down into specific peptides, all the spots are the same except for one crucial spot (shown darkened in the final chromatogram below), which represents the difference between sickle cell and normal hemoglobin. Different steps involved in peptide fingerprinting using paper chromatography are illustrated in Figure 14.10.

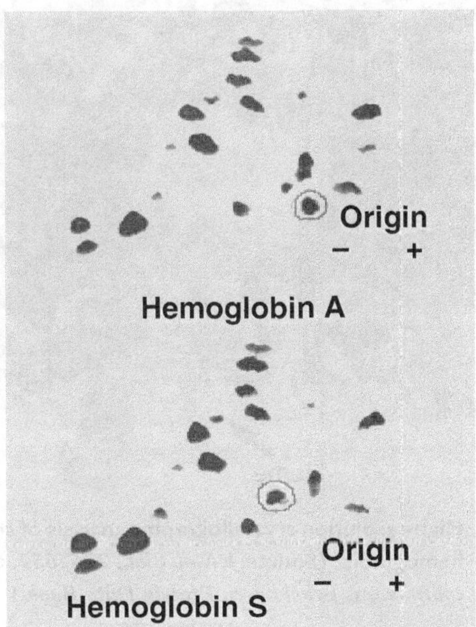

FIGURE 14.10 **Two-dimensional paper chromatography of normal (hemoglobin A) and mutant (sickle cell, hemoglobin S) hemoglobins. The encircled spot represents the position of the peptide. Source: *Stryer, Biochemistry,* 1995**

The fact that the spots migrate to different places on the chromatogram indicates their molecular structures must be somewhat different. Pauling and his colleagues were convinced that the difference might be no more than one or two amino acids, but it was left to biochemist Vernon Ingram at the Medical Research Council in London to demonstrate this directly. Taking the one aberrant peptide and analyzing it one amino acid at a time, Ingram showed that sickle cell hemoglobin differed from normal hemoglobin by a *single* amino acid, the number six position in the beta chain of hemoglobin. The substitution of glutamic acid at the sixth position with valine in sickle cell hemoglobin results in an increase in hydrophobic interaction between the hemoglobin molecules resulting in aggregation and ultimately in the deformation of RBC to a sickle shape. That one small molecular difference makes the enormous difference in people's lives, that of good health and disease.

In overall structure, a complete Hb molecule consists of four separate polypeptide chains (i.e., each a long string, or polymer, of amino acids joined together end-to-end) of two types, designated as the **alpha** and **beta chains.** The two α chains are alike (meaning they have the exact same sequence of amino acids), and the two β chains are also alike.

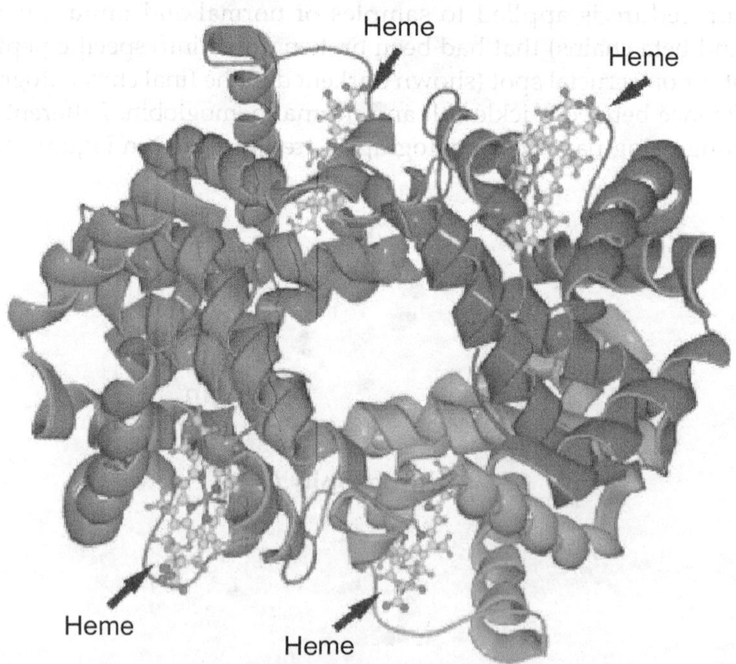

FIGURE 14.11 **High-resolution crystallographic analysis of co-operative dimeric hemoglobin. (Source: *J. Mol. Biol., 235, 657. Oxyhemoglobin PDB coordinates, Brookhaven Protein Data Bank.*)**

The protein fingerprinting or the peptide mapping developed for the molecular studies of sickle cell hemoglobin became a very powerful technique for the identification of protein samples from different sources. The peptide fingerprint of a protein from new sources can be compared with that of the standard protein and thus, the variations can be identified or understood. This simple technique of peptide fingerprinting has given rise to another similar and more powerful technique: **two-dimensional gel electrophoresis.** This is a combination of two electrophoretic techniques—Isoelectric focusing and SDS-PAGE—in a series. First, the protein is subjected to isoelectric focusing, which is followed by SDS-PAGE in a direction perpendicular to the first. This technique was found to be very useful for proteomes studies, expression of protein profiles of cells grown under different conditions (for example, normal cells and diseased cells), and also easy identification of proteins in combination with mass spectrometry. All these technological advancements including amino acid sequencing have provided an enormous quantity of data, and that has given rise to computerized databases and homology searches and protein identification. All these have led to a generation of bioinformatics and computational biology.

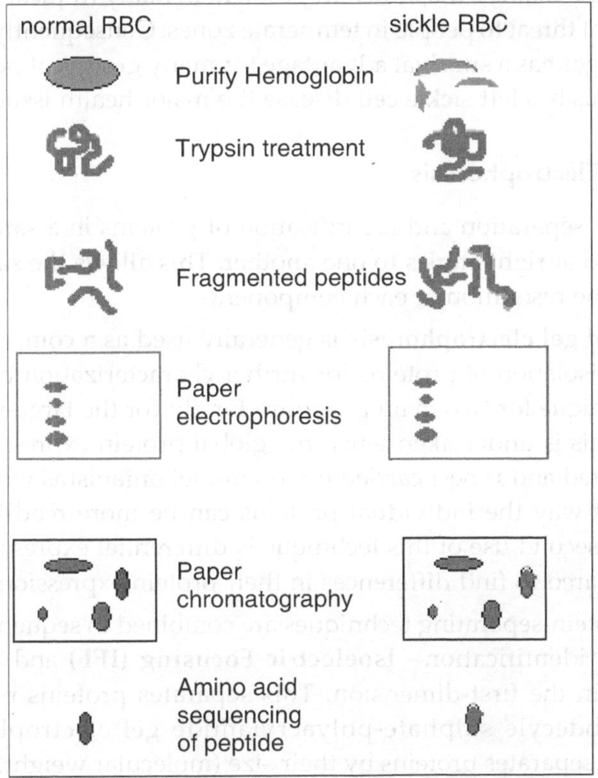

FIGURE 14.12 **Different steps in protein fingerprinting.**

The Sickle Gene and Malaria

The high representation of the hemoglobin S gene in some populations reflect the protection it provides against malaria. The malaria parasite does not survive as well in the erythrocytes of people with the sickle trait as it does in the cells of normal people. The basis of the toxicity of sickle hemoglobin for the parasite is unknown. One possibility is that the malarial parasite produces extreme hypoxia in the red cells of people with the sickle trait. These cells then sickle and are cleared (along with the parasites they harbor) by the reticuloendothelial system. Another possible mechanism is that low levels of hemichromes are formed in sickle trait erythrocytes. Hemichromes are complexes containing heme moieties that have dissociated from the hemoglobin. Hemichromes catalyze the formation of reactive oxygen species, such as the hydroxyl radical, which can injure or even kill malarial parasites.

The malaria hypothesis maintains that during prehistory, on average, people without the sickle gene died of malaria at a high frequency. On the other hand, people with two genes for sickle hemoglobin died of sickle cell disease. In contrast, the heterozygotes (sickle trait) were more resistant to malaria than normal people and yet suffered none of the ill effects of sickle cell disease. This selection for heterozygotes is called **balanced polymorphism.** Support for this concept comes from epidemiological studies in malaria-endemic regions of Africa. The frequency of the sickle cell trait is lower in people coming for treatment to malaria clinics than is seen in the general population. The reasonable assumption is that relative protection from malaria is at work in this situation.

Although malaria remains a major health problem in many tropical regions of the world, the disease is not a significant threat to people in temperate zones. Consequently, the protection afforded by the sickle trait no longer has a survival advantage for many groups of people in whom the sickle cell gene is common. This has left sickle cell disease the major health issue in these populations.

Two-dimensional Gel Electrophoresis

This is a method for the separation and identification of proteins in a sample by displacement in two-dimensions oriented at right angles to one another. This allows the sample to separate over a larger area, increasing the resolution of each component.

Two-dimensional gel electrophoresis is generally used as a component of proteomics and is the step used for the isolation of proteins for further characterization by mass spectroscopy. In the lab we use this technique for two main purposes. Firstly, for the large-scale identification of all proteins in a sample. This is undertaken when the global protein expression of an organism or a tissue is being investigated and is best carried out on model organisms whose genomes have been fully sequenced. In this way the individual proteins can be more readily identified from mass spectrometry data. The second use of this technique is differential expression; this is when two or more samples are compared to find differences in their protein expression.

Two different protein-separating techniques are combined in sequence to achieve the goal of protein separation and identification—**Isoelectric Focusing (IEF)** and **SDS-PAGE.** Isoelectric focusing (IEF) is used in the first-dimension. This separates proteins by their charge (pI) and SDS-PAGE (sodium dodecyle sulphate-polyacrylamide gel electrophoresis) is used in the second-dimension. This separates proteins by their size (molecular weight, MW). The procedure is known as **ISO-DALT** (iso for isoelectric focusing and dalt for molecular weight in dalton).

Isoelectric focusing (IEF). The side chains of amino acid residues of a protein contribute a net charge for protein molecules, which depend on the pH of the medium. In simple electrophoresis the mobility of the protein molecules is dependent on its charge that is controlled by the pH. There is a pH for every protein molecule at which the net charge of the protein becomes zero. This pH is known as **isoelectric pH** or **isoelectric point (PI).** At isoelectric pH the protein loses its mobility in the electric field. The technique of protein separation based on the property of isoelectric pH or PI is known as isoelectric focusing (IEF). A pH gradient is generated on an IEF gel and proteins are allowed to move in an electric field and that results in the separation of an individual protein species according to its isoelectric point.

Isoelectric focusing pH (IEF pH) gradients can be generated by adding ampholytes to an acrylamide gel. These are a mixture of amphoteric species of polymeric buffers with a range of pI values. An ampholyte should have the following properties: even conductivity, high buffering capacity, soluble at isoelectric point, and minimum interaction with focused proteins. Ampholytes, like proteins, are also charged and have their own pI values. These ampholytes gel and may have to be pre-focused before sample application. **Pre-focusing** is carried out by subjecting electrophoretic gels containing the ampholytes to an electric field for three to four hours. The cathode chamber contains a weak base and the anode chamber contains a weak acid. During this process each ampholyte species migrates to a position where its pH is equal to its pI and this generates a pH gradient across the gel. After this pre-focusing of ampholytes, the protein sample is applied and the electrophoresis is continued until all the protein molecules are focused at their isoelectric point, when the net current becomes zero. At this point there won't be any protein molecule with a net charge.

Similar to ampholytes, there are the **immobilines** used for IEF. These are the ampholytes immobilized within the polyacrylamide gel producing an immobilized pH gradient or IPG that does not need to be pre-focused. They are more stable as the gel is plastic packed and the pH gradient is fixed. This leads to improvements in reproducibility as they are mechanically strong and the pH gradient cannot drift.

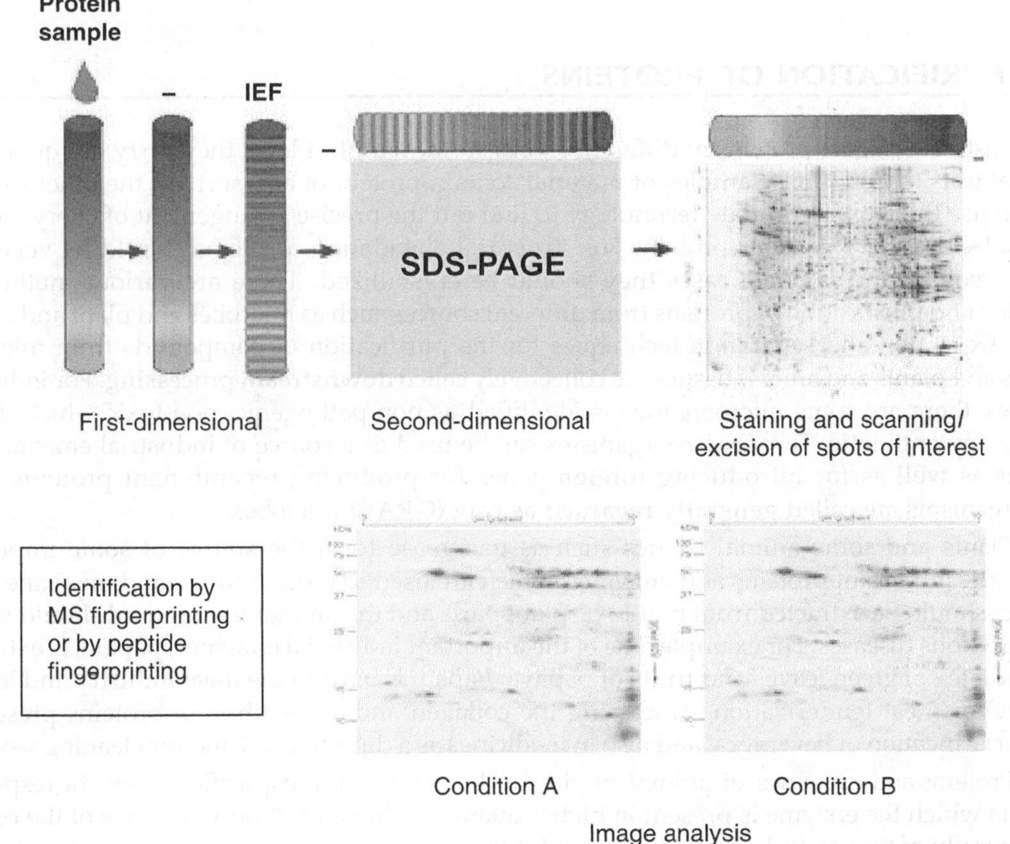

FIGURE 14.13 **Two-dimensional gel electrophoresis.**

SDS-PAGE (second-dimension). Electrophoresis in the second-dimension is by SDS-PAGE. Electrophoresis in the first-dimension is carried out in a tube gel (IEF tube). After running the first-dimensional electrophoresis, the IEF tube gel is placed horizontally across the top of the stacking gel of SDS-PAGE gel. Now the protein already separated on the basis of pI is separated on the basis of molecular size in the second-dimension electrophoresis by SDS-PAGE.

After completing the electrophoresis, the gel is stained with protein-specific dyes such as commasie brilliant blue, amido black, or silver stains. Fluorescent dyes developed recently are more effective and improve protein detection sensitivity. The stained gels can be photographed or scanned by specially designed scanners. These can produce images of high resolution even for a faintly stained protein spot in the gel. These high-quality images are very important for comparing the protein profiles from different cells, tissues, or organisms grown at different environmental or nutritional conditions.

Both IEF and SDS-PAGE are very powerful protein-separating techniques that result in high resolution. Therefore, their combined effect will be more effective and will result in several fold improvement in the resolution of proteins. The chance of having two proteins with the same molecular weight and isoelectric pH is almost absent. Therefore, this type of two-dimensional gel electrophoresis can resolve all the proteins produced by a cell or organism, and thus it is very suitable for studying proteomics.

14.4 PURIFICATION OF PROTEINS

Biochemists investigate proteins at different levels. At the simplest level they carry out qualitative chemical tests to find out if samples of material contain protein of any sort. At the other extreme they can use the most up-to-date technology to find out the precise arrangement of every atom in each molecule of a particular protein. For structural elucidation, proteins should be very pure homogeneously and in some cases they should be crystallized. There are various methods of extraction and purification of proteins from different sources such as microbes and plant and animal tissues. Extraction and separation techniques for the purification of compounds from microbial cultures and plants and animal tissues are collectively called **downstream processing**. For industrial purposes, there are some microorganisms identified as non-pathogenic, non-toxic which do not produce any antibiotic. These microorganisms can be used as a source of industrial enzymes and proteins as well as for introducing foreign genes for producing recombinant proteins. Such microorganisms are called **generally regarded as safe (GRAS)** microbes.

Plants and some animal tissues such as pancrease form the source of some important industrially important proteins and enzymes, which are used in food industry and medicine. Such enzymes should be extracted from non-toxic plant parts and the animal tissues used should be free from infectious diseases. For example, one of the important industrial enzymes, papain, is extracted from the latex of green leaves and fruits of papaya. Papain is used in the meat industry and leather industry for meat tenderization, processing the collagen and other fibrous proteins present in leather, clarification of beverages, and also in medicines as a digestive aid and for cleaning wounds.

Proteins and enzymes of animal origin can be extracted and purified from the respective organs in which the enzyme is present in higher quantities. Slaughterhouses are one of the centers for the supply of tissues and organs necessary for the extraction of certain proteins such as insulin. Traditionally, insulin was obtained from the pancreas of cows and pigs. It requires the slaughtering of about 100 to 150 pigs or 15 to 20 cows to meet the insulin requirement of a single diabetic patient per year. From this we can imagine the number of animals that have to be killed to meet insulin requirements. But modern biotechnology and genetic engineering has come in to help the situation. Nowadays we don't depend on animals for insulin. There are a number of pharmaceutical companies which manufacture and market human insulin produced by genetic engineering. Similarly, a large number of therapeutic proteins such as vaccines and hormones are now produced by genetic engineering. Today, efforts are there to produce transgenic plants and animals capable of producing these therapeutic antibodies and hormones, and other industrially important proteins in specific organs of plants and animals. For example, the production of edible vaccines (edible plant parts such as fruits and tubers containing the vaccine) and expression of certain proteins such as insulin in the milk of cows and goats, and in eggs. This method of genetic engineering for large-scale production of specialized proteins and other molecules is called **molecular farming**. The main advantages of molecular farming is that the cost of production can be reduced, costly and time-consuming fermentation procedures and downstream processes can be avoided, large-scale production of the specific compounds is possible, and ease of production and purification procedures is advantageous.

Edible vaccines have the advantage that they can be stored for a long time without refrigeration, can be easily transported, and can be administered by feeding the fruit or the plant parts having the vaccine. While eating the vaccinated fruit, the vaccine molecules will be absorbed into the bloodstream through the mucous membrane lining the mouth and esophagus. Even animals can be vaccinated by this way. Attempts are being made to develop transgenic fodder grass containing the anthrax vaccine. Cattle can be fed with this transgenic fodder grass and be vaccinated effectively against anthrax.

Whatever the source of protein, whether it is natural sources or genetically-altered organisms, the general principles of protein extraction and purification are almost identical. The procedures depend on whether the protein is extracellular or intracellular. The exact protocol for the extraction and purification has to be designed individually after considering a number of factors, which are given below:

- Source of protein (plant, animal, or microbe)
- Location of protein (organ such as leaf, tuber seeds, fruit, stem, etc., or liver, brain, pancreas, etc.)
- Amount of protein required and the amount of raw material required for the extraction
- Physical, chemical, and biological properties of the desired protein

The extraction procedure has to be designed according to the type of source material. If the protein is extracellular the medium of the culture has to be made cell free by separating the cells from the medium by centrifugation to purify the specific protein. If the protein of interest is intracellular, the cells have to be disrupted to release the protein and an appropriate cell-disruption method, depending upon the type cells or tissue, has to be adopted. Animal cells are easier to break since they don't have a cell wall. Plant and fungal tissues are hard to break and therefore need mechanical devices. Bacterial cells and other types of microbes such as yeast cells are very small and need special techniques that involve high-pressure or sonic vibrations.

Once the cells are broken and proteins are released, a protein purification protocol is designed by applying a number of physico-chemical techniques independently or in combination, depending on the physico-chemical and biological properties of the protein in question. Recombinant proteins produced by genetically-engineered organisms are often attached with certain special peptide molecules such as the maltose-binding protein for applying certain very specialized, single-step purification techniques such as affinity chromatography. For example, recombinant proteins attached with the maltose-binding peptide can be purified by applying maltose-affinity chromatography. In this method the fusion protein having the maltose-binding peptide binds to the affinity column containing the maltose. The protein bound to the maltose can be eluted and recovered later using a buffer of suitable ionic strength and pH. A protein-monitoring technique to detect and quantify the total protein and the target protein levels is very essential for the proper and easy purification and characterization of a protein. The detection and quantity of a specific protein can be monitored by a proper-assay method. By protein-assay method the activity of that protein and its intensity can be measured. The protein activity per quantity of protein present is called the **specific activity**, which is an indicator of protein purity. After each purification step, the enzyme activity and the total protein can be determined and that gives the specific activity and fold of purification as shown below:

Sl. No.	Procedure/Steps	Total Protein (mg)	Activity (Units)
1.	Crude extract	20,000	35,00,000
2.	Ammonium sulphate precipitation	8,000	27,00,000
3.	Size exclusion chromatography	500	10,00,000
4.	Affinity chromatography	50	8,00,000

After purification the protein has to be examined for its biological activity, if it is meant for any biological use, such as in the food industry or in medicine as in the case of insulin and vaccines. But if the protein purification is for studying its structure and other properties there are fast and effective physico-chemical methods.

Extraction of Protein (Initial Recovery of Protein)

In the case of the microbial system, if the protein is intracellular, the cell harvesting is done by centrifugation or by filtration of the fermentation medium. The cells separated can be subjected to an appropriate method of cell disruption in a suitable buffer. If the protein is an extracellular type, the fermented medium itself can be used as the source of protein. Protein can be directly separated by salt precipitation or can be subjected to a suitable chromatographic technique such as gel filtration or affinity chromatography. In the case of multicellular organisms, the first step is the collection of the appropriate organs where the target protein is located in maximum quantity. If you want to get a plasma protein you have to collect the blood, for insulin get the pancreas, for liver specific-proteins you have to collect the liver, for pituitary proteins you have to collect the pituitary glands, and if you want to isolate a photosynthetically-related protein, you have to use green leaves as the starting material. The source tissue has to be homogenated under appropriate conditions to extract the protein and appropriate purification techniques have to be followed after considering various other factors such as the nature of protein, its quantity, and its end use.

The protein of a crude homogenate can be separated or partitioned into two aqueous phases that differ in polarity and density. The crude extract or homogenate is added to a biphasic system consisting of a mixture of dextran and polyethylene glycol (PEG) and subjected to centrifugation. The cell debris and other impurities separate into the lower, more polar and dense phase, the **dextran phase.** The soluble proteins separate into the less dense, less polar and upper layer or phase, the **PEG phase.** Thus, the cellular debris and other insoluble impurities can be separated effectively by this aqueous two-phase partition.

The protein extract contains all other macromolecules such as carbohydrates, lipids, and nucleic acids in varying amounts. These can also interfere with purification protocols and biological activity as well as the functional status of the protein. There are different methods to avoid these dissolved impurities either by separating the protein from the mixture, by precipitation, or by separating the impurities from the protein solution.

The lipids can be removed by passing the solution through a glass wool or a cloth and nucleic acids can be removed by treating with nuclease or by precipitation followed by centrifugation. A convenient method is to precipitate protein either by salting out or by organic solvents such as acetone or by changing the pH or by ion-exchange chromatography. Whatever the techniques adopted for protein purification, there are some basic criteria that should be maintained to keep the protein biologically active and stable and these are given below:

1. The buffer in which the protein is maintained should be around the physiological pH, if the specific pH value of that particular protein is not known.
2. Maintain the temperature below 4°C both during the process of purification and storage.
3. Minimize the extraction and purification time.
4. The extraction buffer should be provided with proteinase inhibitors to avoid possible activity of proteolytic enzymes on the target protein.
5. Processes and chemicals that can destroy the three-dimensional structure of the protein and thereby the biological activity should be avoided.

Obtaining a Pure Sample of Protein

Dialysis

The protein is separated from smaller molecules and ions by enclosing a solution of the impure protein in a partially permeable membrane. Small, solute molecules and ions pass through this, while the larger protein molecules are held back (Figure 14.14).

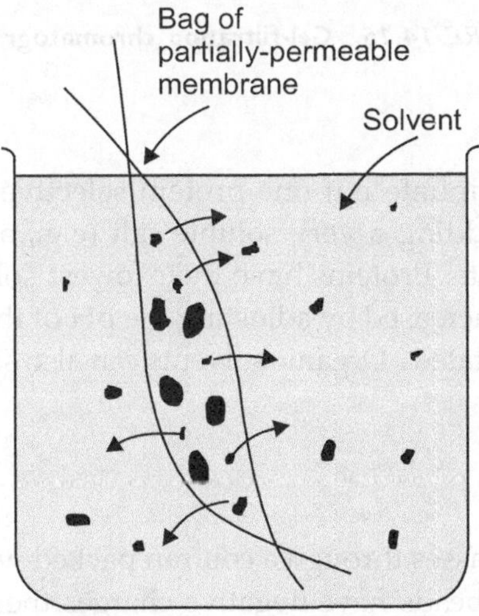

FIGURE 14.14 **Dialysis—small molecules can pass through the membrane; large protein molecules cannot.**

Gel-filtration Chromatography

The solution of impure protein passes through a column packed with beads of a polymer. The large protein molecules can move freely around the beads and pass through quickly. Smaller molecules and ions have to pass through the polymer network inside the beads (Figure 14.15) and so they take longer to travel through the column.

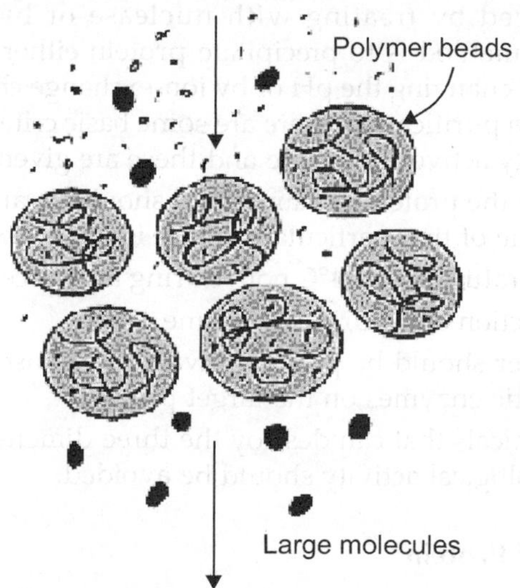

FIGURE 14.15 **Gel-filtration chromatography.**

Precipitation and Salting Out

It is sometimes possible to precipitate out one protein selectively from a solution containing a mixture of different proteins. Adding a very soluble salt (e.g., ammonium sulphate) causes the least soluble protein to precipitate. Proteins have their lowest solubility at their isoelectric point. Selectivity of salting out can be increased by adjusting the pH of the mixture to the isoelectric point of the protein that is to be precipitated. Organic solvents can also selectively precipitate a particular protein.

Ion-exchange Chromatography

The solution of impure protein passes through a column packed with beads of a polymer that have charged groups attached. If the beads have negative charges then the protein molecules with an overall positive charge stick to them, while others wash through. The molecules can be released from the column by changing the pH or salt concentration of the solution washing through. By adjusting the pH at which the column is loaded and the pH or salt concentration at which the column is washed, some of the different protein components of a mixture can be separated.

Affinity Chromatography

Some proteins are able to bind strongly to other specific molecules. These molecules can be attached to a polymer support, packed in a chromatography column. When the column is loaded with an impure sample, one protein binds to the molecules held on the polymer while the impurities pass through. The pure protein can then be displaced from the column (Figure 14.16).

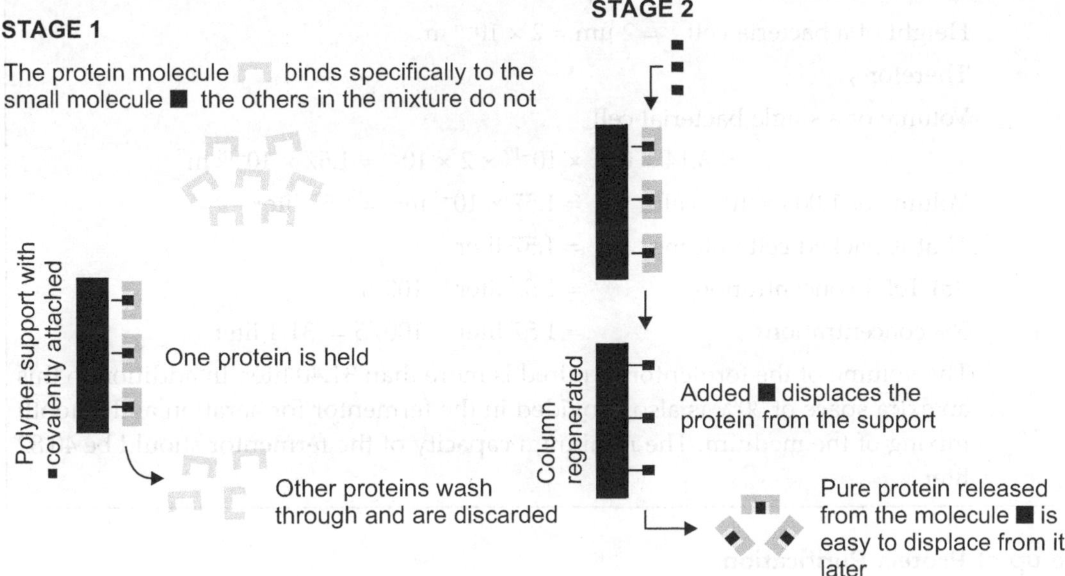

FIGURE 14.16 **Affinity chromatography.**

Box 1

1. A bacterial cell produces at least 3,000 different proteins. One of these proteins is produced at the level of 6,000 molecules per cell under optimal growth conditions. If we want to purify one gm of this intracellular protein, estimate the number of bacterial cells required theoretically. The molecular weight of the protein is 100 kD (kilo Daltons).

   1 Mole of the protein = 100 Kg (1,00,000 g)

   1 mol = 100 kD = 6.023×10^{23} molecules (Avogadro's number)

   Therefore, 1g of protein = $1/1{,}00{,}000 \times 6.023 \times 10^{23} = 6.023 \times 10^{18}$

   6,000 molecules of the protein are present in one bacterial cell and therefore,

   6.023×10^{18} molecules are present in $6.023 \times 10^{18}/6000 = 1.003 \times 10^{15}$ cells.

2. Suppose a bacterial cell is a cylinder of diameter 1 μm and height 2 μm, calculate the following:

   (*a*) The total packed cell volume of the bacteria required to produce 1 gm of intracellular protein.

(*b*) Volume of the fermentor required if the maximum cell concentration inside the fermentor is 5% as the cells need space to multiply.

(*a*) Volume of a cylinder $= \pi r^2 h$

Diameter of bacteria $= 1\ \mu m$

Therefore, radius $= 0.5\ \mu m$

Height of a bacteria cell $= 2\ \mu m = 2 \times 10^{-6}$ m

Therefore,

Volume of a single bacterial cell

$$= 3.14 \times 0.5^2 \times 10^{-12} \times 2 \times 10^{-6} = 1.57 \times 10^{-18}\ m^3$$

Volume of 1.003×10^{15} cells $= 1.57 \times 10^{-3}\ m^3 = 1.57$ liter

That is packed cell volume = 1.57 liter

(*b*) Total concentration = 1.57 liter = 100%

5% concentration = 1.57 liter × 100/5 = 31.4 liter

The volume of the fermentor required is more than 31.40 liter. In addition to this an extra space of 30% is also provided in the fermentor for aeration and smooth mixing of the medium. The minimum capacity of the fermentor should be 40.82 liter.

Scale up of Protein Purification

When the protein purification protocol forms the part of an industrial process for the production of industrially important or pharmaceutically important protein such as vaccines, hormones, etc., the laboratory scale protocol also has to be scaled up. Actually, it forms the process of downstream processing.

Special care should be taken to bring the cost of production of the protein down by purchasing the chemicals in bulk quantities and using equipment and holding vessels, which are permanent or recyclable. The materials and vessels should be made up of cheap materials and should be inert and resistant to corrosion. Leaching of toxic metals and chemicals into the product should be avoided. The process of manufacturing and downstream processing require the approval of the regulatory authority to produce and market a protein that is useful in food industry or as pharmaceuticals. About 80% of production costs is for downstream processing and quality assurance. Therefore, by controlling the various steps and equipment in downstream processing and product recovery, we can bring down the cost of production without compromising the quality of the product.

Bulk-Protein Production

The protein-purification procedure that is applicable for the laboratory scale is applicable for the downstream process of the industrial scale also. The schematic representation of the flow chart given in the Figure 14.17 can be used for the purification of enzymes in bulk quantities not only from microbial fermentation systems but also from plant and animal tissues.

Purification of Proteins Used for Therapeutic and Diagnostic Purposes

Drug proteins such as digestive enzymes, therapeutic proteins such as antibodies or vaccines, hormones, growth factors, etc., and other diagnostic proteins should be purified to a very high degree. Those proteins that are for pericutaneous or intra-venal administration should be highly purified and sterile. Parentral preparations are those that are intended for intra-venal administration or for infusion or for implantation. Such materials should be highly purified and sterile.

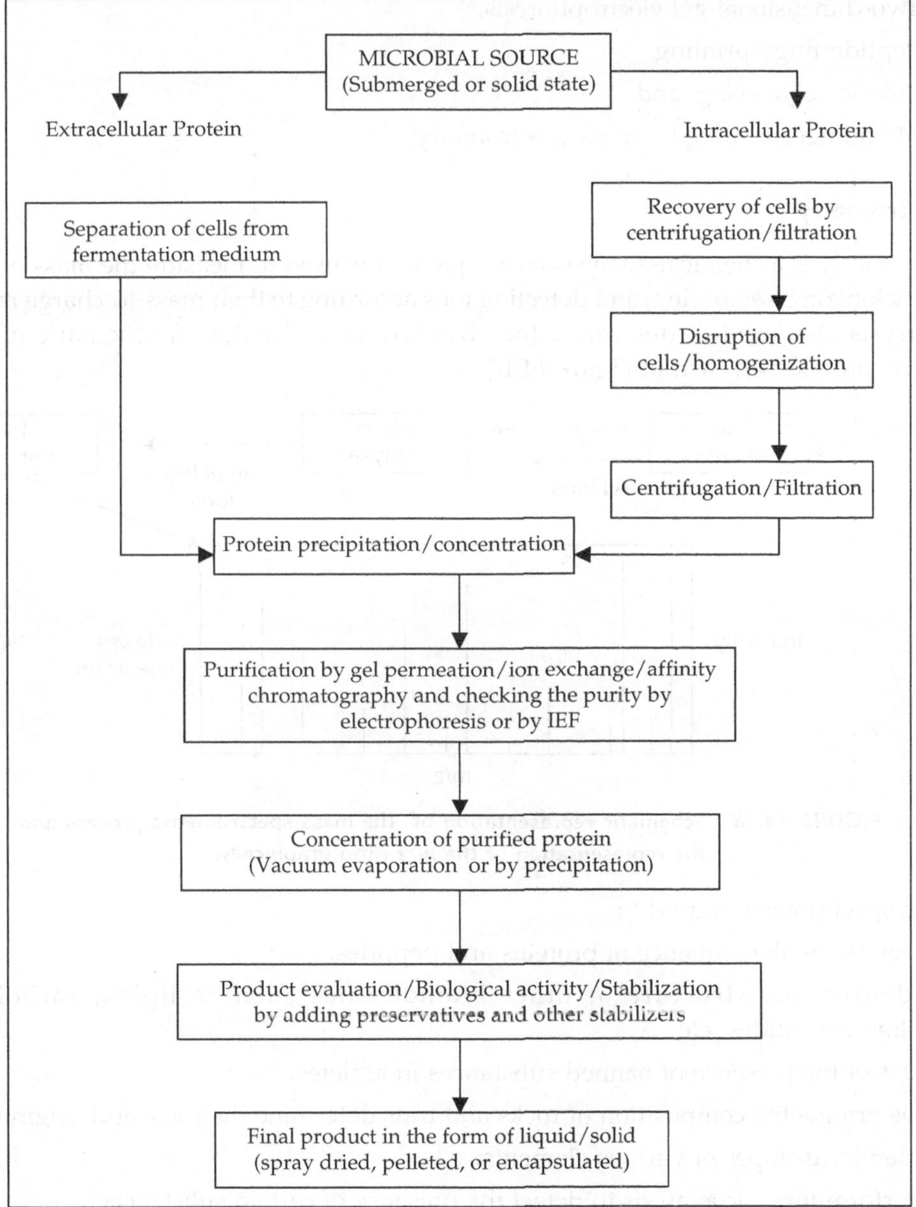

FIGURE 14.17 **A general flow chart for the purification of protein from microbial cultures.**

14.5 CHARACTERIZATION OF PROTEINS

The protein purified has to be identified and characterized. There are various methods to accomplish this. The purity of the protein can be determined by the SDS-PAGE or by isoelectric focusing (IEF) or by two-dimensional gel electrophoresis.

Identification or characterization of protein can be carried out with the help of any one of the following techniques or by a combination of these:

1. Two-dimensional gel electrophoresis,
2. Peptide fingerprinting,
3. Protein sequencing, and
4. Mass fingerprinting by mass spectrometry.

Mass Spectrometry

Mass spectrometry is a chemical-analysis technique that is used to measure the mass of unknown molecules by ionizing, separating, and detecting ions according to their mass-to-charge ratios. Mass spectrometry is also used to determine the structure of molecules. A schematic of the mass spectrometric process is shown in Figure 14.18:

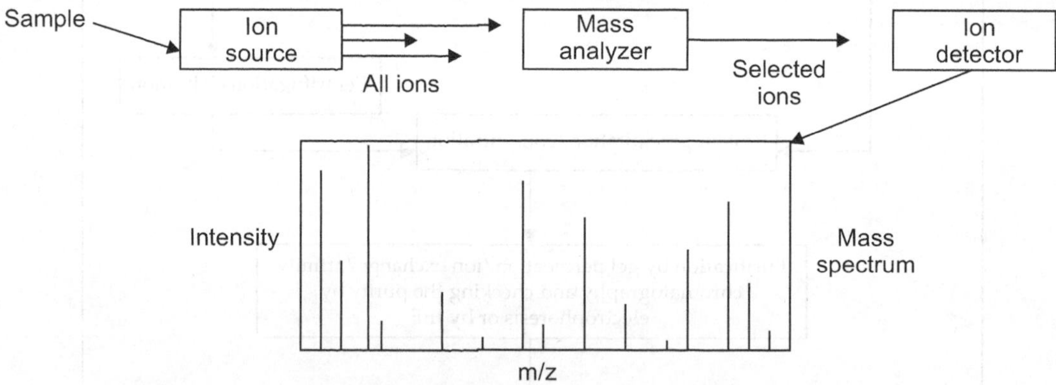

FIGURE 14.18 **Schematic representation of the mass spectrometric process and the representation of the *m/z* ratio graphically.**

Mass spectrometry is used to:

- Determine the sequence of proteins and peptides.
- Identify the structure of other biomolecules such as lipids, carbohydrates, oligonucleotides, etc.
- Detect the presence of banned substances in athletes.
- Determine the composition of rocks and thus determine their age and origin.
- Identify isotopes of various elements.
- Perform forensic analysis to detect the presence of certain substances.

A mass spectrometer creates charged particles (ions) from molecules. It then analyzes those ions to provide information about the molecular weight of the compound and its chemical structure. There are different types of mass spectrometers and sample introduction techniques, which allow a wide range of analyses. This discussion will focus on the principle of mass spectrometry and its use in the study of protein chemistry. The techniques can be used to sequence peptides and proteins and to study their interactions. It is useful in the identification of protein molecules, their accurate molecular weight, the post-translational modifications, and structural functional relationship of proteins.

It is now possible through mass spectrometer to 'see' proteins of molecular weight as high as 100,000 daltons. A big advantage is that a very small sample, as small as picomoles, is required. Thus, it is possible to do two-dimensional gel electrophoresis of proteins of a cell (to separate the few thousand different proteins) and identify them using mass spectrometer. This approach of protein analysis and identification is one of the important techniques of what is called **proteomics**—the study of the complete protein complement of a cell.

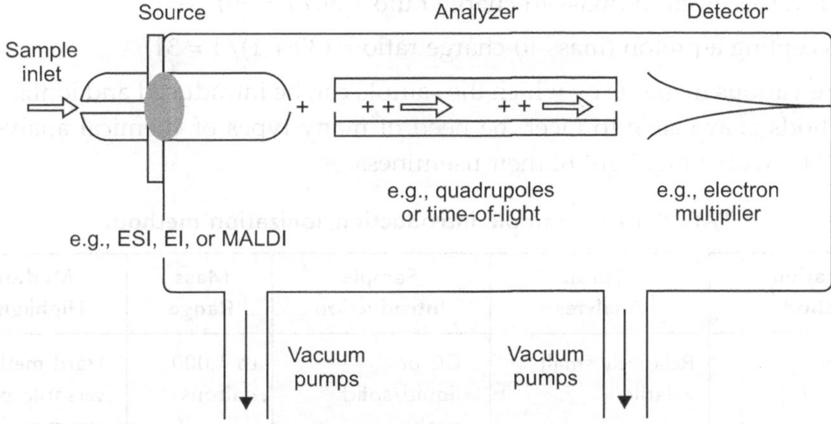

FIGURE 14.19 **Schematic representation of a mass spectrometer.**

A mass spectrometer is an instrument used to separate ions according to their mass-to-charge ratios. The mass spectrometer consists of three components—an ion source, a mass analyzer, and an ion detector. The sample to be analyzed is introduced into the ion source in a liquid or vapor phase where it is fragmented and ionized. These ions are then passed to the mass analyzer where they are separated according to their mass-to-charge ratios (m/z) and then go to the ion detector, where their presence is recorded and the mass spectrum is produced. Figure 14.19 shows the schematic representation of a mass spectrometer.

- Ionization
- Ion Analyzer
- Detector

Ionization

Ionization is a process of producing an electrically charged protein molecule from a neutral protein molecule by either adding protons or removing electrons. Why is ionization of a sample required in mass spectrometry? The sample to be analyzed by mass spectrometry must be ionized in order to separate the ions according to their mass-to-charge ratio and keep the size of the mass spectrometer instrument small. While it is possible to analyze samples that have not been ionized, the size of the instrument required for separating samples by gravity according to their mass alone will be very large.

The 'mass-to-charge' ratio of an ion is the mass of the ion divided by the charge on the ion. The mass of the ion is typically measured in daltons. The charges on an ion are always positive integers (1, 2, 3, ...). Charged ions are produced by ionization. Ions can be charged by adding a proton (in which the total mass of the ion increases by one and the charge on the ion increases by one) or by removing an electron (in which case the mass remains the same, but the charge decreases by one). Mass-to-charge ratios are typically plotted on the X-axis in a mass spectrum.

For instance, consider the molecule C_2H_6. It has a mass of $(2 \times 12) + (6 \times 1) = 30$ Da. This molecule can acquire a charge of one unit in two ways:

- by losing an electron (mass-to-charge ratio = 30/1 = 30)
- by accepting a proton (mass-to-charge ratio = (30 + 1)/1 = 31)

There are various methods by which the sample can be introduced and ionized. An array of ionization methods is available to meet the need of many types of chemical analysis. A few are listed in Table 14.1 with a highlight of their usefulness.

TABLE 14.1 Sample introduction/ionization method.

Ionization Method	Typical Analytes	Sample Introduction	Mass Range	Method Highlights
Electron Impact (EI)	Relatively small volatile	GC or liquid/solid probe	to 1,000 daltons	Hard method versatile provides structure info
Chemical Ionization (CI)	Relatively small volatile	GC or liquid/solid probe	to 1,000 daltons	Soft method molecular ion peak $[M + H]^+$
Electrospray (ESI)	Peptides, proteins non-volatile	Liquid chromatography or syringe	to 200,000 daltons	Soft method ions often multiply charged
Fast Atom Bombardment (FAB)	Carbohydrates, organometallics peptides, non-volatile	Sample mixed in viscous matrix	to 6,000 daltons	Soft method but harder than ESI or MALDI
Matrix Assisted Laser Desorption (MALDI)	Peptides, proteins nucleotides	Sample mixed in solid matrix	to 500,000 daltons	Soft method very high mass

Ion Analyzer

Molecular ions and fragment ions are accelerated by manipulation of the charged particles through the mass spectrometer. Uncharged molecules and fragments are pumped away. The quadrupole mass analyzer in this example uses positive (+) and negative (–) voltages to control the path of the ions. Ions travel down the path based on their mass-to-charge ratio (m/z). EI ionization produces singly charged particles, so the charge (z) is one. Therefore, an ion's path will depend on its mass. If the (+) and (–) rods shown in the mass spectrometer schematic were 'fixed' at a particular rf/dc voltage ratio, then one particular m/z would travel the successful path shown by the solid line to the detector. However, voltages are not fixed, but are scanned so that ever-increasing masses can find a successful path through the rods to the detector.

The following are the various types of mass analyzers used in different types of mass spectrometry (Table 14.2).

TABLE 14.2 Mass analyzers.

Analyzer	System Highlights
Quadrupole	Unit mass resolution, fast scan, low cost
Sector (Magnetic and/or Electrostatic)	High resolution, exact mass
Time-of-Flight (TOF)	Theoretically, no limitation for m/z maximum, high throughput
Ion Cyclotron Resonance (ICR)	Very high resolution, exact mass, perform ion chemistry

Detector

There are many types of detectors, but most work by producing an electronic signal when struck by an ion. Timing mechanisms, which integrate those signals with the scanning voltages, allow the instrument to report which m/z strikes the detector. The mass analyzer sorts the ions according to m/z and the detector records the abundance of each m/z. Regular calibration of the m/z scale is necessary to maintain accuracy in the instrument. Calibration is performed by introducing a well-known compound into the instrument and "tweaking" the circuits so that the compound's molecular ion and fragment ions are reported accurately.

What is a Mass Spectrum?

A **mass spectrum** is a plot that shows the relative abundance of ions of various mass to charge ratios. The X-axis represents the mass-to-charge ratio of the ions and the Y-axis represents the relative abundance of each ion. A typical mass spectrum of carbon dioxide (CO_2) is shown in Figure 14.20.

Mass spectrometry can be in association with or linked to one or more separating techniques. For example, in organic chemistry mass spectrometry is always linked to Gas chromatography and therefore, is called **GC-MS**. Gas chromatography will convert the samples into gaseous phase, which is introduced to the mass spectrometer. In addition to this, GC can be used to separate the components in a mixture and the separated pure compounds can be directly identified and analyzed by a mass spectrometry spectrum. This is the advantage of linking a separating technique to mass spectrometry. Similarly, in the case of protein studies mass spectrometry can be linked to various types of liquid chromatographic systems such as GPC, ion exchange, or affinity chromatography individually or in tandem, or to capillary electrophoretic systems, which can also be operated automatically, like HPLC.

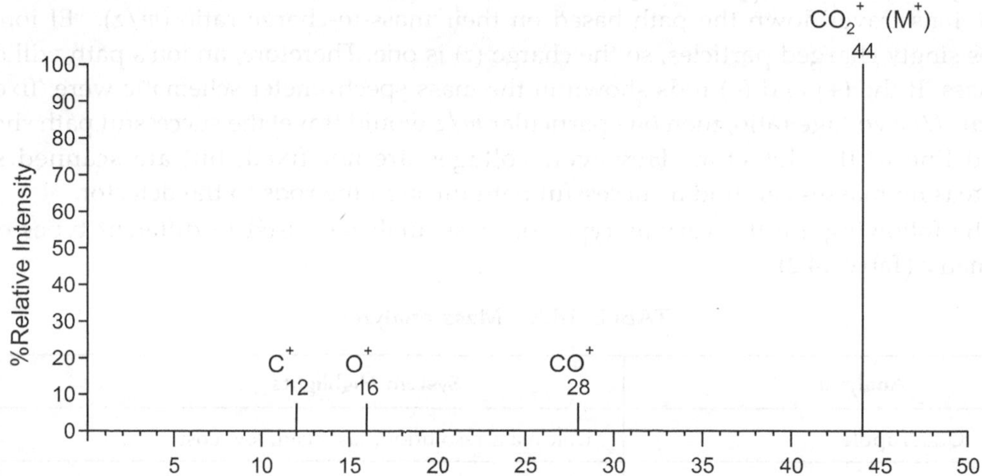

FIGURE 14.20 **Mass spectrum of carbon dioxide (CO_2). Molecular ion is seen at m/z = 44.**

TABLE 14.3 The separation techniques linked to mass spectrometry.

GC/MS	Gas chromatography coupled to mass spectrometry.
LC/MS	Liquid chromatography coupled to electrospray ionization mass spectrometry.
CE/MS	Capillary electrophoresis linked to mass spectrometry such as MALDI-TOF or MS/MS.

There are various types of mass spectrometry depending on the type of ionization source or the ion analyzers. They are used for specialized studies in proteomics such as identification of proteins by generating mass fingerprints or the ionization pattern of the protein molecule, amino acid sequencing of peptides, and thereby the detailed studies of the three-dimensional structure of the protein and modification of the proteins if any can be detected very easily and efficiently. Some of the common types of mass spectrometry are discussed below.

MALDI-TOF Mass Spectrometry

Matrix-Assisted Laser Desorption Ionization—Time of Flight (MALDI-TOF) is another method of ionization that does not require any heating for the production of ions and can be used for heat-sensitive molecules. In this method, the sample to be analyzed is mixed with a matrix compound and crystalized. The compound of this matrix is usually a weak inorganic acid. This mixture (matrix mixed with sample) is then excited with a laser, which results in the evaporation of the matrix compound. The matrix compound also carries the sample molecules into the vapor phase, resulting in indirect vaporization of the sample. Sample ions are formed by the exchange of electrons and protons with the matrix compound. MALDI is especially useful for protein and peptide identification using masses alone, since the masses of ions can be determined with great accuracy. Time of flight is the technique used for the sorting and separation of ions generated by laser ionization.

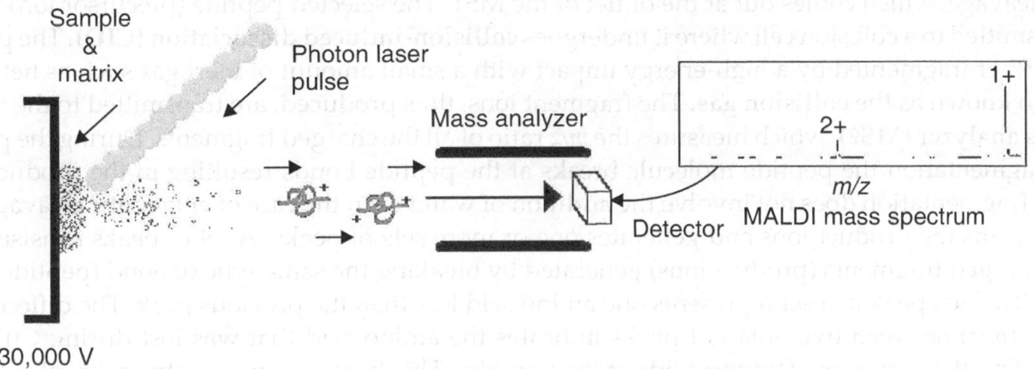

FIGURE 14.21 **Schematic representation of MALDI.**

Electrospray Ionization (ESI)

This is a newer method of ionization that does not cause excessive fragmentation. ESI generates ions directly from the solution without heating. Therefore, this method can be used for heat-sensitive molecules, which cannot be ionized by heating. The sample is finely sprayed in the presence of an

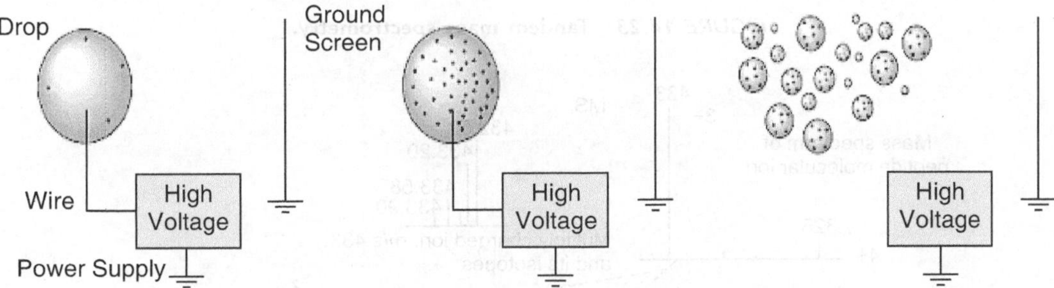

FIGURE 14.22 **Schematic of electron ionization.**

electrical field. Charge accumulates on the sample droplets which then explode due to mutual repulsion of charges leading to the formation of ions. Both single- and multiple-charged ions can be produced in this manner.

Tandem Mass Spectrometry (MS/MS or Two-dimensional MS)

Tandem MS is a procedure in which multiple rounds of mass spectrometry are performed to obtain greater resolution of the ions generated by the unknown sample molecule. Short stretches of polypeptides can be sequenced with tandem MS. In general, a polypeptide is cleaned into a number of fragments enzymatically and the products are taken into the mass spectrometer for analysis one by one to generate the primary structure or the amino acid sequence.

The protein solution is cleaned into short stretches of peptides preferably by an enzyme such as trypsin or chymotrypsin. The mixture of short peptides thus generated is injected into the tandem MS for analysis. The tandem MS is an instrument with two MS in a series. They are designated as MS1 and MS2. In the first mass analyser (MS1) the peptide mixture (precursor ions) is sorted resulting in the selection of only one peptide from the several types of peptides produced by cleavage, which comes out at the outlet of the MS1. The selected peptide (precursor ion) is then transmitted to a collision cell where it undergoes **collision-induced dissociation (CID)**. The peptide is further fragmented by a high-energy impact with a small amount of inert gas such as helium or argon known as the **collision gas**. The fragment ions, thus produced, are transmitted to the **second mass analyzer (MS2)**, which measures the *m/z* ratio of all the charged fragments. During the process of fragmentation the peptide molecule breaks at the peptide bonds resulting in the product ions. This fragmentation does not involve the addition of water as in the case of enzymatic cleavage. The MS2 scans the product ions and generates one or more sets of peaks. A set of peaks consists of all the charged fragments (product ions) generated by breaking the same type of bond (peptide bond) by CID. Each peak in a set represents one amino acid less than the previous peak. The difference in mass (*m/z*) between two adjacent peaks indicates the amino acid that was lost during CID, thus revealing the sequence of amino acids of the peptide. This type of mass spectrometry, also known as 2D MS, is usually used for the analysis of complicated mass spectrum, particularly useful in the amino acid sequencing of peptides and post-translational modifications. The strategy of tandem mass spectrometry using electron spray ionization is shown in Figure 14.24.

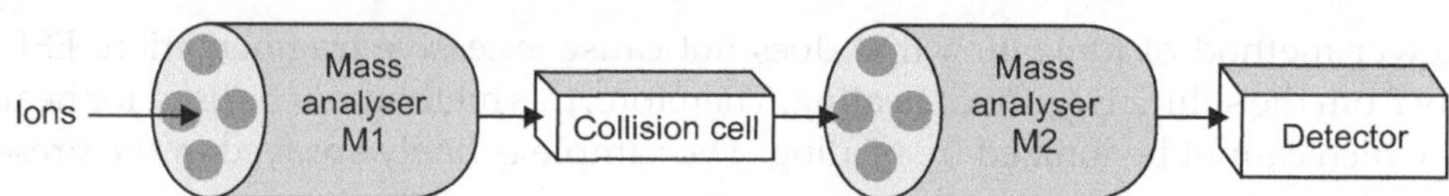

FIGURE 14.23 **Tandem mass spectrometry.**

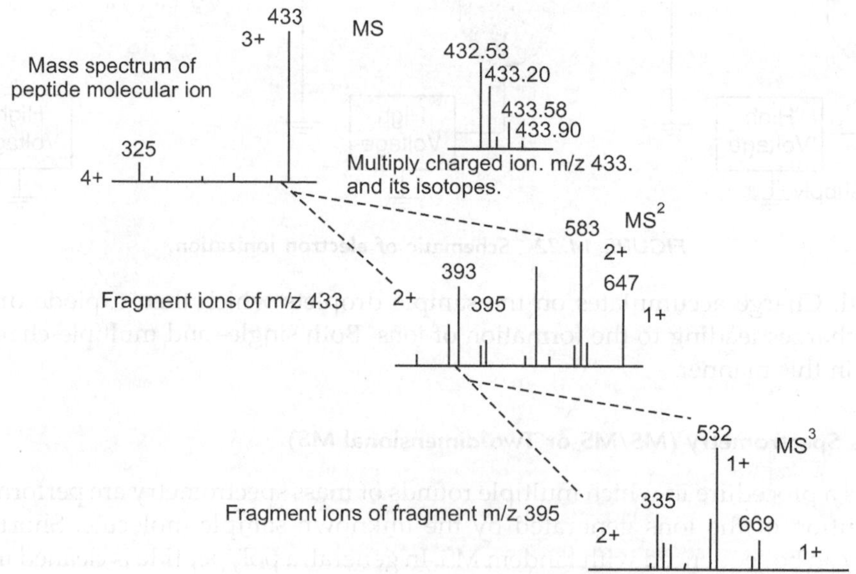

FIGURE 14.24 **Tandem MS spectrum.**

All three mass spectrometry processes are very much useful in the study of proteins—identification and amino acid sequencing. Typically, two-dimensional gel electrophoresis combined with a type of mass spectrometry is a powerful tool for the study of proteomics. A crude protein extract directly from a cell sample is separated on a two-dimensional gel (by IEF followed by SDS-PAGE) and the protein spots are visualized on the gel. The spots that are needed to be analyzed are excised from the gel and introduced into the mass spectrometer along with the gel matrix. Thus, the protein can be identified either by generating the mass fingerprint and comparing it with a standard molecular fragmentation pattern present in the database, or by generating the amino acid sequence of the peptide and comparing it with known sequences. Mass fingerprinting is a very simple and rapid method for the identification of protein expressed by the cells under varying conditions. The molecular fragmentation pattern obtained in the form of a mass spectrum is called the mass fingerprint of a molecule, which is characteristic for that particular compound. Proteins of large size can be cleaved into smaller peptides by treating them with proteases such as trypsin and separating the fragments by column chromatography attached to the MS. The separated peptides can be directly injected into the mass spectrometer.

Other liquid chromatography techniques such as ion exchange, affinity, or reverse-phase column chromatography can be used independently or in tandem for the separation of peptides and proteins from a mixture. In addition to this, capillary electrophoresis also can be attached with mass spectrometer, so that proteins or peptides can be separated very efficiently from a mixture before they are introduced into the mass spectrometer. The peptides that are introduced into the mass spectrometers are either sequenced directly or the mass fingerprint of the peptide is analyzed by comparing it with the database, which is available online in many databases such as the PDB (protein data bank).

14.6 PROTEIN-BASED PRODUCTS

Proteins, as biological macromolecules, are very important for the existence of normal activity in a living system. The biological roles played by various types of proteins have been discussed earlier. Because of the multifaceted role of proteins, they are also important commercially. Proteins may be classified, as follows, based on their commercial importance:

- Blood-derived proteins and vaccines
- Therapeutic antibodies and enzymes
- Hormones and growth factors as therapeutic agents
- Regulatory factors
- Proteins and enzymes used in analytical applications
- Industrial enzymes and proteins
- Non-catalytic functional proteins
- Nutraceutical proteins

Blood-derived Proteins and Vaccines

Blood is the fluid connective tissue of humans. It is the most important medium for the transportation of gas and nutrients between tissues and sources. The blood forms the source for a number of therapeutically important proteins. A detailed understanding of the formation of blood cells (hematopoiesis) and detailed studies regarding the coagulation of blood have shown the presence of several useful proteins. In addition to blood-coagulating factors, whole blood and blood plasma have been used traditionaly as blood products. These products are obtained from human volunteers who donate blood. But today some of the proteins of blood are produced by a recombinant technique. The blood coagulating factor known as factor VIII used in the treatment of hemophilia A and another factor, the factor IX for treating hemophilia B, etc., and the vaccine for hepatitis B are now synthesized by transgenic bacteria in fermentors.

Therapeutic Antibodies and Enzymes

Antibodies and enzymes that are used for clinical applications are known as **therapeutic antibodies**. Polyclonal antibodies and monoclonal antibodies are examples. In tissue and organ transplantation, the body will recognize these as foreign objects and will produce antibodies against these tissues and organs leading to the failure of organ transplantation. Certain specific antibodies can be used against these natural antibodies and can revert the rejection process. For example, OKT-3 is used during kidney transplantation to revert the acute organ rejection in patients. Similarly, ReoPro is used to prevent blood clots. A large number of therapeutic antibodies are now produced by recombinant DNA techniques. Tissue plasminogen activator (t-PA) is an example of therapeutic enzyme used for acute myocardial infarction. There are a number of enzymes that are used as drugs. Asparaginase is used against some types of cancers and DNase is used for the treatment of cystic fibrosis.

Hormones and Growth Factors as Therapeutic Agents

There are a number of proteins and peptides that act as hormones and growth factors. In the case of diseases due to metabolic errors such as diabeties, caused by the absence of these peptides, the concerned molecules can be administered to correct the metabolic errors. Insulin is the best example. In earlier times, insulin was prepared from the pancreas of cows and pigs. But now, human insulin is prepared through genetic engineering, by transferring the gene of insulin into bacteria; the transgenic bacteria will produce the active human protein. This insulin is called **humulin**. The methods of protein engineering have helped to develop altered forms of insulin, which are more active than the normal ones. There are a number of growth factors such as **EGF (Epithelial Growth Factor)** and platelet-derived growth factors that are used for therapeutic purposes. EGF is used for the treatment of burns and injuries, during skin transplantation, for its growth. The platelet-derived growth factors are used for the treatment of diabetic and skin ulcers. A large number of growth factors and hormones are under clinical trials.

Regulatory Factors

Regulatory factors are also like growth factors, which are closely involved in the process of signal transduction and expression of specific genes. These factors are called **cytokines** because they are involved in the growth and proliferation of cells. Cytokines include interferones (INF), interleukins, tumor necrosis factors, colony-stimulating factors, etc. All these factors find use in therapeutic application for the treatment of different diseases, both infectious and non-infectious. For example, α-interferon is used in the treatment of hepatitis C, β-interferon is used for the treatment of multiple sclerosis, and γ-interferon is used for severe granulomatous disease.

Proteins and Enzymes Used in Analytical Applications

In addition to the use of antibodies and enzymes as therapeutic agents, they are also used in the diagnosis of diseases as the components of some confirmatory tests of certain diagnostic procedures. Hexokinase and glucose oxidase are used in the quantification of glucose in the serum and urine. Glucose-oxidase is used in glucose electrodes. Uricase is used for the estimation of uric acid present in urine. Alkaline phosphatase, horseradish peroxidase, and antibodies are used in ELISA (Enzyme Linked Immunosorbent Assay).

Industrial Enzymes and Proteins

Among commercially useful proteins, industrial enzymes have the first place. Industrially useful enzymes include carbohydrate-hydrolyzing enzymes such as amylases, cellulase, invertases, etc., proteolytic enzymes such as papain, trypsin, chymotrypsin, etc., and other bacterial and fungal derived proteolytic enzymes and lipases that can hydrolyze various types of lipids and fats. All these enzymes are important in the food and beverage industries, the textile industry, paper industry, and detergent industry. Proteases have a special use in the beverage industry, meat and leather industries, cheese production, detergent industry, bread and confectionary industry, etc. Various types of lipases are used for the modifications of various types of lipids and fats, production of various organic acids including fatty acids, in detergents, production of coco butter, etc. In addition to all these, enzymes are used in chemical industries as reagents in organic synthesis for carrying out stereospecific reactions.

Non-catalytic Functional Proteins

These commercially important proteins are used in the food industry as emulsifiers, for inducing gelation, water binding, foaming, whipping, etc. These non-catalytic functional proteins are classified as whey proteins. The proteins that remain in solution after the removal of casein are by definition called whey proteins. The following are the major whey proteins present in milk:

Box 2

- Beta-lactoglobulin
- Alpha-lactalbumin
- Immunoglobulins
- Serum albumin
- Lipoproteins
- Minor proteins like lactoferrin, lactoferoxidase, protease, peptones, etc.
- Free amino acids
- Urea
- Lactose, lactic acid, and sialic acid
- Minerals—sodium, potassium, calcium, magnesium, phosphate, chloride, sulphate, citrate, heavy metals
- Membrane, free fatty acids

TABLE 14.4 Functional properties and uses of whey proteins.

Sl. No.	Useful Property	Mode of Action	Types of Food
1.	Emulsification	Formation and stabilization of fat emulsions	Vegetarian sausages, coffee whiteners, infant foods, biscuits, salad dressings, cakes, ice cream, etc.
2.	Water binding	Entrapment of water and hydrogen bonding of water	Cakes, breads, meat, sausages, etc.
3.	Solubility	Making protein solutions and dissolving proteins	Clearing of beverages
4.	Whipping and foaming	Formation of stable film	Whipped toppings, desserts, eggless cakes, etc.
5.	Gelation	Making protein matrix and setting	Meat, baked foods, cheese
6.	Viscosity	Thickening water binding	Salad dressings, soups, gravies, etc.
7.	Flavor and aroma	Lactose reacts with the milk proteins	Baked foods, milk products, biscuits and confectionaries, sauces and soups
8.	Browning	Millard reaction—on heating the amino group of proteins react with aldehyde groups of sugars	Confectionaries, biscuits, cakes, breads, sauces, etc.

Commercially-available whey protein concentrates contain 35 to 95% protein. If they are added to food on a solid's basis, there will be large differences in functionality due to the differences in protein content. Most food formulations call for a certain protein content and thus whey-protein concentrates are generally utilized as a constant protein base. In this case the differences due to protein content as such should be eliminated. As the protein content increases, the composition of other components in the whey-protein concentrate must also change and these changes in composition have an effect on functionality. Table 14.4 gives the various uses of whey proteins.

Nutraceutical Proteins

Nutraceutical proteins represent a class of nutritionally-important proteins having therapeutic activity. The whey-protein concentrates and some of the milk proteins of infant foods contain certain pharmaceutical proteins having high nutritive quality. Infants get the required proteins from the mother's milk, which also contains certain therapeutic proteins that protect the baby from infection and other problems. There are other infant foods, which also have more or less the same composition as that of mother's milk, made up of cow's and buffalo's milk. All these food proteins provide the infants the raw building materials in the form of essential amino acids and at the same time protects them from microbial infections and other diseases. The following table gives a comparative picture of protein, carbohydrates, and fats present in the milk of various animals and humans. Thus, milk is a very good source of nutrition.

TABLE 14.5 Protein fat and carbohydrate content of various types of milk.

Species	Fat %	Protein %	Protein/Fat	Lactose %	Ash %	Total Solids %
Buffalo, Philippine	10.4	5.9	0.6	4.3	0.8	21.5
Cow						
Ayrshire	4.1	3.6	0.9	4.8	0.7	13.1
Brown Swiss	4.0	3.6	0.9	5.0	0.7	13.3
Guernsey	5.0	3.8	0.8	4.9	0.7	14.4
Holstein	3.5	3.1	0.9	4.9	0.7	12.2
Jersey	5.5	3.9	0.7	4.9	0.7	15.0
Zebu	4.9	3.9	0.8	5.1	0.8	14.7
Goat	3.5	3.1	0.9	4.6	0.79	12
Human	4.5	1.2	0.2	6.8	0.2	12.6

Now, with modern analytical techniques and equipment, it is possible to give an explanation for the therapeutic activities of these proteins present in the milk. The research on the therapeutic activity of whey protein has shown that it elevates a tripeptide, glutathione (L-gamma-glutamyl-L - cysteinyl glycine), in the cells of animals when fed with whey concentrates. The tripeptide glutathion is a very reactive reducing compound. In our body the presence of glutathion is very beneficial since it can act as a detoxifying compound. It can detoxify xeno-biotics and protect the cells against the bad effects of oxygen intermediates and free radicals. Table 14.6 gives the composition of milk from human, buffalo and cow.

TABLE 14.6 Composition of milk from buffalos, cows, and humans.

Sl. No.	Nutritional Constituents (In grams)	Buffalo	Cow	Human
1.	Lactose	4.3	4.8	6.8
2.	Total protein	3.8	3.3	1.2
3.	Casein	3.0	2.8	1.4
4.	Lactoglobulin	0.2	0.2	0.2
5.	Lactalbumin	0.4	0.4	0.3
6.	Fat	7.5	3.7	3.8
7.	Calorific value (K Cal)	100.0	71.0	69.0
8.	Calcium (mg)	203.0	125.0	33.0
9.	Phosphorous (mg)	130.0	96.0	15.0
10.	Chloride (mg)	112.0	43.0	103.0

14.7 DESIGNING PROTEINS

Protein design is important because it not only allows us to systematically investigate the forces that determine structure and stability but it also affords the possibility of creating proteins through novel and useful activities. The biotechnological industry is more interested in developing new types of enzymes, which are stable and more active under extreme working conditions such as high temperature and extreme pH such as acidic or alkaline pH.

Biochemists can now use a refinement of the genetic modification technique to redesign proteins. Once they have isolated the gene for a particular protein they can alter its code so that a change occurs in the protein's primary structure. They then incorporate the modified gene into a microorganism where it is decoded as before, but this time a new protein appears. Swapping one α-amino acid in a protein can have a large effect on how the protein behaves. Biochemists have already used this **protein engineering** to modify human insulin in a way that makes it become absorbed more quickly after injection. Multiple changes are needed to 'humanize' antibodies.

Computer-modeling techniques allow protein chemists to make predictions about how proposed α-amino acid changes might change the structure and activity of a protein. Genetic and protein modification have enormous potential. Besides insulin, genetic modification can produce human growth hormone and the blood clotting factor VIII. Genetic engineering, thus, has provided a tool to produce designer proteins having special characteristics by changing the amino acid sequence. In addition, it is already possible to introduce into a plant new genes that enable it to produce its own protein insecticide or that make it resistant to disease. Hepatitis B vaccine, oil-digesting bacteria, and bacteria that produce biodegradable plastic are all recently developed products of protein engineering and genetic-modification techniques.

The following are some examples of protein designing through genetic engineering techniques:

Improving the Detergent Quality of Subtilisin

Subtilisin is a protease enzyme of bacterial origin with a molecular size of 27.4 kD. This protease has a wide range of proteolytic activity unlike other proteases such as trypsin and chymotrypsin. The catalytic activity of subtilisin is due to the presence of a catalytic triad—ser 221, his 64, and asp 32, similar to chymotrypsin. This enzyme is used in the manufacturing of washing powder as an ingredient to improve the quality of detergents; such detergents are called stain cutters or stain removers.

But, the problem is that the native form of subtilisin loses its catalytic activity by 90% in the presence of bleach. Detailed investigations have revealed that loss of activity is due to the oxidation of methionine at position 222 of the polypeptide chain. By site-directed mutagenesis, the methionine at the position 222 of subtilisin molecule was substituted by a variety of other amino acids and the enzyme activity of the modified subtilisin was studied in the presence of detergent (Table 14.7). Finally, it was observed that substitution of Met^{222} with Alanine has imparted stability and activity for the enzyme. Many detergents and washing powders are supplemented with an engineered enzyme produced by the recombinant technique.

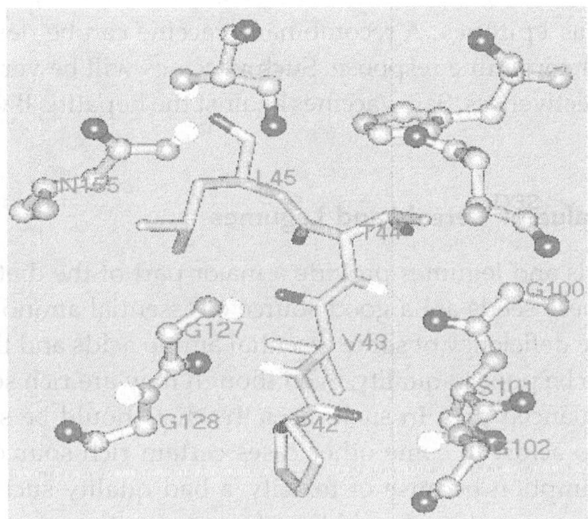

FIGURE 14.25 **The reaction center (catalytic triad) of subtilisin.**

TABLE 14.7 Various site-directed mutagenesis studied at position 222.

Codon-222	% Activity	Codon-222	% Activity
Cys	138.0	Gln	7.2
Met	100.0	Phe	4.9
Ala	53.0	Trp	4.8
Ser	35.0	Asp	4.1
Gly	30.0	Tyr	4.0
Thr	28.0	His	4.0
Asn	15.0	Glu	3.6
Pro	13.0	Ile	2.2
Leu	12.0	Arg	0.5
Val	9.3	Lys	0.3

Creation of Novel Proteins

Vaccines are traditionally prepared from heat-inactivated microbes—bacteria or viruses or the surface proteins of these organisms to immunize against various infectious diseases. But in some cases it was frequently observed that some parts or components of the vaccines were creating some deleterious effects on people. It is known that proteins are the main candidates responsible for imparting the stimulus for immunity. Therefore, efforts have been made to design an altered protein molecule, which can produce only very little deleterious effects. There are some amino acid sequences, which are directly involved in the stimulation of immunity and that part of the protein or polypeptide is known as **epitopes.** A recombinant vaccine can be designed by involving only the essential epitopes in the immune response. Such vaccines will be very simple and will be safe without reducing their effectiveness. Safe vaccines against the hepatitis B virus and anthrax bacteria are under development.

Improving Nutritional Value of Cereals and Legumes

In human nutrition, cereals and legumes provide a major part of the dietary protein requirement. The storage proteins of these seeds are a good source of essential amino acids. But some of these legumes and cereals create deficiency of some essential amino acids and therefore such cereals and legumes are considered to be poor in quality, even though they are rich sources of dietary protein. They cannot provide a balanced diet. In such cases the diet should be supplemented with other sources of essential amino acids. In some other cases certain rich sources of protein may not be suitable for human consumption because of toxicity, a bad quality such as bad smell, taste, etc. Thus, protein for human consumption should be safe and nutritious.

The major parameters for assessing the general effectiveness of a protein as a dietary protein are the **Essential Amino Acids Profile (EAAP), Biological Value (BV),** and **Protein Efficiency Ratio (PER).**

Essential Amino Acids

Proteins are composed of 20 amino acids. Plants and microbes can synthesize all these amino acids, but animals cannot. Such amino acids have to be supplemented through the diet. Such amino acids are the **essential amino acids.** The presence of a high percentage of essential amino acids along with other amino acids increases the nutritive quality of protein for human consumption. Table 14.8 gives the deficiency of essential amino acids for various food materials and their complementary food proteins.

TABLE 14.8 Complementary protein combinations to avoid deficiency of essential amino acids.

Food	Amino Acid Deficient	Complementary Protein Food Combination
Grains	Isoleucine Lysine	Rice + legumes Corn + legumes Wheat + legumes Wheat + peanut + milk Wheat + sesame + soybean Rice + brewer's yeast
Legumes	Tryptophan Methionine	Legumes + rice Beans + wheat Beans + corn Soybeans + rice + wheat Soybeans + corn + milk

		Soybeans + wheat + sesame Soybeans + peanuts + sesame Soybeans + peanuts + wheat + rice Soybeans + sesame + wheat
Nuts and Seeds	Isoleucine Lysine	Peanuts + sesame + soybeans Sesame + beans Sesame + soybeans + wheat Peanuts + sunflower seeds
Vegetables	Isoleucine Methionine	Lima beans + Sesame seeds or Brazil nuts or mushrooms Green beans + Sesame seeds or Brazil nuts or mushrooms Brussel sprouts + Sesame seeds or Brazil nuts or mushrooms Cauliflower + Sesame seeds or Brazil nuts or mushrooms Broccoli + Sesame seeds or Brazil nuts or mushrooms Greens + millet or rice

The nutritive quality of whey proteins is considered to be higher compared to other types of proteins. They are rich in branched chain amino acids—Ile, leu, val, lys, and trp. The **branched chain amino acid (BCAA)** is needed for rapid energy metabolism for muscular activity. BCAA helps in the bioavailability of complex carbohydrates absorbed by muscle cells for anabolic muscle building activity. BCAA plays an important role in the production of metabolic energy during rapid muscular actions. During exercise BCAAs are released from the muscle cells. The carbon chain part of the molecule is used as fuel and the nitrogen part is used to synthesize alanine. Alanine is transported to the liver where it is turned into glucose to meet the energy requirements. So athletes are advised to take BCAA sources before and after exercises to protect their healthy body. Thus, BCAAs are very important for muscle growth and muscular activity.

Biological Value

Biological value or **BV** is a measure of protein nitrogen retained by the body after consuming a particular amount of protein nitrogen. Among various types of dietary proteins whey protein showed the maximum BV compared to rice, soy, egg, and wheat proteins.

Protein Efficiency Ratio

Protein efficiency ratio or **PER** is a measure of the efficiency of a protein on the rate of growth. It is the growth performance in terms of weight gain of an adult by consuming 1 gm of dietary protein. The order of PER for various dietary proteins is given below.

Wheat protein < Rice protein < Soy protein < Casein < Milk protein < Whey protein

The quality of protein can be improved by the modern approach of protein engineering through recombinant DNA technology. The storage protein of cereals and legumes can be modified by introducing essential amino acids. These genes can be transferred to food plants to be expressed in the respective storage parts of the plants such as seeds, tubers, or grains. Modifications in proteins can be made by introducing new amino acids or by substituting the existing amino acids with new ones. Efforts are already being made to enhance the nutritive quality of protein genes in maize and other similar food grains. Scientists are on the look out for new types of dietary protein genes that are rich in essential amino acids, with which the nutritive quality of other plants can be increased.

14.8 PROTEOMICS

A large number of sequencing projects initiated in the late eighties and early nineties have now been completed. The entire genome sequences of numerous prokaryotes, eukaryotes such as *Saccharomyces cerevisiae*, and the first multicellular organism *Caenorabditis elegans* and *Homo sapience* have been deposited into public databases (GenBank,http://www.ncbi.nlm.nih.gov/Entrez/Genome/org.html). Significantly, analysis of the sequence data has revealed a myriad of uncharacterized genes having homologues in many different organisms, yet their functions have not been deciphered in any of these organisms. Statistically, 24% of *Haemophilus influenzae*, 40% of *Escherichia coli*, and 43% of *Saccharomyces cerevisiae* open reading frames (ORFs) remain completely uncharacterized.

A second important realization is that since DNA sequence is a static entity, it is very difficult, if not impossible, to infer the gene function from sequence alone. Therefore, temporal, spatial, and functional properties of dynamic gene products of an organism must be examined. It is expected that the recent development of techniques such as nucleic acid microarrays and proteomics that permit genome-wide expression profiling will provide the vital information required to describe how a cell functions, and perhaps give an insight into the development of diseases.

Of the two methods of genome-wide analysis, proteomics, the examination of the complete set of proteins synthesized by a cell under a given set of physiological or developmental conditions, should be most informative when making functional assignments. This is mainly due to the fact that the proteome is context-dependent and unlike mRNAs, proteins are functional entities within the cell. The proteome is the proteins expressed by a genome.

Proteomics involve the analysis of a large number of proteins with a combination of two-dimensional gel electrophoresis and mass spectrometry. Today, it includes both the identification of a large number of proteins expressed by a genome as well as the characterization of their functional and structural relationships. Figure 14.26 represents the different stages in protein synthesis. It was thought that a single gene would result in the formation of a single protein or polypeptide. But there are different stages in the process of transcription and translation, where the same process can take place but produces a different protein. These processes can take place just after the process of transcription—the post-transcriptional process that can result in different types of mRNAs with the control of splicing mechanisms. The second point where the alteration can occur is after the process of translation or protein synthesis. This is the post-translational modification which can also produce structurally and functionally different proteins from the same gene.

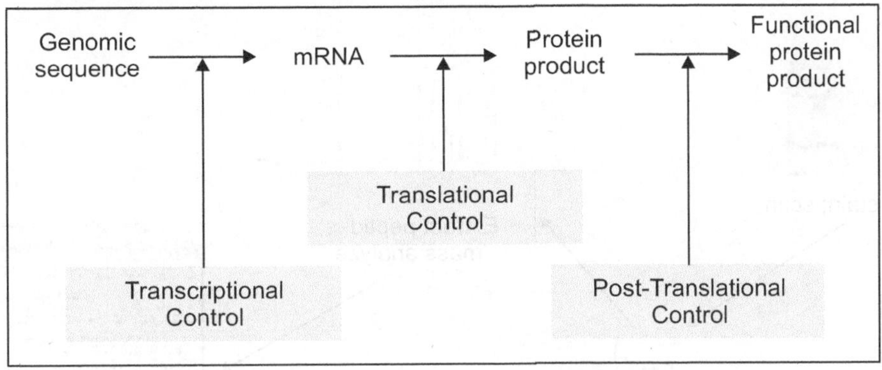

FIGURE 14.26 **Sequence of protein synthesis–a single gene results in the production of more than one protein.**

Proteomics also deals with the use of qualitative and quantitative protein-level measurements to characterize biological processes (e.g., diseases). By comparing the protein profiles from normal cells and diseased or metabolically aberrant cells, the reason for the diseased condition can be ascertained. Therefore, diseases can be treated at the protein level or even at the gene level because the proteins are the drug targets.

The Key Technologies in Proteomics

- Reproducible Two-dimensional Gel Technology
- Staining and Scanning Technology
- Mass Spectrometry for Identification
- Databases (protein and genome)
- Database Searching Algorithms

The crude protein extracted from the tissue sample or cells or from individuals is separated in gel by two-dimensional electrophoresis comprising of IEF and SDS-PAGE. The gel is suitably stained and compared with that of the standard two-dimensional gel profile of protein from the same organisms grown under normal conditions (or healthy individuals). If the formation of any new protein is observed that spot can be excised from the gel along with the gel matrix and washed with sterile distilled water. This protein spot along with the gel can be digested with trypsin and then introduced into the mass spectrometer and (the MS fingerprint can be taken). The MS fingerprint can be used for the identification of the protein by database search or can proceed further to carry out amino acid sequencing by mass spectrometer. Thus, the protein purification and identification become very easy and simple. But the key challenge in proteome research is the automation and integration of these technologies.

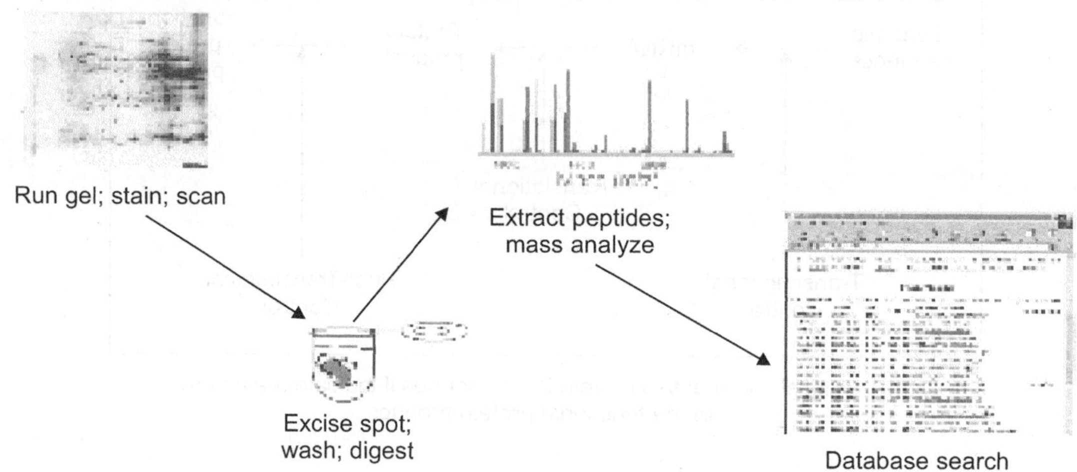

FIGURE 14.27 **A general approach for proteomics.**

Genes and Proteins

Owing to concerted efforts by numerous state and commercial establishments, the human genome had completely been sequenced in 2000. All 24 human chromosomes are mapped, and the defects of hundreds of genes responsible for the development of hereditary diseases have been revealed. A new medical concept has been advanced whereby all diseases can be divided into two major groups: (a) hereditary diseases—a consequence of transmission of defective genes from parents to their children, and (b) non-heritable or so called socially-significant diseases, which make up more than 95% of all human diseases and result from disturbances in normal genes' expression regulation. Thus, all human diseases, one way or another, are associated with the genome; the only difference is that the diseases of the former group are due to a defect(s) in gene's structure while those of the latter group are caused by disturbances in a gene expression's regulation. The cataloging of human genes is an achievement, which can hardly be overestimated: for many years to come this catalogization will serve as a basis for the development of basic biochemistry, molecular biology, and genome-associated applied sciences. The future of this area of research, called **genomics,** is even more brilliant. Currently, the emphasis has gone from sequencing the human genome to sequencing the genomes of animals and microorganisms, particularly pathogenic and plant genomes. This research promises enormous achievements in medicine, especially in the struggle against infectious diseases.

How is the informational structure (gene) connected with the actual working molecular machine (protein)? To answer this question, we have to consider the results of those few works, where the expression map of mRNA was compared with the proteinous map in the same cellular system. It failed to reveal any strict correlation between the two maps. Thus, the informational knowledge cannot be directly converted into the knowledge of actually operating protein molecules. As a result, a new area of research has appeared, called **proteomics,** which deals with inventory of proteins. At first glance, this is an utterly impossible task. While the human genome map is the same for all human cells (24 chromosomes), in the proteomic map each cell is individual. Although the cell may have only 33,000 functional genes, the numerous modification reactions may increase the number of proteins in it up to several million. There are two definitions for proteomics: a narrow one (the so-called structural proteomics) and a broader one, encompassing both the structural and functional proteomics. In a narrow sense of the word, 'proteomics' is a science dealing with the cataloging of proteins based on a combination of several methods: two-dimensional electrophoresis, mass spectrometric analysis of molecular mass, and sequencing of electrophoretically separated proteinacious biological material with subsequent analysis of the results obtained, with the help of bioinformatical and computational methods.

In a broader sense, the terms 'proteome analysis' or 'proteomics' can be used not only for cataloging proteins of a biological subject but also for the monitoring of reversible post-translational modification of proteins by specific enzymes (i.e., phosphorylation, glycosylation, acylation, phrenylation, sulfurization, etc). To date, more than 300 different types of post-translational modification have been characterized with the aid of proteomics. Another aspect of functional proteomics is to clarify the composition of functionally active complexes that constitute different metabolic chains and also, to determine the interactions between various proteins or subunits of oligomeric complexes by a combination of methods to isolate these complexes and subsequent mass-spectrometric analysis. Lately, structural proteomics is often called expressional proteomics while functional proteomics is also designated as cell-mapping proteomics, since it elucidates the interactions of proteins within metabolic pathways. Therefore, in short, the number of proteins and the number of genes are not equal and they are non-linear. The number of proteins easily outnumbers the number of genes.

Types of Proteomics

Even though proteomics can be defined as the study of the protein complement of the genome, the dynamic nature of the proteome made it difficult to study and understand. The total protein expression profile always changes with time and micro- and macro-environmental conditions. Only a small percentage of the total gene is expressed at a particular time. Information about the genome and gene structure can predict the structure and function of its protein complement to a limited extent because of the post-transcriptional modifications that the protein undergoes. These modifications are not represented in the respective gene. This is what makes proteomics difficult to study compared to genomics. The following are the major types of proteomics:

Expression Proteomics

This is the qualitative and quantitative study of the expression of total proteins under two different conditions. For example, expression proteomics of normal cells and diseased cells can be compared to understand the protein that is responsible for the diseased state or the protein that is expressed due to disease. Using this method disease-specific protein can be identified and characterized by comparing the protein-expression profile of the entire proteome or of the subproteome between the two samples.

For example, tumor tissue samples from a cancer patient and the same type of tissue from a normal person can be analyzed for differential protein expression. Using two-dimensional gel electrophoresis, mass spectrometry combined with chromatography and microarray techniques, additional proteins that are expressed in the cancer tissues or the proteins, which are absent, or those, which are over-expressed and under-expressed can be identified and characterized. Identification of these proteins will give valuable information about the molecular biology of tumor formation.

Structural Proteomics

Structural proteomics, as the name indicates, is about the structural aspects, including the three-dimensional shape and structural complexities, of functional proteins. This includes the structural prediction of a protein when its amino acid sequence is determined directly by sequencing or from the gene with a method called **homology modeling**. This can be carried out by doing a homology search and computational methods of protein structural studies and predictions. Apart from this, structural proteomics can map out the structure and function of protein complexes present in a specific cellular organelle. It is possible to identify all the proteins present in a complex system such as ribosomes, membranes, or other cellular organelles and to characterize or predict all the proteins and protein interactions that can be possible between these proteins and protein complexes. Structural proteomics of a specific organelle or protein complex can give information regarding supra-molecular assemblies and their molecular architecture in cells, organelles, and in molecular complexes.

Functional Proteomics

This is an assembly type of proteomic method to analyze and understand the properties of macromolecular networks involved in the life activities of a cell. With these methods it will be possible to identify specific protein molecules and their role in individual metabolic activities and their contribution to the metabolic network that operates in the system. This forms one of the major objectives of functional proteomics. For example, the recent elucidation of the protein network involved in the functioning of a nuclear pore complex has led to the identification of novel proteins involved in the translocation of macromolecules between the cytoplasm and nucleus through these complex pores.

Functional proteomics is yielding large databases of interacting proteins, and extensive pathway maps of these interactions are being scored and deciphered by novel high-throughput technologies. However, traditional methods of screening have not been very successful in identifying **protein-protein interactions** and their inhibitors. The identification and measurement of changes in the concentration of specific proteins that cells make as a result of their genetic response to specific toxicants, and how these proteins are related to each other and to the specific biological condition of the cell, also fall under functional proteomics.

REVIEW QUESTIONS

1. Name two metabolic disorders in humans. Which proteins are responsible?
2. Define zymogen with two examples.
3. The proteolytic enzyme chymotrypsin is produced by the pancreas. At the same time the protein present in that gland is protected from its proteolytic activity. Explain.

4. Chymotrypsin belongs to the group serine protease. Why are they so named? Explain the catalytic triad of serine protease.
5. What is the importance of the three-dimensional shape of proteins in carrying out specific functions? Explain your answer with suitable example.
6. Explain how the serine residue of the catalytic triad in certain enzymes becomes reactive or acidic.
7. What is sickle cell anemia? Explain its physiological basis.
8. Two-dimensional gel electrophoresis is more efficient than other types of electrophoretic techniques. Explain.
9. Differentiate between domain and subunit in protein three-dimensional structure.
10. What is the advantage of linking a separation technique such as chromatography to mass spectrometer?
11. What are the different methods used for characterization of proteins?
12. Explain the principle and application of mass spectrometer in proteomics.
13. Explain how isoelectric focusing is different from SDS-PAGE.
14. What are the different types of mass spectrometer used in proteomics?
15. What does protein engineering mean? How can a protein be designed? Explain with an example.
16. Explain the principle and use of MALDI-TOF.
17. A bacterial cell has a diameter of two μm and a length of three μm. Calculate the total packed cell volume required to produce five gm of intracellular protein.
18. An enzyme produced by a bacterium has a molecular weight of 60 kD. This protein is produced at a level of 2,000 molecules per cell under standard conditions. If we want to produce three gm of the protein, estimate the quantities of cells that we have to grow.
19. What are the different steps for the purification of protein from a plant tissue?
20. What is proteomics? What is its importance in future pharmaceuticals and medical studies?

Chapter 15

RECOMBINANT DNA TECHNOLOGY

In This Chapter

15.1 INTRODUCTION

The field of modern biotechnology started when recombinant human insulin produced by bacteria was first marketed in the United States in 1982. The effort leading up to this landmark event began in the early 1970s when research scientists developed protocols to construct new types of bacterial plasmids or vectors, by cutting out and pasting pieces of DNA together to create a new piece of DNA (recombinant DNA) that could be inserted into host bacterium such as *e. coli*. We have also observed that yeasts can be made to produce vaccines such as hepatitis B, plants having special properties such as resistance to certain diseases, pests, and herbicides and plants with superior nutritive qualities can be generated very efficiently. These excellent goals of genetic engineering were achieved because of the advent of recombinant DNA technology. Recombinant DNA technology is one of the few techniques that made conventional biotechnology into "Modern Biotechnology." Paul Berg, Herbert Boyer, Annie Change, and Stanley Cohen are the team of scientists that made the first recombinant DNA molecule in 1973.

Simply defined, it is the art of cutting and pasting genes. There are, however, many new applications of this technology invented each year, and it is impossible for any textbook to be completely up to date. This technique encompasses a number of methodologies or tools that enable us to construct new combinations of DNA (recombinant DNA or rDNA) in the laboratory for different purposes. The rDNA molecule thus constructed can be introduced into an appropriate host cell, where it can be multiplied and generate many copies. This forms the basic concept of the process known as gene cloning or DNA cloning. In this chapter we will examine the basic tools, methodologies, and applications of recombinant DNA techniques in various fields of biological research.

Gene cloning can generate unlimited copies of a DNA molecule (e.g., recombinant DNA) by replication in a host cell. It was first developed in 1970. The following are some of the major applications of rDNA technique:

- Genetic mapping
- DNA sequencing
- Mutation studies
- Transformation
- Genetic engineering
- Recombinant DNA libraries
- Restriction enzyme site analysis
- Analysis of gene transcripts
- Polymerase chain reaction (PCR)

The goal of rDNA is gene cloning to generate large amounts of pure DNA that can be manipulated and studied. The following are the basic steps involved in the process of the rDNA technique for gene cloning:

1. Isolate DNA from organism (e.g., extraction)
2. Cut DNA with **restriction enzymes** to generate DNA fragments of desired size, which may have a specific type of DNA sequence or gene that has to be cloned.
3. Ligate or splice each piece of DNA into a cloning vector to create a recombinant DNA molecule. A cloning vector is an artificial DNA molecule, capable of replicating in a host organism (e.g., bacteria).

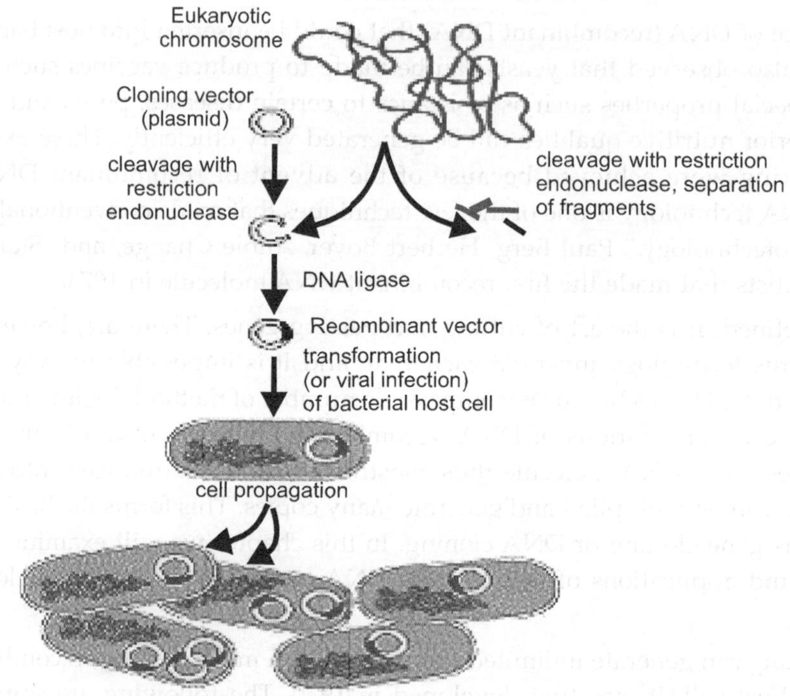

FIGURE 15.1 **Different steps in a typical rDNA experiment.**

4. Transform recombinant DNA (cloning vector + DNA fragment) into a host that will replicate and transfer copies to progeny.

15.2 TOOLS OF RECOMBINANT DNA TECHNOLOGY

For the generation of recombinant DNA molecules, we require that the DNA fragment is cloned (known as the **insert DNA**) and a vehicle DNA (known as the **vector**) to carry the DNA into a host cell and multiply. Since the process of making a recombinant DNA requires the precise cutting and stitching of DNA molecules, it involves a number of molecular tools—the enzymes to cut and modify the DNA, finally resulting in a recombinant DNA molecule. The major enzymes required for the making of an rDNA molecule are the following:

- Type II restriction endonucleases
- DNA ligase
- Alkaline phosphatase

The role of each of these enzymes in the making of an rDNA molecule is explained in the following parts:

Restriction Endonucleases

In 1968 a group of three scientists—H.O. Smith, K.W. Wilcox, and T.J. Kelley—isolated and characterized the first sequence-specific restriction endonuclease. The discovery of these enzymes marked the beginning of recombinant DNA research and sequence-specific modification of DNA molecules. Since these restriction endonucleases are responsible for converting the DNA molecules into a specific number of fragments by cutting the molecule at certain specific sequences known as **restriction sites**, these enzymes are known as **molecular scissors.** These enzymes, along with another equally important DNA modifying enzyme—the DNA ligase—made it possible to create entirely new kinds of DNA molecules and, equally important, manipulate the functioning of the genes located on these new molecules.

Growth Restriction of Bacteriophages

It has clearly been known since the 1950s that phage particles that infect and grow efficiently in one strain of bacteria are unable to infect and grow in other strains of the same bacterial species. A series of studies showed that phage particles that efficiently infect the host cells have DNA molecules that have been chemically modified by the addition of methyl groups to some of their adenine and/or cytosine bases, while the DNA of poorly infecting phage particles does not show this pattern of chemical modification or 'methylation.' Phage particles with unmethylated DNA do not grow and infect efficiently because their DNA molecules are cleaved and degraded by enzymes of the host cell, while methylated DNA is protected from this degradation. This phenomenon of degrading unmethylated DNA destroys the growth ability of the phage, and is responsible for the pattern of growth restriction described above.

In the late 1960s, two scientists—Stewart Linn and Werner Arber—isolated two types of enzymes responsible for phage growth restriction in *e. coli* bacteria. One of these enzymes methylate DNA at specific sequence sites, while the other leave the DNA if it is not methylated at the specific sites along the length of the molecule. The first type of enzyme was named **methylase** and the second type was named **restriction endonuclease**. These restriction nucleases were important to scientists who were gathering the tools needed to cut DNA molecules at specific sites. Because of this, scientists could cut DNA molecules in a predictable and reproducible way and thus laid the foundation for the recombinant DNA technique.

These two enzyme systems—restriction endonucleases and DNA methylases—form the basic components of the defense mechanism in bacterial systems, known as the **restriction-modification system**. The first part of this system is the restriction endonucleases that selectively recognizes specific DNA sequences and fragments the DNA having these sequences, which leads to its complete degradation. Since these enzymes are responsible for the restricted growth and multiplication of bacterial viruses in bacteria they are called **restriction endonucleases.** The second part is the modification system, which recognizes the restriction sites and modifies those sites by adding methyl groups to one or two bases within those restriction sites. Thus, it modifies the restriction sites by methylation, so that the restriction enzymes are not able to cut those sites. This usually happens for the DNA of the host cells. Thus, these modification systems protect the DNA of the host cell from the action of its own restriction systems.

The first sequence specific endonuclease was isolated and characterized by a group of three scientists, H.O. Smith, K.W. Wilcox, and T. J. Kelley, of Johns Hopkins University, in 1968. Working with *Hemophilus influenzae* bacteria, this group isolated an enzyme, called *Hind*-II, which always cuts DNA molecules at a particular point within a specific sequence of six base pairs. This sequence is:

5′ G T (T or C) (A or G) A C 3′
3′ C A (A or G) (T or C) T G 5′

They found that the *Hind*-II enzyme always cuts directly in the center of this sequence. Wherever this particular sequence of six base pairs occurs unmodified in a DNA molecule, *Hind*-II will cleave both DNA backbones between the third and fourth base pairs of the sequence. Moreover, *Hind*-II will only cleave a DNA molecule at this particular site. For this reason, this specific base sequence is known as the 'recognition sequence' for *Hind*-II.

Hind-II is only one example of the class of enzymes known as restriction endonucleases. In fact, more than 900 restriction enzymes, some sequence specific and some not, have been isolated from over 230 strains of bacteria since the initial discovery of *Hind*-II.

These restriction enzymes generally have names that reflect their origin—the first letter of the name comes from the genus and the second two letters come from the species of the prokaryotic cell from which they were isolated. For example, *Eco*RI comes from *escherichia coli* RY13 bacteria, while *Hind*-II comes from hemophilus influenzae strain Rd. Numbers following the nuclease names indicate the order in which the enzymes were isolated from single strains of bacteria. Nucleases are further described by the addition of the prefix 'endo' or 'exo' to the name. The term 'endonuclease' applies to the sequence-specific nucleases that break the nucleic-acid chains somewhere in the interior, rather than at the ends, of the molecule. Nucleases that function by removing nucleotides from the ends of the molecule are called **exonucleases**. There are three classes of restriction endonucleases. They are type I, type II, and type III, and each is characterized by a slightly different mode of action on DNA. Type II restriction endonucleases are used in recombinant DNA techniques as they can recognize specific DNA sequences and cleave at a site that comes within that sequences. These recognition sequences are called **restriction sites.**

The restriction sites or the recognition sequences are usually four to eight nucleotides in length and are palindromic. The palindromic sequences read the same on both the strands of DNA in a 5′–3′ direction. For example, the restriction site of *Eco*RI is 5′GGATCC3′ and the cleave site is between the G and A on the complementary strands, which is demonstrated in Figure 15.2.

The restriction-site recognition sequence for different restriction enzymes is unique and therefore different endonucleases yield different sets of cuts or DNA fragments. But one endonuclease will always cut a particular base sequence the same way, no matter what DNA molecule it is acting on. Once the cuts have been made, the DNA molecule will break into fragments.

Type I and III restriction endonucleases: They have both endonuclease and methylase activities on a single protein. Type I REs cleave the DNA at a random site located at 1,000 base pairs from the recognition site. Type III RE does the same at 24 to 24 bp away from the recognition sites.

Type II RE: They cleave DNA at specific site within the recognition sequence. This property has made type II RE a most important tool in molecular biology.

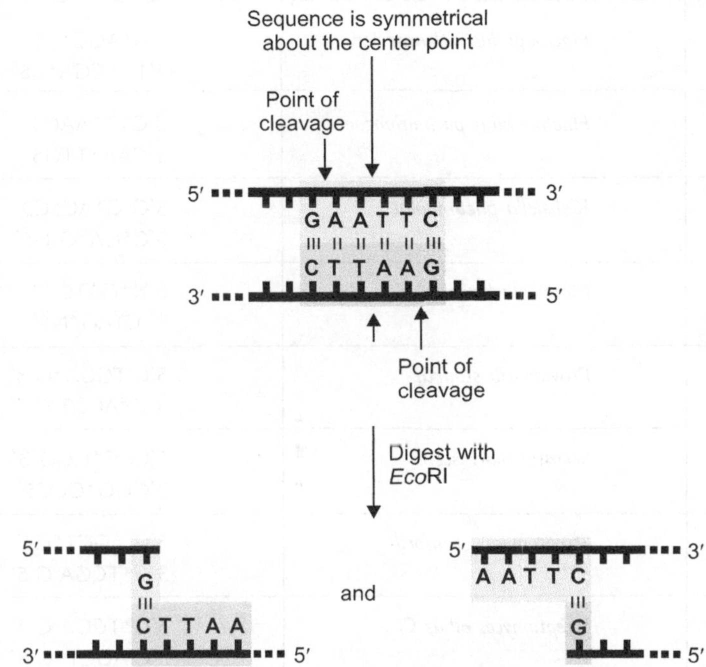

FIGURE 15.2 **The restriction site sequence of *Eco*RI is a palindrome of a hexamer and it cuts between the bases G and A on both the strands resulting in sticky ends or cohesive ends.**

TABLE 15.1 Examples of Type II restriction endonucleases used in recombinant DNA experiments.

Enzyme	Organism from which Derived	Target Sequence (cut at*) 5′ → 3′
Ava I	*Anabaena variabilis*	5′C*C/TCGA/GG 3′ 3′GG/AGCT/C*C5′
Bam HI	*Bacillus amyloliquefaciens*	5′G*GATCC3′ 3′CCTAG*G5′
Bgl II	*Bacillus globigii*	5′A*GATCT3′ 3′TCTAG*A5′
Eco RI	*Escherichia coli* RY 13	5′G*AATTC3′ 3′CTTAA*G
Eco RII	*Escherichia coli* R245	5′*CCA/TGG3′ 3′GGT/ACC*5′
Hae III	*Haemophilus aegyptius*	5′GG*CC3′ 3′CC*GG5′
Hha I	*Haemophilus haemolyticus*	5′GCG*C3′ 3′C*GCG5′

Hind III	*Haemophilus inflenzae* Rd	5′A*AGCT T3′ 3′T TCGA*A5′
Hpa I	*Haemophilus parainflenzae*	5′GTT*AAC3′ 3′CAA*TTG5′
Kpn I	*Klebsiella pneumoniae*	5′G GTAC*C3′ 3′C*CATG G5′
Mbo I	*Moraxella bovis*	5′N*GATC 3′ 3′ CTAG*N5′
Pst I	*Providencia stuartii*	5′C TGCA*G 3′ 3 G*ACGT C 5′
Sma I	*Serratia marcescens*	5′CCC*GGG 3′ 3′GGG*CCC5′
Sst I	*Streptomyces stanford*	5′G AGCT*C3′ 3 C*TCGA G 5′
Sal I	*Streptomyces albus* G	5′G*TCGA C 3′ 3′C AGCT*G 5′
Taq I	*Thermophilus aquaticus*	5′ T*CG A 3′ 3′ A GC*T 5′
Xma I	*Xanthamonas malvacearum*	5′ C*CCGG G 3′ 3′ G GGCC*C 5′

Type II restriction enzymes cut DNA and produce two types of fragments. Some restriction enzymes produce fragments with blunt ends, whereas others produce fragments with sticky (overhanging or staggered) ends. Some enzymes such as *Hind*-II and *Sma*-I cut the DNA symmetrically on both the strands at exactly the same nucleotide position leaving blunt ends as shown in Figure 15.3.

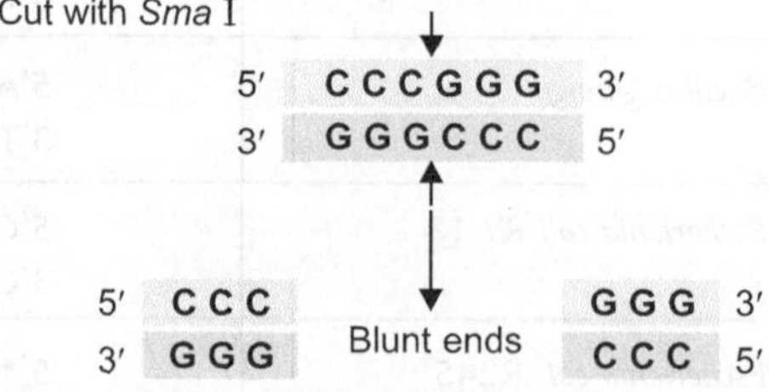

FIGURE 15.3 **Restriction cleavage with *Sma*-I resulting in blunt ends.**

Many endonucleases cleave the DNA backbones in positions that are not directly opposite to each other as in the case of *Eco*RI as illustrated in Figure 15.3.

Restriction Digestion with *Eco*RI

When the restriction endonuclease encounters its respective restriction site sequence (5′GAATTC 3″), it cleaves each backbone between the G and the closest A base residues. Once the cuts have been made, the resulting fragments are held together only by the relatively weak hydrogen bonds that hold the complementary bases to each other. The weakness of these bonds allows the DNA fragments to separate from each other. Each resulting fragment has a protruding 5′ end composed of unpaired bases (Figure 15.2). Other enzymes also create cuts in the DNA backbone in the same manner, which results in protruding 3′ ends. Protruding ends—both 3′ and 5′—are sometimes called 'sticky ends' because they tend to bond with complementary sequences of bases. In other words, if an unpaired length of bases (5′ A A T T 3′) encounters another unpaired length with the sequence (3′ T T A A5′) they will bond to each other—they are 'sticky' for each other. Ligase enzymes are then used to join the phosphate backbones of the two molecules. The cellular origin, or even the species origin, of the sticky ends does not affect their stickiness. Any pair of complementary sequences will tend to bond, even if one of the sequences comes from a length of human DNA, and the other comes from a length of bacterial DNA. In fact, it is this quality of stickiness that allows the production of recombinant DNA molecules (molecules which are composed of DNA from different sources and have given birth to a powerful technology and

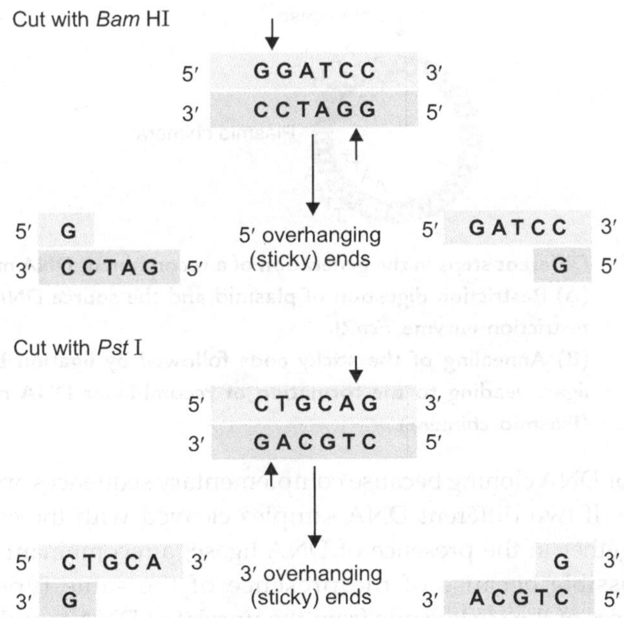

FIGURE 15.4 **Restriction digestion with *Bam* HI resulting in 5′ overhanging ends and with *Pst* -I resulting in 3′ overhanging ends.**

industry) the genetic engineering or the recombinant DNA. Examples of some other restriction enzymes, their mode of cutting and generation of 5′ overhangs and 3′ overhangs, are illustrated in Figure 15.4.

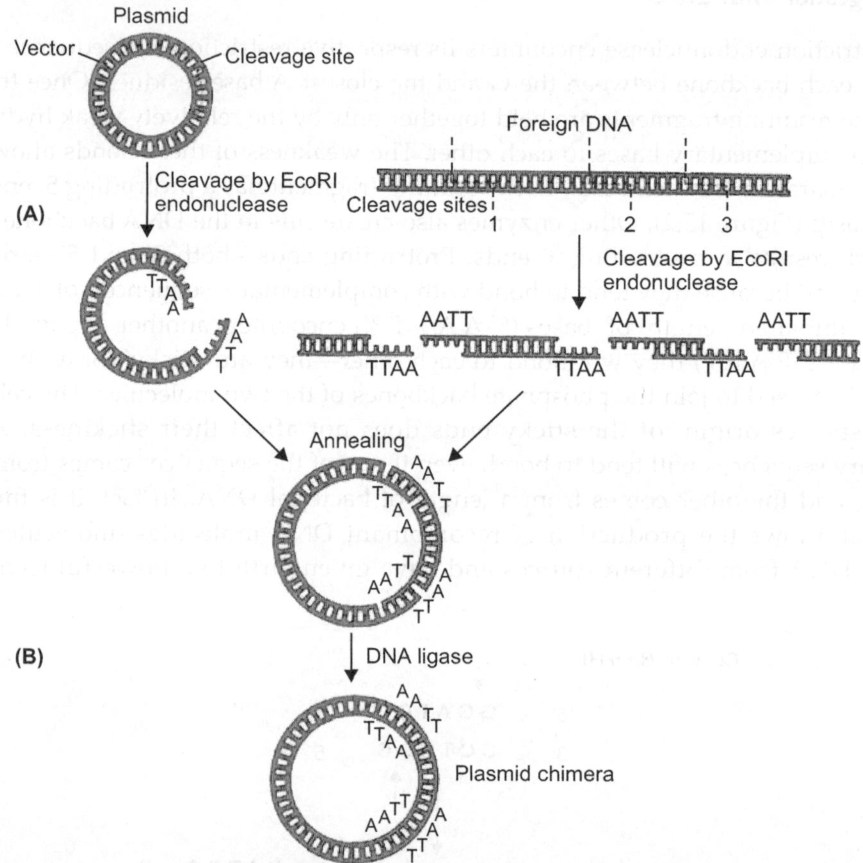

FIGURE 15.5 **Different steps in the generation of a recombinant DNA molecule.**
(A) Restriction digestion of plasmid and the source DNA by the restriction enzyme *Eco*RI.
(B) Annealing of the sticky ends followed by ligation by DNA ligase leading to the formation of recombinant DNA molecule (Plasmid chimera).

Sticky ends are useful for DNA cloning because complementary sequences anneal and can be ligated directly by DNA ligase. If two different DNA samples cleaved with the same type of restriction enzymes are mixed together in the presence of DNA ligase, a recombinant DNA molecule can be generated. This is possible because of the presence of the same type of sticky ends. The complementary sequences of the sticky ends from the unrelated DNA samples will anneal together and are finally joined by the DNA ligase enzyme to form the recombinant DNA molecule. Figure 15.5 illustrates the generation of a recombinant DNA molecule from two different sets of DNA samples.

Restriction Fragment Length Polymorphism (RFLP)

In addition to the use of restriction enzymes in the generation of recombinant DNA molecules, these molecular scissors also have another important application—**restriction fragment length polymorphism (RFLP)** or the **DNA fingerprinting**. Genome-sequencing projects have clearly shown that DNA sequences among the members of a species are almost identical. In humans it has been demonstrated that the DNA sequence is identical up to 99.9%. The remaining 0.1% of sequence variation in the genome is enough for creating all types of genetical and phenotypical variations seen among individuals. This small percentage of sequence variation is reflected also in the distribution of restriction-site sequences for various restriction endonucleases. A specific type of restriction endonuclease can cut the DNA molecule into a set of fragments, which is reproducible for any DNA sample of that organism. The set of DNA fragments of various sizes produced by the restriction digestion of a DNA sample is known as the **restriction pattern**, which is reproducible for that individual and thus forms a molecular marker. The RFLP analysis is based on variations in the DNA sequences even between individuals of a species and will result in the variations in the distribution of restriction sites for one or more types of restriction endonucleases. If two samples of DNA from two different organisms are subjected to restriction digestion with the same type of restriction endonuclease, they can create two entirely different restriction patterns, which depend on the variation in the DNA sequence and number and distribution of the restriction sites of the enzyme used. An example is illustrated in Figure 15.6.

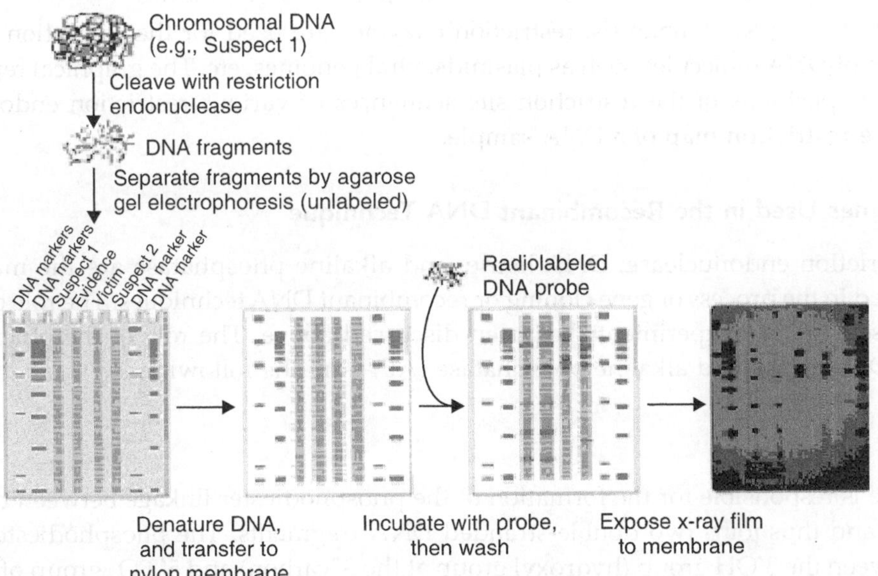

FIGURE 15.6 **Different steps in RFLP analysis in forensic science. Two suspects are screened against a victim along with molecular markers.**

There are two DNA samples, 'A' and 'B', isolated from two different individuals. Sample 'A' has four restriction sites of a restriction endonuclease *Hind*-III. Sample 'B' has three restriction sites for the same enzyme. After restriction digestion with *Hind*-III, both the DNA samples will give a unique restriction fragment pattern. The number of restriction fragments corresponds to the number of restriction sites and the size of the fragments depends on the relative positions of the restriction sites. Thus, individuals can be identified by the number and length of the fragments generated by the restriction digestion of the respective DNA samples by a specific restriction enzyme. Therefore, the restriction length polymorphism is a powerful tool for the identification of organisms to the level of species, strains, varieties, or individuals.

Different steps involved in RFLP analysis or DNA fingerprinting are illustrated in Figure 15.6. It is extensively used for phylogenetic and forensic analysis. The technique of RFLP analysis can be summarized in the following steps. The DNA samples are digested by an appropriate restriction enzyme (single digestion) or by a mixture of two or more enzymes (double digestion, triple digestion, etc.), producing a number of DNA fragments of varying sizes with single-stranded overhangs (sticky ends) or blunt ends. DNA fragments can be separated according to their molecular weight on an agarose gel. In this process of two-dimensional gel electrophoresis, DNA (which has a negative charge) migrates through the gel towards the positive electrode. Large pieces of DNA move through the tangle of agarose carbohydrates slower than small fragments. DNA in the gel can be visualized under UV light after staining with fluorescent dye, ethidium bromide. DNA bands in the gel can be transferred to a more durable nylon filter by capillary action (Southern blotting). DNA can be denatured (made single-stranded), by heat, basic pH, or enzymes. DNA probes are pieces of specific single-stranded DNA fragments that have been tagged (labeled) with radioactivity or photoactive markers. Probes will bind their compliment DNA fragments (DNA bands) and can be visualized either by autoradiography or by UV fluorescence.

In addition to RFLP analysis, restriction enzymes are used for the restriction mapping of new samples of DNA molecules such as plasmids, viral genomes, etc. The graphical representation of the relative positions of the restriction site sequences of various restriction endonucleases is known as the **restriction map** of a DNA sample.

Other Enzymes Used in the Recombinant DNA Technique

Type II restriction endonuclease, DNA ligase, and alkaline phosphatase are the main types of enzymes used in the process of gene cloning or recombinant DNA techniques. The role of restriction endonuclease in rDNA experiments has been discussed above. The role of the other two major enzymes—DNA ligase and alkaline phosphatase (AP)—are the following:

DNA Ligase

This enzyme is responsible for the formation of the phosphodiester linkage between two adjacent nucleotides and thus joins two double-stranded DNA fragments. The phosphodiester linkage is formed between the 3′OH group (hydroxyl group at the 3′ carbon) and 5′ PO_4 group of the pentose sugar of adjacent nucleotides of two different DNA fragments or nicks created in a double-stranded DNA molecule. In rDNA experiments, DNA ligase is used to join two different DNA fragments (plasmid/vector and the foreign DNA) that are annealed by the sticky ends. There are different types of ligase enzymes from different sources. But, the most frequently used one is the T_4 DNA ligase produced by T_4 bacteriophage.

Alkaline Phosphatase

In rDNA experiments there is always the problem of self-ligation and the reformation of the original plasmid or the vector. This reduces the efficiency of the recombinant DNA formation. Self-ligation is the process of annealing the sticky ends of the linearized vector without inserting the foreign DNA fragment. On ligation it will result in the original vector without any insert DNA. This can be prevented by using alkaline phosphatase. We know that the presence of 3′OH group and 5′ PO_4 group is a prerequisite for the ligase enzyme to act. If any one group of this is absent, the DNA ligase cannot function. Alkaline phosphatase can remove the phosphate group from the 5′ end of the DNA molecule resulting in a 5′OH group. Now, the only chance of ligation is between the vector and the foreign DNA resulting in a recombinant DNA molecule as the insert DNA has the 5′PO_4 group. Thus, self-ligation can be prevented and the efficiency of rDNA formation can be increased. This enzyme is isolated from some bacteria **(bacterial alkaline phosphatase—BAP)** or calf intestine **(calf intestine alkaline phosphatase—CAP)**.

In addition to these there are several other enzymes that are used for certain special purposes. Table 15.2 gives a summary of all enzymes that have some role in recombinant DNA experiments.

TABLE 15.2 Various enzymes used in recombinant DNA techniques.

Enzyme(s)	Function
Type II restriction endonucleases	Cleave DNAs at specific base sequences
DNA ligase	Joins two DNA molecules or fragments
DNA polymerase (*e.coli*)	Fill the gaps in duplexes by stepwise addition of nucleotides to 3′ ends
Reverse transcriptase	Makes a DNA copy of an RNA molecule
Polynucleotide kinase	Adds a phosphate to the 5′-OH end of a polynucleotide to label it or permit ligation
Terminal transferase	Adds homopolymer tails to the 3′-OH ends of a linear duplex
Exonuclease III	Removes nucleotide residues from the 3′ ends of a DNA strand
Bacteriophage λ exonuclease	Removes nucleotides from the 5′ ends of a duplex to expose single-stranded 3′ ends
Alkaline phosphatase	Removes terminal phosphates from either the 5′ or 3′ end (or both)

Vectors: The Vehicle for Cloning

One of the major applications of rDNA experiments is the gene cloning or DNA cloning. **Vectors** are another major component required to make an rDNA molecule for gene cloning. Vectors act as a vehicle for carrying foreign DNA into a host cell for multiplication. Usually small circular DNA molecules of bacterial origin are used as cloning vectors. A DNA molecule should possess the following essential characteristics to act as a cloning vector:

Origin of Replication

It is required for autonomous replication of the plasmid using the host's replication machinery. Almost all commonly used plasmids are based on the ColE1 origin of replication (ori). Naturally occurring origins of replication are negatively regulated to keep the copy number down (typically five to ten copies per cell) to reduce the load on the host's replication machinery. While a high copy number is disadvantageous in a natural system, it is a desirable feature in a cloning vector—since the whole idea of gene cloning is to easily isolate substantial quantities of a particular DNA sequence. Therefore, considerable work has gone into engineering the ColE1 ori such that the negative regulatory mechanisms that limit the copy number of plasmids are disabled. Modern plasmid vectors are therefore often called 'runaway replicons' and are present at 100 to 1,000 copies per cell.

Selectable Markers

Selectable markers are essential for the identification of bacteria containing recombinant plasmids. Selection can be divided into two types—**positive selection** and **negative selection**.

Positive selection is used to identify bacteria that contain plasmids. The most common markers used for positive selection are the antibiotic resistance genes carried by the original R factors. While many antibiotics and resistance genes are available, the commonly used ones fall into two general classes: **Antibiotics affecting cell wall synthesis** and **Antibiotics affecting translation.** Ampicillin is a beta-lactam-based antibiotic that acts by inhibiting the synthesis of the bacterial peptidoglycan cell wall. Sensitive bacteria are not actively 'killed,' but on cell division are unable to synthesize the cell wall and suffer from osmotic lysis. The enzyme beta-lactamase is secreted into the periplasmic space where it breaks down the antibiotic, allowing cell wall synthesis to proceed. The antibiotics tetracycline, kanamycin, and chloramphenicol all act by inhibiting translation. The covalent modification (phosphorylation, acetylation) of these antibiotics blocks their interaction with the translation apparatus. Positive selection is particularly important when introducing plasmids into bacteria by transformation. At best, only about 1 in 10,000 bacteria picks up a plasmid that carries the antibiotic resistance. A strong positive selection system is essential to eliminate the 9,999 bacteria that did not pick up a plasmid from the one that did. By plating the transformation products directly on antibiotic plates, all untransformed bacteria die and only those containing the plasmid (and antibiotic-resistance marker) grow to form colonies.

A second selection system is necessary to distinguish between plasmids that are merely recircularized from those that carry a foreign DNA insert. Usually, a negative selection method is used. In order to identify those plasmids carrying a foreign DNA fragment, the site of insertion is chosen such that the insertion disrupts a selectable marker—a phenomenon known as **insertional inactivation**. It can be the insertional inactivation of an antibiotic resistance gene or enzymes such as β-galactosidase, a product of *Lac* Z gene of lac operon. The insertion of the foreign DNA into the vector will disrupt the expression of these genes and will produce a color difference for the colony from that of cells with intact vectors.

Multiple Cloning Sites (MCS) or Polylinker

A vector should have a site specific for cloning the foreign DNA fragment provided with one restriction site for most of the commonly used unique restriction endonucleases. All these unique restriction sites are grouped together in a small region of the vector known as the **multiple cloning site (MCS)** or the **polylinker**. The presence of unique restriction sites at the MCS gives flexibility in the choice of restriction enzymes.

Small Size

Relatively small vectors are more desirable because they increase the transformation efficiency and are easy to manipulate. Small size also helps purification procedures to obtain intact plasmids.

A large number of vectors with the above characteristics have been developed. There are six different types of cloning vectors commonly used in recombinant DNA experiments. They are the following:

- Plasmid-cloning vectors
- Bacteriophage cloning vectors
- Cosmid-cloning vectors
- Yeast artificial chromosomes (YACs)
- Bacterial artificial chromosomes (BACs)
- Animal and plant vectors (Shuttle vectors)

Plasmid-cloning Vectors

Plasmid-cloning vectors are derived from bacterial plasmids and are the most widely used, versatile, and easily manipulated ones. ColE1 of *e. coli* is an example of a naturally occurring plasmid. The *ori* in almost all plasmid vectors is that of ColE1. The following are some of the examples of plasmid-cloning vectors.

pBR322 Vectors

This was the first widely used, purpose built plasmid vector. It has a number of useful features:

Origin of replication. It carries a fragment of the plasmid pMB1 that acts as an origin for DNA replication and thus ensures multiplication of the vector.

Size. It is relatively small at 4,363 bp. This is important because transformation efficiency is inversely proportional to size and above 10 kbp is very low. Thus, there is 'room' in pBR322 for an insert of at least six kbp.

Copy number. Reasonably high copy number (~15 copies per cell), which can be increased 200-fold by treatment with a protein-synthesis inhibitor—chloramphenicol amplification.

Selectable marker. It carries two antibiotic resistance genes—ampicillin and tetracycline.

Cloning sites. It carries a number of unique restriction sites. Some of these are located in one of the antibiotic resistance genes (e.g., sites for *Pst* I, *Pvu* I, and *Sac* I are found in Amp^r and *Bam*HI and *Hind* III in Tet^r). Cloning into one of these sites inactivates the gene allowing recombinants to be differentiated from non-recombinants known as **insertional inactivation.**

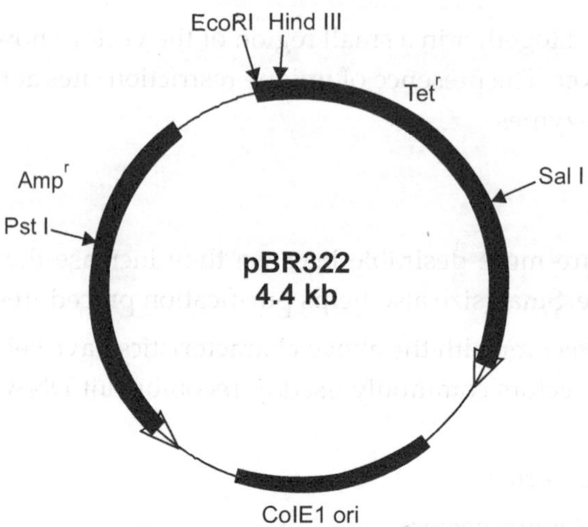

FIGURE 15.7 **Diagram sketch of pBR322.**

pUC Series Vectors

Another popular plasmid vector is the pUC series, which is extensively used as cloning as well as expression vector. These vectors have three important additional features compared to pBR322.

High copy number. A mutation within the origin of replication produces 500 to 600 copies of the plasmid per cell without amplification.

Blue-white screening. This is a special form of insertional activation that can be used during the primary selection of transformants, rather than requiring a second round of screening. It utilizes the N-terminal portion of the *e. coli* beta-galactosidase—the product of lac Z gene (alpha-peptide) encoded by the vector in a form of intermolecular complementation—that restores beta-galactosidase activity to a defective enzyme (omega-peptide) encoded by the host. If a chromogenic substrate (X-gal) and a beta-galactosidase inducer (IPTG) are included in the plates on which the primary transformants are selected, non-recombinant molecules will catabolize the colorless substrate to give blue colonies.

A synthetic polylinker. This is a piece of manmade DNA that contains several unique restriction sites. It has been inserted within the portion of the vector encoding the beta-galactosidase alpha-peptide in such a way that it does not affect its expression. However, inserting a foreign DNA fragment into any one of the polylinker restriction sites almost invariably disrupts the enzyme activity. Thus, recombinant colonies remain white but non-recombinants turn blue. The polylinker sequences differ between different members of the pUC family allowing different choices of cloning enzymes to be used. Moreover, the close proximity of many sites within each plasmid allows the cloning of fragments cut with different enzymes at each end. As well as being more flexible this allows the orientation of the insert within the plasmid to be predetermined known as **forced cloning.**

These vectors can replicate only in *e. coli* cells, since they carry the ColE1 *ori* (origin of replication). There are some other vectors, which are constructed in such a way that they can exist and multiply both in eukaryotic cells and *e. coli* and such vectors are known as **shuttle vectors.** They are attached with two types of *ori* and selectable markers. One functions in the eukaryotic

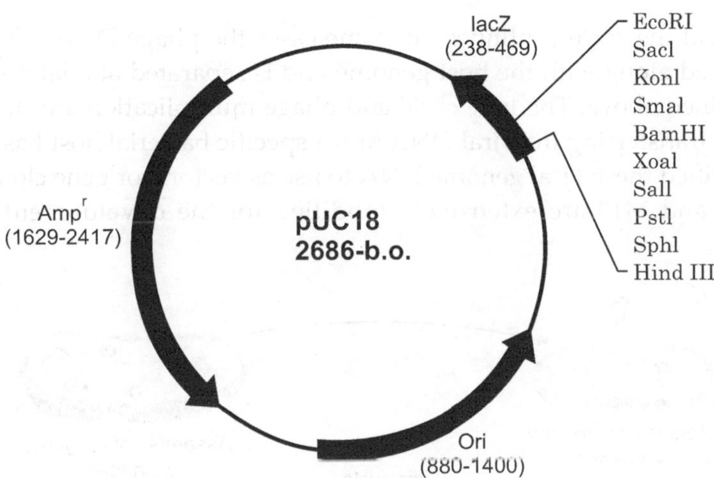

FIGURE 15.8 **Diagram sketch of pUC18. It differs from the pUC19 only in the region of polylinker site in the arrangement of restriction sites.**

cells and the other works in the bacterial cell (*e. coli*). **YEp,** the yeast episomal plasmid, is an example of a shuttle vector. There is a naturally occurring plasmid vector in plants; the plasmid of *agrobacterium tumifaciens* called Ti plasmids. This is suitably modified to function as a shuttle vector, which now carries the genes that have to be introduced into the plant genome instead of Ti (tumor inducing) genes.

Expression Vectors

In most of the cases the main purpose of the rDNA experiment and cloning is to sufficiently multiply or amplify the inserted DNA fragment. But sometimes the aim of the process will be to produce large quantities of protein encoded by the inserted gene. This can be accomplished by incorporating the necessary regulatory elements along with the gene in the vector. Such vectors should be provided with signals necessary for initiation and termination of transcription such as suitable promoters and terminator sequences; and signals for translation initiation such as a start codon and a ribosome binding site into the vector upstream to the multiple cloning site (MCS). These vectors, which have the regulatory elements and other machinery for the expression of the cloned gene, can be used for the production of recombinant proteins and are known as **expression vectors**. Vectors of the pUC series are example of expression vectors.

Bacteriophage-cloning Vectors

These are derived from viruses that specifically infect bacteria. Such viruses usually contain a comparatively small DNA genome surrounded by a protein coat. The viruses that infect bacteria are called **bacteriophages**. Those phages, which most frequently infect *e. coli*, are used for cloning purposes. Bacteriophages infect bacterial cells by injecting their DNA into the host cytoplasm. This DNA in the host cell selectively replicates and expresses the proteins required for the assembly of new phage particles. These processes result in the production of a large number of phages, which break the cells with a process known as the **lytic cycle**. These new viral particles will re-infect the neighboring cells and the cycle continues. In some cases the phage DNA, after entering into the host cell, is integrated along with the host genome and is separated at a later stage and starts the lytic cycle as explained above. The lytic cycle and phage multiplication are shown in Figure 15.9. The natural way of transferring the viral DNA into a specific bacterial host has attracted scientists and they have modified these viral genome DNA to use as vectors for gene cloning. Bacteriophage lambda (λ phage) and M13 are extensively modified for the development of a phage-based cloning vector.

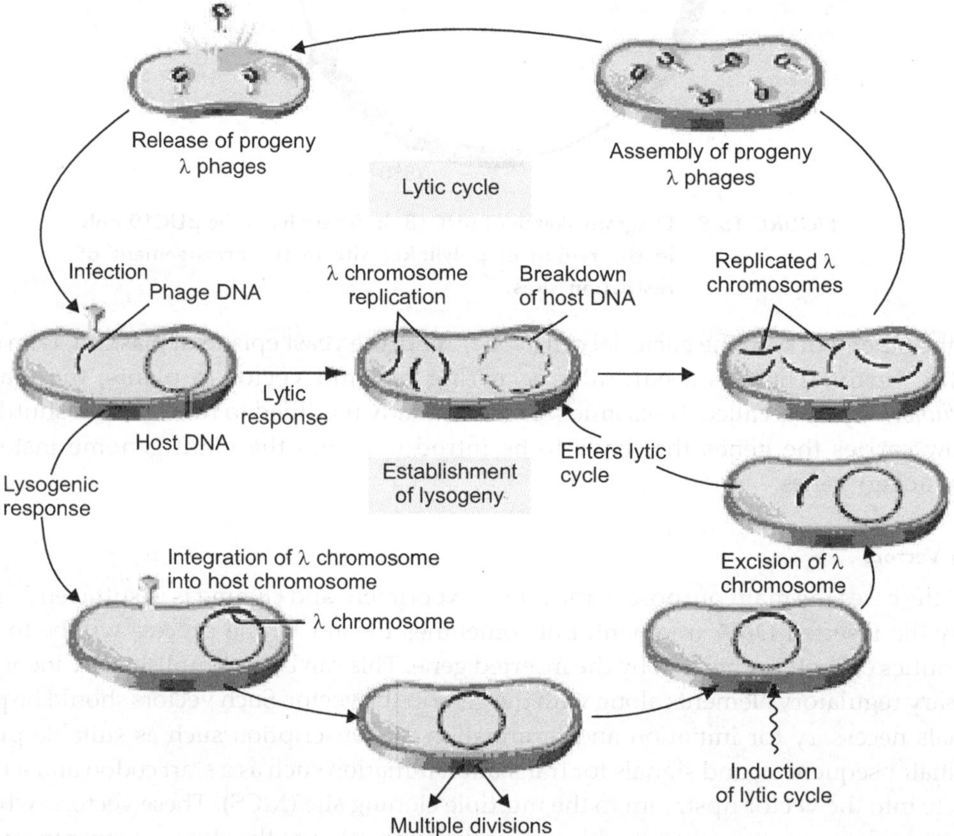

FIGURE 15.9 **The bacteriophage infection and life cycle—lysogeny and lytic cycle.**

Phage-cloning Vectors

These are the engineered version of λ bacteriophage that infects *e. coli.* This vector has a double-stranded linear DNA genome having a size of 48,514 bp (48.5 kb). About 12 bases at the ends of this DNA molecule are unpaired and complementary and therefore are sticky or cohesive and known as **cos sites (cohesive end sites).** These cohesive sites are very essential for packaging the DNA into the viral particle during the lytic cycle. But a major portion of the central region of the viral genome DNA is not required for the infection and lytic cycle in *e. coli* cells. The λ phage vector is designed in such a way that the central region of the λ chromosome (linear) is cut with a restriction enzyme and is replaced with the foreign DNA digested with the same restriction enzyme.

This recombinant DNA is packaged in phage heads to form virus particles (Figure 15.10). Phages with both ends of the λ chromosome and a 37 to 52 kb insert replicate by infecting *e. coli*. Phages replicate using *e. coli* and the lytic cycle (Figure 15.9) and produce large quantities of 37 to 52 kb cloned DNA. A large number of unique restriction sites are also available here.

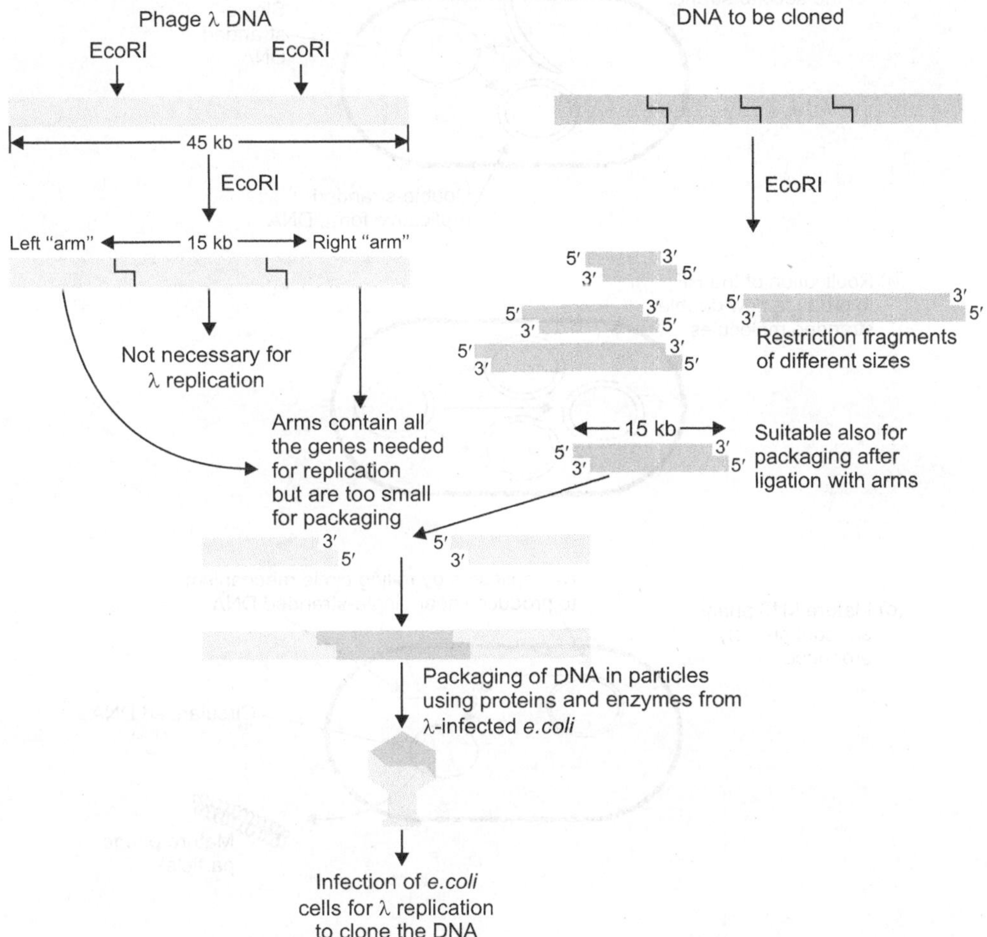

FIGURE 15.10 **Phage λ based cloning vectors for cloning inserts of large sizes.**

M13-based Cloning Vectors

The M13 family of vectors is derived from bacteriophage M13. This is a male-specific (infects *e. coli* having *f. pili*), lysogenic filamentous phage with a circular single stranded DNA genome about 6,407 bp (6.4 kb) in length. On infection, this molecule is transferred to *e. coli* and converted into the double-stranded replicative form (RF). The replication continues and when there are more than 100 copies of DNA in the cell, the DNA replications become asymmetric and produce copies of the original single-stranded molecule, which are packaged into infective particles and extruded from the cell. The host is never lysed but continues to grow throughout the infection, although at a significantly reduced rate. This difference in growth rate between infected and uninfected cells produces characteristic 'plaques' when M13 phages are plated on a suitable host.

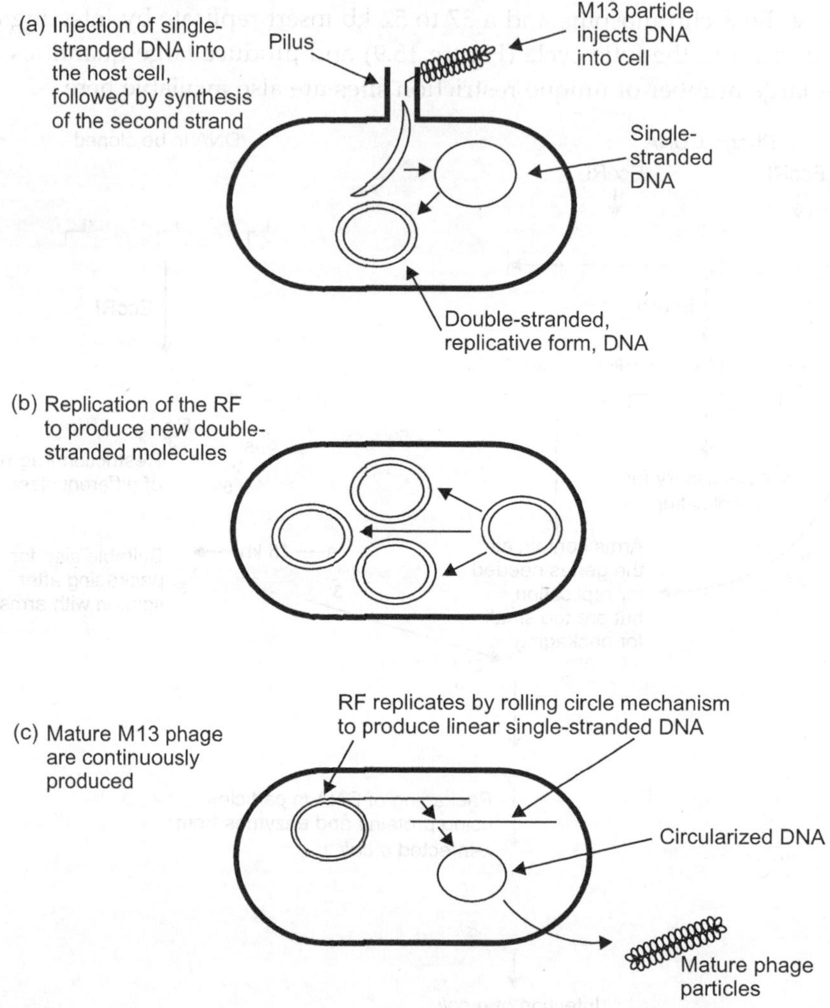

FIGURE 15.11 **Infection and life cycle of M13 phage.**

The advantages of M13-based vectors are that they contain the same polylinker and alpha-peptide fragments as the pUC series and recombinants can be selected by the blue → white color test. The RF form of M13 vectors can be isolated by standard plasmid DNA-preparation procedures, and foreign DNA can be inserted into them as if they were conventional plasmids. The size of the genome is below 10 kb and so is easy to handle. The specific use of M13 is as an aid to DNA sequencing. Once cloned into M13, large amounts of the single-stranded form of any given fragment can be easily isolated from the mature phage that is extruded from infected cells. This is an ideal template for a dideoxy-sequencing reaction. If an oligonucleotide complementary to the region just downstream of the polylinker is used as the primer for the dideoxy reaction, the cloned DNA fragment can be sequenced. With this method it is possible to determine the sequence of any DNA fragment that has been cloned into M13.

Cosmid-cloning Vectors

Cosmid-cloning vectors were among the first large insert cloning vehicles developed. They were constructed by certain features of plasmid and the *cos* site of λ phage. A cosmid can be defined as a plasmid that contains a *cos* site from the lambda phage genome. The vector replicates as a plasmid (it contains a ColE1 origin of replication), and uses Amp^r for positive selection and employs lambda phage packaging to select for recombinant plasmids carrying foreign DNA inserts up to 45 kb in size. The simplest cosmid vector has a ColE1 origin of replication, selectable markers including the antibiotic-resistance gene and β-galactosidase gene (a part of *lac Z* gene), and suitable polylinker sites and lambda cos site (Figure 15.12).

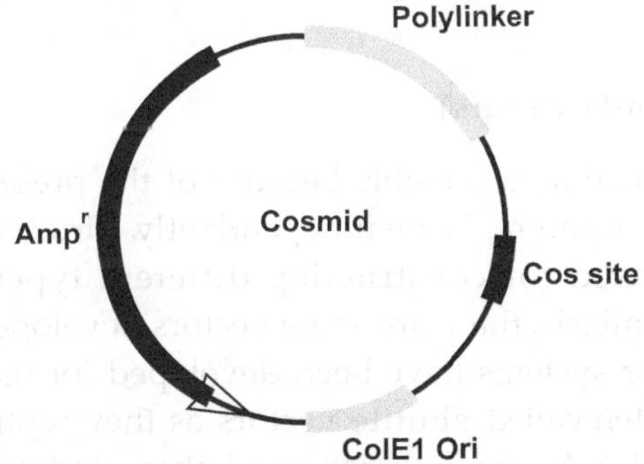

FIGURE 15.12 **A simplest form of cosmid.**

Ligation of the cosmid vector and foreign DNA fragments of sizes upto 45 kb is similar to ligation into a lambda substitution vector. The desired ligation product is a concatemer of 45 kb foreign DNA fragment and five kb cosmid-vector sequences. This concatemer is then packaged into viral particles (remember, packaging is cos site to cos site) and these are used to infect *e. coli* where the cosmid vector replicates using the ColE1 origin of replication (*ori*). Phage packaging serves only to select for recombinant molecules and to transfer these long DNA molecules (50 kb total) into the bacterial host (50 kb fragments transform very inefficiently while phage infection is very efficient).

Yeast Artificial Chromosomes (YACs)

Yeast artificial chromosomes (YACs) are the vectors that enable the formation of artificial chromosomes with the foreign DNA fragments and cloning into yeast. These are used for the cloning of very large DNA fragments in the range of 500 to 1,000 kb. These vectors are constructed with yeast telomere sequences at each ends, yeast centromere sequences, and the yeast **ARS (autonomously replicating sequences)** for replication. In addition to this there are the restriction sites for DNA insertion and selectable markers such as amino acid dependence and antibiotic resistance on each arm. The recombinant DNA is constructed by ligating high-molecular weight to the YAC arms using appropriate restriction enzymes. These recombinant vectors or the artificial chromosomes are used for the transformation of yeast cells by electroporation and the transformed cells are selected on selection growth medium.

Bacterial Artificial Chromosomes (BACs)

Bacterial artificial chromosomes (BACs) are the cloning vectors based on the extra-chromosomal plasmids of *e.coli* called **F factor** or **fertility factor**. These vectors enable the construction of artificial chromosomes, which can be cloned in *e.coli*. This vector is useful for cloning DNA fragments up to 350 kb, but can be handled like regular bacterial plasmid vectors, and is very useful for sequencing large stretches of chromosomal DNA. Like any other vector, BACs contain *ori* sequences derived from *e.coli* plasmid F factor, multiple cloning sites (MCS) having unique restriction sites, and suitable selectable markers.

Both YACs and BACs were very useful in genome-sequencing projects such as the Human Genome Project.

Animal and Plant Vectors (Shuttle Vectors)

The genetic engineering of bacteria is possible because of the presence of special types of extra-chromosomal circular DNA that can replicate independently. Viruses that infect bacteria (M13 and lambda phage) have been used for constructing different types of vectors for the genetic transformation of bacteria. Similarly, there are some vectors developed for transforming plant and animal cells. Numerous vector systems have been developed for use in eukaryotic systems. The most common of these are often called **shuttle vectors** as they replicate in both prokaryotic and eukaryotic hosts. In general, DNA manipulations and characterizations are done in prokaryotic systems, and then the manipulated DNA is reintroduced to eukaryotic systems for functional analysis. The common features of such shuttle vectors or eukaryotic vectors are the following:

- They are capable of replicating in two or more types of hosts including prokaryotic and eukaryotic cells.
- They replicate autonomously, or integrate into the host genome and replicate when the host cell multiplies.
- These vectors are commonly used for transporting genes from one organism to another (i.e., transforming animal and plant cells).

Mammalian Vectors

These are shuttle vectors developed for use in mammalian tissue culture. Like any other vector system, mammalian shuttle vectors require sequences enabling them to replicate in mammalian tissue culture. These eukaryotic origins of replication are typically derived from well-characterized mammalian viruses such as simian virus 40 (SV-40) with sv-40 ori and large tantigen system or Epstein-Barr virus (mononucleosis). SV-40 was the vector used in the first cloning experiment in mammalian cell cultures in 1979. Since then a number of mammalian vectors based on viruses such as adenovirus and pappilomavirus have been developed to transfer genes in mammals for various purposes including gene therapy. At present, retrovirus-based vectors are the most commonly used vectors for cloning genes in mammalian cells. Both systems, eukaryotic and prokaryotic, allow the vector to replicate within the host cell as a plasmid without integrating into the genome. In addition to the origin of replication, these shuttle vectors also carry antibiotic resistance genes, which function in eukaryotic cells (i.e., neomycin (G418) resistance, hygromycin resistance, methotrexate resistance, etc).

Finally, artificial chromosome vectors have been developed for use in mammalian cells known as **mammalian artificial chromosomes (MACs)** such as the YACs and BACs. These eukaryotic vectors contain telomeric sequences (eukaryotic chromosomes are linear not circular molecules) and centromeres to ensure appropriate segregation of the artificial chromosomes as well as selectable markers. Such vector systems may gain in importance as we move toward manipulating the eukaryotic genome.

Plant Vectors

There are a number of shuttle vectors developed for the genetic transformation of plant systems. These shuttle vectors are based on certain viruses and bacteria, which are pathogenic to plants. Plant viruses such as the cauliflower mosaic virus (CMV), tobacco mosaic virus (TMV), and gemini viruses were used for developing vectors for plant cell transformations, but with limited success. The most successful shuttle vectors developed for the plant system are those that are based on the Ti plasmid of *agrobacterium tumifaciens*, a bacterium that causes tumor formation in plants. This has been successfully exploited for the development of *a. tumifaciens*-based vectors for the genetic engineering of plants. The success in plant genetic engineering and the development of a number of transgenic plants is mainly due to the development of vectors based on Ti plasmids and *a. tumifaciens*.

TABLE 15.3 Different types of cloning vectors and the size of the insert DNA that can be accommodated.

Type of Vectors	Insert DNA Size (In kb)
Plasmid	0.5 to 8
Bacteriophage Lambda	9 to 25
Cosmid	30 to 45
YAC	250 to 1000
BAC	50 to 300

Host Cells

The ultimate aim of the development of recombinant DNA molecules or vectors is the multiplication of rDNA by cloning in a suitable host cell. There are many types of host cells available for the purpose of cloning. The kind of host cell used mainly depends on the aim of the cloning experiment. The host cell can be bacteria, yeast, plant or animal cells.

E. coli is the most extensively studied and widely used organism in rDNA experiments. It is very simple, easy to handle, grows rapidly and is able to accept and maintain a range of vectors. The doubling time of *e. coli* under ideal growth conditions is 20 minutes. As the cell undergoes multiplication, the rDNA within the cell also undergoes multiplication independent of its genome. Within a short time the number of cells and the quantity of the rDNA will be very high. If the recombinant plasmids have appropriate promoter signals for expression, a large quantity of protein can also be purified from these systems. There are a large number of genetically defined strains, which can be used freely for genetic engineering experiment and expression of recombinant proteins because they are disarmed and so are non-infective and harmless.

If the recombinant protein that we want to produce is of eukaryotic origin, it is better to opt for a eukaryotic host system such as yeast. There are several reasons for using a eukaryotic host system for the expression of a protein of eukaryotic origin. The polypeptide after its synthesis has to undergo folding in the correct manner to attain its functionally active three-dimensional structure. This may require the removal of some signal peptides or amino acids to initiate the process of folding to the correct three-dimensional structure. The correct three-dimensional structure is very essential for the production of functionally active protein. In addition to this, there may be removal or addition or chemical modification of amino acids or the addition of other macromolecules such as carbohydrates or lipids to make the protein functionally active. All these processes together are known as the post-translational modification of proteins. Prokaryotic systems such as *e. coli* lack all these enzymes and other machinery essential for the post-translational modification of proteins. The eukaryotic protein produced in bacteria may fold incorrectly resulting in the production of a functionally inactive molecule. Prokaryotic systems do not have the molecular machinery necessary for the splicing of introns of eukaryotic genes resulting in the production of correct form of mRNA.

The most convenient eukaryotic system that can be used for the expression of eukaryotic genes is a species of yeast such as *pichea pastoris*. There are several advantages in yeast as an expression system. Yeasts are the simplest of the eukaryotic systems, single-celled, genetically and physiologically well-characterized, and easy to grow and manipulate. Like bacteria it can be cultivated in different volumes, either in flasks, laboratory fermentors, or in industrial scales. Plant and animal cells can also be used as host cells for the expression of cloned genes either in tissue culture or as cultured cells in fermentors or as a genetically-engineered whole organisms—transgenic animals and plants.

15.3 MAKING RECOMBINANT DNA

The first step in the construction of a recombinant DNA is the isolation and purification of vectors and the DNA fragments containing the gene to be cloned. There are a number of standardized procedures for the isolation of plasmids and DNA fragments containing the gene. First, digest the vector DNA with a suitable restriction endonuclease and make it linear with or without sticky ends, which is determined by the type of RE used. Next, isolate the DNA fragment carrying the gene by digesting the genome or the source DNA strand with the same RE used for cutting the vector. The linearized (digested) vector and the target DNA cut with the same restriction enzyme are incubated together in the presence of DNA ligase. During incubation, the sticky ends of the two DNA strands come closer by the base complementation and become hydrogen bonded to each other. At this point, a phosphodiester linkage is established between them, mediated by the DNA ligase enzyme. This results in the formation of recombinant DNA molecule of vector and the target DNA or insert DNA. This is the basic strategy for the construction of a recombinant DNA molecule from a vector and a target DNA fragment.

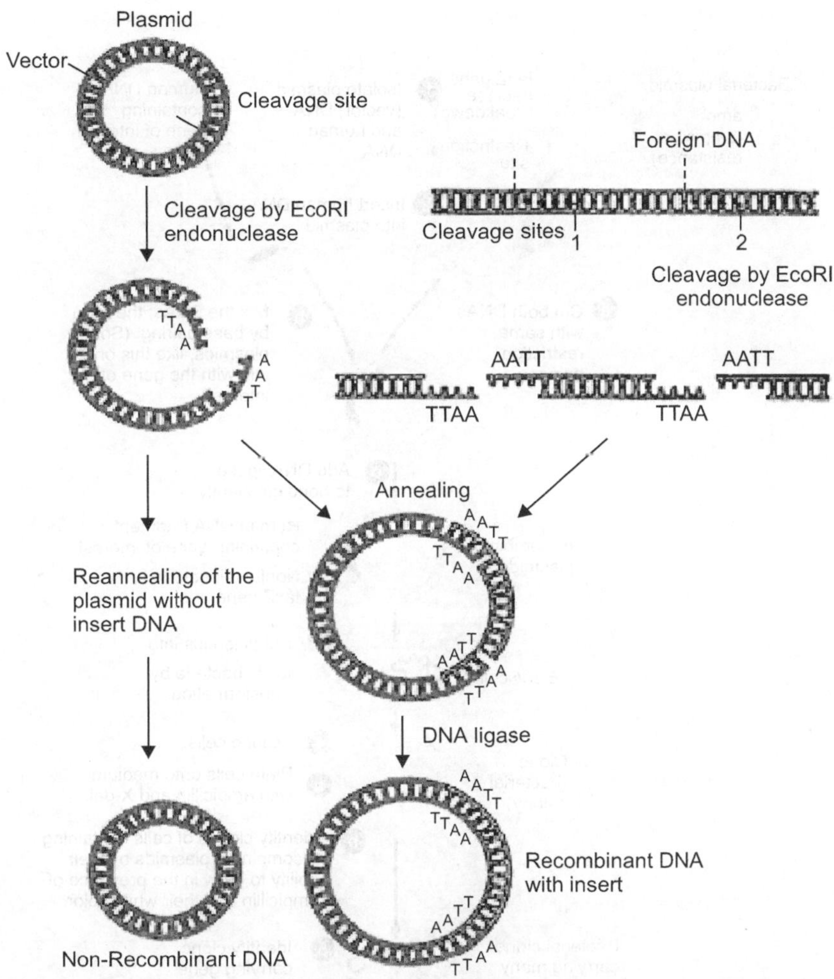

FIGURE 15.13 **A general strategy for constructing a recombinant DNA molecule.**

The disadvantage of this strategy is that there is the possibility of re-annealing of the sticky ends of the linearized vector leading to the formation of the vector itself without any foreign DNA fragment. This problem can be reduced considerably by adopting any one of the following strategies :

- Treating the digested vector with alkaline phosphatase to remove the 5′ phosphate group as described earlier.
- Linearization of the vector (restriction digestion of the vector) and the isolation of the foreign DNA fragment with two restriction enzymes instead of one (double digestion). So that the two sticky ends of the vector will not be identical, which will prevent the re-annealing to some extent.

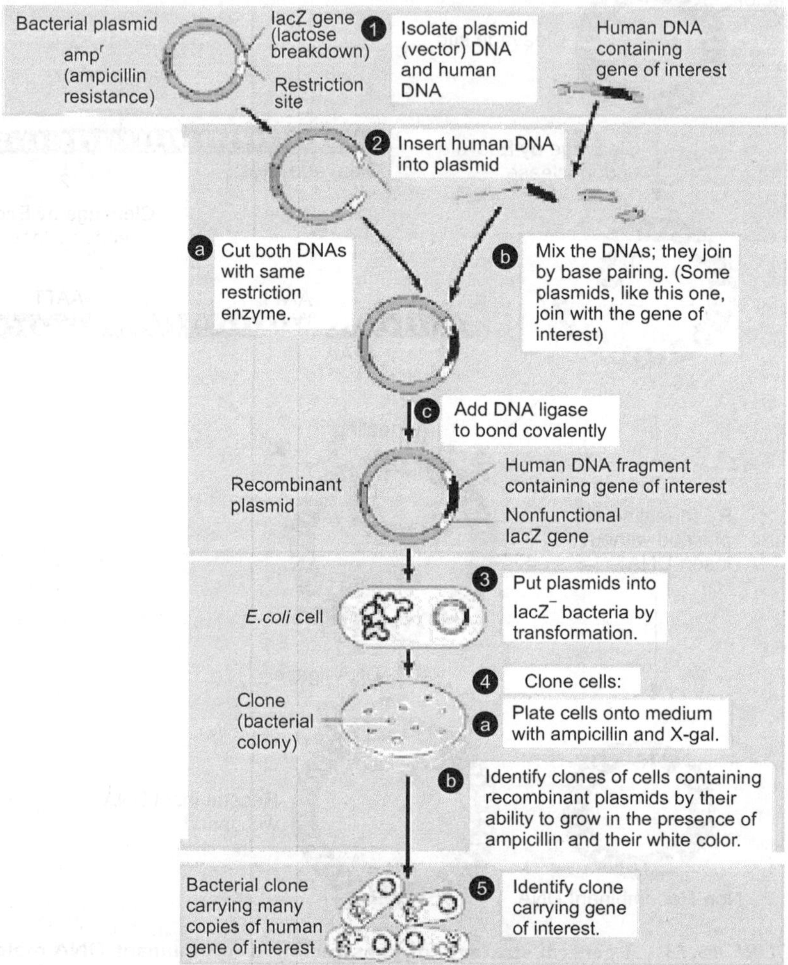

FIGURE 15.14 **General strategy for the construction of a recombinant DNA and its cloning. The figure illustrates the cloning of a human gene in *e. coli* with all the details in each step.**

15.4 DNA LIBRARY

The isolation of a gene or the DNA strand that carries the gene is the first step in a gene cloning experiment. The construction of a DNA library is a prerequisite for this purpose. There are mainly three types of DNA libraries—**genomic library, chromosomal library,** and **cDNA library**. These DNA libraries can be used as the source for screening the target gene. These DNA libraries are generally known as the recombinant DNA libraries.

Genomic Library

A genomic library is a collection of cloned, restriction-enzyme-digested DNA fragments containing at least one copy of every DNA sequence in a genome. The entire genome of an organism is represented as a set of DNA fragments inserted into a vector molecule. These can be propagated or cloned in a suitable organism.

There are three ways to make a genomic library. The genomic DNA of the organism is extracted and is cut into fragments of suitable sizes by any of the following three methods. The genomic DNA is digested completely by a restriction enzyme that converts it into fragments of suitable sizes. The restriction enzyme cuts at all relevant restriction sites and produces a large number of short fragments with sticky ends. The disadvantage of this is that genes containing restriction sites within the reading frame may be cut into two or more fragments and may be cloned separately.

The genomic DNA can be fragmented un-enzymatically by means of mechanical shearing such as sonication, which produces longer DNA fragments. The disadvantage in this case is that the ends of the fragments produced are not uniform and need enzymatic modification for insertion into a cloning vector.

The third method for fragmenting genomic DNA is by partial enzymatic digestion with a restriction enzyme. By limiting the amount of restriction enzyme and the time of exposure of DNA to the active enzyme, all the restriction sites present in the genome will not be acted upon by the restriction enzyme. The restriction digestion will be partial, resulting in long DNA fragments with overlapping sequences. After digestion, the DNA fragments can be size selected by agarose electrophoresis. Like fragments of complete digestion, fragments have sticky ends and can be inserted directly.

The vector is also isolated from a host bacterium and cut with the same restriction enzyme used for the digestion of the genomic DNA. The linearized vector molecule is incubated with the fragments of genomic DNA along with DNA ligase so that each of the genomic DNA fragments is inserted into each vector molecule resulting in recombinant DNA molecules. Each of the recombinant vector molecules will have different fragments of genomic DNA and all these recombinant DNA molecules are introduced into a suitable host organism such as bacterial cells or phage particles.

Chromosomal Library

A chromosomal library is the collection of cloned restriction-enzyme-digested DNA fragments from individual chromosomes. This is very important in the case of genomic sequencing of eukaryotes and physical mapping of chromosomes, which indicate the locus of specific genes in specific chromosomes.

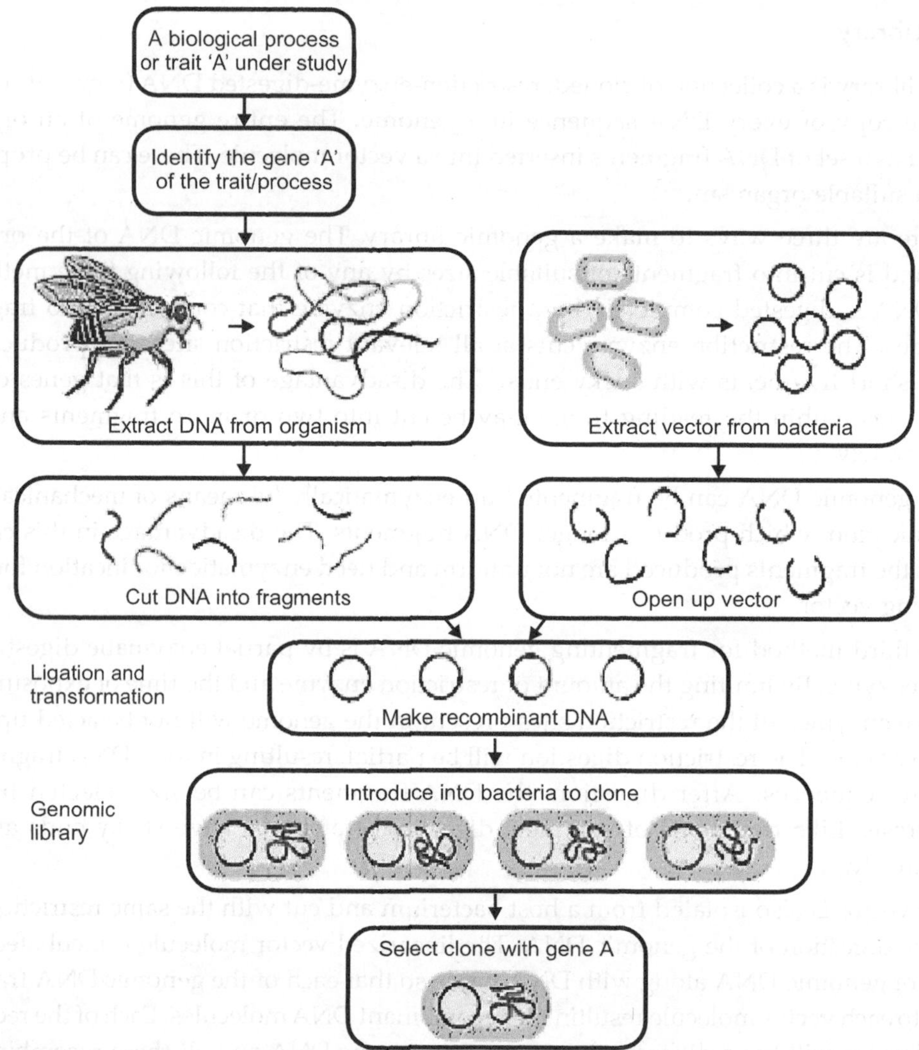

FIGURE 15.15 **Schematic representation of the construction of a genomic library and identification of the gene in the library.**

Complementary DNA Library (cDNA Library)

In the case of the eukaryotic system, the genomic library may not be of much use if the aim is to isolate an intact gene and express the protein in bacterial cells in active form because of the following reasons:

1. Even though the genomic library represents complete genome sequences, it is possible that certain genes may be fragmented and may be present in two different bacterial clones.
2. Eukaryotic genes usually contain intervening non-coding regions known as **introns.** The genes with introns can be expressed in prokaryotic systems as they lack the splicing machinery. So the gene will be expressed along with the introns and the result will be an entirely different protein, which is not active or functional.

3. The regulatory elements of eukaryotic genes may not be properly understood by the prokaryotic systems and that may create problems in the expression of a proper protein in active form.

The cDNA library can solve most of these problems, if our aim of gene cloning is to express the gene in bacterium. The cDNA library is the collection of clones of DNA, which are the complementary copies of messenger RNA (mRNA) isolated from the respective cells. Messenger RNA is the starting material for the construction of cDNA libraries. This DNA library represents only those genes which are being expressed by a group of cells or tissues. Since the mRNAs are produced after splicing, they are devoid of introns. Therefore, the complementary copy of these mRNAs (cDNA) represents only the exons or the coding regions of the actual eukaryotic genes. This cDNA can be directly inserted into an expression vector such as pUC18 or pUC19 and the protein can be expressed in the bacterial systems such as *e. coli*.

The construction of a cDNA library starts with the isolation of mRNA from the specialized cells or tissue in which the particular mRNAs are abundant. The mRNAs are purified from the total RNA by affinity chromatography using oligo dT columns, which selectively bind the poly A tail of mRNA. The purified mRNAs are transcribed back into cDNA by an enzyme known as **reverse transcriptase**. In the first step, a single strand of the cDNA is produced and again by enzymatic method the single-stranded DNA is converted into a double-stranded cDNA molecule. These cDNA molecules are inserted into suitable vectors with the help of appropriate linkers or adapters, which provide the required restriction sites for annealing and ligation. Adapters or linkers are the specialized artificial oligonucleotides carrying restriction sites for a particular restriction enzyme. These can be attached to the ends of the cDNA molecule to provide the restriction site for a particular restriction enzyme. Once the recombinant DNA molecules with the insert cDNA are formed, they are introduced into suitable bacterial host cells for cloning and establishing the cDNA library. These cDNA libraries can be used as the source for the isolation of a specific gene and its expression in a prokaryotic system. The expression vector should be provided with the regulatory sequences such as the correct type of promoters.

15.5 TRANSGENICS—INTRODUCTION OF RECOMBINANT DNA INTO HOST CELLS

After the construction of the recombinant DNA molecule, it has to be introduced into a suitable host cell so that the DNA will multiply. There are several methods available for this, which depend on several factors including the type of vector used and the host cell. The following are common methods used for introducing the rDNA molecule into the host cell:

1. **Transformation** is the most commonly adopted method for introducing a recombinant DNA into a host cell. During the process of transformation, cells take up foreign DNA from their environment. But many cells including *e. coli*, yeast, and mammalian cells are not able to take up the DNA from their environment naturally. Certain chemical treatments can enhance the ability of cells to take up the foreign DNA or make the cell competent for transformation. This enhanced competency for transformation was found out by Mandel and Higa in 1970. They observed that *e. coli* cells become more competent to take up foreign DNA when these cells are incubated in cold calcium chloride solution. This mechanism is still followed for the transformation of *e. coli* cells as a part of gene cloning.

2. **Transfection** is another method used for the transformation of cultured cells. In this method DNA (recombinant vector) is mixed with charged substances such as calcium phosphate, cationic liposomes, or DEAE dextran and overlayered on the host cells. This ultimately results in the uptake of the external DNA by these host cells.
3. **Electroporation** is another efficient method for introducing rDNA into a host cell. During the process of electroporation an electric current is used to create transient microscopic pores in the cell membrane of the host cell. Through these temporary openings foreign DNA enters the cell. This method is especially suitable for cells such as yeast, mammalian cells, and plant protoplasts.
4. **Microinjection** is a specialized technique by which the DNA fragment or gene can be directly injected into the nucleus of plant and animal cells. This method can be carried out without the use of any specialized vector. The procedure involves the direct injection of the DNA into the nucleus of the host cell with the help of a glass microinjection tube or syringe. The instrumentation for micro-injection consists of a microscope of high magnification, attached with fine needles, syringes, and other facilities to carry out the delivery of DNA into the selected cells automatically.
5. **The biolistic method** was developed for introducing foreign DNA into plant cells with the help of a gene or particle gun. During the biolistic method, microscopic particles of gold or tungsten coated with the DNA of interest is bombarded into the cells at a high velocity. The bombardment of the particle is carried out with the help of a mechanical device called a **particle gun**. Normally, the method of introducing foreign genes into the plant cell is done using *agrobacterium tumifaciens*. But this bacterium is host specific and may not infect and transfer the gene in certain cases, particularly in the case of monocots. Biolistic methods can be adopted in such cases.

In the case of **bacteriophage vectors** no special methods or treatments are required for inserting the DNA into the genome because the phage infection and the transfer of DNA into the genome is natural.

15.6 IDENTIFICATION OF RECOMBINANTS

The introduction of recombinant DNA into a suitable host cell is followed by the selection of those cells, which contain the recombinant vectors. During the process of genetic transformation only a very low percentage of the total cell population takes or receives the recombinant DNA. Therefore, it is very important to have an efficient screening method to select the transformed cells or those cells that have taken the foreign DNA. There are various selection methods or screening techniques that are based on the expression or non-expression of some of the traits present in the vector or along with the cloned gene. Some of these traits are the resistance to certain antibiotics. If the antibiotic-resistance gene is present along with the cloned gene, it is very easy to select the recombinant transformants directly on a medium supplemented with the respective antibiotic. In most of the cases there are two stages of selection. First, one is the selection of transformed cells (i.e., the cells that have taken a plasmid). The second one is to identify the transformed cells that have the recombinant plasmid.

The transformed cells contain a plasmid and can be identified by the positive-selection method. The most common markers used for positive selection are the antibiotic-resistance genes present in the cloning vectors. For example, the ampicillin-resistance gene is present in pUC19 and other similar types of vectors. A positive-selection method is very important when the DNA is introduced through a plasmid by transformation. Only a very small percentage of the cells picks up the plasmids and becomes antibiotic resistant. So, a strong positive-selection system can eliminate all the untransformed cells and allow the transformed cells to grow. By plating the transformation products directly on antibiotic plates, all untransformed bacteria die and only those containing the plasmid with the antibiotic-resistance marker gene can grow to form colonies.

By this selection method we will be able to get the transformed cells with the plasmids. But it is not possible to identify the transformed cells or colonies having recombinant plasmids from the transformed cells or colonies with parent plasmid without any insert DNA fragment (non-recombinant vector). A negative-selection method can be adopted for identifying the presence of recombinant vector among the transformed cells. In order to identify those plasmids carrying a foreign DNA fragment, the multiple cloning site in the vector is engineered in such a way that the insertion of a foreign DNA strand disrupts the expression of a selectable marker gene—a phenomenon known as **insertional inactivation.** Two types of selectable markers are used for negative selection. One is the insertional inactivation of antibiotic resistance and the second is the insertional inactivation of an enzymatic activity.

In order to use antibiotic-resistance as a negative- as well as a positive-selection system, the plasmid vector must carry two different antibiotic-resistance genes. An example of such a vector is pBR322. pBR322 carries both an ampicillin and a tetracycline resistance gene. The phenotype of the bacteria containing the intact plasmid is **Amp^r Tet^r** (ampicillin resistant and tetracycline resistant). Insertion of foreign DNA into the Pst I site located in the Amp^r gene results in an **Amp^s Tet^r** (ampicillin susceptible and tetracycline resistant) phenotype. Conversely, insertion of foreign DNA into the *eco*RI, *hind*-III, or *sal* I sites located in the Tet^r gene results in an **Amp^r Tet^s** phenotype. In order to reveal these phenotypes, transformants are first plated on the positive selection media (tet or amp, respectively). Positively selected colonies are then restreaked on the positive selection media (master plate from which we recover our desired vector + insert) and on the negative selection media. (amp or tet media, respectively). The transformed cells with recombinant vector will be either **Amp^s-Tet^r** or **Amp^r-Tet^s** cells or colonies.

While the insertional inactivation of antibiotic-resistance works, it requires a lot of manipulation—picking the positively selected bacteria and replating on negative selection media, etc. In addition to the tedium of picking colonies, vectors such as pBR322 also suffer from a paucity of convenient restriction sites at which to insert the foreign DNA fragments. These limitations promote the development of a set of host-vector systems in which it is possible to positively select for bacteria carrying a plasmid and simultaneously select for insertional inactivation of enzymatic activity. This system is based on the beta-galactosidose gene (lac Z gene) of the *e. coli lac* operon. This method is based on the insertional inactivation of *lac* Z gene. The *lac* Z gene is a component of the lac operon and it produces an enzyme β-galactosidase. This enzyme can cleave a colorless, synthetic substrate, X-gal into a blue-colored product. If the β-galactosidase gene (*lac* Z gene) is disrupted and inactivated by inserting the foreign DNA fragment into it, X-gal will not be cleaved and the development of the blue color will be prevented. Thus, the selection medium is provided with the X-gal and the host cells that carry normal vectors (without insert DNA) will produce blue-colored colonies, and those cells which carry the recombinant plasmids will develop white-colored colonies. Here, the positive selection (expression of antibiotic resistance) and the negative selection (inactivation of an enzyme activity—β-galactosidase) can be combined together and carried out in a single experiment. In addition to X-gal and antibiotics as the means of selection additives, the media also require a compound called **IPTG** as an inducer for the expression of *lac* Z gene.

The selection method just described is primarily used for the screening of colonies with recombinant plasmids in *e. coli*. There are several other methods to identify the clones having the recombinant plasmids, but it depends on the type of vector-host combinations used for the experiment. The presence of the desired insert DNA can be confirmed by isolating the recombinant plasmids and digesting it with the same restriction enzyme used for making the recombinant vectors.

15.7 POLYMERASE CHAIN REACTION (PCR)

Once in a while in the world of science, there comes an idea or a tool so ingenious that it revolutionizes the way people ask questions. **Polymerase chain reaction**, better known as PCR, is one of these technologies. It has not only made a tremendous impact on the scientific community, but it has also affected many aspects of our everyday lives. It was originally invented by Kary Mullis in 1985.

PCR is a process that 'amplifies' or 'copies' a piece of DNA repeatedly until there is an amount which is great enough to observe visually. The following are the components required for carrying out a PCR reaction:

1. Template DNA
2. Nucleoside tri-phosphates
3. Primers
4. Taq DNA polymerase

The **Taq DNA polymerase** is a temperature-tolerant enzyme isolated from *thermus aquaticus*, a bacterium found in hot springs. It catalyzes the synthesis of DNA. This enzyme is stable and active at near-boiling temperatures. Therefore, it is well suited for carrying out PCR reactions. The basic components required for the assembly of a PCR reaction is given below. The part of DNA sequences that has to be amplified can be selected by choosing the correct types of primers and the template. The template carries the DNA sequence that we want to amplify and the two sets of primers—forward primer and the backward primer—select the region of the template by annealing with its complementary sequences.

Genomic DNA	**0.1 to 1 ug**
Primer 1	**20 pmoles**
Primer 2	**20 pmoles**
Tris HCl, pH 8.3	**20 mM**
Mg Cl_2	**1.5 to 2 mM**
dNTPs	**50 uM each**
DNA Polymerase derived from *thermus aquaticus*	

A PCR-amplification cycle consists of three basic steps. They are **denaturation, annealing,** and **primer extension.** Its time duration and sequence of the steps have to be programed in the thermal cycler. In the denaturation step, the template is heated to 94°C for one minute or up to five minutes. At this high temperature the DNA undergoes complete denaturation and the double-stranded DNA (dsDNA) becomes single-stranded (ssDNA). Each single ssDNA can act as the template for the *in vitro* DNA synthesis. The next step is the primer annealing. In this step the two primers, the forward primers and the backward primers, anneal or hybridize to the single-stranded template DNA at its complementary regions. Annealing is usually carried out at a lower temperature depending on the length and sequence of the primers. In standard cases it is in the range of 40 to 50°C.

Denaturation	**94°C, 0.5 min**
Primer Annealing	**55°C, 1.5 min**
Extension	**72°C, 1 min**

The final step in each cycle is the extension in which the Taq DNA polymerase synthesizes the DNA region between the primers using dNTPs and Mg^{2+}. The optimum temperature for carrying out the primer extension reaction or polymerization of dNTPs is standardized at 72°C. The second cycle again starts with the denaturation reaction by heating at 94°C, so that all the newly synthesized DNA are also denatured to single strands, which again act as templates. It will again be followed by the annealing and extension and thus the cycle of denaturation, primer annealing, and extension continues resulting in the amplification of the selected DNA sequence at an exponential phase as shown in Figure 15.16.

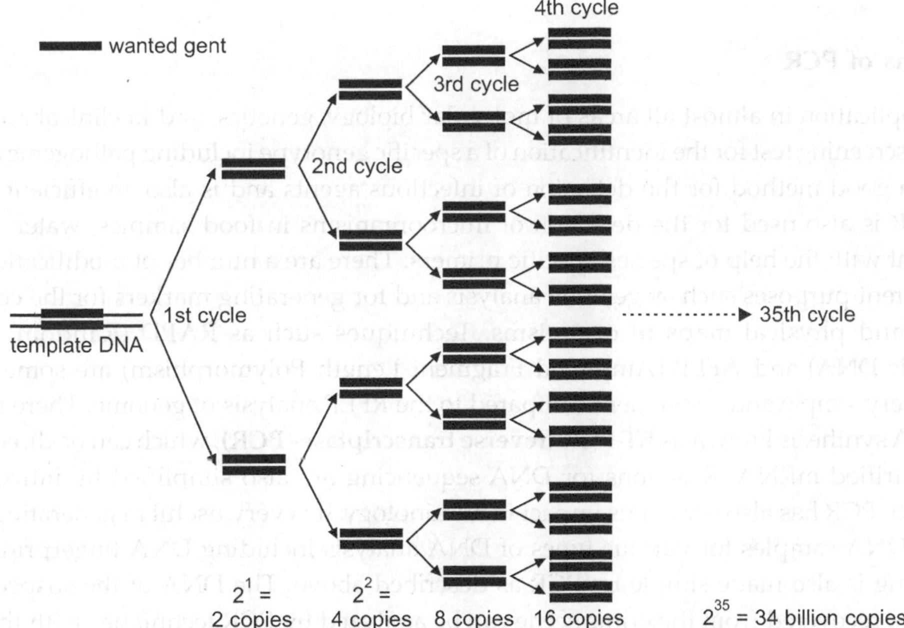

FIGURE 15.16 **Diagram representation of the exponential amplification of selected DNA sequences during PCR reaction at every cycle.**

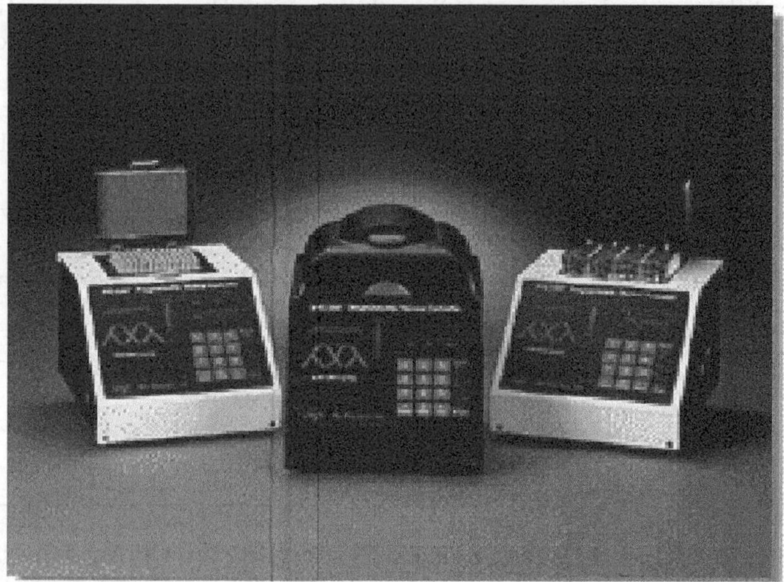

FIGURE 15.17 **Various models of PCR machines or thermo cyclers.**

The extension product of one cycle can serve as the template for the next cycle and in each cycle the DNA content doubles from that of the previous cycle. Thus, it is possible to generate 2^n molecules of DNA if there is '*n*' number of PCR cycles. Thermo cyclers can be programed with the temperature parameters and duration for a specific number of cycles to get a sufficient amount of DNA.

Applications of PCR

PCR has application in almost all areas of molecular biology, genetics, and in clinical areas. PCR is an efficient screening test for the identification of a specific genotype including pathogenic organisms. Thus, it is a good method for the detection of infectious agents and is also an efficient diagnostic technique. It is also used for the detection of microorganisms in food samples, water, and in the environment with the help of species-specific primers. There are a number of modifications of PCR to suit different purposes such as genome analysis and for generating markers for the construction of genetic and physical maps of organisms. Techniques such as RAPD (Random Amplified Polymorphic DNA) and AFLP (Amplified Fragment Length Polymorphism) are some examples, which are very simple and convenient compared to the RFLP analysis of genome. There is the PCR-based cDNA synthesis known as **RT-PCR (reverse transcriptase- PCR)**, which can be directly carried out with purified mRNA. Reactions for DNA sequencing are also simplified by introducing the PCR method. PCR has also shown its impact in criminology. It is very useful in generating sufficient amount of DNA samples for various types of DNA analysis including DNA fingerprinting. DNA fingerprinting is also made simple by PCR as described above. The DNA of the suspects and the DNA sample recovered from the crime scene can be analyzed by PCR techniques with the help of a set of identical random primers or specific primers. The genetic basis of disease can be easily detected using PCR tools. The genetic mutations responsible for certain genetic diseases and cancers can be detected very early and the possibility of causing the disease can be predicted and treated or precautionary measures can even be taken before the disease manifests. Early detection of genetic disease is even possible in embryonic conditions or even in sex cells—sperm and egg.

15.8 DNA PROBES

A DNA probe is a small, fluorescently or radioactively labeled DNA molecule that is used to locate similar or complementary sequences among a long stretch of DNA molecule or bacterial colonies such as genomic or cDNA libraries or in a genome. Such DNA probes are used in hybridization experiments such as Southern hybridization to detect certain specific sequences, which are complementary to the probes. Since the probe is labeled with a fluorescent dye or radioactive isotopes of phosphorous, its binding to specific sequences can be detected. DNA probes labeled with radioactive isotopes or fluorescent dyes can be used for the screening of transformed colonies having the correct recombinant plasmid by Southern hybridization.

15.9 HYBRIDIZATION TECHNIQUES

Any two single-stranded nucleic acid molecules will attempt to base pair with one another under appropriate conditions. In most cases, the hybrid structures thus formed will be very unstable since the total number of H-bonds that formed is very low. However, if there is **significant sequence complementarity** between the two strands, **stable hybrids** will be formed. This phenomenon of **nucleic acid hybridization** can be used to identify a particular recombinant if a suitable probe, complementary to the sequence sought, is available.

Southern Hybridization

Experimentally, the gene library to be screened, or the DNA bands separated by electrophoresis on an agarose gel, is transferred to a nitrocellulose or nylon filters by blotting. These are then processed to release DNA from the bacteria and/or the phage. These are then denatured to separate the complementary strands and are immobilized on to the membrane, in the position occupied by the original clone and in a way that leaves the bases free to interact with the complementary molecules. The probe molecules are then labeled (traditionally with radioactive nucleotides but now more commonly with chromogenic or chemiluminescent labels). Probe and filter are then incubated under conditions that promote hybridization after which the unbound probe is washed off from the filter and the specifically bound probe is visualized. Positive signals reveal the position (and thus identity) of those recombinants carrying the sequences related to the probe.

A fragment of DNA in a genome can also be detected by **Southern hybridization.** This technique was originally devised by Edward Southern in 1975 to identify specific DNA fragments on an agarose gel after the separation by electrophoresis. The technique involves the isolation of the genomic DNA with a suitable procedure. The genomic DNA is digested with a suitable restriction enzyme or with a mixture of different restriction enzymes to cut the long DNA molecule into fragments. Restriction-digested genomic DNA is separated by electrophoresis and the DNA separated on the gel is then transferred to a membrane—a nitrocellulose or nylon membrane—with a process called **blotting**, which is driven by capillary action. The blotting can also be carried out electrophoretically, and then it is known as **electroblotting**. There is another equally effective blotting method known as **vacuum blotting**. The DNA fragments transferred to the membrane are immobilized to its surface by heat or UV mediated cross-linking. Now the membrane is hybridized with a 'labeled' segment of the gene understudy called a probe (the label may be radioactive or fluorescent chemical labeling). When the hybridized filter is exposed to x-ray film, only the band with the labeled probe hybridized to it will be shown on the film. Southern hybridization allows the precise localization of a given DNA sequence in a genome, once a restriction map has been constructed. If the probe is labeled with a fluorescent chemical then the fluorescent band can be visualized after illuminating the nylon or nitrocellulose membrane with UV light after the process of blotting.

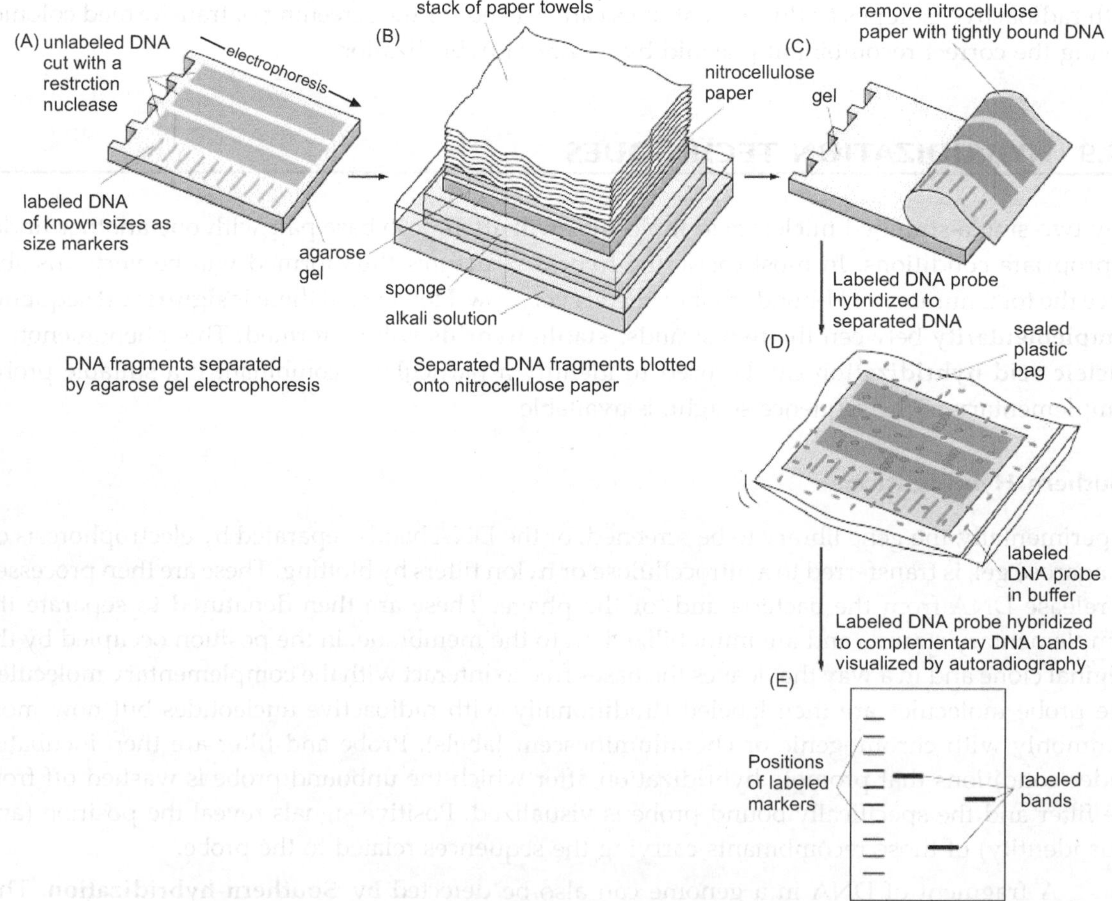

FIGURE 15.18 **Diagram representation of the various steps in the technique of Southern hybridization.**

Since the DNA bands on the membrane will have the same pattern as that present in the gel, the position of a labeled band on the membrane can be traced to the same corresponding position in the gel. There are a number of variations of Southern hybridization or Southern blotting suited for specific types of experiments. Some of them are dot blots, colony hybridization, slot blots, etc., which are comparatively easier. Similar to Southern hybridization there are altered forms of molecular hybridization techniques developed for RNA and proteins. There are the techniques developed in which RNA is hybridized with its complementary DNA sequences or a specific protein such as antibody hybridized to another protein such as its ligand or antigen. Since hybridization of DNA with another DNA molecule is called Southern hybridization or blotting, hybridization between RNA and DNA is called **Northern hybridization** and protein-protein hybridization such as antigen-antibody binding or protein-ligand binding is called **Western hybridization.**

15.10 DNA SEQUENCING

To understand the complexities of gene structure, its expression, its regulation, protein interactions, and molecular mechanisms of genetic diseases—the detailed and exact sequences of the bases in DNA is very essential. During the late 1970s two different sequencing techniques were developed. They are:

1. The **chemical cleavage method** developed by Allan Maxam and Walter Gilbert (the Maxam-Gilbert Method). This method was very important in the beginning but is no longer practiced today.
2. The enzyme-mediated chain termination method developed by Frederick Sanger along with Andrew Coulson is popularly known as the **Sanger sequencing method** or the **enzymatic method of DNA sequencing.**

Both methods involve the labeling of the terminal nucleotide followed by separation and detection of the generated oligonucleotides.

Sanger Method or Dideoxynucleotide Chain Termination DNA Sequencing

DNA sequencing reactions are just like the PCR reactions for replicating DNA or synthesizing a new strand of DNA. The reaction mix includes the template DNA, free 5′deoxyribonucleotide triphosphates, an enzyme, the DNA polymerase (usually a variant of Taq polymerase) and a 'primer'—a small piece of single-stranded DNA about 20 to 30 nucleotides along with a free 3′-hydroxyl group that can hybridize to one strand of the template DNA. The primer with a free 3′OH group can initiate the synthesis of DNA strand with the addition of the free nucleotides to the primer by the polymerase enzyme. **Sanger's chain termination method** utilizes 2′3′-dideoxyribonucleoside triphosphates (ddNTPs). This molecule differs from the normal deoxyribonucleoside triphosphates (dNTPs) by having a hydrogen atom at the 3′ carbon instead of a hydroxyl group (Figure 15.19).

ddNTP

dNTP

FIGURE 15.19 **Chemical structure of ddNTP and dNTP.**

The reaction is initiated with heat until the two strands of DNA (template) separate, then the primer binds to its intended location and the DNA polymerase starts elongating the primer by adding the appropriate dNTPs as indicated in Figure 15.20. If allowed to go to completion, a new strand of DNA would be the result. If we start with a billion identical pieces of template DNA, we will get a billion new copies of one of its strands.

However, we run the sequencing reactions in the presence of a dideoxyribonucleotide. Once this nucleotide, which doesn't have the 3' hydroxyl group, is added to the end of a growing DNA strand, there's no way to continue its elongation. The key point here is that in the reaction mixture that we use for DNA sequencing most of the nucleotides are regular ones (dNTPs), and just a fraction of them are dideoxy nucleotides (ddNTPs). The double-stranded DNA (the template) can be converted into single strands by NaOH or by heating to 94°C as in the case of PCR reaction. A DNA-sequencing reaction following Sanger's method consists of the following components:

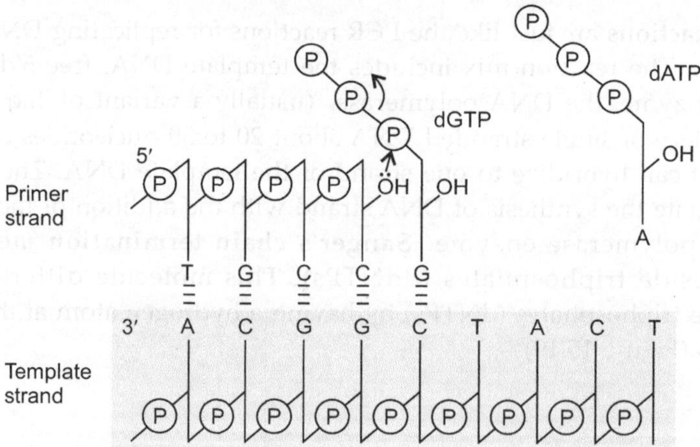

FIGURE 15.20 **Binding of primer to the template and the elongation of the primer with the addition of dNTPs.**

1. DNA sample to be sequenced or the template **2. DNA primers** **3. A mixture of all dNTPs—dATP, dGTP, dCTP, and dTTP** **All this as mixture is distributed among four different reaction tubes and labeled A, T, G, and C.** **4. ddNTP: ddATP, ddTTP, ddGTP, and ddCTP (separately in four tubes)** **5. Taq DNA polymerase**

The DNA sample to be sequenced is the template. The template has to be converted into a single strand by denaturing with NaOH. But if you are carrying out the sequencing reaction using a PCR machine, denaturation of the template occurs as a part of the reaction cycle.

DNA primers. 5′ end radio-labeled DNA primers, which are short fragments of DNA complementary to the template DNA. Primers are labeled with radioactive phosphate at the 5′ end.

A mixture of all dNTPs—dATP, dGTP, dCTP, and dTTP.

All this as a mixture is distributed among four different reaction tubes in appropriate quantities and labeled **'A', 'T', 'G', and 'C'.**

ddNTPs. In each tube a small quantity of the corresponding ddNTP is added. In a tube labeled 'A' a small amount of ddATP is added. In tube 'T' ddTTP is added, in tube 'G' ddGTP is added, and in tube 'C' a small amount of ddCTP is added. The concentration of ddNTP is approximately 1% of the concentration of the dNTPs.

Taq DNA polymerase. When all the components are ready, Taq DNA polymerase is added to all the four tubes and the reaction, the synthesis of DNA or the elongation of the primer, starts.

The elongation of the primer with the addition of dNTPs directed by the template continues until a dideoxynucleotide is incorporated instead of a dNTP. For example, consider the DNA

5′ — TACGCGGTAACGGTATGTTCGACCGTTTAGCTACCGAT
3′ — ATGCGCCATTGCCATACANGCTGGCAAATCGATGGCTAGAGATCCAA — 5
Normal synthesis of DNA strand, when the reaction Mixture contains only dTTPs
5′ — TACGCGGTAACGGTATGTTCGACCGTTTAGCTACCGAT—
5′ — TACGCGGTAACGGTATGTTCGACCGTTTAGCT—
5′ — TACGCGGTAACGGTATGTTCGACCGTTT—
5′ — TACGCGGTAACGGTATGTTCGACCGTT—
5′ — TACGCGGTAACGGTATGTTCGACCGT—
5′ — TACGCGGTAACGGTATGTT—
5′ — TACGCGGTAACGGTATGT—
5′ — TACGCGGTAACGGTAT—
5′ — TACGCGGTAACGGT—
5′ — TACGCGGT—

If 1% of the T nucleotides are ddTTPs then each strand will terminate when it gets a ddTTP on its growing end.

FIGURE 15.21 **Diagram illustrating DNA chain termination due to the incorporation of ddNTPs and the generation of fragments of various types with terminal ddNTP.**

synthesis in the presence of a small amount of ddTTP (the reaction in the tube labeled 'T'). Most of the time when a 'T' is required to make the new strand, the enzyme will get a dTTP and there's no problem. After adding a T, the enzyme will go ahead and add more nucleotides and thus the elongation continues. However, 5% of the time, the enzyme will get a dTTP, but instead it will get a dideoxy T(ddTTP) and that strand can never again be elongated. It eventually breaks away from the enzyme, resulting in a dead-end product. Sooner or later all of the copies will get terminated by a 'T.' But each time the enzyme makes a new strand and the places at which it stops will be random. In millions of starts, there will be strands stopping at every possible T along the way. This is illustrated in Figure 15.21.

The length of the strands thus produced will depend on the position at which the ddNTP is incorporated and can be determined by electrophoresis. When the reaction is completed, the products from each tube labeled 'A', 'T', 'G', and 'C' are taken and electrophoresed under denaturing conditions in four different lanes in a polyacrylamide gel. The gel is dried and exposed to an x-ray film by autoradiography. Only those fragments, which are labeled with the primer, will appear on the autoradiogram. Each fragment or the bands that are present in a lane end with the corresponding ddNTPs. For example, in the lane meant for the base 'T' all the bands or the fragments will have the ddTTP at the terminals. Thus, by reading the four lanes 'A', 'T', 'G' and 'C' from the bottom of the gel to the top, the sequence of the DNA can be elucidated. But it should be noted that whatever sequence you are reading from the gel is the complementary of the template DNA. It is well-illustrated in Figures 15.21 and 15.22.

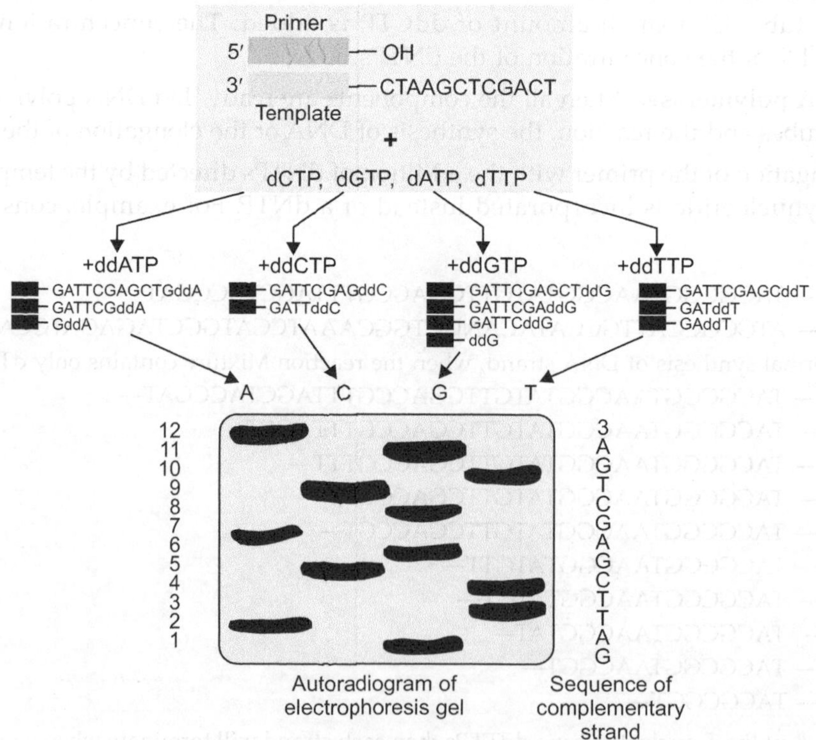

FIGURE 15.22 **Chain termination after incorporating the ddNTPs and its interpretation from an autoradiography of a DNA sequencing gel.**

DNA sequence can be directly read from the autoradiogram. If the labeling is by fluorescent dyes, instead of using radiolabeled primers, fluorescently labeled ddNTPs are used in the reaction mixture. Each type of ddNTPs is labeled with a specific color and can be detected directly from the gel when illuminated with a laser or UV light (see Appendix).

Automated DNA Sequencing

Automatic sequencing machines have greatly improved the quality as well as the speed of the sequencing process. There is no need for radiolabeling and autoradiography. Great improvements in the DNA-sequencing techniques were observed in the 1960s and resulted in **automated DNA sequencing machines.**

The use of fluorescently labeled ddNTPs (dideoxynucleotide triphosphates) has made the reading very easy, convenient, and automatic with the help of UV-laser detectors. Thus, it has greatly improved the speed of sequencing. Each of the four types of ddNTPs can be labeled with a specific dye, so that a specific color can be attributed to the presence of a particular nucleotide or base. For example, if the color of the band is red it represents the base 'T,' because the ddTTP is labeled with a red dye. Similarly, yellow is for 'G,' green is for 'A,' and blue represents the base 'C' (see Appendix).

Imagine that we carried out the sequencing reaction with all four of the dideoxynucleotides (A, G, C, and T) labeled with the appropriate fluorescent colors along with normal deoxynucleotides. The reaction mixture was separated by electrophoresis under standard conditions. Now look at the gel to view the bands in colors under UV illumination. The sequence of the DNA is rather obvious if you know the color codes for each base; just read the colors from bottom to top. The reaction mixture can be electrophoresed on a single lane instead of four.

After electrophoresis, we don't even have to 'read' the sequence from the gel. The computer does that for us. After electrophoresis the colored bands can be monitored using UV-laser beams. The laser beams excite the fluorescent dyes and result in the emission of specific spectral waves (colored light), which are recorded by a specific photoelectric device. The data thus generated is fed to a computer, where the emission data from the gel is converted into a corresponding nucleotide sequence of the DNA sample (see Appendix). The nucleotide sequence is also represented in specific peaks indicating each nitrogen base.

An example of what the sequencer's computer shows us for one sample is given in the figure in the Appendix. This is a plot of the colors detected in one 'lane' of a gel (one sample), scanned from smallest fragments to largest. The computer even interprets the colors by printing the nucleotide sequence across the top of the plot. The sequencer also gives the operator a text file containing just the nucleotide sequence, without the color traces.

15.11 SITE-DIRECTED MUTAGENESIS

Any heritable change in the genome is commonly called a **mutation**. Biochemically, it is a chemical change or alteration in a nitrogen base of a DNA sequence resulting in the production of a defective protein or a truncated protein, which is not functional. These altered proteins can cause serious problems in metabolism leading to changes in the morphology and physiology of the organism. Mutations, in most cases, are spontaneous and may not be dangerous.

Even though the natural mutations are spontaneous and rare, biologists can induce mutations using different methods, which in most cases are not desirable and precise. But, now molecular biologists can alter any amino acid of a protein by changing the corresponding bases in its gene very precisely and accurately resulting in desirable mutations. It is possible to alter properties such as increased stability, temperature resistance, product inhibition, substrate specificity, etc. of any enzyme. The accurate induction of one or more point mutations on selected regions of a gene resulting in amino acid substitutions or deletion or addition is known as **site-directed mutagenesis**. It can be defined as the controlled alteration of selected regions of a DNA molecule. The **principle of site-directed mutagenesis** is that a mismatched oligonucleotide primer is extended, incorporating the 'mutation' into a strand of DNA that can be cloned. This technique of creating desired molecular mutations in a gene has contributed greatly to the basic understanding of functions of genes, DNA-protein interactions, gene regulations, the role of amino acids in the structure and functions of proteins, role of active centres in the enzyme-substrate interactions, etc. A single base change in a gene permits the evaluation of the role of specific amino acids in the function and structure of a protein. This technique also allows one to create or destroy a restriction site at specific locations within a DNA sequence or gene.

Site-directed mutagenesis is actually one of the applications of PCR. The gene, which has to be mutated, should be made into a single-stranded DNA by cloning into a M13 plasmid vector. By following modern PCR methods, it is possible to carry out the site directed mutation without the participation of M13 vector. The designing and chemical synthesis of the primer is the key factor in this technique. The part of the DNA where the mutation has to be introduced should be synthesized as an oligonucleotide primer, which is complementary to the respective region of the DNA except for the nucleotide that has to be changed. In short, the mutation is introduced to the gene in the form of a primer and the primer is extended with a polymerase reaction.

The site-directed mutagenesis is a multistep process that begins with the cloning of the gene in a bacteriophage like M13 to generate single stranded DNA. M13 is a filamentous bacteriophage that specifically infects *e. coli* that expresses sex pili encoded by a plasmid F factor. M13 bacteriophage contains DNA in a single-stranded or replicative form, which is replicated to double-stranded DNA within a bacterial cell. The primer is designed and synthesized, which is an oligonucleotide complementary to the region of the DNA to be mutated except for the nucleotide to be changed. This oligonucleotide with the mismatched base or bases hybridize to the single-stranded DNA and serve as the primer to start synthesizing the complementary DNA strand with the help of a suitable DNA polymerase such as T4 DNA polymerase or Taq polymerase. The resulting double-stranded DNA will be a hybrid of the wild type parent strand and the mutated newly synthesized DNA strand. This DNA molecule can be transformed into an *e. coli* cell, where the mutated DNA strand serves as a template to replicate new strands carrying the mutation along with the wild strands. The bacteriophage plaques containing the mutated DNA can be screened by hybridizing with the labeled probe of the original mutated oligonucleotide. By adjusting the hybridization time and wash temperature of the hybridized probe, only the perfectly matched hybrid will remain and all other mismatched hybrids will dissociate. The presence of the desired mutation in the gene can be checked and confirmed after isolating the plasmid DNA from the single positive plaques and sequencing it.

A single base can be mutated in recombinant DNA plasmids with a process called **inverse PCR.** Two primers are synthesized with their antiparallel 5′ ends complementary to the adjacent bases on the two strands of DNA. One of the two primers carries a specific mismatched base that is faithfully copied during the PCR amplification resulting in a recombinant plasmid with a single mutated base.

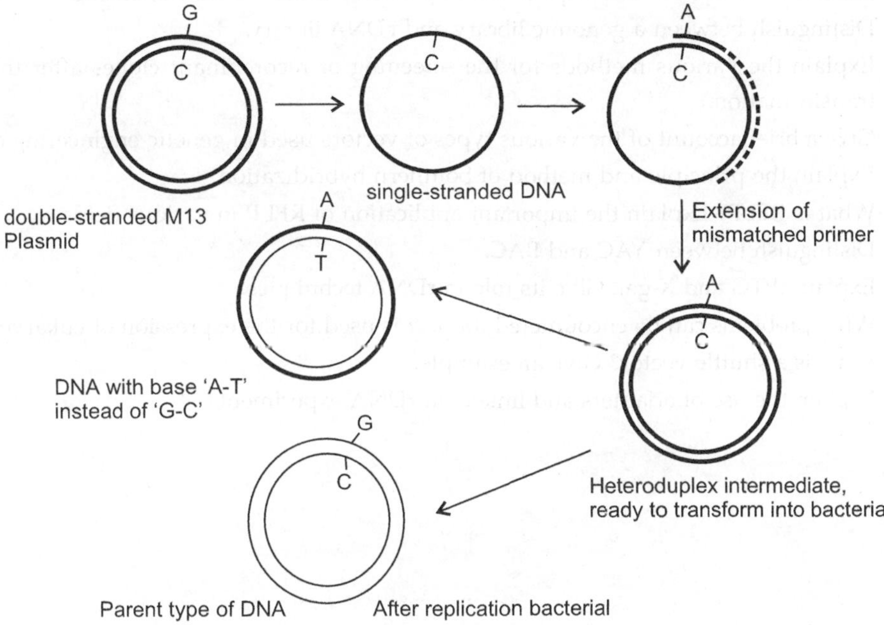

FIGURE 15.23 **Site-directed mutagenesis of a gene mediated through M13 plasmid.**

REVIEW QUESTIONS

1. Give the definition of restriction endonucleases. Explain their practical application in molecular biology and biotechnology.
2. Explain the possible problems that may arise when trying to produce a eukaryotic protein in bacteria.
3. What is meant by site-directed mutagenesis? Explain its various application in genetic and biochemical studies.
4. What is the importance of M13 bacteriophage in site-directed mutagenesis?
5. Give a short account of PCR and its various applications in basic research and applied research.
6. What are Type II restriction enzymes?
7. Distinguish between dNTPs and ddNTPs. Explain the chain-termination method of DNA sequencing.
8. What are the essential characteristics needed for a vector?
9. What are the advantages of fluorescent labeling in DNA sequencing? Explain.
10. Explain insertional inactivation. What is its importance in rDNA techniques?
11. What are the enzymes needed for the recombinant DNA technique? Explain the role of each enzyme.
12. Explain the principle and application of PCR.

13. What are the differences between a cloning vector and an expression vector? Give examples.
14. What is the difference between a probe and a primer?
15. What is a restriction site? Explain the important features of restriction sites.
16. Distinguish between a genomic library and cDNA library.
17. Explain the various methods for the screening of recombinant clones after the process of transformation.
18. Give a brief account of the various types of vectors used in genetic-engineering experiments.
19. Explain the principle and method of Southern hybridization.
20. What is RFLP? Explain the important application of RFLP in various fields.
21. Distinguish between YAC and BAC.
22. Explain IPTG and X-gal. Give its role in rDNA techniques.
23. What problems can be encountered if *e. coli* is used for the expression of eukaryotic proteins?
24. What is a shuttle vector? Give an example.
25. Explain the use of adapters and linkers in rDNA experiments.

Chapter 16 GENOMICS

In This Chapter

16.1 INTRODUCTION

Genomics is simply the study of complete genomes. More recently, genomics has meant the analysis of the complete DNA sequence of an organism's genome. This field was not really possible until the publication of the genome of *haemophilus influenzae* in 1995. Fred Sanger sequenced the first bacteriophage (phiX174, 5,386 pp long), for which he later won the Nobel Prize. Although this was a dramatic improvement over the conventional methods, it was still very slow, compared to the amount of information in a single human cell. About a decade later, the Human Genome Project was launched. This was an international effort, and the U.S. would pay about

$200,000,000 per year for 20 years! Most of this investment was in technology to speed sequencing, which in fact has been realized. Within a few years, it is likely that it will be possible to read the entire DNA sequence of a human cell, in just a few hours.

The completion of the human-genome sequence opens an enormous opportunity for chemistry to understand the complex biological processes of living organisms at a molecular level. This opportunity coincides with a tremendous increase in our ability to design, synthesize, and analyze experiments due to combinatorial methods in biochemistry and genomics, enhanced computational capabilities, and highly sensitive analytical tools. We can apply these new methods to a number of problems in modern genomics, including:

- The synthesis of combinatorial chemical libraries and high throughput cell-based screening of these libraries for molecules with novel biological activities.
- The use of mRNA expression analysis together with yeast genetics and retroviral cDNA libraries to identify novel proteins and biological pathways.
- The development of new tools for identifying peptides and post-translationally modified proteins with novel biological activities.
- High throughput structure determination together with molecular docking to generate molecules that selectively modulates protein activity.

Currently, several hundred prokaryotic genomes have been sequenced, and nearly 100 genomes are publicly available for analysis. Genomics is not a new science, but the term is new and was introduced by Thomas Roderick in 1986. Genomics describes the detailed study of the genome, its structural organization and function, using various modern methods including computational biology. The term genome represents the complete genetic material including both nuclear and cytoplasmic genes present in a cell. The Human Genome Project (HGP), sponsored in the United States by the Department of Energy and the National Institutes of Health, has created the field of genomics and understanding of genetic material on a large scale.

TABLE 16.1 A look at genome sequencing since 1994 (including bacteria, archaea, and eukaryotes).

Year	Genomes Sequenced	Running Total
1994	0	0
1995	2	2
1996	2	4
1997	5	9
1998	8	17
1999	13	30
2000	23	53
2001	42	95
2002	>100	>200

The field of molecular life science is changing rapidly because of the genomic revolution. **High throughput DNA sequencing** and the development of algorithms to analyze the vast amount of data have led to the elucidation of the entire genome sequences. Revolutionary improvements in DNA-sequencing techniques have given rise to a large amount of DNA sequences, which is difficult to manage, particularly for future references and analysis. Technological developments in computer and information technology have helped in managing the huge data of DNA and protein sequences in the form of computerized databases and their access through the Internet. Similarly, a large number of algorithms and software have been developed, which have the simplified DNA-sequencing methods and genome analysis. Now a large number of software is freely available online to analyze DNA sequences for various purposes such as to find out the ORF, promoters, exons, introns, and also for sequence alignment and structural prediction of its product. Computerized sequence databases and their management, retrieval of data from the databases, computational methods of sequence and genome analysis, protein structural prediction and computer-mediated drug development, etc. have given rise to bioinformatics and computational biology. Automatic DNA and protein-sequencing technologies were developed in the 1970s and the use of computerized databases and computational analysis of gene sequences began in the 1980s. Slowly its academic nature has changed into commercial enterprises and thus it has developed into bioinformatics.

Structural Genomics

The science of genomics actually originated in the form of **structural genomics**. It involves the sequencing of genes and genomes using most modern sequencing techniques such as automated fluorescent sequencing of DNA, by adopting shotgun strategies of assembly, organization, and management of DNA sequences. It involves the construction of high-resolution genetic, physical, and gene expression maps of the genome concerned. The physical map of highest resolution of a genome is its complete DNA sequence. Therefore, structural genomics primarily means the structural organization of genes and their arrangement in the genome with all the physical details, including location, in the respective chromosomes. This is known as the **physical map of the genome**. In addition to the structural elucidation and organization of genes and genome, structural genomics also predicts the structure of the possible protein and its metabolic role. If the topic deals with the functional aspects of a gene and its products then it is functional genomics rather than structural genomics.

Functional Genomics

The part of genomics that deals with genome analysis and functional aspects of each gene, its expression and interaction between genes and its gene products including the gene expression network, is **functional genomics**. The information gathered from structural genomics is utilized for the design of global experiments to understand the function of individual genes or groups of genes and their role in the expression of other genes and their interaction. The scope of functional genomics is to expand from the study of a single gene or a single protein to a systematic study of the entire genes of a genome and their proteins as well as their functional role in metabolic activities. To accomplish this, functional genomics experiments are usually characterized by the use of large-scale experimental methodologies combined with analysis by computational tools. Northern hybridization experiments to understand the expression of genes by monitoring the formation of RNAs and the experiments with DNA microarrays are some of the examples.

16.2 GENOME MAPPING

A **genome map** is composed of the relative positions of genes in a genome. Genome maps help scientists locate a specific gene in a genome just like a road map. There are two types of genome maps—**the genetic map** and **the physical map**.

The genetic map and the physical map provide scientists the order and position of the gene in question and its distance from other genes or markers in a genome, just like the address and map of the road that leads to a particular house. The genetic map is just like the road map between cities and the physical map is similar to the map of a city specifying the exact distances between places giving almost complete information to find that place. Both genetic and physical maps provide the likely order of items along a chromosome. One could say that genetic maps serve to guide a scientist toward a gene, just like an interstate map guides a driver from city to city. On the other hand, physical maps mark an estimate of the true distance, in measurements called **base pairs (bp)**, between genes or locations of interest. Depending on the quality of the genetic map and physical map of a genome one could easily locate a specific gene in the crowded city of genome. **Genetic marker** is a broad term, that includes any observable variation that results from an alteration, or mutation, at a single genetic locus. A genetic marker can be used as landmark in a genetic map if it is inherited from the parent to the offspring as per standard genetic laws.

Markers can be a noticeable physical characteristic such as eye color, drug resistance, or a not so noticeable trait such as a disease. **DNA-based reagents** can also serve as markers. Markers for DNA-based reagents are found mostly within the non-coding regions of genes and are used to detect unique regions on a chromosome. DNA markers are especially useful for generating genetic maps when there are occasional, predictable mutations that occur during **meiosis**, which cause a high degree of variations within the DNA content of the marker over many generations from individual to individual.

The following are the most commonly used genetic markers in a genetic map:

1. **RFLPs**, or **Restriction Fragment Length Polymorphisms**, were among the first developed DNA markers. RFLPs are defined by the presence or absence of a specific site, called a restriction site, for a bacterial restriction enzyme. This enzyme breaks apart strands of DNA wherever they contain a certain nucleotide sequence. RFLP also has a number of other uses, particularly in criminology.
2. **VNTRPs**, or **Variable Number of Tandem Repeat Polymorphisms**, occur in non-coding regions of DNA. This type of marker is defined by the presence of a nucleotide sequence that is repeated several times. In each case, the number of times a sequence is repeated may vary.
3. **Microsatellite Polymorphisms,** are defined by a variable number of repetitions of a very small number of base pairs within a sequence. Oftentimes, these repititions consist of the nucleotides, or bases, cytosine and adenosine. The number of repititions for a given microsatellite may differ between individuals, hence the term **polymorphism**—the existence of different forms within a population.

4. **SNPs**, or **Single Nucleotide Polymorphisms**, are individual point mutations, or substitutions of a single nucleotide, that do not change the overall length of the DNA sequence in that region. SNPs occur throughout an individual's genome.

   Genetic maps are also used to generate essential framework needed for the construction of more detailed genome maps. These detailed maps, called **physical maps**, further define the DNA sequence between genetic markers and are essential to the rapid identification of genes.

Physical Map of Genome

Physical maps are detailed genetic maps, which describe the position of genes in terms of base pairs or gene sequences between genetic markers or genes. Physical maps can be divided into three general categories. They are **chromosomal** or **cytogenetic maps**, **radiation hybrid (RH) maps**, and **sequence maps**. The different types of maps vary in their degree of **resolution or clarity.**

The lowest-resolution physical map is the **chromosomal** or **cytogenetic map**, which is based on the distinctive banding patterns observed by light microscopy of stained chromosomes. As with genetic linkage mapping, chromosomal mapping can be used to locate genetic markers defined by traits observable only in whole organisms. Because chromosomal maps are based on estimates of physical distance, they are considered physical maps. Yet, the number of base pairs within a band can only be estimated.

A **linkage map** is created by finding the map distances between a number of traits that are present on the same chromosome, ideally avoiding having significant gaps between traits to avoid the inaccuracies that will occur due to the possibility of multiple recombination events. Linkage mapping is critical for identifying the location of genes that cause genetic diseases. In an ideal population, genetic traits and markers will occur in all possible combinations with the frequencies of combinations determined by the frequencies of the individual genes.

RH maps and **sequence maps**, on the other hand, are more detailed. RH maps are similar to linkage maps in that they show estimates of distance between genetic and physical markers, but that is where the similarity ends. **RH maps** are able to provide more precise information regarding the distance between markers than can a linkage map.

The physical map that provides the most detail is the sequence map. **Sequence maps** show genetic markers, as well as the sequence between the markers measured in base pairs.

RH Mapping (Radiation-Hybrid Maps)

In this technique, scientists, instead of depending on natural recombination to separate two markers, use breaks induced by radiation to determine the distance between two markers. In RH mapping, DNA is exposed to a measured dose of radiation, and in doing so, controls the average distance between breaks in a chromosome. By varying the degree of radiation exposure to the DNA, one can control the induction of breaks between two markers that are very close together. RH mapping, like linkage mapping, shows an estimated distance between genetic markers. The ability to separate closely linked markers allows scientists to produce more detailed maps. RH mapping provides a way to localize almost any genetic marker, as well as other genomic fragments, to a defined map position. Actually, RH maps are an intermediate between linkage maps and sequence maps in clarity or resolution. Therefore, RH maps can be used as a bridge between linkage maps and sequence maps in locating and identifying the position of gene in the genome. This can help in the easy identification of the location(s) of genes involved in certain diseases such as spinal muscular atrophy and hyperekplexia (startle disease). RH maps are extremely useful for ordering markers in regions where highly polymorphic genetic markers are scarce or absent.

STS Mapping

Sequence tagged site, or **STS mapping,** is one of the latest physical mapping techniques, which can give a high degree of resolution. An STS is a short DNA sequence that has been shown to be unique. To qualify as an STS, the exact location and order of the bases of the sequence must be known, and this sequence may occur only once in the chromosome being studied or in the genome as a whole if the DNA fragment set covers the entire genome. The following are some of the most commonly used sequences for the identification of gene locations in a genome.

Expressed sequence tags (ESTs) are short sequences obtained by analysis of complementary DNA (cDNA) clones. Complementary DNA is prepared by converting mRNA into double-stranded DNA and is thought to represent the sequences of the genes being expressed. An EST can be used as a STS if it comes from a unique gene and not from a member of a gene family in which all of the genes have the same, or similar, sequences.

Simple sequence length polymorphisms (SSLPs) are arrays of repeat sequences that display length variations. SSLPs that are polymorphic and have already been mapped by linkage analysis are particularly valuable because they provide a connection between genetic and physical maps.

Random genomic sequences are obtained by sequencing random pieces of cloned genomic DNA or by examining sequences already deposited in a database.

16.3 GENOME-SEQUENCING PROJECTS

The genomes of a large number of organisms have been sequenced and a number of them are in the final stages of different genome project groups. The genome projects are having a profound impact on health care and drug discovery. The main purpose of genome projects is to access the entire genome sequences, which can be used to find out the probable genes and their functions of various organisms. Some important genomes that are partially or completely sequenced include that of *e.coli, mycobacterium tuberculosis, yeast, arabidopsis, drosophila, mice, humans, rice, wheat,* etc. The availability databases for researchers will help them to better understand the gene and its expression and thereby better understand fundamental biological processes. For example, the Human Genome Project would greatly help the development of gene therapy. The final and complete sequence of the human genome would unravel the mystery of heredity and reveal the sources of many diseases such as diabetes, heart disease, cancer, hemophilia, Alzheimer's disease, and cystic fibrosis. It can also help unravel the intricacies of development and differentiation. The genome sequencing of a number of important pathogenic organisms such as the malarial parasite will help to design methodologies to control the disease. Since among the various genome projects, the Human Genome Project is the one that has actually had an impact on science and society, let us see the beginning and the major outcome of the Human Genome Project, in brief.

History and Evolution of Human Genome Project

The history of the Human Genome Project (HGP) can be traced to its roots to an initiative in the U.S. Department of Energy (DOE). Since 1947, the DOE and its predecessor agencies have been encouraged by the U.S. government to develop new energy resources and technologies and to get a deeper understanding of potential health and environmental risks posed by their production and use. In 1986, the DOE took a bold step in announcing the Human Genome Initiative and was convinced that its missions would be well served by a reference human genome sequence. Shortly thereafter, the DOE joined with the National Institutes of Health (NIH) USA to develop a plan for a joint Human Genome Project (HGP) that officially began in 1990. During the early years of the HGP, Welcome Trust, a private charitable institution in the United Kingdom, joined the effort as a major partner. Important contributions also came from other collaborators around the world, including Japan, France, Germany, and China.

The ultimate aim of the HGP was to generate a high-quality reference DNA sequence for the human genome's three billion base pairs and to identify all human genes. Other important goals included sequencing the genomes of model organisms to interpret human DNA, enhancing computational resources to support future research and commercial applications, exploring gene function through mouse-human comparisons, studying human variation, and training future scientists in genomics. The powerful analytic technology and data arising from the HGP raise complex ethical and policy issues for individuals and society. These challenges include privacy, fairness in use and access of genomic information, reproductive and clinical issues, and commercialization. Programs that identify and address these implications have been an integral part of the HGP and have become a model for bioethics programs worldwide.

Scientists announced the completion of the first working draft of the entire Human Genome Project in June 2000. First analysis of the details appeared in the February 2001 issues of the journals *Nature* and *Science*. The high-quality reference sequence was completed in April 2003, marking the end of the Human Genome Project—two years ahead of the original schedule. Coincidentally, this was also the 50th anniversary of Watson and Crick's publication of DNA structure that launched the era of molecular biology.

The human-genome reference sequence provides a magnificent and unprecedented biological resource that will serve as a basis for research and discovery and, ultimately, myriad practical applications. The sequence is already having an impact on finding genes associated with human disease. Hundreds of other genome-sequencing projects—on microbes, plants, and animals—have been completed since the inception of the HGP, and this data now enables detailed comparisons among organisms, including humans. The first set of whole genome sequences released was of the smallest genome *mycoplasma genitalium* and *haemophilus influenzae*. Several groups in Europe and other places started many more sequencing projects during this time. Complete genome-sequencing of *mycobacterium tuberculosis* and *bacillus subtilis* was also initiated and completed by the Pasteur Institute and the Sanger Centre. A large number of genome-sequencing projects are underway or planned because of the research value of the DNA sequence, the tremendous sequencing capacity now available, and continued improvements in technologies. Sequencing projects on the genomes of many microbes, as well as the honeybee, cow, and chicken are in progress.

In addition to the sequencing of genomes, research interests are also developing and focusing on identifying important elements in the DNA sequence responsible for regulating cellular functions and providing the basis of human variation. Possibly the most frightening challenge in functional genomics is to commence to understand how all the biochemical components of cells such as genes, proteins, lipids, and many other similar biomolecules work together to create complex living organisms. Future analysis on this treasury of data will provide a deeper and more comprehensive understanding of the molecular processes underlying life such as how the various metabolic pathways and reactions and gene expressions are interconnected and coordinated to form the metabolic network and how they are regulated at molecular levels.

The Impact of Human Genome Project on Science and Society

The first announcement of the human genetic landscape revealed a wealth of information and also some surprises. A few highlights from the first publications analyzing the sequence are the following:

- The human genome contains three billion nucleotide bases (A, C, T, and G).
- The average gene consists of 3,000 bases, but sizes vary greatly, with the largest known human gene being dystrophin at 2.4 million bases.
- Functions are unknown for more than 50% of discovered genes.
- The human genome sequence is almost (99.9%) exactly the same in all people.
- About 2% of the genome encodes instructions for the synthesis of proteins.
- Repeat sequences that do not code for proteins make up at least 50% of the human genome.
- Repeat sequences are thought to have no direct functions, but they shed light on chromosome structure and dynamics. Over time, these repeats reshape the genome by rearranging it, thereby creating entirely new genes or modifying and reshuffling existing genes.
- The human genome has a much greater portion (50%) of repeat sequences than the mustard weed (11%), the worm (7%), and the fly (3%).
- Over 40% of the predicted human proteins share similarity with fruitfly or worm proteins.
- Genes appear to be concentrated in random areas along the genome, with vast expanses of non-coding DNA in between.
- Chromosome 1 (the largest human chromosome) has the most genes (2,968), and the Y chromosome has the fewest (231).
- Genes have been pinpointed and the particular sequences in those genes associated with numerous diseases and disorders including breast cancer, muscle disease, deafness, and blindness.
- Scientists have identified about three million locations where single-base DNA differences occur in humans. This information promises to revolutionize the processes of finding DNA sequences associated with such common diseases as cardiovascular disease, diabetes, arthritis, and cancer.

The estimated number of human genes is only one-third as great as previously thought, although the numbers may be revised as more computational and experimental analyses are performed. Scientists suggest that the genetic key to human complexity lies not in gene number but in how gene parts are used to build different products in a process called alternative splicing.

TABLE 16.2 Genome size and estimated gene number of organisms whose genome has been sequenced completely.

Organism	Genome Size (Bases)	Estimated Genes
Human (*Homo sapiens*)	3 billion	30,000
Laboratory mouse (*M. musculus*)	2.6 billion	30,000
Mustard weed (*A. thaliana*)	100 million	25,000
Roundworm (*C. elegans*)	97 million	19,000
Fruit fly (*D. melanogaster*)	137 million	13,000
Yeast (*S. cerevisiae*)	12.1 million	6,000
Bacterium (*E. coli*)	4.6 million	3,200
Human immunodeficiency virus (HIV)	9,700	9

Other underlying reasons for greater complexity are the thousands of chemical modifications made to proteins and the repertoire of regulatory mechanisms controlling these processes.

Genome-Cloning Strategies

There are several strategies based on Sanger's chain termination reaction for sequencing small stretches of DNA strands. But the sequencing of a complete genome requires special strategies. There are two main strategies used for the sequencing of large genomes such as human and other eukaryotes. These were also discussed in the introduction.

Direct Sequencing of YACs and BACs Contigs

This strategy developed by NIH starts from the top to the bottom. That is, it starts from the bigger fragments of cloned genomic DNA fragments that are adjacent to each other in the genome or have overlapped fragments. Yeast-artificial chromosomes and bacterial-artificial chromosomes are very suitable for cloning of large DNA inserts of genomic origin in *e. coli*. These vectors are usually used for the construction of genomic libraries with large DNA inserts having a size range of 100 to 150 kb. The most common approach to sequence the whole genome uses a three-stage divide and conquer strategy. It involves the construction of three different libraries of chromosomal DNA. This is done by randomly cutting (with restriction enzymes) high molecular DNA into fragments, separating these fragments into size classes by electrophoresis, including PFGE (pulse field gel electrophoresis), for larger fragments. Then inserting the fragments into vectors like BACs or YACs capable of propagating them into appropriate hosts (bacteria or yeast). A clone comprises a vector with a single insert of genomic DNA, whereas a clone library or genomic library comprises the entire collection of genomic DNA fragments, each inserted into a vector. Different numbers of clones constitute a clone library depending on the average size of the insert and the size of the donor DNA.

The sequencing depends on retrieving clones that have overlapping segments from a genomic DNA library. On the basis of the overlaps, a continuum of clones can be established for a chromosome region, a whole chromosome, or the entire genome. Sets of **contiguously ordered clones (contigs)** are generated at every phase. The construction of an array of contigs involves the construction of a genomic library and its screening with a common marker such as a probe. The process can be summarized as follows.

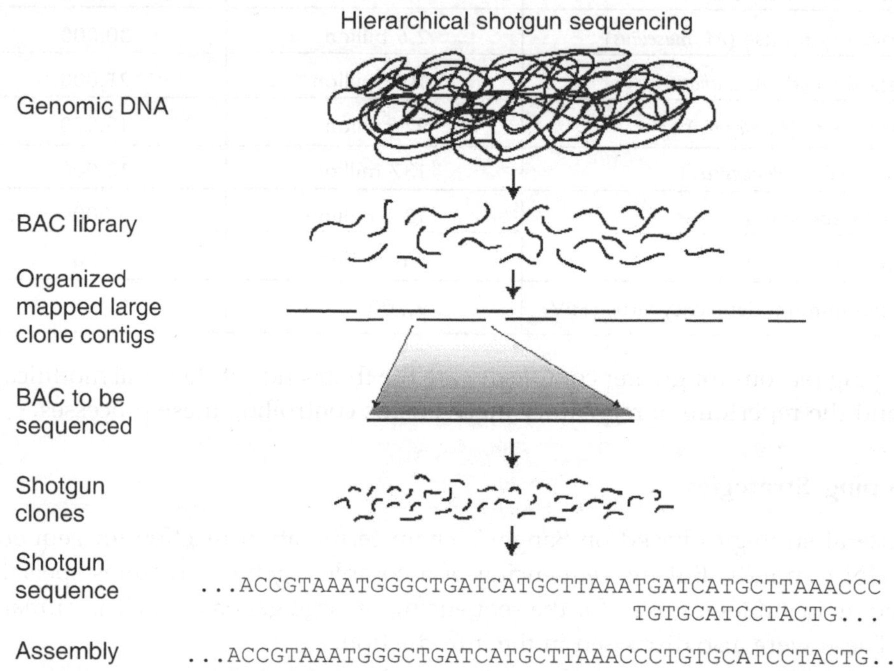

FIGURE 16.1 **Different stages of genome sequencing by direct-sequencing strategy.**

First a genomic library is prepared with the help of high-capacity cloning vectors such as BAC and YAC in suitable host cells such as *e.coli* or yeast. These libraries screened with shared landmarks (molecular markers) such as common restriction-site sequences such as probes to identify clones with overlapping DNA fragments. These overlapping arrays of contiguous clones are known as **contigs.** The rationale of this approach is to assemble large contigs, corresponding to whole chromosomes or chromosomal regions in which the entire DNA has been cloned as a series of overlapping fragments in identified clones. Genomic DNA libraries normally contain clones with overlapping inserts because they are made from many copies of each chromosome that have been cut at random, or by partial restriction enzyme digests. When the individual DNA fragments are inserted into recipient cells during construction of the library, the overlapping fragments end up in separate cells. The clones with overlapping fragments can be fragmented randomly and sub-cloned in suitable vectors such as cosmids and can be sequenced by shotgun method. Therefore, in this method of sequencing adjacent DNA fragments of the genome can be sequenced and will be able to make the high-resolution physical map of the genome.

Strategies for Identifying Overlapping Inserts in a Library

- Hybridization of short oligonucleotide probes with isolated clones to test for sharing of a specific sequence.

- Fingerprinting by restriction mapping—identifying a diagnostic pattern of restriction sites, which is common to independent DNA clones.
- PCR amplification—in this case a short stretch of a DNA clone must be first sequenced in order to establish a sequence-tagged site (STS) from which primers can be designed that will amplify in all DNA clones that contain this site, but not in other clones.

Shotgun-Sequencing Method

This strategy of genome sequencing was actually developed by Celera Genomics, the commercial partner of the Human Genome Project. In this strategy there is no involvement of mapping for the identification of overlapping fragments among the clones of the genomic library. The genomic clones of large sizes are further fragmented into smaller sizes of defined length, 2kb to 10kb, and cloned into a convenient plasmid randomly. Each of these plasmids having inserts is sequenced and the sequences are arranged to find the overlap. On sequencing, there is every chance of getting the same type of sequences in different clones, entirely different sequences without overlapping, and at the same time different lengths of overlapping sequences may be present in the inserts of different clones. All these sequences may be arranged and the overlapping sequences and the continuity of sequences can be determined with the help of suitable computer programs. For example, if a number of inserts that have been sequenced don't show any overlapping, it means that they are present at different locations of the genome. But if the sequences show a different degree of overlapping, then it shows that they are represented in the same location of the genome and are continuous. But a single type of genomic library of an organism and a single sampling may not be sufficient to generate the genomic sequence completely without any gap. Different types of genomic libraries of the organism (libraries constructed with different cloning strategies such as restriction digestions with different restriction enzymes and cloning using different cloning vectors, etc.) can provide more overlapping sequences that can fill the gap in the genome sequences. As the overlapping of sequences among the inserts of the clones increase the continuity, the sequences of the genome can be ascertained.

16.4 GENE PREDICTION AND COUNTING

The completion of the genome sequencing of an organism doesn't mean anything. It is just like a storybook in which there are only letters without any demarcation into words, sentences, and paragraphs. It has to be properly annotated. The genome sequence has to be analyzed to find out which genes are immersed in it. The gene prediction or gene finding can be carried out with the help of known genes, whose structural components and organizations are very clear. There are a number of markers that can be used for locating a gene in a DNA sequence.

The DNA strand has to be read on all six possible **open reading frames (ORF)** and characteristics, by which the gene can be marked. The presence of the start codon and stop codon and the presence of the minimum number of nucleotides in between these codons should be defined. The presence of the stop codon frequently in an ORF shows the absence of a valid gene. Identification of intron-exon junctions, the presence of Shine Delgarno sequence near to the start signal, and the presence of relevant promoter sequences upstream to the start codons, etc. are some other identification marks by which the presence of a valid gene in an ORF can be predicted. A number of computer algorithms incorporating all the above characteristics are available for the prediction of genes and also for estimating the possible number of genes present in a DNA sequence or genome. But there are problems in the computational methods of genomics and gene prediction and analysis. The existence of overlapping genes, variation in the splicing junctions due to the special feature of the organisms, shifting in the reading frame, and there is the possibility of the presence of more than one gene in the same fragment of DNA. Even then we can make some computer programs allowing some experimental error, to predict the possible number of genes that can be accommodated in a genome.

Genome sequencing and computational analysis of the genome sequences to carry out the gene prediction and counting has given quite surprising results in the total number of genes against the expected genes based on genome size. The genome size and the predicted number of genes in the corresponding genomes of the completely sequenced organisms are given in Table 16.3.

Caenorhabditis elegans is a worm in which there are 18,000 genes for a genome size of one million bp, whereas a human in spite of his huge genome size (three billion bp) has only 30,000 genes. But, there is another view that the number of genes in a human genome can still be more and may come between 40,000 and 50,000. Even then all these results indicate that there is no simple correlation between the size and complexity of the genome and the number of total genes present in the genome.

TABLE 16.3 Comparison of genome size and predicted genes in different organisms.

Organisms	Number of Chromosomes	Genome Size (bp)	Total Number of Predicted Genes	The Percentage of Genome that Codes for Protein
Escherichia coli	1	500,000	5,000	90%
Saccharomyces cerevisiae	12	14,000,000	6,000	70%
Caenorhabditis elegans	6	100,000,000	18,000	27%
Drosophila melanogaster	4	300,000,000	14,000	20%
Arabidopsis pthaliana	5	125,000,000	25,500	20%
Homo sapiens	23	3,000,000,000	30,000	2-5%

16.5 GENOME SIMILARITY, SNPs, AND COMPARATIVE GENOMICS

The Human Genome Project has proved beyond doubt that the genetic sequences of all humans are identical to about 99.8%. That means the DNA sequences in all individuals are identical. But this 0.2% difference is more than enough to make unique differences between individuals. But it has become evident from recent genome analysis and sequencing that an average one out of one thousand nucleotide sequences is different in the genome of two individuals. **Single Nucleotide Polymorphism** or **SNPs** are an important variation observed between two individuals.

A single nucleotide polymorphism, or SNP ('snip'), is a small genetic change, or variation, that can occur within a person's DNA sequence. Among the four **nucleotide** 'letters' are A (adenine), C (cytosine), T (thymine), and G (guanine), which specify the gene sequence SNP variation that occurs when a single nucleotide, such as an A, replaces one of the other three nucleotide letters—C, G, or T. For example, an SNP is the alteration of the DNA segment GTATTAGGTTA to GTATTACGTTA, where the seventh 'G' in the first snippet is replaced with a 'C.' On average, SNPs occur in the human population more than one percent of the time. Snips can be in the coding and non-coding regions of the genome. But since only about three to five percent of a person's DNA sequence code for proteins, most SNPs are found in the non-coding sequences. Interestingly, those SNPs found within a coding sequence are very important because they are more likely to alter the biological function of a protein. Because of recent advances in technology, coupled with the unique ability of these genetic variations to facilitate gene identification, there has been a recent flurry of SNP discovery and detection. This comparative genomics is very important in various fields, particularly in the pharmaceutical industry and in disease diagnosis. For example, a single base difference in the ApoEi gene is associated with the Alzheimer's disease. A simple deletion within the chemokine-receptor gene CCR5 leads to resistance to AIDS. Therefore, SNPs are very important in diagnosis and prescription of medicines.

Each person's genetic material contains a unique SNP pattern that is made up of many different genetic variations. Researchers have found that most SNPs are not responsible for a disease state. Instead, they serve as biological markers for pinpointing a disease on the human genome map, because they are usually located near a gene found to be associated with a certain disease. Occasionally, an SNP may actually cause a disease and, therefore, can be used to search for and isolate the disease-causing gene. It will only be a matter of time before physicians can screen patients for susceptibility to a disease by analyzing their DNA for specific SNP profiles.

Another very important surprise that has been revealed through genome-sequencing projects of various organisms is the similarity of the genome sequences of unrelated organisms. For example it was observed that 12% of the 18,000 genes in the worm *caenorhabditis* are similar in sequence to the yeast genome. The biological roles of the proteins coded by these genes were inferred by the sequence similarity with yeast genes. Approximately three percent of the 6,000 yeast genes are functionally equivalent to the worm genes. Another important point evident from the Human Genome Project is that the difference between the genome of human and chimpanzee is only one to three percent. Similarly, human and mouse genomes are identical to about 97.5% of their functional DNA. All these similarity and homology data shared by unrelated organisms show that they have evolved from a common ancestor some 100 million years ago and their genomes have not undergone much change during this period of evolution.

16.6 PHARMACOGENOMICS

The way a person responds to a drug (both positive and negative reactions) is a complex trait that is influenced by many different genes. Without knowing all of the genes involved in drug response, scientists have found it difficult to develop genetic tests that could predict a person's response to a particular drug. Once scientists discovered that people's genes show small **variations** (or changes) in their **nucleotide** (DNA base) content, all of that changed—genetic testing for predicting drug response is now possible. **Pharmacogenomics** is a science that examines the inherited variations in genes that dictate drug response and explores the ways these variations can be used to predict whether a patient will have a good response to a drug, a bad response to a drug, or no response at all.

16.7 FUNCTIONAL GENOMICS AND MICROARRAYS

Functional genomics as stated earlier is the part of genomics that deals with the function and expression of individual genes and interaction of the gene products with each other and with the gene itself. Since it deals with gene products and their interactions, it also involves proteomics. As genomics indicates the structural and functional aspects of the total genome and genes of an organism, proteomics is the structural and functional aspects of total protein complement of the genomes. New techniques such as DNA microarrays and newer techniques of proteomics such as mass spectrum (MS) can provide a wealth of information about the identity of the gene, its product and function, its interaction with other genes, etc. Microarray analysis is specially suited for the study of gene expression profiles of tissues and cells under different environmental conditions. It is very efficient to study the expression of thousands of genes at a time.

Microarray Technology

The use of microarrays, small spots of DNA fixed to glass slides or nylon membranes, promise to provide new and fundamental information for cell and molecular biology. **Microarray technology** provides a tool to potentially identify and quantitate levels of gene expression for all genes in an organism. DNA microarrays are perfectly suited for comparing gene expression in different populations of cells. The hows and whys of such an experiment provide insight into the power of microarrays, their limitations, and the kinds of biological questions, which they can help to answer. The illustration below shows the steps that make up a comparative cDNA hybridization experiment.

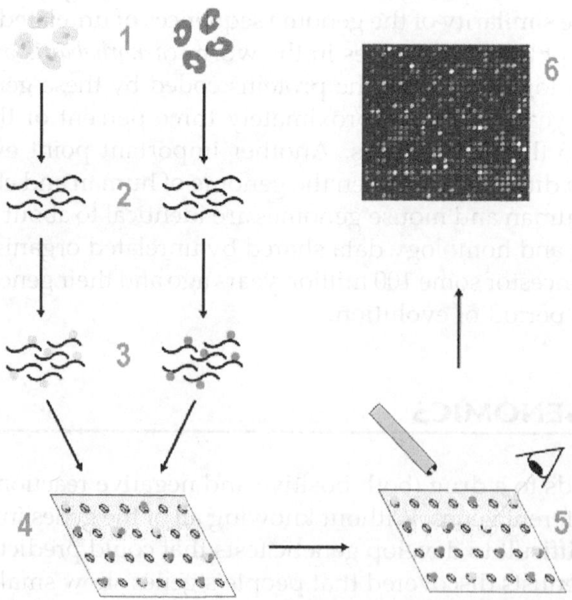

FIGURE 16.2 **A comparative hybridization experiment with microarrays.**

The major steps of a comparative cDNA hybridization experiment are:

1. Choosing cell populations
2. mRNA extraction and cDNA synthesis
3. Fluorescent labeling of cDNAs
4. Hybridization to a DNA microarray
5. Scanning the hybridized array
6. Interpreting the scanned image

Choosing Cell Populations

The goal of comparative cDNA hybridization is to compare gene transcription in two or more different kinds of cells. We will describe some experiments of particular interest; however, the possibilities for informative comparative transcription studies are limited only by the investigator's imagination.

Tissue-specific Genes

Cells from two different tissues (say, cardiac muscle and prostate epithelium) are specialized for performing different functions in an organism. Although we can recognize cells from different tissues by their phenotypes, it is not known just what makes one cell function as smooth muscle, another as a neuron, and still another as a prostate. Ultimately, a cell's role is determined by the proteins it produces, which in turn depend on its expressed genes. Comparative hybridization experiments can reveal genes, which are preferentially expressed, in specific tissues. Some of these genes implement the behaviors that distinguish the cell's tissue type, while other controlling genes make sure that the cell *only* performs the functions for its type.

Regulatory Gene Defects in Cancer

Genetic disease is often caused by genes, which are inappropriately transcribed—either too much or too little—or which are missing altogether. Such defects are especially common in cancers, which can occur when regulatory genes are deleted, inactivated, or become constitutively active. Unlike some genetic diseases (e.g., cystic fibrosis in which a single defective gene is always responsible), cancers that appear clinically similar can be genetically heterogenous. For example, prostate cancer (prostatic adenocarcinoma) may be caused by several different, independent regulatory gene defects even in a single patient. In a group of prostate cancer patients, every one may have a different set of missing or damaged genes, with differing implications for prognosis and treatment of the disease.

Comparative hybridization can serve two purposes in studying cancer: it can pinpoint the transcription differences responsible for the change from normal to cancerous cells, and it can distinguish different patterns of abnormal transcription in heterogeneous cancers. Understanding the diverse basis of cancer is crucial for inventing therapies targeted to the different varieties of the disease, so that each patient receives the most appropriate and effective treatment.

Cancers are common examples of genetically heterogeneous diseases, but they are by no means the only ones. Diabetes, heart disease, and multiple sclerosis are among the diseases for which genetic risk factors are known to be heterogeneous.

Cellular Responses to the Environment

How does a cell adapt to changes in its environment? Cells survive in the face of changes in temperature and pH, changing nutrient availability, and the presence of environmental toxins and ionizing radiation. Usually, a change in environment requires that expression of some genes be turned up or down so that the organism can respond appropriately. For example, common yeast has been extensively studied to understand how it switches from metabolizing sugars into ethanol which, in turn, changes into acetic acid (this is why wine with active yeast eventually becomes vinegar). The move from one metabolic state to the other, called **diauxic shift,** involves shutting down genes for processing sugars and activating others for processing ethanol, as well as a general stress response due to the greater difficulty of deriving energy from ethanol.

Comparative hybridization experiments can point out genes whose transcription changes in response to an environmental stimulus. In the simplest experiment, a population of cells is subjected to the stimulus and allowed to reach a steady state of transcription. Transcription levels in the altered cells can then be compared to those in a control population. A more informative experiment subjects cells to a change, then takes samples of the cell population at successive points in time. In this way, the experimenter can watch as the gene transcription patterns change from the old to the new steady state. Temporal studies can identify not only genes whose transcription changes but also the *order* of the changes, providing evidence about which genes control the response directly and which are only indirectly affected by it.

Environmental changes of interest also include the introduction of signaling molecules, such as hormones, interleukins, and interferons, as well as the actions of drugs. All these molecules stimulate a change in a cell's behavior (including possibly its death). While some of the changes may be mediated purely at the protein level, others require new transcription, which can be detected with comparative hybridization.

Cell-cycle Variations

Even in a stable environment, cells undergo DNA replication, mitosis, and eventually death. These activities require quite different gene products, such as DNA polymerases for genome replication or microtubule spindle proteins for mitosis. A cell's genes encode the 'programs' for these activities, and gene transcription is required to execute those programs. Comparative hybridization can be used to distinguish genes that are expressed at different times in the cell cycle. In this way, the pathways responsible for controlling basic life processes can be uncovered.

mRNA Extraction and cDNA Synthesis

Genes that code for protein are transcribed into messenger RNAs (mRNAs) in the cell nucleus. The mRNAs in turn are translated into proteins by ribosomes in the cytoplasm. The transcription level of a gene is taken to be the amount of its corresponding mRNA present in the cell. Comparative hybridization experiments compare the amounts of many different mRNAs in two cell populations.

To prepare mRNA for use in a microarray assay, it must be purified from total cellular contents. The total amount of mRNAs accounts for only about three percent of all RNA in a cell. So isolating it in sufficient quantity for an experiment (one to two micrograms) can be a challenge. Common mRNA isolation methods take advantage of the fact that most mRNAs have a poly-adenine (poly(A)) tail. These poly$(A)^+$ mRNAs can be purified by capturing them using complementary oligodeoxythymidine (oligodT) molecules bound to a solid support, such as a chromatographic column or a collection of magnetic beads. Captured mRNAs are still difficult to work with because they are prone to being destroyed. The environment is full of RNA-digesting enzymes (there are some on your fingers, your work table, glassware, pipettes, and every other exposed surfaces around you) and so free RNA is quickly degraded. To prevent the experimental samples from being lost, they are reverse transcribed back into more stable DNA form. The products of this reaction are called **complementary DNA** or **cDNAs,** because their sequences are the complements of the original mRNA sequences. The reverse transcription reaction usually starts from the poly(A) tail of the mRNA and move toward its head; such a reaction is called **oligo(dT)-primed.**

Fluorescent Labeling of cDNAs

In order to detect cDNAs bound to the microarray, we must label them with a reporter molecule that identifies their presence. The reporters currently used in comparative hybridization to microarrays are fluorescent dyes (fluors), represented by the circles attached to the cDNAs in Figure 16.2. A different colored fluor is used for each sample so that we can distinguish the two samples apart on the array. The labeled cDNA samples are called probes because they are used to probe the collection of spots on the array.

The number of fluor molecules that label each cDNA, depends on its length and possibly its sequence composition, both of which are often unknown. There is one more reason that fluorescent intensities for different cDNAs cannot be quantitatively compared. However, identical cDNAs from the two probes are still comparable as long as the same number of label molecules are added to the same DNA sequence in each probe.

To equalize the total concentration of the two cDNA probes before applying them to the array, the probe solutions are diluted to have the same overall fluorescent intensity. This procedure makes two possibly unjustified assumptions: first, that the total amount of mRNA in each cell type being tested is identical; and second, that each fluor emits the same amount of light relative to its concentration. The second assumption can be eliminated by suitable calibration, but the first one is difficult to check. It may therefore be considered that cells with more abundant mRNA are made into a probe with artificially low mRNA concentrations.

Hybridization to a DNA Microarray

The two cDNA probes synthesized from mRNA produced by two different cell populations are tested by hybridizing them to a DNA microarray. The array holds hundreds or thousands of spots, each of which contain a different DNA sequence. If a probe contains a cDNA whose sequence is complementary to the DNA on a given spot, that cDNA will hybridize to the spot, where it will be detectable by its fluorescence. In this way, every spot on an array is an independent assay for the presence of a different cDNA. There is enough DNA on each spot that both probes can hybridize to it rapidly without interference.

Microarrays are made from a collection of purified DNAs. A drop of each type of DNA in solution is placed on to a specially prepared glass microscope slide or silicon chip by an arraying machine. The arraying machine can quickly produce a regular grid of thousands of spots in a square about 2 cm on a slide, small enough to fit under a standard slide coverslip. The DNA in the spots is bonded to the glass to keep it from washing off during the hybridization reaction.

The choice of DNAs to be used in the spots on a microarray determines which are the genes that can be detected in a comparative hybridization assay. For organisms whose genomes have been completely sequenced it is possible to array genomic DNA from every known gene or suspected open reading frame (ORF) in the organism. Each gene or ORF is amplified from the total genomic DNA by PCR, producing enough DNA to make unlimited number of arrays. The Pat Brown Lab at Stanford University has arrayed all known or suspected genes of *s. cerevisiae* (roughly 6,100) on a single microarray. The number of human genes has been estimated at somewhere between 30,000 and 50,000, so several arrays will probably be required to hold them all. Despite these limitations, several strategies can be used to make arrays for studying human genes. We do know the location and sequence of quite a few human genes now, so the same method used to array yeast genes will produce at least a partial human genome array. There are two other ways to produce arrayable DNA even for unknown genes. Amplify clone inserts from human cDNA libraries, or synthesize oligonucleotides directly from known expressed sequence information such as ESTs. While neither of these methods will produce DNA for every human gene, both can yield enough different expressed sequences to make substantial arrays. Both types of DNA have been used before in array-like applications: cDNA libraries were used for comparative hybridization before the advent of fluorescent microarrays, while oligonucleotide arrays are available commercially today from Affymetrix Corporation for rapid resequencing of a few genes important to AIDS and some cancers. There are several names for this microarray technology. Different manufacturers give the names. It includes, DNA arrays, DNA chips, biochips, gene arrays, etc. In the beginning of this technology there were distinctions between these names, but today they are all synonyms. Now this microarray technology has been extended to protein and peptide arrays, carbohydrate arrays, nucleotide arrays, etc. in which the concerned biomolecules are immobilized on a microscopic glass slides as spots and used for different types of studies and analysis.

Scanning the Hybridized Array

Once the cDNA probes have been hybridized to the array and any loose probe has been washed off, the array must be scanned to determine how much of each probe is bound to each spot. The probes are tagged with fluorescent reporter molecules, which emit detectable light when stimulated by a laser. The emitted light is captured by a detector, either a **charge-coupled device (CCD)** or a **confocal microscope,** which records its intensity. Spots having more bound probes will have more reporters and will therefore fluoresce more intensely.

Each of the two fluorescent reporters (fluors) used has a characteristic excitation wavelength; only light of this wavelength will cause the molecule to fluoresce. The emitted light has a characteristic emission wavelength, which is different from the excitation wavelength. The detector for the emitted fluorescence from the array is sensitive to the emission wavelength but filters out the excitation wavelength; in this way, the fluorescent light of interest can be separated from the laser light scattered off the slide.

A good pair of fluors for a comparative hybridization experiment should have very different emission or excitation wavelengths. If the emission wavelengths are different, light emitted from the two fluors can be selectively filtered to measure the amount emitted by each fluor separately. If the excitation wavelengths are different, the two fluors can be stimulated and scanned one at a time. If one of these conditions is not met, the scanned intensities can be contaminated by crosstalk between the two fluorescent channels.

Although the purpose of the scanner is to pick up the light emitted by probe cDNAs bound to their complementary spots, it also records light from a few molecules that hybridized either to the wrong spot or nonspecifically to the glass slide. This extra light becomes the background of the scanned array image. One advantage of current microarray technology over its predecessors is that the background is extremely low; consequently, the signal-to-noise ratio of the scanned data can be quite high.

Interpreting the Scanned Image

The end product of a comparative hybridization experiment is a scanned array image. A small piece of such an image is shown in the figure in the Appendix. Interpreting the data from a microarray experiment can be challenging. Quantitation of the intensities on each spot is subjected to noise from irregular spots, dust on the slide, and non-specific hybridization. Deciding the intensity threshold between spots and the background can be difficult, especially when the spots fade gradually around their edges. Detection efficiency might not be uniform across the slide, leading to excessive red intensity on one side of the array and excessive green on the other side. Even after overcoming detection and calibration problems, the measured intensities for each spot only represent the ratio of cDNAs in each cell population. Low levels of cDNA due to reverse transcription bias, sample loss, or an inherently rare mRNA can cause large uncertainties in these ratios.

Numerous software packages, both free and commercial, exist for quantitating microarray data. The interpreted array data will highlight a relatively small number of spots representing very differentially expressed mRNAs whose genes deserve further investigation. Alternatively, the overall pattern of expression can be used as a "fingerprint" to characterize specific cell types (e.g., different classes of tumors), even if not all the differentially expressed genes on an array have been identified.

16.8 MOLECULAR PHYLOGENY

Phylogenetic systematics is the field of biology that deals with identifying and understanding the evolutionary relationships among the many different kinds of life on earth, both living (**extant**) and dead (**extinct**). Evolutionary theory states that similarity among individuals or species is attributable to common descent, or inheritance from a common ancestor. Thus, the relationships established by phylogenetic systematics often describe a species' evolutionary history and, hence, its **phylogeny**, the historical relationships among lineages or organisms or their parts, such as their genes.

Evolution is not always discrete with clearly defined boundaries that pinpoint the origin of a new species, nor is it a steady continuum. Evolution requires genetic variation and genetic variation results from changes within a **gene pool**, the genetic make-up of a specific population. A gene pool is the combination of all the **alleles**—alternative forms of a genetic locus—for all traits that population may exhibit. Changes in a gene pool can result from **mutation**—variation within a particular gene—or from changes in **gene frequency**—a measure of the proportion of an allele in a given population.

Macromolecular data, meaning gene (DNA) and protein sequences, are accumulating at an increasing rate because of recent advances in molecular biology. For the evolutionary biologist, the rapid accumulation of sequence data from whole genomes has been a major advance, because the very nature of DNA allows it to be used as a 'document' of evolutionary history. Comparisons of the DNA sequences of various genes between different organisms can tell a scientist a lot about the relationships of organisms that cannot otherwise be inferred from **morphology**, or an organism's outer form and inner structure. Because genomes evolve by the gradual accumulation of mutations, the amount of nucleotide sequence difference between a pair of genomes from different organisms should indicate how recently those two genomes shared a common ancestor. Two genomes that diverged in the recent past should have fewer differences than two genomes whose common ancestor is more ancient. Therefore, by comparing different genomes with each other, it should be possible to derive evolutionary relationships between them, the major objective of molecular phylogenetics.

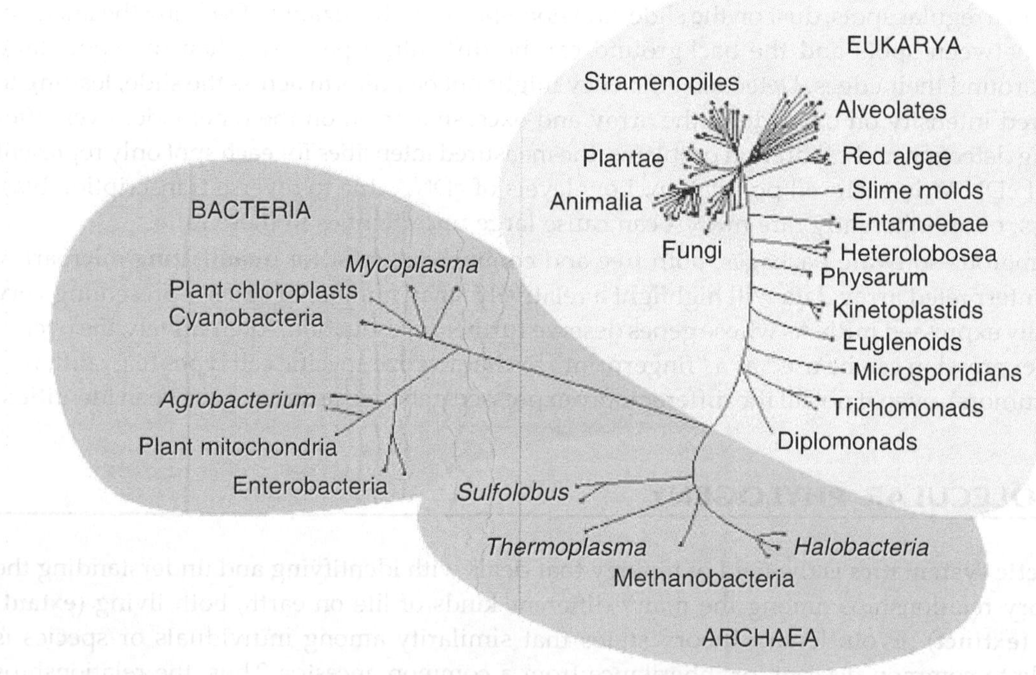

FIGURE 16.3 **Molecular phylogeny.**

Molecular phylogenetics attempts to determine the rates and patterns of change occurring in DNA and proteins and to reconstruct the evolutionary history of genes and organisms. Two general approaches may be taken to obtain this information. In the first approach, scientists use DNA to study the evolution of an organism. In the second approach, different organisms are used to study the evolution of DNA. Whatever the approach, the general goal is to **infer process from pattern**: the processes of organismal evolution deduced from patterns of DNA variation and processes of molecular evolution inferred from the patterns of variations in the DNA itself.

As we just discussed, macromolecules, especially gene and protein sequences, have surpassed morphological and other organismal characters as the most popular forms of data for phylogenetic analyses. Therefore, this next section will concentrate only on molecular data.

It is important to point out that a single, all-purpose recipe does not exist for phylogenetic analysis of this type of data. Although numerous algorithms, procedures, and computer programs have been developed, their reliability and practicality are, in all cases, dependent on the size and structure of the data set under analysis. The merits and shortfalls of these various methods are subject to much scientific debate, because the danger of generating incorrect results is greater in computational molecular phylogenetics than in many other fields of science. Occasionally, the limiting factor in such analyses is not so much the computational method used, but the users' understanding of what the method is actually doing with the data. Therefore, the goal of this section is to demonstrate to the reader that practical analysis should be thought of both as a search for a correct model (analysis) as well as a search for the correct tree (outcome).

Phylogenetic tree-building models presume particular evolutionary models. For any given set of data, these models may be violated because of various occurrences, such as the transfer of genetic material between organisms. Therefore, when interpreting a given analysis, a person should always consider the model used and entertain possible explanations for the results obtained. For example, models used in molecular phylogenetic analysis methods make 'default' assumptions, including:

The sequence is correct and originates from the specified source. The sequences are homologous—are all descended in some way from a shared ancestral sequence. Each position in a sequence alignment is homologous with every other in that alignment. Each of the multiple sequences included in a common analysis has a common phylogenetic history with the other sequences. The sampling of taxa is adequate to resolve the problem under study. Sequence variation among the samples is representative of the broader group. The sequence variability in the sample contains phylogenetic signal adequate to resolve the problem under study. A straightforward phylogenetic analysis consists of four steps:

1. Alignment—building the data model and extracting a dataset
2. Determining the substitution model—consider sequence variation
3. Tree building
4. Tree evaluation

Studies of gene and protein evolution often involve the comparison of **homologs**, sequences that have common origins but may or may not have common activity. Sequences that share an arbitrary level of similarity determined by alignment of matching bases are **homologous**. These sequences are inherited from a common ancestor that possessed similar structure, although the ancestor may be difficult to determine because it has been modified through descent.

The field of molecular phylogenetics has grown, both in size and in importance, since its inception in the early 1990s, attributable mostly to advances in molecular biology and more rigorous methods for phylogenetic tree building. The importance of phylogenetics has also been greatly enhanced by the successful application of tree reconstruction, as well as other phylogenetic techniques, to more diverse and perplexing issues in biology. Today, a survey of the scientific literature will show that molecular biology, genetics, evolution, development, behavior, epidemiology, ecology, systematics, conservation biology, and forensics are but a few examples of the many disparate fields conceptually united by the methods and theories of molecular phylogenetics. Phylogenies are used essentially the same way in all of these fields, either by drawing inferences from the structure of the tree or from the way the character states map on to the tree. Biologists can then use these clues to build hypotheses and models of important events in history. Broadly speaking, the relationships established by phylogenetic trees often describe a species' evolutionary history and, hence, its phylogeny—the historical relationships among lineages or organisms or their parts, such as their genes. Phylogenies may be thought of as a natural and meaningful way to order data, with an enormous amount of evolutionary information contained within their branches. Scientists working in these different areas can then use these phylogenies to study and elucidate the biological processes occurring at many levels of life's hierarchy.

REVIEW QUESTIONS

1. Explain Phylogenetic systematics.
2. Explain the principle and the applications of Microarray Technology.
3. Name a technology that can be used for the comparison of gene expression in different population of cells. Explain the principle.
4. What are the various steps in the comparative cDNA hybridization experiment?
5. Explain the technique of labeling mRNA in Microarray experiments.
6. Explain the technique of mRNA extraction and its purification.
7. What is functional genomics? Explain how Microarrays can be used for functional genomics.
8. Explain how the information of SNPs can be used in Pharmaceutical Industry.
9. Explain how one can predict the presence of a valid gene in an ORF.
10. Explain the Shotgun Sequencing Method.
11. What is STS? Explain its use in PCR.
12. What are the types of cloning vectors used for making genomic libraries for Genome sequencing?
13. Explain various Genome cloning strategies.
14. Explain the impact of human genome project on science and society.
15. Explain the history of Human Genome Project.
16. What is the advantage of RH Mapping?
17. Explain the major differences between Physical Map and genetic map of Genomes.
18. Explain VNTRPs and its application.
19. Differentiate structural genomics and functional genomics.
20. Explain the principle and method of DNA sequencing by chain termination method.

Chapter 17 BIOINFORMATICS

In This Chapter

17.1 HISTORY OF BIOINFORMATICS

Bioinformatics can be described as the science of collecting, modeling, storing, searching, annotating, and analyzing biological information. It involves a range of activities from data handling and publication to data mining and analysis. Bioinformatics is the field of science in which biology, computer science, and information technology merge to form a single discipline. An essential part of bioinformatics is the creation of new algorithms for the analysis of complex and/or large data sets. Bioinformatics deals with the issues created by the massive amounts of new types of data obtained through novel biological experiments. The most well-known example is, of course, the determination of the complete nucleotide sequence of the human genome, and other organisms. Actually, computational biology began as a part of bioinformatics to solve the problems of bioinformatics and to simplify the data storage, management, retrieval, and classifications. But now bioinformatics is a part of computational biology, which deals with molecular modeling, drug designing, protein structural and functional prediction, phylogenetic analysis, metabolic networks, etc.

Bioinformatics started as the databases for sequences of both protein and DNA. As the sequencing of more and more genes and proteins was carried out, sequence data accumulated in the database and it became very difficult to manage the sequences and their comparisons. Thus, computer-based databases evolved, which simplified the management and classification of data, and also its comparisons.

It was the National Biomedical Research Foundation that compiled the first comprehensive database of proteins and other macromolecules in the **Atlas of Protein Sequence and Structure** published from 1965–1978 by Margaret O. Dayhoff. This printed database became a computational database for analysis of sequences, comparison and alignment of protein sequences, detection of functionally-related proteins, to detect duplication of sequences within the database, and to study the molecular evolution of proteins. It is this Atlas of Protein Sequence and Structure that has given rise to the famous protein-structure database known as the **Protein Information Resource (PIR)**, established in 1984 by the **National Biomedical Research Foundation (NBRF)**. PIR acts as a resource to assist researchers in the identification and interpretation of protein sequences based on homology searching and modeling. In 1962, Zuckerkandl and Pauling observed that functionally-related proteins have similar amino acid sequences. This has initiated a new field known as **molecular evolution**. Later, Margaret Dayhoff observed that amino acid sequences are similar for functionally-related proteins and they undergo variation during evolution following certain patterns. From the protein sequence comparison and functional analysis, she observed the following points:

- During molecular evolution of proteins, amino acids were not replaced or changed at random, but followed with a specific preference. One amino acid will be replaced with another one, which is very close to the previous one in physical and chemical characteristics.
- Certain amino acids such as tryptophan were never replaced by other amino acids.

Based on the systematic analysis of a large number of homologous sequences researchers could develop a matrix called **point accepted mutation (PAM)** for computational homology analysis among the protein sequences.

A data library was started as a part of the **European Molecular Biology Laboratory (EMBL)** to collect, classify, and distribute nucleotide sequences and other information regarding molecular biology research in 1980. Later, this was developed into the **European Bioinformatics Institute (EBI)**, located in Hinxton. Now it is an important database for proteomics and genomics. More or less at the same time, the National Institute of Health started the **National Center for Biotechnology Information** as a databank for protein and gene sequences including software development for conducting various studies related to this data. **DNA databank** is the database started by Japan just after this. Today, all these databases are in close collaboration and share the information and update their database content and developments on a regular basis. Most of their services to the scientific world are free of cost and anyone can access their databases and freely available softwares for research and academic purposes.

The tremendous improvements in the techniques of molecular biology, biotechnology, and computational biology for the easy and rapid sequencing of nucleic acids and proteins have resulted in the rapid growth of databases. This has further forced the development of improved, powerful software for genome analysis, gene finding, gene annotation, protein structural analysis, homology searching, modeling, etc. With all this input from computer science, mathematics, statistics, physics, etc., bioinformatics has developed into a fast growing branch of learning, which in addition to the primary function of spreading scientific knowledge for research purposes, also provides the essential requirements for the development and testing of newer software for newly developed analytical techniques and data management. For example, databases for x-ray crystallography data, NMR structural databases and their involvement in the homology modeling of new proteins, microarray techniques and their databases, two-dimensional gel electrophoresis, mass spectrum analysis of proteins, etc., require efficient software for the analysis as well as the storage and use of their results.

The basic data has so far been the sequence data (nucleotide or protein), but other types of data (microarray, functional analysis, interactions, metabolic pathways and networks) are now rapidly coming into focus. The advantage of bioinformatics is that it mainly deals with biopolymers such as proteins, nucleic acids, carbohydrates, etc. Their molecular sequences can be transformed into digital symbols. Their monomers can be represented by a limited number of alphabets. In the case of DNA and RNA, they require only four letters and protein sequences can be represented by 20 types of letters. It is this digital nature of data that differentiates genetic data from other types of biological data. This has made bioinformatics unique among information science. Another important point in protein- and DNA-sequence information is that sequence implies the structure of the molecule and the structure in turn implies function. Therefore, the sequence data can be treated as context free, because the prediction of the biological importance of a sequence, DNA sequence or protein sequence, can be understood without the involvement of any other biological factors.

Based on the involvement of bioinformatics in the various areas of biological sciences it can be classified into the following three levels. They are:

1. Analysis of a single gene (protein) sequence. For example:

- Similarity with other known genes
- Phylogenetic trees; evolutionary relationships
- Identification of well-defined domains in the sequence
- Sequence features (physical properties, binding sites, modification sites)
- Prediction of subcellular localization
- Prediction of secondary and tertiary structure

2. Analysis of complete genomes. For example:

- Which gene families are present, which are missing?
- Location of genes on the chromosomes, correlation with function or evolution
- Expansion/duplication of gene families
- Presence or absence of biochemical pathways
- Identification of "missing" enzymes
- Large-scale events in the evolution of organisms

3. Analysis of genes and genomes with respect to functional data. For example:

- Expression analysis; microarray data; mRNA concentration measurements
- Proteomics; protein concentration measurements, covalent modifications
- Comparison and analysis of biochemical pathways
- Deletion or mutant genotypes vs phenotypes
- Identification of essential genes, or genes involved in specific processes

Legal Issues of Bioinformatics

Bioinformatics databases, algorithms, and software have widely different levels of copyrights, occasionally patents, and license conditions attached to them. This sometimes causes considerable problems for academic research. For example, a novel data modeling scheme may use data from several sources and combine them into an entire novel framework. This new database becomes difficult to distribute, since it builds on other databases. Of course, such legal protection may also be very important for the academic researcher. Software patents in bioinformatics are not very common, thus far. Large sequence databases and a few others are also publicly accessible and redistributable. But, most other databases have conditions that make development and subsequent redistribution problematic.

Commercial Uses of Bioinformatics

Bioinformatics is of direct relevance to the pharmaceutical and agricultural industries. It is necessary for analysis of experimental data. Bioinformatics can suggest hypothesis to be tested by experiment. There is a very close connection between basic bioinformatics science and its technological applications.

Most of the large pharmaceutical companies do not have any large groups doing original science research in bioinformatics. They usually have enough expertise to use and apply academic software and commercial products, but not much more. Most bioinformatics efforts in industries are directed at finding interesting genes, or helping experimental scientists with handling and analyzing their data. The relationship between academia and companies is basically good. There is much collaboration and communication. However, the very fast progress and movement of the field makes some kinds of long-term collaborations, such as Ph.D. projects, difficult. A company (especially a small company) is reluctant to commit to a four-year project; much can happen during that time.

It must be noted that different parts of the industry have different interests: some companies sell information (databases), algorithms or analysis tools, while other companies use these tools in pharmaceutical and agricultural research. For example, large pharmaceutical companies have no strong interest in software patents, while a small bioinfo company may build its entire business case on proprietary software.

17.2 SEQUENCE AND NOMENCLATURE

In science there should be some approved formats and nomenclature systems for getting uniformity in scientific studies and comparisons. In bioinformatics, we also adopt a nomenclature system as per the recommendations of the **International Union of Pure and Applied Chemistry (IUPAC)**. Because of this uniform nomenclature and symbols, it is very important to access the database and results from all over the world. All databases of the major institutes and journals that publish the research articles and reports follow the same nomenclature and symbols as per the recommendations of IUPAC. This ensures uniformity and rapid reproducibility. The basic nomenclature systems for nucleic acids and proteins are discussed here.

IUPAC Nomenclature for DNA and Proteins

In the case of DNA, the first letter of the corresponding nitrogen base present in the nucleotide is taken as the symbol for the four nucleotides. They are the G, A, T, and C, and this system is very clearly understood. But during experiments to determine the sequence of nucleic-acid strands, some nucleotides may not be clearly understood due to problems related to compression artifacts or secondary structures. Such problems, in most cases, are solved by repeating the experiments two or more times. Another approach to solve these types of problems is to sequence the complementary strand. But these problems may not be resolved properly and may persist even after this. In such cases, we adopt certain uniform codes or symbols to represent such nucleotides or nitrogen bases. For example, there may be some points in DNA sequences at which it is unclear whether the nucleotide is G or C even after repeated sequencing. But the experimental data shows that there is no chance of A or T at that position. It is then recommended that the symbol 'S' be used. The letter 'S' was chosen because it denotes the strong linkage between G and C due to the presence of three hydrogen bonds.

TABLE 17.1 Summary of single-letter code recommendations by IUPAC.

Symbol	Meaning	Base(s)
G	G	Guanine
A	A	Adenine
T	T	Thymine
C	C	Cytosine
R	G or A	PuRine
Y	T or C	PYrimidine
M	A or C	Amino
K	G or T	Keto
S	G or C	Strong (3 Hydrogen bonds)

W	A or T	Weak (2 Hydrogen bonds)
H	A or C or T	Not-G, H follows G in the alphabet
B	G or T or C	Not-A, B follows A in the alphabet
V	G or C or A	Not- T (not-U), V follows U in the alphabet
D	G or A or T	Not-C, D follows C in the alphabet
N	G or A or T or C	Any

Since the DNA molecule is a double-stranded helix in most organisms except in certain viruses, the strands are complementary and antiparallel to each other. By sequencing one strand we can predict the opposite or the complementary strand. But in some cases, the problem comes when we start encountering the symbols that mean more than one base at a single position. IUPAC has solved that problem and the symbols that are recommended in such situations are given in Table 17.2.

TABLE 17.2 Symbols and their complementary symbols, according to IUPAC.

Symbol	A	B	C	D	G	H	K	M	S	T	V	W	N	R	Y
Complement	T	V	G	H	C	D	M	K	S	A	B	W	N	Y	R

Similarly, in the case of protein sequences, each of the 20 amino acids is represented by a symbol recommended by the IUPAC. Each amino acid has a single-letter code and a three-letter code, which are given in Table 17.3.

TABLE 17.3 IUPAC symbols—Single-letter and three-letter symbols for amino acids.

Amino Acids	Three-letter Code	Single-letter Code
Alanine	Ala	A
Arginine	Arg	R
Asparagine	Asn	N
Aspartate	Asp	D
Cysteine	Cys	C
Glutamine	Gln	Q
Glutamate	Glu	E
Glycine	Gly	G
Histidine	His	H
Isoleucine	Ile	I

Leucine	Leu	L
Lysine	Lys	K
Methionine	Met	M
Phenylalanine	Phe	F
Proline	Pro	P
Serine	Ser	S
Threonine	Thr	T
Tryptophan	Trp	W
Tyrosine	Tyr	Y
Valine	Val	V
*	Asx	B
*	Glx	Z
*	Xaa	X

The Concept of Directionality

Nucleic-acid synthesis in any biological system proceeds in 5′–3′ direction and this is universal for both DNA and RNA. Considering this fact, it became a way to collect and store DNA and RNA sequence data in a 5′–3′ pattern even if it is published in the other way. The nucleotide-sequence data in the databases is generally present in the same form as submitted except for the facts mentioned above, subjected to some conventions which have been adopted for databases as a whole. Bases are numbered sequentially beginning with 1 at the 5′ end and the complementary strand is marked by 'c.' Usually, only one strand of the DNA molecule is given in the database entry and if you want to get its complementary strand you can get it by using various software packages available in the database.

In the cases of proteins, the translation process starts from the N-terminus to the C-terminus. This convention is followed in the database entries of proteins. Thus, the directionality that is present in biological systems is very helpful in describing the conventions to be adopted in the database systems.

Different Types of Sequences

The DNA sequences that are deposited in the gene bank of various bioinformatics centers and institutes include the following types:

Genomic DNA. Genomic-DNA sequences are sequences of the DNA directly isolated from the chromosomes or the total genome. They include both coding parts and non-coding parts. Coding parts of the genome are the genes of proteins and other types of RNAs.

cDNA. These sequences are derived from the complementary sequences of mRNA that are expressed in cells and tissues. These are the expressed parts of the genome. It can be very useful for the detection of actual genes in the genome as a probe or primer, and also for the expression of certain valued proteins.

ESTs. These are similar to cDNA, but may not be a complete complementary copy of an mRNA. ESTs or Expressed Sequence Tags are the sequenced ends of cDNAs used as probes for the detection and isolation of complete genes and also for the sequencing of genome fragments of the genome library. There are separate databases for ESTs known as dbESTs. Specific ESTs can be used for studying the expression patterns of genes and can be used for the manufacturing of DNA chips and microarrays.

GSTs. These are similar to ESTs, but are unspecific; that is, they may not be a part of a gene or expressed parts of the genome. They are Genome sequence tags ESTs, any small random fragment of the genome that can be sequenced and used as a probe for the identification of genome parts for sequencing. Genomic DNA is fragmented into bits by restriction digestion or randomly by Mung Bean nuclease (Mnase) and the fragments are used for the construction of the genomic library. A small fragment or a small part of a fragment is picked up randomly and sequenced. This sequence (GSTs) is used for picking up further fragments, which are continuous with the tags, and can be sequenced. These fresh sequences can be used as new GSTs for further studies. Thus, sequencing of the genome or the DNA can be carried out continuously to get the complete genome sequenced or to get into a specific region of the genome or chromosomes.

Organelle DNA. This database is specialized for the genome of organelles such as chloroplasts and mitochondria. These are the cytoplasmic genome. They have a circular DNA, which makes them semi-autonomous, and code for important proteins needed for their specialized functions, in addition to the proteins coded by the nuclear genome. If the mitochondrial and chloroplasts DNA sequences are deposited in the general genomic database, it should be indicated in the format when submitting the data.

Other molecules. The genome database also contains other types of sequences such as tRNA, rRNA, and snRNAs, which may be present as genes or complementary sequences or as RNA sequences.

17.3 INFORMATION SOURCES

The major information sources in biotechnology and molecular biology are located in different types of databases attached with the major research centers of the world. The following are the major bioinformatics centers:

National Center for Biotechnology Information (NCBI)

The National Library of Medicine (NLM) of the National Institute of Health started an information center for Biotechnology in 1988, and it has developed into a separate institute for Bioinformatics, the National Center for Biotechnology Information. In addition to maintaining the various types of databases such as pubMed, GeneBank, etc., it provides a data-retrieval facility and other computational resources and tools for analyzing the data obtained from the databases. In addition to the basic databank there are various other types of biological information databases. Since it was started, the database, services, and information given by NCBI have grown enormously and are still growing. The resources, both databases and computational tools for the analysis of sequence data, molecular modeling and structural predictions available with the NCBI, include the following:

- Database retrieval tools such as ENTREZ.
- Computational tools for sequence homology searching such as BLAST family.
- Organism level genome sequences.
- Chromosome level sequences.
- Gene-level sequences.
- Database of microarrays and gene-expression patterns.
- Programs for sequence alignment structural predictions.
- Homology, modeling, etc.

Database-Retrieval Tools

NCBI has a number of retrieval tools for different types of databases. For example, PubMed is the tool for searching Medline, the literature database of NCBI. The searching of DNA sequences and protein sequences can be carried out by the powerful database retrieval tool, the Entrez. Entrez allows a user to not only access and retrieve specific information from a single database, but also to access the integrated information from many NCBI databases. For example, the Entrez protein database is cross-linked to the Entrez taxonomy database. This allows a researcher to find taxonomic information for the species from which a protein sequence was derived. **Taxonomy** is a division of the natural sciences that deals with the classification of animals and plants.

Map Viewer is a recent computational tool that allows a user to view an organism's complete genome or integrated map for each chromosome, if it is available. It can also be used for visualizing sequence data for a genomic region of interest. With the help of Map Viewer researchers can find the following information:

- Position of a particular gene within an organism's genome.
- The location of gene within a specific chromosome.
- The types of genes located in a particular chromosome or chromosome region.
- The distance between two genes within a chromosome, etc., which can be obtained from a gene or genome map.

The Locus Link carries information on the official name and other essential descriptive information about selected genes, which are studied extensively. Through this tool one can access the homologous genes. It is very easy to obtain the information about the rabbit or mouse hemoglobin gene if the information about the human hemoglobin gene is known.

BLAST (Basic Local Alignment Search Tool)

BLAST is a collection of computational tools for carrying out sequence homology searching and alignment. Using this tool one can find out the other DNA or protein sequences present in the database that is similar to the DNA or protein that has been isolated. If it is similar, this tool will give the percentage homology or similarity between the sequences. This is very helpful in identifying the gene and its function.

The basic principle behind the BLAST is as follows:

- A sequence of your choice—DNA or protein is aligned with the other DNA or protein sequences present in the database. The database can be genome specific or chromosome specific. Alignment and comparison is carried out using scoring matrices, a statistical method, in which a score will be rewarded (positive value for correct matching) and penalized for a mismatch (negative value for bases or amino acid mismatch) while doing the alignment.

- Top scoring matches are ranked according to the set of criteria that serve to distinguish between similarities due to genetical or evolutionary relationship or by random chance.
- Sequences having homology or similarity are further analyzed extensively with other possible sequences with the help of other tools that are available in NCBI or other databases.

The terms homology and similarity, even though used as synonyms today, actually have very different meanings. Homology is a type of similarity. All similarities cannot be considered as having homology. If two sequences have homology, the similarity in the sequence is due to a common ancestry. If two proteins, 'A' and 'B,' have sequence homology they might have descended from a common protein, 'C', by evolution. A duplicated gene within a genome may also have sequence homology, but its function may be different. Such duplicated sequences within a genome are called **paralogs**. Homologous sequences of protein or DNA will have functional similarities but the paralogs may differ in their functions.

Computational Resources for Gene-Level Analysis

These programs are very helpful in the identification of genes within a DNA sequence or to locate a gene in a genome. UniGene, HomoloGene, RefSeq, etc., are examples of computational tools for gene-level analysis. UniGene databases were created for the classification and grouping of ESTs of the same or identical cDNAs. ESTs are produced by the sequencing of randomly chosen cDNA clones and so there can be ESTs of the same cDNAs represented in a cDNA library. In this database, the ESTs representing the same gene are grouped into sets known as clusters. HomoloGene is another specialized database developed for the orthologs and homologs for the human, rat, cow, mouse, zebrafish, etc., represented in the UniGene and Locus Link, and this database is found to be very easy to understand the homologous relationship between genes with the help of this database.

TABLE 17.4 Important databases of bioinformatics.

Database	Information Available
NCBI (National Center for Biotechnology Information)	Nucleic acid sequences, amino acids sequences of proteins
EMBL (European Molecular Biology Laboratory)	Nucleotide and protein sequences
SWISS-PORT	Protein sequences
PDB (Protein Database)	Three-dimensional structures of proteins elucidated by NMR and X-ray crystallography techniques
Ribosomal RNA Database	Ribosomal RNA sequences
PALI Database	Phylogenetic analysis and alignment of proteins

There is a database for mRNA and proteins of humans, mice, and rats and is known as RefSeq. This database is very useful for cases such as designing of DNA chips and microarrays for gene expression analysis in humans, mice, and rats. This type of database was also developed for microorganisms such as yeast and bacteria. In addition to these, there are a number of computational tools and resources available in NCBI for various molecular biology and biotechnology applications.

Just like NCBI there are a number of equally good centers for biotechnology information having specialized and general applications and computational tools for biotechnology research. Some of the important ones are listed in Table 17.4. The databases are managed and controlled by a person or team of persons known as **curators.** They review and check the newly submitted data for all the essential requirements such as biological features, translation of coding regions, use of universal rules, etc., needed by the format of the concerned database. **Curation** is the term used for this process of review and checking.

17.4 ANALYSIS USING BIOINFORMATICS TOOLS

As mentioned earlier, sequences of DNA and proteins and other biological data can be analyzed by various computational tools. Some of the data analysis that can be carried out by the computational tools available in the NCBI and other sites are the following:

1. Gene Annotation (processing of raw information)—Experimentally-derived genomic sequences are analyzed using specialized programs to find out the possible genes and their coding regions or ORF. The protein that can be encoded by the gene and its biological functions can be predicted. Regulatory elements such as promoters and demarcation of the ORF into exons and introns can be carried out using specialized computer programs. The phylogenetic relationship of the protein or gene with other known molecules of the same type from other organisms can also be elucidated.
2. Genes present in DNA sequences can be predicted with the help of computer tools such as GenScan, GeneFinder in the case of eukaryotes, and GeneMark for bacterial genomes.
3. There are a number of efficient computer tools available for structural and functional genomic analysis. The DNA sequences can be translated into the corresponding amino acid sequences to find the possible polypeptide. This polypeptide sequence can be compared with the sequences that are present in the database, the biological function of which is clear, and thus biological function of the new protein can be predicted because of the structural and functional relationship.
4. Regulatory sequences present in genome sequences can be determined using computational tools specialized for that purpose.
5. Phylogenetic relationships can be understood very clearly by studying the homology and similarity between sequences present in the database and also with the new sequence elucidated. The computational methods for comparing the sequences can give valuable insights into the evolutionary sequences of macromolecules present in unrelated organisms.

Thus, with the help of databases and bioinformatics tools it is possible to predict the functions of unknown genes and the relationship with the functioning of other genes present in the genome.

REVIEW QUESTIONS

1. Describe how bioinformatics tools are used to analyze DNA sequence to find out a gene and its possible protein.
2. What are the various types of databases available for bioinformatics?
3. What are the DNA-sequence databases present in NCBI?
4. What are the conventions adopted for the storage of DNA-sequence data and protein-sequence data with regard to the direction of the sequences? Explain.
5. What are the major bioinformatics institutes that maintain public databases and computational tools for bioinformatics analysis?
6. What is the ESTs database? Explain its use in bioinformatics.
7. What are the single-letter IUPAC codes for arginine, lysine, tyrosine, phenylalanine, and glycine?
8. What are the IUPAC codes for (1) 'G' or 'C', (2) 'C' or 'T', (3) 'A' or 'G', (4) 'G' or 'A' or 'T', and (5) 'G' or 'A' or 'T' or 'C'?
9. What is the need for an IUPAC system in bioinformatics?
10. Describe the contribution made by Margaret Dayhoff toward the development of bioinformatics.
11. What is meant by PIR? Describe its origin.
12. What is BLAST? How is it used in bioinformatics analysis?

Part 6 CELL-CULTURE TECHNOLOGY

Biotechnology is not just about recombinant DNA technology, cloning, and genetics. But these are some of the techniques that changed the face of biotechnology forever. Whatever may be the techniques used in biotechnology, the ultimate aim is to make use of the ability of cells to produce certain valuable compounds or the genetic improvements of plants and animals for the benefit of mankind. But these cannot be accomplished without the most fundamental technique of biotechnology—cell-culture technology. In fact, classical biotechnology or fermentation technology is wholly dependent on cell-culture technology.

Cell-culture technology is the core of any biotechnology-based production process—be it vaccines, proteins, alcohol, organic acids, amino acids, vitamins, antibiotics, enzymatic- or cell-mediated biotransformation, biodegradation, bioremediation, or bio-fertilizers. The production process involves the respective cell-culture in a specialized bioreactor or fermentor and the application of the principles of biochemical engineering. The creation of transgenic animals and plants also involves *in vitro* culturing of the respective types of cells and tissues.

This part of the textbook discusses the various types and techniques of cell cultures and their applications in three chapters—Chapter 18, Chapter 19, and Chapter 20. Microbial culture and application is the subject of Chapter 18. It discusses various microbial culture techniques and growth kinetics, industrial application of microbial cultures and isolation of microbial products, strain isolation and improvements, and finally, bioethics in microbial technology. Chapter 19 is about the technique of plant cell and tissue cultures and their application in agriculture, industry, and other areas. This chapter discusses various methods of genetic engineering on plant systems systematically. This also includes the bioethics of plant biotechnology. The last chapter is about animal cell cultures and their various applications in medicines and animal biotechnology. It also discusses monoclonal-antibody production by hybridoma technology and stem-cell technology and its possible impact in medical biotechnology and gene therapy. It ends with a discussion of bioethics in animal-cell technology and genetic engineering.

Chapter 18

MICROBIAL CULTURE AND APPLICATIONS

In This Chapter

18.1 INTRODUCTION

Microbial populations dominate the biosphere in terms of metabolic impact and numbers. Among the various types of microbes, prokaryotes are the most pervasive lifeform on the planet, often tolerating extremes in pH, temperature, salt concentration, etc. Metabolic diversity is greater among prokaryotes than all eukaryotes combined. Men have long been utilizing the bacterial and yeast and fungal population for the manufacturing of various chemicals, biochemicals, antibiotics, beverages, etc. Production of antibiotics, alcohols, vinegar, amino acids, vitamins, therapeutic antibodies, acetone and other solvents, and recombinant proteins is accomplished by the large-scale cultivation of microbial cells such as bacteria, algae, yeast, and fungus on industrial scale. In all these industrial applications the metabolic activities or the biochemical pathways are used for the production of specific chemicals with the consumption of the substrates or a carbon source such as sucrose. Here, the microbial culture acts as a factory, where the substrate is the raw material. It is converted into the product and secreted into the media. The product can be recovered from the media with a process called downstream processing. There is a limitation for a single cell to convert the raw material into products in a given period of time. It is possible to calculate the rate of product formation by a single cell under a specific metabolic condition, if we know the quantity of product formed over a period of time and the number of cells in the culture. If we want to produce a specific quantity of the product over a period of time it is possible to calculate the number of bacterial or microbial cells required to operate the bioprocess on an industrial scale.

Like any other chemical equation this microbial-mediated biotransformation can also be considered a chemical reaction and can be expressed by a chemical equation:

$$C_wH_xO_yN_z + aO_2 + bH_gO_hN \longrightarrow cCHaObNd + dCO_2 + eH_2O$$

Even though the microbial conversion of a substrate can be compared to a chemical reaction, where the reactant is converted into a product, the efficiency of conversion will be comparatively less. A chemical reaction requires an appropriate temperature, pressure, pH, and solvent system for maximum product output. The microbial system also has to be provided with the optimum environmental and nutritional conditions such as temperature, pH, and the correct substrate for converting it into the product. The efficiency of microbial conversion of substrate into product is comparatively less because a major part of the metabolic energy is utilized for the generation of biomass by cell growth and multiplication.

18.2 MICROBIAL CULTURE TECHNIQUES

As stated previously, microbial cultures should be provided with the required chemical and physical environment for proper multiplication and physiological state, so that the cells can carry out the required bioconversion satisfactorily. The chemical environment of a microbial cell is its nutritional conditions in which it is growing. It also includes the correct pH and temperature.

Nutrients for Microbial Culture

Like any other living system, microorganisms also require a source of energy, carbon, nitrogen, oxygen, iron and other minerals, micronutrients, and water for growth, and multiplication. All these nutrients that are essential for the growth and multiplication of microbial organisms are supplied in the form of nutrient media. For laboratory-scale cultivation we may use certain costly media components, but for industrial purposes they should be economical and readily available. The media that we use for growing microbes may be synthetic, semi-synthetic, or completely natural. If the nutritive components of the media are not of natural origin, such nutritive media are known as **synthetic media**, which we can synthesize in the laboratory following certain recipes, mixing the required salts, minerals, and carbon source. There are a large number of commercially-available nutrient media, which contain both salts and minerals. Such nutrient media are known as **semi-**

synthetic media. For example, commercially available nutrient broth, trypticase soya broth (TSB), brain-heart infusion (BHI) broth, yeast extract, potato dextrose agar, casein digest, etc., are some examples of semi-synthetic media. For laboratory-scale cultivation of bacteria and other microorganisms, these synthetic or semi-synthetic media are preferred, but for industrial-scale cultivation these media are not recommended from an economical point of view. For commercial purposes, the recommended media should be cheap and available year round. The following are the minimum components required in a microbial medium for cultivation of microbes in a laboratory:

Carbon source. A simple carbon source, which is simple to use and easily available, can be used. Sugars such as glucose, lactose, sucrose, and complex polysaccharides such as starch, glycogen cellulose, a mixture of various carbohydrates, and other compounds such as cereal grain powders, cane molasses, etc., are usually used as carbon sources in microbial culture media. The main purpose of the carbon source is to provide energy and carbon skeleton for the synthesis of various other biological compounds.

Nitrogen sources. The major types of nitrogen sources used in culture media are ammonium salts, urea, animal tissue extracts, amino acid mixtures, and plant-tissue extracts.

Microelements or trace elements. Elements required in small amounts or in traces are to be added into the medium as salts in required amounts. The elements such as copper, cobalt, iron, zinc, manganese, magnesium, etc., are the microelements.

Growth factors. Growth factors are certain organic compounds that are essential for the growth and multiplication of cells, but cannot be synthesized by the cells. Such compounds should be supplemented in the medium. Certain amino acids and vitamins are also included in this category.

Anti-foams. This is not a nutritive component of the media. Media rich in nutritive components such as starch, protein, and other organic material and also the proteins and other products secreted by the growing cells can result in excessive foaming while the culture media is agitated for aeration. To prevent the formation of foam some anti-foaming agents are included in the media. Certain types of fatty acids such as olive oil and sunflower oil and silicones are commonly used in cell cultures as anti-foam agents.

Energy sources. The carbon sources used in culture media such as carbohydrates, sugars, proteins, lipids, etc., can work as energy sources for the growth and metabolism of the microbial cells.

Water. Water is the base of any culture media, whether it is liquid or solid. In solid culture media such as the media of solid state fermentation or agar media the quantity of water is comparatively less than liquid media. In laboratory experiments, single-distilled water or double-distilled water is usually used. But in large-scale microbial cultivation for industrial purposes, the pH and the dissolved salts present should be considered when formulating the media requirements and its concentration. Water is also required for a large number of other services in the laboratory such as cooling, heating, steaming, etc. Therefore, any laboratory should be provided with a source of clean water of consistent quality.

Culture Procedures

1. **Sterilization.** The media and culture vessel have to be sterilized to prevent the growth of unwanted microorganisms and thus contamination. If laboratory-scale experiments are carried out in 100 to 1,000 ml flasks, or in lesser volumes such as 50 ml or 10 ml, the media along with the culture flasks or vials can be steam-sterilized with an autoclave. Depending on the quantity of the materials autoclaved, the sterilization can be carried

out alternatively in a pressure cooker of convenient size. Steam-sterilization with an autoclave or pressure cooker is carried out at 120°C for 15 to 20 minutes under 15 psi pressure.

When microbes are cultivated in a fermentor for large-scale operation, it is convenient to sterilize the fermentor as a whole with or without media. Media may be sterilized separately or *in situ*, in the fermentor itself. Steam is used for the sterilization of the media and fermentor, by passing the steam through the sterilization jacket or the coil around the fermentor. When the fermentor is sterilized without media in it, steam can be sparged into the vessel through all openings, allowing it to exit very slowly. Sparging is a process by which sterile air or steam is allowed to pass through the medium in the vessel with the help of a sparging device placed at the bottom of the fermentor. The steam pressure is held at 15 psi for 20 to 30 minutes while circulating or holding the steam within the vessel or in the jacket (Figure 18.1).

2. **Environment for microbial growth.** The nutrient composition of the medium, the ionic concentration of salts, pH, and temperature influence the growth of microorganisms in the culture and its metabolic state. Most of the bacteria grow at neutral pH, where as yeast and fungi prefer acidic pH. Similarly, different organisms prefer different optimum temperatures for active growth and multiplication. The optimum temperature has to be maintained in the culture with the help of an incubator in the case of small-scale cultures and circulating water of the appropriate temperature through the jacket of the fermentor.
3. **Aeration and mixing.** Mixing of the broth is essential for the uniform distribution of the nutrients and the microbial population in the culture. Aeration is needed for the easy gas exchange between the medium and the environment. Aerated medium will be rich in oxygen. Aeration and mixing can be easily achieved by shaking the medium on a shaker in the case of small-scale cultures (shake flasks cultures). In large-scale cultivation in bioreactors the transfer of oxygen to organisms is very difficult because it requires proper mixing. In fermentors, the proper mixing of cells, media components, and oxygen is achieved by stirring the medium with the help of a mechanical stirrer with baffles attached to it. Baffles help in maintaining turbulence. Microbial-free air passed through the media ensures proper aeration, and this forced aeration also helps in the mixing of media, cells, and oxygen.

Microbial Culture Equipment

In the laboratory, microbial cells can be grown in tubes and vials, when the volume is five to ten ml, and in Erlenmeyer flasks when the volume is 100 to 1,000 ml. Improvements in the culturing of microbes can be done by making improvements in the design of the flasks and also by using shakers.

Baffle Flasks

Baffle flasks are the modified flasks for microbial cultivation, in which there are v-shaped notches or indentations in the sides of the flasks. The presence of baffles improves the efficiency of oxygen transfer and thereby the growth of microbes because the baffles increase the turbulence while the media is agitated on a shaker.

Shakers

Shakers are the special equipment designed for rotating a platform orbitaly, so that the culture flasks with media kept on the platform of the shaker will be continuously agitated. This agitation helps the medium to be homogeneous in cell-mass distribution, media components, and efficient oxygen transfer.

Fermentors

These are bioreactors used for the cultivation of microbial cells on large scale under controlled conditions for industrial purposes. This closed metallic or glass vessel has the adequate arrangement for aeration, mixing of media by agitation, temperature control, pH control, anti-foaming, control of overflow, sterilization of media and vessel, cooling, and sampling (removal of sample, while the fermentor is on). Agitation of the media in the bioreactor may be through stirring or aeration or both. This equipment is convenient for operation continuously for a number of days. The essential parts of a laboratory fermentor are given in Figure 18.1.

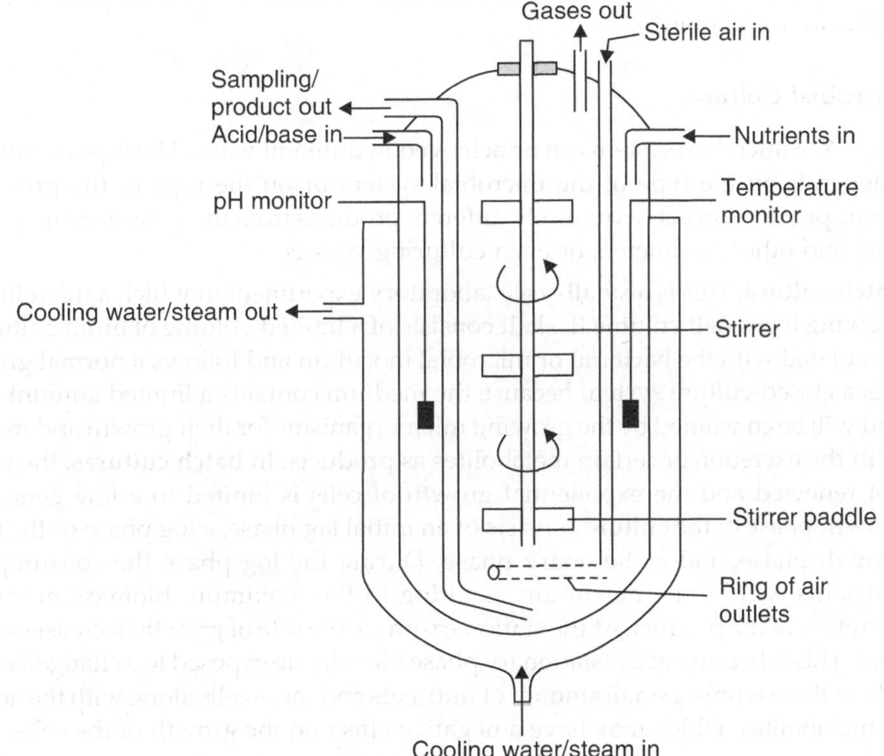

***FIGURE 18.1* Diagram sketch of a laboratory fermentor showing the essential components.**

As indicated in the figure, the bioreactors are provided with controls for monitoring and adjusting the many physical and chemical parameters such as temperature, pH, nutrient composition, foaming, etc. Maximum cell growth and product formation can be achieved by controlling these parameters that assist cell growth and metabolism leading to high output of the product. A stirred tank bioreactor is the most commonly used bioreactor for microbial cultivation, in which the microbial medium is stirred with an impeller. A high density of metabolically active cells in the medium can result in sudden depletion of dissolved oxygen creating an anaerobic condition in the medium. This can result in serious consequences in the quality of the product or even in the type of products formed in the fermentation reaction. Similarly, the cell growth and product formation can alter pH of the medium, which can also create problems in the further growth and metabolism of the cell cultures. Rapid growth also results in the depletion of essential nutrients that directly link to the growth and metabolism that causes the production of the product. All these changes are monitored by the accessories of the fermentor or bioreactor and are accordingly indicated or rectified automatically. For example, whenever there is a change in pH from the optimum value, automatically a sufficient amount of acid or alkali is added to the media to keep the optimum pH constant. Similarly, if there is foaming in the media the sensor will detect the foam formation and accordingly, the antifoam agent is delivered into the medium to prevent the foaming.

In addition to the industrial type of bioreactors or fermentors, there are fermentors of small volumes suitable for operating in the laboratories, known as **laboratory fermentors.** These laboratory fermentors are for 10 to 100 liters of volume and are used for optimizing culture conditions and nutritional parameters for better growth of cells and production of metabolites for conducting research studies in the laboratory.

Types of Microbial Cultures

The culturing of the microbial system can be achieved in different ways. The type of culture method sometimes depends on the type of the microbial system or on the type of the product that we expect. For example, one can get two entirely different products from the same organism by changing the nutritional and other parameters or even culturing vessels.

1. **Batch culture.** This is a small-scale laboratory experiment in which a microbial culture is growing in a small volume flask. It consists of a limited volume of broth culture in a flask inoculated with the bacterial or microbial inoculum and follows a normal growth phase. It is a closed-culture system because the medium contains a limited amount of nutrients and will be consumed by the growing microorganisms for their growth and multiplication with the excretion of certain metabolites as products. In **batch cultures**, the nutrients are not renewed and the exponential growth of cells is limited to a few generations. The growth phase of the culture consists of an initial lag phase, a log phase or the exponential growth phase, and a stationary phase. During the log phase the consumption of the nutrients will be the maximum resulting in the maximum biomass output with the excretion of the product. At the stationary phase the rate of growth decreases and becomes zero. This is because at the stationary phase the cells are exposed to a changed environment where there is only a small amount of nutrients and more cells along with the accumulation of metabolites, which may have a negative effect on the growth of the cells.
2. **Fed-batch culture.** The batch culture can be made into a semi-continuous culture or fed-batch culture by feeding it with fresh media sequentially at the end of the log phase or in the beginning of the stationary phase without removing cells. Because of this the volume of the culture will go on increasing as fresh media is added. This method is specially suited for cultures in which a high concentration of substrate is inhibitory to cell multiplication and biomass formation. In such situations the substrate can be fed at

low concentrations to achieve cell growth. This method can easily produce a high cell density in the culture medium, which may not be possible in a batch fermentor or shake flask culture. This is especially important when the product formation is intracellular to achieve maximum product output per biomass.

3. **Continuous culture.** Bacterial cultures can be maintained in a state of exponential growth over long periods of time using a system of **continuous culture,** designed to relieve the conditions that stop exponential growth in batch cultures. Continuous culture, in a device called a **chemostat**, can be used to maintain a bacterial population at a constant density, a situation that is, in many ways, more similar to bacterial growth in natural environments. This is a very convenient method to get continuous cell growth and product formation over a long period of time. In continuous culture, the nutrient medium including the raw material is supplied at a rate that is equal to the volume of media with cells and product displaced or removed from the culture. The volume removed and the volume added is the same. In effect there is no change in the net volume as well as the chemical environment of the culture.

   In a **chemostat**, the growth chamber is connected to a reservoir of sterile medium. Once the growth is initiated, fresh medium is continuously supplied from the reservoir (Figure 18.2).

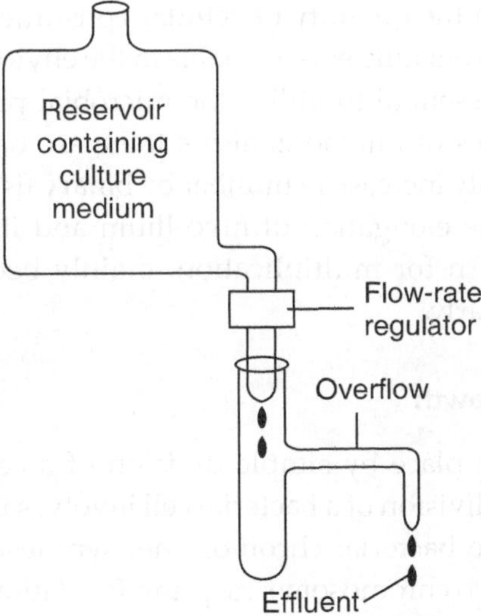

FIGURE 18.2 **Schematic diagram of a chemostat, a device used for the continuous culture of microbes. The chemostat relieves the environmental conditions that restrict growth by continuously supplying nutrients to cells and removing waste substances and spent cells from the culture medium.**

The volume of fluid in the growth chamber is maintained at a constant level by some sort of overflow drain. Fresh medium is allowed to enter into the growth chamber at a rate that limits the growth of the bacteria. The bacterial cells grow (cells are formed) at the same rate at which bacterial cells (and spent medium) are removed by the overflow. The rate of addition of the fresh medium determines the rate of growth because the fresh medium always contains a limiting amount of an essential nutrient. Thus, the chemostat relieves the insufficiency of nutrients, the accumulation of toxic substances, and the accumulation of excess cells in the culture, which are the parameters that initiate the stationary phase of the growth cycle. The bacterial culture can be grown and maintained at relatively constant conditions, depending on the flow rate of the nutrients.

If the chemical environment is constant in a chemostat continuous culture, the cell density is constant in a turbidostat culture, which is also a continuous culture. Since the culture is fed with the fresh medium at specific rate, a steady state of growth and metabolism is achieved. At a steady state, the cell multiplication and substrate consumption for growth and product formation occur at a fixed rate. The growth rate is maintained constantly. The formation of new biomass is balanced with the removal of cells from the outlet. Continuous culture is very suitable for the production of cell biomass and products, if it is excreted into the medium. It is widely used for the production of single-cell protein from liquid effluents as a byproduct of the waste treatment. The organic waste present in the effluent is converted into microbial biomass, which is known as single-cell proteins.

18.3 MEASUREMENT AND KINETICS OF MICROBIAL GROWTH

Growth is an orderly increase in the quantity of cellular constituents. It depends on the ability of the cell to form new protoplasm from nutrients available in the environment. A proper understanding regarding microbial growth is essential to utilize the microbial process to get maximum product output. Among the various types of microorganisms there are basically four general patterns of cell multiplication. Bacteria mainly increase in number by binary fission, yeast multiply by budding, fungi increas the biomass by the elongation of mycellium and its branching, and in the case of virus, there is no regular pattern for multiplication, mainly because its multiplication is host dependent and grow intracellularly.

Measurement of Microbial Growth

In bacteria, multiplication takes place by simple division of a cell into two by a process called **binary fission.** The growth and division of a bacterial cell involves increase in cell mass and number of ribosomes, duplication of the bacterial chromosome, synthesis of new cell wall and plasma membrane, partitioning of the two chromosomes, septum formation, and cell division. This asexual process of reproduction is called binary fission.

During this process, there is an orderly increase in cellular structures and components, replication and segregation of the bacterial DNA, and formation of a septum or cross wall, which divides the cell into two progeny cells. The process is coordinated by the bacterial membrane, perhaps by the mesosomes. The DNA molecule is believed to be attached to a point on the membrane where it is replicated. The two DNA molecules remain attached at points side-by-side on the membrane while new membrane material is synthesized between the two points. This draws the DNA molecules in opposite directions while new cell wall and membrane are laid down as a septum between the two chromosomal compartments. When septum formation is complete, the cell splits into two progeny cells. The time interval required for a bacterial cell to divide or for a population of bacterial cells to double is called the **generation time.** Generation time for bacterial species growing in nature may be as short as 15 minutes or as long as several days. For unicellular organisms such as bacteria, growth can be measured in terms of two different parameters: changes in **cell mass** and changes in **cell numbers**.

Growth Kinetics and Specific Growth Rate

When bacteria are grown in a closed system (also called a batch culture) such as a test tube, the population of cells almost always exhibits these growth dynamics: cells initially adjust to the medium (lag phase) until they can start dividing regularly in the exponential phase. When their growth becomes limited, the cells stop dividing (stationary phase), until eventually they show loss of viability (death phase). Growth is expressed as change in the number of viable cells versus time or the cell biomass versus time.

Growth of cells in a microbial culture under a steady state or balanced growth can be compared to a chemical reaction, in which the substrate is getting converted into products. In a microbial culture under a steady state the substrate is the nutrients and the product is the cell biomass. In such a reaction the rate of growth will be proportional to the cell biomass present in the culture. The cell culture behaves like an autocatalytic reaction. When the microbial culture is under a steady state the rate of increase in cell biomass dX/dt is equal to the product of specific growth and cell concentration (biomass concentration).

$$DX/dt = \mu X \tag{1}$$

where X is the cell concentration (gm/L) and μ is the specific growth rate (in $hour^{-1}$).

The specific growth rate $\mu = dX/dt \times 1/X$, which is an index of rate of growth of cells in those particular conditions. Specific growth rate can be determined by plotting dX/dt against X, the cell concentration, and determining the slope of the straight line. It is possible to calculate the generation time or the doubling time of the organism or bacterial cell, if we know the initial and final cell concentration of the culture and the specific growth rate.

The cell biomass at the starting of exponential growth is X and after time, t, it is 2X. Time required for doubling the biomass or the generation time of the cell can be calculated by following the above equation (Equation No. 1).

$$\ln 2X/t = \mu X$$

$$\ln 2X/X = \mu t$$

$$\ln 2 = \mu t$$

Therefore,

$$t = \ln 2/\mu$$

$$t = 0.693/\mu \tag{2}$$

Here t is the generation time or the doubling time of the cells. If the specific growth 'μ' of the cells is calculated it can be substituted in the above equation to determine the generation time or the doubling time. From this it is very clear that the doubling time and specific growth rate are inversely related. As the doubling time increases, the specific growth rate decreases. The microbial cell cultures usually have a high specific growth rate, because they have short doubling time or generation time.

The cell numbers of the microbial culture at different time intervals can be counted and this data can be used for calculating the specific growth rate as follows:

$$\ln 2X/X = \mu t \text{ or } \ln 2X - X = \mu (t - t_0)$$

on converting natural logarithm to logarithm to the base 10

$$\log_{10} 2X - \log_{10} X = \mu/2.303\ (t - t_0)$$

2X and X represent the amount of microbial cells at time t and t_0.

Suppose if the culture medium contains 10^4 cells/ml at time t_0, and after 4 hours (t) the number of cells is 10^8 cells/ml, then the specific growth rate of the culture is

$$\mu = (t - t_0)\ 2.303/\log_{10} (10^8 - 10^4) = (8 - 4)\ 2.303/4$$

$$= 4 \times 2.303/4 = 2.303 \text{ hr}^{-1}$$

Thus, the specific growth rate is 2.303 hr^{-1}

From equation 2, $t = 0.693/\mu$

$$t = 0.693/2.303$$

$$t = 0.3 \text{ hrs.}$$

$$0.3 \text{ hrs.} = 20 \text{ minutes}$$

This clearly illustrates that when the specific growth rate increases, the doubling time of the culture will decrease. The specific growth rate of microbial cells changes according to the change in the growth phase. Usually the specific growth of a culture is calculated during the steady state of growth, which is during the exponential phase of growth. Specific growth during the exponential phase or the log phase indicates the growth capacity of the culture in that particular environment because the specific growth of a culture is also influenced by environmental factors such as temperature, pH, nutrient composition of the medium, and other factors that affect the growth of the cells in the culture.

18.4 SCALE UP OF A MICROBIAL PROCESS

The scaling up of a small laboratory experiment into an industrial process has to address several issues, which have to be solved satisfactorily. The direct conversion of a laboratory-scale process into an industrial process is not advisable if it has to be operated economically. When a product is produced commercially with microbial cultures, large financial investments will be needed. In addition to this, there is the risk of creating products that cannot be commercialized because of the problems associated with quality control. This can be avoided by including an intermediate step or a validation step to validate the laboratory scale before attempting production on a commercial scale. The validation step is carried out in a pilot plant.

The pilot plant is a mini-version of the commercial scale plant in which the commercial and economical aspects, capital equipments, utilities such as steam, electricity consumption, transportation, fittings, labor, and market survey can be studied in a very systematic way. The data from these studies can be analyzed and its economical feasibility can be evaluated before going into the commercial scale.

The following example is given to illustrate the importance of scale-up calculations for the commercial application of a bioprocess in terms of reactor size.

In a pilot plant study it was observed that *e. coli* at a cell concentration of 1 g/L can produce 100 mg/L of insulin. Calculate the reactor size needed for producing 1 Kg of insulin in the following conditions:

(*a*) Cell concentration is 1 g/L and insulin formation is 100 mg/gm of cells.
(*b*) Cell concentration is 50 g/L and insulin production is 100 mg/gm of cells.
(*c*) Cell concentration is 50 g/L and insulin production is 500 mg/gm of cells.

The answers are as follows:

(*a*) Cell concentration is 1g/L. Insulin production is 100 mg/L. Target volume of insulin is 1 Kg.

The volume required for the production of 1 Kg of insulin is 1 Kg/100 mg/L 1000,000 mg/100 mg/L = 10,000 L.

(*b*) Cell biomass is 50 g/L. Therefore, insulin production is 50 × 100 = 5000 mg/L (5g/L).

The volume of cell culture required to produce 1 Kg of insulin is 1 Kg/ 5g/L 1000 g/5g/ L = 200 L.

So, the reactor volume required is 200 L.

(*c*) Cell biomass concentration is 50 g/L and the insulin production is 500 mg/L.

Insulin yield = 500 × 50 = 25,000 mg/L i.e., 25 g/L
Volume required to produce 1 Kg of insulin is 1 Kg/ 25 g = 1000g/25g/L = 40 L.
So we need a 40 L reactor or fermentor to produce 1 Kg of insulin if the cell concentration is 50 g/L and insulin production is 500 mg /gm of cells.

From the above examples it is evident that the reactor volume decreases as we have high cell density and high rate of insulin production per cell biomass or per volume of the culture. Based on these data and calculations we can design the reactor or fermentor of appropriate volume for industrial-scale production of any product.

18.5 ISOLATION OF MICROBIAL PRODUCTS

The process of separation and purification of the desired product from the microbial culture of a bioprocess is called **downstream processing.** This involves the separation of cells from the medium, if the product is present in the medium and includes purification techniques such as chromatography.

If the metabolites are produced intracellularly, then the procedure includes the disruption of the cell and extraction of the metabolites. The major and common steps of a typical downstream process can be classified as follows:

1. Separation of cells from the fermentation media.
2. Disruption of the cells and extraction of the product if the product is intracellular and extracted directly from the culture broth if the product is extracellular (present in the media).
3. Initial purification of metabolites, such as precipitation or a general chromatographic separation.
4. Product specific purification procedure or technique such as affinity chromatography or ion-exchange chromatography.
5. Polishing of the product, which involves the purification or concentration of the specific product to a high level of purity (98 to 100% purity) and making it ready for commercialization.

Figure 18.3 illustrates the different steps involved in downstream processing for the isolation and purification of a product from the media of a fermentation process. Depending on the type of product, the isolation and purification steps may vary involving more steps. But it is advisable to include fewer purification steps so that the cost of production of the product can be reduced and also the product loss during downstream processing can be reduced.

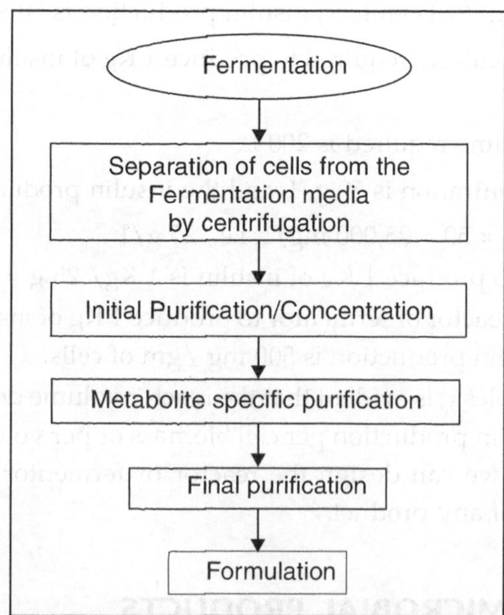

FIGURE 18.3 **Flow chart illustrating the different steps involved in the isolation and purification of a microbial metabolite.**

The first step of any downstream process is the separation of cells from the media of fermentation. This is generally achieved by centrifugation or by ultra-filtration. In some cultures, the cells rapidly settle down when aeration and agitation of the fermentation broth is stopped. This sedimentation of cells can be achieved rapidly by adding flocculating agents. In cultures where the cell settling doesn't occur, cell separation can be achieved by centrifugation. Alternatively, cells can be separated by ultra-filtration. In this method the culture is forcibly passed through a membrane filter of pore size 0.5 mm. The cells will be retained in the filter and the broth will be free of cells.

The cell-free media or broth contains microbial metabolites, extracellular enzymes, and other compounds. There are many methods to deal with the separation and purification of the desired product, whether it is an enzyme, a sugar, or an antibiotic. The methods include precipitation, solvent extraction, ion-exchange chromatography, etc. If the desired product is localized internally the procedure will also change. Figures 18.4 and 18.5 give the flow chart showing the major steps involved in the isolation and purification of products both from the media and cells. For example, the recovery of antibiotics (streptomycin, for example), in most cases, is secreted into the medium and follows the steps as illustrated in Figure 18.4.

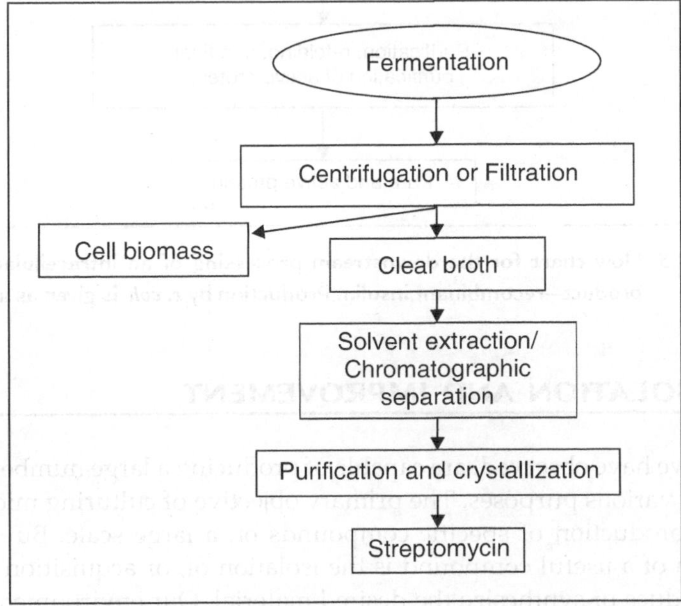

FIGURE 18.4 **Flow chart for the purification of extracellular microbial product. Production of streptomycin by *s.gresius* is given as an example.**

Some proteins produced by bacteria are intracellular and are not secreted into the culture media, particularly recombinant proteins. In such cases the downstream processing involves the breaking of cells followed by extraction and purification. Production of recombinant insulin by *e. coli* cell cultures is illustrated in Figure 18.5. Once the product is purified, it is concentrated to the required purity and stabilized by the required ingredients, known as **excipients,** and is ready for marketing after the quality evaluation.

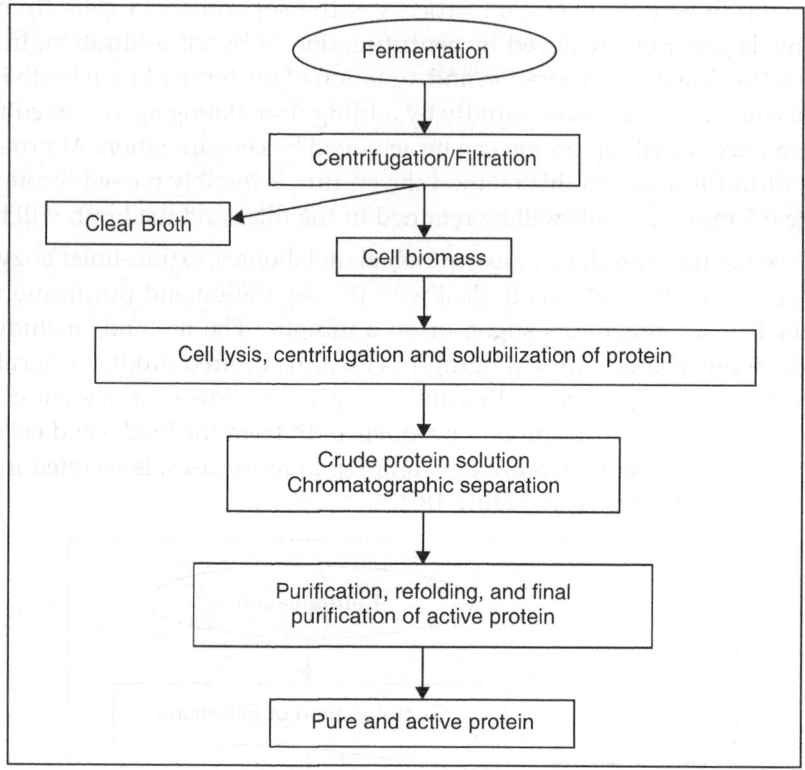

FIGURE 18.5 **Flow chart for the downstream processing of an intracellular microbial product—recombinant insulin. Production by *e. coli* is given as an example.**

18.6 STRAIN ISOLATION AND IMPROVEMENT

Microorganisms, as we have observed, are capable of producing a large number of useful chemicals and biochemicals for various purposes. The primary objective of culturing microbes is to use them industrially for the production of specific compounds on a large scale. But the first step in the microbial production of a useful compound is the isolation of, or acquisition of, the microbe that has the ability to produce or synthesize the desired material. Our environment, including the soil, air, and water is a rich source of various types of microorganisms. Therefore, to isolate a microorganism having a particular enzyme activity or capable of producing a compound, we have to search in the natural habitats and isolate them. Microorganisms, especially bacteria, also live in extreme habitats such as very cold environments, hot springs, low pH, and also in alkaline conditions. Thus, microbes of desired quality can be isolated from their natural environments. The desired quality of the organisms that we have isolated can be improved by a variety of methods including induced mutation.

Acquisition or Isolation

A bacterial strain can be obtained either by purchasing a commercial sample or from a known source. Otherwise, a new strain has to be isolated from its natural habitat using different screening techniques. The screening technique for isolating a specific microbe depends on the type of metabolic activity or the type of product. In some cases, the microbe can produce characteristic color change in the medium when an indicator compound is supplemented in the medium. To screen out an organism we have to use a selection medium, in which a specific component or condition is introduced to identify the required microorganism. In addition to this **direct-screening technique**, there is another technique for isolating microorganisms known as the **enrichment technique**. In this method the sample containing the microbes such as soil or water is inoculated in a nutrient liquid medium and allowed to grow on a shaker. The specific media composition and growth conditions such as temperature are provided in such a way that it promotes the growth or enrichment of the desired microorganism. The enriched culture can be sub-cultured in fresh medium using a small inoculum taken from the enriched culture. In this way the desired organism can be enriched to a higher degree and ultimately the culture will be of the desired organism. This enriched culture can be used for the isolation of pure culture by screening methods using its specific properties.

Screening of microorganisms can also be carried out directly with the help of antibodies that are specific for the microorganism. The antibody may be against some organism-specific compounds present in the cell wall or cell membrane or may be against some proteins or other compounds produced and secreted into the media. Instead of antibodies, organism-specific DNA probes or markers can be designed and used for the detection of specific organisms. Now this is one of the diagnostic techniques for the detection of certain diseases by detecting the presence of the pathogenic organism. Microarray techniques can be used for the detection and isolation of specific known microorganisms or organisms producing one or more desired products with the help of product (protein) specific DNA probes.

Strain Improvement

The isolation methods described above will provide a strain that is capable of producing a compound of interest. But it may not be able to produce the product in sufficient quantities so it may be difficult to develop an economically viable process. But the strain can be modified genetically to get the product in sufficient quantities. The techniques of molecular biology and classic genetics can be used for strain improvements to get the desired characteristics. The following are some of the techniques routinely used for the strain improvement:

Mutation

Selection of mutated strains or mutants is a widely used technique to get strains of improved qualities. Bacteria that can tolerate high temperature are an example. There is a natural occurrence of mutated strains in the habitat and the desired strain can be isolated directly. But mutations can be induced on exposing the cultures to some chemicals or physical mutagens and the mutated strains can be screened out from the cultures. Chemicals such as nitrosoguanidine (or NTG) or physical mutagens such as UV rays are the frequently used methods for creating mutation in a strain of bacteria or fungus. In most cases, it is necessary to carry out multiple successive mutations before getting the right kind of genetic variation among the organisms in a culture. After inducing the mutation, the right kind of organism with the desired characteristic is isolated with a screening technique. A high-yielding strain of *pencillium chrysogenum* has been developed from the wild strain by inducing several successive mutations. The new strain thus developed is capable of producing penicillin 100 times more concentrated than that of the wild strain. This high-yielding strain has helped to develop an economically viable process for commercial purposes.

Genetic Recombination

These are a group of natural processes that bring genetic elements from two different genomes into one unit resulting in a new genotype in the absence of mutations. There are generally three mechanisms of gene transfer between organisms. They are the following:

- Transformation
- Transduction—including generalized transduction and specialized transduction
- Conjugation

Once the foreign DNA is in the cell, a homologous region of donor DNA is cut out and ligated or joined to the recipient DNA. Pieces of DNA that the host cell recognizes as foreign are degraded by restriction endonucleases.

Genetic-Engineering Techniques

This is a set of tools (not a science) involving the manipulation of DNA outside the cell to create artificial genes or novel combinations of genes with predesigned control elements. A recombinant DNA generally has the following components—the gene of interest along with the appropriate type of control elements or promoters, the replicons or the origin of replication, and selection markers.

The gene of interest can be obtained through any one of the following methods— shotgun cloning, hybridization, reverse transcription of the mRNA, chemical synthesis of the gene if the sequence of the gene is known, or by polymerase chain reaction (PCR) if the gene specific primer is available.

Strain Preservation

Once you have isolated a strain of desired characteristics, it has to be properly preserved or stored. If the strain is not properly preserved, it may change its unique properties or they may decline over the course of time. The strain can be stored or preserved for future use through any of the following methods:

Storage on Agar

This is a routinely used method in all microbiology and molecular biology laboratories for storing cultures for a short duration. Cultures are grown and **stored on agar** slants or stabs and stored at 5 to 20°C. This type of culture can be stored for six months and has to be sub-cultured every six months. Sometimes it is advisable to over layer the culture with mineral oil and such cultures can be stored at least for one year. In this case, the cultures have to be sub-cultured every year.

Cryopreservation

Another very effective and simple method of preservation is **cryopreservation**. In this case the cultures are mixed with cryoprotective agents such as glycerol (10 to 30%) and are dispensed in small tubes or ampoules. These glycerinated cultures are stored in liquid nitrogen (– 176 to 196°C) for a very long time without any change. These can also be stored at – 80°C for a short duration of six months or one year.

Lyophilization

This is a method in which the cultures are dried at very low temperature. This process is called **freeze-drying or lyophilization**. This involves the sudden freezing of the culture followed by drying under a vacuum. In a vacuum, the water molecules will be sublimed and the dried cultures will remain viable for many years. Under this condition the cells are dormant and so can be stored even for ten years or more without any loss of viability.

Culture Collection Centers

These are some official centers that maintain strains of microorganisms that have been isolated and identified there or by others and deposited into these centers for maintaining the strains. Once a new strain has been isolated from the natural habitat or a new strain has been developed by mutation, it can be deposited into some of these official centers for evaluation or proper identification and preservation. These centers also provide strains of organisms for scientists who are involved in research activities. But considering the commercial potential of these cultures, there is stringent regulations for the use of culture strains for commercial purposes and the depositors of the cultures will be protected through intellectual property rights.

The major centers of microbial culture collection around the world are the American Type Culture Collection (ATCC), National Collection of Industrial bacteria of Britain (NCIB), and DSM, Germany (Deutsche Sammiung von Mikroorganismen and Zelkulturen, Germany).

18.7 APPLICATIONS OF MICROBIAL CULTURE TECHNOLOGY

As we discussed earlier, microbial cultures have a large number of applications. In our daily life they play an important role in different ways. Microbial cells can be used for the production of various substances, depending on the metabolic activities of the cells. The following are the major applications of microbial cultures for basic research and industrial applications:

1. Production of whole microbial cells such as food and vaccines.
2. Production of primary metabolites such as acids, alcohol, enzymes, and microbial polysaccharides.
3. Production of secondary metabolites such as antibiotics and biodegradable plastics.
4. Microbial leaching of metals, effluent and waste treatment.
5. Microbial cells such as agents of biotransformation of organic compounds.
6. As host cells for the production of recombinant proteins, gene cloning, and other molecular biology research.

The oldest use of microbial cultures is to produce fermented foods such as cheese and wine. Here, the whole cells are used. Whole cells are also used for the production of single-cell proteins and certain vaccines such as tetanus vaccines, typhoid vaccines, and tuberculosis vaccines. For the production of single-cell proteins selected microorganisms such as yeast and spirulina are allowed to grow in the culture (may be starch factory effluents) and the whole microbial cells are dried and used as food material both for human consumption and cattle feed. Production of organic acids, alcohol, various types of enzymes, and polysaccharides are examples for primary metabolites. Antibiotics and other types of organic molecules such as hydrocarbons are examples of secondary metabolites. Microbial extraction of metals such as copper and iron and treatment of liquid waste or effluents using microbial systems are examples where microbial metabolism is used to convert the waste into useful products and thereby the disposal of the waste without creating any environmental pollution. In modern biology, particularly in molecular biology and recombinant techniques, certain microbial cells such as *e.coli* are used as host cells for the cloning and expression of certain genes and the production of recombinant proteins. Production of human insulin in *e.coli* and the hepatitis B vaccine in yeast are good examples. Microbial production of vitamin C is another example of exploiting the primary metabolism for the production of useful compounds.

TABLE 18.1 Some microbial products and their organisms.

Microorganisms	Products
Saccharomyces cerevisiae	Ethanol and production recombinant proteins like hepatitis B surface antigens
Aspergillus niger	Citric acid
Aspergillus oryzae	Amylase
Pencillium chrysogenum	Penicillin
Streptomyces griseus	Streptomycin
Propionibacterium shormani	Vitamin B_{12}
Escherichia coli	Recombinant proteins
Alcaligenes eutrophus	Poly 3-hydroxybutyrate
Leuconostoc mesenteroides	Dextran

18.8 BIOETHICS IN MICROBIAL TECHNOLOGY

Biotechnology, like any other scientific research and development, has to be closely monitored and regulated to avoid its negative impacts on society and the environment. Biotechnology is a double-edged sword. This is also true in the case of microbial biotechnology. So the products and processes should be safe, and the required safety measures have to be implemented for industrial applications as well as in the laboratory. Some important safety considerations specific to microbial biotechnology are the following:

1. The microorganisms used in the laboratory and for industrial purposes may be pathogenic and may infect humans and animals. Special safety measures have to be implemented while working with such organisms. When handling genetically-altered organisms for

industrial purposes, special care should be taken to make sure that it is not allowed to leak into the environment. Therefore, clear information about the possible pathogenisity of microorganisms used in the studies should be available.

2. The microbial product should be evaluated for its toxicity and allergic activities before it is released into the market.
3. Disposal of waste, which also includes the spent microorganisms, should not cause any harm to the environment and lifeforms.
4. The strains of bacteria or any other microorganism should be maintained without causing any contamination or mutations.
5. The antibiotic-resistant bacterial cultures can increase the environmental pool of antibiotic-resistant organisms. This should not be promoted.

If a genetically-modified organism escapes, it can cause serious genetic changes in other organisms, including humans, and affect normal metabolisms. Generally Recognized As Safe (GRAS) organisms found in nature can change into potential hazards if genetically modified, which can create unpredictable and undesirable changes in the environment. Guidelines have been established to address these safety concerns of biotechnology. The genetically-engineered organism that is released into the environment for environmental cleaning has to be monitored regularly to see if it creates any problems in the environmental balance. The Environmental Protection Agency (EPA) is engaged with the safety aspects of genetically-modified organisms released into the environment. If a new genetically-modified organism has to be released into the environment, it has to be approved by the EPA.

REVIEW QUESTIONS

1. What is lyophilization? What is the advantage of using this technique?
2. How can microbial cultures be stored for a long time without affecting viability?
3. What are the different types of bioreactors used in industrial biotechnology?
4. How can large-volume fermentors be sterilized?
5. Differentiate between cryopreservation and lyophilization.
6. Explain different methods for improving the quality of an industrial microorganism.
7. What are the services provided by microbial culture collection centers?
8. What are the different methods available for the measurement of microbial growth in a culture?
9. What do GMP and GRAS mean?
10. Differentiate between batch culture and fed-batch culture.
11. How does the airlift reactor differ from a stirred-tank reactor?
12. Lyophilized cells are viable for several years. Explain.
13. What is the use of anti-foams in microbial media while using fermentors?
14. Draw a diagram of an industrial fermentor and mark the essential parts. Give the function or use of individual parts.
15. What is a chemostat culture?

Chapter 19

Plant Cell Culture and Applications

In This Chapter

19.1 INTRODUCTION

Tissue culture is the culture and maintenance of plant cells or organs in sterile, nutritionally and environmentally supportive conditions (*in vitro*). It has applications in research and commerce. In commercial settings, tissue culture is often referred to as micropropagation, which is really only one form of a set of techniques. Micropropagation refers to the production of whole plants from cell cultures derived from explants (the initial piece of tissue put into culture) of (usually) meristem cells. The basic principle of tissue culture is the totipotency of cells. A plant cell has all the genetic information for developing into a whole plant. Most of the early tissue culture experiments failed, at least in part, because isolated plant tissues are extremely prone to microbial attack by the inborne microbial population that compete with the plant tissue in the culture medium, without any resistance from the plant tissue. Bacteria, fungi, and other organisms, which can be resisted to some degree by a whole plant, can easily outcompete an isolated fragment of tissue from the plant in the relatively nutrient-rich environment of a culture flask. Therefore, it is necessary to remove competitor organisms from the culture and isolate it in aseptic conditions, most usually in a small flask or test tube. This is usually done by chemical surface sterilization of the explant with an agent such as bleach at a concentration and for a duration that will kill or remove pathogens without injuring the plant cells beyond recovery. The medium and culture flasks used must also be sterile.

It was G. Haberlandt who first tried, in 1902, to culture the isolated plant cell on an artificial nutrient medium. Even though he was not successful in growing and differentiating the cells, he did maintain the cells in living condition for a long time. So, he is considered the father of tissue culture. After that, during the period 1902 to 1930, attempts were made to culture isolated plant tissues and organs such as the root, shoot, apex, and embryos known as organ culture. During this period, different types of nutrient media were evaluated for the culture and differentiation of organs and tissues. Suitable nutrient media were developed for culturing the isolated embryos, anther, pollen, cells, and protoplast and regenerating the whole plants from the cultured cells, tissues, and organs for different plant species. During 1940 to 1970, lots of experiments were carried out to formulate suitable nutrient media and standardize culture conditions including various plant-hormone combinations on *in vitro* morphogenesis. With the advent of methods of genetic engineering in the 1980s, it became possible to transfer specific genes from other organisms including microbes and animals into plant cells and protoplasts and regenerate the whole plant— the transgenic plants. The introduction of **transgenic plants** or **genetically-modified (GM)** crop plants with improved agronomic characteristics is a real boost to agriculture and may be responsible for the second green revolution. In addition to the applications in agriculture, genetic engineering of plants has also helped to understand the basic mechanism of gene expression and its involvement in various activities such as morphogenesis and differentiation.

19.2 CELL- AND TISSUE-CULTURE TECHNIQUES

Basic Technique

Plant-tissue culture (micropropagation) is a technique that can be used for the mass multiplication of exotic plants. Plant-tissue culture is a tool which is used extensively in the nursery business and in plant biotechnology for the rapid production of many genetically-identical plants using relatively small amounts of space, supplies, and time.

Basically, the technique consists of taking a piece of a plant (such as a stem tip, node, meristem, embryo, or even a seed) and placing it in a sterile, (usually gel-based) nutrient medium where it multiplies. The formulation of the growth medium is changed depending on whether one is trying to produce somatic undifferentiated callus tissue, multiply the number of plantlets, grow roots, or multiply embryos for 'artificial seed.' The following are the major steps involved in plant-tissue culture:

1. Selection of explant or tissue.
2. Surface sterilization of the explant to remove the microflora present on the surface using disinfectants such as sodium hypochlorite or mercuric chloride, followed by repeated washing with sterile distilled water.
3. Inoculation of the surface sterilized explant of suitable size into the culture media under aseptic conditions inside a laminar air flow chamber.
4. The culture media should contain suitable nutrient components in right composition including the growth regulators.
5. Inoculated cultures are transferred to growth chambers or tissue-culture rooms having appropriate physical conditions such as the right temperature (26 to 28°C), relative humidity (50 to 60%), and fluorescent light (16 h photoperiod).
6. Plantlets are regenerated from the cultured tissues and cells.
7. Fully regenerated plants (with shoot and root) are transferred to the greenhouse and then to fields after proper acclimatization.

This is a standard method or protocol for the micropropagation of a plant species. But all these conditions have to be standardized for each and every plant species. Micropropagation allows the production of large number of genetically-identical plants from small pieces of the stock plant in relatively short periods of time. Depending on the species in question, the original tissue piece may be taken from shoot tip, leaf, lateral bud, stem, or root tissue (Figure 19.1 (A)). Once the plant is placed in tissue culture, proliferation of lateral buds and adventitious shoots (Figure 19.1 (B)) or the differentiation of shoots directly from callus (unorganized mass of cells) (Figure 19.2), results in the tremendous increase in the number of shoots available for rooting. Rooted **plantlets** of many species have been established in production situations and have been successfully grown either in containers or in field plantings. The two most important points that we have to bear in mind is that this methodology is a means of accelerated asexual propagation and that plants produced by these techniques respond similarly to any vegetatively propagated plant.

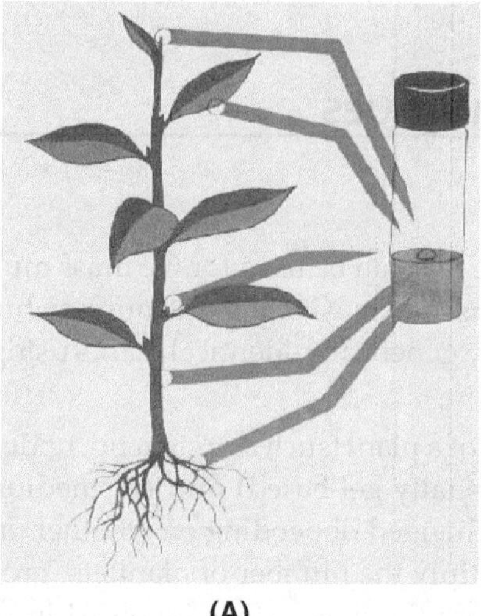
(A)

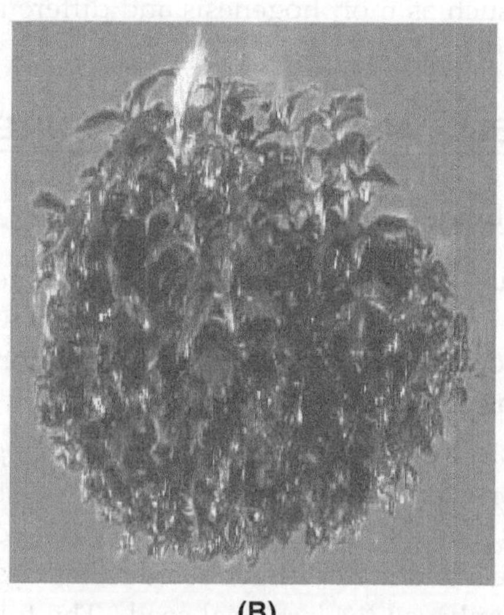
(B)

FIGURE 19.1 (A) The sources of explants for tissue culture. (B) Regeneration of plantlets.

Micropropagation offers several distinct advantages, which are not possible with conventional propagation techniques. A single explant can be multiplied into several thousand plants within a very short period of time. With most species, the taking of the original tissue explant does not destroy the parent plant. Once established, actively dividing cultures are a continuous source of microcuttings, which can result in plant production under greenhouse conditions without seasonal interruption. Using methods of micropropagation, the nursery can rapidly introduce selected superior clones of ornamental plants in sufficient quantities to have an impact on the landscape plant market.

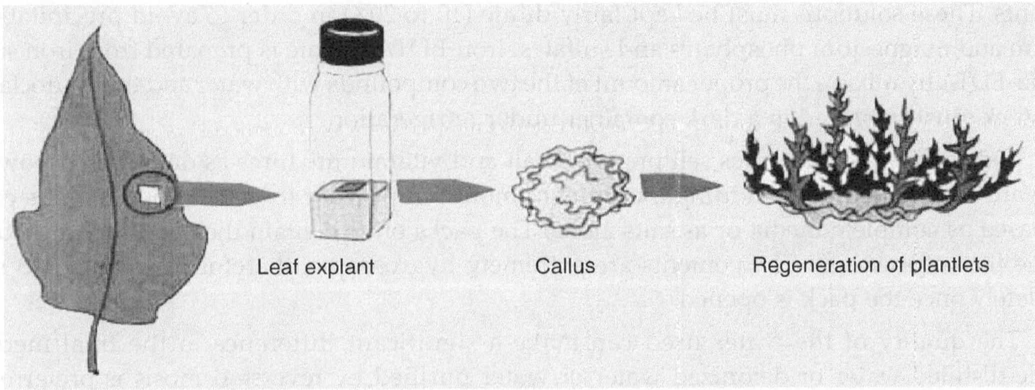

FIGURE 19.2 **Regeneration of plantlets from callus cultures.**

Nutrient Media

When a small portion of a plant is isolated, it is no longer able to receive nutrients or hormones from the plant, and these must be provided to allow growth *in vitro*. In addition, the culture must be provided with the ability to excrete the waste products of cell metabolism. This is accomplished by culturing on or in a defined culture medium. The medium must be periodically replenished. A controlled environment is needed for the growth of cells *in vitro*. Tissue cultures, sustained by the nutritive medium and confined in a protective vessel, require a stable and suitable climate. Thus, light and temperature must be more carefully regulated than in the case of the whole plant.

A typical plant-tissue culture media consists of all the macro- and micro-nutrients in addition to a carbon source, vitamins, and growth regulators. The macro- and micro-nutrients are generally supplemented as inorganic salts. The usual carbon source in tissue culture media is sucrose and the important vitamins included are nicotinic acid, thiamine, and pyridoxine. In certain media, amino acids such as glycine or arginine are also supplemented. These ingredients constitute the **basal medium**. The growth regulators are supplemented to the basal medium as per the requirements of the culture conditions and the aim of the experiments. The important growth regulators supplemented in tissue-culture media are auxins such as indole acetic acid (IAA), indole butyric acid (IBA), and 2, 4–D, cytokinins such as kinetin and gibberellins such as gibberellic acid. These growth regulators are supplemented in appropriate concentrations individually or in combinations. The concentrations and combinations of growth regulators have key roles in the growth and differentiation of cultured cells and tissues. Another important organic compound present in all tissue culture media is *myo*-inositol. In some cases, the media will be supplemented with tissue extracts such as coconut milk, tomato juice, yeast extract, casein hydrolysate, or malt extracts to meet special purposes. The media compositions and the special supplements including the hormonal composition for a particular plant species or tissue type are determined by trial and error methods. There are a number of basal-media compositions available for plant-tissue cultures, specified for different types of plants. The most extensively used basal media is the **Murashige and Skoog (MS)** medium developed by Murashige and Skoog in 1962.

In the old days, mineral-salt mixtures were prepared as stock solutions ranging from 10 to 100 times the final concentrations used in the medium. Stock solutions can be prepared as two solutions, one containing all of the macro-nutrients and the other containing all of the micro-nutrients. These solutions must be kept fairly dilute (10 to 20X) in order to avoid precipitation of calcium and magnesium phosphates and sulfates. Iron-EDTA chelate is prepared from iron sulfate and Na-EDTA by mixing the proper amount of the two compounds with water and then autoclaving. This stock must be stored in a dark container under refrigeration.

Today, several companies sell prepared salt and vitamin mixtures as dehydrated powders. These are easily handled by adding the proper amount of powder to water. The mixtures can be purchased as complete media or as salts alone. The packs often contain the necessary ingredients for one liter of medium. The contents are extremely hygroscopic, therefore it is better to use it completely once the pack is opened.

The quality of the water used can make a significant difference in the final medium. Double-distilled water or deionized water or water purified by reverse osmosis is preferred for making media. The pH of most culture media is adjusted to 5.7 ± 0.1 before autoclaving. The pH can influence the solubility of ions in nutrient media.

Pouring and Storing Media

Once media is made and the pH adjusted, it may be heated to dissolve the gelling agent and dispensed into culture vessels, such as tubes, baby food jars, or magenta boxes, and then autoclaved, if no filter-sterilized components need to be added. Agar is the gelling agent or the support for the solid medium. It is a polysaccharide obtained from a red algae, *gelidium amansii*. Media should be stored in the refrigerator or at least in the dark (light causes some reagents to break down). Under good conditions, most media can be stored for at least a month.

Types of Cultures

Organ cultures. Culturing isolated organs or tissues such as roots, stem, or leaf in an artificial media under controlled conditions are known as **organ culture**. Depending on the type of organs or tissue used for establishing the culture, organ cultures are named accordingly. The following are the various types of organ culture and its specific purpose:

Seed culture. Increasing the efficiency of germination of seeds that are difficult to germinate *in vivo*, precocious germination by application of plant-growth regulators, and production of clean seedlings for explants or meristem culture.

Embryo culture. Overcoming embryo abortion due to incompatibility barriers, overcoming seed dormancy and self-sterility of seeds, and embryo rescue in distant (interspecific or intergeneric) hybridization where endosperm development is poor, shortening of breeding cycle, etc.

Ovary or ovule culture. A common explant for the initiation of somatic embryogenic cultures, for the production of haploid plants, overcoming abortion of embryos of wide hybrids at very early stages of development due to incompatibility barriers, and *in vitro* fertilization for the production of distant hybrids avoiding style and stigmatic incompatibility that inhibits pollen germination and pollen tube growth.

Anther and microspore culture. Production of haploid plants, production of homozygous diploid lines through chromosome doubling, thus reducing the time required to produce inbred lines, and for uncovering mutations or recessive phenotypes.

Explant Culture

Explant culture is actually the tissue culture. Culturing of any excised tissue or plant parts such as leaf tissue, stem parts, cotyledon, hypocotyls, root parts, etc., is called explant culture. The primary purpose of explant culturing is to induce callus cultures or to regenerate whole plantlets directly from it without the formation of callus. Shoot apical meristem culture is an example, and its important uses are the following: Production of virus-free germplasm or plantlets, mass production of desirable genotypes, facilitation of exchange between locations (production of clean material), and cryopreservation (cold storage) or *in vitro* conservation of germplasm, etc., are the main purposes of meristem or shoot apex culture.

Callus Culture

Callus represents an unorganized or undifferentiated mass of cells. They are generally composed of parenchymatous cells and usually undergo division. When an explant is cultured in a medium supplemented with sufficient amount of auxins, it starts producing mass of cells from the surface of the explant (Figure 19.3). The concentration of auxins required for each type of explant will be different and is mainly dependent on the physiological state of the explant tissue. Callus cultures can be maintained for a very long time by intermittent sub-culturing to a fresh medium. The callus cultures can be manipulated for different purposes by changing the hormone concentrations in the media. Callus cultures can be used for regeneration of plantlets, preparation of single cells or suspension cultures, or for protoplasts preparation. Callus cultures can also be used for genetic transformation studies. In some instances, it is necessary to go through a callus phase prior to regeneration via somatic embryogenesis or organogenesis. Callus cultures are suitable for the generation of useful somaclonal variants (genetic or epigenetic) and can be used for *in vitro* selection of cells and tissue variants.

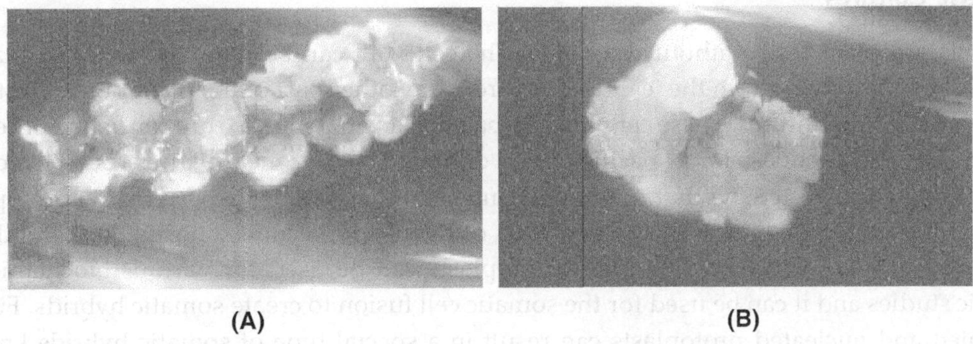

(A) (B)

FIGURE 19.3 **(A) Callus initiation on an explant. (B) Callus initiation of a growing callus culture.**

Cell-Suspension Cultures

Single-cell cultures and suspension cultures can be established from callus cultures by transferring a piece of callus tissue into liquid medium and subjecting it to continuous shaking. The growth rate of the suspension-cultured cells is generally higher than that of the solid culture. The former is more desirable, particularly for the production of useful metabolites on a large scale. A piece of the callus is transferred to a liquid medium in a vessel such as an Erlenmeyer flask and the vessel placed on a rotary or reciprocal shaker. The culture conditions depend on plant species and other factors, but in general, the cells are cultivated at 100 rpm on a rotary shaker at 25°C. By subculturing for several generations, a fine cell suspension culture containing small-cell aggregates and single cells is established. The time required to establish the cell-suspension culture varies greatly and depends on the tissue of the plant species and the medium composition. The cells in suspension are also used for a large-scale culture with jar-fermentors and tanks. The suspension cultures can be grown either as batch cultures or as continuous cultures for producing phytochemicals.

Enzymatic methods can also be adopted for establishing a fine cell-suspension culture. This is based on the use of certain pectin digesting enzymes in the culture medium, such as pectinase or macerozyme. These enzymes act on the pectin, which joins two adjacent cells in plant tissues, so that the cells become independent and grow freely as single cells. The cell-suspension cultures can be used for inducing somatic embryogenesis and the preparation of artificial seeds, induction of somatic mutation, and selection of mutants by screening the cells just like microbial cultures. The main application of plant cell-suspension cultures is that it can be used for the bioproduction of certain important phytochemicals or secondary metabolites by applying the principle of biochemical engineering. The suspension cultures can be cultivated in specially designed bioreactors known as **airlift bioreactors** for the **mass-cell cultures** for the production of plant secondary metabolites on industrial scale. Normal bioreactors with mechanical stirrer cannot be used in plant-cell cultures because it can result in the breaking of cells and thereby the cell viability can be drastically reduced. Whereas the airlift bioreactor can provide both stirring and air inflow to meet the high demand of oxygen.

The cells can also be used for genetic transformation experiments to produce transgenic plants.

Protoplast Cultures

Protoplasts are plant cells without cell walls. The cell wall can be removed with an enzymatic method. The cells may be from the leaf tissue or from any other part of the plant or may be the cells from the suspension cultures. These cells are incubated in an enzyme mixture consisting of cellulase, hemicellulase, and pectinase for a specific period of time. The enzyme mixture acts on the cell wall and is completely digested, so that the underlying cell membrane is exposed. This protoplast on culturing in a proper medium will regenerate its cell wall and become a normal cell and then can regenerate into a whole plant. The plant protoplasts can be used for various biochemical and metabolic studies and it can be used for the somatic cell fusion to create somatic hybrids. Fusion of aenucleated and nucleated protoplasts can result in a special type of somatic hybrids known as cybrids, in which there is no fusion of nucleus, but fusion in between the cytoplasm. Protoplasts can also be used for genetic transformation studies with biolistic methods, electroporation techniques, by PEG-mediated DNA transfer or by direct injection of DNA into the nucleus of the protoplast using micro-syringes.

Plant Regeneration Pathways

There are two methods of whole plant regeneration: **organogenesis** and **somatic embryogenesis**.

Organogenesis

This is a major path of regeneration that involves the differentiation of culture cells or callus tissue into organs such as shoot and roots. Plant regeneration through the formation of shoots and roots is known as plant regeneration through organogenesis. Organogenesis can occur directly from the explants depending on the hormonal combination of the medium and the physiological state of the explants (Figure 19.4). Miller and Skoog demonstrated that the initial formation of roots or shoots on the cultured callus or explant tissue depends on the relative concentration of auxins and cytokinins in the culture media. Medium supplemented with relatively high auxin concentration will promote root formation on the explants and high cytokinin concentration will promote shoot differentiation. In tissue-culture practices there may be three types of medium in relative combinations of auxins and cytokinins, which promote either the shoot formation or root formation or both simultaneously. In the latter case, you can get the complete plantlets, having both shoot and roots, which can be directly transferred to the pots in the greenhouse. Whereas in other cases, after the formation of shoots, individual shoots are transferred to the rooting medium, which promote root formation. The rooted plantlets can be transferred to a greenhouse for acclimatization. Plant regeneration through organogenesis is commonly used for mass multiplication, for micropropagation, and for conservation of germplasm at either normal or sub-zero temperatures (cryopreservation).

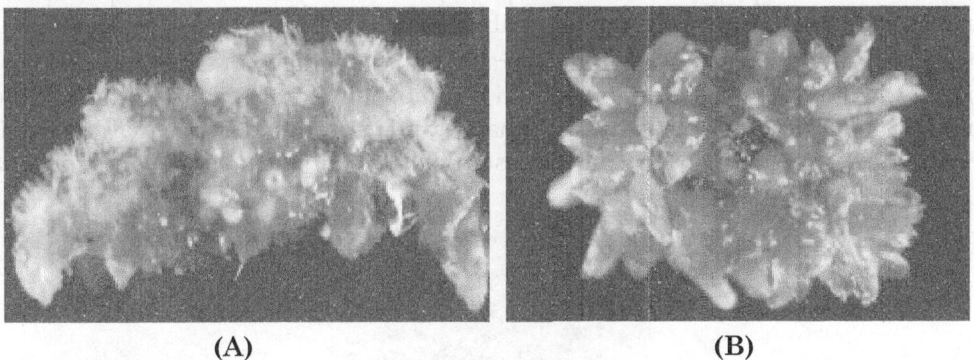

FIGURE 19.4 **(A) Shoot regeneration in callus culture . (B) Root regeneration (organogenesis) in callus culture.**

Somatic Embryogenesis

This is another major path of regeneration and development of plantlets for micropropagation or mass multiplication of specific plants. The cells, under a particular hormonal combination, change into the physiological state similar to zygotes (somatic zygotes) and follow an embryonic path of development to form somatic embryos (Figure 19.5 (A)). These somatic embryos are similar to normal embryos (seed embryos) developed from zygotes formed by sexual fertilization (Figure 19.5 (B)). The somatic embryos can develop into a complete plant. Since somatic embryos can germinate into a complete plant, these can be used for the production of artificial seeds. Somatic embryos developed by tissue or cell cultures can be entrapped in certain inert polymers such as calcium alginate and used as artificial seeds. Since the production of artificial seed is amenable to mechanization and for bioreactors, it can be produced in large numbers.

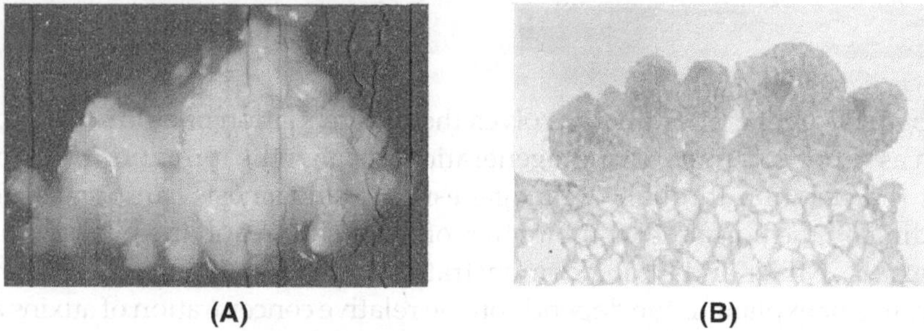

(A) (B)

FIGURE 19.5 **Somatic embryogenesis in callus cultures. Embryoids are seen all over the surface of the callus (A). To the right is a microscopic preparation, a cross-section through an embryoid (B).**

19.3 APPLICATIONS OF CELL AND TISSUE CULTURE

Micropropagation

This is a tissue-culture method developed for the mass multiplication and propagation of agricultural and horticultural plants as per the demand. The plant parts usually used for this type of propagation are apical meristem or auxiliary buds or the seedlings obtained by the *in vitro* germination of seeds.

Since this method of micropropagation is rapid and convenient for generating a large number of genetically-identical plantlets it has been used commercially for the mass multiplication of crop plants such as cardamom, apples, eucalyptus, bananas, and many other ornamental and horticultural plants (Figure 19.6). The following are the main steps involved in the method of micropropagation:

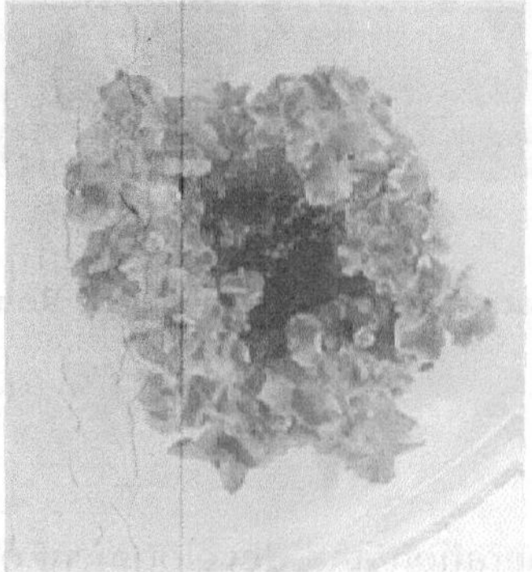

FIGURE 19.6 **Regenerated plantlets ready for transfer to the greenhouse.**

Step 1. Selection of the Leaves

Leaves, stems, or any other suitable plant part may be cut from healthy plants. The explants should be cut to suitable sizes. Old tissue parts should be avoided. Wash the dust off the tissue in a beaker of distilled water, which should be surface sterilized.

Step 2. Surface Sterilization and Preparation of the Explants

This part of the procedure should be carried out in a sterile working area, or with a meticulous aseptic technique. The leaf or any other tissue should be immersed in 70% ethanol for 30 seconds, and then transferred to a sterile petri dish. Sterile scissors and forceps are then used to cut the tissues into small suitable sizes of explants. The explants are transferred into a 10% hypochlorite bleach solution or mercuric chloride for five minutes, gently agitating once or twice during this time. They are then washed thoroughly by immersing in four successive beakers of sterile distilled water, leaving them for two to three minutes in each.

Step 3. Initiation of Culture

A suitable explant such as terminal meristem or auxiliary is used for initiating the culture. The surface sterilized explant is implanted into a suitable nutrient medium with the correct type of growth hormones in appropriate concentrations. Depending on the nutrient conditions and hormone combination, the explant either starts developing callus or directly produces shoot buds and roots to form a large number of plantlets. If callus is formed, it is transferred to another medium having the hormone combination favoring multiple shoot formation. The shoots are then transferred to another medium, which favors the root initiation. The rooted shoots are now ready for transfer to the pots for hardening under greenhouse conditions.

Shoot formation. If the cultures develop multiple shoots, they are excised and transferred to a rooting medium.

Root formation. Shoots that are transferred to the rooting medium develop roots and are ready for being transferred to the greenhouse for hardening.

Step 4. Transplantation and Hardening

The plantlets that develop roots can be transferred to special pots with sterilized sand for hardening under greenhouse conditions. **Hardening** is a process by which the *in vitro* generated plants are acclimatized to the greenhouse conditions. Then they can be transferred to the field.

Recently, the mass multiplication of plants through micropropagation has been automated and robotized to increase the efficiency and speed of the methods of micropropagation of agricultural and horticultural plants. Micropropagation methods for a large number of useful plants including forest trees and other medicinal herbs are now available.

(A)

(B)

FIGURE 19.7 **Plants regenerated from calli. (A) The plantlets with roots are still on agar. (B) The same plant kept for hardening in a pot.**

Virus-Free Plants

One of the main purposes of propagation through tissue culture is to produce virus-free plants. This is very important to increase yield and quality. Since meristems such as apical meristem and auxiliary meristem are free from viral infections these tissues can be used as explants for mass multiplication to get a large number of virus-free plants. Agricultural plants, which have a high risk of damage due to viral diseases include plants such as bananas, sugarcane, potatoes, tapioca, apples, etc.

Meristem culture is usually practiced to eliminate viruses and to make virus-free plants. Tissue culture using meristem is sometimes referred to as 'meristem culture' or 'meristemming,' and the plantlets produced are known as 'mericlones.'

Artificial Seeds

Artificial seeds are the somatic embryos encapsulated by certain inert polymeric compounds such as alginate. They are also very useful in the mass propagation of agricultural and hybrid varieties.

Embryo Rescue

Tissue culture also provides the means to overcome reproductive isolating barriers between distantly related wild relatives of crops through embryo rescue and *in vitro* fertilization. It is very difficult to produce hybrid plants by interspecific and intergeneric crosses. Such hybridizations can result in abnormal endosperm, which causes premature death of embryos resulting in the formation of sterile seeds. The embryo from such hybrid plants can be excised and cultured on a suitable nutrient to produce new hybrids, which may not be possible by other methods. This method of protecting the embryo viability and its development *in vitro* is called **embryo rescue**.

Haploids and Triploids

The culture of immature anthers and pollen grains is done to induce the pollen grains to develop into multicellular forms, particularly into embryos, with half the normal number of chromosomes for the species. When such haploid embryos are treated with chromosome-doubling agents (e.g., colchicine) their normal chromosome number is restored (and thus their fertility). Thus, it is a very easy method for the development of homozygous genotypes or pure lines. Pure line formation is a natural tendency for self-fertilizing species. In the case of cross-fertilizing species pure lines can be obtained by repeated in-breeding for 10 or more generations. The generation of pure lines and haploids is very helpful for analyzing the expression of recessive characteristics, which cannot be expressed in heterozygous conditions.

Triploids are raised by culturing endosperm tissues and are useful in generating triploid plants, which are sterile. Therefore, it is very useful for creating seedless plants that are desirable in fruit crops such as grapes, citrus fruits, apples, pears, etc.

Somatic Hybrids and Cybrids

Somatic hybrids are the hybrid plants developed from the fusion products of two somatic cells, which is very helpful for producing distant hybrids with desirable characteristics. Protoplasts

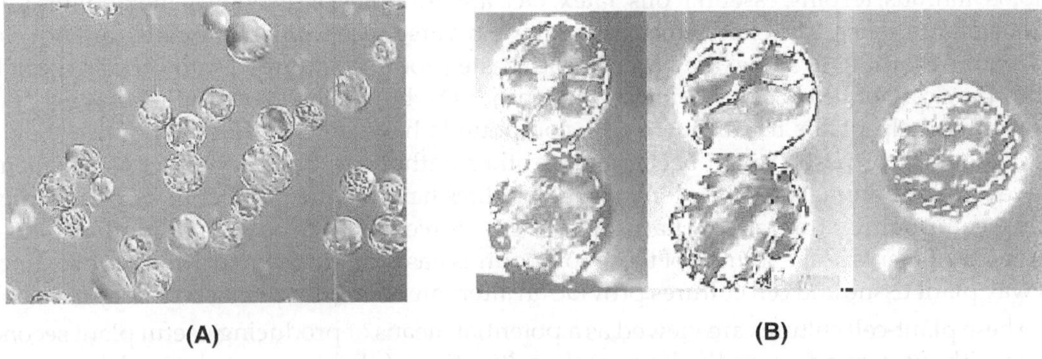

FIGURE 19.8 **(A) Protoplasts isolated from leaves and (B) different stages of protoplast fusion mediated by electric force.**

isolated from two distantly-related plants can be fused together to produce a cybrid or a hybrid protoplast. The hybrid protoplasts can be selected visually and cultured separately to develop into a callus and then differentiated into a complete plant, resulting in a somatic hybrid. This type of hybridization of distantly-related plants through sexual fertilization is never possible in nature.

In the beginning, there were two steps for the preparation of protoplasts from plant tissues.

First, the middle lamellae that cement the two adjacent cells in a tissue was dissolved by pectinase, and followed by the removal of the cell wall by cellulase. Nowadays, the method is shortened to just one operation in which a mixture of enzymes is applied. Due to the high osmotic pressure within the protoplast, isotonic media are used for the protoplast isolation. The enzyme mixture in the isotonic solution consists of pectinase, cellulase, and hemicellulase. When the protoplasts are cultured, a new wall is synthesized, thus restoring the original state of the plant cell. These protoplasts can be used for introducing foreign genetic materials and also for somatic hybridization, as stated earlier. Somatic hybridization is also known as **parasexual hybridization**. The fusion of isolated protoplasts can be carried out with chemical agents such as PEG (polyethylene glycol) or by electrical force, a method called electrofusion (Figure 19.8 (B)). Somatic hybridization can also be carried out between plants that cannot be hybridized because of some sexual incompatibility problems, even though they are closely related.

Carlson and his group[4] produced the first somatic hybrid of two different species—*nicotiana glauca* and *nicotiana langsdorfii*—in 1972. In 1978, Melchers[5], et al developed for the first time a hybrid between plants of two different genus—*solanum tuberosum* (potato) and *lycopersicon esculentum* (tomato). The somatic hybrid thus developed was named 'pomatoes' or 'topatoes'.

Plant Secondary Metabolites

Plants are the source of a large number of chemicals belonging to different classes and uses. Plant-derived chemicals can be divided mainly into two classes—primary metabolites and secondary metabolites. Chemicals such as proteins, amino acids, sugars, lipids, nucleic acids, and their derivatives essential for the basic metabolic activities are called **primary metabolites**. **Secondary metabolites** are certain chemicals that are formed as by-products of the primary metabolites, but are not essential for keeping the normal state of functioning of the cells. Chemicals such as alkaloids, flavanoids, tannins, terpins, essential oils, latex, etc., are examples of secondary metabolites present in various plant tissues. The exact functions of these diverse secondary metabolites are not clearly understood. According to some, they are simple waste products, dumped into certain specialized cells. But they cannot be considered mere waste products, but instead compounds having specific functions, that help plants to survive. Some compounds have defense functions since they fight against the attack of pests, fungi, bacteria, and other pathogens. Some even act as repellents to keep away grazing animals. Plant secondary metabolites have a large number of uses as medicines, flavoring compounds, fragrances, natural rubber, sources of hydrocarbons, biofuels, etc. The extensive use of plants as the sources of these compounds has brought them to the verge of extinction. In this way plant tissue and cell cultures provide an alternate source for plant secondary metabolites.

These plant-cell cultures are viewed as a potential means of producing useful plant secondary products without depending on the large-scale cultivation of these source plants. A large number of problems associated with the cultivation of these plants can also be overcome by adopting cell-culture techniques for the production of secondary metabolites. These problems include: environmental factors (drought, floods, etc.), disease, political and labor instabilities in the producing countries, uncontrollable variations in the crop quality, inability of authorities to prevent crop adulteration, losses in storage and handling, etc. Thus, the production of useful and valuable secondary metabolites in large bioreactors by cell cultures of specific plant species is an attractive alternative.

However, this technology is still being developed and despite the advantages outlined above, there are a variety of problems to be overcome before it can be adopted on a wide scale for the production of useful plant secondary metabolites. The success of Mitsui Petrochemical Industry Co. Ltd., in Japan, in producing shikonin on a commercial scale from *lithospermum erythrorhizon* cultivations and that of Nitto Denko Co. Ltd., also in Japan, in mass production of *panax ginseng* or ginseng cells using 20 kL tanks have demonstrated that many of the problems can be overcome with perseverance. The economic feasibility of these processes is another question that has to be handled carefully. In theory, it is anticipated that large-scale suspension cultures will be suitable for industrial production of useful plant chemicals such as pharmaceuticals and food additives, in a manner similar to that of microbial fermentation. Nevertheless, there are some significant differences between microbial- and plant-cell cultures that must be considered when attempting to apply them to the available technology. Table 19.1 shows a comparison of some of the characteristics of plant and microbial cultures of relevance to fermentation.

TABLE 19.1 Characteristics of microbial and plant cells relevant to fermentation.

Characteristics	Microorganism	Plant Cell
Size	2 μ	>10 μ
Shear stress	Insensitive	Sensitive
Water content	75%	>90%
Duplication time	<1 hour	days
Aeration	1 to 2 vvm	0.3 vvm
Fermentation time	Days	Weeks
Product accumulation	Medium	Vacuole
Production phase	Uncoupled	Often growth-linked
Mutation	Possible	Requires haploids

Table 19.2 gives the important plant secondary metabolites produced through plant cell suspension cultures.

TABLE 19.2 Secondary metabolites produced in high levels by plant-cell cultures.

Compound	Plant Species	Uses
Shikonin	*Lithospermum erythrorhizon*	Red pigment, used as lipstick, anti-microbial
Ginsenoside	*Panax ginseng*	Stress tonic
Anthraquinones	*Morinda citrifolia*	Anti-bacterial anti-inflammatory
Ajmalicine	*Catharanthus roseus*	Anti-carcinogenic
Scopalamine	*Datura stramonium*	Anti-hypertensive
Taxol	*Taxus buccata*	Anti-carcinogenic
Diosgenin	*Dioscorea deltoides*	Anti-fertility
Berberine	*Coptis japonica*	Anti-inflammatory, anti-bacterial
Berberine	*Thalictrum minor*	Anti-inflammatory, anti-bacterial
Berberine	*Coptis japonica*	Anti-inflammatory, anti-bacterial
Vincristine	*Cathranthus roseus*	Anti-carcinogenic
Digoxin	*Digitalis lanata*	Cardiac tonic
Azadirachtin	*Azadirachta indica*	Insecticide, anti-bacterial
Quinine	*Cinchona officinalis*	Anti-malarial
Artemisin	*Artemisia spp.*	Anti-malarial
Codeine	*Papaver somniferum*	Analgesic

Somaclonal Variation

The advantage of micropropagation for mass multiplication of a plant species is that it can provide a large number of genetically-identical plants in large numbers. But it has been observed that it is not true in the case of plants which are regenerated from the callus cultures maintained for very long periods and also from cell-suspension cultures. Such plants show variations and such variations observed during these asexual means of propagation are called **somaclonal variations**. These variations are often associated with chromosomal changes, and therefore are due to genetic changes occurring during the callus cultures. The somaclonal variations have potential applications in agriculture and crop improvement by producing somaclonal mutants with desirable characters.

The term **somaclones** was proposed by Larkin and Scowcroft[6] in 1981, for the plants that were produced from a tissue (somatic cells) isolated from a plant variety and the variation observed among the somaclones was termed **somaclonal variation.** If the cell or tissue used for the generation of clones has a gametic origin such as pollen or ovule, the clones are termed **gametoclones** and the variation seen among them is termed **gametoclonal variation.** Since long-term cell-suspension cultures are prone to somaclonal variation, they are used for the isolation of cell variants just like microbial cultures.

In Vitro Plant Germplasm Conservation

Germplasm collection and conservation is one of the major tasks of many research establishments as a means for the conservation of biological diversity. Tissue-culture methods are used for the conservation of plant varieties for future studies and plant-breeding purposes. Germplasm is the plant genotype representing whole genetic materials and that of its related species. Primitive crops and their wild varieties are very important as they may have unique and valuable genes, which are agriculturally very important. Because of the extensive use of modern agricultural varieties and hybrids, primitive crops and their wild relatives are on the path to extinction. The best method of protecting them and to prevent the erosion of plant genetic resources is to conserve them through tissue culture. Germplasm conservation can be carried out by collecting and preserving genetic resources by conventional methods in the form of seeds, vegetative propagules, etc. This collection of plant seeds and propagules is done in ***in vivo* gene banks**. Collection and conservation of any plant parts or tissue through tissue culture is done in ***in vitro* gene banks**. The conventional method of germplasm conservation has several limitations. Germplasm cannot be conserved for a very long time. The seeds are short lived and have demerits of seed dormancy, seed-borne diseases, and germplasm needs high input of labor, space, and cost. The *in vitro* approaches through cell and tissue-culture can overcome many of these inconveniences and problems, with the following tissue-culture-mediated conservation methods:

Cryopreservation. This method is the free storage of plant tissues and cells under frozen conditions. It is a long-term preservation of selected cells and tissues at very low temperature, equal to the temperature of liquid nitrogen (– 196°C). The tissues can be stored at this temperature indefinitely with the help of cryoprotectants such as glycerol, mannitol, dimethylsulfoxide, proline, etc. Cryopreservation has proved to be an efficient conservation method for the long-term storage of genetic stocks without any loss of viability and mutations.

Cold Storage. This method of storage of plant parts and tissues for conservation is suited for short term or medium term. The germplasm can be stored in the form of meristems such as shoot tips or auxiliary buds or other similar parts under low temperatures ranging from 4 to 15°C under nutrient limitations. The disadvantage is the tissue cannot be stored indefinitely.

19.4 GENE-TRANSFER METHODS IN PLANTS

The totipotency of plant cells and the ability of isolated single plant cells to develop into a complete plant, along with the tremendous progress that has been made in molecular biology to isolate and characterize a specific gene and its cloning have helped the development of plant genetic transformation methods. Isolation of the desired genes and construction of a suitable vector consisting of the gene of interest attached to the right type of promoter elements for its proper expression in the respective plant tissue is a prerequisite for the genetic transformation of crop plants. There are a number of genetic transformation methods developed for the production of transgenic plants. Genetic transformation methods in plants can be divided mainly into two classes—vector-mediated gene transfer and direct gene transfer or vector-less gene transfer.

Vector-Mediated or Indirect Gene Transfer

Among the various types of gene transfer methods, agrobacterium-mediated gene transfer is the most efficient, because it is a natural method of genetic engineering. The bacterium *agrobacterium tumefaciens* (AT) causes certain plants to develop a tumor, called a **gall**, located most commonly at the crown of the plant—the junction of root and stem. This "crown gall disease" affects broad-leafed plants such as fruit trees, grapes, and ornamental plants.

The disease has been known since the turn of the last century, when scientists determined that AT isolated from a crown gall and inoculated into a wound in a healthy plant would cause the latter to develop a tumor as well. Scientists suspected by the early 1970s that the causative agent was nucleic acid in some mobile form that could be transferred from the bacterium to the plant and incorporated into the plant cells, probably carried by a virus or a plasmid. A plasmid is a small piece of "auxiliary" double-stranded DNA, separate from the regular host chromosomes, carrying special genes that may confer survival advantages or other characteristic traits such as antibiotic resistance, pathogenisity, or the ability to metabolize rare compounds.

In the late 1970s, microbiologists Mary-Dell Chilton[7], Eugene W. Nester, and Milton P. Gordon were the first to prove that it was DNA from a plasmid in *agrobacterium* that was transferred and integrated into the plant DNA to cause crown gall tumors. Furthermore, they identified the parts of the plasmid DNA involved in the tumor development. The plasmid that was responsible for the tumor was named **Ti plasmid**. The unique biological characteristics of this plasmid make it a natural agent for gene transfer. The plasmid consists of certain group of genes, which were cut from the plasmid and inserted into the plant genome. The host plant then maintains that foreign genetic material, replicating it along with the host plant's own genes, which are transcribed to produce proteins ultimately. These proteins are responsible for the synthesis of plant hormones that cause the uncontrolled cell division leading to the formation of the tumor. T-DNA also has the genes for the synthesis of metabolites for the growth and multiplication of *agrobacterium* living in the plant tissue. This part of the plasmid is known as **T-DNA**. Its borders are marked by certain specific DNA sequences, which can be recognized by the enzyme system that cleaves the T-DNA and inserts it into the plant chromosomes resulting in the formation of the crown gall tumor. This natural case of genetic engineering of plants is of major importance to biology because it demonstrates that DNA can be transferred between biological kingdoms. This knowledge is now exploited for the transfer of new valuable characteristics from other organisms such as bacteria and animals to plants, apart from transferring genes between plants. Thus, *agrobacterium* has become the major vehicle for genetic engineering of plants. The cycle of infection by agrobacterium and the formation of the crown gall disease are explained in Figure 19.9.

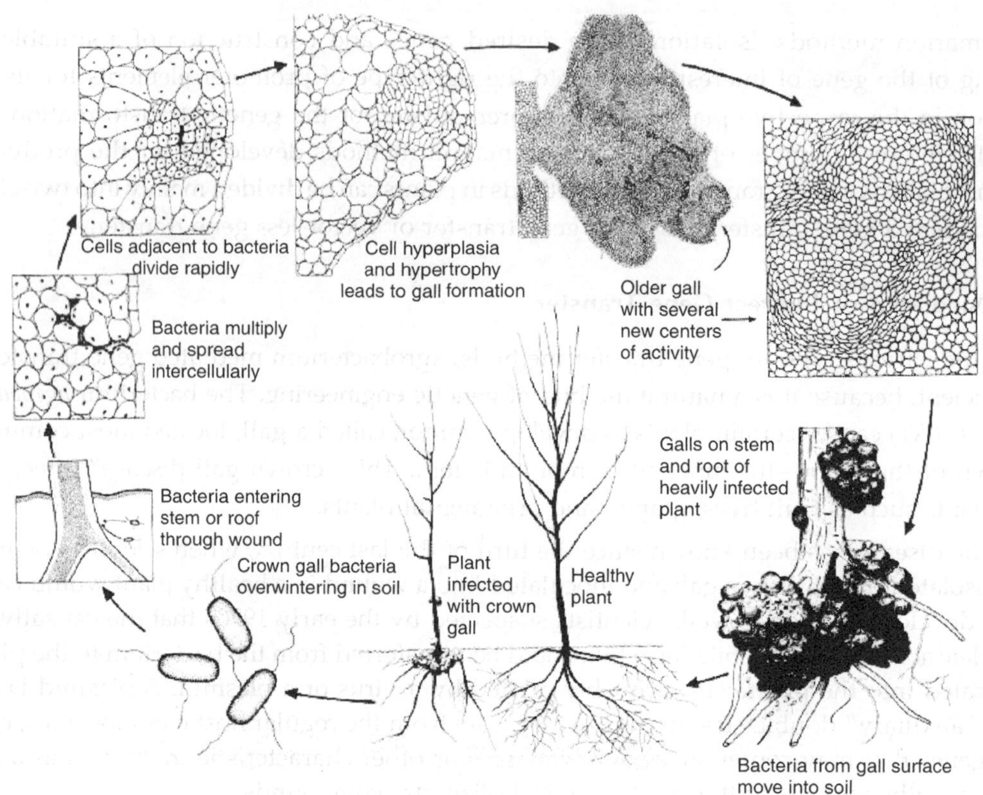

FIGURE 19.9 **Cycle of infection by agrobacterium tumefaciens and formation of crown gall tumor formation in plants.**

There is another species of agrobacterium, *a. rhizogenes* that induces extensive hairy root formation at the site of its infection. They harbor root-inducing plasmids (Ri plasmids) instead of Ti plasmids. The Ti plasmids (tumor-inducing plasmid) can be disarmed by removing the *t*-DNA parts keeping the borders intact and can also be used as vectors. Genes of interest can be cloned in place of *t*-DNA, and agrobacterium can now introduce this foreign DNA into the plant gene mistaking it for *t*-DNA. The molecular basis of crown gall disease and its modification for genetic engineering in plants is illustrated in Figure 19.10.

The general strategy would therefore be to design a 'recombinant' plasmid containing the desired gene that codes for a desired trait that is to be transferred to the plant. The gene may be for resistance to herbicides, for example, or the ability to synthesize a specific protein of commercial interest, such as an antibody or hormone. Next, the recombinant plasmid is transferred into the *agrobacterium*. Finally, the bacteria containing these modified Ti plasmids (without t-DNA) are inoculated into the host plant and these bacteria go about their business of inserting plasmid DNA fragments into the host plant genome. The first genetically engineered food now available in supermarkets, the *'Flavor-Savr'* tomato, owes its existence to gene transfer by *agrobacterium*. Plant tissue such as terminal meristems, or leaf discs, or cultured plant cells are co-cultivated or incubated with the *a. tumefaciens* having the recombinant plasmids, allowing the bacterium to infect the tissue or cells. The infected tissues are allowed to grow on shoot regeneration medium. During this time the transfer of genes into the plant genome takes place. The transformed tissues are now transferred to another selection medium, which promotes only the growth of transformed tissues and prevents the growth of *agrobacterium*. This medium is usually a shoot-inducing medium supplemented with one or two antibiotics that eliminate both bacteria and untransformed plant cells. The transformation can also be detected by the expression of some reporter genes linked along with the gene of interest like that of luciferase. The transformed tissues will develop shoots and can be transferred to a root-inducing medium, in which it develops roots. The completely developed transgenic plants can be transferred to soil after proper acclimatization or hardening. The transgenic plants can be evaluated

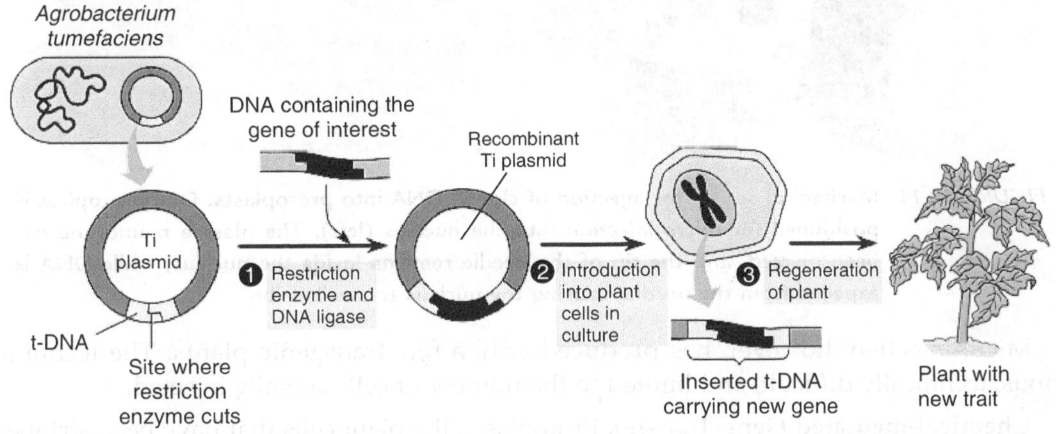

FIGURE 19.10 **Diagram explaining the Ti plasmid mediated (Agrobacterium mediated) genetic transformation of plants.**

for the presence of the foreign gene and its proper expression can be analyzed by different sets of molecular techniques such as Southern hybridization, Northern hybridization, PCR, and also by enzyme assays. Once the transgenic plant is completely evaluated for its possible biological activities and its impact on environment and other ethical and safety issues, it can be recommended for release to farmers or companies.

Direct Gene Transfer (Physical Methods of Gene Transfer)

One of the major problems in agrobacterium- or vector-mediated gene transfer is the host specificity shown by the concerned infectious agent and the difficulty in plant regeneration after the gene transfer through infection. Agrobacterium cannot infect most of monocot plants. In such cases alternate gene transfer techniques have been developed, which don't involve infection by any organism. The following are some of the direct gene transfer methods developed for the genetic engineering of specific plants:

Microinjection. The desired gene in the form of plasmid or alone is injected directly into the plant protoplast. The DNA may be directly delivered into the nucleus or in the cytoplasm. Injection into the nucleus is more efficient than that into the cytoplasm. Micro-injection is carried out with automatic equipments (robotics) using a micro-needle (Figure 19.11).

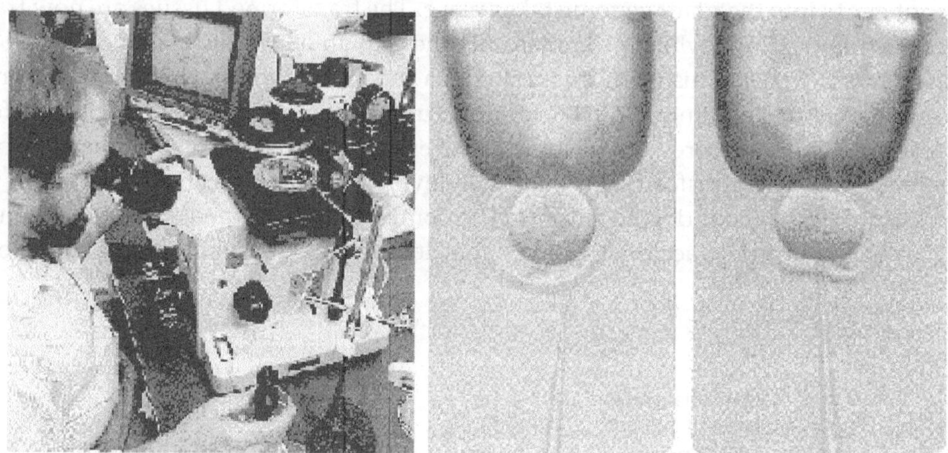

FIGURE 19.11 **Mechanical set up for injection of cloned DNA into protoplasts. One protoplast is positioned for micro-injection into the nucleus (left). The plasma membrane has been pierced, and the tip of the needle remains inside the nucleus, while DNA is expelled from the needle, causing the nucleus to swell visibly.**

Micro-injection, however, has produced only a few transgenic plants. The technique is laborious, technically difficult, and limited to the number of cells actually injected.

Chemical-mediated Gene Transfer. Protoplasts, the plant cells that have been stripped off their protective cell walls using a cocktail of enzymes, will take up pure DNA when treated with certain membrane-active agents such as polyethylene glycol (PEG) and dextran sulphate. Once inside the cell, the DNA is integrated into the genome and the foreign gene will express. This technique largely depends on the development of protoplast systems that retain the capacity to regenerate intact plants.

Electroporation. This is a physical technique developed for DNA delivery into plant protoplasts through membranes under a rapid pulse of high-voltage direct current. When an electric pulse of high voltage is applied in the presence of protoplasts, it makes a large number of temporary pores in the cell membrane. These temporary openings facilitate the DNA uptake by the protoplasts. Once inside the cell, the DNA is integrated and the foreign gene will express.

All these three techniques—chemical-mediated DNA uptake, micro-injection, and electroporation, are largely dependent on the development of protoplast systems that retain the capacity to regenerate intact plants. Transgenic corn, rice, and soybeans have been produced with these techniques, especially electroporation. Success rates, however, are low, and the techniques are not very reproducible in some cases.

Biolistic Method (Microprojectile or Particle-gun Method). This is a new method that involves accelerating very small particles of tungsten or gold coated with DNA into cells using an electrostatic pulse, air pressure, or gunpowder percussion. The DNA molecules containing the desired gene are coated on to microscopic gold or tungsten particles. As the particles pass through the cell, the DNA dissolves and becomes free to integrate into the plant-cell genome. This improbable technique actually works quite well and has become, along with electroporation, one of the methodologies of choice. Biolistics has the advantage of being applicable to whole cells in suspension or to intact or sliced plant tissues. For example, plant meristems or tissues capable of regeneration can be targeted directly. Unlike electroporation and microinjection, this technique does not require protoplasts or even single-cell isolations. Using biolistics, transgenic corn and soybean plants have been produced that contain heritable copies of the inserted gene.

This is also a highly mechanized or robotic mediated technique in which the speed of the micro-projectile particles is controlled by certain specific mechanisms such as high voltage electric pulse, air pressure, or gunpowder percussion.

Transgenic Analysis

The selection of transformed cells, whether plant or animal or microbial cells, is an important step in genetic-engineering procedures. Usually it is carried out by including one or more marker genes, for example, antibiotic-resistance genes along with the desired gene. The transformed cells can be selected on a special medium known as **selection medium** supplemented with the right type of antibiotics. The expression of the transferred gene (desired gene) can be monitored with the help of a reporter gene such as luciferase attached with the gene of interest. The gene integration and its expression can also be monitored and studied by Southern hybridization, Northern hybridization, Western blot analysis, PCR, and assay of the respective enzymes.

19.5 TRANSGENIC PLANTS WITH BENEFICIAL TRAITS

Only a few genes of agronomic importance have been inserted into plants: genes conferring resistance to certain insects and viruses and also those conferring tolerance to broad-spectrum herbicides. The latter result in increased herbicide specificity, allowing the farmer to use more effective, environmentally safe chemical agents. More recently, a gene has been introduced into a tomato that delays over- ripening and prolongs the shelf life of the fruit. Other traits of interest include those associated with grain quality. Genes to increase the content of amino acids such as lysine, methionine, and tryptophan in seeds will increase the nutritional value, thereby decreasing the need for amending grains with costly feed supplements. Table 19.3 gives the list of transgenic crops developed in the United States and approved by the United States departments of food, agriculture, and drugs.

TABLE 19.3 Examples of transgenic crops developed by different Agro-industrial companies.

Transgenic Crop	Genes Introduced	New Trait Developed	Company
Soybeans	EPSP Synthase	Weed control	Monsanto
Cotton	Acetolactate synthase Nitrilase EPSP synthase Bt Cryl A(c)	Weed control Weed control Weed control Insect control	DuPont Calgene Monsanto Monsanto
Corn	EPSP synthase Bt Ceyl A(c)	Weed control Insect resistance	Monsanto, Ciba Geigy, Monsanto, Dekalb genetics

Canola	Thioesterase EPSP synthase PAT & Barnase	High laurate oil Weed control Weed control	Calgene Monsanto AgrEvo
Rice	Genes of pro vitamin A synthesis	Pro vitamin A in the grains	Monsanto
Papaya	Coat protein	Virus resistance	Cornell University University of Hawaii
Potatoes	Bt Cryl IIA Coat protein	Insect resistance Viral resistance	Monsanto
Tomatoes	Bt Cryl A(b) Antisense PG	Insect resistance Delayed ripening	Monsanto Calgene
Sweet potatoes	Gmfad2-1	Improved oil	DuPont

Genetic Engineering for Stress Tolerance

Agricultural crops are always under various types of biotic and abiotic stress, which results in a great loss of productivity. By controlling these biotic and abiotic stresses, an ideal agricultural condition can be provided with substantial increases in agricultural productivity. Biotechnological approaches can reduce the use of toxic chemicals as pesticides and herbicides and thereby reduce environmental problems substantially.

Herbicide Tolerance

One of the major problems in agriculture is the growth of weeds. Weeds are the unwanted herbs growing along with crop plants. These plants compete with the crop plants for space, light, water, and nutrients, and thus crop productivity is reduced. There are a number of strategies for weed control. The most common approach is to use herbicides to control the weeds. But these herbicides can reduce the crop productivity if the crop is not resistant to the herbicide used. Agricultural plants can be made herbicide-tolerant through genetic engineering. Herbicide-degrading enzymes are introduced into the crop plants either in the chloroplasts or the enzymes toward the chloroplasts. These plants become insensitive to herbicides. **Herbicide-tolerant crops** allow farmers to apply a specific herbicide to control weeds without any harm to the crop. It gives farmers greater flexibility in pest management and promotes conservation effectively.

Insect Resistance

Various types of insects and pests always attack plants because leaves form their food source. These pests can damage agricultural plants and reduce productivity substantially. One common pest management strategy is to apply pesticides to kill the insects and pests. These chemicals may also be toxic to humans and other animals and can also pollute water and the environment. These toxic chemicals enter into the food chain and cause ecological magnification in those organisms, which are present at the terminals of the food chain. Ultimately, it can create deleterious health problems for human and animals. In these circumstances, transgenic plants provide an effective solution for pest management and environmental protection by creating insect-resistant crops. Pest management through insect-resistant plants is an eco-friendly, sustainable, and effective approach.

The first insect-resistant genes used for making insect resistant crops were the *Cry* genes or the Bt genes isolated from *bacillus thuringiensis*, a pathogen that kills certain group of insects. These Bt proteins or the bacteria are not harmful to humans or any other animals including flies, butterflies, silkworms, etc. A large number of crops such as cotton, potatoes, rice, maize, wheat, tomatoes, soybeans, wheat, brinjal, cauliflower, cabbage, etc., have been made transgenic with the Bt toxin gene. These transgenic crops were found to be very effective in controlling pests and it significantly increased productivity along with a considerable reduction in pesticide contamination in the environment. The latest example of cultivating an insect-resistant crop is the Bt cotton plant, the cotton plant transformed with the Bt toxin genes, the *CryI A*(c) gene (Figure 19.12). It is resistant to the ballworm (*helicoperpa armigera*), which causes great damage to cotton cultivation all over the world.

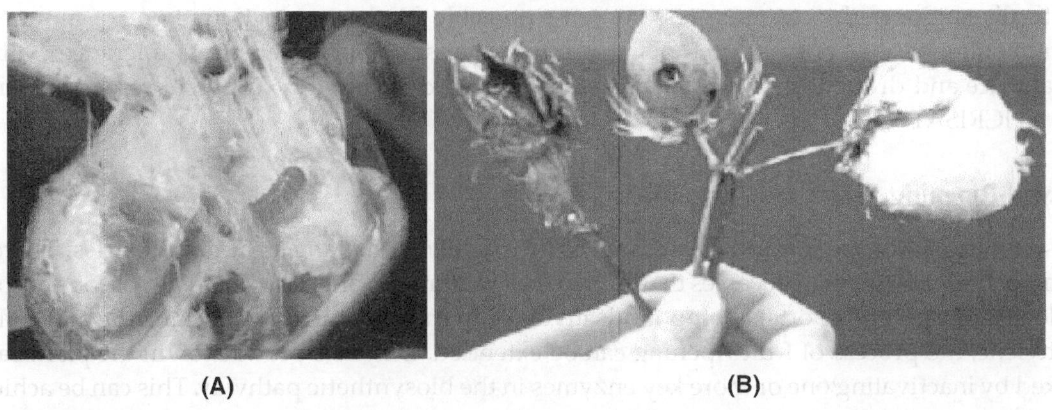

(A) **(B)**

FIGURE 19.12 **(A) Cotton infected with pink ballworm. (B) Results of insect infestation on Bt (right) and non-Bt (left) cotton.**

Disease Resistance

Plants are prone to all types of disease that include both pathogenic and deficiency diseases. Deficiency diseases can be managed by supplementing the correct type of manures. But the diseases due to pathogens cause destruction of crops and financial loss. A major pathogenic attack is by fungi, viruses, and bacteria resulting in the reduction of productivity or complete damage of the crops.

Biotechnological approaches can be used for solving this problem. There are several disease-resistant crops engineered to fight the attack of a virus. These plants are engineered with viral coat proteins. This virus-derived resistance has shown promising results in crops such as potatoes, tobacco, tomatoes, papayas, etc. Transgenic papayas, resistant to the ring spot virus have been commercialized in some countries.

When a plant is attacked by a bacterium or fungus it will respond by producing certain compounds including proteins and enzymes, which can act against the invading pathogens. The proteins produced in plants due to a pathogenic attack are known as **pathogenesis-related proteins or PR proteins**. They may include enzymes such as chitinase and other antifungal proteins. The organic compounds produced in plants due to the attack of pathogens and other injuries developed due to insect bites are known as **phytoalexins**. A large number of transgenic crops with increased resistance toward various fungal diseases have been raised with genes responsible for producing any of the above-mentioned anti-fungal compounds.

Abiotic Stress Tolerance

Abiotic stress is also a major factor that influences the growth and productivity of plants directly. For example, drought, salinity, snow, temperature, etc., have a direct link to plant growth and crop productivity. There are metabolites produced in plants due to the stress response of high temperature, drought, or salinity. These metabolites include stress-related osmolytes such as sugars and sugar derivatives such as trahalose, fructans, mannitol, etc., and amino acids such as proline, glycine, betaine, and proteins such as heat-shock proteins and anti-freeze proteins. Crop plants have been genetically engineered to over-express genes responsible for the production of one or more of the above-mentioned metabolites. Such plants are found to show increased stress tolerance. Salinity-tolerant rice and drought-resistant sugarcane have been developed by different research agencies such as ICRISAT.

Delayed-Ripening Fruits and Vegetables

The shelf life of fruits and vegetables is one of the problems that affects agro-industries. There are a number of key metabolic processes that lead to fruit maturation and ripening. The gaseous hormone ethylene is one of the important compounds involved in fruit ripening. By blocking the biosynthesis of ethylene, the process of fruit ripening can be extended. Ethylene production in the plant can be blocked by inactivating one or more key enzymes in the biosynthetic pathway. This can be achieved by anti-sense mRNA technology, in which a reverse copy (a complementary gene) is introduced with all the necessary regulatory elements present in the normal gene. In such plants, ethylene production may not occur and so the fruit ripening will be very slow under normal conditions. Fruit ripening can occur artificially by supplementing ethylene. This delayed ripening helps in exporting tender fruits such as tomatoes.

Male Sterility

In plant breeding, male sterility is an important characteristic that prevents self-pollination in plants and promotes cross-pollination. If the plant is a sterile male, there is no need for emasculation, which is usually carried out in bisexual flowers to avoid self-pollination during plant breeding. Therefore, the creation of male sterility in agricultural plants is important as far as plant breeding is concerned. Male sterility is introduced into crop plants by transferring gene coding for the enzyme barnase, an RNA-hydrolyzing enzyme. This enzyme inhibits pollen formation and the plant becomes male sterile. This gene is made to express in the tapetal cells of the anther with the help of the

tapetal specific promoter (TA29), so that the enzyme activity is restricted only to the cells involved in pollen production. Male fertility can be restored through the introduction of another gene, barstar, that inhibits or suppress the activity of barnase.

Transgenic Plants with Improved Nutrient Qualities

Crop plants can be genetically engineered for the improvement of nutritive qualities of the edible parts such as grains, fruits, tubers, etc., for protein content, starch qualities, etc. Some examples are described below:

Rice grain with vitamin A. Rice doesn't normally contain vitamin A. A genetically-engineered form of rice, **golden rice**, does make vitamin A. The deficiency of vitamin A can result in a condition called night blindness. This condition is severe in third-world countries. The hope is that, in countries where rice is the staple of the diet, introducing golden rice will save millions of people from the consequences of vitamin A deficiency.

Protein quality of seeds and tubers. Nutritionally-enhanced foods will offer an increased level of nutrients such as proteins, vitamins, and other healthful phytochemicals. Introduction of new types of seed-storage proteins from other sources can increase protein qualities of cereals, legumes, pulses, and tubers such as potatoes. The genes of protein molecules rich in essential amino acids can be isolated and transferred to food crops with the right type of promoters to express it in the edible part of the plant. Another strategy for the quality improvement of storage proteins through genetic engineering is by the modification of the protein already present in the seeds or tubers so as to increase the quantity of essential amino acids.

Protein-enhanced sweet potatoes, rice, wheat, and maize, high-vitamin-A canola oil, and increased antioxidant fruits and vegetables, etc., are some other examples of increasing the nutritive quality of food materials through genetic engineering.

Transgenic Plants as Bioreactors (Molecular Farming)

The development of modern techniques for creating transgenic plants and the isolation of new genes responsible for the production of various industrial proteins and enzymes and pharmaceuticals for humans and other organisms has created an entirely new and amazing technology. Drugs and animal proteins can be cultivated in the field. Plants are very cheap and the raw materials required for the cultivation of plants, their harvesting and extraction of the products, are relatively simple and economical. If the plants are transformed with the right kind of gene along with the proper promoters, they can function like a bioreactor to produce the required compounds such as proteins (therapeutic peptides and proteins) and antibodies or vaccines, enzymes for food, amino acids, vitamins, drugs, etc., in sufficient quantities. Some examples are described below.

Diagnostic and Therapeutic Proteins

Plants that are easy to grow can be genetically engineered to produce a variety of proteins that are used as drugs, vaccines, and diagnostic agents in humans and animals. By using plants as the bioreactor or factory against the use of conventional fermentation techniques, which include costly equipments, upstream and downstream processing, the cost of production can be brought down substantially. Proteins such as plasma proteins, peptide hormones, interferons, and cytokinins can be produced in specific organs of transgenic plants such as the leaves in the case of tobacco, tubers in the case of the potato and the stem in sugarcane effectively.

A recombinant form of the natural hormone—rBST (bovine somatotropin) which causes cows to produce milk—can be synthesized through transgenic plants. rBST increases milk production by as much as 10 to 15%. It is used to treat over 30% of U.S. cows. Food enzymes, including a purer, more stable form of chymosin used to curdle milk in cheese production, are now being synthesized by transgenic plants. It is used to make 60% of hard cheeses. This can replace the chymosin of rennet from slaughtered calves stomachs. Rennet is the extract from the membrane living in the stomach of calves.

Edible Vaccines

Transgenic plants engineered for vaccine production, in addition to their use as a bioreactor, can also be used as edible vaccines if the vaccines are directly expressed in the edible parts of the plant. If the plants are transformed with the gene responsible for vaccine production attached with the organ-specific promoter, it will express, and the raw, edible part can be eaten for vaccination. The vaccine molecules present in the raw food (tuber, fruit, or leaves) will be absorbed into the cell lining of the mouth and food canal while it reaches the stomach. This type of vaccination is applicable not only in humans but also in animals. For example, fodder plants can be genetically engineered to produce the protective antigen anthrax bacteria, so that cattle can be automatically vaccinated against deadly anthrax.

The expression of antigenic protein in the edible parts of crops such as tomatoes, bananas, and apples, is useful for the immunization of humans. The edible vaccines have a number of advantages against conventional methods. Edible vaccines (edible parts of plants) are easy to store and transport without refrigeration. The delivery method is also very simple. The cost of the product is comparatively low since the costly equipments and downstream processing techniques are not required for the industrial production by fermentation. Vaccination against diseases such as cholera, hepatitis, small pox, tuberculosis, etc., with transgenic crops will soon be a reality. Research and field trials are in progress at many of international and national laboratories and universities.

Biodegradable Plastics

The use of plastics is creating many environmental problems, as it is not biologically degradable. There is some bacteria, that can produce plastics, polyhydroxy butyrate (PHB), that are biodegradable. It has been demonstrated that arabidopsis plants can produce these PHB globules in the chloroplasts, when genetically-engineered with the genes responsible for the biosynthesis of PHB. The production and accumulation of PHB did not create any problem with the normal metabolism and growth of the plant in the field trials. This bioplastic can be easily extracted and purified from the leaves of the plants. A number of companies are already engaged in the process of developing transgenic plants for exploring the possibility of producing biodegradable plastics industrially.

Metabolic Engineering and Secondary Products

Development of transgenic medicinal plants can improve the production of specific secondary metabolites. With metabolic engineering it is possible to enhance a particular secondary metabolite or to produce a new compound by a medicinal plant. Over- expression of a key enzyme gene in the biosynthetic pathway generally results in the enhanced production of the desired product. Introduction of one or more genes of enzymes or silencing the expression of a specific enzyme in the biosynthetic pathway can alter or divert the biosynthetic pathway in a new direction resulting in the production of a new desired compound, which may be absent in the plant under normal conditions. For example, the production of the anti-cancer drug taxol has been enhanced considerably by the methods of metabolic engineering in tissue cultures of *taxus buccata*.

Another *in vitro* method for enhancing the production of metabolites, particularly those compounds produced in roots, involves the induction of hairy root cultures with *agrobacterium rhizogenes*. The transfer and integration of the root-inducing DNA of the root-inducing plasmid (Ri plasmid) can result in the profuse production of hairy roots, which can be cultured and maintained indefinitely. These root cultures can act as the source for root-specific metabolites.

19.6 DIAGNOSTICS IN AGRICULTURE AND MOLECULAR BREEDING

There are a number of biotechnological methods available to produce plants that are either free of pathogens or resistant to microbial infections. But these plants must be tested. Today, a number of biochemical and molecular tests are available as diagnostic methods to test or to detect the presence of pathogens in the plant or plant tissues. There are two types of diagnostic methods. One is based on proteins or immunological techniques and the second is based on DNA probes. Pathogen-specific antibodies or DNA fragments as probes can be used for the detection of specific pathogens in plantlets or plant tissues that are used for plantations or as explants for micropropagation. Similarly, using molecular markers—protein-based or DNA-based—it is possible to ascertain whether the plant is disease-resistant or not. This gives the freedom to farmers to adopt right type of control measures if required.

Molecular Breeding

Plant breeding assisted by a genetic marker or a molecular marker is called **molecular breeding**. Molecular markers or genetic (linkage) markers are used to identify or to tag certain traits that are important for breeding. If the genes that are located in a chromosome are very close together, they are inseparable by crossing over. Markers that are close together always inherit together and the chance of crossing over is very low. Molecular markers or genetic markers can be identified by restriction analysis such as RFLP analysis or by genetic hybridization experiments. Such linked traits can be taken as observable markers for the gene that we are trying to locate or isolate. With restriction analysis of the genome, DNA markers can be assigned to the locations of these genes in the chromosomes. Based on these markers, the genes can be cloned and sequenced. In the same way a plant variety can be selected for a particular trait, for example drought resistance, just by looking for its marker, which may be a molecular marker or another observable trait that is linked to the trait of drought resistance. This type of cloning of genes is called as map-based cloning or **marker-based cloning**.

There are three types of markers available for different types of screening and breeding:

Morphological Markers. These are the observable phenotypic expression of genes. These morphological characteristics can be linked to some other traits, which cannot be detected phenotypically or morphologically.

Biochemical Markers. This marker is based on the isoenzyme (isozymes) pattern shown by the plant varieties or genotypes on native gel electrophoresis. Plant genotypes are identified by the specific isozyme profile of one or more enzymes present.

Molecular Markers. These are based on DNA fragments. The restriction fragment length polymorphism can be detected by known DNA sequences and that polymorphism forms the molecular markers. The DNA probes are the known sequences, which may be generated by PCR or may be VNTR (variable number tandem repeats), or microsatellites sequences or may be a random fragment isolated from genomic libraries or may be EST (expressed sequence tags). All these DNA probes can be used for RFLP analysis to get a meaningful marker. The above-mentioned DNA fragments can also be used as primers for generating PCR-based markers such as AFLP (amplification fingerprinting) or RAPD (randomly amplified polymorphic DNA).

19.7 BIOETHICS IN PLANT GENETIC ENGINEERING

Almost all of the crops that we cultivate today are much different than their wild ancestors. Breeding by selection and saving the best seed for the next generation has been in progress for many thousands of years. In most crops, the incorporation of traits compatible with agriculture, such as free threshing in cereals, was achieved centuries ago. Scientific breeding, however, which followed the rediscovery of Mendel's Laws, has been under way only for over 90 years. For example, in the 1950s, a type of maize was used for F1 hybrid production in the United States, which was subsequently discovered to be associated with susceptibility to southern corn leaf blight. During the 1970s, wheat varieties with single major gene resistance to fungal diseases were released in the UK, one after another. The best varieties were usable, on average, for only 14 months as the fungus populations repeatedly overcame the disease-resistance genes. The supply of resistance genes became limiting and farmers increasingly had to rely on chemical fungicides to keep their crops free of disease. These instances, although costly and apparently very serious at the time, did not produce significant, long-lasting effects.

'Transformation', 'genetic modification', 'genetic engineering', and 'transgenesis' are all synonyms for the transfer of isolated and cloned genes into the DNA, usually the chromosomal DNA, of another organism. Genetically-modified plants (GM plants) and GM foods are now becoming a part of agriculture and the the food industry. The advancement that has been made in plant biotechnology is not contributing to the promotion of industries based on agriculture, and the food industry, unlike the contributions made in the pharmaceutical industry. The main problem in GM crops and GM foods is that it is not accepted by the public because of the wrong and misleading information given by the media, environmentalists, and anti-scientific people in society. However, like any genetic-engineering techniques and transgenic organisms, care should be taken in handling GM crops. The main public concerns about GM crops and GM foods are the following:

- Safety of GM foods for human and animal consumption.
- The impact of GM crops on the environment, that is, on other organisms, and how GM crops will influence biodiversity.
- What may be the influence of insect-resistant crops on non-target insects and microbes, which are beneficial to the environment. For example, nitrogen-fixing bacteria and insects that help pollination.
- Transgenic plants may undergo natural pollination with the wild relative of the crop plants, which may result in the escape of the gene that has been transferred to the crop plant. This may result in the so-called gene pollution and the formation of super weeds.
- Vegetarians fear that if the plants are transformed with some animal genes, the vegetarian nature of the plants will change.
- The antibiotic-resistant markers that are transferred to the plants can escape into microbes in the soil and can create the problem of antibiotic-resistant pathogens in humans and animals.

All these fears and objections can be explained or solved satisfactorily. But unfortunately, public debate on the benefits and risks of plant biotechnology suffers a large number of unexpected objections mainly because of the misinformation, misunderstanding, and manipulative analysis of the data of surveys. But it appears that it is a short phenomenon and as time passes, with the support of new convincing evidence on safety and efficiency and at the same time the complete absence of evidence that shows the contrary, more and more consumers will accept GM foods and crops.

REVIEW QUESTIONS

1. What is micropropagation? How does it differ from clonal propagation?
2. Compare somatic embryogenesis with organogenesis.
3. Describe the various steps in plant-tissue culture.
4. What is meant by embryo rescue? How is it useful in producing new hybrids?
5. What is a protoplast? Describe the method of its isolation.
6. Explain the application of protoplasts in plant breeding.
7. What are the various genetic markers used for plant breeding?
8. What is meant by molecular breeding?
9. Explain the natural genetic engineering method by *agrobacterium tumefaciens*.
10. What are the diagnostic methods available for crop plants?
11. Explain the strategies adopted for creating transgenic plants with the following traits:

    (*a*) Disease resistance (*b*) Herbicide tolerance

    (*c*) Insect resistance.
12. What is golden rice?

13. What are biodegradable plastics? How is plant genetic engineering used for producing biodegradable plastics?
14. Explain how can plant biotechnology be used for enhancing the nutritive qualities of edible parts. Give examples.
15. What are the social and ethical problems associated with plant genetic engineering? How can these be addressed?

REFERENCES

[4] Peter S.Carlson, Harold H. Smith, Rosemarie D. Dearing, *Proceedings of the National Academy of Sciences of the United States of America,* Vol. 69, No. 8 (Aug., 1972), pp. 2292-2294

[5] Melchers, G., M.D. Sacristan and A.A. Holder: Somatic hybrid plants of potato and tomato regenerated from fused protoplasts. Carlsberg Res. Commun. 43, 203-218 (1978)

[6] Larkin, P.J. and W.R. Scowcroft. 1981, Somaclonal variation: A novel source of variability from cell cultures for plant improvement. Theor. Appl. Genet. 60:197-214.

[7] Chilton MD., *Agrobacterium Ti plasmids as a tool for genetic engineering in plants,* Basic Life Sci. 1979;14:23-31.

Chapter 20 ANIMAL-CELL CULTURE AND APPLICATIONS

In This Chapter

20.1 INTRODUCTION

Cell culture is defined as growth of cells dissociated from the parent tissue by spontaneous migration or mechanical or enzymatic dispersal. Primary cell culture refers to the cell culture initially derived from the parent tissue prior to any subsequent culture *in vitro*. Tissue culture has been known since the beginning of this century as a method of studying animal cell behavior. The history of animal cell culture dates back to 1856 when Ludwig maintained animal organs alive outside the body by pumping blood through them. Actually, this can be considered the start of animal tissue culture or organ culture. After some years, in 1885, Roux removed a portion of the medullary plate of a chick embryo and maintained it in a warm saline solution for some days. It was Ross Harrison who demonstrated the *in vitro* growth of living animal tissue by explanting a fragment of nerve cord from a frog tadpole and placing it in a drop of frog lymph, in 1907. He observed that within hours the nerve fibers began to grow from the cells in the tissue fragment.

During 1911 to 1912 Burrows and Alexis Carrell succeeded in growing tissue fragments from adult dogs, cats, rats, and guinea pigs and also in growing malignant (cancerous) tissues. Carrell started a cell line from a tissue taken from the heart of a chick embryo, which was propagated by Albert H. Ebeling for 34 years. Ebeling used the culture to test the toxicity of germicides.

This technique, which used aggregated fragments of tissue from which cells were restricted to migrate outwards, dominated the field for more than 50 years. However, it was the work of Enders and colleagues in 1948 that gave tissue culture a major boost. These researchers demonstrated, using dispersed cells, that the polio virus could be grown *in vitro* in the absence of nerve tissue. Hence, the origin of the term "cell culture." This work stimulated the development of cell-culture technology, which has been adapted successfully into many routine applications both in medicine and industry (e.g., production of antiviral vaccines, antigen and monoclonal antibody production, chromosomal analysis of embryonic cells (from amniocentesis) skin grafting, etc). The cultivation of viruses using animal cell culture has actually showed the way to the discovery of interferons in 1966 by Alec Isaacs. He took the filtrate from virus-infected cell cultures and made it virus-free. This filtrate was supplemented to a cell-culture media and cultured fresh cells. He observed that these cells could not be infected with the same virus. From these observations he concluded that some compounds were secreted into the media by the cells in response to the viral infection. This compound, when supplemented into fresh cell cultures, prevented the viral infection. The compounds that interfered with the viral infection were named **interferons**. Interferons were purified from the virus-infected cell cultures, and today different types of interferons such as alpha, beta, and gamma forms are some of the valued products of biotechnology available in the market.

Chinese hamster ovary cell lines, an important cell line used extensively in basic and applied research were developed in the 1980s to produce erythropoietin, another important product of biotechnology introduced to the pharmaceutical market in the beginning. Now, CHO cell lines are one of the approved cell lines for the industrial production of many recombinant proteins. There was a tremendous improvement in media formulations and culture techniques adopted for the culturing of animal cells for various purposes. Unlike plant and microbial cells, animal cells are very fragile and require very complex and special nutrients in the medium. A number of synthetic media are now available in addition to the conventional culture media in which blood serum is an important and common ingredient. The serum is a very complex mixture of very essential nutrients such as vitamins, hormones, and essential amino acids. Different types of culture vessels and bioreactors were designed and developed for the large-scale cultivation of various types of cell lines and cultures. Large-scale cultivation of cells mediated by micro-carrier beads, roller bottles, and spinner baskets, etc., are examples of culture techniques used commercially. The advent of hybridoma technology and advances in immunology has made the production of therapeutic antibodies comparatively less expensive.

It has been repeatedly observed that cell cultures, particularly primary cell cultures, can be used as a model to study cellular responses to a stimulus or stress. These models can provide supplementary information to the larger models of live experimental organisms. Modern political and social situations also demand models as an alternative to live animals. Because of the wide availability of embryos such as chicken embryos, and the ease of obtaining tissues from these embryos without resorting to sophisticated surgical procedures, embryos are the ideal sources for a variety of primary cell cultures. Also, chicken or mice embryos are not subjected to tight regulations as is the use of live animals and are not surrounded by controversy, as is the use of human fetus tissues in research. This circumstance allows the researcher to be devoted more to the research project itself. Using chicken embryos to prepare primary cell cultures also offers an enormous economic advantage. Fertile eggs can readily be purchased from commercial sources, relieving the investigator of the expenses associated with live animal care. Incubation of fertile eggs is simple and inexpensive.

New insights into the genetical and cellular basis of embryology and developments in stem-cell technology has given the hope of culturing the tissues of different organs such as smooth muscles of the heart and blood vessels, tissues of the kidney and liver, blood cells, bone-marrow cells, muscle, skin, and brain or neurons. It may be possible to grow the organs in culture developed from the stem cells and the defective organs may be replaced or repaired by the cultures, organs, or tissues. This may be particularly important for the treatment of leukemia, neuro-degenerative disease, and problems related to kidneys, liver, heart, etc.

20.2 ANIMAL CELL CULTURE TECHNIQUES

Eukaryotic cells are much more difficult to culture than most prokaryotes. They demand complex media and are very susceptible to contamination and overgrowth by microbes such as bacteria, yeast, and fungi. Before dealing with the details let us discuss some of the important characteristics of animal cells and their growth in culture.

Features of Animal Cell Growth in Culture

Animal-tissue culture can be divided into organ culture, tissue culture, and cell culture. In organ culture, entire embryos, organs, or tissues are excised from the body either by vivisection or shortly after brain death. In such cases, the normal physiological functions of organs or tissues are maintained. The cells remain fully differentiated and scale-up is not recommended in organ cultures. It is characterized by slow growth and fresh explantation is required for every experiment.

When fragments of excised tissue are grown in a defined medium it is called tissue culture in a true sense. In these cases, most of the normal functions of the cultured tissues may be maintained even though the original organization of the tissue is lost. Compared to organ cultures, tissue cultures are better for scale-up but not ideal. When the excised tissues or tissue outgrowths are dispersed, mostly enzymatically, into a cell suspension, which may then be cultured as a monolayer or suspension culture, it is called **animal cell cultures**. The cell cultures can be maintained for a very long time as a particular type of cell line and are very suitable for scale-up studies. But these cells may lose some of their differentiated characteristics over time.

HISTORICAL TIMELINE OF TISSUE CULTURE RESEARCH

1801	Bichat investigates and names the tissues of the body and stresses the study of the different tissues that make up the organs.
1856	Ludwig maintains animal organs alive outside the body by pumping blood through them.
1885	Roux removes a portion of the medullary plate of a chick embryo and maintains it in a warm saline solution for a few days.
1907	Harrison demonstrates the *in vitro* growth of living animal tissue by explanting a fragment of nerve cord from a frog tadpole and placing it in a drop of frog lymph. Within hours nerve fibers begin to sprout from the cells in the cord.
1911–12	Burrows and Carrell succeed in growing explants from adult dogs, cats, rats, and guinea pigs and also in growing malignant tissues. Carrell begins a cell line, taken from the heart of a chick embryo, which is propagated by Albert H. Ebeling for 34 years. Ebeling uses the culture to test toxicity of germicides.
1913	Steinhardt, Israeli, and Lambert show that vaccinia virus can survive for several weeks in fragments of rabbit cornea.
1914	David Thompson begins to experiment with organ culture, explanting toes, feather germs, optic lens, and tail bud from embryonic chicks.
1922	Ebeling succeeds in culturing epithelial cells.
1923	Warburg develops a method of studying respiration in thin slices of tissue.
1925	Parker and Nye inoculate fragments of rabbit testes with vaccinia virus and cultivate them in plasma.
1929	Fell studies the development of bones and joints *in vitro* using fragments of developing organs from chick embryos.
1931	Goodpasture and associates note the possibilities for research and practical application in the use of developing chick embryos. Use of the chorioallantoic-membrane technique for the study of fowl-pox virus is described.
1946	Beveridge and Burnet describe the cultivation of viruses and rickettsiae in the chick embryo.
1947	The Tissue Culture Conference held in Hershey, Pennsylvania initiates the founding of the American Tissue Culture Association. Enders, Weller, and Robbins grow mump viruses in chick tissue using penicillin to prevent bacterial contamination.
1948	Enders, Weller, and Robbins prove that poliomyelitis virus can be cultivated *in vitro* in the absence of nerve tissue. Sanford, Earle, and Likely establish clone cultures from single cells.
1954	Pomerat and Leake propose the use of tissue culture for toxicity studies.
1958	Harris induces hybridization of cells on a large scale, with cells of widely different species as well as cells of widely differing types in the same species.

Animal cells, unlike plant and microbial cell cultures, show limited growth even if the best of nutritive and physical conditions are maintained. In plant and microbial cell cultures, the cells continue to grow if the adequate physical and chemical environments are maintained in the medium. The behavior of animal cells in culture depends on the source organs or the tissue from which the cells are excised. In most cases, it grows and divides only to a limited number of generations and then it dies or stops its growth, even if all the appropriate physical and nutritive factors are provided. In some cases, the cells will not divide in culture as in the case of neurons. But cells such as fibroblasts or skin cells can undergo cell division for some generations and then they come to a static phase and eventually die. Thus, mortality is associated with all animal cell cultures unless it is an immortal cell line or cancer cell line. Cells derived from tissues such as kidneys, liver, muscles, and embryos grow and spread as a monolayer and fill the surface of the container in which they are growing. Once the cells spread to the entire surface and come in contact with the walls of the plate or the container, the cells stop further growth. This phenomenon is called **contact inhibition.**

Another important feature of cell cultures is the artificial environment, which can be altered and controlled. A number of features applicable for cells *in vivo* are lacking in *in vitro* cultures. Cell-cell interaction and cell-matrix interactions are reduced or almost absent in *in vitro* cultures. Nutritional milieu and cell proliferation are also changed *in vitro*. The cells in culture may have an entirely altered hormonal condition, which can change the physiological state of the cells. The three-dimensional architecture that was present in the tissue is totally absent in cell cultures. The changed physiological conditions of the cultured cells result in the way the cells adhere to the support and increased spreading and migration of the cells as well as the shape. These features are mainly applicable to normal cells, which have a limited life span. But cancer cells behave differently than normal cells. They lose the phenomenon contact inhibition and instead of spreading as a monolayer, they pile up one above the other to form multilayered cell cultures to the uncontrolled cell divisions. Such cells are round in shape and show continuous cell multiplications if proper nutritional conditions are provided. Oncologists utilize these properties of cancer cells, particularly the growth pattern, to confirm the diagnosis of cancerous nature of tumors.

Animal cell cultures in simple terms can be classified into three types. They are primary cell cultures, secondary cell cultures, and immortalized cell cultures.

Primary Cell Cultures

Primary cell cultures are the cells explanted directly from a tissue or organ of a donor organism. The cell cultures of white blood cells, fibroblasts derived from skin, smooth muscle cells of heart, neurons, etc., are some examples of primary cell cultures. They may be capable of one or two divisions in culture, and if given the right conditions, can survive for some time, but they do not continue to grow and eventually senescence and die. So, the main purpose of growing these demanding cells is that they represent the best experimental models for *in vivo* situations, and they may express characteristics that are not seen in cultured cells. Cultures of epithelial cells with beating cilia is a good example of this.

The type of cell culture mainly depends on the source tissue of the organism used for generating the cell. The cells can be grown as **suspension cultures** or as **anchored cells** adhered to a support. The cultures of lymphoid cells (white blood cells) can be cultivated as a suspension culture and made to undergo a limited number of divisions before senescence by adding cytokines and **lectins** to the growth media. There is no need for any proteinase or collagenase to establish a cell culture of lymphocytes from lymph or blood. Lectins are proteins, which react with specific sugar residues on cell surfaces, cross-linking them and often causing intracellular signalling. Different lectins stimulate the outgrowth of different types of lymphoid cells. For example, phytohaemagglutinin (PHA) stimulates T-cells only, while conconavalin-A has mitogenic activity (inducing cell division) for both T-cells and B-cells. Figure 20.1 shows primary human peripheral blood lymphocytes in culture.

The primary cultures derived from tissues other than blood or lymph have to be dispersed mechanically or enzymatically and should be maintained in appropriate nutrient media in suitable culture containers. Since these cells are anchorage dependent, the surface of these containers serve as the support on which cells can grow and spread as a monolayer. The tissue used for establishing primary cell cultures are derived from a specific organ such as the kidneys or liver. The cells are not mobile and are embedded in the connective tissues. Enzyme treatment can separate the cells from each other by digesting the protenacious cementing materials that joins cells. The enzymes used for this purpose is a mixture of proteases such as trypsin and collagenase.

The primary cell cultures provide a system, which is not much different from the *in vivo* system. It preserves most of the special characteristic features of the cells in the respective organ. At the same time, a number of physical and physiological factors can be regulated by altering the culture conditions. The main drawback of primary cultures is that it takes a lot of time to establish a specific type of primary cell culture. Each time it requires a live organism or a fresh organ or tissue to establish a culture. The quality of the cell cultures can vary if two different people prepare it at different times. The primary cell cultures can be made viable by subculturing for a very long time with or without cell division. Once the primary cell culture is split or subcultured, it becomes a **secondary cell culture**, and can lead to the formation of a cell line with special physiological or genetical characteristics.

Secondary Cells

Secondary cells are originally explanted from a donor organism and maintained by subculturing for a long time. If these cell cultures are given the correct physical and nutritive culture conditions including the correct hormones continuously by subculturing, they can divide and grow for some time *in vitro*, at least 50 to 100 generations. However, they do not continue to divide indefinitely and eventually, and their physical characteristics may change, after which the cells will eventually senescence and die. The factors which control the replication of such cells *in vitro* are related to the degree of differentiation of the cell—in general, **terminally differentiated cells** are harder to maintain than less specialized cells as in the case of embryonic cells. For example, MRC5 cells are secondary human lung fibroblasts, which undergo 60 to 70 doublings before senescence. These cells are widely used to study viruses *in vitro* and for vaccine production.

The frequency of subculturing depends on the cell type and its source. If the source is differentiated cells or terminally differentiated cells, its rate of multiplication may be very slow and therefore it demands less frequency of subculturing. Whereas if the cell source is embryo cells, it will multiply and spread very fast, and therefore it needs frequent subculturing to avoid confluency, to remove the products and excretory materials, which can adversely affect the stability of the cells, and to replenish the nutrients. Otherwise, the cells may stop their growth and may die over time due to insufficient nutrients and space. But the frequent subculturing or splitting of the slow-growing cultures may lead to low cell density and eventually loss of the cell lines.

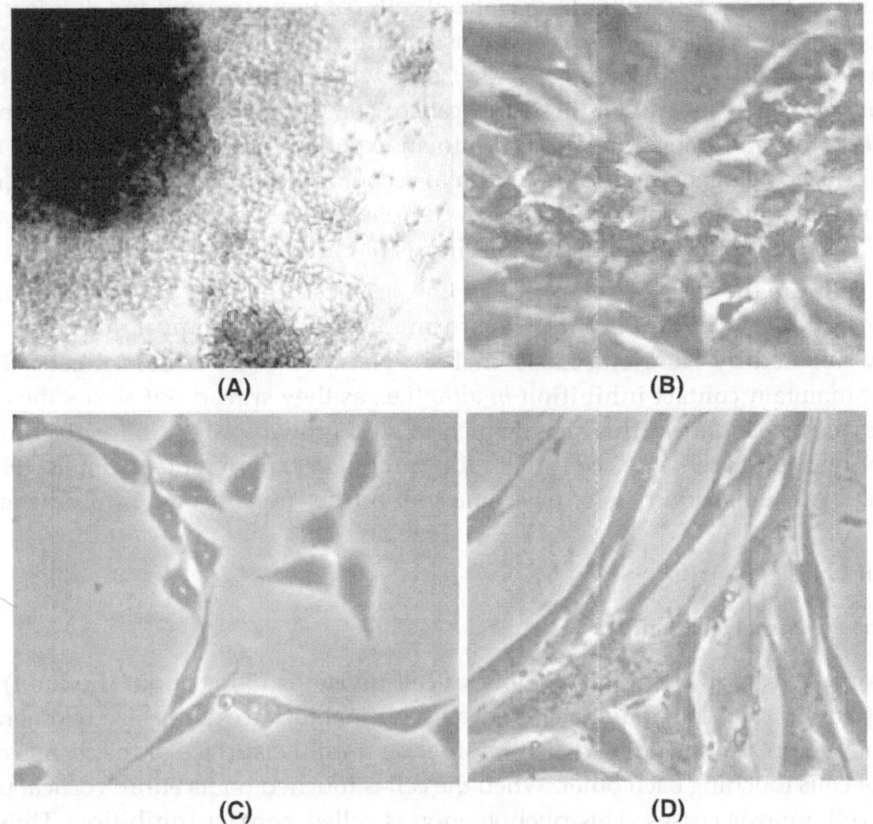

FIGURE 20.1 **Various types of cell lines. (A) Primary human lymphocytes. (B) Ciliated nasal epithelium. (C) MRC5 cells—secondary human lung fibroblasts. (D) BEAS-28, human bronchial epithelial cells.**

Subculturing of suspension culture is very easy. It involves the dilution of the cultures having high cell density with fresh nutrient media or by separating the cells from the old culture to a fresh media with sufficient quantity of cells to maintain the initial cell density. Subculturing of the anchor-dependent cultures involves a special procedure. The growth media is removed from the cultures. The culture plate is washed repeatedly with fresh media. The anchored cells are removed or dissociated from the plate by enzymatic treatment with proteases and collagenase. These cells are transferred to the fresh culture plates supplemented with fresh media. In some cases, the cells can be removed from the plates mechanically with the help of a plastic spatula or by gently pipetting.

Sometimes these secondary cell cultures may spontaneously become transformed resulting in the formation of cell lines having indefinite growth or immortality such as cancer cell lines. Such cell lines are called immortal cell lines.

Immortalized Cells

Unlike primary and secondary cells, immortalized cells continue to grow and divide indefinitely *in vitro* for as long as the correct culture conditions and nutrients are maintained. **Immortalized cell lines** are also known as **transformed cells**—i.e., cells whose growth properties have been altered. This does not necessarily mean that these are "cancer" or "tumor" cells. The tumor cells or cancer cells are able to form a tumor if introduced into an experimental animal. But a true transformed cell or immortalized cell generated from a normal secondary cell culture cannot generate a tumor in an experimental animal. Transformation is a complex process and can occur by many different routes (e.g., infection by transforming tumor viruses or chromosomal changes).

HeLa cells are the classic example of an immortalized cell line. These are human epithelial cells from a fatal cervical carcinoma transformed by human pappilomavirus 18 (HPV18) involuntarily donated by Henrietta Lacks in 1951. HeLa cells are **adherent cells** (they stick to surfaces) that maintain **contact inhibition** *in vitro* (i.e., as they spread out across the culture flask, when two adjacent cells touch, this signals then to stop growing). Loss of contact inhibition is a classic sign of **oncogenic cells** (i.e., cells that form tumors in experimental animals). HeLa cells are not oncogenic in animals, but they may become so if further transformed by a virus **oncogene.**

Types of Cell Lines

Finite Cell Lines (Normal Cell Lines)

Normal cells pass through a limited number of cell divisions (50 is about the limit) before they decline in vigor and die. This may be caused by their inability to synthesize telomerase. Normal cells, when placed on a tissue culture dish, proliferate until the surface of the dish is covered by a single layer of cells touching each other. When the cell is touched on its entire vertical planes by the neighboring cell, mitosis ceases. This phenomenon is called **contact inhibition.** Thus, these cells are limited in their growth and finally die. The growth rate of each cell line may vary depending on the source organ, its age, and degree of differentiation. Normal cells are exceedingly fussy about the nutrients that must be supplied to them in their tissue culture medium. These cells ordinarily have the normal set of chromosomes of the respective species. That is, they have a normal karyotype. But due to differentiation and other *in vitro* variations, the cells may lose some of their original characteristics or may gain some special features that make them a special cell line.

Continuous Cell Lines (Immortal Cell Lines)

Cancer cells may be immortal. They proliferate indefinitely in culture. Example: HeLa cells are cultured in laboratories around the world. They are all descended from cells removed from a cancer (of the cervix) of a woman who died of her disease almost a half century ago. Cancer cells in culture produce telomerase and so can divide continuously and never cease their division capacity. Cancer cells show no contact inhibition. Once the surface of the dish is covered, the cells continue to divide, piling up into mounds.

The photographs in Figure 20.2 show mouse fibroblasts (connective tissue cells) growing in culture. The cells on the left (A) show contact inhibition. Those to the right do not exhibit the contact inhibition even though they are immortal cells (transformed but not derived from cancer cells). The cells to the far right are said to be **transformed.** These cells (called 3T3 cells) were not derived from a mouse cancer but were produced by laboratory treatment of normal cells. Radiation, certain chemicals, and certain viruses are capable of transforming cells. Although transformed cells are not derived from cancers, they can often develop into malignant tumors when injected into an appropriate test animal such as a mouse. Cancer cell lines (and transformed cells) can usually grow in much simpler culture mediums. Cancer cells almost always have an abnormal

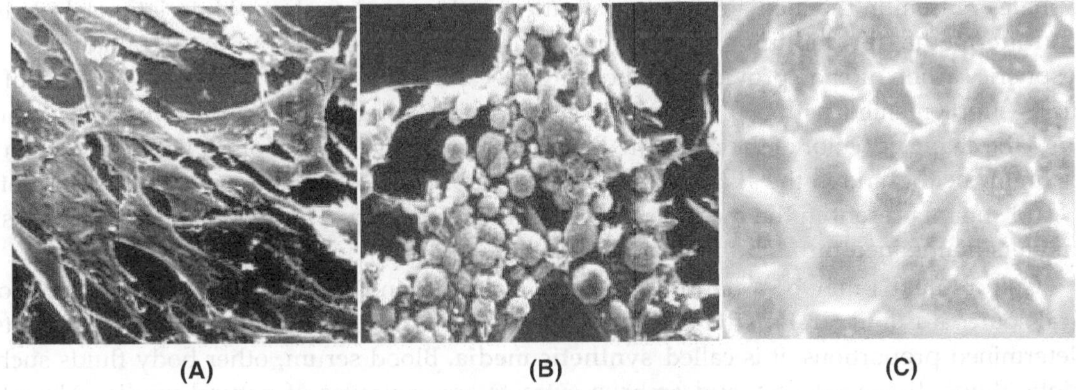

FIGURE 20.2 **(A) Normal mouse fibroblast cell lines showing contact inhibition.**
(B) Transformed mouse fibroblast cell lines.
(C) HeLa cell lines.

karyotype with abnormal numbers of chromosomes (polyploid or aneuploid), or chromosomes with abnormal structures such as translocations, deletions, duplications, or inversions, the different forms of chromosomal mutations responsible for variations among the cell lines.

Physical Environment for Culturing Animal Cells

The term cell culture or tissue culture refers to the culturing of whole organs, tissue fragments, or dispersed cells in a defined nutrient medium providing all the necessary physical and chemical environments. Animal cells are much more difficult to culture than most prokaryotes. They demand complex media and are very susceptible to contamination and overgrowth by microbes such as bacteria, yeast, and fungi. They are also very sensitive to environmental factors created by the special *in vitro* conditions. The culture media has to provide the cells all the important physical and chemical environments that keep the cells intact and help in maintaining the special physiological activities or functions. The major physical environment includes the temperature, pH, osmolarity, and the gaseous environment. The media, in addition to providing the necessary nutrients and hormonal requirements, also protects the cells from all types of chemical, physical, and mechanical stresses.

Media

The media used for culturing animal cells is usually very complex in its composition and is predetermined. A large number of different types of media have been formulated for the cultivation of various types of animal cells. Certain cell types require very specialized nutritional compositions for growth and metabolism or to carry out their function *in vitro*. A media is a mixture of balanced salt solution containing the metabolically important micro- and macro-elements in correct concentration to satisfy nutritive needs and to maintain the osmotic pressure in the media throughout the culture period. Other metabolically important nutritive components include all amino acids, fatty acids, vitamins, cofactors, sugar as a carbon source, and the essential hormones and growth factors needed to maintain the physiological state of the cells. Certain media components may have some other roles in addition to their primary nutritive role. For example, the carbonates and phosphates present in the media can act as a buffer system to normalize pH variations. In animal cell culture media, the sodium or potassium bicarbonate present in the media along with the atmospheric carbon dioxide provided in the culture vessel can minimize pH variation. Most of the media is supplemented with a pH indicator such as phenol red, which shows the pH variation through the change in color of the medium. Phenol red gives a yellow color for the media when the pH is highly acidic, and in the alkaline range it gives a pink color. The mineral components, other salts, and sugars present in the media, can control the osmolarity. These are some of the physical parameters maintained through the nutritive media.

Culture media may be a natural type or a synthetic one. If the media is made from some natural source it is a **natural media** and if it is made from salt solutions and other components in predetermined proportions, it is called **synthetic media**. Blood serum, other body fluids such as amniotic fluids, tissue extracts, and embryo extracts are examples of natural media. Almost all types of media have serum as one of the ingredients, even if the media is a synthetic type. Recently, some serum-free media has been developed for special purposes.

pH

Intracellular pH is very critical for the life activities of a cell in culture. Intracellular pH is controlled by the extracellular pH of the media. PH controls the metabolic reactions through the control of cellular enzymes and binding of hormones, growth factors, and signal molecules to their respective receptors present on the surface of the cell membranes. Thus, the state of pH is closely related to the physiological state of the cells in culture. Even a transient change in pH can alter or block important metabolic activities such as respiration and can result in cell death. The optimal pH in most culture media is maintained between 7 and 7.4. Metabolism of the cultured cells can excrete various types of metabolites into the media including the desired product, which can alter the pH of the medium. This change in media can adversely affect the growth and metabolism of the cells resulting in death and reduction in product formation. Therefore, the pH should be controlled within the range through a proper buffer system. The most frequently used buffering system in cell culture media is the bicarbonate—a carbon dioxide system. Sodium or potassium bicarbonate is a nutrient component of the medium and carbon dioxide is a major component of the filtered air present in the atmosphere above the medium in the container and in the incubator (CO_2 incubator) in addition to the CO_2 evolved due to cellular respiration.

$$H_2O + CO_2 \rightleftharpoons H_2CO_3 \rightleftharpoons H^+ + HCO_3$$

$$NaHCO_3 \rightleftharpoons Na^+ + HCO_3$$

A shift in the pH of the media is rectified either by the formation of carbonate ions due to the ionization of carbonic acid formed by CO_2 dissolved in water or by the formation of $NaHCO_3$ from the excess carbonate.

This equilibrium between the carbonate and sodium carbonate is maintained in the medium by the CO_2 present in the atmosphere above the media in the culture vessels. Because of this buffering activity the pH is maintained between 7 and 7.4 in the medium. This kind of buffer system is known as an **open system**. The choice of buffer system and pH of the medium affects both solubility of media components and osmolarity, in addition to the functions of the cellular enzymes and other aspects of cell physiology controlled by the ionic concentration of the medium.

Osmolarity

The concentration of salts and other water-soluble materials such as glucose, amino acids, etc., present in the media determine the **osmolarity**. Osmolarity of a media can be monitored by an **osmometer**. The recommended osmolarity of cell culture media is about 300 mOsm. All the commercially available cell culture media compositions are adjusted so as to get 300 mOsm. Alteration in the osmolarity of the medium can affect cell growth and metabolism very significantly and can even result in the death of the cells. Change in salt concentration can affect the membrane integrity. High salt concentration outside the cell (of the media) can result in exosmosis, and finally that can lead to the plasmolysis or shrinkage of the cell. When the salt concentration of the media is less than the cell matrix, it can result in endosmosis, and finally an increase in the turbidity of the cells, resulting in lysis.

Temperature

Temperature is another important physical factor that directly controls the physiology of a living cell. Cells in the *in vivo* condition grow at about 37°C, and this temperature is considered the physiological temperature. This is the body temperature of *homo sapiens*. Therefore, the optimum temperature for the *in vitro* culturing of cells is 37°C and most of the mammalian cells and those from warm-blooded animals can be grown at this temperature. In other cases, the optimum temperature has to be determined by trial and error methods.

Cryopreservation

Cells in culture undergo continuous cell division and may differentiate resulting in genetic and physiological variations. There are several factors that affect the properties of the cells during prolonged storage through subculturing. If the cells have to be stored for a long time, the better choice is to freeze the cells and store them. Cryopreservation is the storing of cells and tissues at very low temperatures (liquid nitrogen temperature –180°C to –196°C). By freezing the cells at liquid nitrogen temperature, it is possible to keep the cells intact without any change in the cells for a long duration.

Freezing of cells and tissues can result in serious damage, which is mainly caused by intracellular ice crystals, osmotic effects leading to severe dehydration resulting in pH changes, denaturation of proteins, etc. These damages can be controlled by the use of preservative agents and slow cooling, which allows water to move out of the cells before freezing. Freezing at very low temperature—at 180°C—can retard the growth of ice crystals within the cells. The formation of ice crystals within the cells can result in the breaking of cells. It is very important that the cells be cooled slowly, to prevent the formation of ice crystals within the cells. This can be accomplished in one of two ways. One way is to freeze the cells at – 20°C for an hour, and then transfer them to – 70°C overnight. The next day, move them to liquid nitrogen, where they can be stored for very long time.

Cryopreservative agents can prevent the formation and growth of ice crystals within the cells. The media components such as sugars, serum, and some solvents can act as the cryopreservative. Most common compounds used as cryopreservatives are DMSO (dimethyl sulfoxide) and glycerol. These agents lower the freezing point and have low toxicity. Compounds such as DMSO are soluble in lipids, and therefore can easily pass through the lipid membranes. These antifreeze agents or cryopreservatives strongly bind water to prevent it from freezing at 0°C. Lack of ice crystal formation keeps the membrane integrity intact. High serum concentration contributes to the cell integrity by marinating the intracellular protein concentration by preventing its loss into the medium by diffusion.

The main reasons for freezing cells are to prevent the genotypic drift due to genetic instability, senescence and cell death, transformation of the cells leading to the formation of cancer cells or immortal cells, and phenotypic instability due to selection and redifferentiation. The technique of cryopreservation can also prevent microbial contaminations, cross contamination with other cell lines, during some technical problems like incubator failure, etc. This technique of preservation can help to save time and material, when cells are not in immediate use, for distribution to other users, and this is very helpful for cryobiology—the study of life at low temperature.

Thawing

Cryopreserved cells have to be brought to room temperature or to incubator temperature to carry out the culture experiments. This process is called **thawing**, which has to be carried out carefully. The frozen cells should be thawed rapidly to avoid the formation of ice crystals, which can create adverse effect on cells. After thawing, the cells can be directly transferred to the culture media in suitable containers. The vial containing frozen cells removed from the liquid nitrogen has to be placed immediately in a water bath kept at 37°C. Agitate the vial very slowly and finally transfer the contents to a flask or petri plate or any other suitable container having the appropriate media and incubate at its optimum culture temperature. Change the media after 8 to 12 hours.

Equipment Required for Animal-Cell Culture

A cell-culture laboratory needs some special instruments and equipment in addition to the usual laboratory equipment needed in a biotechnology laboratory. The major equipment essential for an animal cell culture laboratory and its uses are described here.

Laminar Air Flow Chamber (LAF)

A laboratory that operates aseptic works requires a sterile work area. The sterile work area should provide an environment free of dust and microbial contaminations. Contamination by bacteria or fungi can completely ruin the cell culture, because the growth rate of microbes—both bacteria and fungi— is much faster and immediately it can over-grow the animal cell cultures. A **laminar airflow chamber (LAF)** can provide a microbial-free work environment by constantly pumping filtered air over the workbench. All the aseptic works such as handling the cells, tissues, implanting of the cell and tissue into the media, filter, sterilization of the media, and all such experiments and works can be carried out in this chamber. In addition to the protection given to the sterile materials and cell cultures from microbial contamination, a laminar air flow chamber also protects the technician or operator from hazardous materials such as infected tissues and pathogenic organisms. The LAF mainly provides two types of protection:

- Protection of the cell culture from microbial contamination.
- Protection of the researcher from infectious cell cultures.

A laminar flow chamber is provided with a UV lamp, which can be used as a germicidal agent. Before operating the LAF, UV light is switched on for 15 to 20 minutes for sterilizing the interior of the workspace.

Types of Laminar Air Flow Chambers

There are various types of laminar air flow chambers designed for special situations. The choice of LAF depends on the nature of the cell cultures and the hazardous nature of the work. There are mainly three types of LAFs used in cell-culture and microbiological laboratories.

Class I LAFs or bio-cabinets are the ordinary type of chambers that can be used for protecting cells from other microbial or cell contamination. This type of cabinet has a horizontal air flow within the sterile area. The researcher is less protected and therefore he or she has to use other protective measures while working with infectious organisms. Air is sucked from the top or bottom of the hood and is filtered and flown into the workspace as a gentle horizontal stream toward the operator. It is most useful for plant tissue and cell culture and also ordinary microbiological works.

Class II LAFs have a vertical air flow within the sterile area directed from the roof to the floor of the cabinet. Therefore, it can give protection to both the cells and the user. Air from the room is sucked in underneath the work area and is filtered and flown over the work area vertically downward from the top. This operation does not create any health hazard for the researcher as the air is flowing vertically downward. This type of laminar air flow chamber is most frequently used in animal cell-culture laboratories and microbiological and genetic-engineering laboratories, where there is risk of infection or other types of health hazards.

Class III LAFs are similar to class II with respect to the direction of air flow, but are provided with special arrangements to protect the cells from escaping to outside and also to protect the operator. The user is completely separated from the workspace by transparent physical barriers such as glass plastic shields with two openings attached with gloves, through which the user can carry out the aseptic operations inside the sterile work area. This type of LAF is recommended for experiments with highly pathogenic microorganisms or for achieving stringent sterile conditions as some experiments demand.

Carbon-dioxide Incubator

A carbon-dioxide incubator is specially made for incubating animal cell cultures. It provides constant temperature, sterility, humidity, and adequate carbon-dioxide concentration in the atmosphere, which maintains the pH and reproduces the environmental conditions of a living cell acting as an open buffer system. Maintenance of relatively high humidity inside the chamber prevent the media from desiccation changes in the salt concentration. It helps in maintaining the correct osmolarity. Since sterility is maintained inside the CO_2 incubator, open culture dishes and multiwell plates can be incubated without contamination. The chamber of the incubator is made airtight by a silicon gasket on the inner door, and therefore sterility and humidity can be maintained. But frequent cleaning and humidification is essential to keep the chamber sterile. Air containing high percentage of carbon dioxide is known as **high efficiency particulate air (HEPA)** and is filtered to make it bacteria-free.

Inverted Microscope

A microscope is an important instrument in all cell-culture laboratories to examine the cultured cells, whether microbial, animal, or plant cell cultures. It helps to see morphological changes and can monitor the early phases of contamination. In animal cell-culture labs inverted microscopes are used, because it is essential to see the cells *in situ*. The cells can be observed while they are inside the container, without disturbing their state. In cell cultures the cells are usually adhered to the bottom or on the lower surface (floor) of the container. Because of this, ordinary microscopes are not useful for visualizing the cells. The inverted microscope can be used for seeing the cells adhered to the bottom of the plates as a monolayer. In an inverted microscope the optical systems and light sources are located below, just opposite to that of an ordinary microscope. Examination of the cells while they are in the culture can give a direct idea of morphology, growth, and health. But an ordinary microscope with a movable slide holder may be used for cell counting, and for the determination of growth parameters with the help of a counting chamber.

Centrifuges

This is another essential piece of equipment required in a cell-culture laboratory. A low-speed, table-top refrigerated centrifuge is recommended for the isolation of cells from suspension cultures or for separating the cells from a liquid phase. Spinning of cells at high rpm can result in the breakage of cells. Centrifugation in an ordinary centrifuge can increase the temperature of the rotor because of the heat generated from friction and also by the motor. Exposure of cells to this high temperature can damage the cells. The uncontrolled rise in temperature can be effectively controlled in refrigerated centrifuges.

Other Instruments

There are a large number of other ordinary instruments that are required in a cell-culture laboratory. Some of the most important instruments are listed as follows.

Sterilizing equipment—autoclave to sterilize non-heat labile liquids, sterilizing oven used for glassware and metallic instruments.

Vacuum pump and sterilization filters—many fluids are heat labile and cannot use autoclave. Such fluids can be filter sterilized.

Refrigerators and freezers—to store media and frozen serum.

Cryostorage for storing cell stocks for very long durations.

Fluid-handling systems—pipettors and micropipettes, pipette aids (electric and mechanical), multichannel pipettors, peristaltic pumps, etc.

In addition to this are the instruments that are needed for any biological or biochemical laboratory such as balances, hemocytometer, pH meter, conductivity meter for ion concentration, osmometer, magnetic stir plate, dishwasher, water-purification systems, etc.

20.3 CHARACTERIZATION OF CELL LINES

Cell cultures and cell lines have to be characterized properly and all the details regarding their identity and physiological as well as genetical properties should be available along with the cell lines. It is very important to keep all these records related to their origin and specific features for future use. A primary cell culture established initially may not be very useful as a special cell line, but it may develop into a very useful cell line. The records regarding its characterization are of particular importance if cells are used for the production of some compounds requiring government validation.

A cell line has to be characterized for a number of biological parameters. The origin of the cell line—that is, the species of organism from which the cell cultures were established and confirmation of the tissue of origin—is required for the characterization of a cell line. The species of organism and the type of organ or tissue to some extent can be identified by chromosomal analysis. Cells should be characterized by their lineage with the help of enzymes and surface (receptor) markers. For example, cells can be identified by the type of stem cells, or precursor cells of stem cells, or specific type of differentiated cells or immortalized or transformed cells or cancer cells. Cells can be examined for cross-contamination and its absence can be confirmed. And also the cell lines should be well-characterized for genetic instability and mutations or transformations.

The cross-contamination of one cell line with another is a very serious problem in cell-culture studies. Therefore, cell lines must be checked for cross-contamination by other cell lines. For example, many cell lines in the 60s were found to be contaminated with HeLa cells. In such cases of cross-contaminations, the rapid growers will overgrow slow growers and still today it is a major problem in cell-culture studies. Before starting any experiment with *in vitro* cell cultures it should be confirmed that the cells are in fact the original cell type. The cell lineage can be characterized or studied with the help of cell-specific markers such as surface antigens, receptor proteins, specific enzymes, and other unique markers including DNA fingerprinting. Visual characteristics such as morphological features such as cell shape and structures are the easiest and most direct way to identify cell lineage. These characteristics can be studied and analyzed by special staining and microscopy combined with microphotography. But the main drawback of this method is that cells may acquire different shapes and properties under different culture conditions.

Chromosome analysis is another important method to characterize and identify a cell line. Karyotyping or chromosome analysis can be used to identify cell lines regarding species and sex of original tissues. With the help of karyotyping it is possible to distinguish normal from malignant cells. The cells under mitosis are arrested during metaphase with colchicine and the chromosomes are stained with a suitable stain or with specific labeled probes, which gives a banded appearance to the chromosomes. The chromosome numbers can be counted, and, can also give details regarding the structure as well as possible chromosome mutations such as deletion, duplication, inversion, etc. When different types of fluorescently-labeled DNA probes are used for the *in situ* hybridization of chromosomes it is called FISH or fluorescent *in situ* hybridization (chromosome painting).

Total DNA content is another data for the characterization and identification of a cell line. Normal cell lines have stable DNA content and transformed cell lines are aneuploid or heteroploid with more DNA content than standard cell type of that species. The DNA content can be analyzed visually or via flow cytometry. Analysis of RNA and protein content of a cell line is another method of characterizing the cell lines for future use and identification. Examine the cell culture for the characteristic gene product via Northern blotting or get the characteristic protein fingerprint or profile of the cell type and that will indicate identity of the cell line. Antigenic markers can also be used for screening the cells for specific surface proteins with the help of a monoclonal antibody, immunostaining, or ELISA.

The growth pattern or growth characteristics of a cell line is very important to predict the growth phases and the duration of the cultures and doubling time of the cells. All this information is very essential when dealing with an industrial process or when setting up an experiment with animal cell culture. To analyze the growth pattern of a cell type it is necessary to develop a growth curve by following the rate of growth at regular intervals of time. The growth curve tells all the important factors regarding cell growth such as the doubling time, time taken by the cell cultures to approach the log phase and stationary phase, etc. A typical growth curve of a cell culture shows the lag phase, log phase, the stationary phase, and the expected cell density after a specific duration of cell culture. The lag phase of a growth curve represents the initial stage of the cell culture in which the cells take some time to recover from subculture and acclimatize with the new environmental conditions and begin to multiply or enter into the growth phase or log phase. During the lag phase, cells start attaching to the substratum and slowly spread to the position in which they are able to take nutrients and multiply. In the log phase, active cell growth starts by cell division and cell enlargement. In this phase cell multiplication occurs exponentially and therefore this phase is called the exponential phase. This phase is very limited and is closely related to the media availability and space. This is followed by the next phase, the stationary phase. In this phase, rate of growth declines and finally stops due to exhaustion of nutrients or confluency. When cell density increases to a particular level that will adversely affect the rate of growth and finally result in the stationary phase of cell growth. The rate of cell growth is an efficient technique to assess the effect of certain growth regulators, hormones, cytotoxic compounds, etc., on specific cell lines. The cell growth can be assessed primarily by cell counting using a hemocytometer or a cell counter.

A Colorimetric Method to Measure Cell Viability: MTT Assay

In order to compare the quantitative differences of a given response to any chemical or physical stimuli, the cell population under various treatments should be the same. But it has been observed that at certain times, a stimulus may kill the cells instead of giving them less or non-responsiveness. It is, therefore, essential to ensure that the response measurement is based on equal size of cell population. Tritiated thymidine (radioactive) has been traditionally used to determine cell proliferation, which requires the use and disposal of radioactive isotopes. A substitution based on measuring an enzymatic color formation is now widely used, thus bypassing the problem with isotope handling. The water soluble, yellow-colored MTT tetrazolium salt is converted by functional mitochondrial dehydrogenases in living cells to a non-water soluble blue-colored MTT formazan. The formazan color is proportional to the number of live cells, and thus can be used to measure relative cell viability and cytotoxicity.

20.4 SCALE-UP OF ANIMAL CELL CULTURE PROCESS

Animal cell-culture technology is used for the industrial production of a large number of therapeutic proteins, hormones, antibodies etc. Some of the important products of animal cell cultures include,

- Vaccine production
- Monoclonal antibody production
- Enzymes and hormones production
- *In vitro* skin cell growth and other organ tissues by stem-cell culture.

All these products are produced by the scale-up of laboratory-scale cell cultures capable of production of the respective compounds. Scale-up of a laboratory experiment to an industrial process requires the application of the principles of biochemical engineering. The scale up of animal cell cultures faces a number of special problems, which are much different from the microbial and plant cell cultures. Animal cells are comparatively larger and are very fragile. They cannot resist the shear stress and it results in mechanical damage. Most of the animal cell cultures are anchor-dependent and need a special type of culture vessel or bioreactor that support anchor-dependent growth. Fermentors have been used for the cultivation of bacteria and yeasts for a long time. Initially, fermentation was synonymous with alcohol production. Later, bacteriologists learned to use the same principles for the preparation of vitamins, organic acids, antibiotics, etc. This led to the rapid development of a variety of fermentors and methodologies.

Naturally, the same principles are applied for the mass cultivation of animal and plant cells. However, adaptation of these processes is required. Cultivation of plant and animal cells is difficult, mainly due to the slow metabolism of these cells, which is reflected by slow cell growth. Animal cells have complex nutritional requirements compared to bacteria and yeast. Thus, the media are complex and costly and prone to contamination. Animal cells lack the cell wall present in bacterial cells, making them fragile and shear sensitive. Therefore, the agitation and aeration systems have to be designed differently. Cell densities achieved are low resulting in low concentration of products. Downstream-processing procedures required to concentrate and purify the product lead to increased costs. In spite of these drawbacks, fermentors have been used for growing animal cells for the last few decades. Different cell lines such as BHK-21, namalwa cells, etc., have been grown in fermentors as submerged cultures for the production of viral vaccines and other products. The shear sensitivity and fragility of animal cells are overcome by the introduction of novel impellers, which are paddle shaped. Direct-gas sparging produces bubbles capable of rupturing cells, and hence the supply of gas needs to be done via diffusion through silicone tubing. The medium contains plenty of serum proteins capable of generating froth. Therefore, agitation has to be slow and gentle. For high-density cultivation, oxygen supply becomes critical. The silicone-tubing method of aeration has many advantages. No bubbles are formed and the oxygen transfer rate has been found to be satisfactory. However, the winding of the fragile tubing is difficult and is limited to small-scale reactors used in laboratories. Thus, suitably modified fermentors can be used for the mass culture of animal cells growing as suspension cultures. If an anchorage-dependent cell line needs to be grown, it becomes necessary to use a carrier system such as a microcarrier.

Roller bottles and microcarriers are the techniques adopted for the scale-up of anchor-dependent cell cultures and spinner flasks are used for the scale-up of suspension cultures. But recently it was shown that spinner flasks can also be used for the culturing of anchored cells with the help of microcarriers or other types of supports that can be accommodated within the flask. Spinner baskets, a modified version of spinner flasks, are an example for this.

Anchor-dependent Culture

Anchor-dependent cell lines can be grown with any of the following conventional methods. They include MD bottles, t-flasks, roux bottles, and roller bottles. If production is to be increased, the number of units has to be increased making the process time consuming and laborious. The result is that the product may not be cost effective. Roller bottles are one of the convenient methods for the culture of anchor-dependent cell cultures.

Roller Bottles

Roller bottles are cylindrical bottles made up of glass or special plastics with a curved internal surface area in which the cells can adhere as a monolayer and grow. These culture bottles can be

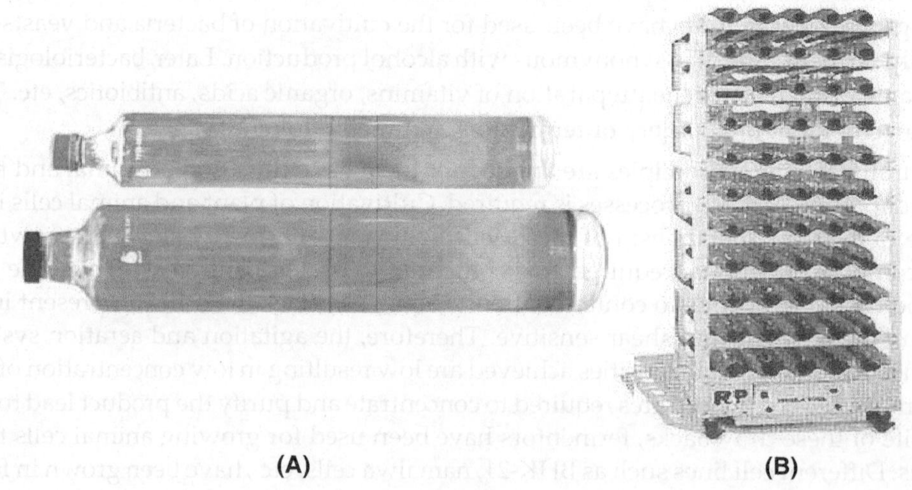

(A) (B)

FIGURE 20.3 **(A) Roller bottles for the cultivation of anchor-dependent cells. (B) Attachments for rotating the roller bottles in CO_2 incubator.**

incubated in special types of carbon-dioxide incubators provided with special attachments for slowly rotating the bottles along the long axis continuously. During each rotation the monolayer cells are transiently exposed to the medium, which is required in very low quantity to form a shallow covering over the cells attached to the inner surface of the culture bottle. But these conventional methods have limitations in the volume and cost of production. There are a number of modifications that can produce a larger quantity of cells and can reduce the cost of production.

The following are the new trends in the large-scale culturing of anchor-dependent cells for industrial purposes:

1. **Multiple-surface tissue-culture propagators.** This consists of stacks of glass plates maintained in a supporting rack within a glass or acrylic culture vessel. These serve as the substratum on which cells can grow. There is an airlift pump, which circulates the medium at a predetermined rate. A sparger is provided to pass air and CO_2.
2. **Capillaries.** Semi-permeable membranes, in the shape of capillaries 200 to 300 μ in diameter, are bundled together in a glass culture vessel and medium is circulated continuously. The cells grow along the walls of the capillary. The nutrients can diffuse in and products can diffuse out of the semi-permeable membrane. Thus, it is possible to operate such a system continuously.
3. **As aggregates.** This is possible mainly with insect cell lines. Enzyme treatments can release cells growing as a monolayer, and they form aggregates ranging from 20 to 500 μ in diameter. These aggregates are then grown as a suspension culture. The cells in the aggregates multiply and release products in the surrounding medium.
4. **Encapsulation.** Cells can be trapped inside beads made up of agarose, karageenan, calcium, alginate, etc. These beads are then introduced into stirred or airlift reactors. The products released remain localized in the material, which we use to make the beads thereby making downstream processing very easy.
5. **Microcarriers.** These are bead-like structures ranging from 100 to 180 μ in diameter that can be held in homogeneous suspension in stirred or airlift reactors. They are capable of providing a large surface area to cells in small volumes of media. Commercially-available microcarriers have been made from a variety of materials such as DEAE, Sephadex, cellulose, glass, gelatin, etc. Cells attach and spread on the microcarriers and gradually grow into confluent monolayers. The microcarriers are held in a medium that is constantly stirred. The features of both suspension and monolayer culture are brought together in this system. Porous microcarriers, which allow growth of cells inside them, are called as macrocarriers (Figure 20.4). There are different qualities of **macrocarriers** available commercially (e.g., Cultispher-G). These types of macrocarriers will produce better yields; the cells growing in them are not exposed to shearing forces in fermentors. Such microcarriers and macrocarriers have been used for cultivating different cells such as vero and CHO and also for viruses such as VSV.

   This system has the following advantages over other methods of large-scale cultivation:

   - High surface area to volume ratio can be achieved, which can be varied by changing the microcarrier concentration. This leads to high cell densities per unit volume with a potential for obtaining highly concentrated cell products.
   - Cell propagation can be carried out in a single high productivity vessel instead of using many low productivity units, thus achieving better utilization and a considerable savings of medium.

- Since the microcarrier culture is well mixed, it is easy to monitor and control different environmental conditions such as pH, PO_2, PCO_2, etc.
- Cell sampling is easy.
- Since the beads settle down easily, cell harvesting and downstream processing of products is easy.
- Microcarrier cultures can be relatively easily scaled-up using conventional equipment such as fermentors that have been suitably modified.

Because of the many advantages of the technique itself, it has gained great popularity. Thus, a large variety of microcarriers are available on the market.

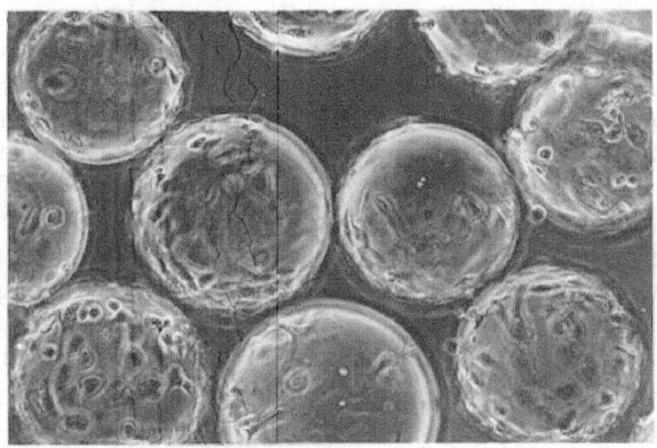

FIGURE 20.4 **Vero cells cultured on cytodex microcarriers.**

6. **Fixed-bed reactors.** Microcarriers, macrocarriers, or encapsulated beads could be used in fixed-bed reactors. The cells are immobilized in a matrix and the culture fluid is circulated in a closed loop. There is no agitation system. If the bed of immobilized cells is disturbed by the circulating medium, it is said to be a fluidized-bed reactor. Such a process achieves a high degree of aeration and agitation.

Suspension Culture

The above-mentioned culture techniques and bioreactors are meant for anchor-dependent cells, and the following methods are available for growing cells as a suspension.

Hollow fiber reactor. These consist of semi-permeable membranes packed together. In the lumen of these 'fibers', cells can grow. The semi-permeable membrane has a definite cut-off and hence nutrients can diffuse into the lumen and products and by-products of metabolism can diffuse out. Since the medium is constantly circulated, products can be obtained continuously. Large surface area to volume is provided for cell growth. Such systems are now available as highly automated systems.

Spinner flasks. These culture vessels or bioreactors are very suitable for the large-scale cultivation of suspension cells such as lymphocytes. They consist of a flask made up of glass or polymer plastics with a suspended Teflon paddle in the center, which is magnetic and can be agitated when placed on a magnetic stirrer. The spinner bottle may be provided with two or more side arms to facilitate aeration with CO_2 and oxygen, for sampling and removal of medium for product recovery. This can be used for both suspension cells and also for anchor-dependent cells attached to microcarriers or macrocarriers (Figure 20.5).

FIGURE 20.5 **Different types of spinner flasks.**

Air-lift fermentors. These have a cylinder inside the culture vessel through which air is passed for circulating the medium. It is a very well-aerated system. This type of fermentor is used mainly for monoclonal antibody production.

Chemostats. This is a continuously stirred tank reactor with an inlet and outlet so that the medium can be passed continuously at a fixed rate keeping the volume constant. The theory for chemostats was established by Monod[8] and Novick and Szillard, independently, in 1942. The culture is first grown in batch mode, and when cells reach the log phase they are grown in continuous mode. The medium employed has one growth-limiting nutrient and the rest are in excess. Hence, cell growth is proportional to the growth-limiting nutrient. A steady state is obtained by a continuous and constant feed and harvesting rate where the dilution rate is determined by growth rate.

Suspension Culture Technology

The following modes of cultivation are available for cells grown as a suspension in bioreactors:

Batch culture. Until a few years ago, most biologicals were produced as a single batch. Growth is determined by the rate of nutrient depletion and toxic metabolite accumulation. Vaccines and interferons are still produced in this mode. The yields are low compared to other modes. In this type of culture, cells are inoculated into a fixed volume of medium. They consume nutrients as they grow and release toxic metabolites into the surrounding medium. This results in the continuous variation in the chemical and physical environments such as osmolarity, pH, and depletion of nutrients. All these variations eventually cause cells to cease multiplication and die.

[8] Monod J."Recherches sur la croissance des cultures bacteriennes" Hermann, 1942.

Fed-batch culture. The culture is fed intermittently either with the limiting nutrient or with the whole growth medium. Cell growth is limited not by nutrients but by the accumulation of waste products. In this mode, a higher cell concentration is achieved and cell viability is maintained for larger periods compared to the batch mode.

Continuous culture. This mode is based on the principle established by Monod in 1942. Homoestatic culture conditions are maintained in chemostats with no fluctuations of nutrients, metabolites, or cell numbers. Once established they can be maintained at high cell density and high product yield for a long period.

Perfusion culture. In contrast to continuous culture where a steady state is maintained by constant dilution of the culture, in perfusion cultures, the cells are physically retained in the vessel. Spent medium is withdrawn continuously from the culture system and an equal volume of fresh medium is added. Cell concentration increases constantly until it becomes limited at high cell population densities.

20.5 APPLICATIONS OF ANIMAL-CELL CULTURES

Animal-cell cultures are used for various basic research as well as for commercial purposes as the source of valuable therapeutic chemicals and agents. There is an ever-increasing list of proteins and other biologicals that can be produced by culturing mammalian and other types of cells. Some of the major services and different products obtained from mammalian cells have been summarized below. In addition to the natural compounds produced by the cultured cells, genetically-altered cells can be used for the production of recombinant proteins.

- Vaccine production
- Monoclonal antibody production
- Enzymes and hormones production
- *In vitro* skin and other tissues and organs by stem culturing
- Viral cultivation

Some of the products, both recombinant and natural proteins, produced by animal-cell cultures are tabulated below:

TABLE 20.1 Products of animal-cell cultures available on the market or under clinical trials.

Products	Medical Application
Urokinase	Heart attacks
Tissue plasminogen activator	Heart attacks
FSH, Factor VIII	Hemophilia A
Factor IX	Hemophilia B

Human-rabies, mumps, rubella etc. Veterinary-FMD vaccine, New Castle's Disease etc.	Vaccines
Monoclonal antibodies Hybridoma technology	Various diagnostic tools
Baculovirus Bioinsecticides	Virus cultivation
Interferons and interleukins	Used as immunoregulators
Toxicological testing	Cultured cells as experimental models
Stem cells	Stem cell culturing various basic research for clinical applications

Tissue Plasminogen Activator (t-PA)

This is an anti-clot drug developed to dissolve blood clots selectively. It is a natural protease secreted by mammalian cells and is used against heart attacks and other problems due to the formation of blocks in the blood flow. Cultured mammalian cells are transformed with the gene of t-PA and the cells capable of secreting high levels of t-PA are selected and cultivated in a suitable bioreactor for commercial production as a pharmaceutical compound (Figure 20.6). It is one of the first products of mammalian cell culture that hit the market first.

Plasminogen acts on blood clots and dissolves the clot resulting in the removal of block in the blood vessels and capillaries. The different steps in the commercial production of action of t-plasminogen activator are shown in Figure 20.6.

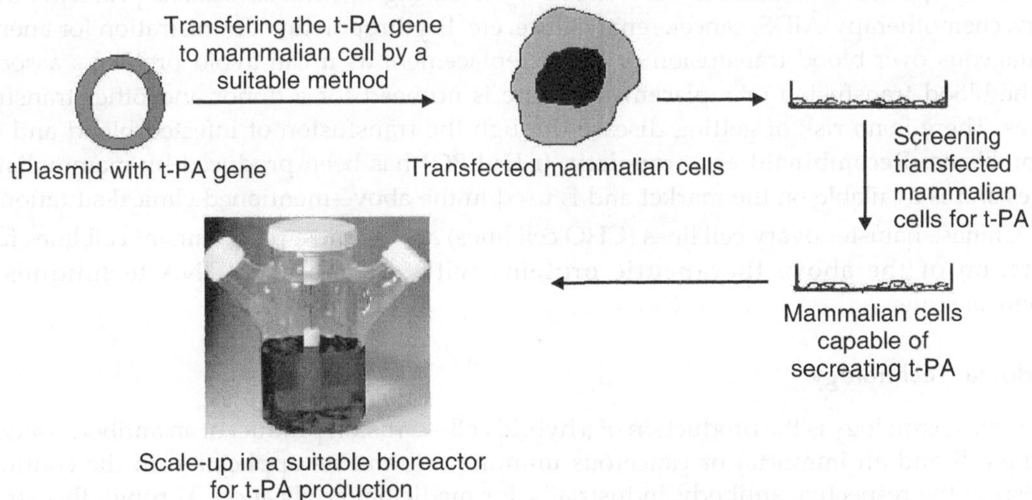

FIGURE 20.6 **Genetic transformation and culturing of mammalian cells for t-PA production on a commercial scale.**

Factor VIII and Factor IX

These are also the first of the few successful products of animal cell culture that have come to the market. Factor VIII and IX are the complex proteins required for the normal clotting of blood, which is activated at the time of injuries and causes the rupture of the blood vessels. The malfunctioning or absence of any one of these protein factors results in the failure of clot formation and causes bleeding. This condition is known as hemophilia.

Hemophilia A is due to the absence or malfunctioning of factor VIII in the blood, an important factor essential for the clotting of blood. This is the most commonly inherited X-linked genetic disorder. With the help of recombinant DNA technology and mammalian cell cultures, factor VIII has reached the market and is now accessible. Factor VIII is a very complex glycoprotein with at least 25 sites with glycosilation. Because of this structural complexity, which is created by very specific post-translation modification, transgenic bacteria cannot produce biologically active factor VIII.

Hemophilia B is another genetic defect caused by the absence of a protein factor in the blood-factor IX due to a defective gene or a mutated gene. This product is also available on the pharmaceutical market as a product of animal cell culture and genetic engineering.

Erythropoietin (EPO)

This is an important growth factor or hormone-like protein responsible for the formation of blood cells from hematopoietic stem cells in the bone marrow. During the conditions of hypoxia (shortage of oxygen) or anoxia (lack of oxygen), a condition caused by anemia, kidney cells release erythropoietin into the bloodstream and activate the formation of blood cells, particularly erythrocytes.

Erythropoietin treatment is very effective for curing anemia-associated problems during surgery, chemotherapy, AIDS, cancer, renal failure, etc. Erythropoietin administration for anemia is advantageous over blood transfusion or blood replacement as it can avoid problems associated with the blood transfusion or replacement. There is no need for a donor and other transfusion facilities. There is no risk of getting disease through the transfusion of infected blood and other such problems. **Recombinant erythropoietin (r-HuEPO)** has been produced by mammalian cell cultures and is available on the market and is used in the above-mentioned clinical situations.

Chinese hamster ovary cell lines (CHO cell lines) are the most predominant cell lines for the production of the above therapeutic proteins with recombinant DNA techniques in a commercial scale.

Hybridoma Technology

Hybridoma technology is the production of a hybrid cell—a fusion product of an antibody-secreting immune cell and an immortal or cancerous immune cell and its application in the continuous synthesis of the respective antibody, industrially, for medicinal application. Through this strategy scientists are now able to synthesize substances secreted by cells of the immune system—both antibodies and lymphokines—for commercial purposes. The ready supply of these materials has not only revolutionized immunology but has also created a resounding impact throughout medicine industry.

A hybridoma is created by fusing two cells, a secreting cell from the immune system and a long-lived (immortal) cancerous immune cell, within a single membrane. The resulting hybrid cell can be cloned, producing many identical cells. Each of these daughter clones can secrete the immune cell product over a long period of time (Figure 20.7).

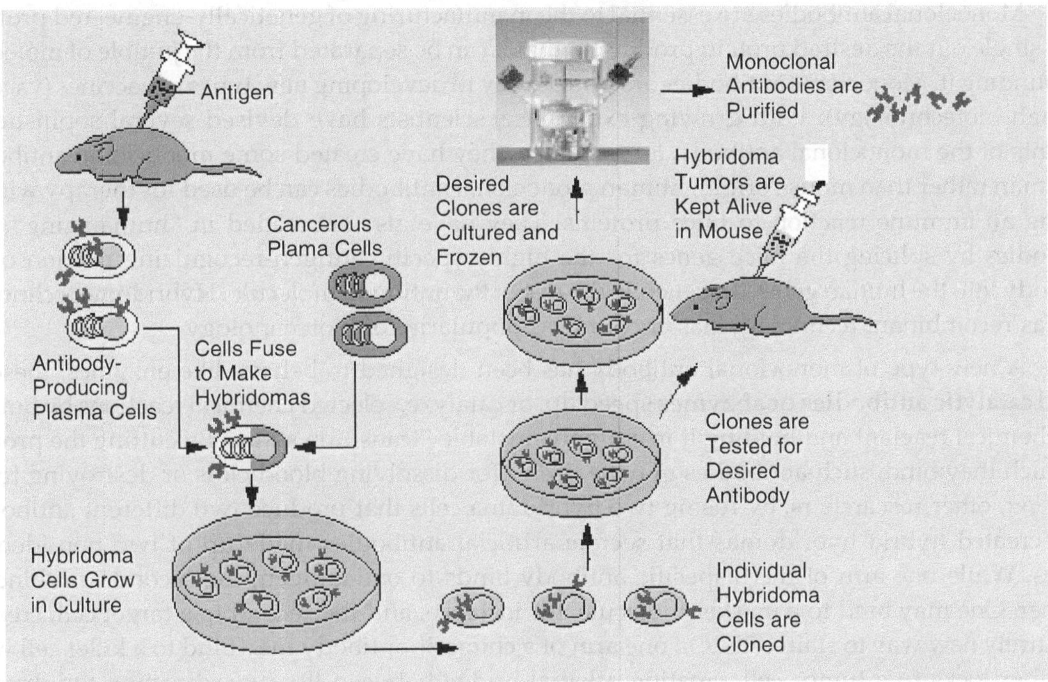

FIGURE 20.7 **Hybridoma technology. Different steps in the production of monoclonal antibody production.**

Monoclonal Antibodies

It was observed that when foreign substances enter the body, such as disease-causing bacteria and viruses and other infectious agents known as antigens, the body's immune system recognizes them as invaders. In response, our natural defense mechanisms produce certain proteins known as **antibodies**. These proteins are very specific to the antigens and neutralize the toxins or the invading foreign particles by binding with them. This process helps to destroy them. The antibodies produced by the immune system bind to certain specific sites or domains on the antigen macromolecules. These antigen sites are known as **epitopes**.

There are two basic and very useful characteristics of the antibodies produced by the immune cells of the system. First, they are extremely specific; that is, each antibody binds to and attacks one particular antigen. Second, most of the antibodies, once activated by the antigen (occurrence of a disease), continue to confer resistance against that disease or pathogen. Antibodies to childhood disease such as chickenpox and measles are very good examples. The second characteristic of antibodies makes it possible to develop vaccines. A vaccine is a preparation of killed or weakened bacteria or viruses that, when introduced into the body, stimulate the production of antibodies against the antigens it contains. It is the first trait of antibodies, their specificity, which makes monoclonal antibody technology so valuable. Thus, specific antibodies can not only act as therapeutic agents to protect against disease, they can also help to diagnose a wide variety of illnesses, and can detect the presence of drugs, viral and bacterial products, and other unusual or abnormal substances in the blood.

Monoclonal antibodies are essential to the manufacturing of genetically-engineered proteins. They single out the desired protein product so that it can be separated from the jumble of molecules surrounding it. Monoclonal antibodies are also the key to developing new types of vaccines (vaccines through biotechnology). With growing experience, scientists have devised several sophisticated variants of the monoclonal antibody. For instance, they have created some monoclonal antibodies of human rather than mouse origin; human monoclonal antibodies can be used for therapy without risking an immune reaction to mice proteins. They have also succeeded in "humanizing" mice antibodies by splicing the mice genes for the highly specific antigen-recognizing portion of the antibody into the human genes that encode the rest of the antibody molecule. Hybridoma technology such as recombinant technology has encouraged popularity of biotechnology.

A new type of monoclonal antibody has been designed to behave like enzymes; these so-called **catalytic antibodies** or **abzymes** speed up, or catalyze, selected chemical reactions by binding to a chemical reactant and holding it in a highly unstable "transition state." By cutting the proteins to which they bind, such antibodies may be useful for dissolving blood clots or destroying tumor cells. Yet, other researchers, by fusing two hybridoma cells that produce two different antibodies, have created hybrid hybridomas that secrete artificial antibodies made up of two non-identical halves. While one arm of the bispecific antibody binds to one antigen, the second arm binds to another. One may bind to a marker molecule, for instance, and the second to a target cell, creating an entirely new way to stain cells. Or, one arm of a chimeric antibody may bind to a killer cell while the other locks to a tumor cell, creating a lethal bridge between the two, directing the drugs to tumor cells.

But the antibodies produced in a living system are a heterogenous population originated from different types of immune cells (B-lymphocytes), and therefore are known as **polyclonal antibodies**. They can bind to more than one epitope of the antigen. The conventional method for the production of antibodies was to inject a laboratory animal with an antigen and then, after antibodies had been formed, collect those antibodies from the blood serum (antibody-containing blood serum is called antiserum). There are two problems with this method: It yields antiserum that contain undesired substances; that is, antibodies produced by different lymphocytes; and it provides a very small amount of usable antibody. These problems were solved by the development of techniques for the production of antibody secreted by a single type of immune cell. These antibodies are known as **monoclonal antibodies**.

Monoclonal antibodies can be produced with the hybridoma technology. A hybridoma can be produced by injecting a specific antigen into a mouse, collecting antibody-producing cells from the mouse's spleen, and then fusing them with long-lived cancerous immune cells. This hybrid cell of antigen-activated B-lymphocyte and cancerous lymphocyte or myeloma cell is known as **hybridoma**. Individual hybridoma cells are cloned and screened to find hybridoma that produces the desired antibody. The selected hybridoma is cloned and identical daughter clones will secrete the monoclonal antibody over a long period of time, and can be used for commercial application.

The different steps in the production of a hybridoma and production of monoclonal antibodies are illustrated in Figure 20.7. It was Cesar Milstein and George Kohler[9] who developed the hybridoma by fusing antibody-producing B-lymphocytes having limited lifespan with myeloma or cancerous B-cells in the presence of polyethylene glycol (PEG). They received the Nobel Prize for developing the hybridoma technology for producing monoclonal antibodies.

Therapeutic Antibodies

The development of monoclonal antibodies has resulted in remarkable new approaches for preventing, diagnosing, and treating disease. Monoclonal antibodies are used, for instance, to distinguish subsets of B-cells and T-cells. This knowledge is helpful not only for basic research but also for identifying different types of leukemias and lymphomas and allowing physicians to tailor treatment accordingly. Quantitating the number of B-cells and helper T-cells is all-important in immune disorders such as AIDS. Monoclonal antibodies are being used to track cancer antigens, alone or linked to anticancer agents, and to attack cancer metastases. The monoclonal antibody known as OKT3 is saving organ transplant patients threatened with rejection, and preventing bone-marrow transplant patients from setting off graft-versus-host disease. T-cells have an important role in rejecting the foreign tissues in organ and tissue transplantation. Therefore, as a prelude to the organ transplantation, the patient has to be treated to suppress the T-cells. Otherwise the organ or tissue transplantation will end in failure. Antibodies are raised against some surface antigen present on the membranes of the T-cells. The antibodies raised in such a way can bind to the T-cell and inactivate it effectively. The surface antigen CD3 is exposed to develop its specific antibodies (anti-CD3MoAb). When the patient is treated with anti-CD3 monoclonal antibodies known as OKT3 it will bind to the T-cell specific CD3 surface antigens and remove it from circulation in the organ. This can prevent organ or tissue rejection during transplantation or grafting.

20.6 STEM-CELL TECHNOLOGY

Advances in the uses of traditional tissue culture for human applications are now focused on the culture of stem cells. Stem cells are special types of cells that provide tissue renewal and may be used to grow tissue or organs for replacements. A stem cell is a primitive type of cell capable of developing into most of the 220 types of cells found in the human body. Stem cells can be isolated from blood, brain, or muscle tissue, but can now also be cultured from early human embryos. The advantage of embryonic stem cells is that they are 'master' cells, able to develop into any sort of tissue type (e.g., bone, muscle, nerve tissue, blood cells, heart cells, etc.). Some researchers regard them as offering the greatest potential for the alleviation of human suffering since the development of antibiotics.

Stem cells are cells that display two properties. When they divide by mitosis the daughter cells can either:

- Enter a path leading to a fully-differentiated cell or
- Remain a stem cell thus ensuring that the pool of stem cells is not 'used up.'

[9] Kohler G & Milstein C. 1975. Continuous cultures of fused cells secreting antibody of predefined specificity. *Nature* 256: 495-497.

Several adjectives are used to describe the developmental potential of stem cells; that is, the number of different kinds of differentiated cells that they can become.

Totipotent cells. In mammals, totipotent cells have the potential to become any type of cell in the adult body; any cell of the extraembryonic membrane (e.g., placenta). The only totipotent cell is the **fertilized egg** and the first four or so cells produced by its cleavage, as shown by the ability of mammals to produce identical twins, triplets, etc. In mammals, the expression of totipotent stem cells is a misnomer because these cells fail to meet the second criterion—they cannot make more of themselves.

Pluripotent stem cells. These are true stem cells, with the potential to make any differentiated cell in the body, but they cannot contribute to making the extra embryonic membranes (which are derived from the trophoblast). Three types of pluripotent stem cells have been found in mammals:

- **Embryonic Stem (ES) Cells.** These can be isolated from the inner cell mass (ICM) of the blastocyst—the stage of embryonic development when implantation occurs.
- **Embryonic Germ (EG) Cells.** These can be isolated from the precursor to the gonads in aborted fetuses.
- **Embryonic Carcinoma (EC) Cells.** These can be isolated from teratocarcinomas, a tumor that occasionally occurs in a gonad of a fetus. Unlike the other two, they are usually aneuploid.

All these three types of pluripotent stem cells can only be isolated from embryonic or fetal tissue and can be grown in culture, but only with special methods to prevent them from differentiating.

Multipotent stem cells. These are true stem cells but can only differentiate into a limited number of types. For example, bone marrow contains multipotent stem cells that give rise to all the cells of the blood but not to other types of cells. Multipotent stem cells are found in adult animals; perhaps every organ in the body (brain, liver) contains them where they can replace dead or damaged cells.

The most well-studied multipotent stem cells are the hematopoietic stem cells present in bone marrow and the spleen. The formation of blood cells from hematopoietic stem cells is known as **hematopoiesis.** In humans it is estimated that about one million hematopoietic stem cells produce one billion RBC, one billion platelets, one million T-cells, and the same number of B-cells per Kg of body weight. Although different types of human stem cells are yet to be isolated and purified, scientists have discovered that progenitor cells capable of giving rise to an array of blood cells if not of actually reproducing themselves carry the cell surface marker CD34. These cells can be sorted out from marrow and blood with monoclonal antibodies that recognize CD34. In experimental programs, CD34 cells are being tested as long-lived vehicles for gene therapy and as an alternative to bone-marrow transplants.

Identification of Stem Cells: The Morphological Approach

Different types of stem cells can be identified by different methods. Identification of stem cells by microscopic examination of morphological characteristics is one method. But this method is not very effective as there are a large number of different types of stem cells with the same or similar morphological features. For example, the mammalian hematopoietic system is charged with the life-long daily production of numerically large mature cell populations, which collectively represent a wide variety of peripheral blood cell lineages. Attempts to identify the different hematopoietic stem cells present in bone-marrow tissue by microscopic examination were ineffective due to similar morphology for different types of cells. There evolved an indirect method to identify or to understand the type of hematopoietic stem cells that continuously renew the blood cells. The method involved the study of the repopulation ability of blood cells in an irradiated mouse.

A genetically-identical mice population was raised through inbreeding; that is, by sibling mating for more than 20 generations. These genetically-identical mice populations were used for the studies. Two groups of mice were lethally irradiated to destroy the ability to produce the blood cells in the circulatory system. Of these two groups, one was injected with bone-marrow cells, the hematopoietic cells from the normal genetically identical mice. These mice were able to survive whereas the other group of irradiated mice, which did not receive the bone-marrow cells, died. On examining the spleen of the survived irradiated mice it was observed that they developed cells that produced colonies on culturing. These colonies were named **colony-forming units of the spleen** or **CFU-S**, and this technique is known as **repopulation assay**. When these colony-forming cells of the spleen were injected into another similar group of irradiated mice they also survived. This study and its observations has given rise to the present-day understanding of stem cells, in general, and hematopoiesis, in particular.

Hematopoiesis is self-renewing and resembles other developmental systems such as the small intestine, the epidermis, and hair follicles of the skin, as well as the male germ line. Other tissue types such as the central nervous system, the liver, and muscle seem to replenish mature cell types much more slowly or in response to injury. Greater than four decades of elegant *in vivo* transplantation studies have defined the activities of a rare bone-marrow stem cells that are both self-renewing and multipotential (pluripotent) in their abilities to produce all blood cell types in engrafted hosts permanently and clonally. Eventually, a large number of strategies were developed to purify the hematopoietic stem cell physically from adult bone marrow and from other sources such as a fetal liver. Individual hematopoietic stem-cell purification schemes come in different forms. There is general agreement that the degree of enrichment is quantitatively similar in most strategies and that the only reliable measure of stem-cell activity in any physically purified population is by *in vivo* transplantation. One particular purification strategy that is relevant to the present discussion employs vital dye uptake and efflux properties together with flow cytometry to define a subset of bone-marrow cells called the **side population (SP)**. These cells are highly enriched in transplantable stem cell activity. The complexity of the whole-animal assay system does not permit accurate estimates of the absolute homogeneity of any purification protocol; that is, it is not possible to obtain a quantitatively rigorous one-to-one 'mapping' of stem-cell activity on to individual physically purified cells. Remarkably, however, in some cases, it has been possible to show that a single transplanted stem cell is both necessary and sufficient to transfer an intact, normal hematopoietic system to a recipient host. Although less extensive, other studies have physically identified candidate stem cells from a number of other tissues.

The *In vitro* Clonal Assay

This is based on the *in vitro* culturing of stem cells such as hematopoietic cells and their assay by different methods. Stem-cells proliferate in culture media to form the clones or colonies of different types of cells. This has helped researchers carry out a large number of assays on stem cells to identify them and to study the differentiation of these cells. These *in vitro* assays also helped to study the hormonal requirements that initiate or promote a specific type of cell formation or differentiation. For example, it was possible to identify the growth factors required for the formation of various types blood cells such as RBC and different types of lymphocytes from the hematopoietic cells of bone marrow. These types of stem-cell cultures form a very good system to study the influence of certain hormones and growth factors on cell differentiation and related processes. The role of erythropoietin, one of the first biotechnology products of animal cell culture marketed, was assayed by this procedure.

Long-term Bone Marrow Cells

The *in vitro* culturing of bone-marrow cells on plastic and glass surfaces has helped to study the process of hematopoiesis under *in vitro* conditions resulting in the development of various types of blood cells. Figure 20.8 shows the different stages in the differentiation of stem cells of bone marrow into erythrocytes *in vitro*. The process of hematopoiesis can be divided into four stages:

1. Stem cells that can retain the capacity for self-renewing. These cells are pluripotent cells or multipotent cells.
2. Stem cells with limited self-renewing capacity, which are uni- or bi-potential cells. Their differentiation capacity is very limited, and may develop into one or two types of mature or differentiated cells.
3. Precursor cells that do not have any self-renewing ability, and therefore cannot divide, but can undergo differentiation to form mature highly differentiated cells.
4. The last stage of the process of hematopoiesis, which results in the mature and completely differentiated cells.

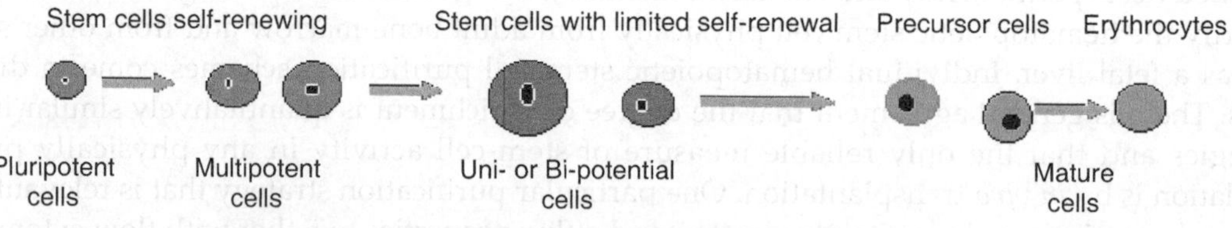

FIGURE 20.8 **Hematopoiesis—differentiation of pluripotent hematopoietic stem cells leading to the formation of mature blood cells through uni- or bi-potent cells.**

This technique is also very useful to understand the formation of various blood cells from the pluripotent stem cells of bone marrow with the process of hematopoiesis and also to study the effect or role of various hormones and growth factors on cell differentiation. The information gathered from these studies has helped to carry out bone-marrow transplantation as treatment for leukemia or blood cancers. Bone-marrow transplantation or grafting is carried out with bone-marrow tissues containing a high population of purified hematopoietic stem cells.

Embryonic Stem-cell Cultures

A zygote, just after its formation by fertilization, develops into an eight-celled stage and then into a group of cells by mitotic cleavage. The cells in the eight-celled stage (morula) are *totipotent* and are called **embryonic cells**. It can produce all types of tissue. Thus, it is the very early stage of a recently fertilized egg. When it becomes a group of cells in the next stage known as blastula, the inner cells are concentrated to a side and become *pluripotent*. These embryonic stem cells can become almost all types of cells. On further development, the cells become more differentiated and become adult stem cells. Adult stem cells are typically called *multipotent* cells such as bone marrow cells, which can produce a wide range of different blood cells.

Pluripotent embryonic stem cells are 'harvested' or collected from the very early stages of a fertilized egg called a blastocyst and are maintained in tissue cultures along with fibroblasts cells. These embryonic totipotent and pluripotent cells can also be derived by *in vitro* fertilization and culturing of the zygote to the eight-celled stage and blastocyst. If the cells are derived from the eight-celled stage they are totipotent embryonic cells, which are not stable and may get differentiated into pluripotent stem cells in culture. If the cells in culture are derived from the inner cells of a blastocyst, they are pluripotent embryonic stem cells. Adult stem cells or multipotent cells are collected from a limited number of cell types in the body. Typically, these are bone-marrow cells or cells from the very early stages of tissue development.

The studies conducted on mice embryonic stem cells has revealed that:

- These cells were able to maintain the embryonic nature and their pluripotency for a very long time in culture without any decline in properties.
- The cells were able to reintegrate into the process of embryogenesis if returned into the early embryonic condition such as the blastula stage. This was proved correct by creating chimeric mice.
- The pluripotent embryonic cell cultures were found to maintain a constant and stable euploid (actual diploid chromosome number of a species) karyotype.
- The most important point is that they could divide and produce the same type of embryonic cells without undergoing any further differentiation in specific *in vitro* conditions.

These embryonic stem cells can be manipulated genetically by introducing foreign genes or selectively removing genes and can be used for creating various types of mouse models to study various genetic diseases. The experimental models created by removing or silencing a specific gene or genes is known as **gene knockouts.**

Development of Chimeric Mouse

Cultured pluripotent embryonic stem cells (ES cells) were used for creating chimeric mice. ES cells taken from a black mouse were maintained in culture. This cell was implanted into an embryo of an albino mouse at the blastula stage as illustrated in Figure 20.9. The progeny developed from this embryo had skin color with white and black patches known as chimera.

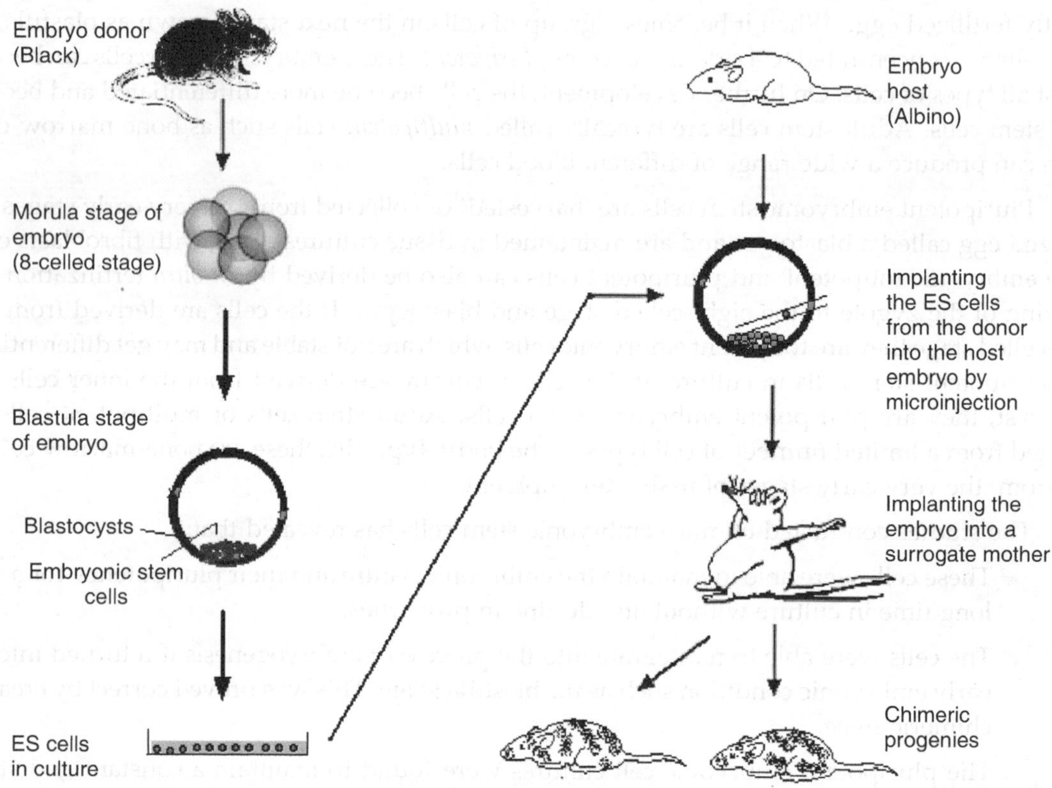

FIGURE 20.9 **Creation of chimeric mouse by injecting embryonic stem cells into the developing embryo at the blastula stage.**

Human Embryonic Cells

It has been demonstrated that human embryonic stem cells can be cultured and maintained just like mice ES cells. The inner cell mass of a blastocyst can be derived from embryos developed from in vitro fertilization or from germ-cell precursors. These pluripotent stem cells or embryonic stem cells can be maintained in a medium along with feeder fibroblasts cells. These stem cells can be cultured and specific cell lines or tissues may be derived or created from these with the help of specific growth factors such as **fibroblast growth factor (FGF)** or **platelet derived growth factor (PDGF)**. The studies on the manipulation of embryonic stem cells and their *in vitro* differentiation studies along with genetic transformation techniques have opened up new avenues in treatments based on cell and tissue engineering and gene therapy (Figure 20.10).

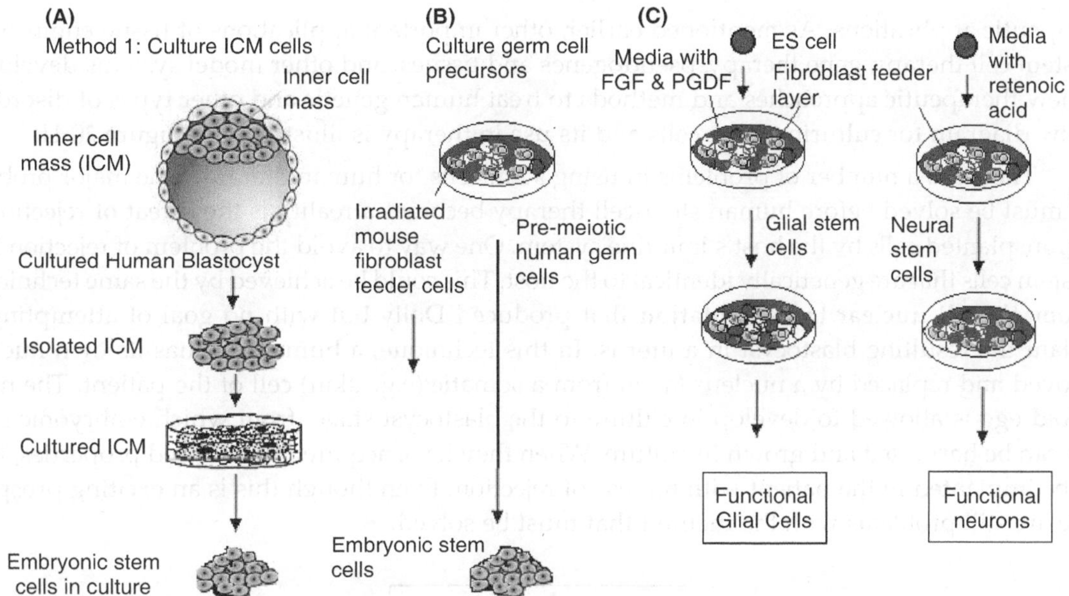

FIGURE 20.10 **Use of embryonic stem cells (ES cells) as therapeutics. (A) Development and *in vitro* culture of human embryonic stem cells from inner cell mass (ICM) of blastula stage of embryo or from (B) germ-line cells before meiosis. These stem cells can differentiate in culture to form the more restricted stem cells such as neural, blood, muscle, etc. (C) Differentiation of ES cells in culture into lineage restricted cells. The same type of ES cells differentiate into glial cell and neuronal cells by altering the components in the culture media.**

Cell and Tissue Engineering: Application of Stem Cells

The dream of using stem cells for human therapy surely will be a reality in the near future. It may be used for the treatment of many medical problems arising from damage to differentiated cells. For example, insulin-dependent diabetes mellitus (IDDM) where the beta cells of the pancreas have been destroyed by an autoimmune attack; Parkinson's disease, where dopamine secreting cells of the brain have been destroyed; spinal cord injuries leading to paralysis of the skeletal muscles; ischemic stroke, where a blood clot in the brain has caused neurons to die from oxygen starvation; multiple sclerosis with its loss of myelin sheaths around axons, etc., are some of the diseases caused by tissue damages, which may be cured by stem-cell therapy combined with gene therapy. The great developmental potential of stem cells has created intense research into enlisting them to aid in replacing the lost cells of such disorders. Even though some success has been achieved with laboratory animals, not much has yet been achieved with humans.

Biotechnologists are able to grow skin cells, cardiac tissues and cells, blood cells and hematopoietic cells, etc., *in vitro*, which can be used for therapeutic purposes. Sheets of cultured skin tissues are already available for grafting on to the burn victims. *In vitro* cell cultures of different types of stem cells and tissue engineering can supply the required tissues or cells for repair of damaged tissues and organs, without causing any immune response, infection, or damaging other body parts of the patient. Primary cell types such as keratinocytes, epithelial hepatocytes, and endothelial cells have great commercial potential in a wide range of emerging technologies and therapeutic applications. As mentioned earlier, other important applications of tissue engineering are stem cell therapy, gene therapy, pseudogenes and tissues, and other model systems developed for new therapeutic approaches and methods to treat human genetic and other types of disorders. A flow diagram for culturing stem cells and its use in therapy is illustrated in Figure 20.11.

There are a number of problems in using stem cells for human therapy. One major problem that must be solved before human stem-cell therapy becomes a reality is the threat of rejection of the transplanted cells by the host's immune system. One way to avoid the problem of rejection is to use stem cells that are genetically identical to the host. This could be achieved by the same techniques of **somatic cell nuclear transplantation** that produced Dolly but with no goal of attempting to implant the resulting blastocyst in a uterus. In this technique, a human egg has its own nucleus removed and replaced by a nucleus taken from a somatic (e.g., skin) cell of the patient. The now-diploid egg is allowed to develop in culture to the blastocyst stage, from which embryonic stem cells can be harvested and grown in culture. When they have acquired the desired properties, they can be implanted in the patient with no fear of rejection. Even though this is an exciting prospect, there are still problems with the method that must be solved.

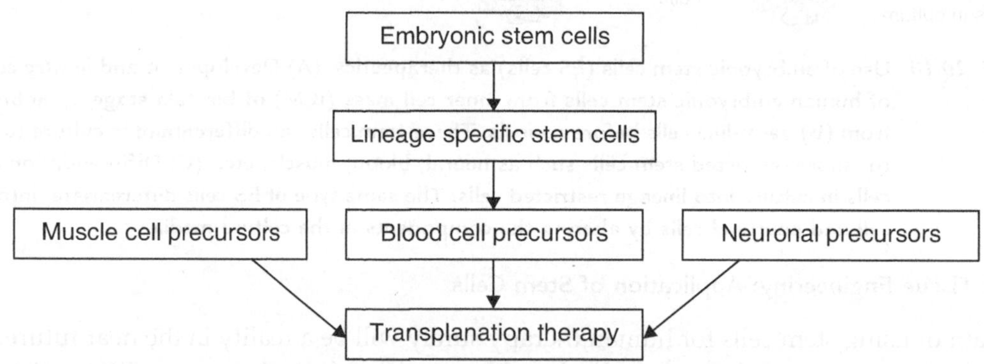

FIGURE 20.11 **Stem cell therapy. Use of stem-cell cultures for the development of specific stem-cell lineages and their use in transplantation therapy.**

20.7 BIOETHICS IN ANIMAL GENETIC ENGINEERING

The use of animal biotechnology, particularly genetic transformation techniques, human cell culture, hybridoma technology, and advances in cell culture techniques have developed new opportunities in medicine to develop new, more effective drugs and other therapeutics and approaches in the treatment of deadly genetic disorders. Studying the genetics of humans is allowing us to understand what happens when genes go wrong in inherited diseases, or cancers, and to start to develop new therapies that treat the genetic cause, not the symptoms. Development of transgenic farm animals that are resistant to disease has already changed the course of animal farming. But there are variety of ethical problems and concerns in creation of transgenic farm animals, stem cell research, and cloning and gene therapy.

Some of the major ethical concerns associated with transgenic animals are the following:

1. Creation of transgenic animals with human genes. Certain animals that are used for food such as sheep and pigs were genetically engineered with human genes to produce certain therapeutics such as factor VIII, factor IX, t-PA, etc.
2. Transfer of genes of animal proteins to food plants; some vegetarians are of the opinion that eating this plant part is equivalent to eating non-vegetarian food.
3. Transfer of genes from certain sources, which is considered forbidden by certain religious groups, to food animals they normally use.
4. The use of animal feeds containing human genes. Yeast single-cell protein is a very good cattle feed. Yeast is also used as a host organism to produce recombinant human proteins of therapeutic value. The spent recombinant yeast cells may be used as the component of the animal feed.

Based on the above concerns and objections it was recommended by the concerned authorities to adopt some precautions or to discourage the use of transgenic organisms as food, when the normal foods with the same nutritive value are available. Food materials containing the foreign genes have to be properly labeled with all details such as the type of gene, its protein, and the source organism, so that people will have the choice of avoiding foods containing genes from the unethical sources according to their religious practices and beliefs.

The advances made in cell-culture and tissue-culture technology could replace the animal models for specific experiments related to drug development and drug metabolism. This has resulted in substantial reduction in the use of different types of animals for studying cancer biology and other aspects of genetics and metabolism. But transgenic animals created for special purposes like production of vaccines or some other proteins or enzymes, which are normally not present in their system, may adversely affect the normal behavior and health of the organism. Considering these facts, there is severe moral opposition against the creation of transgenic animals as it may result in the suffering of animals. For example, there is a report that when animals were treated with transgenic growth hormones that enhance meat quality, they developed acute arthritis. All genetic engineering studies whether it involves animals, plants, or microbes should consider and evaluate the potential risk of its misuse or accidental faults and errors directly to humans, other animals, and also to the environment in a very systematic way. All the experiments and development of transgenic animals or gene therapy or cell culture should meet certain basic safety measures in the laboratory as well as with the organisms to minimize possible risk factors. The following are some of the risks that have to be addressed satisfactorily:

1. Transgenic animals may escape into the environment and breed with the wild types, which can result in the transfer of that gene into other populations. So transgenic animals have to be protected from other populations.
2. The use of pathogenic viruses and retroviruses as vectors for delivering DNA into genomes may be very risky in the sense that these viruses might be mutated to pathogenic form and might cause disease or other problems.
3. Public concern over the possible risk to humans and other animals due to the consumption of genetically-modified food materials. The public should be properly informed of the actual facts.

4. Application of drug-resistant genes as markers of genetic transformation is a usual practice in genetic engineering. There is the risk of transferring this gene into other undesired organisms or pathogens, which may pose a threat to humans and other organisms.
5. Uncontrolled production of disease-resistant organisms, both animals and plants, can create an imbalance in nature due to the destruction of other organisms.
6. At times of scarcity of healthy human organs or tissue for transplantation, it is recommended to use the tissues and organs from suitable compatible animals such as pigs. This method of transplantation of organs or tissue grafting from animals is called **xenotransplantation**. It also poses concerns about the possible infection through the organs and tissues if the animals are infected or have disease in the dormant form. But, the use of healthy animals without any type of infections or diseases can solve this problem to a greater extent.

Human blood, blood products, body fluids, and tissues are listed as potentially hazardous biological materials. Biosafety practices and procedures must be followed when handling human blood, blood products, body fluids, and tissues because of the infectious agents they may contain. Biosafety practices and procedures are consistent with the concept of "Universal Precautions," which requires all specimens of human blood, blood products, body fluids, and tissues to be treated as if they are infectious. Occupational exposure to blood-borne pathogens mandates a combination of work-practice controls and training for laboratory employees to prevent infection by pathogens and other risks. Vaccination against known pathogens such as hepatitis B has to be carried out among laboratory employees.

Most of the public is scientifically illiterate and should be made aware of the actual scientific facts related to the problems of GM foods and GM animals and plants. They should be given proper information about the scientific facts related to the so-called danger and moral ethics in the practice of some of the techniques of modern biotechnology. Modern molecular approaches in the treatment of genetic diseases such as IVF (*in vitro* fertilization), stem-cell research, organ transplantation, gene therapy, tissue engineering, etc., should be made available to the public so that they will appreciate it in the correct manner without any mystification. It is very essential for the expansion and commercial application of biotechnology.

REVIEW QUESTIONS

1. Describe the principle and application of animal-cell culture.
2. Explain hematopoiesis and its application in therapy.
3. Name two important products of animal-cell culture. Explain their use and mode of action.
4. What is the difference between primary and secondary cell cultures?
5. What is HeLa?
6. What are immortalized cells? Give a practical method to make a cell culture immortal.
7. How can we differentiate an immortalized cell line from a cancer cell line?
8. Usually animal cell cultures are incubated in a carbon-dioxide incubator. Explain.
9. What are the major ethical and moral issues related to animal cell culture and its various applications?

10. What is stem-cell culture? Explain its therapeutic uses.
11. What are embryonic stem cells and adult stem cells? Illustrate with examples.
12. Explain the procedure used to make a chimeric mouse.
13. What are the types of fermentors used for the large-scale cultivation of animal cells for industrial purposes?
14. What is a tissue plasminogen activator? How can it be produced commercially for pharmaceutical purposes?
15. Explain the mode of activity of t-PA.
16. What are factor VIII and IX? Explain their importance as a therapeutic protein. How can these proteins be produced by animal cell cultures?
17. Certain therapeutic proteins, even though they are direct gene products, cannot be produced in *e. coli* with recombinant DNA methods. Explain the reasons.
18. What is the difference between pluripotent and unipotent stem cells? Give examples.
19. Most animal cell culture media contains serum as one of the major ingredients. Explain the importance of serum in a culture medium.
20. What is meant by cryopreservation? What are the difficulties and how can they interfere with the viability of the preserved cells? How can they be solved? Give examples of cryopreservatives.

Chapter 21 APPLICATIONS OF BIOTECHNOLOGY (SUMMARY)

In This Chapter

21.1 BIOLOGICAL FUEL GENERATION

We are slowly depleting our fossil fuel energy resulting in the need to seek out alternative sources of energy. So far, these have included the harnessing of hydro, tidal, wave and wind power, the capture of solar and geothermal energy supplies, and nuclear power. There is now a growing appreciation of biological solar energy systems and biotechnological advances in this area will soon bring economic reality to selected processes. As fossil-fuel resources are depleted and become increasingly more expensive, conversion of organic residues to liquid fuels will become a more economically-attractive consideration.

There are three main directions that can be followed to achieve biomass supplies: (1) cultivation of so-called energy crops, (2) harvesting of natural vegetation, and (3) utilization of agricultural and other organic wastes. The conversion of the resulting biomass to usable fuels can be accomplished by either biological or chemical means or by a combination of both. The two main end products that will be formed, will be either methane or ethanol although other products may arise depending on initial biomass and the processes utilized, for example, solid fuels, hydrogen, low-energy gases, methanol and longer-chain hydrocarbons.

Although biomass may ultimately only supply a relatively small amount of the world's energy requirements, it will nevertheless be of immense overall value. In some parts of the world, such as Brazil and countries of similar climatic conditions, biomass will surely attain wider exploitation and utilization. There may still be some disadvantages when comparing it with coal or oil, but the very fact that it is renewable and they are not must spur further research. In time, biomass will become much more easily available and economically useful as a source of energy for mankind.

21.2 SINGLE-CELL PROTEIN

One of the biggest problems facing the world today is population growth, especially in developing nations. Conventional agriculture may not be able to provide sufficient supply of food and, in particular, protein. Today, new agricultural practices are widespread, high-protein cereals have been developed, the cultivation of soybeans and groundnuts is ever expanding, and so on. The use of microbes as protein producers has also gained wide experimental success. This field of study has become known as **single-cell protein production (SCP)** and reflects the fact that most microorganisms used as producers grow as single or filamentous individuals rather than as complex multicellular organisms such as plants or animals.

There are many reasons microbes are the prime candidate for SCP production. Some of the reasons include: (1) microorganisms can grow at remarkably rapid rates under optimum conditions (some microbes can double their mass every 0.5 to 1 hour); (2) microorganisms are more easily modified genetically than plants and animals (they are more amenable to large-scale screening programs to select for higher growth rate, improved amino acid content, etc., and can be more easily subjected to gene transfer technology); (3) microorganisms have relatively high protein content, and the nutritional value of the protein is good; (4) microorganisms can be grown in vast numbers in relatively small continuous fermentation processes using a relatively small land area and are also independent of climate; and (5) microorganisms can grow on a wide range of raw materials, in particular low value wastes, and can also use plant-derived cellulose.

The acceptability of SCP, when presented as human food, depends not only on its safety and nutritional value, but also on other factors. People do not usually take to the idea of eating food derived from microbes. In many cultures, there are guidelines of what you can and cannot eat. Also, odor, color, taste, and texture need to be taken into consideration when dealing with people's desires. Thus, if SCP is to be used as direct food for humans, then the skills of the food technologist will be greatly challenged.

21.3 SEWAGE TREATMENT

Waste is, as any material or energy form that cannot be economically used, recovered or recycled at a given time and place. Growth in human populations has generally been matched by a greater formation of a wider range of waste products, many of which cause serious environmental pollution if allowed to accumulate in the ecosystem. In rural communities, recycling of human, animal, and vegetable waste has been practiced for centuries, providing in many cases valuable fertilizers or fuel. In urban communities where most of the deleterious wastes accumulate, efficient waste collection and specific treatment processes have been developed because it is impractical to discharge high volumes of waste into natural land and waters. The development of these practices in the last century was one of the main reasons for the spectacular improvement in health and well being of humans.

Mainly by empirical means, a variety of biological treatment systems have been developed, ranging from cesspits, septic tanks, and sewage farms to gravel beds, percolating filters, and activated sludge processes coupled with anaerobic digestion. The primary aim of all of these systems or biotreaters is to alleviate health hazards and to reduce the amount of oxidizable organic compounds and thus produce a final effluent or outflow, which can be discharged into the natural environment without producing any adverse affect. Bioreactors rely on the metabolic versatility of mixed microbial populations for their efficiency. The fundamental feature of biotreaters is that they should contain a range of microorganisms with the overall metabolic capacity to degrade any compound entering the system. Controlled use of microorganisms has lead to the virtual elimination of such water-borne diseases as typhoid, cholera, and dysentery in industrialized communities.

21.4 ENVIRONMENTAL BIOTECHNOLOGY

In industrialized countries, there is increasing public concern over the impact of human activities on the environment and the legacy we leave for future generations. Attention is being given to minimizing environmental damage, and to cleaning up past environmental damage, while trying to ensure that the standard of living to which we have been accustomed is maintained.

As an example, in the United States, $100 billion worth of crops are destroyed annually by soil-dwelling nematodes damaging crop plants. Chemical nematicides are the only current option for crop protection, but these are among the most toxic and environmentally-damaging pesticides in widespread use.

It is obviously far better not to pollute the environment in the first place, than have to develop ingenious ways of cleaning up the environment. For this reason, many agricultural industries are investigating the potential of plants and microorganisms to provide cleaner processes and products.

The benefits of this research include:

- An increase in the productivity of crops, without an increase in the dependency on environmentally-damaging agrochemicals.
- As a result of increased productivity, a reduced pressure to exploit the remaining uncultivated habitats.
- As a result of increased productivity, a reduction in energy inputs (mostly from reduced agrochemical manufacture).
- The creation of alternative, renewable, sources of energy (e.g., biodiesel).
- The creation of new more environment-friendly raw materials for industry (e.g., biodegradable plastics from plant starches, or high-value speciality chemicals).

- As a result of the development of genetically-modified crops (if properly used), a reduction in the amount of agrochemical (e.g., pesticides and herbicides) released into the environment.

Environmental Concerns

Many environmental groups, and some biotechnologists, have concerns about the use of genetically-modified crops with respect to their effect on the environment. They warn that the environmental impact of their use might not become apparent until after their large-scale commercial use. These concerns include:

Herbicide use. The use of herbicide-resistant crops may increase rather than decrease herbicide use.

Genetic pollution and superweeds. Genes that have been copied from one species and inserted into another might 'escape' and spread to other organisms, thus causing 'genetic pollution.' For example, herbicide-resistant crops might cross-breed with related weeds and produce herbicide-resistant 'superweeds.'

Antibiotic resistance. Genes that code for antibiotic resistance are sometimes inserted into plant cells together with the 'useful' gene; these 'marker' genes enable scientists to select cells that have been successfully modified. Such antibiotic-resistance markers in crops might spread to animals or humans, rendering medical or veterinary use of the antibiotic ineffective.

Unexpected effects. Some genes that are inserted into genetically-modified plants might be unstable (there is a high probability that they will be lost from the plant cells), or might show unexpected effects because it is difficult to predict where in the plant genome the genes will insert.

Pest resistance. Genetically-modified crops that are designed to be resistant to a particular pest might have unintentional (and unpredictable) effects on harmless or beneficial organisms (e.g., ladybirds and bees).

Persistence and weediness. Genetically-modified plants might, intentionally or unintentionally, be more vigorous than their non-modified relatives. They can therefore effectively become 'weeds.' If plants are more 'persistent' (e.g., survive over winter better), they could rapidly dominate ecosystems at the expense of other plants. If they show 'weediness' characteristics, they could also spread to new habitats.

Damage to wildlife and biodiversity. Monocultures of pest-resistant crops may also harm beneficial insects and non-target organisms, with knock-on effects on the food chain.

Addressing Environmental Concerns

Because it is difficult to assess the risks associated with the potential environmental impact of using genetically-modified crops, precautionary environmental legislation is in place in many countries. Before allowing experimental or commercial release of genetically-modified organisms, most countries use risk-assessment procedures that are based on estimating the impact and interactions of genetically-modified crops and their environment. The following counter-views might also be considered:

Genetic Modification of Plants Achieves Essentially the Same Result as Conventional Plant Breeding

The ability to move single genes, or sets of genes, from one species to another allows 'step changes' in genetic composition to be made quickly. However, in some cases, the end product is effectively the same as might be achieved eventually by conventional breeding.

Wide-crossing

'Wide-crossing' is regarded as a conventional breeding technique, but is used to create new varieties of crops between plants that would not normally interbreed. For example, some varieties of wheat that are used to make bread contain some genes from rye. This is the result of a plant-breeding experiment that introduced the rye genes to the wheat that confers disease resistance.

In Nature, Genes are Exchanged between Species

Bacteria that are not related can exchange genes quite readily (this is one mechanism for antibiotic-resistance spreading). *Agrobacterium tumefaciens* is a bacterium known as 'nature's own genetic engineer,' which routinely transfers some of its genes into plant cells.

Gene Instability in Conventional Crops

The genes in new varieties of crops that have been developed by conventional breeding can also be unstable, particularly when they are produced as a result of wide-crosses. As with GM crops, selective breeding and variety evaluation 'weed out' undesirable gene combinations.

Conventional Breeding is not 'Natural' Either

Plant breeders have for many years used techniques to introduce more variation than nature produces. Bombardment of a plant with chemicals or radiation causes mutations randomly throughout its genes. Among the plants that survive might be tens or thousands of mutations—perhaps including ones that plant breeders want.

Gene Transfer from Crops to their 'Wild' Relatives

Little is known about the transfer of genes from any crop to the environment. 'Escaped' genes are most likely to spread to the crops' nearest relatives. Many, but not all, crops are grown in environments in which their 'wild' relatives do not exist. For example, in Europe, genes are much less likely to transfer from genetically-modified potatoes than from those grown in Latin American countries where there are wild relatives of the potato. There is no evidence to suggest the genes in genetically-modified plants 'escape' any more easily than genes in other plants. Herbicide-tolerant crops—produced by conventional breeding—already exist, as does the possibility of gene transfer and 'superweeds.' To date this has not been a problem.

Novel, Sustainable, Agricultural Practices Needed

There are concerns that pest-resistant plants will remove important food sources for wildlife and affect non-target organisms. The ecological impact of such crops must be evaluated, particularly for large-scale use. But we must also remember that many pesticides currently in use are not highly selective for pests, harming non-target organisms. Agriculture, as currently practiced, is the apparent cause of some disastrous effects on wildlife biodiversity—we must look for new options.

Application of biotechnology that we discussed in biological fuel generation, single-cell protein, single-cell oils, sewage treatment, etc., can also be covered under environmental biotechnology and can be used for the above purposes while cleaning the environment using biotechnological methods.

21.5 MEDICAL BIOTECHNOLOGY

The use of biotechnology in medicine is growing rapidly and is providing us with opportunities to develop new, more effective drugs and other therapeutics. Studying the genetics of humans is allowing us to understand what happens when genes go wrong in inherited diseases, or cancers, and to start to develop new therapies that treat the genetic cause, not the symptoms. By studying the genetics of viruses, fungi or bacteria, we can understand how they cause disease and develop better drugs and antibiotics that target them more specifically.

Safer, Cheaper Medicines from Biotechnology

Many medicines have traditionally been extracted from plants, microorganisms, or animals. However, in many cases, it is not possible to produce the medicines in large enough amounts, safely, or cheaply enough to treat large numbers of people. Valuable drugs may be available only from endangered plant species (e.g., the anti-cancer drug tactual from the Pacific yew), or produced naturally in such small amounts that extraction is just too expensive. Biotechnology offers ways to produce valuable medicines in larger amounts by 'cell culture,' or using genetic modification to increase the amounts produced, or to make entirely new medicines.

Medicines from 'Cultured' Cells

Many different types of plant, animal, and human cells can be grown 'in culture,' that is, as a suspension of cells in a large fermentor vessel of liquid that contains all of the nutrients they need, just as though they were yeasts or other microorganisms. These cells can be treated in various ways to make them produce large amounts of valuable compounds: microorganisms grown in this way can produce antibiotics. Plant cells grown in this way have been used to produce the anti-leukemia drugs vinblastine and vincristine: these are natural products of the pink periwinkle. Sedatives and heart drugs (e.g., digoxin from foxglove) have also been produced in plant culture.

Genetic Modification for Medicine Production

Organisms can be genetically modified either to increase the amount of medically useful products, such as antibiotics that they produce, or to produce entirely new medicines. Microorganisms and plants are both being genetically modified to produce large amounts of medicines safely and cheaply: genetically-modified microorganisms containing the gene for human insulin produce a safe and plentiful supply of insulin for diabetics; while genetically-modified plants are starting to be used to produce vaccines for a variety of human and animal diseases.

Research into finding new antibiotics is probably more urgent now than at any time since the discovery of penicillin some 50 years ago. Why? Many antibiotics that were once highly effective are now proving to be useless as the disease-causing microorganisms develop resistance. A new approach is being used to try to make entirely new antibiotics. A soil bacterium—*streptomyces*—is a source of antibiotics, immunosuppressants, anti-cancer and anti-parasitic agents, and natural herbicides. Many of these valuable products are made by the bacterium from smaller building blocks according to the 'blueprint' determined by the bacterium's genes. It is possible to generate new compounds by changing the blueprint—switching some of the genes, and thus some of the units, around. A whole range of new compounds can be made using this approach, potentially leading to new antibiotics. This area of research is extremely important, considering the rising number of diseases caused by antibiotic-resistant microorganisms.

21.6 AGRICULTURE AND FOREST BIOTECHNOLOGY

Genetic engineering can be used to modify the genetic compositions of plants, animals, and microorganisms. The number of genes that have been isolated and are available for transfer is growing daily. Currently, the technology is used primarily to modify crops, although a number of other applications are in the wings.

Like other products, genetically-engineered products undergo a period of research and development before they are ready for commercial release. Many products never emerge from the research and development pipeline. While this is true for almost any technology, genetic engineering is turning out to be more difficult and more expensive than early proponents expected. Although in the early 1980s biotechnology was touted as a miracle technology that was going to usher in a new era of agricultural abundance with minimal harm to the environment, the initial set of products have proven modest. Some of the most important commercial applications of biotechnology include:

- Crops that are resistant to pests or herbicides, which will allow reduced or more selective use of agrochemicals.
- Crops that are better able to tolerate hostile environmental conditions such as frost or drought.
- New and more efficient crop-breeding systems that allow production of hybrids that cannot be produced by conventional plant breeding.
- New crop varieties that are qualitatively different (e.g., plants that produce seeds or tubers that have an altered starch composition, and thus represent new valuable products).
- New crop varieties that have different growth characteristics (e.g., altered flowering time or rate of growth) to increase yield.

These tools fall into two major categories: improving crop performance in the field and developing new products with enhanced values.

Genetic Improvement through Traditional Crop Breeding

The cultivation and improvement of food crops began centuries ago, when seeds from the most successful plants of one season were used to sow the next year's crop. Improved crop varieties have been developed by selection and breeding, and few of today's crops resemble their wild ancestors. Some crops were developed by these conventional plant-breeding processes to be disease resistant or to grow faster, for example; others to make the edible parts of the plant larger or tastier. However, developing new varieties of crops by traditional methods is not only slow but also limits plant breeders to transferring genes between crops that normally cross-fertilize.

FIGURE 21.1 **Many of our modern cereals bear little resemblance to their 'wild' grass ancestors. Through breeding and selection, the characteristics of these crops have been greatly altered. Left to right: pearl millet, barley, wheat, maize, rice, and sorghum.**

Plant Genetics

Over the past few years, scientists have been able to identify the individual genes that determine particular characteristics in a plant. Just as importantly, we can now transfer these genes between plants, which means that novel features can be introduced into crop species. It is unlikely that biotechnology will solve every agronomic problem, but it can offer farmers new options.

Plant Physiology and Biochemistry

Studies of plant physiology, and biochemistry, have given us an understanding of how plants sense their environment, and respond by germinating, growing, flowering, or setting seed. All these processes are determined by their genes. How they photosynthesize, and make proteins, oils, or carbohydrates is now well understood. How they recognize insect pests or disease-causing fungi, bacteria or viruses (pathogens), and mount their own defense responses is also much better understood.

Genetic Modification of Crop Plants

The discovery that genes are naturally transferred from the bacterial pathogen *agrobacterium tumefaciens* when it infects plants allowed a breakthrough in introducing new genes and desirable traits into crop plants. The subsequent development of other genetic modification methods means that it is now quite straightforward to introduce a gene, or set of genes, into most of the world's major crops.

Advanced breeding program means that genes introduced by genetic modification can often be moved into 'elite varieties' (highly productive and commercially successful varieties) within economically viable timeframes. A better understanding of how genes work is allowing the introduced genes to be controlled in quite sophisticated ways: switched 'on' or 'off' at a particular time during plant development or in response to environmental signals. For example, research is underway to make plants switch from using their energy supplies for 'growing tall' to storing energy in the form of starch and sugars. Other studies are aimed at getting plants to switch on the genes for defence mechanisms in response to attack by pests or diseases, or at stages of growth when they are most likely to be susceptible to attack.

Engineered Crops

The most widespread application of genetic engineering in agriculture by far is in engineered crops. Thousands of such products have been field-tested and over a dozen have been approved for commercial use. The traits most commonly introduced into crops are herbicide tolerance, insect tolerance, and virus tolerance.

Herbicide Tolerance

Herbicide tolerance allows crops to withstand otherwise lethal doses of herbicides, which are chemicals that kill plants. Some herbicides kill virtually all plants and cannot be used on crops. By offering crops tolerant to herbicides, chemical companies can expand the market for their products. And indeed, the major developers of herbicide-tolerant plants are companies with herbicides to sell. The current set of commercially-available herbicide-tolerant crops is tolerant to three herbicides based on three active ingredients: bromoxynil, glyphosate, and glufosinate.

Insect Tolerance

All of the commercially-available insect-tolerant plants contain a version of the toxin *bacillus thuringiensis* (Bt), which is found in nature in soil bacteria. Bt toxins are highly effective for many pest organisms, such as beetles and moth larva, but not toxic to mammals and most other non-target organisms. A major concern among farmers and environmentalists is that wide use of Bt crops will lead to the rapid development (over the course of perhaps as few as three to five years) of resistance to the toxin. If resistance develops, the Bt toxin will be as useless as a pesticide. In this case, the environmental benefits of the product will be short lived.

Loss of Bt will affect those who currently use the engineered Bt crops, but also many other farmers who use Bt in its natural bacterial form, usually as a spray. These other farmers include those who grow food organically and those who use Bt as part of integrated pest management (IPM). Natural Bt sprays are a valuable mode of pest control for these farmers. Organic farmers and others who rely on Bt question whether the companies who sell the Bt crops have the right to use up this resource guided only by commercial calculations.

Virus Tolerance

The third major application of biotechnology to crops is virus tolerance. These crops contain a gene taken from a virus. By a process that is not well understood, plants that produce certain viral proteins are able to fend off infections by the viruses from which the proteins were taken. Two virus-tolerant crops are currently approved for commercial use: papaya and squash. The squash, which is resistant to two viruses, is currently off the market. Although it is difficult to get information on why products are not on the market, it is possible that the squash did not perform well enough in the field to capture market share.

Other Engineered Products

Many other genetically-engineered products have been envisioned, but only a few have come to market so far.

Several tomatoes engineered to delay ripening have been approved for commercial use. In some cases, delayed ripening just prolongs shelf life. But, for the Flavr Savr™, the objective was to increase the time on the vine without softening, producing a transportable, tasty winter tomato. After a highly publicized rollout, the Flavr Savr™ too is off the market. The problem appears to have been with transportability rather than taste.

On the livestock side, a drug has been produced for dairy cows—Bovine Growth Hormone (BGH) or Bovine Somatotropin (BST)—by engineering a bacterium to contain the gene for the hormone. The drug is administered to cows to increase milk production, despite the chronic oversupply of milk in the United States. A highly controversial product when it was first introduced, BGH is currently used on about 10 percent of U.S. dairy herds.

An interesting product not related to agriculture is a rabies vaccine intended for use on wild raccoons. In this case, genetic engineering was used to construct a 'hybrid' virus made up of a component of the rabies virus inserted into an unrelated 'carrier' virus. The resulting virus confers immunity to rabies but poses no danger of causing the disease. Baits laced with the vaccine have been distributed in many parts of the eastern United States in attempts to combat rabies in wild raccoon populations. The vaccine has been approved by the USDA, despite suspicions that it has been only marginally, if at all, effective. Early studies on efficacy failed to demonstrate that the product could control rabies in wild raccoon populations. Data from more recent studies are being withheld from the public as confidential business information.

Agricultural Biotechnology Research Projects

The following lists some of the organisms being engineered by agricultural researchers and indicates which products are commercially available and which are not.

Genetically-Engineered Livestock and Poultry

Animals Engineered for Leaner Meat

No livestock engineered for leaner meat is currently near commercialization. Research done early in the 1980s to genetically engineer leaner pigs failed because of unacceptable side effects, including low fertility, arthritis, and impaired immune systems. Some low level of research activity may still be underway.

Animals Engineered as Drug-Production Facilities

Status: Goats and sheep have been engineered to secrete bioactive molecules into their blood, urine, or milk. Companies are in the process of developing commercial enterprises based on these animals. So far, none of the drugs is on the market. It is likely that producers will want to slaughter the animals for food after they are no longer useful for drug production.

Animals Engineered as Sources of Transplant Organs

Status: Commercial entities are engineering pigs so that their organs will not be rejected by human transplant recipients. So far, the organs are not commercially available. It is likely that producers will want to use the carcasses of donor pigs as food.

Animals Engineered for Disease Resistance

Status: Chickens and turkeys have been engineered to resist avian diseases. None have been commercialized.

Genetically-Engineered Fish and Shellfish

Status: Fish and shellfish have been engineered to cause changes in hormones that accelerate growth in several laboratories. So far, none have been commercialized in the United States.

Genetically-Engineered Plants Eaten Whole as Food

Status: Many plants have been commercialized, including tomatoes and squash and commodity crops such as corn and soybeans. Most have been engineered for one of three traits: herbicide tolerance, insect resistance, or virus tolerance.

Genetically-Engineered Fiber Plants

Status: Genetically-engineered cotton has been approved for commercial use.

Engineered Insects Used in Agricultural Systems

Status: No engineered insects have been approved for commercial use. An engineered predatory mite has been field tested in Florida. Researchers have produced honeybees and other beneficial insects engineered to tolerate pesticides.

Engineered Microorganisms Used as Pesticides

Status: Several bacteria engineered to enhance their ability to kill or repel pests have been approved for commercial use. These products are used as pesticides in agricultural fields and gardens.

Food Processing Aids Made from Engineered Bacteria

Status: Bacteria have been genetically engineered to produce rennet, an enzyme important in making cheese. Genetically engineered rennet (chymosin) is approved for commercial use and widely used by US cheese processors.

Animal Drugs Made by Engineered Bacteria

Status: Bovine Growth Hormone (BGH) made from engineered bacteria is used commercially to stimulate milk production in cows. Currently this hormone is being used in about 10 percent of the nation's dairy herds.

21.7 FOOD AND BEVERAGE BIOTECHNOLOGY

Application of Modern Biotechnology in the Food Industry

The techniques of modern biotechnology are becoming an increasingly important part of the overall effort to improve methods of food production and to increase the variety and quality of foods. Modern biotechnology has potential applications in the production of food, food processing, and also in the assurance of food quality and safety.

Although many genetically-modified food crops are under development, very few have so far reached the supermarket shelf. The first genetically-modified whole food to reach the shops was the Flavr Savr™ tomato in the USA, and the first product in the UK was tomato puree. The recent introduction of genetically-modified soybeans has caused controversy. Up to 60% of processed foods contain soy or soy derivatives, such as starch, protein, or oils.

Modern biotechnology is also being used to improve microorganisms, processing aids (e.g., enzymes) or ingredients for use in the production and processing of food. The enzyme chymosin, which is used in the manufacturing of 'vegetarian cheese,' is an example of this type of genetically-modified product.

21.8 SAFETY IN BIOTECHNOLOGY

Based on international guidelines, governments in each country frame rules to ensure not only the safety of the product but also for those who conduct the genetic-engineering experiments. When the first rDNA experiments were carried out, scientists themselves imposed a moratorium on further genetic experiments until proper guidelines were structured. In 1977, the Asilomar Conference formulated some guidelines and the major points are the following:

- Recombinant DNA experiments should be carried out in laminar flow-chamber kept in a clean room, so that the recombinant organisms do not escape.
- The host organism used to carry out the gene cloning or the cloning of rDNA should be specially made for such experiments. Even if the organism escapes the laboratory it will

not survive and the recombinant DNA molecule will be lost. There are special safe vectors that can be used for rDNA experiments without fear. There is also non-pathogenic *e. coli* developed to use in rDNA experiments as host cells.

The products, in most cases, of modern biotechnology will be therapeutic proteins such as vaccines, hormones, and enzymes. The proteins that are to be used as drugs should meet all the safety requirements and regulations imposed by the regulatory authorities. They are very strict about the data regarding the experiments on animal models, terminal patience, non-terminal patience, and independent clinical trials by separate agencies. If a transgenic organism is released into the environment, in addition to the ordinary safety norms it should also monitor its environmental impact. The clinical trials of drugs produced by rDNA methods have to be carried out at least in three phases.

Phase I. In the first phase of study, the side effects and dose of tolerance by the patients is systematically carried out on selected patients.

Phase II. In the second phase, collecting data from pharmacological, pharmacokinetic, metabolic, and toxicological studies on a selected number of patients optimizes the use of the drug.

Phase III. In the last phase, the studies focus on the safety aspects of the drugs. During these studies the harmful effects of the drug, if any, are monitored in addition to the effectiveness as a drug. In dosage range, the interactions with other drugs are also investigated.

A new biotechnology product, whether it is a drug or a food material, has to undergo trial studies, and only after the regulatory agencies are fully satisfied by the data, will they give permission or license it for mass production and marketing.

REVIEW QUESTIONS

1. Explain the importance of biotechnology in food and agriculture.
2. Explain the contribution made by biotechnology and molecular biology to medicine.
3. Explain the role of biotechnology in the production of fuel.
4. What is single cell protein ? Explain how can it be related to environmental biotechnology?

INDEX

A

B

C

D

E

F

G

H

I

K

L

M

N

O

P

Q

R

S

T

U

V

W

X

Y

Z

T

U

V

W

X

Y

Z

APPENDIX
4-COLOR FIGURES

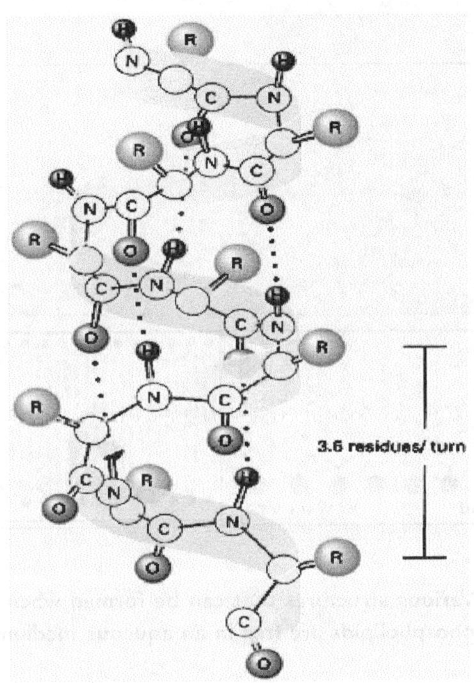

PAGE 165 **Right-handed α-helix.**

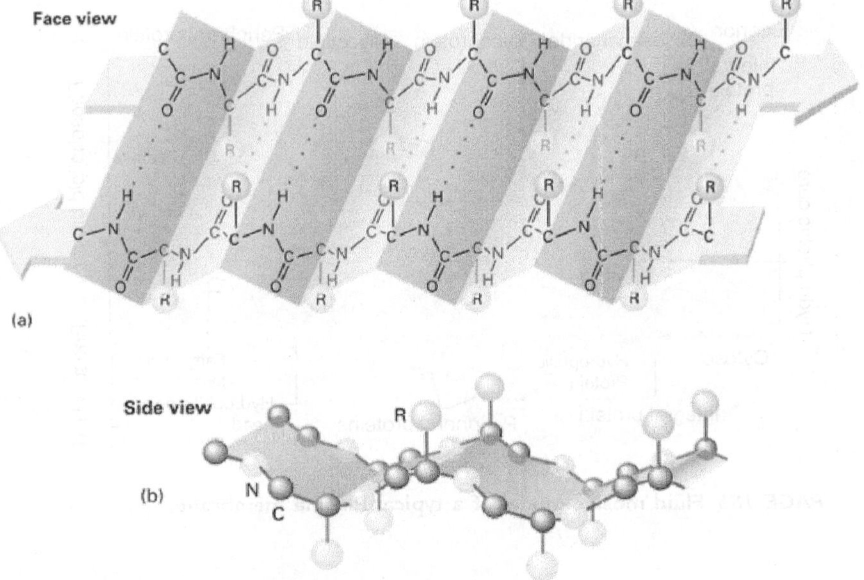

PAGE 166 **Diagram sketch of beta-pleated sheat.**
(a) View from the top and (b) view from the side.

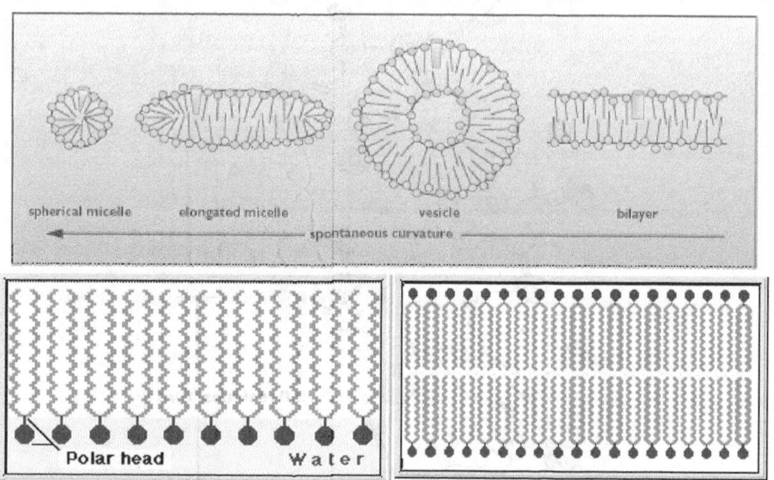

PAGE 184 **Various structures that can be formed when hydrophobic phospholipids are free in an aqueous medium.**

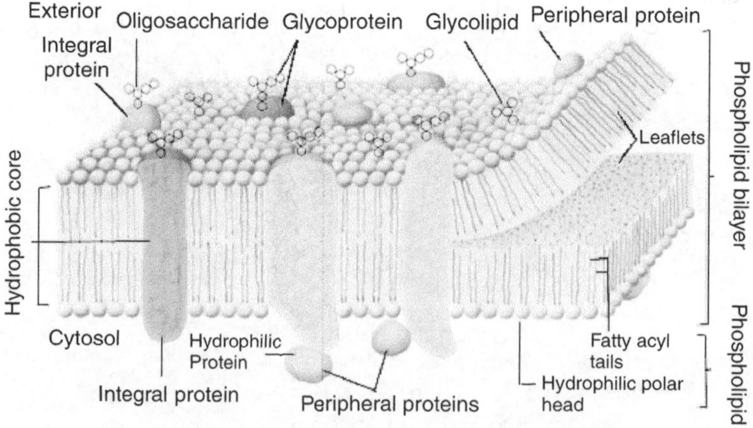

PAGE 185 **Fluid mosaic model of a typical Plasma membrane.**

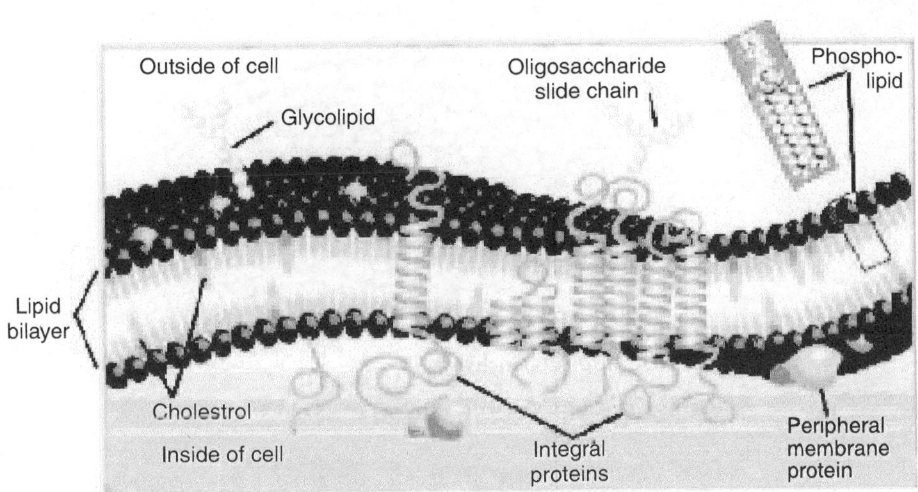

PAGE 218 **Fluid mosaic model for cell membranes.**

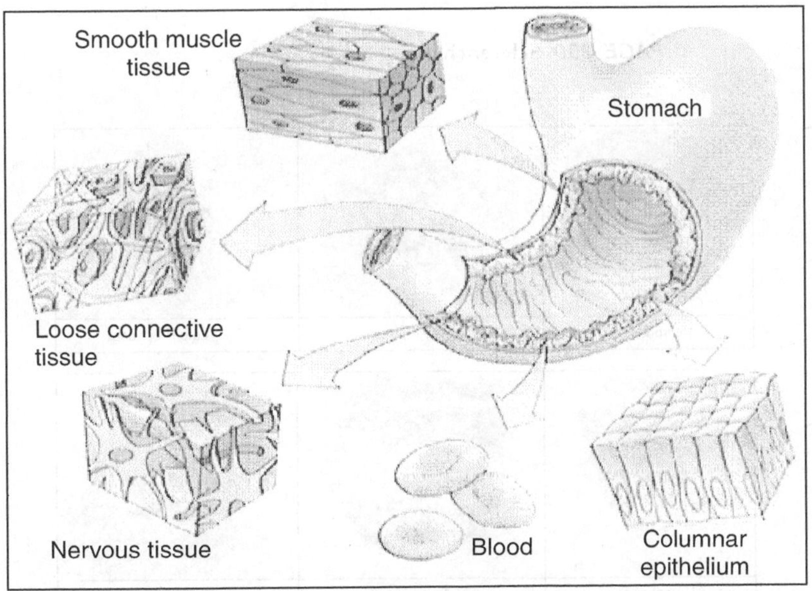

PAGE 226 **Various types of tissues present in the stomach.**

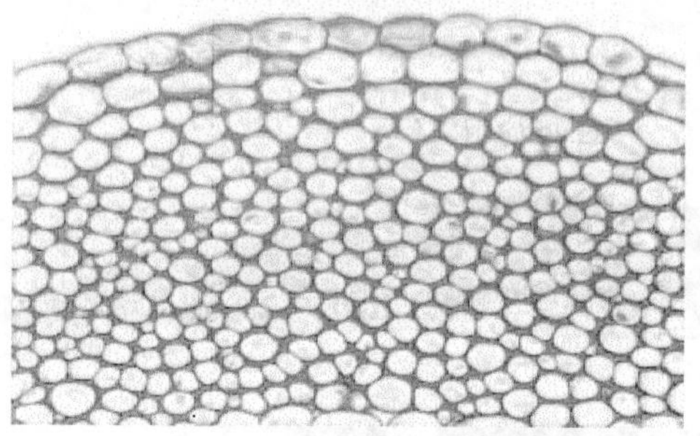

PAGE 230 **Collenchyma.**

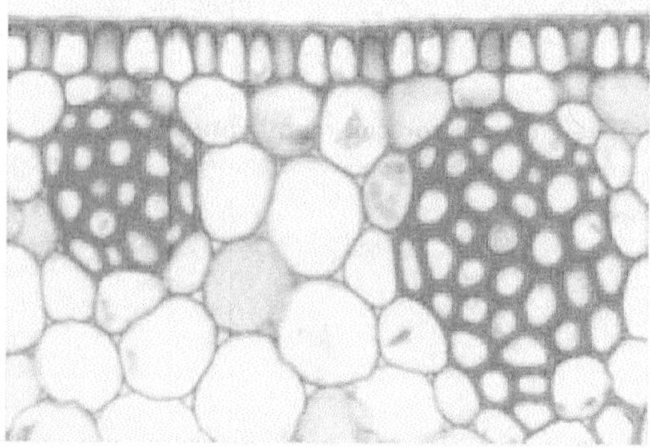

PAGE 230 **Sclerenchyma.**

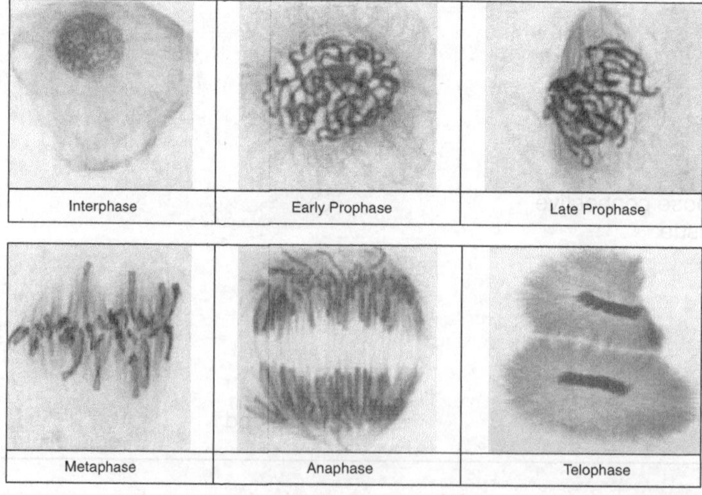

PAGE 243 **The stages of mitosis.**

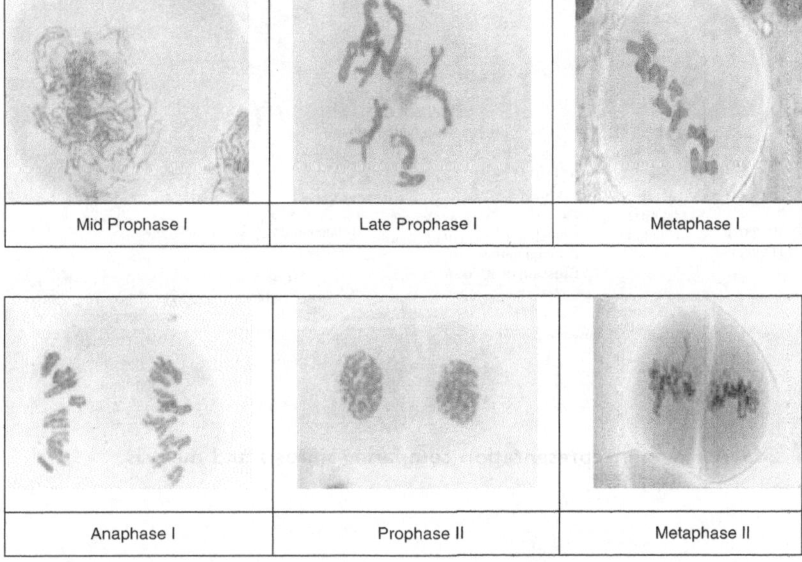

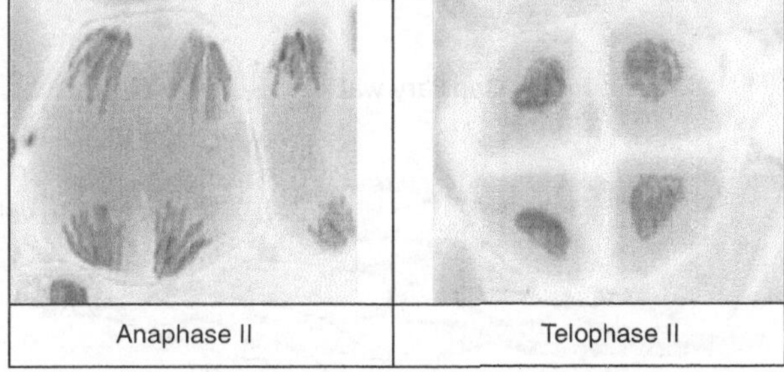

PAGE 244 **The Stages of Meiosis (The photographs show meiosis in *Lilium longiformum* (lilly). Note that as in mitosis the process is dynamic and that the terms do not refer to static stages.**

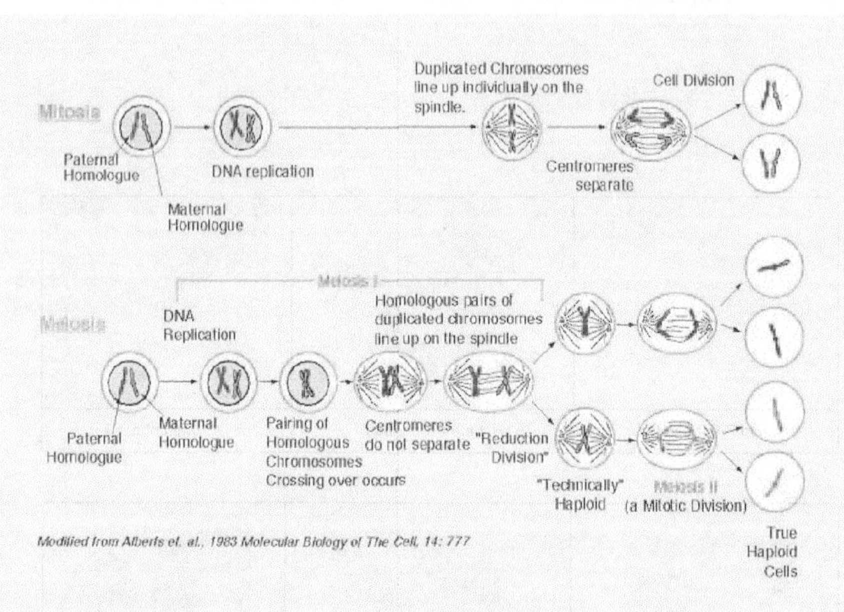

PAGE 245 **A diagram representation comparing mitosis and meiosis.**

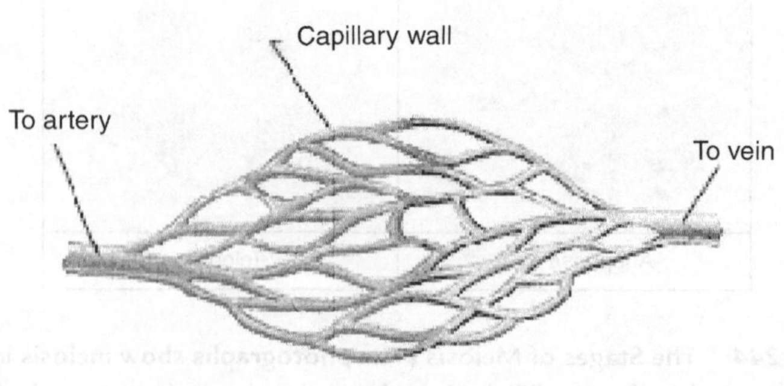

PAGE 258 **Capillary beds.**

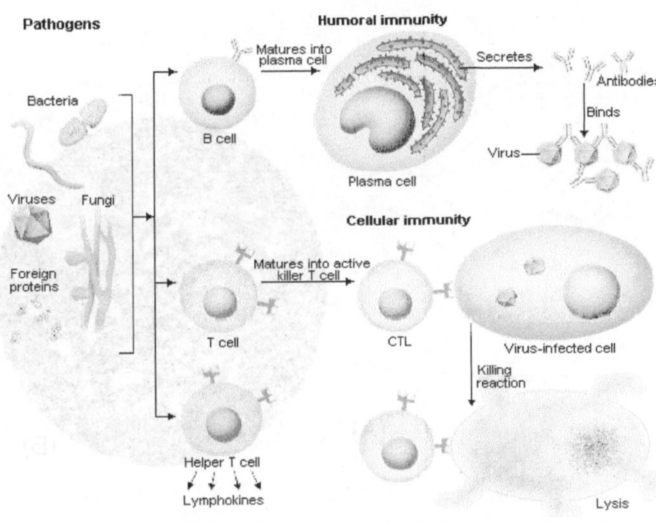

PAGE 272 **Pathway of immune response.**

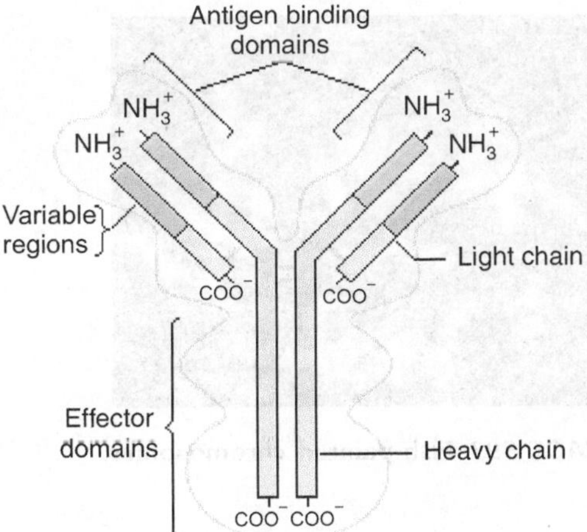

PAGE 273 **Structure of an antibody.**

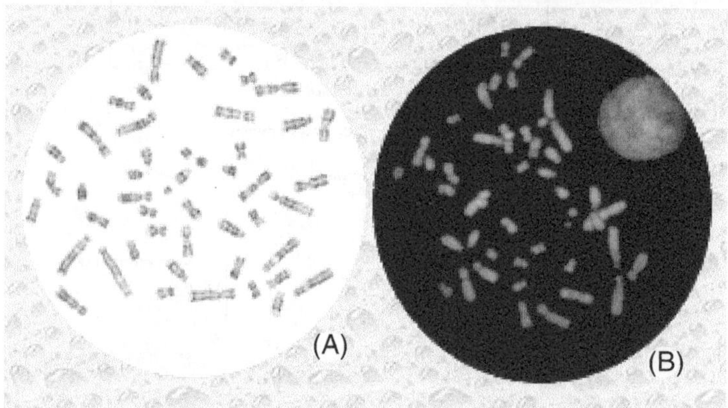

PAGE 391 **(A) G-banded metaphase cell at 1000x;**
(B) Q-banded metaphase cell at 1000x.

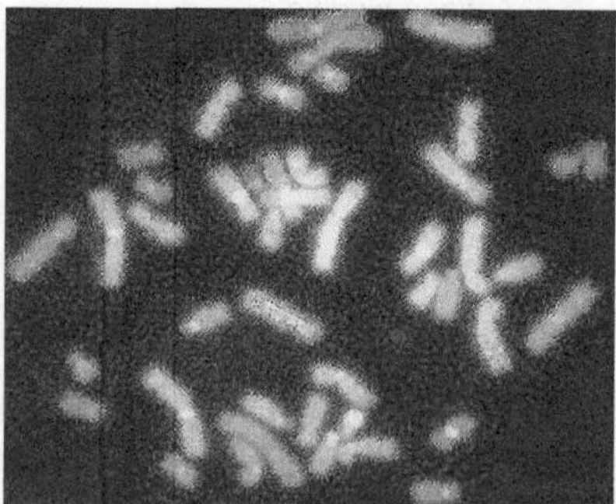

PAGE 393 **Fish-painted chromosomes.**

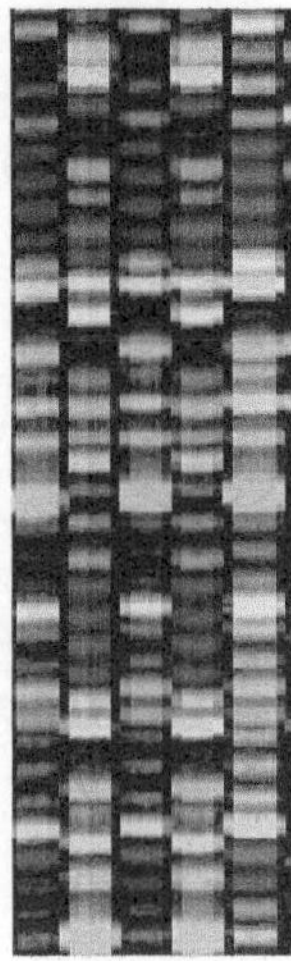

PAGE 505 **A sequencing gel after electrophoresis, illuminated with ultraviolet laser. Here, the primers are not radiolabeled. Instead, each type of ddNTPs are fluorescently labeled and therefore are visible in the gel itself under UV. This can be read automatically with a UV laser scanner or reader. The introduction of fluorescent labeling and the automatic reading of the bands became an integral part of the automatic DNA sequencing machine or the automated DNA sequencer, which made the technique of DNA sequencing very simple and rapid.**

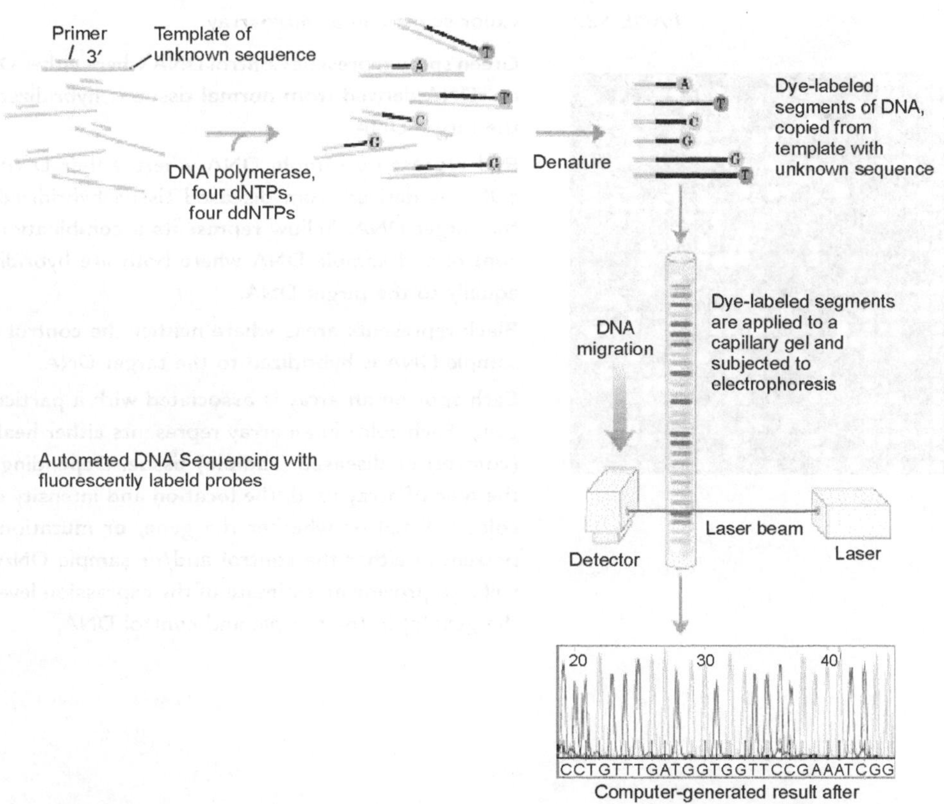

PAGE 505 **Different steps of automated DNA sequencing using fluorescently-labeled probes.**

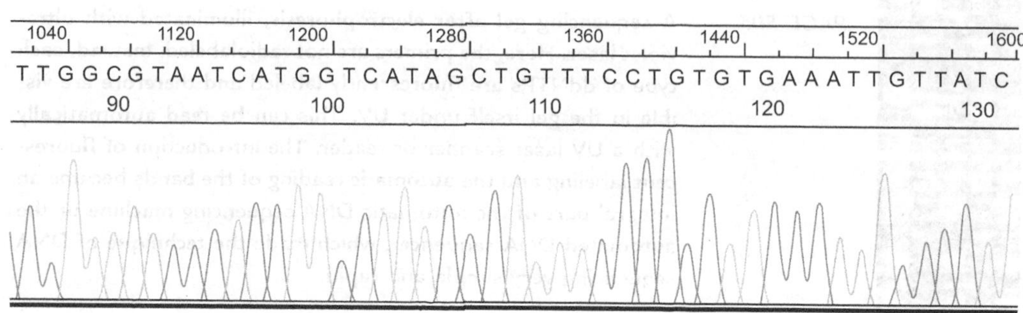

PAGE 505 **Data of DNA sequence obtained from an automatic DNA sequencer. The black color is instead of yellow, which represents the base 'G'.**

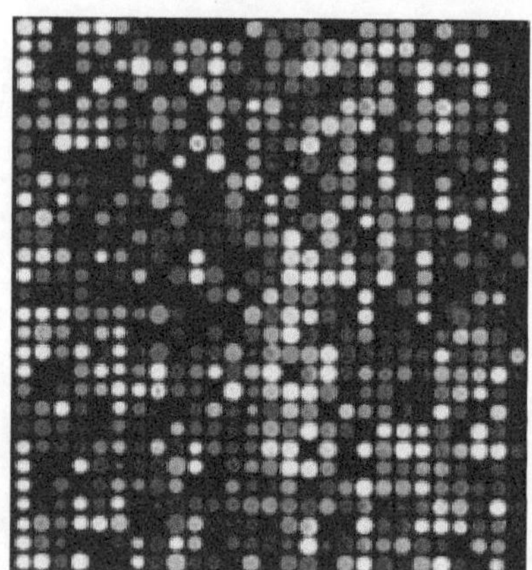

PAGE 527 **Color scheme in a Microarray.**

Green spots represent Control DNA where either DNA or cDNA derived from normal tissue is hybridized to the target DNA.

Red represents sample DNA where either DNA or cDNA is derived from diseased tissue hybridized to the target DNA. Yellow represents a combination of control and sample DNA where both are hybridized equally to the target DNA.

Black represents areas where neither the control nor sample DNA is hybridized to the target DNA.

Each spot on an array is associated with a particular gene. Each color in an array represents either healthy (control) or diseased (sample) tissue. Depending on the type of array used, the location and intensity of a color will tell us whether the gene, or mutation, is present in either the control and/or sample DNA. It will also provide an estimate of the expression level of the gene(s) in the sample and control DNA.